Unmanned Aerial Vehicles:

Breakthroughs in Research and Practice

Information Resources Management Association
USA

IGI Global
DISSEMINATOR OF KNOWLEDGE

Published in the United States of America by
IGI Global
Engineering Science Reference (an imprint of IGI Global)
701 E. Chocolate Avenue
Hershey PA, USA 17033
Tel: 717-533-8845
Fax: 717-533-8661
E-mail: cust@igi-global.com
Web site: http://www.igi-global.com

Library of Congress Cataloging-in-Publication Data

Names: Information Resources Management Association.
Title: Unmanned aerial vehicles : breakthroughs in research and practice / Information Resources Management Association, editor.
Description: Hershey, PA : Engineering Science Reference, an imprint of IGI Global, [2019]
Identifiers: LCCN 2018053028| ISBN 9781522583653 (hardcover) | ISBN 9781522583660 (ebook)
Subjects: LCSH: Drone aircraft--Design and construction. | Drone aircraft--Maintenance and repair.
Classification: LCC TL685.35 .U56 2019 | DDC 629.133/39--dc23 LC record available at https://lccn.loc.gov/2018053028

British Cataloguing in Publication Data
A Cataloguing in Publication record for this book is available from the British Library.

For electronic access to this publication, please contact: eresources@igi-global.com.

List of Contributors

Table of Contents

Detailed Table of Contents

which the developed model is verified. Four control algorithms are developed to control the quadrotor's degrees of freedom: a linear PID controller, Gain Scheduling-based PID controller, nonlinear Sliding Mode, and Backstepping controllers. The performances of these controllers are compared through the developed simulation environment in terms of their dynamic performance, stability, and the effect of possible disturbances.

In computer science, the study of mimicking nature has given rise to Swarm Intelligence, a distributed system of autonomous agents interacting with each other to collectively perform intelligent tasks. This chapter investigates how groups of holonomic flying robots such as quad copters can seek shelter autonomously when encountering bad weather. In this context three alternative autonomous shelter seeking techniques that address the unsolved plateau-problem had to be implemented. The methods were inspired by cockroaches and hunting strategies observed in apex predators. Previous studies on cockroaches have provided facts about their behaviour and resulted in algorithms that can be used for robotic systems. This research builds on these previous studies by formulating three alternative techniques and carrying out a comprehensive analysis of their performance. Simulation results confirm a scalable system where swarms of flying robots successfully find shelters in 3-D environments.

Recently, Flying Ad-hoc Networks (FANETs), enabling ad-hoc networking between highly mobile Unmanned Aerial Vehicles (UAVs), are gaining importance in several military, commercial and civilian applications. The sensitivity of these missions requires precise and prompt data delivery. Thus, the most important communication requirements that need to be addressed while designing FANETs are of high reliability and low latency. Considering these demands, this chapter focusses on mobility models, MAC protocols and routing protocols.

One of the fundamental challenges faced by an inexperienced user in portable unmanned aerial vehicle (UAV) such as quadcopters is flight control, often leading to crashes. Addressing this challenge, and leveraging upon the technological advancement in perceptual computing and computer vision, this research presents a modular system that allows for hand gesture based flight control of UAV, alongside a transport mechanism for portable objects. In addition to ascertain smooth flight control by avoiding obstacles in navigation path, real-time video feedback is relayed from the UAV to user, thus allowing him/her to take appropriate actions. This paper presents the design implementation by discussing the various sub-systems involved, inter system communication, and field tests to ascertain operation. As presented from testing results, the proposed system provides efficient communication between the subsystems for smooth flight control, while allowing for safe transport of portable objects.

This chapter discusses HCI interfaces used in controlling aerial robotic systems (otherwise known as aerobots). The autonomy control level of aerobot is also discussed. However, due to the limitations of existing models, a novel classification model of autonomy, specifically designed for multirotor aerial robots, called the navigation control autonomy (nCA) model is also developed. Unlike the existing models such as the AFRL and ONR, this model is presented in tiers and has a two-dimensional pyramidal structure. This model is able to identify the control void existing beyond tier-one autonomy components modes and to map the upper and lower limits of control interfaces. Two solutions are suggested for dealing with the existing control void and the limitations of the RC joystick controller –the multimodal HHI-like interface and the unimodal BCI interface. In addition to these, some human factors based performance measurement is recommended, and the plans for further works presented.

This chapter highlights important aspects of an intelligent fighter pilot support for distributed unmanned and manned decision making. First the background is described including current trends within the domain, and characteristics of a decision support system are discussed. After that a scenario and example situations are presented. The chapter also includes reflections of an intelligent fighter pilot support for distributed unmanned and manned decision making from the joint cognitive systems view, regarding human interoperability, and function allocation.

In this chapter, the authors present conceptual models of decision support systems (DSS) and Expert systems (ES) for Human-Operator (H-O) of Air Navigation System (ANS), such as air traffic controller (ATC) in flight emergencies, Unmanned Aerial Vehicles (UAV) operator, Safety Management System (SMS), etc. The authors made an analysis of the International civil aviation organization (ICAO) documents on risk assessment and the impact of the social environment on the aviation system. Automated system of pre-flight information preparation with intelligent module for support the decision making (DM) about aircraft departure are presented and program realization of systems are shown. The authors obtained algorithm of determining the optimal aerodrome for the forced landing of aircraft is provided. Expert systems of aviation enterprise's estimation were developed. Inhomogeneous factors of internal and external management environment of aviation enterprise were generalized using set-theoretical approach. This gave possibilities to define that the level of safety of aviation activity has the greatest influence among factors of internal environment and global economic situation – among factors of external environment. The authors demonstrate some interesting applications of DM in Socio-Technical Systems (STS). The DSS for H-O of ANS in the emergency situational were developed. Examples of DSS: DSS of ATC in the emergency situational; the flight dispatcher's DSS; DSS of UAV's operator, etc. In addition, the chapter presents some cases of DSS developed by the authors and colleagues at National Aviation University, Ukraine.

Dismount skin detection from an aerial platform has posed challenges compared to ground-based platforms. A small, area scanning multispectral imager was constructed and tested on a Small Unmanned Aerial System (SUAS). Computer vision registration, stereo camera calibration, and geolocation from autopilot telemetry were utilized to design a dismount detection platform. The test expedient prototype was 2kg and exhibited skin detection performance similar to a larger line scan hyperspectral imager (HSI). Outdoor tests with a line scan HSI and the prototype resulted in an average 5.112% difference in Receiver Operating Characteristic (ROC) Area Under Curve (AUC). This research indicated that SUAS-based Spectral Imagers are capable tools in dismount detection protocols.

Unmanned aerial vehicles (UAVs), commonly known as drones, are a robotic form of military aircraft that are remotely operated by humans. Due to lack of situation awareness, such technology has led to the deaths of civilians through the inaccurate targeting of missile or gun attacks. This chapter presents the case for how a patented invention can be used to reduce civilian casualties through attaching an affect recognition sensor to a UAV that uses a database of strategies, tactics and commands to better instruct fighter pilots on how to respond while in combat so as to avoid misinterpreting civilians as combatants. The chapter discusses how this system, called VoisJet, can reduce many of the difficulties that come about for UAV pilots, including reducing cognitive load and opportunity for missing data. The chapter concludes that using UAVs fitted with VoisJet could allow for the reduction of the size of standing armies so that defence budgets are not overstretched outside of peacetime.

This chapter addresses the key issues of chemical plume mapping and tracing via swarm robots. First, the authors present the models of turbulent odor plumes with both non-buoyant and buoyant features, which can efficiently evaluate strategies for tracing plumes, identifying their sources in two or three-dimensions. Second, the authors use the Monte Carlo technique to optimize moth-inspired plume tracing via swarm robots under formation control, which includes a leader to perform plume tracing maneuvers and non-leaders to follow the leader during plume tracing missions. Third, the authors introduce a variety of robot-based plume tracers, including ground-based robots, autonomous underwater vehicles, or unmanned aerial vehicles. Finally, the authors prospect the further research in this area, e.g., applying swarm robots to detect oil or gas leak, or to investigate subsea chemical pollution and greenhouse gases.

This article proposes an approach which deals with the problem of monitoring ocean pollution and cleaning dirty zones using autonomous unmanned vehicles. The authors present a cooperative agent-based planning approach for heterogeneous unmanned vehicles with different roles. Unmanned aerial vehicles (UAVs) monitor multiple ocean regions and unmanned surface vehicles (USVs) tackle the cleaning of dirty zones. Due to the rapid deployment of these unmanned vehicles, and the increase of ocean pollution, it is convenient to use a fleet of unmanned vehicles. Thus, most of the existing studies deal with the monitoring of different zones, the detection of the polluted zones and then the cleaning of the zones. In order to optimize this process, the authors' solution aims to use one UAV and one USV to reduce the pollution level of the ocean zones while taking into account the problem of fault tolerance related to these vehicles.

This article describes how a healthcare delivery drone has the potential for developing countries to leapfrog the development of traditional transportation infrastructure. Inaccessible roads no longer will prevent urgent delivery of blood, medications or other healthcare items. This article reviews the current status of innovative drone delivery with a particular emphasis on healthcare. The leading companies in this field and their different strategies are studied. Further, this article reviews the latest decision models that facilitate management decision making for operating a drone fleet. The contribution in this article of two new models associated with the design of a drone healthcare delivery networks will facilitate a more timely, efficient, and economical drone healthcare delivery service to potentially save lives.

Although most unmanned systems that militaries use today are still unarmed and predominantly used for surveillance, it is especially the proliferation of armed military robots that raises some serious ethical questions. One of the most pressing concerns the moral responsibility in case a military robot uses violence in a way that would normally qualify as a war crime. In this chapter, the authors critically assess the chain of responsibility with respect to the deployment of both semi-autonomous and (learning) autonomous lethal military robots. They start by looking at military commanders because they are the ones with whom responsibility normally lies. The authors argue that this is typically still the case when lethal robots kill wrongly – even if these robots act autonomously. Nonetheless, they next look into the possible moral responsibility of the actors at the beginning and the end of the causal chain: those who design and manufacture armed military robots, and those who, far from the battlefield, remotely control them.

While some are reluctant to admit it, we are witnessing a fundamental shift in the way that advanced militaries conduct their core business of fighting. Increasingly autonomous 'unmanned' systems are taking on the 'dull, dirty and dangerous' roles in the military, leaving human war fighters to assume an oversight role or focus on what are often more cognitively demanding tasks. To this end, many military forces already hold unmanned systems that crawl, swim and fly, performing mine disposal, surveillance and more direct combat roles. Having found their way into the military force structure quite rapidly, especially in the United States, there has been extensive debate concerning the legality and ethicality of their use. These topics often converge, but what is legal will not necessarily be moral, and vice versa. The authors' contribution comes in clearly separating the two parts. In this paper, they provide a detailed survey of the legality of employing autonomous weapons systems in a military context.

This is the second paper of two on the role of autonomy in the unmanned systems revolution currently underway and affecting military forces around the globe. In the last paper, the authors considered the implications of autonomy on the legal obligations of military forces and their ability to meet these obligations, primarily through a survey of the domestic law of a number of drone wielding nations and relevant international legal regimes, including the law of armed conflict, arms control law, international human rights law, and others. However, the impact of autonomy in the military context extends well beyond the law and also encompasses philosophy and morality. Therefore, this paper addresses perennial problems concerning autonomous systems and their impact on what justifies the initial resort to war, who may be legitimately targeted in warfare, the collateral effects of military weaponry and the methods of determining and dealing with violations of the laws of just war theory.

The United States is now relying on Reaper and Predator drone strikes as its primary strategy in the continuing War on Terrorism. This paper argues for the rational scrutiny drone warfare has yet to receive. It is argued that drone warfare is immoral as it fails both the jus in bello and the jus ad bellum conditions of Just War theory. Drone warfare cannot be accepted on utilitarian grounds either, as it is very probable that terrorists will acquire drones capable of lethal strikes and deploy them against defenseless civilians. Moreover, by examining the psychological bases for reliance on drone warfare, as well as the message the United States is sending adversaries, we need to be concerned that, rather than reduce the likelihood of terrorists strikes, the U.S. reliance on drones strikes threatens to institutionalize terrorism as the status quo for the foreseeable future.

Technology has always allowed agents of war to separate themselves from the harm that they or their armed forces inflict, with spears, bows and arrows, trebuchets, cannons, firearms and other modern weaponry, all serving as examples of technologies that have increased the distance between belligerents and supposedly made warfare less sickening than the close-quarters combat of the past. However, this paper calls into question the extent to which new military technologies actually mitigate the savagery of war. It contends that with the introduction of technologies that eliminate the need for a human presence on the battlefield, we are the cusp of a major revolution in warfare that presents new challenges and questions for military technoethics, namely as to how soldiers should conduct themselves and fight justly, if they are to do so at all. Ultimately, it argues that only way to address these issues is through the design of the mediating technologies themselves, which is by no means an easy task.

We are at the cusp of a revolution in the development of autonomous weapons, yet current arguments both for and against such weapons are insufficient to the task at hand. In the context of Just war theory, arguments for and against the use of autonomous weapons focus on Jus in bello and in doing so miss addressing the implications of these weapons for the two other aspects of that theory- Jus ad bellum and Jus post bellum. This paper argues that fully autonomous weapons would likely undermine adherence to the Jus ad bellum and Jus post bellum prescriptions of Just war theory, but remote controlled weapons, if designed with ethical concerns in mind, might improve adherence to all of the theory's prescriptions compared to war as currently waged from a distance, as well as help to undo the occlusion of violence which has been a fundamental characteristic of all forms of modern war.

In 2012, the U.S. Congress passed the FAA Revitalization and Reform Act, which among other provisions called for the integration of drones into the U.S. national airspace by September 2015. While the statutory provision was an attempt to meet emerging industry needs which includes the defense sector, Congress inadvertently failed to examine many of the potential problems relating to domestic drone integration. In spite of industry and government efforts to mitigate these problems, four primary issues exist for UAS integration: the FAA is behind schedule, inadequate rule making, inadequate threat analyses, and an incomplete drone categorization system. Each problem is discussed in detail along with proposed solutions.

This article describes how small unmanned aircraft systems (sUAS) are growing at a rapid pace. They are inexpensive and widely available for both hobbyist and commercial use. However, with this rapid growth, regulations are having a difficult time keeping pace to safely incorporate them into the United States National Airspace. Recent regulations requiring the registration of all sUAS have been overturned by the United States Courts of Appeals. This research provides a statistical analysis of the effectiveness of the registration regulation in the reduction of unauthorized and careless sUAS operation prior to being

Preface

We are living in an age of automation and as the result are seeing a steady decrease in the level of human activity, monitoring, and control behind a broad range of services, devices, and applications.

Autonomous aircraft, especially unmanned aerial vehicles (UAVs; also known as “drones”), are becoming increasingly popular, especially for government and recreational use. Whether for wedding videography and photography, data collection, or military use, there is a need for discourse pertaining to the adoption, policy, and ethics surrounding UAVs.

In order to provide a well-rounded overview of this technology, IGI Global is pleased to offer this one-volume all-inclusive reference that will empower aviation professionals, engineers, computer scientists, military administrators, government officials, researchers, practitioners, academicians, and students with a single reference source covering the conceptual, methodological, and technical aspects, and will provide insight into emerging topics including but not limited to intelligent drones, human aerobatic integration, decision support systems, and airspace regulation.

The chapters within this publication are sure to provide readers the tools necessary for further research and discovery in their respective industries and/or fields.

Unmanned Aerial Vehicles: Breakthroughs in Research and Practice is organized into seven sections that provide comprehensive coverage of important topics. The sections are:

1. Algorithms and Mathematical Models;
2. Control, Design, and Simulations;
3. Decision Support Systems;
4. Detection, Imaging, and Sensor Technologies;
5. Government and Public Welfare Use;
6. Military Use and Ethics; and
7. Regulations and Privacy.

The following paragraphs provide a summary of what to expect from this invaluable reference source:

Section 1, “Algorithms and Mathematical Models,” opens this extensive reference source by highlighting the latest trends in drone modeling and control techniques. The first chapter in this section, “Intelligent Drones Improved With Algae Algorithm,” explores the importance of optimal flight routes of various aircrafts in improving natural disaster responses. The second chapter in this section, “Dynamic Modeling and Control Techniques for a Quadrotor,” presents the detailed dynamic model of a vertical take-off and landing (VTOL) type unmanned aerial vehicle (UAV) known as the quadrotor and the algorithms developed to control the quadrotor’s freedom. The final chapter in this section, “Cockroach-Inspired

Shelter-Seeking for Holonomic Swarms of Flying Robots," investigates how groups of holonomic flying robots such as quadcopters can seek shelter autonomously when encountering bad weather.

Section 2, "Control, Design, and Simulations," includes chapters on flying ad-hoc networks, aerobotic interaction, and gesture controllers. The first chapter in this section, "Cross-Layer Scheme for Meeting QoS Requirements of Flying Ad-Hoc Networks: QoS Requirements of Flying Ad-Hoc Networks," addresses the important communication requirements needed when designing flying ad-hoc networks to be highly reliable and have low latency. The following chapter, "A Perceptual Computing-Based Gesture-Controlled Quadcopter for Visual Tracking and Transportation," presents unmanned aerial vehicle design implementation by discussing the various sub-systems involved, intersystem communication, and field tests to ascertain operation to avoid vehicle collisions and promote intersystem communication. The concluding chapter, "Multimodal Human Aerobotic Interaction," discusses HCI interfaces used in controlling aerial robotic systems and the autonomy control level of aerobot.

Section 3, "Decision Support Systems," presents coverage on intelligent pilots and pilot support systems. The first chapter in this section, "Intelligent Fighter Pilot Support for Distributed Unmanned and Manned Decision Making," highlights important aspects of an intelligent fighter pilot support for distributed unmanned and manned decision making. The final chapter in this section, "Applications of Decision Support Systems in Sociotechnical Systems," presents conceptual models of decision support systems (DSS) and expert systems (ES) for human-operator (H-O) of air navigation system (ANS), such as air traffic controller (ATC) in flight emergencies, unmanned aerial vehicles (UAV) operator, and safety management system (SMS) while analyzing the International Civil Aviation Organization.

Section 4, "Detection, Imaging, and Sensor Technologies," discusses coverage and research perspectives on multispectral imagers and sensor technologies for effective drone action. The first chapter in this section, "Skin Detection With Small Unmanned Aerial Systems by Integration of Area Scan Multispectral Imagers and Factors Affecting their Design and Operation," examines the challenges skin detection from an aerial platform poses compared to ground-based platforms through constructing and testing small unmanned aerial systems. The second chapter in this section, "Compute-Efficient Geo-Localization of Targets From UAV Videos: Real-Time Processing in Unknown Territory," contributes to the computation of location of target anywhere in the image even at larger slant ranges, optimized algorithm to compute terrain elevation at target point, and the use of a visual simulation tool to validate the unmanned aerial vehicle. The following chapter, "Alternative Methods for Developing and Assessing the Accuracy of UAV-Derived DEMs," investigates the utility of UAV acquired imagery for generating high resolution terrain models and provides a detailed accuracy assessment according to recommended protocols. Another chapter in this section, "A Practical UAV Remote Sensing Methodology to Generate Multispectral Orthophotos for Vineyards: Estimation of Spectral Reflectance Using Compact Digital Cameras," explores the use of compact digital cameras to remotely estimate spectral reflectance based on unmanned aerial vehicle imagery. One of the concluding chapters, "The Role of Affective Computing for Improving Situation Awareness in Unmanned Aerial Vehicle Operations," presents the case for how a patented invention can be used to reduce civilian casualties through attaching an affect recognition sensor to a UAV that uses a database of strategies, tactics, and commands to better instruct fighter pilots on how to respond while in combat so as to avoid misinterpreting civilians as combatants. The last chapter in this section, "Chemical Plume Tracing and Mapping via Swarm Robots," addresses the key issues of chemical plume mapping and tracing via swarm robots.

Section 5, "Government and Public Welfare Use," explores the impact unmanned aerial vehicles and drones may have on society at large and government uses. The first chapter of this section, "A Centralized Autonomous System of Cooperation for UAVs-Monitoring and USVs-Cleaning," presents a cooperative agent-based planning approach for heterogeneous unmanned vehicles with different roles. Unmanned aerial vehicles (UAVs) monitor multiple ocean regions and unmanned surface vehicles (USVs) tackle the cleaning of dirty zones. The last chapter of this section, "Models for Drone Delivery of Medications and Other Healthcare Items," describes how a healthcare delivery drone has the potential for developing countries to leapfrog the development of traditional transportation infrastructure.

Section 6, "Military Use and Ethics," covers the various applications of drones in military endeavors and the impact they may have on target outcomes. The initial chapter in this section, "Lethal Military Robots: Who Is Responsible When Things Go Wrong?" assesses the chain of responsibility with respect to the deployment of both semi-autonomous and (learning) autonomous lethal military robots. The second chapter, "Autonomous Systems in a Military Context (Part 1): A Survey of the Legal Issues," discusses the ethical and legal concerns surrounding the use of unmanned systems that crawl, swim, fly, perform mine dispersal, and survey in military contexts. The following chapter, "Autonomous Systems in a Military Context," addresses perennial problems concerning autonomous systems and their impact on what justifies the initial resort to war, who may be legitimately targeted in warfare, the collateral effects of military weaponry and the methods of determining and dealing with violations of the laws of just war theory. Another noteworthy chapter in this section, "Drone Warfare: Ethical and Psychological Issues," argues for the rational scrutiny drone warfare has yet to receive by examining psychological bases for reliance on drone warfare. One of the final chapters of this section, "War 2.0: Drones, Distance, and Death," calls into question the extent to which new military technologies actually mitigate the savagery of war. The concluding chapter, "Robots in Warfare and the Occultation of the Existential Nature of Violence," argues that fully autonomous weapons would likely undermine adherence to the Jus ad bellum and Jus post bellum prescriptions of Just war theory, but remote controlled weapons, if designed with ethical concerns in mind, might improve adherence to all of the theory's prescriptions compared to war as currently waged from a distance, as well as help to undo the occlusion of violence which has been a fundamental characteristic of all forms of modern war.

Section 7, "Regulations and Privacy," explores the ramifications of drone surveillance on the public at large. The first chapter in this section, "Drones in the U.S. National Airspace System," discusses the four primary issues existing for UAS integration: the FAA is behind schedule, inadequate rule making, inadequate threat analyses, and an incomplete drone categorization system. The second chapter in this section, "An Analysis of Unmanned Aircraft Registration Effectiveness," provides a statistical analysis of the effectiveness of the registration regulation in the reduction of unauthorized and careless small unmanned aircraft systems (sUAS) operation prior to being overturned by the courts. The final chapter in this section, "Drones and Privacy," investigates drone and UAV technology capabilities and how they could and are currently affecting privacy laws globally in comparison to those currently in the Republic of Ireland.

Although the primary organization of the contents in this work is based on its seven sections, offering a progression of coverage of the important concepts, methodologies, technologies, applications, social issues, and emerging trends, the reader can also identify specific contents by utilizing the extensive indexing system listed at the end.

Section 1

Algorithms and Mathematical Models

Chapter 1
Intelligent Drones Improved With Algae Algorithm

Alberto Ochoa
Juarez City University, Mexico

Tania Olivier
Juarez City University, Mexico

Raymundo Camarena
Juarez City University, Mexico

Guadalupe Gutiérrez
Universidad Politécnica de Aguascalientes, Mexico

Daniel Axpeitia
Juarez City University, Mexico

Irving Vázque
Juarez City University, Mexico

ABSTRACT

Implement an optimal arrangement of equipment, instrumentation and medical personnel based on the weight and balance of the aircraft and transfer humanitarian aid in a Drone, by implementing artificial intelligence algorithms. This due to the problems presented by geographical layout of human settlements in southeast of the state of Chihuahua. The importance of this research is to understand from a Multivariable optimization associated with the path of a group of airplanes associated with different kind of aerial improve the evaluate flooding and send medical support and goods to determine the optimal flight route involve speed, storage and travel resources for determining the cost benefit have partnered with a travel plan to rescue people, which has as principal basis the orography airstrip restriction, although this problem has been studied on several occasions by the literature failed to establish by supporting ubiquitous computing for interacting with the various values associated with the achievement of the group of drones and their cost-benefit of each issue of the company and comparing their individual trips for the rest of group. There are several factors that can influence in the achievement of a group of Drones for our research we propose to use a Bioinspired Algorithm.

DOI: 10.4018/978-1-5225-8365-3.ch001

INTRODUCTION

The Cessna 208 Caravan, also known as Cargo master, is a regional jet / turboprop short-range utility manufactured in the United States by the company Cessna. The standard version has 10 places (9 passengers and a pilot), although a subsequent design according to new regulations of the Federal Aviation Administration (FAA) can carry up to 14 passengers. The aircraft is also widely used to make connections in freight services, so that goods arrive at smaller airports are transported to major hubs for distribution as in Figure 1.

The concept of the Cessna 208 appeared in early 1980, the first prototype flew on 8 December 1982. After two years of testing and review, in October 1984, the FAA certified the model for flight. Since then, the Caravan has experienced many evolutions. Hand international logistics company FedEx; Cessna produced first the Cargo master, which was followed by an improved and extended version called Super Cargo master and other passengers called Grand Caravan. Practitioners will be free fall boarding a Cessna 208 in the Dutch island of Texel. Currently Cessna 208B offers different configurations to meet the varied market demand. The core 208 can be supplemented with different types of landing gear and can operate in a variety of terrains. Some adaptations include skis, larger tires for unprepared runways or floats with wheels in the case of the Caravan Amphibian. In cabin seats, can be placed or leave room for cargo in various configurations. The standard configuration of airline consists of 4 rows of seats 1-2 after two seats in the cockpit. This variant is capable of carrying up to 13 passengers; albeit only lead to 4 longer an operation rentable 1. The cabin can also be configured to a low density of passengers, in combination or alone as a freighter. Some versions include an additional compartment at the bottom to increase the capacity or luggage. In the cockpit, the 208 has standard analog gauges with some modern digital avionics equipped with autopilot and GPS, modern radio and transponder. Cessna currently offers two different packages avionics manufacturers, one of Garmin and another Bendix / King, a subsidiary of Honeywell. Routing problems vehicle (Vehicle Routing Problem - VRP) are actually a broad range of variants and customizations problems. From those that is simplest to some that remain today research as in Barbucha (2013). They generally were trying to figure out the routes of a transportation fleet to service a customer, nowadays including aerial transportation. This type of problem belongs to the combinatorial optimization problems. In the scientific literature, Dantzig and Ramser were the first authors

Figure 1. Type of drone used to our research in Southwestern Chihuahua

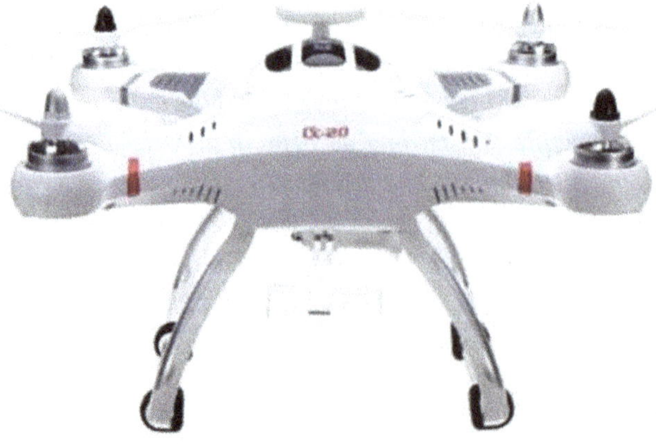

in 1959, when they studied the actual application in the distribution of gasoline for fuel stations. The objective function depends on the type and characteristics of the problem. The most common is to try: minimize the total cost of ownership, minimize total transportation time, minimize the total distance traveled, minimize waiting time, maximize profit, maximize customer service, and minimize the use of vehicles, balance of the resource utilization.

Another consideration to this problem:

Optimization of Spaces Associated With Medical Emergencies for Patients Without Mobility in a Cessna 208

The southeastern state of Chihuahua is an area whose geographical characteristics confer greater difficulty in moving the terrestrial inter hospitalary for that reason is to develop an intelligent system that can consider the transfer of patients using aircraft adapted for it.

The Cessna 208 is a short-range regional aircraft turboprop that meets the expectations routes to cover the landing strips of wood, Batopilas, Temoris, Balleza Morelos and all these with connection to the airport in Hidalgo del Parral, estimated to last approximately 45 minutes.

The planes that travel the state of Chihuahua in time and date will not cause any problems regarding the accumulation of air traffic in that area so it is contemplating that. It is determining the departures and arrivals of each Cessna aircraft through logistics.

Identifying and Optimizing Response Times for Aerial Medical Emergencies for Climbers in Southwestern Chihuahua

Today, technological advances have allowed new devices are at our disposal and make use of them for commercial or personal purposes.

Among them is the Dron which is a device able to fly and reach heights that human needs for tasks such as multimedia Taking pictures and videos.

Figure 2. Views fuselage dimensions associated with Cessna 208

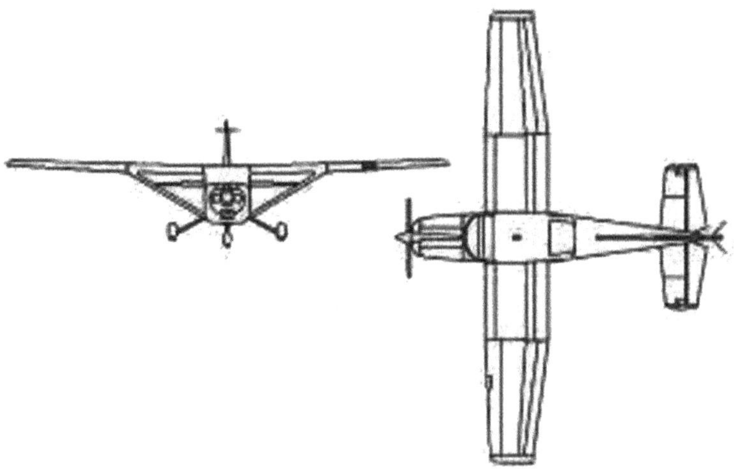

The proposal is focused on how we can use these devices to search for people in need or in distress and can go to various support centers more quickly and effectively.

As part of Cisco initiative under the competition Dron, we develop how to apply for technological purposes. The main idea of this project is based on the use of the same for searches of people in situations of natural disasters or accidents in inaccessible areas such as the mountains of the country, forests and jungles where access by any means of transport is difficult and it takes longer.

As statistics in southern Chihuahua an accident of this type only the transfer of the person to an urban area takes about an hour, this presents a time or a major brand if it is a risk that threatens occurs against life

Originality

Currently the use of drones is primarily for panoramic photographs, personal recreational use.

The originality of this project lies in the fact of taking as a starting point the features and capabilities of the Drones and take them to an area of difficult access where current means of transportation are limited.

Thus the use of these technologies would impact significantly on search methods support brigades such as civil protection and the relevant state government considering that this type of environment complicates its location and signal decreases.

Benefit

Profit and main objective of this proposal is efficiently and quickly meet in accidents in natural areas as follows:

- Saving Lives
- Improve time search and identification of people at risk
- Optimize search and rescue operations.
- Recognition of land
- To minimize as much as possible the use of rescue personnel.

Figure 3. Implement the study regions

Figure 4. One of the areas where the draft Aeronautical Logistics apply to medical emergencies

The priority is to safeguard the integrity and welfare of a person in a situation of risk as quickly as possible.

The viability lies mainly in the fact that this technology is within reach and, tested and currently in use, which is why it is intended to implement such artifacts in a focused prone community around, suffer any type of natural disaster in your area.

The facts that this initiative is implemented as such involve limited financial resources and that all he is concerned as such is the device and the operating personnel.

Figure 5. The priority is to safeguard human lives in alpine rescues

Analysis of Data

The process of data analysis is first to launch the drone in the nearest part where he last saw the person, taking that as a starting point then the operator would be installed to search starting with the closest areas and gradually move forward walking along the area. The drone activates your camera at all times wirelessly transmit the video which will be seen by the operator in progress. If the main battery is exhausted, enter in seconds the spare battery to enhance the autonomy Travel Drone. Once the person is located at the current coordinates are taken to rescue personnel to approach and take presence.

Likewise, the Dron guide you to the destination that is the person to rescue.

The drone that is to implement a tool that would represent and again indispensable important because it increases the organization, monitoring and response times for rescue either air or ground.

Simply save and secure something as valuable as human life gives this project the importance of implementing drones in risk areas. The cost of drone rises above 60,000 pesos and this considering the implementation of the best technologies that mind current account market.

This project would bring great benefit of technology and customer growth considering that the loss of a rescuer represent a major negative economic impact, which can be avoided with the implementation of this technology.

Determine Charges

The charges are the distances that require conveys travel from the airstrip, to the point where there is demand and return to the airstrip of origin. What you have to take into account a factor: the rise, which will cause the vehicle height change and help (with decreasing elevation) or give more load (with increasing elevation).

To simplify the determination of the load with altitude by this is considering the altitude start and final altitude and each way between two points is considered. A simple formulation was used to take the load:

$$c = r_{horizontal} * \left(1 + \frac{h_f - h_i}{r_{horizontal}}\right).$$

where $r_{horizontal}$ is the horizontal path, h_f is the final altitude and h_i is the altitude at first time. This formulation is shown in Figure 4. If the final altitude is increased if the load increases and the load decreases is smaller.

Implementation of an Optimization Model for the Transfer of Organs in a Non-Commercial Aircraft

The importance of this research is the implementation organ in a non-commercial aircraft, it is necessary because in cities of Chihuahua distances are very long or very damaged roads to take organs and some have very little time preservation, as is the heart and lung. To preserve the organs associated hypothermia at 4 ° C for short distances is commonly used ambulance, while the aircraft is used for two hours above

Figure 6. Topography of Chihuahua

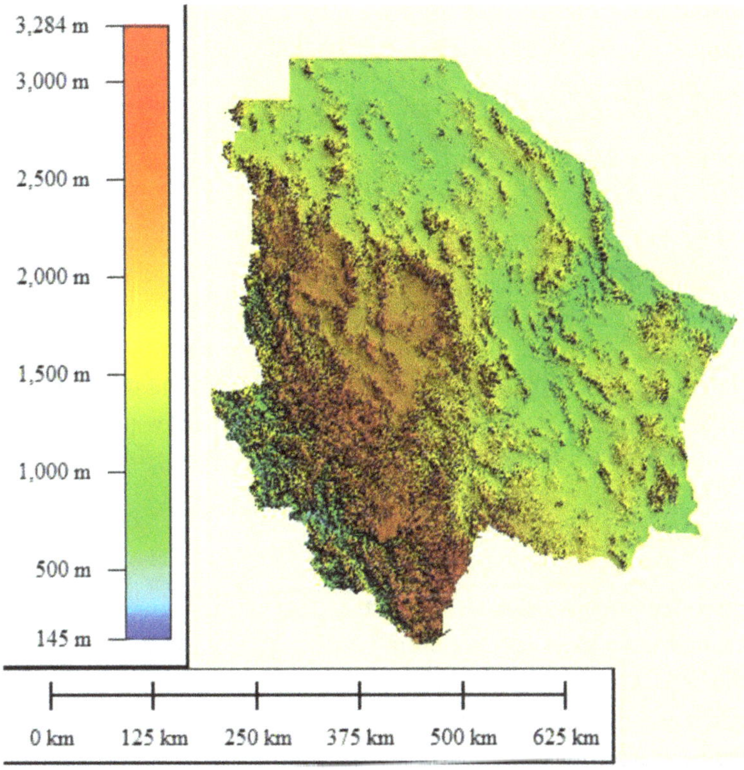

Figure 7. Graphically representation associated with charge

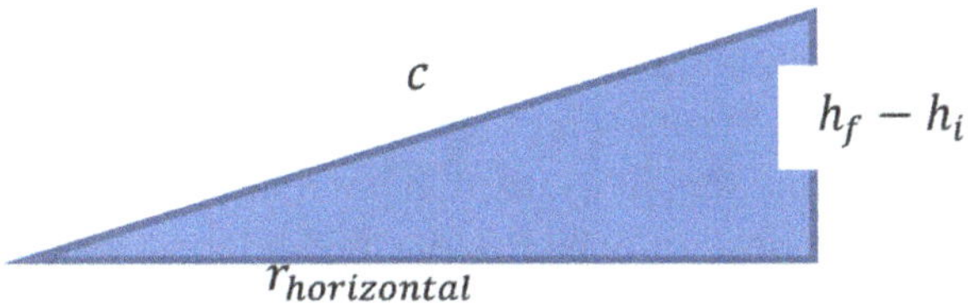

routes. It will approach the Cessna 208 Caravan, using container and determining optimal space and energy required for the cooling system inside the aircraft.

The type of aircraft to be used is a Cessna 208 Caravan turboprop aircraft used was manufactured by the United States, the traditional version has 10 seats, nine passengers and one pilot, later design can reach up to 14 passengers.

- Electric Power (ATA-24)
- Diesel GPU 28.5 (28.5 VDC, 800 continuous, 2,000 amp. peak) (CE)
- Power plant

The PT6A-114A turboprop Pratt & Whitney Canada engine is flat-rated 675 horsepower at 1,900 rpm shaft. Motor operation is controlled by the engine indication system, showing numerical readouts of engine fuel and critical electric signs for the pair, propeller speed, fuel flow and others. Torque meter was also installed pending wet type.

- **Manufacturer**: Pratt & Whitney Canada
- **Model:** (1) PT6A-114A
- **Output Power:** 675 shp
- **Propeller Manufacturer:** McCauley
- **Description:** 3 sheet metal, constant speed, full plumage

For the transportation of bodies shall be used hypothermic perfusion machines that are new tools for the conservation of the organ: the organ perfused constantly and allow at the same time, monitor and add perfusion liquid drug substances that will improve their viability.

Storage Organ Preservation and Transport

The organ preservation techniques used to reduce the damage, improve function and survival of the transplanted organ. Conditions may vary according to the type of organ. To preserve solid organs associated hypothermia at 4 ° C

Hypothermic preservation techniques most commonly used today are cold storage and preservation in hypothermic machine perfusion.

Preservation by Cold Storage (CF)

The most common preservation also the least expensive method, and is to perfuse the body internally or wash with cold preservation solution immediately after removal in the operating room itself. Later that preservation solution or the like to bathe and keep it stored in a refrigerator at 4 ° C until the time of implantation is used. For its extreme simplicity, the CF has a number of advantages, such as its almost universal availability and ease of transport, which is the method most used preservation.

Preservation Hypothermic Machine Perfusion (MPH)

The concern of the transplant to improve the quality of donated organs and the need for expanded criteria donors given the shortage of organs and the inability to assess the viability of organs before transplantation, has promoted in recent years developing methods better than the simple cold storage preservation.

The combination of continuous perfusion and hypothermic storage used by Belzer in 1967 represented a new paradigm in organ preservation, as it managed to successfully preserve canine kidneys for 72 h. According to this technique, after the initial washing is performed in the operating room during perfusion, the organ is introduced into a device that maintains a controlled flow continuously or pulsatile preservation solution of cold (0-4 ° C). This flow allows full organ perfusion and clean microthrombi the bloodstream and facilitate the removal of metabolic end products. Its beneficial effects are a lower incidence of delayed initial graft function, the possibility of evaluating in real time viability and the ability to provide metabolic support (oxygen and substrates) or drug during the infusion. The continuous flow MPH has shown advantages over the pulsatile flow.

Due to the size that occupies a hypothermic machine perfusion and the energy required is more feasible in a Cessna 208 Caravan carry out a cooler where

Transfer Process for the Transfer of Organs

When required a transplant over long distances and is now a donor in the hospital this is brought into contact with the National Organization of Transportation, which is responsible for setting for a short ambulance and a superior ride two hours the plane

If you are going to perform an organ transplant by a cooler this requires special labeling to indicate the type of organ and that patient should attend.

Large distances. In these cases, given the short cold ischemia time (time since being placed in transport solution time and the start of disinfection) that tolerate bodies, private aircraft aviation companies are hired and occasionally uses to aircraft of the Air Force. The preparation of a flight needs not less than 2 hours (check the aircraft, notify the crew, flight plan preparation, etc.), so it is important to communicate the existence of the donor to the ONT to most as soon as possible. Once hired the flight and the scheduled times, the coordinators of the hospitals involved is notified. Most domestic airports are not operating 24 hours, so the coordinator of the ONT must be taken into account because if not 24 hours will implement the mechanisms necessary for opening or to keep it operating after hours.

Possible incidents when transporting organs.

Project Development

To do this research project was developed by dividing into three sections which are modules of application development, implementation of the server and the intelligent module associated with Algae Algorithm and Data Mining. Android is the operating system that is growing in to Streak 5 from Dell, for this reason we select this mobile dispositive along with other manufacturers are propelling the Latin American landing on Android with inexpensive equipment, and on the other hand, some complain about the fragmentation of the platform due to the different versions. Android is free software, so any developer can download the SDK (development kit) that contains your API (Alejandro, 2011). This research tries

Figure 8. This image shows the distribution of passengers by special missions

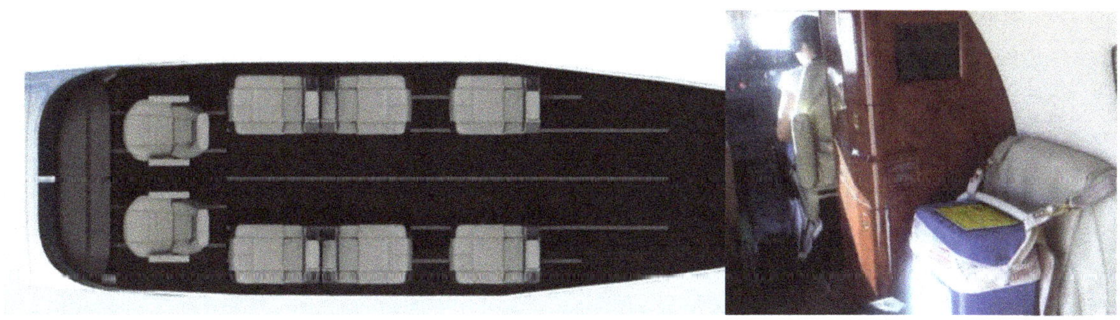

to improve group travel related with Recreational Vehicles in Chihuahua where 7500 people conforms the Caravan Range Community.

Components of the Application

Artificial algae algorithm AAA has been proposed by Uymaz et al. (2015). This algorithm is inspired by the living behaviors of microalgae. In the algorithm, population is composed of algal colonies. An algal colony is a group of algal cells living together. AAA is based on 3 basic parts called "evolutionary process", "adaptation" and "helical movement". In evolutionary process, if the algal colony receives enough light, algal cells in algal colonies grows and reproduces itself to generate two new algal cells similar to the real mitotic division. In adaptation process, Algal colonies, which cannot grow sufficiently in an environment, try to adapt itself to the environment. Helical movement is a process which update of algal colonies. In this process, only three algal cells of each algal colony are modified. The steps of AAA can be described as follows.

Step 1: Initialization.

1.1 Parameters of problems (number of dimension (D), maximum and minimum values for each dimension (UB, LB)).

1.2 Parameters of algorithm (Shear force (_), energy loss (e), adaptation (Ap)), population number (N) and maximum fitness evaluations number (MaxFES) are initialized.

1.3 Initialize algal colonies with random solutions. Define the size of each algal colony as (1). (1)

1.4 Evaluate fitness of each algal colony.

1.5 Evaluate size (G) of algal colonies. (2) where K is the substrate half saturation constant of the algal colony and at time t Ki will be equal to the half of Gi. (3)

1.6 Evaluate friction surface (K) of algal colonies (4)

1.7 Evaluate energy (E) of algal colonies from normalized G Et+1=norm((rank(Gt))2)

Step 2: Main section. This section is iterated until reach MaxFES.

2.1 Helical movement phase for every algal colony.

2.1.1. Select another colony with tournament selection.

2.1.2. Select three algal cells (k, l and m) in the colony randomly.

2.1.3. Modification the colony. (5) (6) (7)

2.1.4. Decrease energy caused by movement.

2.1.5. If new solution is better, move new position else decrease energy by metabolism.

2.1.6. If energy of the colony did not finish go to 2.1.1.

2.1.7. If colony did not find better solution increase starvation of colony.

2.2 Reproduction phase.

2.2.1 Select smallest and biggest colonies. (8) (9)

2.2.2 Select randomly algal cell (m).

2.2.3 Algal cell is replicated from biggest to smallest. (10)

2.3 Adaptation phase.

2.3.1 Select most hungry colony. (11)

2.3.2 Modification the colony.

Step 3: Finally, report best solution.

Modified artificial algae algorithm is a vanguards proposal to increase the opportunity that obtain found adequately the best option in a search space.

Artificial algae algorithm with multi-light source movement (AAAML)In AAA, exploration and exploitation is provided with a helical movement, in other words, with the modification of the algal colonies. The evolutionary process and adaptation process help exploitation. Balance of local and global search ability is important in the solution of optimization problems. If the distribution of

xij = LBj + (UBj − LBj) × Rand i = 1, . . ., N; j = 1, . . ., D _i = S Ki + S Gt+1 i

= Gti + (Gti × _i) _(xi) = 2_ _ 3 _3Gi 4_ _2 xt+1 im

= xtim + (xtjm − xtim)(_ − _t (xi))p xt+1 ik

= xtik + (xtjk − xtik)(_ − _t (xi))cos ˛ xt+1 il = xtil + (xtjl − xtil)(_ − _t (xi)) sin ˇ biggest t

= max Gti i = 1, 2, . . ., N smallest t = min Gti i = 1, 2, . . ., N smallest tm

= biggest m m = 1, 2, . . ., D starving t = max Atii = 1, 2, . . ., N starvingt+1

= starving t + (biggest t − starving t) × rand (12)

Figure 9. Flowchart of modified AAA

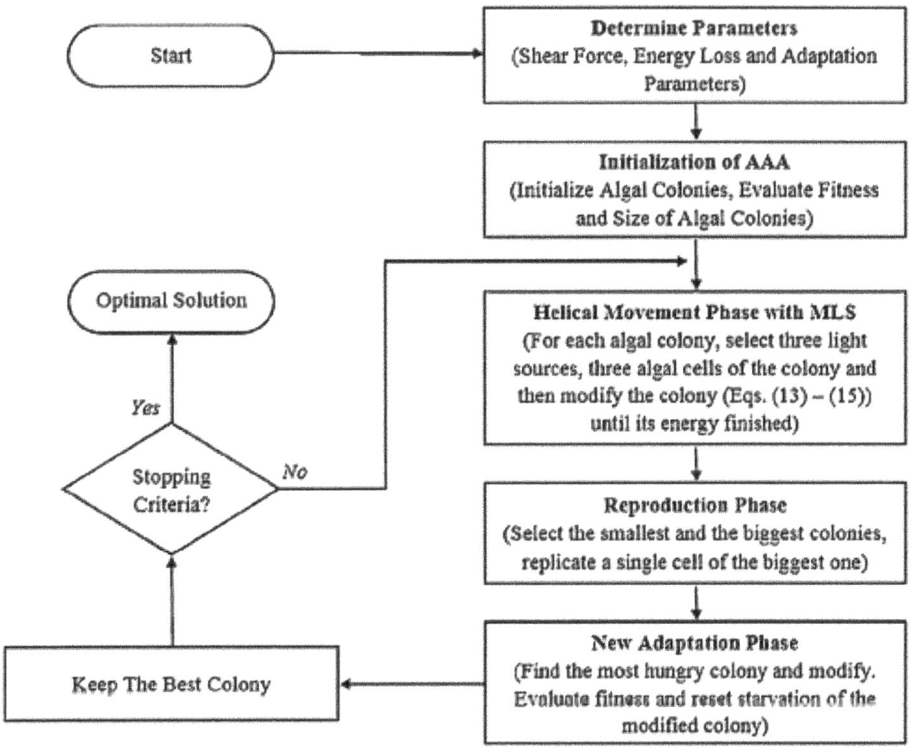

Algal colonies in the search space is too wide, the global search ability increases and the speed of convergence to the global optimum decreases (Elsayed et al., 2011). On the contrary, if this distribution is too narrow, the local search ability increases and premature convergence occurs (Elsayed et al., 2011). Thus, the balance between local and global search capability of the algorithm is disrupted. To create a stronger balance, the selection of a different light source for each dimension that is modified with the helical movement is proposed (Figure 9). These new helical movement equations are as follows: (13) (14) (15) where

i = 1, 2, . . ., N, i /= j /= r /= v; m, k, l = 1, 2, . . ., D, m /= k /= l;˛, ˇ ∈ [0, 2_]; p ∈ [−1, 1]; _

is shear force; _t(x, j, r, v) is the friction surface area of ith algal cell. Equations related with Algae Algorithm have been adapted from multi-parent recombination studies used in evolutionary algorithms (Tsutsui et al., 1999; Elfeky et al., 2008; Kita et al., 1999; Elsayed et al., 2011). Light sources used are selected by tournament method and each light source is different from each other. This presents some solutions that are different from each other in the search space. An orientation is provided toward to the better region of search space with the one of three light sources of which has the closest optimum value in the modified dimension. Also, the diversity in the source space is obtained with the worst light source. In addition, the other light source increases the balance. This approach can be called as multi-light source movement (MLS) or multi-light source helical movement like multi-parent recombination in evolutionary algorithms.

New adaptation operator Adaptation process helps algal colony which has maximum starvation level go to a better part of the search space. In AAA, the starvation levels of each algal colony are kept and these starvation levels of algal colonies which could not afford good solutions increase. With this new proposed adaptation operator, the starvation level of algal colony which undergoes adaptation process is reset and more local search is provided at the part of search space where it goes. In addition, the starvation level of algal colony which goes to the better region in search space with helical movement is reset. With this new approach, with the difference of pure AAA, the starvation levels of algal colonies indicate the starvation level in the region where they are.

The steps of new adaptation process are organized to improve the result.

1. Find the hungriest colony (same with AAA).
2. Modify the colony (best colony is used as the biggest; same with AAA).
3. Evaluate the modified colony.
4. Reset starvation of the modified colony.

xt+1 im = xtjm + (xtrm − xtvm) (_ − _t (xj)) p xt+1 ik

= xtrk + (xtvk − xtjk)(_ − _t (xr)) cos ˛ xt+1 il = xtvl + (xtjl − xtrl)(_ − _t (xv)) sin ˇ starving t

= max Atii = 1, 2, . . ., N starvingt+1 = starvingt + _biggestt – starving t _ × rand

IMPLEMENTATION OF AN INTELIGENT APPLICATION

When designing an interface for mobile devices has to take into account that the space is very small screen, plus there are many resolutions and screen sizes so it is necessary to design an interface that suits most devices. This module explains how to work with different layout provided by the Android API. The programming interface is through XML. Obtaining the geographical position of a device can be made by different suppliers; the most commonly used in this project through GPS and using access points (Wi-Fi) nearby, and perform the same action but differ in some as accuracy, speed and resource consumption. Data Server Communication is the module most important because it allows communication with the server, allowing you to send the GPS position obtained by receiving the processed image and map of our location, thus showing the outcome of your application that is the indicator of insecurity. To communicate to a server requires a HTTP client which can send parameters and to establish a connection using TCP / IP, client for HTTP, can access to any server or service as this is able to get response from the server and interpreted by a stream of data. The Android SDK has two classes with which we can achieve this, HttpClient and HttpPost. With the class HttpClient is done to connect to a remote server, it needs HttpPost class will have the URI or URL of the remote server. This method receives a URL as a parameter and using classes HttpPost HttpClient and the result is obtained and received from the server, in this specific case is only text, which can be JSON or XML format. Here, the server responds with a JSON object which would give the indicator is then used to create the map, as is shown in the Figure 10.

It intercepts the reconnaissance or current conditions facilitating the accident rescue work in inhospitable places where man himself would not, achieving save hundreds of lives in both rescue and prevention.

What kind of connection generated?

- **Processes:** The current process in contrast to this innovation represents a waste of time, human capital and financial resources as well as transportation and staff
- **People:** It generates confidence in people to know that there are more efficient and technologies at the forefront methods.

Figure 10. Intelligent Tool recommend a group travel associated with limited resources and optimize energy (oil), time and comfort

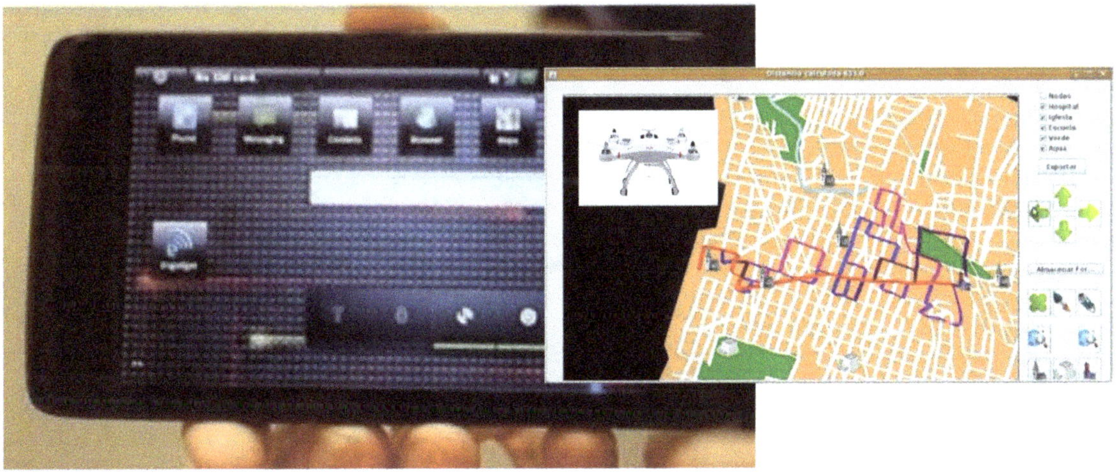

- **Facts:** Quantifiable data such as geo-spatial location of human lives, weather conditions in the region and its longitude, latitude, quantification of people in case of natural disasters, facilitating better planning for civilian protection.
    - Technical description of this solution.
    - It seeks to develop a Drone (INSERT PATTERN) which will have two batteries Nano Tech 4.0 (specified), will alternate means of an electronic circuit and recharging the battery out of use, with solar being self-sustaining.
    - Assuming a lifetime of flight, and data about 2 hours.
    - It will provide environmental data to facilitate human search and reconnaissance, as are the following.
    - Live streaming or GoPro HD Live broadcast, for reconnaissance.
- **Air Pressure:** Or silicon MXP2010DP transducer measuring air model provides linear analog voltage that varies according to the pressure, this will provide the operator with a wind analysis, giving estimated rescue if necessary at high altitudes, such parachute in rocky areas in another.
- **Thermal Vision:** Or in situations where visibility is a problem, for example night environments, or is difficult to implement an alternate vision camera, THERMAL-EYE 36000AS model.
- **GPS Triangulation:** GPS location or returning the location of rescue human lives, using a sensor Dualav xGPS-150A transmission triangulation and GPS location.

To implement the application is installed in operating system devices with Android 2.2 or higher, which tests the system in different areas of the three different parks and natural reserve in Chihuahua based on the previously research related with Cultural Algorithms on Urban Transport (Cruz et al, 2010), by answering a questionnaire of seven questions to all users related with the Caravan Range Community

Figure 11. Statistics graphics related with the solution proposed by algae algorithm

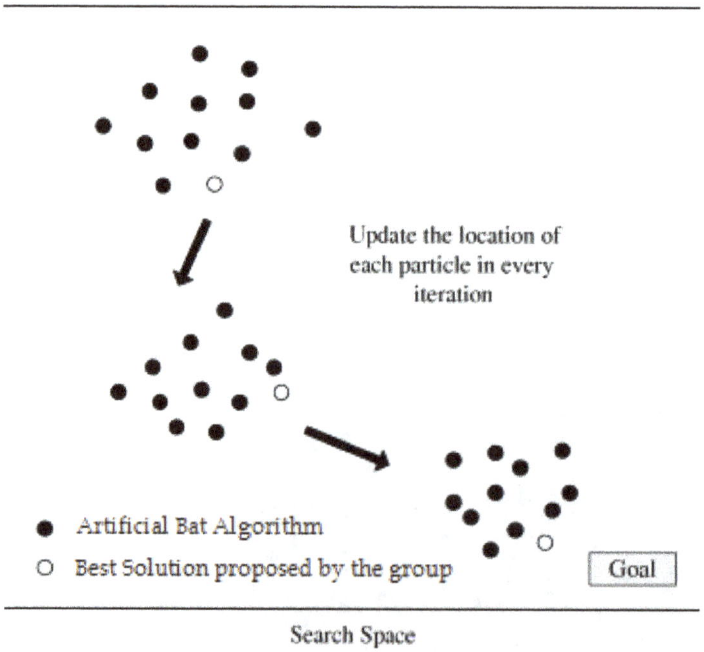

have elapsed since installing the application, the questions are to raise awareness of the performance, functionality and usability of the system, the demonstrate use of this application is shown in Figure 12. To understand in an adequate way the functionality of this Intelligent Tool, we proposed evaluate our hybrid approach and compare with only data mining analysis and random select activities to protect in the city, we analyze this information based on the unit named "époques" used in Algae Algorithm Algorithms, which is a variable time to determine if exist a change in the proposed solution according at different situation of different routes with better use of restricted resources.

We consider different scenarios to analyze during different time, as is possible see in the Figure 13, and apply a questionnaire to a sample of users to decide search a specific number of places to travel, when the users receive information of another past travels (Data Mining Analysis) try to improve their space of solution but when we send solutions amalgam our proposal with an Algae Algorithm and Data Mining was possible determine solutions to improve the resources of the group, the use of our proposal solution improve in 91.72% against a randomly action and 14.56% against only use Data Mining analysis the possibilities of recommend arrive in a specific time to use less energy, these messages permits in the future decrease the possibility of deplete the supply of food and spend time rerouted in an orography terrain with more obstacles and uncertainty of the weather conditions and the use of limited resources.

Experimentation

In order to be able similar, the most efficient arrangement of individuals in a social network, we developed an atmosphere able to store the data of each one of the representing individuals of each society, this with the purpose of distributing of an optimal form to each one of the evaluated societies. One of the most interesting characteristics observed in this experiment was the diversity of the cultural patterns established by each community. The scenes structured associated with the agents cannot be reproduced in general, since they only represent a little while dice in the space and time of the different societies. These represent a unique form and innovating of adaptive behavior which solves a computational problem that it does not try to clustering the societies only with a factor associated with his external appearance (attributes of each society), trying to solve a computational problem that involves a complex change

Figure 12. Hybrid intelligent application based on algae algorithm and data mining

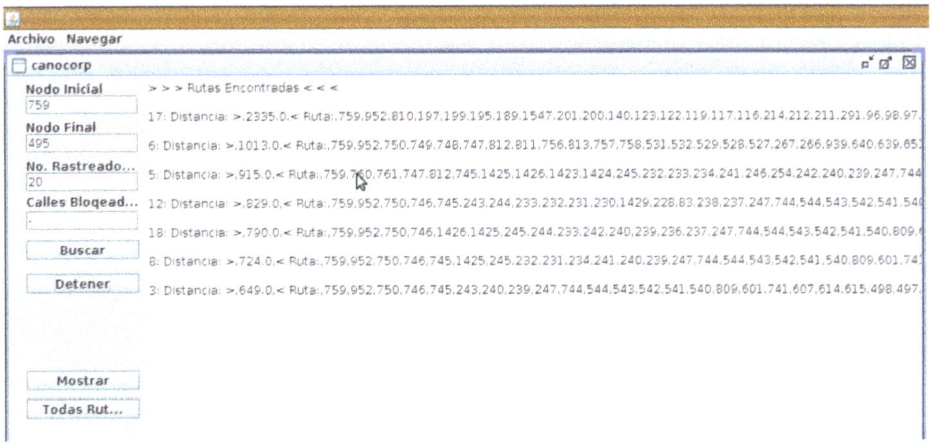

Figure 13. Solutions proposed to each minority group in Chihuahua considering the behavior under floods

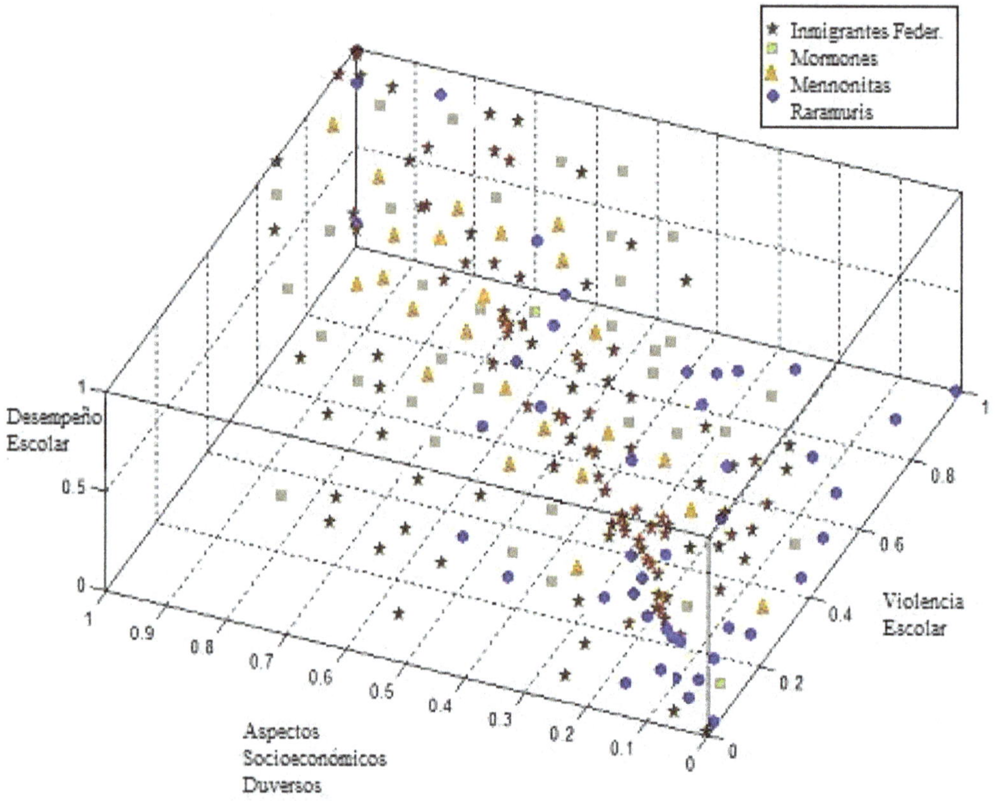

between the existing relations. The generated configurations can be metaphorically related to the knowledge of the behavior of the community with respect to an optimization problem (to select culturally 47 similar societies, without being of the same quadrant (Yang, 2013). The main experiment consisted of detailing each one of the 1087 communities, with 500 agents, and one condition of unemployment of 50 époques, this allowed us to generate the best selection of each Quadrant and their possible location in a Diorama, which was obtained after comparing the different cultural and social similarities from each community, and to evaluate with Multiple Matching Model each one of them (Ochoa, Garci, & Yanez, 2011). The developed tool classified each one of the societies pertaining to each quadrant, with different wardrobe for societies that included linguistic identity and for societies only with cultural identity, this permit identifies changes in the time respect at other societies (To see Figure 4).

The design of the experiment consists in an orthogonal array test, with the interactions between the variables: emotional control, ability to fight, intelligence, agility, force, resistance, social leadership, and speed. These variables are studied in a range of color (1 to 64).

The orthogonal array is L-N(2**8), in other words, 8 factors in N executions, N is defined by the combination of possible values of the 8 variables an the possible range of color (To see Figure 14).

Figure 14. Orthogonal array

Variable Value								
A	**B**	**C**	**D**	**E**	**F**	**G**	**H**	**Color**
H	H	H	H	H	H	H	L	1
H	H	H	H	H	H	L	H	2
H	H	H	H	H	L	H	H	3
H	H	H	H	L	H	H	H	3
...	...	...	...	...	...	...	...	...

Experiment Design for the Developed Tool

Although CAs dates back to the 90s, their use has shown erratic behavior, and only a little more than 40 publications of this topic.

As there is no Benchmarking of tests, the researchers in the subject have developed their own DOEs according to what is intended to be done. For this reason we consider more closely to modelling our AAA.

Ochoa et al., 2010 (ASNA'2010, Switzerland) propose a CA capable of realizing the successful accommodation of 1047 Societies in a Mega Diorama, each of them possessing up to 47 different attributes that make them unique with respect to the others, to And based on the existing literature it was proposed to use a Model of 500 individuals and 27 Communities to carry out the Vote-Inherit-Promote and a condition of unemployment of 50 epochs was established.

Based on the literature, this kind of parameters associated with the development of individuals. The intention was to tune the AAA, it was decided to perform 500 experiments each with the intention of achieving the most successful accommodation, it should be noted that there was no more than 100 objects in the literature using this evolutionary algorithm.

Once the 500 experiments were carried out, it was found that only 22.4% of them were over 80% of the existing accommodations, and the value with the greatest reach was 84.7% of the 1047 Companies in the Mega Diorama, the rest of the societies Were accommodated using a similarity function and following a process of Restrictions Propagation.

The computational cost was excessive since the ACs used up to 7 hours to reach the 50th period, this can be explained due to the complexity of the proposed VIP.

The interpretation that is determined is that the AAA is viable for an Application Domain such as Museography, where other techniques such as PSO, ACO or Forage Bacteria have not been useful. If the objects are the same and their accommodation is based on restrictions of size, color and shape the PSO and Forage Bacteria present a better performance.

CONCLUSION

With this work the implementation of drones is looking to optimize and monitor rescues people on land or dangerous situations. For public servants as rescue workers, Red Cross and civil protection is a challenge to locate and remove people in a natural disaster or accident because the field is unknown and no one has an aerial perspective that does not involve the risk of loss of life or materials. The drone to be constructed satisfies multiple needs in a single device with the different tools that mind current account. This can make aerial reconnaissance of the area and the terrain in real time, current mind is possible to use high-definition digital cameras to take and transmit video, so the drone can locate people who are at risk and save your life or their lives in the case of multiple affected by monitoring the ground support equipment for medical care and immediate removal.

Drafts and uneven ground may affect the operation of a helicopter. In most cases estimated conditions are taken, and operate in areas of collapse, hurricanes or uneven ground pose a risk to the operation of a helicopter aboard lives doing very risky air support. It is possible to monitor environmental conditions with different devices and sensors that are on the market, the digital technology allows devices are more accurate and light, then you might census a particular area to check is navigable air space to attempt an air extraction. It seeks to achieve the implementation of these tools and seek union of the same to reduce costs and not exceed the weight limit of operation has the drone, however the implementation of these devices present mind is used in situations where the use of drones can make a big difference in saving lives and synchronization of current technologies to people who need them.

ACKNOWLEDGMENT

The authors were supported with funds from Social Research Center in Juarez City University and used data from a department specialized on disasters in Chihuahua whom permits compare the simulation with real travels realized by them. The Dell mobile device was bought with funds from a Prodep project supported by SEP.

REFERENCES

Andreu Ropero, A. (2011). *Estudio del desarrollo de aplicaciones RA para Android. In Trabajo de fin de Carrera*. Catalunya: España.

Barbucha, D. (2013). Experimental Study of the Population Parameters Settings in Cooperative Multi-agent System Solving Instances of the VRP. *Transactions on Computational Collective intelligence*, *9*, 1-28.

Glass, S., Vallipuram, M., & Portmann, M. (2010). The Insecurity of Time-of-Arrival Distance-Ranging in IEEE 802.11 Wireless Networks. In Proceedings of the 2010 IEEE 30th International Conference on Distributed Computing Systems Workshops (ICDCSW) (pp. 227–233).

Griffin, D. R., Webster, F. A., & Michael, C. R. (1960). The echolocation of flying insects by bats. *Animal Behaviour*, *8*(34), 141–154. doi:10.1016/0003-3472(60)90022-1

Metzner, W. (1991). Echolocation behaviour in bats. *Science Progress Edinburgh*, *75*(298), 453–465.

Ochoa, A., Garcí, Y., & Yañez, J. (2011). Logistics Optimization Service Improved with Artificial Intelligence. *Soft Computing for Intelligent Control and Mobile Robotics*, *2011*, 57–65.

Reyes, L., Zezzatti, C., Santillán, C., Hernández, P., & Fuerte, M. (2010). A cultural algorithm for the urban public transportation. In Hybrid Artificial Intelligence Systems (pp. 135-142).

Schnitzler, H.-U., & Kalko, E. K. V. (2001, July). Echolocation by insect-eating bats. *Bioscience*, *51*(7), 557–569. doi:10.1641/0006-3568(2001)051[0557:EBIEB]2.0.CO;2

Souffiau, W., Maervoet, J., Vansteenwegen, P., Vanden Berghe, G., & Van Oudheusden, D. (2009). A Mobile Tourist Decision Support System for Small Footprint Devices. In *Bio-Inspired Systems: Computational and Ambient Intelligence* (pp. 1248-1255).

Yang, X.-S. (2013). Bat algorithm for multi-objective optimisation. *International Journal of Bio-inspired Computation*, *3*(5), 267–274. doi:10.1504/IJBIC.2011.042259

This research was previously published in the Handbook of Research on Emergent Applications of Optimization Algorithms edited by Pandian Vasant, Sirma Zeynep Alparslan-Gok, and Gerhard-Wilhelm Weber, pages 279-297, copyright year 2018 by Business Science Reference (an imprint of IGI Global).

Chapter 2
Dynamic Modeling and Control Techniques for a Quadrotor

Heba Elkholy
American University in Cairo, Egypt

Maki K. Habib
American University in Cairo, Egypt

ABSTRACT

This chapter presents the detailed dynamic model of a Vertical Take-Off and Landing (VTOL) type Unmanned Aerial Vehicle (UAV) known as the quadrotor. The mathematical model is derived based on Newton Euler formalism. This is followed by the development of a simulation environment on which the developed model is verified. Four control algorithms are developed to control the quadrotor's degrees of freedom: a linear PID controller, Gain Scheduling-based PID controller, nonlinear Sliding Mode, and Backstepping controllers. The performances of these controllers are compared through the developed simulation environment in terms of their dynamic performance, stability, and the effect of possible disturbances.

INTRODUCTION

In the past decade, a lot of researchers are now focusing on developing miniature flying objects due to the recent advances in technologies and the emergence of miniature sensors and actuators depending on MEMS and NEMS. The developed miniature flying objects can be used in a broad set of applications ranging from military to civil ones. The reason for choosing quadrotors over other UAVs to be the focus of our work is their advantages over their counterparts due to the presence of four separately powered propellers thus giving the quadrotors a higher payload and better maneuverability. Also, their VTOL and hovering capabilities make them good candidates for surveillance and monitoring tasks and for the use in small spaces. The quadrotors control research field is still facing a lot of challenges; this is due to the fact that the quadrotor is a highly nonlinear, multivariable and underactuated system [Hou et al. (2010)]. Underactuated systems are those having a less number of control inputs compared to the system's degrees of freedom. They are very difficult to control due to the nonlinear coupling between

DOI: 10.4018/978-1-5225-8365-3.ch002

the actuators and the degrees of freedom [Kim et al. (2010)]. Although the most common flight control algorithms found in literature are linear flight controllers, these controllers can only perform when the quadrotor is flying around hover, they suffer from huge performance degradation whenever the quadrotor leaves the nominal conditions or performs aggressive maneuvers [Kendoul (2012)].

STATE OF THE ART

Controlling the degrees of freedom of the quadrotor can be done through various control algorithms which vary from the classical linear Proportional-Integral-Derivative (PID) or Proportional-Derivative (PD) controller to more complex nonlinear schemes such as backstepping or sliding mode controllers. Starting with the linear control algorithms; Bouabdallah et al. applied a PID and LQ controllers on an indoor micro quadrotor. The performance of the two controllers was comparable in stabilizing the attitude of the quadrotor around its hover position and under the effect of little disturbances [Bouabdallah et al. (2004)]. Li and Li used the classical PID to control the position and orientation of a quadrotor and it was able to stabilize in a low speed wind environment [Li & Li (2011)]. Simulation based results showed that Yang et al. were able to control the attitude and heading of a quadrotor using a self-tuning PID controller based on adaptive pole placement [Yang et al. (2013)]. Raffo et al. used an H_∞ controller to stabilize the rotational angles and a Model Predictive Controller (MPC) to track the desired position of a quadrotor. The effect of wind and model uncertainties was added to the simulated model and it performed robustly with a zero steady-state error [Raffo et al. (2010)].

In order to employ a linear controller to control a nonlinear system like that of the quadrotor, the system's nonlinearities can be modeled as a collection of simplified linear systems and for each system a separate controller can be designed, this is the concept of gain scheduling and it is commonly used in flight controllers. Gillula et al. divided the state space model of a STARMAC quadrotor to a set of simple hybrid modes and this approach enabled the quadrotor to carry out aerobatic maneuvers [Gillula et al. (2011)]. Ataka et al. used gain scheduling on a linearized model of the quadrotor around some equilibrium points and tested the controllability and observability of the resulting system [Ataka et al. (2013)]. Amoozgar et al. compared the performance of a conventional PID controller to that of a gain scheduled PID controller with its parameters tuned using a fuzzy logic based inference scheme. The gain scheduled PID controller outperformed the conventional PID controller when the system was tested under actuator fault conditions [Amoozgar et al. (2012)]. In load dropping applications, Sadeghzadeh et al. found that a gain scheduled PID controller was able to stabilize the system during the dropping operation [Sadeghzadeh et al. (2012)].

Moving to the nonlinear flight control algorithms, Bouabdallah and Siegwart compared the performance of a backstepping and sliding mode control algorithms to the performance of that of linear PID an LQ controllers in their prior work. They found that nonlinear controllers gave better performance in the presence of disturbances [Bouabdallah & Siegwart (2005)]. Waslander et al. compared the performance of an integral sliding mode controller to that of a reinforcement learning controller to stabilize a quadrotor in an outdoors environment. It was found that, both of the implemented control techniques were able to stabilize the quadrotor and gave a better performance over the classical control algorithms [Waslander et al. (2005)]. Madani and Benallegue used a backstepping controller based on Lyapunov stability theory to track desired values for the quadrotor's position and orientation. They divided the quadrotor model into 3 subsystems: underactuated, fully-actuated and propeller subsystems. Their proposed algorithm was able to stabilize the system under no disturbances [Madani & Benallegue (2006)]. Fang and Gao merged a backstepping controller with an adaptive controller thus resulting in an integral backstepping algorithm to overcome the problems of model uncertainties and external disturbances. The controller reduced the system's overshoot and response time and eliminated the steady state error [Fang & Gao (2012)]. Lee et al. used a backstepping controller to control the position and attitude of a quadrotor, the proposed controller was tested in a noisy environment and gave a satisfactory performance [Lee et al. (2013)]. Zhen et al. combined a backstepping controller with an adaptive algorithm to control the attitude of a quadrotor. A robust adaptive function is used to approximate the external disturbances and modeling errors of the system. Simulations showed the success of the proposed controller in overcoming disturbances and uncertainties [Zhen et al. (2013)]. Gonzalez et al. proposed using a chattering free sliding mode controller to control the altitude of a quadrotor. The proposed controller performed well in both simulations and on a real system in the presence of disturbances [González et al. (2014)]. Kendoul et al. was able to control a quadrotor in several flight tests based on the concept of feedback linearization [Kendoul et al. (2010)]. Alexis et al. relied on a MPC to control the attitude of a quadrotor in the presence of atmospheric disturbances. MPC relies on predicting the future states of the system and tracking the error to give an improved performance [Kendoul (2012)]. The proposed algorithm behaved well in performing rough maneuvers in a wind induced environment as was able to accurately track the desired attitude [Alexis et al. (2011)]. Sadeghzadeh et al. also used a MPC applied to a quadrotor in dropping a carried payload, the MPC was able to stabilize the system with a promising performance [Sadeghzadeh et al. (2012)].

Opposing the previous control techniques, learning based flight control systems do not need a precise and accurate dynamic model of the system to be controlled. On the other hand, several trials are carried out and flight data are used to "train" the system. Efe used a Neural Network to simplify the design of a PID controller and decrease the computational time and complexity [Efe (2011)]. Another approach is using fuzzy logic based control systems, where the experience of a skilled pilot is employed to train the controller [Kendoul (2012)].

In more recent literature, it was found that using more than one type of control algorithms together generated a better performance, especially when the quadrotor is not flying near hover. Nagaty et al. proposed the usage of a nested loop control algorithm; the outer loop consists of a PID controller responsible for generating the desired attitude angles that would achieve the desired position. These attitude angles are then fed to the inner loop. The inner loop stabilization controller relies on the backstepping algorithm to track the desired altitude, attitude and heading [Nagaty et al. (2013)]. Azzam and Wang used a PD controller for altitude and yaw rotation and a PID controller integrated with a backstepping controller for the pitch and roll control. An optimization algorithm was used instead of the pole placement technique to overcome the difficulty of pole placement in a nonlinear time variant system. The system was divided into rotational and translation subsystems where the translation subsystem stabilizes the quadrotor position in flight and generates the needed roll and pitch angles to be fed to the rotational subsystem [Azzam & Wang (2010)].

Research Challenges

After an intensive literature review of the control algorithms applied on quadrotors, it was found out that one of the research challenges facing researchers in this field is flying the quadrotor outside the linear or hovering region and having it perform acrobatic maneuvers. Also, there is no work focused on having a comparative study that compares between employing different types of controllers on the quadrotor system and comparing their performances in stabilizing the quadrotor under different flight conditions and this will be the focus of this work.

THE QUADROTOR CONCEPT AND STRUCTURE

The quadrotor is a rotary wing UAV consisting of four rotors, each fitted on one end of a cross-like structure as shown in Figure 1. Each rotor consists of a propeller fitted to a separately powered DC motor. Propellers 1 and 3 rotate in the same direction while propellers 2 and 4 rotate in an opposite direction leading to balancing the total system torque and cancelling the gyroscopic and aerodynamics torques in stationary flights [Bouabdallah et al. (2004), Hou et al. (2010)].

A quadrotor is a 6 DOF object, thus to express its position in space, 6 variables are used $(x, y, z, \varphi, \theta, \psi)$. x, y and z represent the distances of the quadrotor's center of mass along the x,y and z axes respectively from an Earth fixed inertial frame. φ, θ and ψ are the three Euler angles representing the orientation of the quadrotor. φ is called the roll angle which is the angle about the x-axis, θ is the pitch angle about the y-axis, while ψ is the yaw angle about the z-axis. Figure 2 clearly explains the Euler Angles. The roll and pitch angles are usually called the attitude of the quadrotor, while the yaw angle is referred to as the heading of the quadrotor. For position, the distance from the ground is referred to as the altitude and the x and y position in space is often called the position of the quadrotor.

Figure 1. Quadrotor configuration

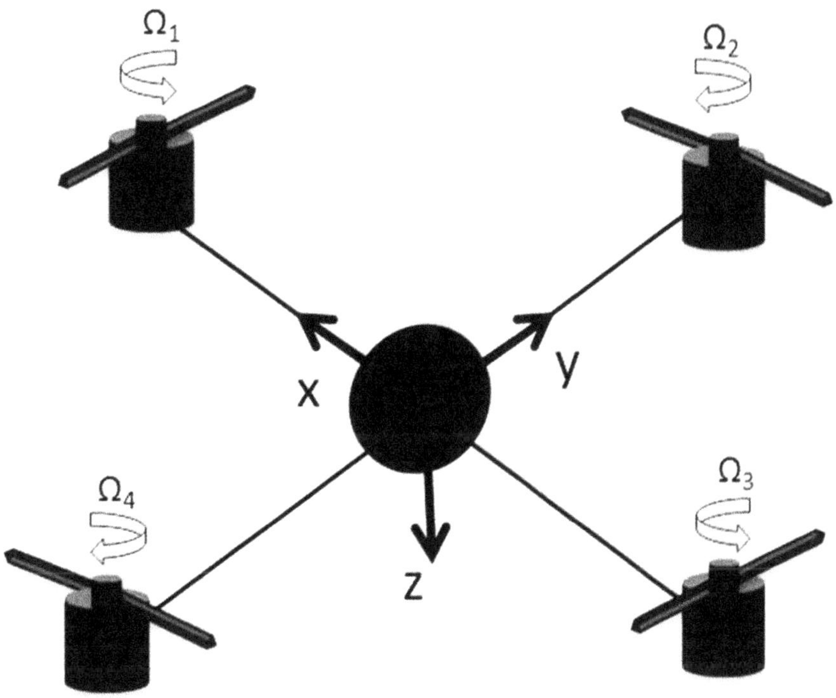

To generate vertical upwards motion, the speed of the four propellers is increased together whereas the speed is decreased to generate vertical downwards motion. To produce roll rotation coupled with motion along the y-axis, the second and fourth propellers speeds are changed while for the pitch rotation coupled with motion along the x-axis, it is the first and third propellers speeds need to be changed.

One problem with the quadrotor configuration is that to produce yaw rotation, one needs to have a difference in the opposite torque produced by each propeller pair. For instance, for a positive yaw rotation, the speed of the two clockwise turning rotors needs to be increased while the speed of the two counterclockwise turning rotors needs to be decreased [Bouabdallah et al. (2004), Mistler et al. (2001)].

Advantages and Drawbacks of Quadrotors

Some advantages of the quadrotor over helicopters is that their rotor mechanics are simplified as it depends on four fixed pitch rotors unlike the variable pitch rotor in the helicopter, thus leading to easier manufacturing and maintenance. Moreover, due to the symmetry in the configuration, the gyroscopic effects are reduced leading to simpler control.

Stationary hovering can be more stable in quadrotors than in helicopters due to the presence of four propellers providing four thrust forces shifted a fixed distance from the center of gravity instead of only one propeller centered in the middle as in the helicopters structure [Hou et al. (2010)].

More advantages are the vertical take-off and landing capabilities, better maneuverability and smaller size due to the absence of a tail [Li & Li (2011)], these capabilities make quadrotors useful in small area monitoring and buildings exploration [Bouabdallah & Siegwart (2005)].

Figure 2. Euler angles for a quadrotor

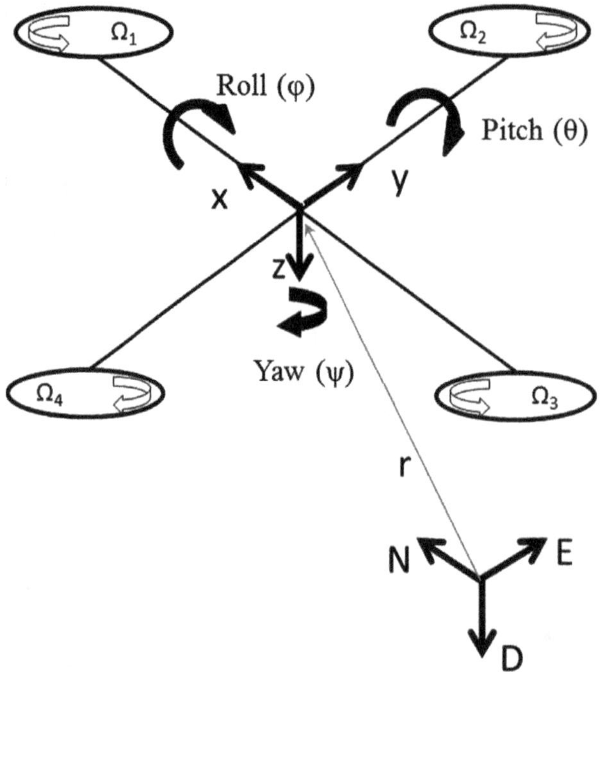

Moreover, quadrotors have higher payload capacities due to the presence of four motors thus providing higher thrust [Hou et al. (2010)].

On the other hand, quadrotors consume a lot of energy due to the presence of four separate propellers [Bouabdallah & Siegwart (2005)]. Also, they have a large size and heavier than some of their counterparts again to the fact that there is four separate propellers [Bouabdallah (2007), Bouabdallah & Siegwart (2005)].

SYSTEM MODELING

The kinematics and dynamics models of a quadrotor will be derived based on the Newton-Euler formalism with the following assumptions:

- The structure is rigid and symmetrical.
- The center of gravity of the quadrotor coincides with the body fixed frame origin.
- The propellers are rigid.
- Thrust and drag are proportional to the square of propeller's speed.

Kinematic Model

Two sets of coordinate frames are defined in order to discuss the modeling of the quadrotor. Figure 2 shows the Earth reference frame, which is fixed on a specific place at ground level with its N, E and D axes positing towards North, East and Downwards respectively. The second coordinate frame is the body frame which is fixed at the center of the quadrotor body with its x, y and z axes pointing towards propeller 1, propeller 2 and to the ground respectively.

The distance between the Earth frame and the body frame describes the absolute position of the center of mass of the quadrotor $r = [x\ y\ z]^T$. The rotation R from the body frame to the inertial frame describes the orientation of the quadrotor. The orientation of the quadrotor is described using roll, pitch and yaw angles (φ, θ and ψ) representing rotations about the x, y and z-axes respectively. Assuming the order of rotation to be roll (φ), pitch (θ) then yaw (ψ), the rotation matrix R which is derived based on the sequence of principle rotations is:

$$R = \begin{bmatrix} c\theta c\psi & s\phi s\theta c\psi & c\phi s\theta c\psi + s\phi s\psi \\ c\theta s\psi & s\phi s\theta s\psi + c\theta c\psi & c\phi s\theta s\psi - s\theta c\psi \\ -s\theta & s\phi c\theta & c\phi c\theta \end{bmatrix} \tag{1.1}$$

where c and s denote cos and sin respectively.

The rotation matrix R will be used in formulating the dynamics model of the quadrotor, its significance is due to the fact that some states are measured in the body frame (e.g. the thrust forces produced by the propellers) while some others are measured in the inertial frame (e.g. the gravitational forces and the quadrotor's position). Thus, to have a relation between both types of states, a transformation from one frame to the other is needed.

To acquire information about the angular velocity of the quadrotor, typically an on-board Inertial Measurement Unit (IMU) is used which will in turn give the velocity in the body coordinate frame. To relate the Euler rates $\dot{\eta} = [\dot{\phi}\ \dot{\theta}\ \dot{\psi}]^T$ that are measured in the inertial frame and angular body rates $\omega = [p\ q\ r]^T$, a transformation is needed as follows:

$$\omega = R_r \dot{\eta} \tag{1.2}$$

where

$$R_r = \begin{bmatrix} 1 & 0 & -\sin\theta \\ 0 & \cos\phi & \sin\phi\cos\theta \\ 0 & -\sin\phi & \cos\phi\cos\theta \end{bmatrix}$$

Around the hover position, small angle assumption is made where $\cos\phi \equiv 1$, $\cos\theta \equiv 1$ and $\sin\phi = \sin\theta = 0$ thus R_r can be simplified to the identity matrix I [Nagaty et al. (2013)].

Dynamics Model

The motion of the quadrotor can be divided into two subsystems; rotational subsystem (roll, pitch, and yaw) and translational subsystem (altitude and x and y position). The rotational subsystem is fully actuated while the translational subsystem is underactuated [Nagaty et al. (2013)].

Rotational Equations of Motion

The rotational equations of motion are derived in the body frame using the Newton-Euler method with the following general formalism,

$$J\dot{\omega} + \omega \times J\omega + M_G + M_a = M_B \tag{1.3}$$

where J represents the quadrotor's diagonal inertia matrix, ω represents the angular body rates. M_G represent the gyroscopic moments due to the rotors' inertia. M_a is the aerodynamic moments acting on the quadrotor body while M_B are the moments acting on the quadrotor in the body frame.

The first two terms of Equation 1.3, $J\dot{\omega}$ and $\omega \times J\omega$, represent the rate of change of angular momentum in the body frame. The reason behind deriving the rotational equations of motion in the body frame and not in the inertial frame, is to have the inertia matrix independent on time.

Inertia Matrix

The inertia matrix for the quadrotor is a diagonal matrix, the off-diagonal elements, which are the product of inertia, are zero due to the symmetry of the quadrotor.

$$J = \begin{bmatrix} I_{xx} & 0 & 0 \\ 0 & I_{yy} & 0 \\ 0 & 0 & I_{zz} \end{bmatrix} \tag{1.4}$$

where I_{xx}, I_{yy} and I_{zz} are the area moments of inertia about the principle axes in the body frame.

Gyroscopic Moment

The gyroscopic moment of a rotor is a physical effect in which gyroscopic torques or moments attempt to align the spin axis of the rotor along the inertial z-axis [Derafa et al. (2006)]. It is defined to be,

$$M_G = \omega \times [0\ 0\ J_r \Omega_r]^T \tag{1.5}$$

where J_r is the rotors' inertia and Ωr is the rotors' relative speed which is defined to be, $\Omega_r = -\Omega_1 + \Omega_2 - \Omega_3 + \Omega_4$

Aerodynamic Moments

Due to the friction of the moving quadrotor body with air, a moment acts on the body of the quadrotor resisting its motion. As the angular velocity of the quadrotor increases, the drag moments in turn increase. The drag moments M_a can be approximated by,

$$M_a = K_r \dot{\eta} \tag{1.6}$$

where K_r is a constant matrix called the aerodynamic rotation coefficient matrix and $\dot{\eta}$ is the Euler rates.

Moments Acting on the Quadrotor (M_B)

For the last term of Equation 1.3, there is a need to define two physical effects which are the aerodynamic forces and moments produced by a rotor. As an effect of rotation, there is a generated force called the aerodynamic force or the lift force and there is a generated moment called the aerodynamic moment. Equations 1.6 and 1.7 show the aerodynamic force F_i and moment M_i produced by the ith rotor [Azzam & Wang (2010)].

$$F_i = \frac{1}{2}\rho A C_T r^2 \Omega_i^2 \tag{1.7}$$

$$M_i = \frac{1}{2}\rho A C_D r^2 \Omega_i^2 \tag{1.8}$$

where ρ is the air density, A is the blade area, C_T and C_D are aerodynamic coefficients, r is the radius of the blade and Ω_i is the angular velocity of rotor i.

Clearly, the aerodynamic forces and moments depend on the geometry of the propeller and the air density. Since for the case of quadrotors, the maximum altitude is usually limited, thus the air density can be considered constant, Equations 1.6 and 1.7 can be simplified to [Nagaty et al. (2013)],

$$F_i = K_f \Omega_i^2 \tag{1.9}$$

$$M_i = K_M \Omega_i^2 \tag{1.10}$$

where K_f and K_M are the aerodynamic force and moment constants respectively. The aerodynamic force and moment constants can be determined experimentally for each propeller type.

By identifying the forces and moments generated by the propellers, we can study the moments M_B acting on the quadrotor. Figure 3 shows the forces and moments acting on the quadrotor. Each rotor causes an upwards thrust force F_i and generates a moment M_i with direction opposite to the direction of rotation of the corresponding rotor i.

Figure 3. Forces and moments acting on quadrotor

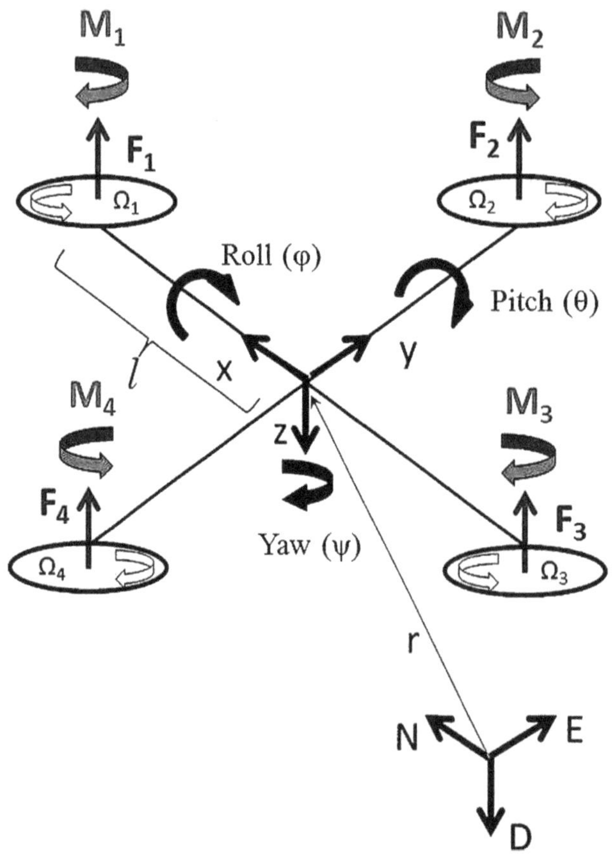

Starting with the moments about the body frame's x-axis, by using the right-hand-rule in association with the axes of the body frame, F_2 multiplied by the moment arm l generates a negative moment about the y-axis, while in the same manner, F_4 generates a positive moment. Thus the total moment about the x-axis can be expressed as,

$$\begin{aligned} M_x &= -F_2 l + F_4 l \\ &= -(K_f \Omega_2^2) l + (K_f \Omega_4^2) l \\ &= l K_f (-\Omega_2^2 + \Omega_4^2) \end{aligned} \tag{1.11}$$

For the moments about the body frame's y-axis, also using the right-hand-rule, the thrust of rotor 1 generates a positive moment, while the thrust of rotor 3 generates a negative moment about the y-axis. The total moment can be expressed as,

$$\begin{aligned} M_y &= F_1 l - F_3 l \\ &= -(K_f \Omega_1^2) l - (K_f \Omega_3^2) l \\ &= l K_f (\Omega_1^2 - \Omega_3^2) \end{aligned} \tag{1.12}$$

For the moments about the body frame's z-axis, the thrust of the rotors does not cause a moment. On the other hand, moments are caused by the rotors' rotation as per Equation 1.9. By using the right-hand-rule, the moment about the body frame's z-axis can be expressed as,

$$\begin{aligned} M_z &= M_1 - M_2 + M_3 - M_4 \\ &= (K_M \Omega_1^2) - (K_M \Omega_2^2) + (K_M \Omega_3^2) - (K_M \Omega_4^2) \\ &= (K_M \Omega_1^2 - \Omega_2^2 + \Omega_3^2 - \Omega_4^2) \end{aligned} \tag{1.13}$$

Combining Equations 1.10, 1.11 and 1.12 in vector form, we get,

$$M_B = \begin{bmatrix} l K_f (-\Omega_2^2 + \Omega_4^2) \\ l K_f (\Omega_1^2 - \Omega_2^2) \\ K_M (\Omega_1^2 - \Omega_2^2 + \Omega_3^2 - \Omega_4^2) \end{bmatrix} \tag{1.14}$$

where l is the moment arm, which is the distance between the axis of rotation of each rotor to the origin of the body reference frame which should coincide with the center of the quadrotor.

Translational Equations of Motion

The translation equations of motion for the quadrotor are based on Newton's second law and they are derived in the Earth inertial frame [Nagaty et al. (2013)],

$$m\ddot{r} + F_a = \begin{bmatrix} 0 \\ 0 \\ mg \end{bmatrix} + RF_B \tag{1.15}$$

where $r = [x\ y\ z]^T$ is the quadrotor's distance from the inertial frame, m is the quadrotor's mass. F_a is the aerodynamic forces. g is the gravitational acceleration and F_B is the nongravitational forces acting on the quadrotor in the body frame.

Aerodynamic Forces

Similar to the aerodynamic moments, there are aerodynamic forces acting on the quadrotor that oppose its motion. The drag forces F_a can be approximated by,

$$F_a = K_t \dot{r} \tag{1.16}$$

where K_t is a constant matrix called the aerodynamic translation coefficient matrix and $\dot{r}$ is the time derivative of the position vector r.

Nongravitational Forces Acting on the Quadrotor

When the quadrotor is in a horizontal orientation (i.e. it is not rolling or pitching), the only nongravitational force acting on it is the thrust produced by the rotation of the propellers which is proportional to the square of the angular velocity of the propeller as per Equation 1.9. Thus, the nongravitational forces acting on the quadrotor, F_B, can be expressed as,

$$F_B = \begin{bmatrix} 0 \\ 0 \\ -K_f(\Omega_1^2 + \Omega_2^2 + \Omega_3^2 + \Omega_4^2) \end{bmatrix} \tag{1.17}$$

The first two rows of the force vector are zeros as there is no forces in the x and y directions, the last row is simply an addition of the thrust forces produced by the four propellers. The negative sign is due to the fact that the thrust is upwards while the positive z-axis in the body framed is pointing downwards.

F_B is multiplied by the rotation matrix R to transform the thrust forces of the rotors from the body frame to the inertial frame, so that the equation can be applied in any orientation of the quadrotor.

State Space Model

Formulating the acquired mathematical model for the quadrotor into a state space model will help make the control problem easier to tackle. Note that the aerodynamic forces and moments (F_a and M_a) were neglected due to their minor effects in simulation.

State Vector X

Defining the state vector of the quadrotor to be,

$$X = \begin{bmatrix} x_1 & x_2 & x_3 & x_4 & x_5 & x_6 & x_7 & x_8 & x_9 & x_{10} & x_{11} & x_{12} \end{bmatrix}^T \tag{1.18}$$

which is mapped to the degrees of freedom of the quadrotor in the following manner,

$$X = \begin{bmatrix} \phi & \dot{\phi} & \theta & \dot{\theta} & \psi & \dot{\psi} & z & \dot{z} & x & \dot{x} & y & \dot{y} \end{bmatrix}^T \tag{1.19}$$

The state vector defines the position of the quadrotor in space and its angular and linear velocities.

Control Input Vector *U*

A control input vector, *U*, consisting of four inputs; U_1 through U_4 is defined as,

$$U = \begin{bmatrix} U_1 & U_2 & U_3 & U_4 \end{bmatrix} \tag{1.20}$$

where

$$\begin{aligned} U_1 &= K_f(\Omega_1^2 + \Omega_2^2 + \Omega_3^2 + \Omega_4^2) \\ U_2 &= K_f(-\Omega_2^2 + \Omega_4^2) \\ U_3 &= K_f(\Omega_1^2 - \Omega_3^2) \\ U_4 &= K_M(\Omega_1^2 - \Omega_2^2 + \Omega_3^2 - \Omega_4^2) \end{aligned} \tag{1.21}$$

Equations 1.21 can be arranged in a matrix form to result in,

$$\begin{bmatrix} U_1 \\ U_2 \\ U_3 \\ U_4 \end{bmatrix} = \begin{bmatrix} K_f & K_f & K_f & K_f \\ 0 & -K_f & 0 & K_f \\ K_f & 0 & -K_f & 0 \\ K_M & -K_M & K_M & -K_M \end{bmatrix} \begin{bmatrix} \Omega_1^2 \\ \Omega_2^2 \\ \Omega_3^2 \\ \Omega_4^2 \end{bmatrix} \tag{1.22}$$

U_1 is the resulting upwards force of the four rotors which is responsible for the altitude of the quadrotor and its rate of change $(z, \dot{z})$. U_2 is the difference in thrust between rotors 2 and 4 which is responsible for the roll rotation and its rate of change $(\phi, \dot{\phi})$. U_3 on the other hand represents the difference in thrust between rotors 1 and 3 thus generating the pitch rotation and its rate of change $(\theta, \dot{\theta})$. Finally U_4 is the difference in torque between the two clockwise turning rotors and the two counterclockwise turning rotors generating the yaw rotation and ultimately its rate of change $(\psi, \dot{\psi})$. This choice of the control vector U decouples the rotational system, where U_1 will generate the desired altitude of the quadrotor, U_2 will generate the desired roll angle, the desired pitch angle will be generated by U_3 whereas U_4 will generate the desired heading.

If the rotor velocities are needed to be calculated from the control inputs, an inverse relationship between the control inputs and the rotors' velocities is needed, which can be acquired by inverting the matrix in Equation 1.22 to give,

$$\begin{bmatrix} \Omega_1^2 \\ \Omega_2^2 \\ \Omega_3^2 \\ \Omega_4^2 \end{bmatrix} = \begin{bmatrix} \frac{1}{4K_f} & 0 & \frac{1}{2K_f} & \frac{1}{4K_M} \\ \frac{1}{4K_f} & -\frac{1}{2K_f} & 0 & -\frac{1}{4K_M} \\ \frac{1}{4K_f} & 0 & -\frac{1}{2K_f} & \frac{1}{4K_M} \\ \frac{1}{4K_f} & \frac{1}{2K_f} & 0 & -\frac{1}{4K_M} \end{bmatrix} \begin{bmatrix} U_1 \\ U_2 \\ U_3 \\ U_4 \end{bmatrix} \tag{1.23}$$

Taking the square root of that, the rotors' velocities can be calculated from the control inputs as follows,

$$\begin{aligned} \Omega_1 &= \sqrt{\frac{1}{4K_f}U_1 + \frac{1}{2K_f}U_3 + \frac{1}{4K_M}U_4} \\ \Omega_2 &= \sqrt{\frac{1}{4K_f}U_1 - \frac{1}{2K_f}U_2 - \frac{1}{4K_M}U_4} \\ \Omega_3 &= \sqrt{\frac{1}{4K_f}U_1 - \frac{1}{2K_f}U_3 + \frac{1}{4K_M}U_4} \\ \Omega_4 &= \sqrt{\frac{1}{4K_f}U_1 + \frac{1}{2K_f}U_2 - \frac{1}{4K_M}U_4} \end{aligned} \tag{1.24}$$

Rotational Equation of Motion

Substituting equations 1.21 in Equation 1.14, the equation of the total moments acting on the quadrotor becomes,

$$M_B = \begin{bmatrix} l\,U_2 \\ l\,U_3 \\ U4 \end{bmatrix} \tag{1.25}$$

Substituting Equation 1.25 into the rotational equation of motion and expanding each term with their prior definition, the following relation can be derived,

$$\begin{bmatrix} I_{xx} & 0 & 0 \\ 0 & I_{yy} & 0 \\ 0 & 0 & I_{zz} \end{bmatrix} \begin{bmatrix} \ddot{\phi} \\ \ddot{\theta} \\ \ddot{\psi} \end{bmatrix} + \begin{bmatrix} \dot{\phi} \\ \dot{\theta} \\ \dot{\psi} \end{bmatrix} \times \begin{bmatrix} I_{xx} & 0 & 0 \\ 0 & I_{yy} & 0 \\ 0 & 0 & I_{zz} \end{bmatrix} \begin{bmatrix} \dot{\phi} \\ \dot{\theta} \\ \dot{\psi} \end{bmatrix} + \begin{bmatrix} \dot{\phi} \\ \dot{\theta} \\ \dot{\psi} \end{bmatrix} \times \begin{bmatrix} 0 \\ 0 \\ J_r\Omega_r \end{bmatrix} = \begin{bmatrix} lU_2 \\ lU_3 \\ U_4 \end{bmatrix} \tag{1.26}$$

Expanding that, leads to,

$$\begin{bmatrix} I_{xx}\ddot{\phi} \\ I_{yy}\ddot{\theta} \\ I_{zz}\ddot{\psi} \end{bmatrix} + \begin{bmatrix} \dot{\theta}I_{zz}\dot{\psi} - \dot{\psi}I_{yy}\dot{\theta} \\ \dot{\psi}I_{xx}\dot{\phi} - \dot{\phi}I_{zz}\dot{\psi} \\ \dot{\phi}I_{yy}\dot{\theta} - \dot{\theta}I_{xx}\dot{\phi} \end{bmatrix} + \begin{bmatrix} \dot{\theta}J_r\Omega_r \\ -\dot{\phi}J_r\Omega_r \\ 0 \end{bmatrix} = \begin{bmatrix} lU_2 \\ lU_3 \\ U_4 \end{bmatrix} \tag{1.27}$$

Rewriting the last equation to have the angular accelerations in terms of the other variables,

$$\begin{aligned} \ddot{\phi} &= \frac{l}{I_{xx}}U_2 - \frac{J_r}{I_{xx}}\dot{\theta}\Omega_r + \frac{I_{yy}}{I_{xx}}\dot{\psi}\dot{\theta} - \frac{I_{zz}}{I_{xx}}\dot{\theta}\dot{\psi} \\ \ddot{\theta} &= \frac{l}{I_{yy}}U_3 - \frac{J_r}{I_{yy}}\dot{\phi}\Omega_r + \frac{I_{zz}}{I_{yy}}\dot{\phi}\dot{\psi} - \frac{I_{xx}}{I_{yy}}\dot{\psi}\dot{\phi} \\ \ddot{\psi} &= \frac{1}{I_{zz}}U_4 + \frac{I_{xx}}{I_{zz}}\dot{\theta}\dot{\phi} - \frac{I_{yy}}{I_{zz}}\dot{\phi}\dot{\theta} \end{aligned} \tag{1.28}$$

To simplify, define,

$$a_1 = \frac{I_{yy} - I_{zz}}{I_{xx}}$$

$$a_2 = \frac{J_r}{I_{xx}} \qquad b_1 = \frac{l}{I_{xx}}$$

$$a_3 = \frac{I_{zz} - I_{xx}}{I_{yy}} \qquad b_2 = \frac{l}{I_{yy}}$$

$$a_4 = \frac{J_r}{I_{yy}} \qquad b_3 = \frac{l}{I_{zz}}$$

$$a_5 = \frac{I_{xx} - I_{yy}}{I_{zz}}$$

Using the above definition of $a_1 \rightarrow a_5$ and $b_1 \rightarrow b_3$, equations 1.28 can then be rewritten in a simpler form in terms of the system states,

$$\begin{aligned}\ddot{\phi} &= b_1U_2 - a_2x_4\Omega_r + a_1x_4x_6 \\ \ddot{\theta} &= b_2U_3 + a_4x_2\Omega_r + a_3x_2x_6 \\ \ddot{\psi} &= b_3U_4 + a_5x_2x_4\end{aligned} \tag{1.29}$$

With the choice of the control input vector U, it is clear that the rotational subsystem is fully-actuated, it is only dependent on the rotational state variables $x_1 \rightarrow x_6$ that correspond to $\phi, \dot{\phi}, \theta, \dot{\theta}, \psi, \dot{\psi}$ respectively.

Translational Equation of Motion

Substituting Equation 1.21 in Equation 1.17, the equation of the total moments acting on the quadrotor becomes,

$$F_B = \begin{bmatrix} 0 \\ 0 \\ -U_1 \end{bmatrix} \tag{1.30}$$

Embedding that into the translational equation of motion and expanding the terms, we get,

$$\begin{aligned} m\begin{bmatrix}\ddot{x}\\ \ddot{y}\\ \ddot{z}\end{bmatrix} &= \begin{bmatrix}0\\0\\mg\end{bmatrix} + \begin{bmatrix} c\psi c\theta & c\psi s\phi s\theta & s\phi s\psi + c\phi c\psi s\theta \\ c\theta s\psi & c\theta c\psi + s\phi s\psi s\theta & c\phi s\psi s\theta - c\psi s\theta \\ -s\theta & c\theta s\phi & c\phi c\theta \end{bmatrix}\begin{bmatrix}0\\0\\-U_1\end{bmatrix} \\ m\begin{bmatrix}\ddot{x}\\ \ddot{y}\\ \ddot{z}\end{bmatrix} &= \begin{bmatrix}0\\0\\mg\end{bmatrix} + \begin{bmatrix} (s\phi s\psi + c\phi c\psi s\theta)(-U_1) \\ (c\phi s\psi s\theta - c\psi s\phi)(-U_1) \\ (c\phi c\theta)(-U_1) \end{bmatrix}\end{aligned} \tag{1.31}$$

Rewriting Equation 1.31 to have the accelerations in terms of the other variables, we get,

$$\begin{aligned}\ddot{x} &= \frac{-U_1}{m}(\sin\phi\sin\psi + \cos\phi\cos\psi\sin\theta) \\ \ddot{y} &= \frac{-U_1}{m}(\cos\phi\sin\psi\sin\theta - \cos\psi\sin\phi) \\ \ddot{z} &= g - \frac{U_1}{m}(\cos\phi\cos\theta)\end{aligned} \tag{1.32}$$

Rewriting in terms of the state variable X,

$$\begin{aligned}
\ddot{x} &= \frac{-U_1}{m}(\sin x_1 \sin x_5 + \cos x_1 \cos x_5 \sin x_3) \\
\ddot{y} &= \frac{-U_1}{m}(\cos x_1 \sin x_5 \sin x_3 - \cos x_5 \sin x_1) \\
\ddot{z} &= g - \frac{U_1}{m}(\cos x_1 \cos x_3)
\end{aligned} \tag{1.33}$$

It is clear here that the translational subsystem is underactuated as it dependent on both the translational state variables and the rotational ones.

State Space Representation

Using the equations of the rotational angular acceleration and those of translation, the complete mathematical model of the quadrotor can be written in a state space representation as follows,

$$\begin{aligned}
\dot{x}_1 &= \dot{\phi} = x_2 \\
\dot{x}_2 &= \ddot{\phi} = x_4 x_6 a_1 - x_4 \Omega_r a_2 + b_1 U_2 \\
\dot{x}_3 &= \dot{\theta} = x_4 \\
\dot{x}_4 &= \ddot{\theta} = x_2 x_6 a_3 + x_2 \Omega_r a_4 + b_2 U_3 \\
\dot{x}_5 &= \dot{\psi} = x_6 \\
\dot{x}_6 &= \ddot{\psi} = x_2 x_4 a_5 + b_3 U_4 \\
\dot{x}_7 &= \dot{z} = x_8 \\
\dot{x}_8 &= \ddot{z} = g - \frac{U_1}{m}(\cos x_1 \cos x_3) \\
\dot{x}_9 &= \dot{x} = x_{10} \\
\dot{x}_{10} &= \ddot{x} = \frac{-U_1}{m}(\sin x_1 \sin x_5 + \cos x_1 \sin x_3 \cos x_5) \\
\dot{x}_{11} &= \dot{y} = x_{12} \\
\dot{x}_{12} &= \ddot{y} = \frac{U_1}{m}(\sin x_1 \cos x_5 - \cos x_1 \sin x_3 \sin x_5)
\end{aligned} \tag{1.34}$$

or

$$f(X,U) = \begin{bmatrix} x_2 \\ x_4 x_6 a_1 - x_4 \Omega_r a_2 + b_1 U_2 \\ x_4 \\ x_2 x_6 a_3 + x_2 \Omega_r a_4 + b_2 U_3 \\ x_6 \\ x_2 x_4 a_5 + b_3 U_4 \\ x_8 \\ g - \frac{U_1}{m}(\cos x_1 \cos x_3) \\ x_{10} \\ \frac{-U_1}{m}(\sin x_1 \sin x_5 + \cos x_1 \sin x_3 \cos x_5) \\ x_{12} \\ \frac{U_1}{m}(\sin x_1 \cos x_5 - \cos x_1 \sin x_3 \sin x_5) \end{bmatrix} \tag{1.35}$$

SYSTEM CONTROL

The mathematical model of the quadrotor implies a nested structure controller, where the outer controller will control the position of the quadrotor in space by generating the required attitude that in turn will be fed to the inner controller that will directly control the attitude by generating the required control inputs to be fed to the quadrotor dynamics as shown in Figure 4.

Unlike the altitude and orientation of the quadrotor, its *x* and *y* position is not decoupled and cannot be directly controlled using one of the four control laws U_1 through U_4. On the other hand, the *x* and *y* position can be controlled through the roll and pitch angles. The desired roll and pitch angles ϕ_d and θ_d can be calculated from the translational equations of motion as follows,

$$\begin{aligned} \ddot{x} &= \frac{-U_1}{m}(\sin\phi_d \sin\psi + \cos\phi_d \sin\theta_d \cos\psi) \\ \ddot{y} &= \frac{-U_1}{m}(\cos\phi_d \sin\theta_d \sin\psi - \sin\phi_d \cos\psi) \end{aligned} \tag{1.36}$$

Since the quadrotor is operating around hover, which means small values for the roll and pitch angles, we can use the small angle assumption ($\sin\phi_d \equiv \phi_d, \sin\theta_d \equiv \theta_d$ and $\cos\phi_d = \cos\theta_d = 1$) to simplify the above equations,

Figure 4. Closed loop control system of a quadrotor

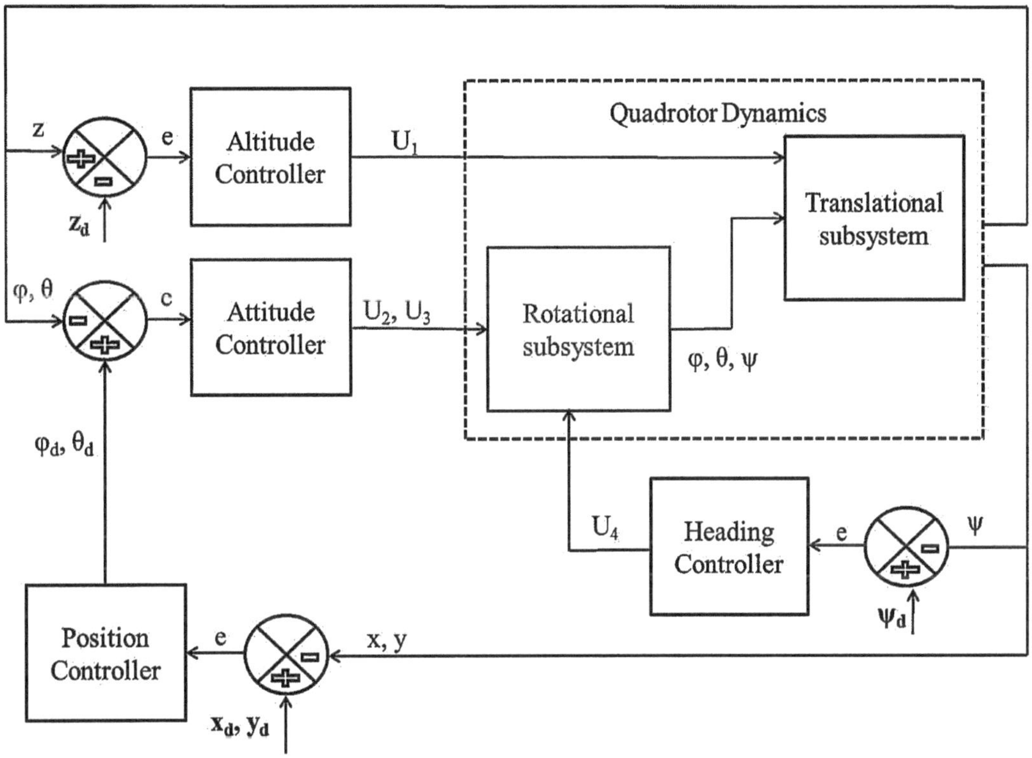

$$\begin{aligned} \ddot{x} &= \frac{-U_1}{m}(\phi_d \sin\psi + \theta_d \cos\psi) \\ \ddot{y} &= \frac{-U_1}{m}(\theta_d \sin\psi - \phi_d \cos\psi) \end{aligned} \tag{1.37}$$

which can be written in a matrix form as,

$$\begin{bmatrix} -\sin\psi & -\cos\psi \\ \cos\psi & -sin\psi \end{bmatrix}\begin{bmatrix} \phi_d \\ \theta_d \end{bmatrix} = \frac{m}{U_1}\begin{bmatrix} \ddot{x}_d \\ \ddot{y}_d \end{bmatrix} \tag{1.38}$$

which can be inverted to get

$$\begin{aligned} \begin{bmatrix} \phi_d \\ \theta_d \end{bmatrix} &= \begin{bmatrix} -\sin\psi & -\cos\psi \\ \cos\psi & -sin\psi \end{bmatrix}^{-1} \frac{m}{U_1}\begin{bmatrix} \ddot{x}_d \\ \ddot{y}_d \end{bmatrix} \\ &= \frac{m}{U_1}\begin{bmatrix} -\sin\psi & \cos\psi \\ -\cos\psi & -sin\psi \end{bmatrix}\begin{bmatrix} \ddot{x}_d \\ \ddot{y}_d \end{bmatrix} \\ &= \frac{m}{U_1}\begin{bmatrix} -\ddot{x}_d \sin\psi + \ddot{y}_d \cos\psi \\ -\ddot{x}_d \cos\psi - \ddot{y}_d sin\psi \end{bmatrix} \end{aligned} \tag{1.39}$$

The calculated ϕ_d and θ_d have to be limited to the range between -20° and 20° to fulfill the small angle assumption used in the derivation and this can be done via a saturation function in the simulation.

The controller blocks in the block diagram in Figure 4 can contain any type of control algorithm, whether linear or nonlinear. All controllers input(s) are the error related to some of the quadrotor's states and produce an output which is either one or several control inputs U_1 through U_4 or ϕ_d and θ_d if it is the position controller.

Parameters Tuning Using GA

For the proceeding control algorithms, tuning the controller constants (gains and different parameters) was done using GA. The objective function of the GA was set to be the settling time of the response of the system. The GA is an iterative optimization algorithm that works in the following way; first it generates a random "population" consisting of many individuals, which in our case will be a vector of values for the controller gains. The fitness of the individuals of the population is evaluated using an objective function, which is the settling time of the response of the system with these values set as the control gains. Another population is then generated from the current one using genetic operations like evolution, mutation and crossover and their fitness is also evaluated. The "elite" of the two populations are then selected to form a third population. The term "elite" indicates those individuals having the best fitness or the least value of the objective function (settling time in our case). The algorithm keeps on iterating until it reaches a population where all (or most of) its individuals are elite individuals and returns the individual (the value for the control gains) that has the least possible fitness (produces the least possible settling time for the system). In this work, we have not gone through the process of implementing a GA from scratch as this is out of our scope. Instead, the optimization toolbox in MATLAB was used and it includes a built-in command for GA optimization. The Block Diagram for the GA is shown in Figure 5.

PID Controller

After the mathematical model of the quadrotor along with its open loop simulation is verified, a PID controller was developed. The PID controller generates the desired control inputs for the quadrotor. The block diagram for a PID controller is shown in Figure 6.

Figure 5. Controller tuning using genetic algorithm

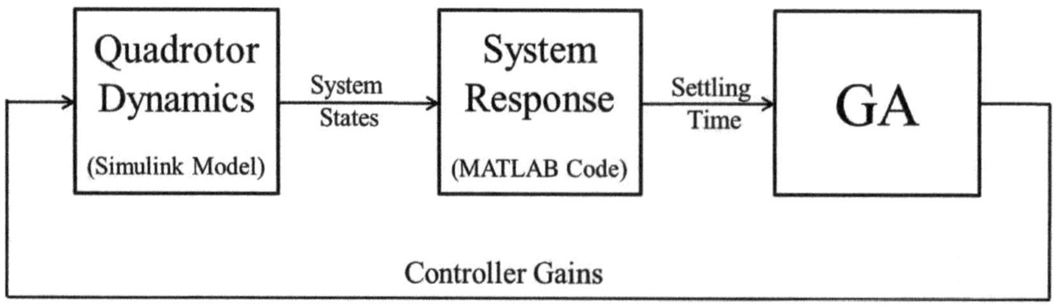

Figure 6. PID controller block diagram

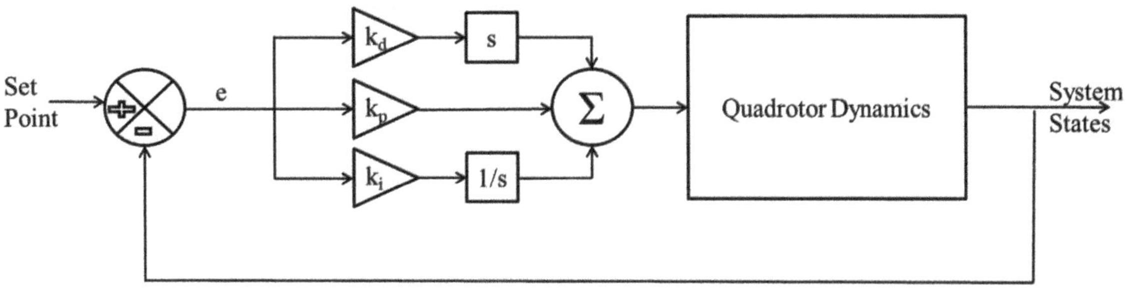

A PID controller is used to track user defined trajectories for the altitude, attitude and heading of the quadrotor. The PID controllers generated the control inputs U_1 through U_4 based on the following control laws,

$$\begin{aligned}
U_1 &= k_p(z - z_d) + k_d(\dot{z} - \dot{z}_d) + k_i \int (z - z_d)dt \\
U_2 &= k_p(\phi_d - \phi) + k_d(\dot{\phi}_d - \dot{\phi}) + k_i \int (\phi_d - \phi)dt \\
U_3 &= k_p(\theta_d - \theta) + k_d(\dot{\theta}_d - \dot{\theta}) + k_i \int (\theta_d - \theta)dt \\
U_4 &= k_p(\psi_d - \psi) + k_d(\dot{\psi}_d - \dot{\psi}) + k_i \int (\psi_d - \psi)dt
\end{aligned} \tag{1.40}$$

where k_p, k_d and k_i are the proportional, derivative and integral gains respectively and z_d, ϕ_d, θ_d and ψ_d are the desired altitude, roll, pitch and heading respectively and $\dot{z}_d$, $\dot{\phi}_d$, $\dot{\theta}_d$ and $\dot{\psi}_d$ are their desired rate of change.

After acquiring stable controllers for the altitude and the attitude of the quadrotor, a complete position controller is developed. PID controllers are used to calculate the desired accelerations $\ddot{x}_d$ and $\ddot{y}_d$,

$$\begin{aligned}
\ddot{x}_d &= k_p(x_d - x) + k_d(\dot{x}_d - \dot{x}) + k_i \int (x_d - x)dt \\
\ddot{y}_d &= k_p(y_d - y) + k_d(\dot{y}_d - \dot{y}) + k_i \int (y_d - y)dt
\end{aligned} \tag{1.41}$$

where again, k_p, k_d and k_i are the proportional, derivative and integral gains respectively and x_d and y_d are the desired *x* and *y* position and $\dot{x}_d$ and $\dot{y}_d$ are their desired rate of change.

Plugging the values of the desired accelerations $\ddot{x}_d$ and $\ddot{y}_d$ into Equation 1.39, the desired roll and pitch angles ϕ_d and θ_d can be calculated which are in turn fed to the attitude controller previously expressed in Equation 1.40.

PID Controller Simulation

For the altitude controller, GA was used to choose the control gains for the PID controller with a desired altitude z_d of 2 m. No steady state error was observed, so the PID controller was simplified to a PD controller by settling the integral gain k_i to zero. The control gains produced by the GA were k_p=5.2 and k_d=1.3. The objective function used to evaluate the GA was the settling time of the system. GA works to find control gains that would result in the least possible settling time for the altitude of the quadrotor. Running the closed loop simulation with the acquired gains resulted in a settling time of 1.3 sec and an overshoot of 1.4%.

Similarly, attitude, heading and position controllers gains were optimized using GA, Table 1 shows a summary of the optimized control gains and the performance of the system in terms of its settling time and overshoot. The response of the system is shown in Figure 7. Note that the reason the altitude response is in the negative z-axis is our previously assigned N-E-D axes for the quadrotor that the z-axis points downwards.

Due to the symmetry of the quadrotor, the controller for the pitch rotation is equivalent to that of the roll rotation. This theory was verified and proved using the closed loop simulation and their results is shown in the attitude row in Table 1.As with the roll and pitch, the performance of the y position controller was exactly the same as the performance of the x position controller due to the symmetry of the quadrotor.

Thus, a complete position and altitude PD controller was developed for the quadrotor. This controller is able to perform well near hovering. The controller was also tested in commanding the quadrotor to follow a circular trajectory as shown in Figure 8.

Gain Scheduling Based PD Controller

To overcome the shortcomings of the linear PD controller in its ability to only operate in the linear near hover region, a gain scheduling based PD controller is proposed. The theory behind gain scheduling is developing a set of controllers for different operating points and switching between these controllers depending on the operating point of the system [McNichols & Fadali (2003)].In this work, a family of PD controllers will be developed, each PD controller having different controller gains and will be able to stabilize the quadrotor system in a certain range of operation. Gain scheduling will then be used to choose an appropriate controller from the family of developed PD controllers. This approach renders the classical PD controller an adaptive controller since the controller's parameters are adapting to different operating conditions. Similar to the previously implemented controllers, GA was used to acquire the control gains, that would result in the least possible settling time, for the family of PD controllers

Table 1. PD controller results

	Desired Value	k_p	k_d	Settling Time	Overshoot
Altitude (z)	2 m	5.2	1.3	1.3 sec	1.4%
Attitude (φ and θ)	5°	4.5	0.5	0.3 sec	2%
Heading (ψ)	5°	3.9	0.7	0.42 sec	1.9%
Position (x and y)	1 m	7.5	4.2	1.4 sec	1.9%

Figure 7. PD controller simulation response

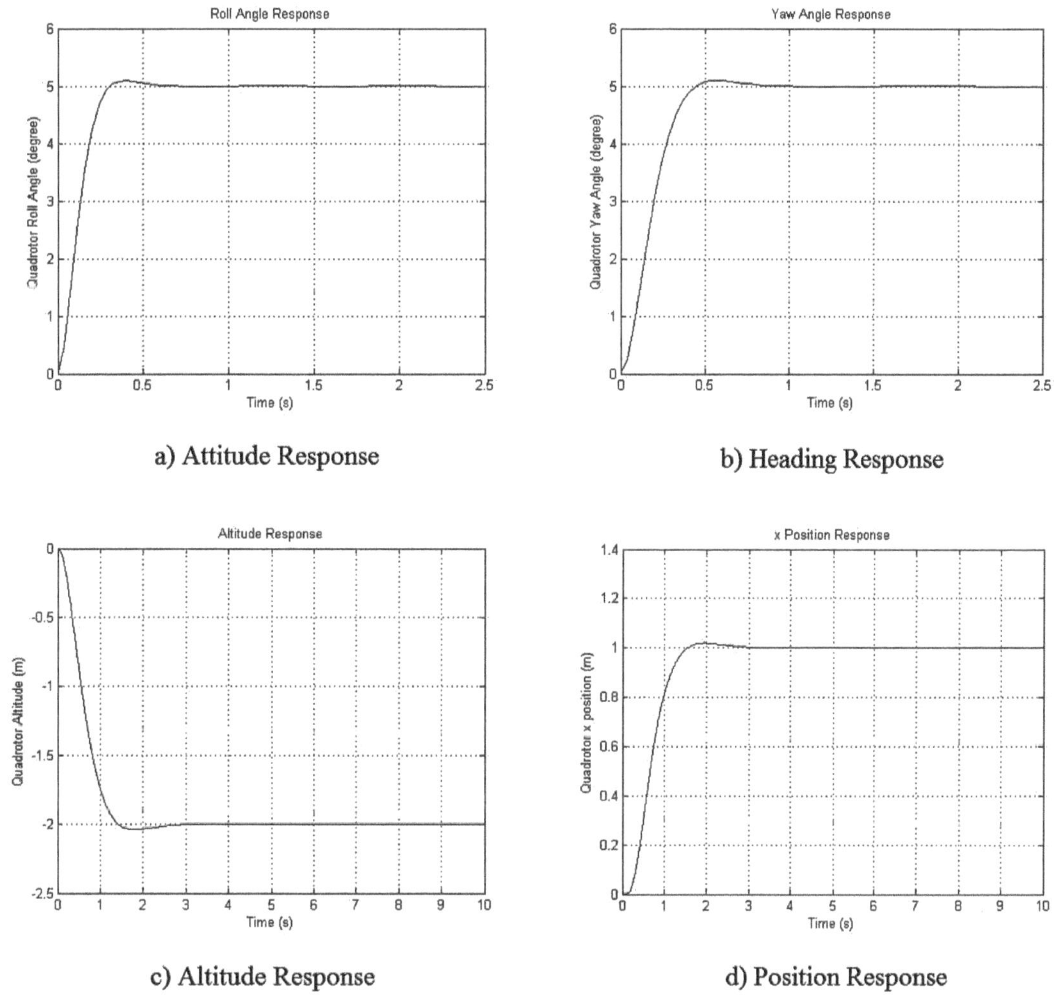

at different operating points. The acquired gains were used in a look up table fashion in the developed MATLAB/Simulink shown in Figure 9.

Gain Scheduling Based PD Controller Simulation Results

Altitude Controller

GA was used to tune the parameters of the PD controller for the system, the parameters for different desired altitudes are shown in Table 2 together with the resulting settling time for the system.

In order to show the strength of the developed gain scheduling based PD controller, it was tested to follow a wide range trajectory unlike the step input that was used in the classical PD. The response shown in Figure 10 compares the performance of the classical PD to the gain scheduled PD.

Figure 8. Trajectory response under PD

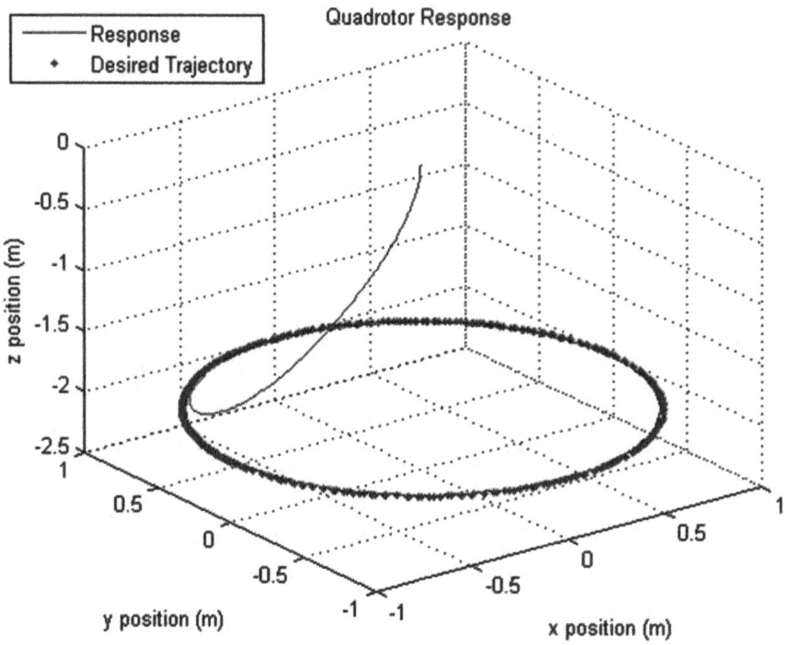

Figure 9. Gain scheduling block diagram

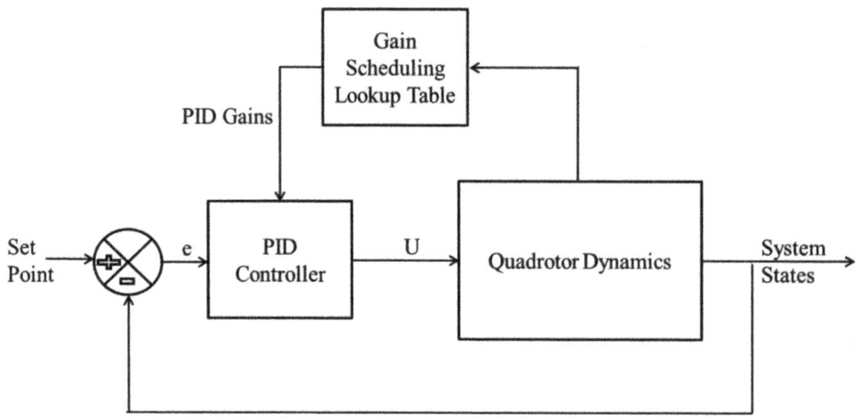

Attitude Controller

For the roll and pitch control, GA was also used to find the controller gains for a set of operating points. Table 3 shows the operating points together with their controller gains and performance.

Figure 11 shows a comparison between the performance of the gain scheduled PD controller and the classical PD controller in following a varying trajectory.

Table 2. Altitude gain scheduling based PD controller gains and results

Desired Altitude	k_p	k_d	Settling Time
1 m	8.91	3.75	0.98 sec
2 m	5.97	3.07	1.24 sec
3 m	4.67	2.71	1.42 sec
4 m	8.77	3.64	1.25 sec
5 m	5.06	2.79	1.51 sec
6 m	6.10	3.04	1.51 sec
7 m	5.14	2.79	1.64 sec
8 m	6.24	3.21	1.69 sec
9 m	4.64	2.67	1.82 sec
10 m	5.69	3.13	1.82 sec

Figure 10. Altitude response

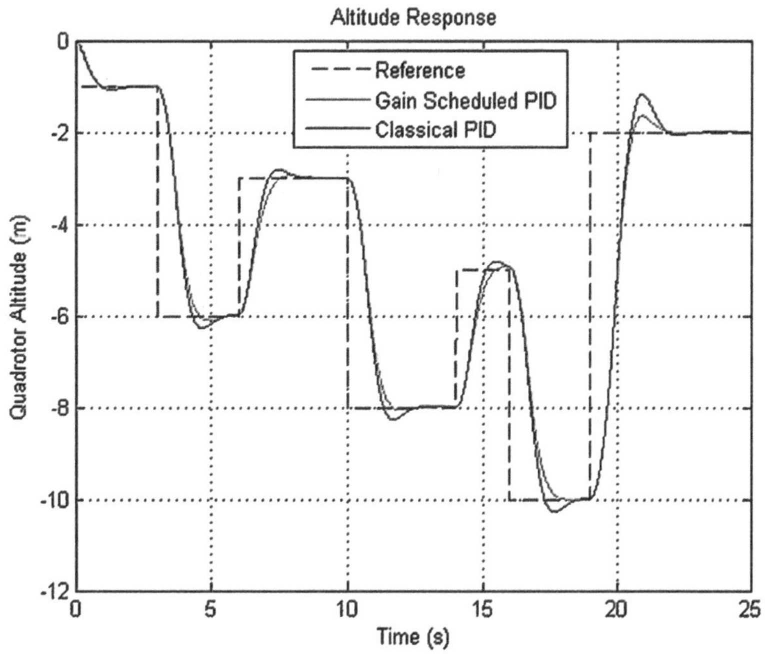

Heading Controller

Similar to the attitude controller, a look up table was synthesized for the heading controller. The controller gains and their respective performances at multiple operating points are shown in Table 4 and the response is shown in Figure 12.

Table 3. Attitude gain scheduling based PD controller gains and results

Desired Attitude	k_p	k_d	Settling Time
2°	6.29	0.694	0.26 sec
4°	5.89	0.675	0.27 sec
6°	7.10	0.737	0.24 sec
8°	7.04	0.742	0.25 sec
10°	4.25	0.573	0.31 sec
12°	5.69	0.661	0.27 sec
14°	5.90	0.678	0.27 sec
16°	5.24	0.637	0.28 sec
18°	3.05	0.486	0.37 sec
20°	5.40	0.657	0.29 sec

Figure 11. Attitude response

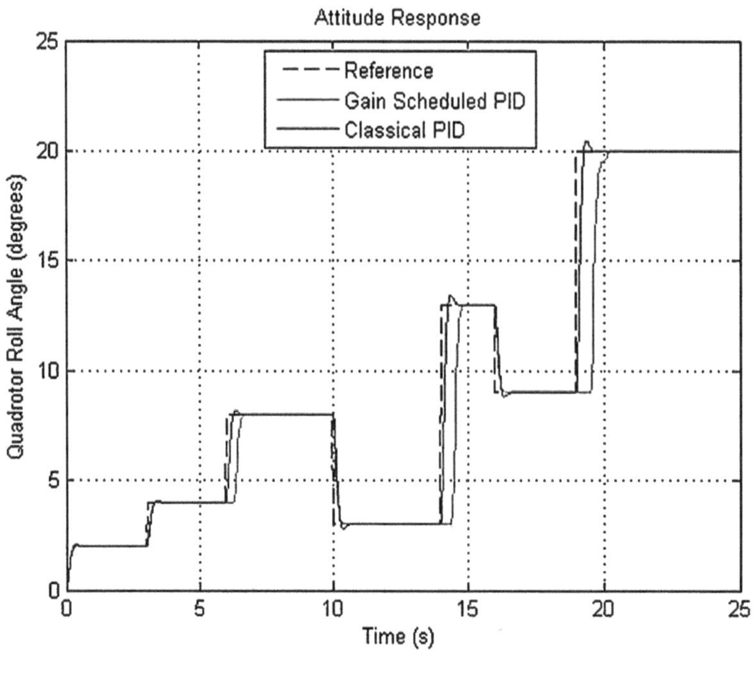

Table 4. Heading gain scheduling based PD controller gains and results

Desired Heading	k_p	k_d	Settling Time
2°	6.25	0.921	0.38 sec
4°	3.28	0.669	0.53 sec
6°	4.96	0.809	0.52 sec
8°	3.75	0.697	0.60 sec
10°	4.00	0.775	0.64 sec
12°	3.94	0.816	0.69 sec
14°	4.51	0.991	0.75 sec
16°	2.27	0.570	0.825 sec
18°	3.31	0.821	0.84 sec
20°	4.70	1.20	0.88

Sliding Mode Controller

Since the quadrotor system is a nonlinear type system, we proposed using a nonlinear Sliding Mode Controller (SMC) to control the states of the quadrotor.

Introduction to SMC

A SMC is a type of Variable Structure Control (VSC). It uses a high speed switching control law to force the state trajectories to follow a specified, user defined surface in the state space and to maintain the state trajectories on this surface [Liu & Wang (2012)].The control law for a SMC consists of two parts as per Equation 1.42; a corrective control part and an equivalent control part. The corrective control function is to compensate any variations in the state trajectories from the sliding surface in order to reach it. The equivalent control on the other hand, makes sure the time derivative of the surface is maintained to zero, so that the state trajectories would stay on the sliding surface.

$$U(t) = U_c(t) + U_{eq}(t) \tag{1.42}$$

where $U(t)$ is the control law, $U_c(t)$ is the corrective control and $U_{eq}(t)$ is the equivalent control.

A block diagram showing the SMC is shown in Figure 13.

The developed sliding mode controller is based on the approach used by [Bouabdallah & Siegwart (2005)]. The SMC is used to track a reference trajectory for the roll angle. The error in the roll is defined as,

Figure 12. Heading response

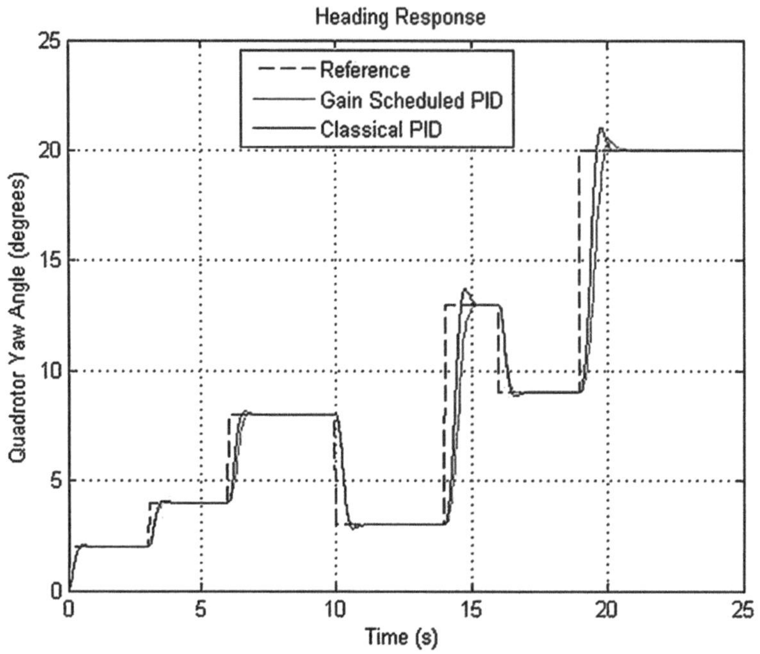

Figure 13. SMC block diagram

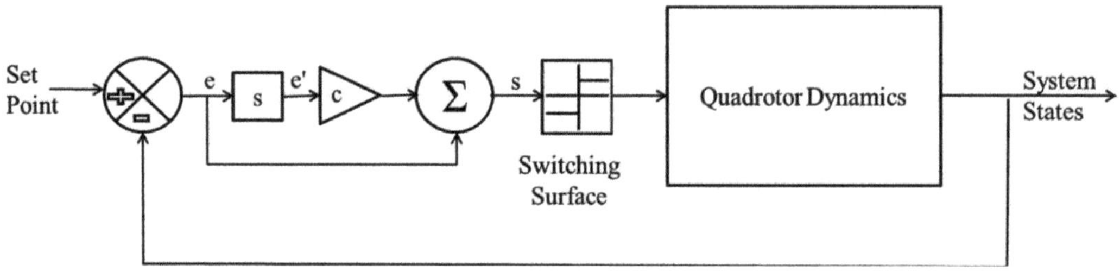

$$e = \phi_d - \phi \tag{1.43}$$

The sliding surface is defined as,

$$s = ce + \dot{e} \tag{1.44}$$

where c is a constant that has to be greater than zero. This format is a common format for the sliding surface in tracking problems. The derivative of the sliding surface defined in Equation 1.44 with the substitution of Equation 1.43 is formulated as the following,

$$\begin{aligned} \dot{s} &= c_1\dot{e} + \ddot{e} \\ &= c_1(\dot{\phi}_d - \dot{\phi}) + \ddot{\phi}_d - \ddot{\phi} \end{aligned} \tag{1.45}$$

A Lyapunov function is then defined to be,

$$V(e,s)=\frac{1}{2}(e^2+s^2) \tag{1.46}$$

Based on the Lyapunov function, an exponential reaching law is proposed for the sliding mode controller as follows,

$$\dot{s}=-k_1 sgn(s)-k_2 s \tag{1.47}$$

where

$$sgn(s)=\begin{cases}-1 & \text{if } s<0;\\ 1 & \text{if } s>0.\end{cases} \tag{1.48}$$

and k_1 and k_2 are design constants. To satisfy the sliding mode condition $s\dot{s}<0$, limits has to be set on k_1 and k_2 such as $k_1>0$ and $k_2>0$. By equating the proposed reaching law 1.47 to the derivative of the sliding surface in Equation 1.45 and substituting $\ddot{\phi}$ by its definition from the rotational equations of motion, the control input U_2 is calculated to be,

$$U_2=\frac{1}{b_1}[k_1 sgn(s)+k_2 s+c_1(\dot{\phi}_d-\dot{\phi})+\ddot{\phi}_d+a_2\dot{\theta}\Omega_r-a_1\dot{\theta}\dot{\psi}] \tag{1.49}$$

Similarly, if the same steps are repeated for altitude, pitch and heading; the control inputs U_1, U_3 and U_4 are calculated to be,

$$\begin{aligned}U_1&=\frac{m}{\cos\phi\cos\theta}[k_1 sgn(s)+k_2 s+c_1(\dot{z}-\dot{z}_d)+g-\ddot{z}_d]\\ U_3&=\frac{1}{b_2}[k_1 sgn(s)+k_2 s+c_1(\dot{\theta}_d-\dot{\theta})+\ddot{\theta}_d-a_4\dot{\phi}\Omega_r-a_3\dot{\phi}\dot{\psi}]\\ U_4&=\frac{1}{b_3}[k_1 sgn(s)+k_2 s+c_1(\dot{\psi}_d-\dot{\psi})+\ddot{\psi}_d-a_5\dot{\phi}\dot{\theta}]\end{aligned} \tag{1.50}$$

SMC Simulation Results

As for the previously implemented PD controller, GA was used to find the design parameters (c_1, k_1 and k_2) for the implemented SMCs to achieve the least settling time for the system. Table 5 shows the GA generated design parameters and the resulting settling time and overshoot of the system. The response is shown in Figure 14. The problem of chattering is clear in the response plots and the problem to overcome it will be addressed in the next section.

Due to the symmetry of the quadrotor, the results for the pitch controller were exactly the same as that of the roll and accordingly the GA produced the same control gains.

SMC Chattering Reduction

As shown in Figure 14, there is a clear chattering effect which is a common outcome of a SMC due to its switching nature.

The presence of the *sgn* term in the SMC's control law makes it a discontinuous controller. Figure 15 shows that whenever the value of the surface s is positive, the control law works to decrease the trajectory to reach the sliding surface $s=0$ at point a. Ideally it should continue sliding on the surface once hitting it, but due to the delay between the change of sign of s and the change in the control action, the trajectory passes the surface to the side ($s<0$). Accordingly, the control law works to drive the trajectory again to ($s=0$), yet it passes it and this causes the famous chattering effect. The main drawbacks of chattering are that it causes the excitation of unmodeled system dynamics that yields a possible instability of the system. In addition to that it is associated with a high power consumption and possible actuator damage. These drawbacks make the SMC hard to be implemented on real systems [González et al. (2014)].

Re-Tuning Using GA

In a trial to eliminate chattering effect, we proposed a mathematical formula to have a numerical value for the chattering based on calculating the difference in areas under the desired and actual response graphs starting from the settling time until the end of the simulation time.

$$chat = \int_{T_s}^{T_{end}} \left\| \phi - \phi_d \right\| dt \tag{1.51}$$

where T_s is the settling time for the system response and T_{end} is the simulation end time.

By visual inspection, Equation 1.51 did not provide accurate means of numerically measuring the chattering of the SMC. Hence, another method was proposed and verified for measuring the chattering which depends on the time derivative of the sliding surface s. The lower $\dot{s}$ is, the lower the chattering,

$$chat = \max_{T_s \rightarrow T_{end}} \dot{s} \tag{1.52}$$

Table 5. SMC results

	Desired Value	c	k_1	k_2	Setting Time	Overshoot
Altitude (z)	2 m	6.98	2.66	6.64	0.57 sec	2%
Attitude(φ and θ)	5°	4.68	1.99	1.80	0.8 sec	1.9%
Heading (ψ)	5°	5.24	1.72	4.33	0.74 sec	1.7%

Figure 14. SMC controller simulation response

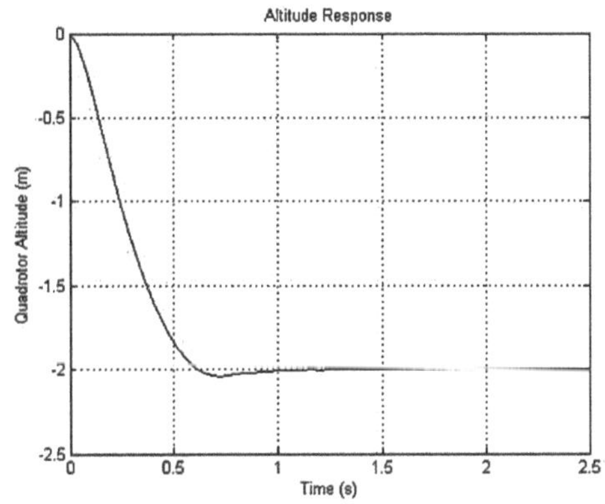

a) Altitude Response

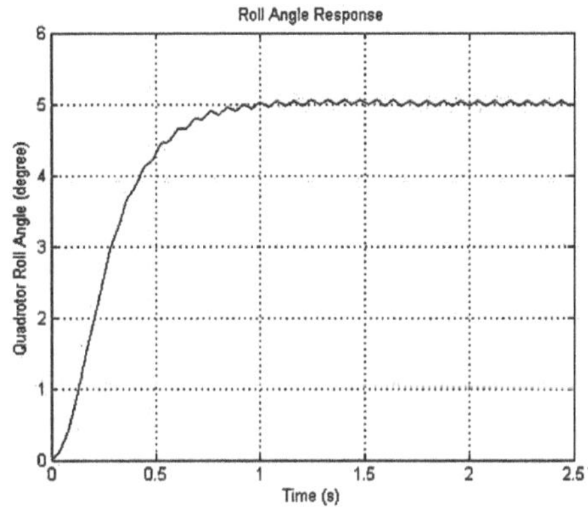

b) Attitude Response

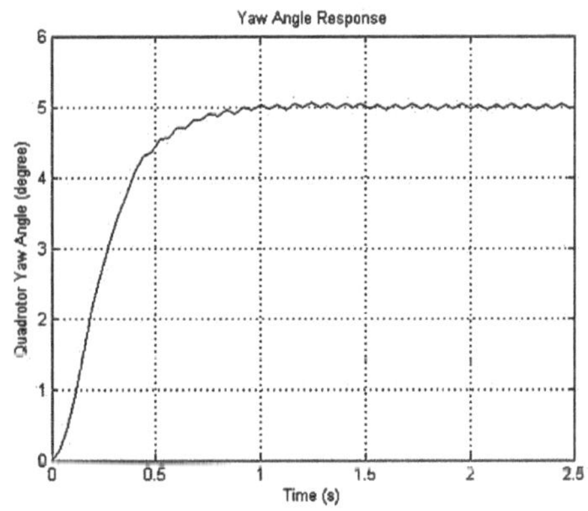

c) Heading Response

Figure 15. The chattering effect
[González et al. (2014)].

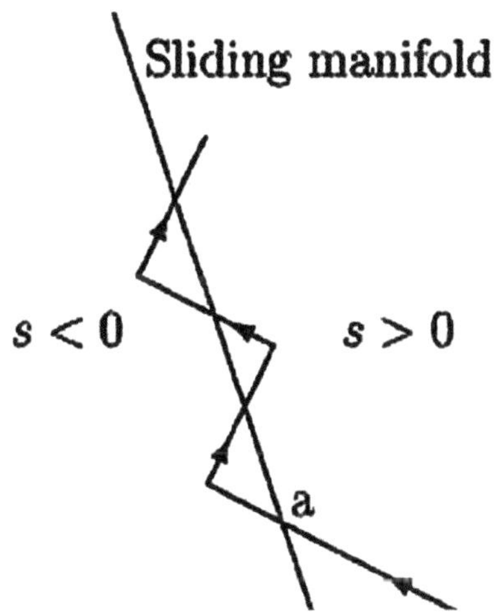

Simulations

The result of Equation 1.52 was used as an objective function for the GA to calculate the design gains. Table 6 shows the GA generated design parameters that result in the least chattering and based on these parameters the resulting settling time and overshoot of the system. The response is shown in Figure 16.

Using the chattering as the objective function of the GA led to the desired reduction of chattering but as a counter effect, the settling time of the system was largely affected. It resulted in an increase in the altitude's settling time from 0.57 sec to 3.11 sec. The attitude's settling time increased from 0.8 sec to 15 sec. While that of the heading increased from 0.74 to 5.25 sec.

Discontinuous to Continuous Control Law

Another approach was proposed to get rid of the chattering effect. As stated earlier, the reason behind chattering is the discontinuity of the control law due to the presence of *sgn* function which is a discontinuous function. Replacing *sgn* with the saturation function *sat* will change the control law to be a continuous one instead of discontinuous. Figure 17 shows graphs of both functions. The *sat* function is defined by,

$$sat(s) = \begin{cases} s & \text{if } \mid s \mid\leq 1; \\ sgn(s) & \text{if } \mid s \mid> 1. \end{cases} \tag{1.53}$$

Thus, to implement this modification on the SMCs for our system, the *sgn(s)* terms in the control laws should be replaced by $sat(s \,/\, \epsilon)$. Where ϵ is a constant that determines the slope of the line between 1 and -1, this region is called the boundary region or boundary layer. Higher values ϵ means a thicker boundary layer and thus an increase in the error [Fuh et al. (2008)].

Simulations

Using the obtained control parameters in Table 5 with the modified control laws (replacing *sgn(s)* by $sat(s \,/\, \epsilon)$ in the control laws), chattering was eliminated as shown in Figure 18, the response graphs for altitude, attitude and heading respectively. Moreover, the quadrotor was commanded to follow a circular trajectory and its response graph is shown in Figure 19.

Backstepping Controller

In this section, a Backstepping controller is used to control the attitude, heading and altitude of the quadrotor. The Backstepping controller is based on the state space model previously derived.

Table 6. SMC results with minimal chattering

	Desired Value	c	k_1	k_2	Setting Time
Altitude (z)	2 m	2.84	0.0011	1.49	3.11 sec
Attitude(φ and θ)	5°	1.49	0.0062	0.0474	15 sec
Heading (ψ)	5°	4.94	0.0016	0.7334	5.25 sec

Figure 16. SMC controller simulation response with chattering reduction

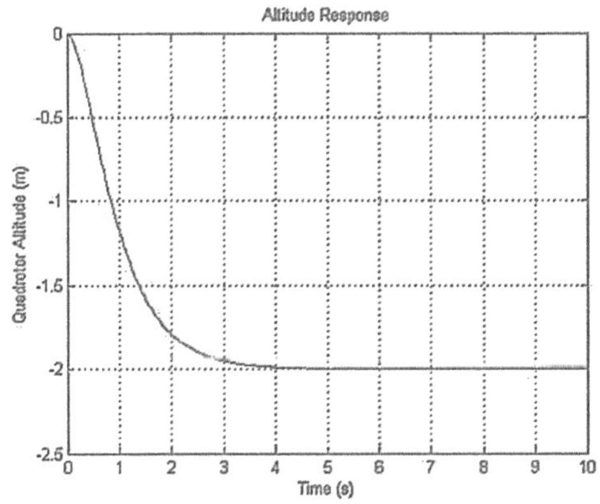

a) Altitude Response

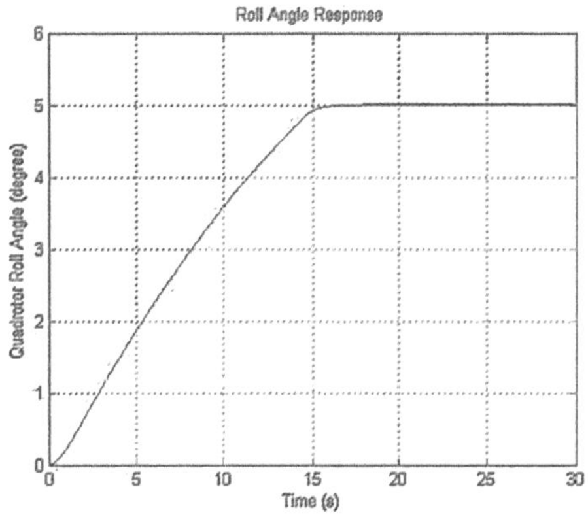

b) Attitude Response

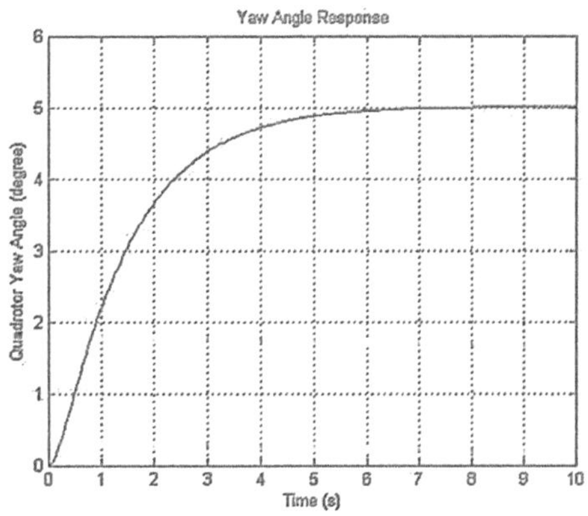

c) Heading Response

Introduction to Backstepping

Backstepping is a recursive control algorithm that works by designing intermediate control laws for some of the state variables. These state variables are called "virtual controls" for the system [Krstic et al. (1995)]. Unlike other control algorithms that tend to linearize nonlinear systems such as the feedback linearization algorithm, backstepping does not work to cancel the nonlinearities in the system. This leads to more flexible designs since some of the nonlinear terms can contribute to the stability of the system. An example of such terms that add to the stability of the system are state variables taking the form of negative terms with odd powers (e.g. $-x^3$), they provide damping for large values of x [Marquez (2003), Krstic et al. (1995)].

Figure 17. Sgn vs. sat functions
[González et al. (2014)].

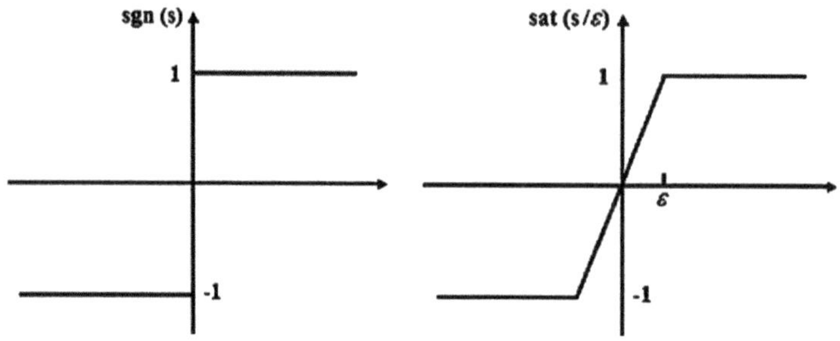

Figure 18. Modified SMC controller simulation

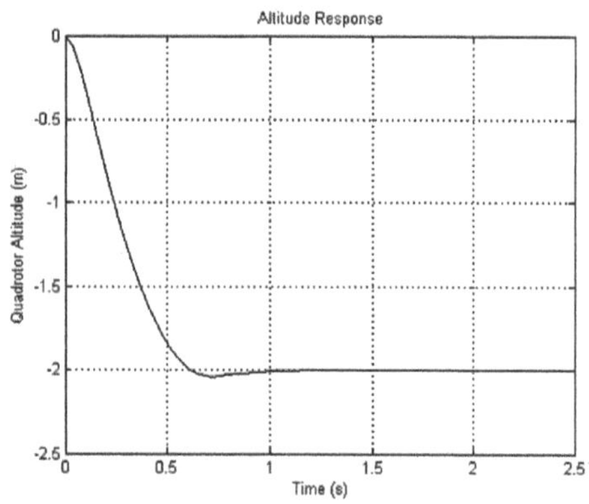

a) Altitude Response

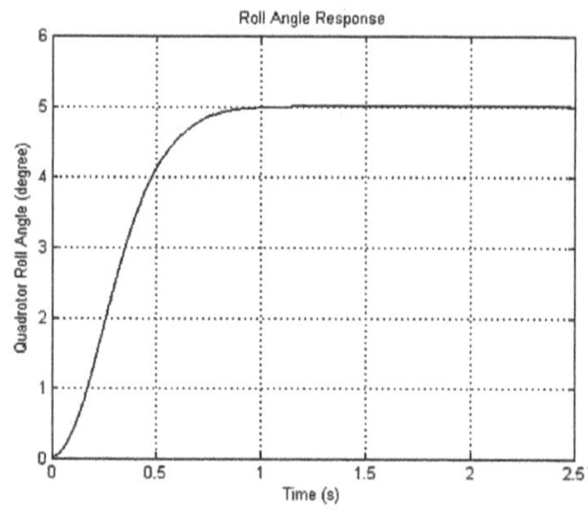

b) Attitude Response

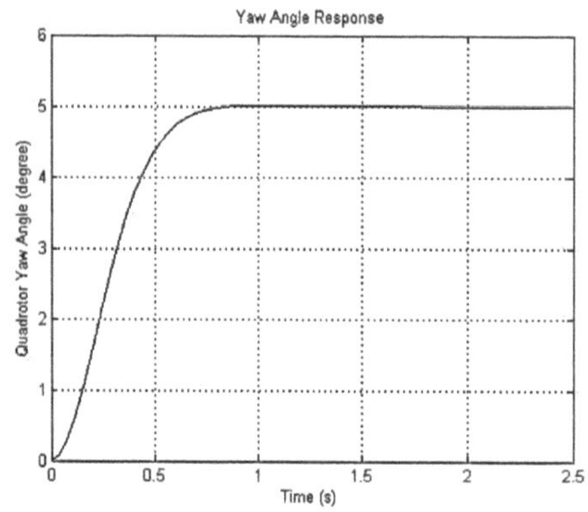

c) Heading Response

Figure 19. Trajectory response under SMC

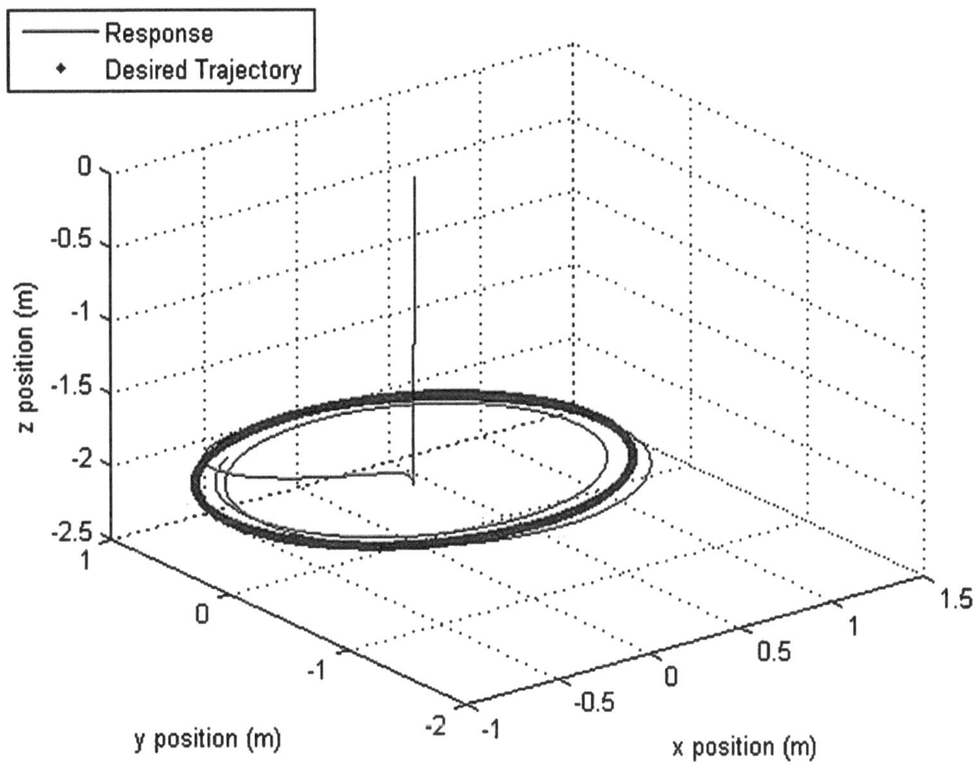

Attitude and Heading Control

The backstepping controller implemented to control the quadrotor's orientation is based on the control approaches proposed in [Nagaty et al. (2013)] and [Bouabdallah & Siegwart (2005)]. For the roll controller, the first two states of the state space model are used which are the roll angle and its rate of change. Extracting those we get,

$$\begin{aligned} \dot{x}_1 &= x_2 \\ \dot{x}_2 &= x_4 x_6 a_1 - x_4 \Omega_r a_2 + b_1 U_2 \end{aligned} \tag{1.54}$$

The roll angle subsystem is in the strict feedback form (only the last state is a function of the control input U_2) which makes it easy to pick a positive definite Lyapunov function for it,

$$V_1 = \frac{1}{2} z_1^2 \tag{1.55}$$

where z_1 is the error between the desired and actual roll angle defined as follows,

$$z_1 = x_{1d} - x_1 \tag{1.56}$$

The time derivative of the Lyapunov function defined in Equation1.52 is derived to be,

$$\begin{aligned} \dot{V}_1 &= z_1 \dot{z}_1 \\ &= z_1(\dot{x}_{1d} - \dot{x}_1) \end{aligned} \tag{1.57}$$

and from Equation 1.54 this can be rewritten as,

$$\dot{V}_1 = z_1(\dot{x}_{1d} - x_2) \tag{1.58}$$

According to Krasovskii--LaSalle principle, the system is guaranteed to be a stable system if the time derivative of a positive definite Lyapunov function is negative semi-definite [Krstic et al. (1995)]. To achieve this, we choose a positive definite bounding function $W_1(z) = c_1 z_1^2$ to bound $\dot{V}_1$ as in Equation 1.59. This choice of $W_1(z)$ is also a common choice for a bounding function for strict feedback systems [Krstic et al. (1995)].

$$\dot{V}_1 = z_1(\dot{x}_{1d} - x_2) \leq -c_1 z_1^2 \tag{1.59}$$

where c_1 is a positive constant. To satisfy this inequality the virtual control input can be chosen to be,

$$(x_2)_{desired} = \dot{x}_{1d} + c_1 z_1 \tag{1.60}$$

Defining a new error variable z_2 to be the deviation of the state x_2 from its desired value,

$$z_2 = x_2 - \dot{x}_{1d} - c_1 z_1 \tag{1.61}$$

Rewriting Lyapunov's function time derivative $\dot{V}_1$ in the new coordinate (z_1, z_2) we get,

$$\begin{aligned} \dot{V}_1 &= z_1 \dot{z}_1 \\ &= z_1(\dot{x}_{1d} - x_2) \\ &= z_1(\dot{x}_{1d} - (z_2 + \dot{x}_{1d} + c_1 z_1)) \\ &= -z_1 z_2 - c_1 z_1^2 \end{aligned} \tag{1.62}$$

Note that the presence of the term $z_1 z_2$ in $\dot{V}_1$ may not lead to a negative semi-definite time derivative but this will be taken care of in the next iteration of the backstepping algorithm. The next step is to augment the first Lyapunov function V_1 with a quadratic term in the second error variable z_2 to get a positive definite V_2,

$$V_2 = V_1 + \frac{1}{2} z_2^2 \tag{1.63}$$

with time derivative,

$$\begin{aligned} \dot{V}_2 &= \dot{V}_1 + z_2 \dot{z}_2 \\ &= -z_1 z_2 - c_1 z_1^2 + z_2(\dot{x}_2 - \ddot{x}_{1d} - c_1 \dot{z}_1) \end{aligned} \tag{1.64}$$

Choosing the positive definite bounding function to be $W_2(z) = -c_1 z_1^2 - c_2 z_2^2$ where c_2 is a positive definite and substituting by the value of $\dot{x}_2$ from Equation 1.54 leads to the following inequality,

$$\dot{V}_2 = -z_1 z_2 - c_1 z_1^2 + z_2(x_4 x_6 a_1 - x_4 \Omega_r a_2 + b_1 U_2 - \ddot{x}_{1d} - c_1 \dot{z}_1) \leq -c_1 z_1^2 - c_2 z_2^2 \tag{1.65}$$

Solving the last inequality, the control input U_2 can be written as,

$$U_2 = \frac{1}{b_1}(-c_2 z_2 + z_1 - x_4 x_6 a_1 + x_4 \Omega_r a_2 + \ddot{x}_{1d} + c_1 \dot{x}_{1d} - c_1 x_2) \tag{1.66}$$

Repeating exactly the same steps for the pitch, heading and altitude, control laws are found to be,

$$\begin{aligned} U_1 &= \frac{m}{\cos x_1 \cos x_3}(-z_7 + g - \ddot{x}_{7d} - c_7 \dot{x}_{7d} + c_7 x_8 + c_8 z_8) \\ U_3 &= \frac{1}{b_2}(-c_4 z_4 + z_3 - x_2 x_6 a_3 - x_2 \Omega_r a_4 + \ddot{x}_{3d} + c_3 \dot{x}_3 d - c_3 x_4) \\ U_4 &= \frac{1}{b_3}(-c_6 z_6 + z_5 - x_2 x_4 a_5 + \ddot{x}_{5d} + c_5 \dot{x}_{5d} - c_5 x_6) \end{aligned} \tag{1.67}$$

where

$$\begin{aligned} z_3 &= x_{3d} - x_3 \\ z_4 &= x_4 - \dot{x}_{3d} - c_3 z_3 \\ z_5 &= x_{5d} - x_5 \\ z_6 &= x_6 - \dot{x}_{5d} - c_5 z_5 \\ z_7 &= x_{7d} - x_7 \\ z_8 &= x_8 - \dot{x}_{7d} - c_7 z_7 \end{aligned} \tag{1.68}$$

Backstepping Controller Simulation Results

The control inputs U_1 through U_4 derived in the previous section were added to the previously implemented simulation model and similar to the PD and the SMC controllers, GA was used to tune the parameters of the Backstepping controllers. The parameters to be tuned are c_1 through c_8. The objective function used for the GA was the settling time of the system. Table 7 shows the optimized parameters acquired from the GA and the resulting settling time for the attitude, heading and altitude of the quadrotor. Figure 20 shows the response when running the simulation with the constants in Table 7.

RESULTS DISCUSSION

Varying Trajectory

To be able to compare fairly between the four implemented control techniques, the response graph of the system under the effect of each the four controllers was plotted superimposed on one another. Figure 21 shows the altitude response while Figure 22 and Figure 23 show the attitude and heading responses respectively.

Performance in a Windy Environment

Disturbance was then added to the quadrotor model in the form of additional forces and moments to give the effect of operating the quadrotor in a windy environment. The forces were added to the right hand side of the system's translational equation of motion as Gaussian noise with zero mean and with a maximum value of 1 N. The added moments were also added to the right hand side of the system's rotational equation of motion as Gaussian noise with zero mean and a maximum value of 0.5 Nm. The system was commanded to follow a certain desired altitude and attitude. The performance of the system under the effect of wind is shown in Figure 24 for PD, SMC and Backstepping.

Nonhovering Operation

In order to operate the system outside its linear region (the hovering condition) a nonlinear controller has to be used. Applying the SMC and the Backstepping controller, the system response for the altitude, attitude and heading is shown in Figure 25. Since the SMC and the Backstepping controller are nonlinear controllers, their tuning is not affected by the operating region. Thus, the control gains acquired in Table 5 and Table 7 were used.

Table 7. Backstepping controller constants and results

	Desired Value	$c_1/c_5/c_7$	$c_2/c_6/c_8$	Setting Time	Overshoot
Attitude (φ and θ)	5°	5.52	3.40	0.80 sec	1.9%
Heading (ψ)	5°	3.07	4.71	0.86 sec	1.6%
Altitude (z)	2 m	6.11	7.96	0.58 sec	2%

Figure 20. Backstepping controller simulation response

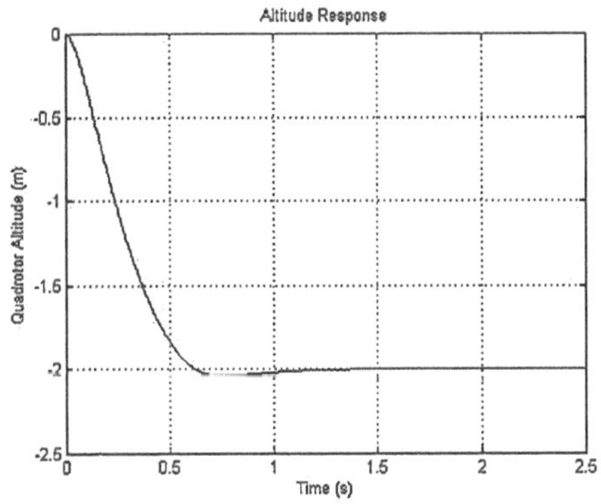

a) Altitude Response

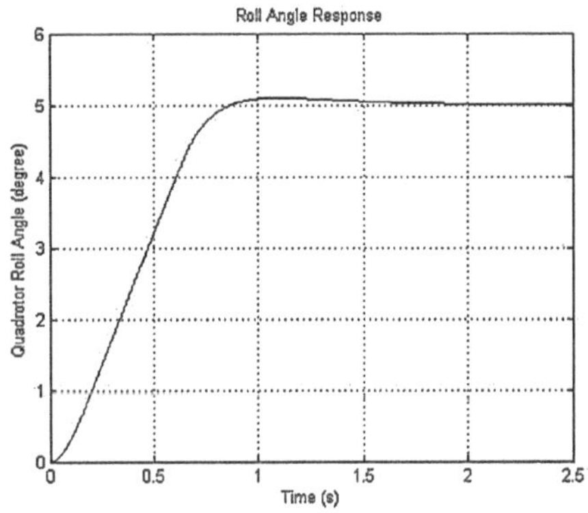

b) Attitude Response

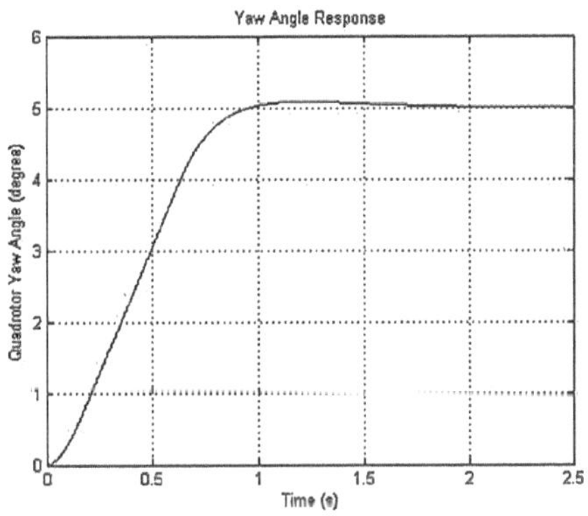

c) Heading Response

Figure 21. Altitude response

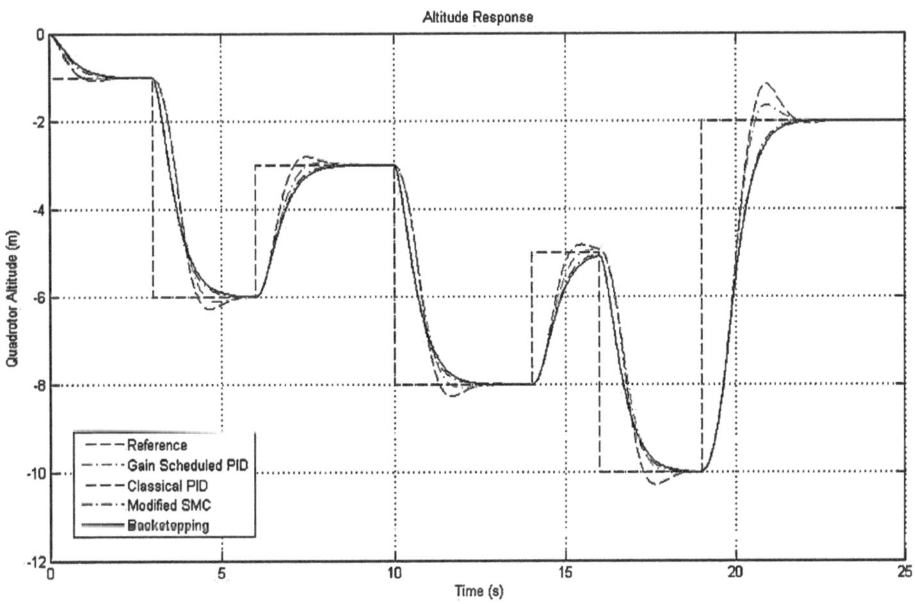

Figure 22. Attitude response

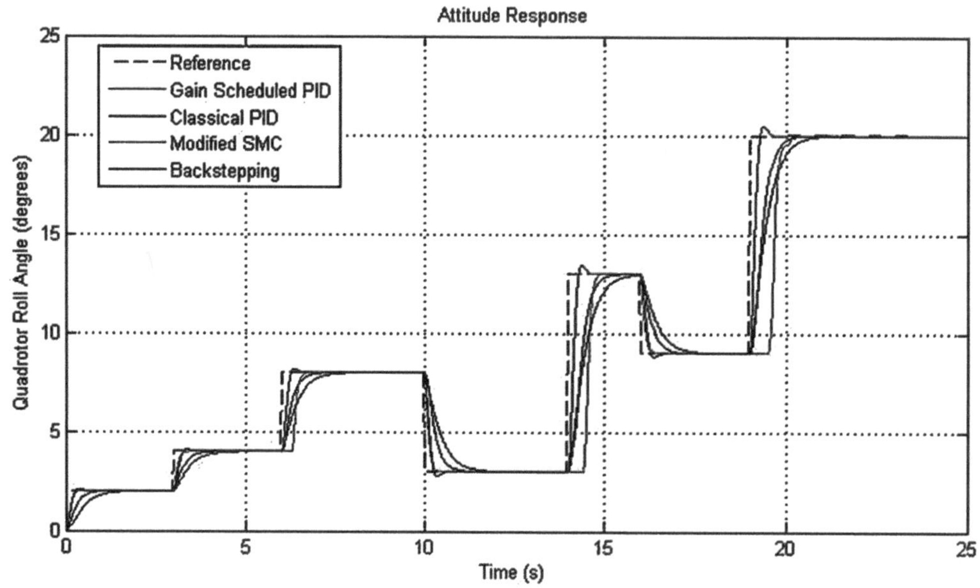

Figure 23. Heading response

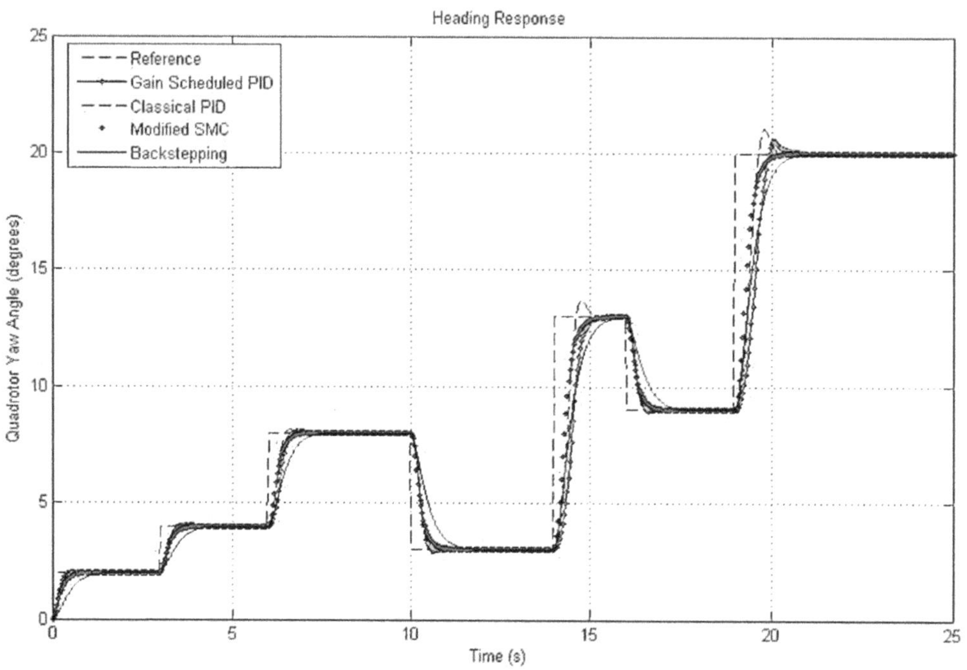

Figure 24. System response in a windy environment

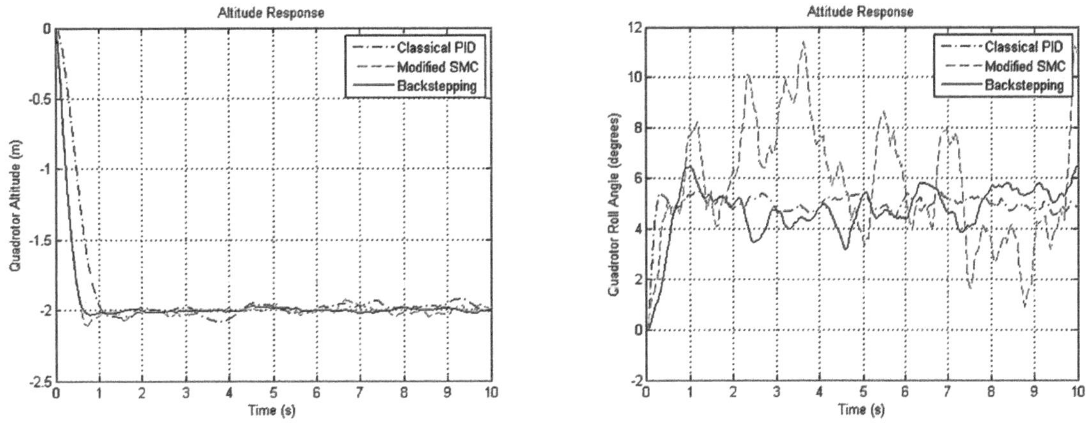

a) Altitude Response with a desired altitude of 2m

b) Attitude Response with a desired attitude of 5 degrees

When the PD controller was used to operate the system outside its linear region (more than 20°) of orientation, the system went unstable.

Summary of Findings

Linear Operation

The four employed controllers developed to control the quadrotor model under consideration gave comparable dynamic performances in terms of settling time and overshoot when they were deployed in near hover stabilization of the quadrotor. When first implemented, the SMC resulted in an undesirable chattering effect which was very notable in the attitude response unlike the altitude's. This chattering effect was then eliminated by using a modified version for the control law rendering the response chattering free.

Nonlinear Operation

When the controllers were used outside of the linear region (away from hover), the PD controller failed to stabilize the system due to the fact that PD comes from a family of linear controllers. On the other hand, the SMC and the Backstepping controller were able to stabilize the system with a good dynamic performance. The developed gain scheduled PD controller was also not able to stabilize the system as it is also based on the linear PD controller.

PD

One notable advantage of the PD controller over the other implemented controllers is that its control law is not a function of the system parameters; it is only a function of the state error and its derivative

Figure 25. System response when operated outside the linear region

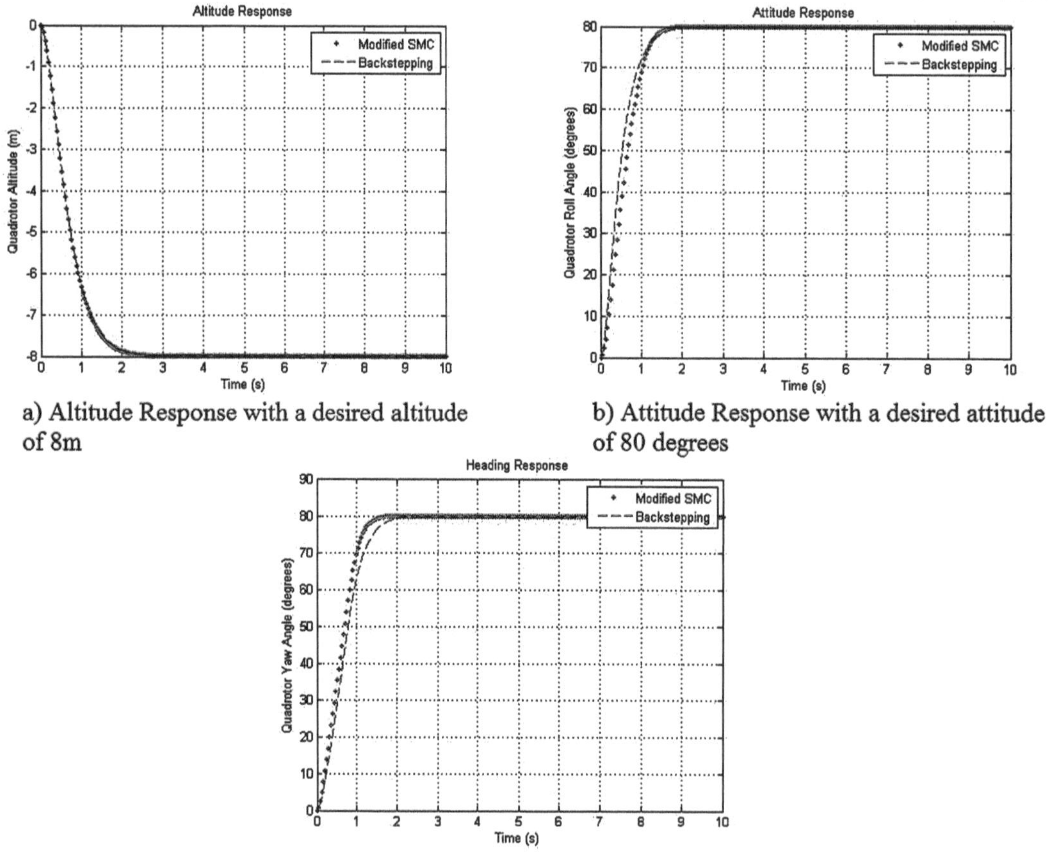

a) Altitude Response with a desired altitude of 8m

b) Attitude Response with a desired attitude of 80 degrees

c) Heading Response with a desired heading of 80 degrees

making the control law less computationally intensive and easier to implement. Also, the controller is less prone to slight variations or uncertainties in system parameters.

Gain Scheduled PD

A gain scheduling based PD controller is essentially a PD controller with its gains tuned for a different set of operating conditions, thus its performance at following a fixed value trajectory is the exactly the same as the performance of the classical PD. On the other hand, the gain scheduled PD controller performs better than the traditional PD controller when following a varying trajectory, which is a more practical or realistic application for a quadrotor UAV. A quadrotor is more likely to be commanded follow a changing trajectory rather than flying to fixed place in space and hovering or maintaining its position there.

A worth mentioning drawback of the Gain Scheduling algorithm is the criticality of the switching time, the switching from a set of controller gains to the other has to be done in infinitesimally small time to guarantee a good performance. This is a critical issue in some quadrotor applications that mainly rely on Gain Scheduling such as the load drop applications, if the switching is not done once the load is dropped, the quadrotor might overshoot and go unstable.

Windy Environment

When operating in a windy environment, which was simulated by Gaussian noise of zero mean, the performance of the modified SMC suffered a huge degradation. This is due to two reasons; the first of them is that the presence of noise excites the chattering phenomenon and cancels the effect of the boundary layer. The second reason for the degradation of the performance of the SMC is that the system model is changed by the added forces and moments that simulate the windy environment. As a consequence to that change, the time derivative of the Lyapunov functions is not guaranteed to be negative semi-definite anymore thus causing the system to be unstable. While the performance of the PD and the Backstepping controllers was comparable in controlling the quadrotor's altitude, yet Backstepping also suffered slight performance degradation in stabilizing the quadrotor's attitude. This is due to the fact that the Backstepping's reaching law design also depends on a Lyapunov function and the added forces and moments cause its time derivative not to be guaranteed negative semi-definite. Figure 26 shows the time derivative of the Lyapunov function under the effect of wind, while Figure 27 shows it in the no wind condition.

Choice of Controller

The choice of the controller to be used will depend mainly on the application, if the quadrotor is to be operating near a hovering condition, a PD controller will be sufficient to stabilize it. On the other hand, if it will be performing tough acrobatic maneuvers thus operating outside its linear region, a SMC or a Backstepping controller should be employed. The environment too will help make the choice, for example simulations showed that PD and Backstepping controllers were more robust to disturbances which might come in the form of a windy environment.

Figure 26. Lyapunov function derivative under the effect of wind

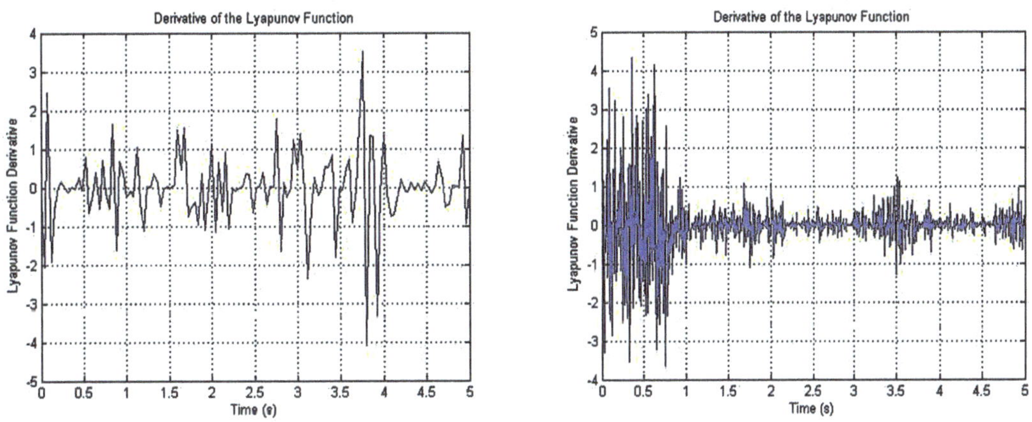

a) SMC

b) Backstepping

Figure 27. Lyapunov function derivative in no wind condition

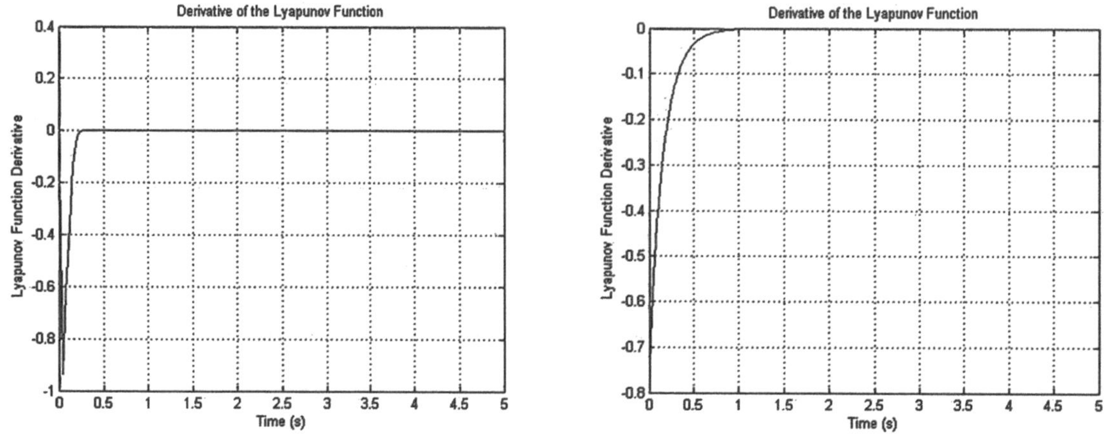

a) SMC

b) Backstepping

CONCLUSION AND FUTURE WORK

The goal of this work was to derive a mathematical model for the quadrotor UAV and develop linear and nonlinear control algorithms to stabilize the states of the quadrotor, which include its altitude, attitude, heading and position in space and to verify the performance of these controllers with comparisons via computer simulations.

The mathematical model of a quadrotor UAV was developed in details including its aerodynamic effects. Four control techniques were then developed and synthesized; a linear Proportional-Integral-Derivative controller, a Gain Scheduling based PD controller, a nonlinear SMC and a nonlinear Backstepping controller. A complete simulation was then implemented on MATLAB/Simulink relying on the derived mathematical model of the quadrotor. The simulation environment was used to evaluate the mentioned controllers and compare their dynamic performances under different types of input conditions.

Tuning the parameters and constants of the four used controllers was done using GA where the objective function was the dynamic response of the system in terms of its settling time and/or overshoot. The four controllers performed comparably in near hovering operation of the quadrotor in the range of $0 \sim 20^{\circ}$ of attitude and heading. The Gain Scheduling based PD controller gave a better performance than the traditional PD controller when the quadrotor was commanded to follow a varying trajectory. The SMC and Backstepping controllers gave better performance outside the linear hovering region due to their nonlinear nature. The PD and Backstepping controllers gave better performance than all the other controllers when the effect of wind was added to the system. The wind effect was modeled as extra forces and moments on the quadrotor body.

For future work, we recommend testing the Gain Scheduled PD controller in load drop applications and actuators failure conditions and compare it to the three other controllers. Also, changing the Gain Scheduled PD to have the switching to be a function of the error and its derivative instead of the desired final value. This might enhance its performance in following a varying trajectory. One valuable addition would be the robustification of the developed control techniques against wind as this is a common problem with quadrotors control and our simulation results showed a huge degradation of the performance of the controllers when the system was exposed to wind. Moreover, in our work it was assumed that all the model parameters are known accurately without any uncertainties, which is not the case in reality, thus, developing adaptive control algorithms to count for the system uncertainties would enhance the performance of the quadrotor when operating in a real environment. Adding an integral action to the developed Backstepping controller will lead to the formulation of an adaptive control algorithm robust to system uncertainties. Moreover, sensors were assumed to be perfect which is not the case in reality, sensors modeling and noise need to be taken into consideration and checking the effect on system stability under the effect of the developed controllers. Also, it is a good idea to study and implement global linearization methods including differential geometric approaches and differential flatness theory. Approximate linearization methods should also be tested, an example to these methods is the nonlinear H∞ control and the Takagi-Sugeno fuzzy control. Last but not least, implementing the developed control techniques on a real quadrotor hardware to give a more fair comparison between their performances.

REFERENCES

Alexis, K., Nikolakopoulos, G., & Tzes, A. (2011). Switching model predictive attitude control for a quadrotor helicopter subject to atmospheric disturbances. *Control Engineering Practice*, *19*(10), 1195–1207. doi:10.1016/j.conengprac.2011.06.010

Amoozgar, M. H., Chamseddine, A., & Zhang, Y. (2012). Fault-tolerant fuzzy gain scheduled pid for a quadrotor helicopter testbed in the presence of actuator faults. In *Proceedings of IFAC Conference on Advances in PID Control*. Academic Press.

Ataka, A., Tnunay, H., Inovan, R., Abdurrohman, M., Preastianto, H., Cahyadi, A., & Yamamoto, Y. (2013). Controllability and observability analysis of the gain scheduling based linearization for uav quadrotor. In *Proceedings of Robotics, Biomimetics, and Intelligent Computational Systems* (ROBIONETICS) (pp. 212–218). IEEE. 10.1109/ROBIONETICS.2013.6743606

Azzam, A., & Wang, X. (2010), Quad rotor arial robot dynamic modeling and configuration stabilization. In *Proceedings of Informatics in Control, Automation and Robotics* (CAR) (Vol. 1, pp. 438–444). Academic Press. 10.1109/CAR.2010.5456804

Bouabdallah, S. (2007). *Design and control of quadrotors with application to autonomous flying*. (Phd. thesis). Ecole Polytechnique Federale de Lausanne.

Bouabdallah, S., Noth, A., & Siegwart, R. (2004). PID vs LQ control techniques applied to an indoor micro quadrotor. In *Proceedings of Intelligent Robots and Systems, 2004* (IROS 2004) (Vol. 3, pp. 2451–2456). Academic Press. 10.1109/IROS.2004.1389776

Bouabdallah, S., & Siegwart, R. (2005). Backstepping and sliding-mode techniques applied to an indoor micro quadrotor. In *Proceedings of Robotics and Automation* (pp. 2247–2252). Academic Press. 10.1109/ROBOT.2005.1570447

Derafa, L., Madani, T., & Benallegue, A. (2006). Dynamic modelling and experimental identification of four rotors helicopter parameters. In *Proceedings of Industrial Technology* (pp. 1834–1839). Academic Press. 10.1109/ICIT.2006.372515

Efe, M. (2011). Neural network assisted computationally simple PID control of a quadrotor UAV. *IEEE Transactions on Industrial Informatics*, *7*(2), 354–361.

Fang, Z. & Gao, W. (2012). Adaptive backstepping control of an indoor micro-quadrotor. *Research Journal of Applied Sciences, 4*.

Fuh, C.-C., & .... (2008). Variable-thickness boundary layers for sliding mode control. *Journal of Marine Science and Technology*, *16*(4), 286–292.

Gillula, J. H., Hoffmann, G. M., Huang, H., Vitus, M. P., & Tomlin, C. J. (2011). Applications of hybrid reachability analysis to robotic aerial vehicles. *The International Journal of Robotics Research*, *30*(3), 335–354. doi:10.1177/0278364910387173

González, I., Salazar, S., & Lozano, R. (2014). Chattering-free sliding mode altitude control for a quad-rotor aircraft: Real-time application. *Journal of Intelligent & Robotic Systems*, *73*(1-4), 137–155. doi:10.100710846-013-9913-8

Hou, H., Zhuang, J., Xia, H., Wang, G., & Yu, D. (2010). A simple controller of minisize quad-rotor vehicle. In *Proceedings of Mechatronics and Automation* (ICMA) (pp. 1701–1706). Academic Press. 10.1109/ICMA.2010.5588802

Kendoul, F. (2012). Survey of advances in guidance, navigation, and control of unmanned rotorcraft systems. *Journal of Field Robotics*, *29*(2), 315–378. doi:10.1002/rob.20414

Kendoul, F., Yu, Z., & Nonami, K. (2010). Guidance and nonlinear control system for autonomous flight of minirotorcraft unmanned aerial vehicles. *Journal of Field Robotics*, *27*(3), 311–334.

Kim, J., Kang, M.-S., & Park, S. (2010). Accurate modeling and robust hovering control for a quadâ€“rotor vtol aircraft. *Journal of Intelligent & Robotic Systems*, *57*(1-4), 9–26. doi:10.100710846-009-9369-z

Krstic, M., Kokotovic, P. V., & Kanellakopoulos, I. (1995). *Nonlinear and adaptive control design*. John Wiley & Sons, Inc.

Lee, H., Kim, S., Ryan, T., & Kim, H. J. (2013), Backstepping control on se (3) of a micro quadrotor for stable trajectory tracking. In *Proceedings of Systems, Man, and Cybernetics* (SMC) (pp. 4522–4527). IEEE.

Li, J., & Li, Y. (2011). Dynamic analysis and PID control for a quadrotor. In *Proceedings of Mechatronics and Automation* (ICMA) (pp. 573–578). Academic Press. 10.1109/ICMA.2011.5985724

Liu, J., & Wang, X. (2012). *Advanced sliding mode control for mechanical systems: Design, analysis and MATLAB simulation*. Springer.

Madani, T., & Benallegue, A. (2006). Backstepping control for a quadrotor helicopter. In *Proceedings of Intelligent Robots and Systems* (pp. 3255–3260). Academic Press. 10.1109/IROS.2006.282433

Marquez, H. J. (2003). *Nonlinear control systems: Analysis and design*. John Wiley.

McNichols, K. H., & Fadali, M. S. (2003). Selecting operating points for discrete-time gain scheduling. *Computers & Electrical Engineering*, *29*(2), 289–301. doi:10.1016/S0045-7906(01)00031-3

Mistler, V., Benallegue, A., & M'Sirdi, N. (2001). Exact linearization and noninteracting control of a 4 rotors helicopter via dynamic feedback. In *Proceedings of Robot and Human Interactive Communication* (pp. 586–593). Academic Press. 10.1109/ROMAN.2001.981968

Nagaty, A., Saeedi, S., Thibault, C., Seto, M., & Li, H. (2013). Control and navigation framework for quadrotor helicopters. *Journal of Intelligent & Robotic Systems*, *70*(1-4), 1–12. doi:10.100710846-012-9789-z

Raffo, G. V., Ortega, M. G., & Rubio, F. R. (2010). An integral predictive/nonlinear hâ^ž control structure for a quadrotor helicopter. *Automatica*, *46*(1), 29–39. doi:10.1016/j.automatica.2009.10.018

Sadeghzadeh, I., Abdolhosseini, M., & Zhang, Y. M. (2012). Payload drop application of unmanned quadrotor helicopter using gain-scheduled PID and model predictive control techniques. In *Intelligent robotics and applications* (pp. 386–395). Springer. doi:10.1007/978-3-642-33509-9_38

Waslander, S., Hoffmann, G., Jang, J. S., & Tomlin, C. (2005). Multi-agent quadrotor testbed control design: integral sliding mode vs. reinforcement learning. In *Proceedings of Intelligent Robots and Systems* (IROS 2005). IEEE. 10.1109/IROS.2005.1545025

Yang, J., Cai, Z., Lin, Q., & Wang, Y. (2013), Self-tuning PID control design for quadrotor UAV based on adaptive pole placement control. In *Proceedings of Chinese Automation Congress (CAC)*. IEEE. 10.1109/CAC.2013.6775734

Zhen, H., Qi, X. & Dong, H. (2013). An adaptive block backstepping controller for attitude stabilization of a quadrotor helicopter. *WSEAS Transactions on Systems & Control, 8*(2).

KEY TERMS AND DEFINITIONS

Backstepping Controller: A recursive nonlinear controller that can be used to control a dynamic system given its model.

Gain Scheduling Control: An adaptive control approach that is used to control nonlinear systems by deploying a set of linear controllers for different linear operating regions.

Genetic Algorithm: An iterative optimization algorithm that works to minimize a given objective function by generating a random population and performing genetic operations to generate a new population.

Proportional Derivative (PD) Controller: A simplification of a PID controller, consisting only of a proportional and a derivative term.

Proportional Integral Derivative (PID) Controller: A linear controller that can be used to control any dynamic system given its model, it consists of a proportional, an integral and a derivative term.

Quadrotor: A class of UAV that flies with the aid of four separately powered rotors.

Sliding Mode Controller (SMC): A nonlinear controller of a switching nature that can be used to control any dynamic system given its model.

Unmanned Aerial Vehicle (UAV): A flying vehicle that is flown without a pilot which can be controlled autonomously or remotely.

This research was previously published in the Handbook of Research on Advancements in Robotics and Mechatronics edited by Maki K. Habib, pages 408-454, copyright year 2015 by Engineering Science Reference (an imprint of IGI Global).

Chapter 3
Cockroach Inspired Shelter Seeking for Holonomic Swarms of Flying Robots

Hamoon Shahbazi
Luleå University of Technology, Sweden

Jan Carlo Barca
Monash University, Australia

ABSTRACT

In computer science, the study of mimicking nature has given rise to Swarm Intelligence, a distributed system of autonomous agents interacting with each other to collectively perform intelligent tasks. This chapter investigates how groups of holonomic flying robots such as quad copters can seek shelter autonomously when encountering bad weather. In this context three alternative autonomous shelter seeking techniques that address the unsolved plateau-problem had to be implemented. The methods were inspired by cockroaches and hunting strategies observed in apex predators. Previous studies on cockroaches have provided facts about their behaviour and resulted in algorithms that can be used for robotic systems. This research builds on these previous studies by formulating three alternative techniques and carrying out a comprehensive analysis of their performance. Simulation results confirm a scalable system where swarms of flying robots successfully find shelters in 3-D environments.

INTRODUCTION

Robotic systems have received significant attention over the last decades. We are on the verge of entering a new phase where the next generation of robotic technology is integrated in our daily lives. This may be beneficial in many ways. For instance, when humans are not prone to take on a certain task, robots can be used to avoid exposure to danger. In some cases robots replace humans for repetitive tasks or simply because they are more efficient in strength, speed or accuracy (Campo, 2010). Different types of robots (e.g. crawling, climbing or flying robots) engender their own unique benefits. For example, one of the many distinctive advantages of flying robots is their broad view coverage. In fact, these types of robots

DOI: 10.4018/978-1-5225-8365-3.ch003

are used today to help Japanese farmers monitor their crops and spray pesticides (Greiner, 2013). Other UAS protect the wildlife in Africa by tracking poachers. The future of flying robots looks promising. For instance, swarms of tiny flying robots could quickly, safely and cost efficiently sweep and visually inspect infrastructures such as bridges and dams. The possible applications are almost endless, e.g. firefighting, police and border observation, search and rescue, mapping, radiation detection, damage observation and assessment (UAV Design, 2013). Some of these applications are already in the development stage (Liu et al., 2010). David G. Green (2014) mentions a prosperous future for swarm robotics with the evolution of nanotechnology. He mentions that small nano-bots can extract contaminants from different mixtures in food or even in our bodies (tracking and removing viruses). As the technology in this particular field will continue to evolve, surely the area of application can be extended even further. Perhaps even to planetary exploration, where flying robots can replace ground moving rovers or even satellites in search for life or close mapping on other earth-like planets (Påhlsson et al., 2011).

Before we can send robot swarms to other planets, we need to make sure they can survive on their own, i.e. make them completely autonomous. Robots used for outdoor missions are exposed to a world of harsh environments. Shelter seeking is for that reason very important if the robots shall remain independent. It is a self-defensive mechanism that may protect them in situations where there is a risk for mission, or even system failure. Shelters are not always easy to reach, and in some cases the capacity of a shelter is low, forcing the swarm formation to morph, by reducing the distance between each robot. For UAVs the minimum distance to objects is crucial to avoid inter-robot collisions and collisions with environmental objects. On top of that, flight instability may occur as a consequence of turbulent air, from e.g. the down-wash of a neighbouring rotor craft.

In cases where the shelter size is not sufficient to hold the entire swarm it may have to be split into multiple groups. A maximum distance has to be defined in order to prevent the robots from losing contact. Here the term connectivity becomes a key component in a dynamic swarm. A connected swarm ensures the continuous flow of information throughout the whole swarm. It may be interpreted as a condition on how scattered a swarm can become without breaking up into sub-swarms. This is important for each individual within the group to keep improving its own current position by using its neighbours as references.

We can now see that shelter seeking is a complex behaviour to implement in a swarm of robots. The lack of research in the field extends the difficulty even further. The problem is solvable though. In fact it has already been solved, by nature itself. For example, a cockroach has the ability to effectively seek shelter with the help of its friends and its love for darkness. Dark areas are usually signs of potentially good shelters against bad weather, e.g. rain and harsh winds, which roaches dislike and try to avoid (Ganihar et al., 1994). We can relate this to flying robots that may experience flight instability or electrical malfunctions due to turbulence, rain, lightning and other harsh weather conditions.

Because biological systems are so complex, biomimetics is applicable for a vast variety of fields. Over extremely long time and through natural selection, nature has evolved and adapted to solve engineering problems. Seeking inspiration from nature is a great place to start, because in most cases life has tackled similar problems which we are facing today. With traditional engineering methods it may become extremely difficult to design a new system because all possible scenarios have to be taken into consideration. In space engineering there is no room for mistakes. In 1961, during the Americas space race with the Soviet Union, an astronaut named Alan B. Shepard was sitting in the Mercury capsule in hope of becoming America's first man in space (Green, 2014). There had previously been little success and the space programme was therefore hanging by a thread. Because of the complexity of the thousand parts of the rocket system, the technicians had to check and recheck everything to make sure there was not going to be any failure. Due to all the system analysing the coffee that Shepard had been drinking four hours earlier, made his bladder full. He had to relieve himself in the suit, which could have caused some problems with the electrical equipment that was attached to his suit for medical purposes. The rocket was launched and in the end everything went fine. It still needs to be emphasized that it could have ended badly had the equipment short circuited and Shepard's suit caught fire. The engineers were positive that they had thought of everything, but they had not. The traditional engineering technique almost failed because of an unexpected changed scenario. Nature inspired techniques are great because nature has the ability to adapt to changing circumstances. It is therefore both beneficial and convenient to take advantage of living organisms to help solve our own challenges, or at least guide us in the right direction.

Studying insect swarm behaviour is intriguing and has been helpful to find simple solutions to complicated problems before. As Deborah Gordon, a Stanford Professor, said in a CBS news segment in 2011, regarding the study of ants: "Ants are not smart. But colonies are smart…" Simple individual rules are in the collective group amplified, thus a higher intelligence is reached which is otherwise unknown to the single individual (CBS News, 2011; Collins English Dictionary n.d.). A swarm robot does not need to be too complex and mass production of the same robot can be made. Hence, by exploiting the properties of swarm systems one may use less resource. Other benefits may involve using multiple agents to carry payloads exceeding the capacity of one individual agent (Kumar & Michael, 2012). One of the most significant benefits of being able to use a large number of robots is faster exploration performances. A big swarm can cover a greater search area than a small one, although this may be at the cost of group coordination (Bjerknes & Winfield, 2013).

One major challenge for autonomous flying swarms of robots is being able to navigate, detect and avoid any obstacle in the shelter finding trajectory. Outdoor environments are difficult to operate in and affect the robots with limited sight, wind and direct sunlight. If the robots are using navigation systems such as GPS, there is a great chance that they may be shadowed. All of these issues are severe and may in fact lead to a total failure of an individual robot. One may think to add additional on-board sensors as a safety precaution, but that takes its toll on both power consumption and computational resources.

As previously mentioned, the technology of autonomous UAS is constantly evolving, leading to reductions in size, weight and power consumption. While these challenges are obvious and addressed, other issues such as scalability, are easily forgotten. It is a term that refers to the systems limit of attaining coordination amongst the group. It has been found that larger groups need considerably more interactions between the group members to coordinate, than a group with less number of entities (Klavins, 2004). Since interactions require sharing information received from different sensors, this process exerts a lot of time and resources which in turn affects the coordination of the group. Self-organizing systems involve minimal selection of interactions, hence is not as dependent on the group size. While coordination can and has been proven successful in centralized systems, it is intrinsically weak. For example, a leader receives information from its subordinates, processes the data and makes a decision which is ordered out to the whole team. Since this has a single point of failure, it is also unreliable should e.g. communication with the leader be lost. A distributed system is more redundant in the sense that it does not require any central processor to collect all data or perform iterations. This type of systems is preferable for shelter seeking, as we want a more robust system than that of a centralized architecture.

The importance of shelter seeking, obstacle avoidance, connectivity and scalability is clear. That is why this chapter focuses on finding an effective shelter seeking method that is compatible with obstacle avoidance techniques while keeping the inter-robot connectivity intact, regardless of the number of robots. The result is a self-governed system of flying robots. The problems we will tackle are nonlinear decision-making processes that may introduce difficulties in the form of confusion and uncertainty of the local best position known to the swarm. A way to counter this complexity is by utilizing the self-organization pattern found in insects, in particular cockroaches, and animal predator societies.

Through interactions with neighbours and without being governed by any leader, this stochastic system results in a unified decision made by a group with limited information (Campo, 2010). The collective decision-making enables effective ways for a swarm of cockroaches or robots to agree upon appropriate shelters, by exploiting the sensing and processing abilities of the group.

The reason for choosing roaches over any other insect, besides their great survival abilities, is that they are harmless, easily accessible and cheap. This makes them popular to use by researchers and as a consequence of that, there are a lot of available papers on experiments and analysis on these creatures.

RELATED WORK

Although nature inspired optimization techniques resemble one another, cockroach based algorithms seek direct inspiration from natural shelter seeking behaviour. This means that they seek temporary shelters instead of permanent ones as in Ant Colony Optimization and the Artificial Bee Colony Algorithm.

In 2005 Jeanson et al. published their experiment with the cockroach *Blattella germanica*. They investigated how the interaction between larvae, at the individual level, influenced the formation of aggregates. The results showed that cockroaches seek shelter in dark places and that they much rather form groups than walk around by themselves. Amé et al. (2006) carried out a similar test where they found that the interaction between individual agents result in a social amplification that leads to optimal formation of groups. A larger population in a shelter, lowers the probability to enter and to leave a shelter. If multiple shelters are present, rather than forming groups by filling up each shelter to its maximum and leaving the surplus individuals in other shelters, they would fill the shelters more evenly. E.g. in an experiment with 50 cockroaches having three shelters with the capacity of holding 40 roaches each, the solution will be 25 in one shelter, 25 in the second shelter and none in the third. Their result support evidence that without elaborate communication and information about their environment, the cockroaches are able to adapt and form groups. Halloy et al. (2007) introduced cockroach-like robots into a group of cockroaches and managed to somewhat change the behaviour/choice making of the whole swarm. This is in accordance with the fact that the roaches act in a distributed system with no obvious leader to guide them.

There have been tests done where the cockroach's individual behaviour has been translated to robots such as in and Garnier et al. (2005). Their experiments showed that the collective behaviour of cockroaches could be simulated by robots programmed to follow simple rules, and the result was a close match between the artificial and biological system. Halloy et al. (2007) conducted an experiment by implementing and using the cockroach aggregation behaviour in micro-robots. These robots then managed to change the decision making of a shelter seeking swarm of roaches.

RIO

This made Timothy C. Havens et al. (2008) choose cockroaches as a model for their swarm intelligence (SI) algorithm called *Roach Infestation Optimization* (RIO). It is an algorithm based on the previous stochastic optimization technique from 1995, called *Particle Swarm Optimization* (PSO). That is, a computational model that guides particles in a system towards the best known location continuously found by new particles (Kennedy & Eberhart, 1995). These kinds of models are convenient to use when navigating swarms of robots through unknown environments without the aid of an external pilot. RIO follows three basic rules inspired by previous research on cockroaches:

1. Cockroaches *randomly search for the darkest location* in the search space. The level of darkness at a location $\vec{r} \in \mathbb{R}^D$ is directly proportional to the value of the fitness function at that location $f(\vec{r})$.
2. Cockroaches *enjoy the company of friends* and socialize with nearby cockroaches with a probability equal to the numerical results of cockroach aggregation behaviour studied by Jeanson et al. (2005).
3. Cockroaches periodically *become hungry and leave* the comfort of darkness or friendship to search for food.

Just like PSO, RIO is a meta-heuristic and does not make any presumptions about the problem that is being optimized (Beheshti & Shamsuddin, 2013). It has a simple structure, with few adjustable parameters which makes it easy to implement and fast in computation. Consequently, RIO inherits the same limitations as its predecessor, namely that it does not guarantee that an optimal solution is ever found. In some complex problems it simply gets stuck in a local optimum due to lack of exploration. We say that the solution prematurely converges (more on this in the "Connectivity" section). RIO has somewhat countered this effect by adding a "find food" behaviour. This means that even if a particle is stuck in an optimum, it will eventually get hungry and leave the place to find food. When the particle is no longer hungry it will once again search for an optimum and hopefully find the true global best.

Although there are only a few parameters involved, it may still be necessary to tune these to achieve fast and correct convergence for a specific search space. This technique has not been tested on real robots. Although PSO has, there is no record of it being used in a 3-dimensional scenario.

RAMBLER

Daltorio et al. (2013) presented an experiment done with the cockroach Blaberus discoidalis, where they chose to investigate what Jeanson et al. did not consider. They wanted to know if there was any shelter-seeking bias model (i.e. if they somehow adjusted their path upon visually observing a shelter) that could fit. The resulting model became the *Randomized Algorithm Mimicking Biased Lone Exploration in Roaches* (RAMBLER), a more improved randomized walk model than previously made. The main concept of this search is for a robot to walk along walls until it can visualize a shelter with e.g. camera sensors. The technique is not made for three dimensions as ground moving robots with antennas were used in these experiments. The science behind the algorithm only considers a lone cockroach's navigation in an unknown environment. This precludes any swarm behaviour.

Lévy Flights

Lévy flight, just like the swarm optimization algorithms, has random walk behaviour. In contrast to Gaussian or Brownian diffusion, a Lévy distribution has a probability to make the random walking unit jump a step length l_j drawn from a probability density function which follows a power law in its tail. This gives it a fat-tailed distribution, or a large skewness compared to a normal distribution and the density function decays at large l (see Figure 1).

Figure 1. The left side plots show the Lévy probability density function with different scale parameters. Observe that the long fat exponentially decaying tail is only on one side of the function, causing a skewness in the graph. This can be compared with the standard normal distribution (Gaussian distribution) plotted on the right side.

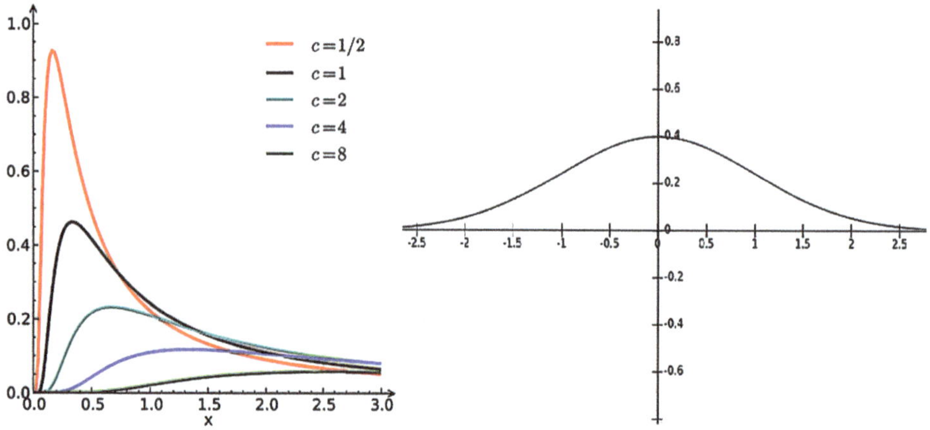

The Lévy distribution has the majority of its probability mass under higher values. Consequently there is a chance for a super-long step to occur (this is the so called Levy flight) as we expect a sample from this distribution to return a relatively high value.

This leads to super-diffusive search patterns where a particle is more likely to explore a larger area of the search space (Figure 2).

Viswanathan et al. (1999) state the hypothesis that since Lévy flights optimize random searches, biological organisms must therefore have evolved to exploit Lévy flights. While Lévy flight has yet to be proved to actually be used by living creatures, extensive studies on animals such as, eagles, spider monkeys, sharks, honey bees, wolfs etc. show tendencies towards Levy flight or at least to a process with similar behaviour. Another good example that describes the process of Lévy flight is the fast spreading

Figure 2. A particle moves randomly in a 2-dimensional search space. In a) the particle draws its movement from the Lévy distribution and therefore sometimes experiences a large jump to a new area. In b) the particle exhibits a Brownian movement drawn from a normal distribution. In this case the particle covers less ground as it usually has smaller step sizes.

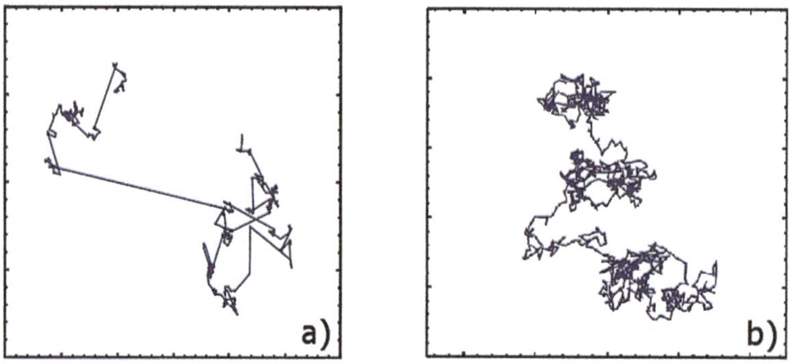

of diseases due to high air traffic connectivity, where viruses and bacteria literally make a "flight" (large step size) to a remote location.

Lévy flight is still an idealization of actual natural foraging. This is because the jump steps are straight vector-like, while in real scenarios a curvature of some degree should exist.

As with search algorithms using normal distribution, Lévy flights inherit the property of occasionally overshooting a target. In fact, in some situations the large jumps can be more troublesome than beneficial. An example of this is when a Lévy flight is made so that the target is lost because the distance is too great to detect with any sensor (see Figure 3). Because of this the scale parameter of the probability density function needs to be tuned to fit the search space.

In an article by Lin et al. (2012) it is presented how a chaotic sequence and a Lévy random process combined and successfully merged into a bat swarm search algorithm, can be used as an effective optimization model.

Plateau Problem

The fitness function incorporated in meta-heuristics such as RIO and PSO is used to evaluate the fitness of each particle, i.e. how well it is doing in its translation from previous position to the current one. For example, PSO can be used to minimize the commonly used test function; the *Rosenbrock* function:

$$f(x, y) = (1 - x)^2 + 100(y - x^2)^2.$$

As a particle randomly walk in the function domain it constantly checks if the fitness function $f(x, y)$ in the new position is lower than it was in the previous position. If true, that position becomes the new best known position. Whenever the particle moves further away from the target it is biased towards the latest best known location and is more likely to be dragged back to that position and try moving in another direction. By continuing to examine its own fitness, the particle moves in a vortex like manner (Figure 4) towards the lowest point in function $f(x, y)$, which is at the point $f(1, 1)$.

The solvable problems for this type are restricted to those where it is possible to compare two different points in the search space. We will henceforth refer to this as *gradient problems*, since wherever the particle traverses there is always a gradient or a force pulling it towards a better known area. A search space where there is no such gradient is from here on referred to as a *plateau problem*. Every route the particle takes will yield the same fitness, thus the particle has no idea if it did good or bad. Obviously the particle cannot share its experience with the neighbours around it since they are on the same plateau level. Figure 5 below illustrates the gradient and plateau problem.

Figure 3. The target (star) is overshot by a distance D. In some unfortunate cases, the distance D may become so large that the target is left unexplored.

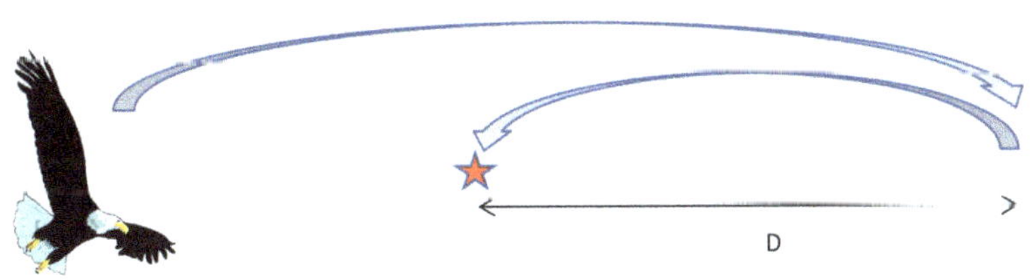

Figure 4. One particle is moving about the search space. Sometimes the particle overshoots the target but due to the fact that it "feels" less fit, it is pulled back towards its personal best (and also towards its neighbour's best position). The result is a spiral around the target.

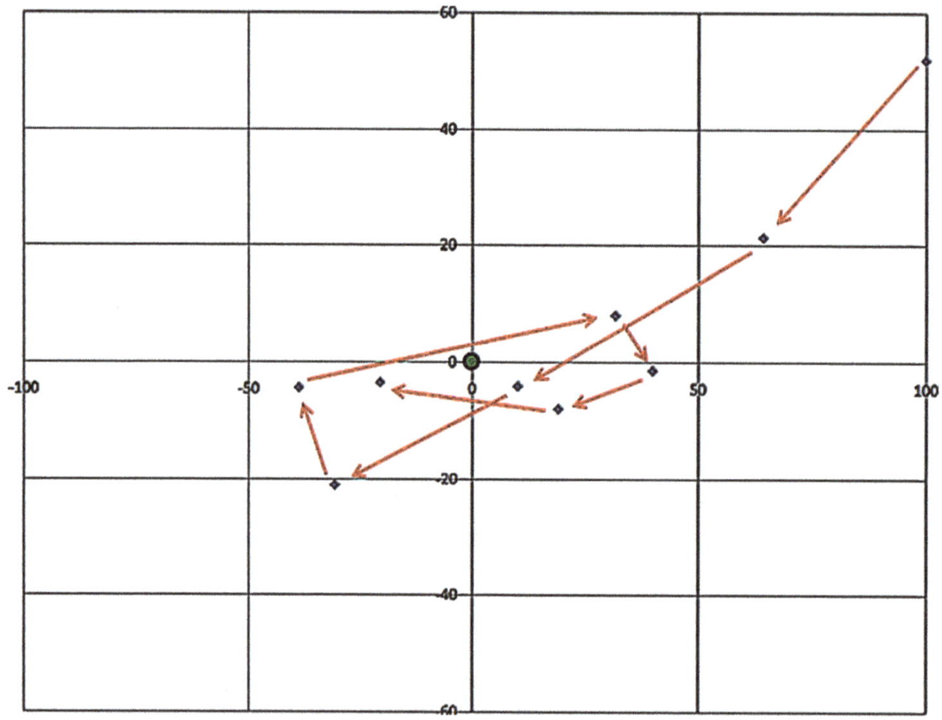

Figure 5. The particles (dots) move around freely in the search space. The goal is to reach as low as possible, hence if a particle moves in the wrong direction it notices this and can correct its trajectory. On the right side the particles are trapped, they cannot decide which way is the best. A particle cannot even rely on its neighbours since they are in the same situation.

Connectivity

The most important feature of functional self-governed multi-robot system is perhaps the ability to communicate between each robot. Gazi and Passino (2011) present two different neighbourhood topologies; *local* and *global neighbourhoods*. In a local neighbourhood every particle has its own neighbourhood best position, since a particle is only connected to a subset of neighbours in the swarm. A global neighbourhood topology on the other hand shares only one neighbourhood best position, a global best position. This is because every particle is in direct contact with all other individuals in the swarm. It has been stated that a search algorithm using PSO converges faster if the neighbourhood topology is that of a global type, rather than that of a local one. However, with a global neighbourhood approach, the solution is somewhat susceptible to converge in a local optimum instead of a global. Since this point would not represent the actual best point in a given search space, we say that the solution has prematurely converged (Figure 6).

It is possible to use both topologies to get their respective benefits. We call that using a *dynamic neighbourhood*, which means that the topology changes from one to another with time or as the optimization process evolves. For instance, if we want to reduce the risk for premature convergence, we can simply start off with a local neighbourhood topology. Since the particles are not directly connected to every other swarm particle, they are less prone to be pulled in by a neighbour that has found a local best position. This results in a more exploratory start of the search, enabling optima that are difficult to detect in some search spaces, to be reached. The number of neighbours can then gradually be increased until we have a fully connected topology and one global best solution (or best solution found by the swarm).

One can measure the connectivity in a swarm by introducing the concepts of *strong connection* and *weak connection*. To achieve a strong connection within a swarm it is necessary for each individual to get information from every other friend in the swarm. This does not necessarily mean that only a global neighbourhood can achieve a strong connection. A formation which allows a node to be indirectly connected to other nodes (via in-between neighbours) is also considered to be strong.

Without a strong connectivity there is a risk that the swarm fails to improve its estimates because some particles lack access to the knowledge of other particles. Thus we can say that a swarm must be strongly connected to stay together, otherwise it will break down into sub-swarms (which is desirable in some cases). Figure 7 below shows examples of both weak and strong connectivities.

Figure 6. One of the particles in the swarm was exploring and heading towards the best solution in the search space when a connected neighbour found an optima. Because the particle's current position is less fit than its neighbour's, an attractive force pulls this particle towards its neighbour. Not knowing that this is only a local optimum, the swarm has now prematurely converged.

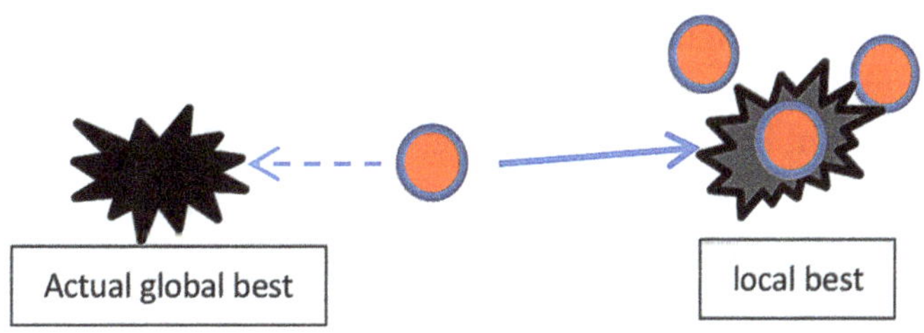

Figure 7. The group on the left side is strongly connected since all nodes can get information from everyone else through in-between robots. The right side group is weakly connected because the link from robot "A" to robot "E" was lost. This means that robot "E" can no longer get any information from the rest of the swarm (Figure adapted by author from Gazi & Passino, 2011).

Scalability

The ability to work at a variety of scales, i.e. different number of robots, degree of cooperativity, loads, etc. without losing an excessive amount of reliability or performance, is considerably desired in a system of swarm robots. Scalability is the general ability of the system to maintain a specific relationship between such scaling parameters and the collective performance (Kernbach, 2013). Roughly there are four different types of scalability, namely:

- Unscalable
- Scalable
- Super-scalable
- Hyper-scalable

In an *unscalable* system the collective communication and cooperation between the robots is weak, resulting in system failure or performances with significantly weaker efficiency as the swarm grows in number. Normally one would notice a marginally reduced performance with increasing number of units, a so called *scalable* system. *Super-scalable* systems are ideal to produce. The characteristic of such a type is the almost constant performance regardless of the quantity of robots. *Hyper-scalability* is defined by the properties of enhanced performance while additional robots are introduced to the system. The scalability types are illustrated in Figure 8.

It is quite rare to find swarm systems or any other collective systems that are super- or hyper-scalable. Those systems, sooner or later, exhibit a bottleneck effect with higher scaling parameters. Decent engineering can provide for high scalability, by keeping the system simple and efficient. Often one has to sacrifice some system functionalities, or hardware, to achieve this.

Figure 8. The relation between scaling parameters and the performance can tell how scalable a system is. (Figure adapted by author from Kernbach, 2013).

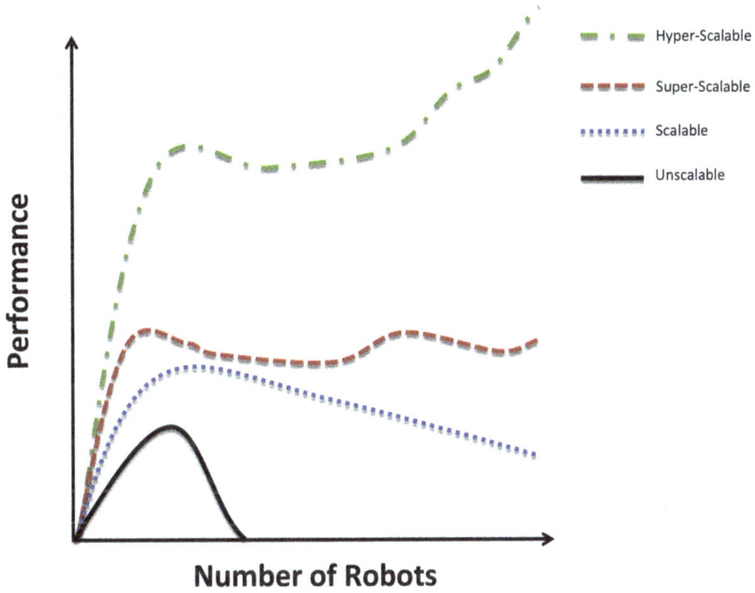

Limitations of Current Techniques

When summing up the weaknesses one can conclude that none of the reviewed shelter seeking techniques have been tested on quad copters and some of them are restricted to two dimensions and single robot searches, instead of swarms. Meta-heuristics such as RIO are vulnerable to the plateau problem, while RAMBLER is adapted to navigate through them.

RAMBLER has been tested on real robots but is only optimized for a single individual, while RIO considers an entire swarm.

It is clear that the research comes with several of challenges. To address these issues a number of requirements were devised. From many requirements two main challenges were defined to support the objectives of this research.

1. The system shall be able to navigate autonomously through an unknown 3-dimensional environment to find the nearest best known location, even in flat regions of the search space.
2. The system should be scalable with up to 25 robots.

None of the techniques previously described can fulfil these requirements by themselves. Even so, these techniques have (with modification or combination) the potential of fulfilling the requirements. Since there are alternate proposed solutions to this system, we need to evaluate each of them to gain the knowledge of how well each solution works, what modifications need to be made and which ones may be combined. The most propitious solution will be the one used in the system design.

The remaining parts of this chapter are outlined as follows; first, a presentation of the shelter mechanism is presented together with the architecture of three different shelter seeking methods. This is followed by the experiment design and result section which describes the experiment and simulation tools used. As a final point, a conclusion is given together with suggestions for future work.

SHELTER SEEKING MECHANISM

Three alternative techniques that can be used to address the shelter seeking problem are described in this section. The hybrid systems, henceforth referred to as *ModRIO*, *GaussRIO* and *StdRIO* allow a homogeneous swarm of robots to follow three basic instincts:

- Random Walks
- Hang on to "good friends"
- Leave shelter

The systems have to address the issue of finding a proper shelter and at the same time being able to navigate through flat regions in the search space as well as escaping possible local optima. Each system's strength shall be measured by comparisons of the two other similar devised techniques.

A flowchart, relevant to all three techniques, showing the core processes of the system is presented in Figure 9.

Figure 9. System process in the form of a flowchart showing how the system continuously runs in a loop.

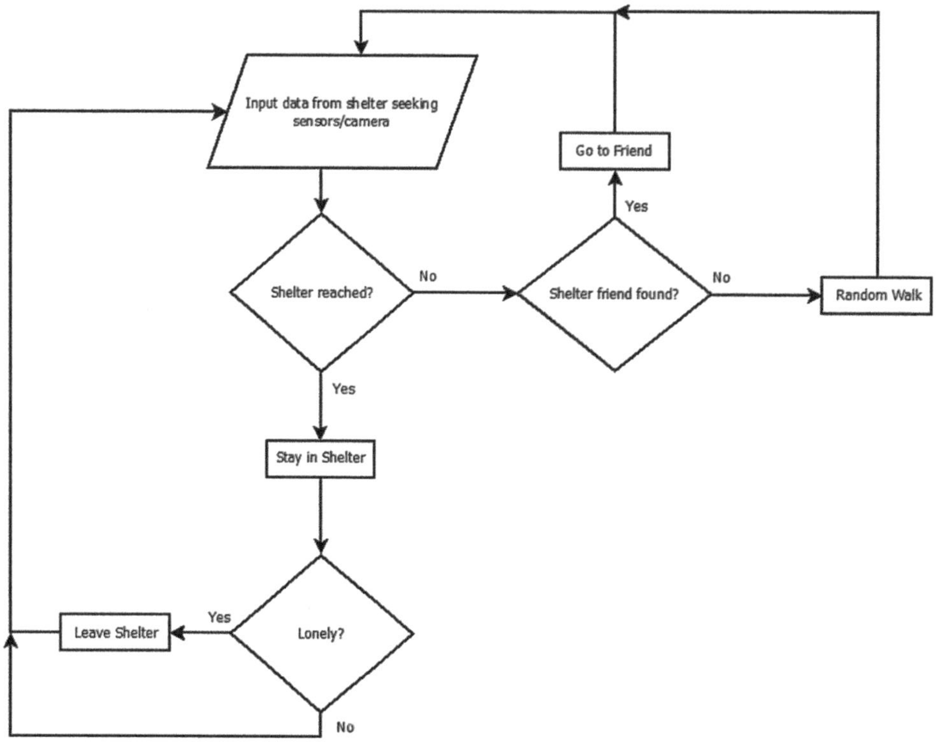

System Architectures

In an ideal case we would use the RIO algorithm without making any alternations to the algorithm, but due to the plateau problem and the fact that we preferably want to be independent of positioning sensors such as GPS, this is not possible.

The three modified RIO systems devised to fit this research disregard the fitness during their search, as the robots will only experience two discrete types of fitness; fit or not fit. The robot is either inside or outside a shelter, there is no in-between. To determine whether or not a robot finds itself inside a shelter, a darkness threshold, $dark_{th}$ is defined. Depending on the desired quality of the shelter, the threshold can be adjusted to lower values (darker areas). If the input data from the light sensor is higher than the threshold (row 18, Algorithm 2) each neighbour message in the first bit is evaluated.

If the input data on the other hand is below the selected threshold level (row 28, Algorithm 2), the robots stops and becomes an attracting beacon. The robot then counts how many robots that are currently inside the shelter and calculates a probability to leave the shelter.

The robot chooses to leave or stay after comparing the value with a random number drawn from the uniform distribution. When the robot leaves, it resets to its initial state, travels to a random "food" location and runs the algorithm all over again to find a shelter.

The maximum number of robots relies on both the desired performance of the swarm and also on the size of the search space. Experiment results that show a trend of significantly fast convergence or hyper scalability can be used to decide the quantity of robots that saturate the selected search volume.

Standard RIO

Standard RIO, or StdRIO, is designed to mimic the original RIO algorithm. Here an exception is made, where positioning sensors (i.e. GPS) are allowed to be used by the system. The technique uses memory to store the position of the personal and local best known shelter location. The leave shelter behaviour is replaced with a find food behaviour, which sends out hungry robots to search for food in different intervals. Because this technique is specially dependent on memory it breaks for robots that are alone and have no best known position. For these cases it is therefore altered to follow the design of blind random walk that is drawn from the normal distribution, until shelter positions can be established.

The parameters for StdRIO are setup according to the original RIO presented in Algorithm 1 (Havens et al., 2008).

Modified RIO

To describe the Modified RIO system, also known as ModRIO, we will divide it into four parts. Each part represents the basic instincts of the robots as previously mentioned in the beginning of this head section. Algorithm 2 shows the structure of the system and how to achieve the flow illustrated in Figure 9.

Shelter Detection

This technique only requires light and relative positioning sensors mounted on the robots to remain functional. The data from the light sensors I_s varies with light intensity according to the inverse square law $(r) = I_{source}/r^2$. Input from these sensors is used to determine if the robot is inside a shelter or not,

Algorithm 1. Roach infestation optimization

Input: Fitness Function $F(\vec{x}) \in \mathbb{R}^D$
Parameters:
N_a
t_{max}
C_0, C_{max}
A_1, A_2, A_3
t_{hunger}
Initialization:
Set hungers $hunger_i = rand\{0, t_{hunger-1}\}$
Set population $\vec{x}_i$ and $\vec{v}_i$, randomly
Set food location $\vec{b}$ randomly
For $1 \le t \le t_{max}$ **do**
    $M = [M_{jk}] = [\| \vec{x}_j - \vec{x}_k \|_2]$
    $d_g = \text{median}\{M_{jk} \in \boldsymbol{M}: 1 \le j < k \le N_a\}$
    For $1 \le i \le N_a$ **do**
        If $F(\vec{x}_i) \le F(\vec{p}_i)$ then $\vec{p}_i = \vec{x}_i$
            Compute the neighbours of roach *i* as,
            $\{j\} = \{k: 1 \le k \le N_a, k \ne i, M_{ik} < d_g\}$
            N_i = number of neighbours $|\{j\}|$
        For $1 \le q \le N_i$ **do**
            If $rand[0, 1] < A_{min\{N_i,3\}}$ **then**
                $\vec{l}_i = \vec{l}_{j_q} = argmin_k \{F(\vec{p}_k)\}, k = \{i, j_q\}$
        If $hunger_i < t_{hunger}$ **then**
            $\vec{v}_i = C_0 \vec{v}_i + C_{max} \vec{R}_1 (\vec{p}_i - \vec{x}_i) + C_{max} \vec{R}_2 (\vec{l}_i - \vec{x}_i)$
            $\vec{x}_i = \vec{x}_i + \vec{v}_i$
        Else
            $\vec{x}_i$ = random food location $\vec{b}$
            Relocate eaten food randomly
            $hunger_i = 0$
    If *Hungry* **then**
        Increment $hunger_i$ counters

i.e. if the darkness level is high enough. If a shelter is not detected the robot looks for any neighbouring robot that already resides in a shelter (hereafter referred to as shelter robots). This is done by analysing incoming messages from nearby robots via the relative positioning system. When a robot eventually reaches an area that is dark enough it stops and becomes an attracting beacon by transmitting a shelter message. If several shelter robots are discovered the robots chooses to move towards the nearest one.

Algorithm 2. Modified roach infestation optimization

Input: Data from light sensors, I_s
Parameters:
N_a
t_{max}
C_0, C_{max}
P_s
υ_{max}
Initialization:
Set population $\vec{x}_i$ and $\vec{v}_i$, randomly
Set food location $\vec{b}$ randomly
Set all robot messages, $msg_s = false$, $msg_f = false$
Set darkness threshold, $dark_{th}$, to minimum shelter darkness quality
for $1 \le t \le t_{max}$ **do**
  $M = [M_{jk}] = [\| \vec{x}_j - \vec{x}_k \|_2]$
  $d_g = \text{median}\{M_{jk} \in \boldsymbol{M}: 1 \le j < k \le N_a\}$
  For $1 \le i \le N_a$ **do**
    Compute the neighbours of roach i as,
    $\{j\} = \{k: 1 \le k \le N_a,\ k \ne i,\ M_{ik} < d_q\}$
    N_i = number of neighbours $|\{j\}|$
    If $I_s > dark_{th}$ **then** $msg_s = false$
      For $1 \le q \le N_i$ **do**
        If $msg_{sq} = true$ **then**
          $\vec{x}_i = \vec{x}_{j_q}$
          msg_f = true
        Else if $msg_{fq} = true$ **and** $msg_f = false$ **then**
          $\vec{x}_i = \vec{x}_{j_q}$
        Else
          msg_f = false
          $\vec{v}_i = C_0 \vec{v}_i + C_{max} \vec{R}(\vec{x}_i + Lévy)$
      If $|\vec{v}_i| > v_{max}$ **then**
        $\vec{v}_i = v_{max} \cdot \hat{v}_i$
    Else
      $\vec{v}_i = 0$
      msg_s = true
      $P = P_s \times ([N_i \in \mathcal{S}] + 1)$
      If $rand[0,1] > prob$ **then**
      $msg_s = false$
      msg_f = false

continued on following page

Algorithm 2. Continued

```
x_i = random food location b
Relocate eaten food randomly
```

Robots that have no sheltered friends can instead follow neighbours that are in contact with a robot inside a shelter. Those robots transmit a different message (row 23 in algorithm 2).

Sheltered Friends Pull In Neighbours

In the case that a sheltered friend is discovered the action is simply to travel to the source of the received message. When using optimization techniques to solve mathematical problems, the algorithms infer that particles "teleport" or travel with infinite velocity to any position. This is not plausible for real life implementations, thus a speed limit has to be introduced. By setting a constraint on the velocity update equation, it might take several system loops for a robot to reach a spotted friend unlike when performing instantaneous travel. During this time a robot may also lose contact with the target and return to a blind random walk again. The range and bearing system provides enough information for the robots to create a three-dimensional vector pointing to the transmitting robot. The robot could then add thrust to its rotors and make the necessary flight trajectory. Eventually the robot will be pulled into a shelter, start sending out shelter messages and thus becomes a beacon itself for other passing robots (Figure 10). The networking can be extended even further by letting other robots know when a sheltered friend has been found, or even when a robot has discovered another robot that has seen a sheltered neighbour, etc. An endless chain reaction can alert the whole swarm. This type of networking may become problematic since it can decrease the exploration of the swarm by frequently pulling individuals towards already discovered shelters. Additionally, in case of lost connections in a distributed system, robots were fooled to think that a good leading robot still existed and started to follow each other as a consequence.

That behaviour was avoided when running the system with a centralized or a decentralized communication network (as presented in Figure 11). That is because the hierarchy system does not allow inter slave communication. Thus, the message chain immediately breaks when a master neighbour loses contact with a shelter robot for instance. A node length of maximum two decentralized robots was tested (i.e. one shelter robot pulling an outside friend who in turn attract its own share of blind robots). This solution was still capable of achieving strong connectivity within the swarm without splitting the swarm into several sub swarms, and was for that reason accepted as a solution for ModRIO. For very large search areas the node length needs to be adjusted to accept larger chains.

Blind Search with Random Walks

In large search areas, it is likely that some robots are unable to sense a shelter or a sheltered friend. Such a robot is completely "blind", and has no idea which way is the most suitable way to go. Here we suggest that the robot takes a random walk either from a Lévy distribution or a Gaussian distribution.

Figure 10. A simulation where each robot performs a randomized search for a shelter. Whenever a robot finds itself in a shelter, it immediately becomes a beacon for other connected robots. Robots who are not in direct line of sight are attracted to their neighbours who themselves have discovered a sheltered robot. Thus, the swarm converges into one shelter.

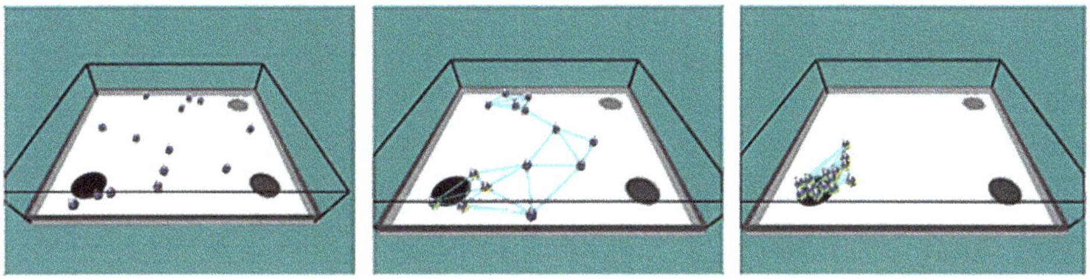

Figure 11. In the decentralized communication model slaves do not influence each other. When a central robot loses contact with its master, its slaves will stop following it. The chain can be made infinitely long, but for each new central level more communication data is required.

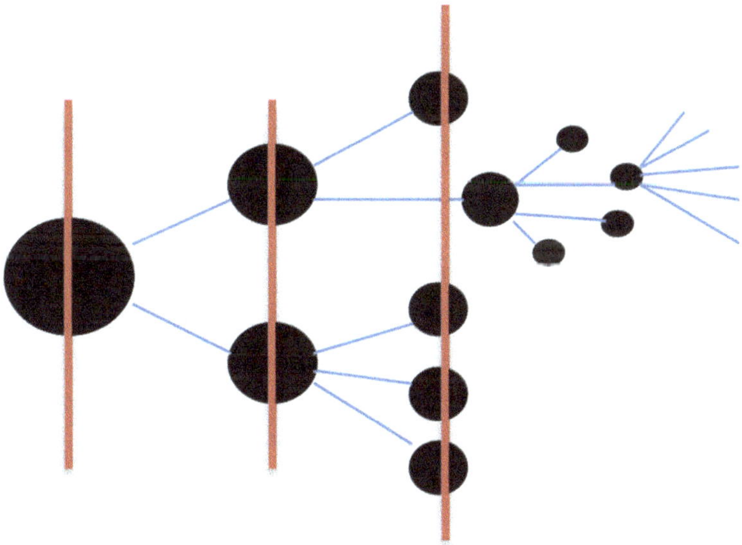

When a robot finds itself in a situation where no surrounding friends can help, or if it is all alone, a random walk is performed by setting the velocity vector to

$$\vec{v}_i = C_0\vec{v}_i + C_{max}\vec{R}\left(\vec{x}_i + L\acute{e}vy\right)$$

$\vec{v}_i$ and $\vec{x}_i$ are the current velocity and position of the robot *i*, C_0 and C_{max} are both weight parameters where C_0 is the inertia weight that can be seen as a heavy weight slowing down the movement of the current speed. In previous experiments inertia weights of 0.9 to 0.4 have shown to produce good results (Bansal et al., 2011). Such algorithms normally provide good results with total $C_{max} \leq 4$ (Carlisle & Dozier, 2011).

$\vec{R}$ and Lévy are vectors of uniform random numbers drawn from the normal and Lévy distribution respectively. To draw a random number from the distribution a method called rejection sampling is used. In short, a uniform random coordinate from a box covering the Lévy probability density function (PDF) is generated. The dimensions of the box can be set to a height that is equal to the amplitude of the chosen PDF so that it encases all possible values. To cover the horizontal size of the Lévy PDF is impossible since its fat tail stretches to infinity. For these scenarios it is sufficient to cut off the tail at any point where the rate of change can be regarded as 0.

If the point happens to belong to the area beneath the Lévy PDF, the corresponding value from the PDF is chosen. In a way this technique can be compared to throwing dart blindfolded at a target being the Lévy PDF

$$\sqrt{\frac{c}{2\pi}} \times \sqrt{\frac{e^{-c/2(x-\mu)}}{\left(x-\mu\right)^{2/3}}},$$

where the shift parameter, μ, is zero in the standard form and *c* is the scale parameter. In Figure 1 the different forms of the function with varied scale parameters is displayed.

The scale parameter was chosen so that the maximum value drawn from the distribution does not exceed the 1/10th of the arena side length or so that

$$C_{max}\vec{R}\left(L\acute{e}vy\right) \leq C_{max}\vec{R}\left(\vec{x}_i\right)$$

This way the maximum jump may be up to twice as large as that of the Brownian motion component.

The blind fumbling will continue until either a shelter is reached or a sheltered friend is spotted, as described in Algorithm 2. Yang & Deb (2010) called the method of combining Lévy walk with stochastic optimization techniques “Eagle Strategy”. The most essential benefits are robots with higher chance to make larger jumps and less probable to wander around the same area (as in Figure 12).

Figure 12. Two simulations showing a single robot’s path. The left side uses ModRIO, were it is obvious that the Lévy flights contribute to occasional large jumps, thus making the robot cover more ground. On the right side example the GaussRIO has a more dense search and often searches in areas it has already visited before.

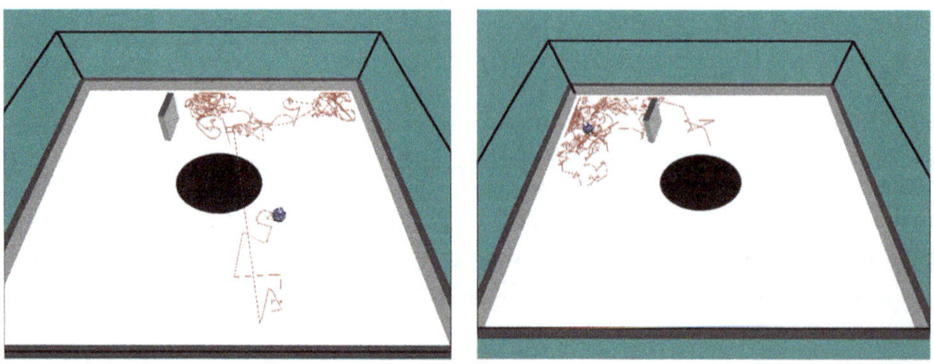

Lonely Robots Leave Shelters

When a robot ultimately finds itself inside a shelter it stops. Similar to cockroaches, the next move is for the individual robot to decide whether to leave the shelter or not. The decision is made based upon how many robots (including itself) that are currently confined in the same shelter

$$P = P_s \times ([N_i \in \mathcal{S}] + 1)$$

P is the probability for a robot to stay depending on the base probability, P_s, and the number of robot neighbours, N_i, currently inside the same shelter $\mathcal{S}$.

For every friend the robot senses there is a slightly less chance for the robot to leave, because it feels comfortable around its friends. For a robot that is completely alone in a shelter it is more likely to move on to find a shelter with other robots hiding in it. Empirical results showed that a 100% chance to stay for approximately 1/5th of the total swarm population inside one singe shelter was the most time efficient method for the scale of robots used in this research. For a swarm population of up to 25 robots $P_s = 0.20$, thus a robot will still consider leaving a shelter with up to three other robots currently inside it. Experiments with a large number of shelters with various sizes may require P_s to be tuned differently, since swarms have shown to be able to "sense" different aggregation proportions (Garnier et al., 2005). This loneliness behaviour is comparable to the "find food" behaviour of the original RIO, where a cockroach over time becomes hungry to the point that it willingly leaves the comfort of its shelter to look for food.

Gaussian RIO

Gaussian RIO, or GaussRIO, is identical to ModRIO in every way except that the Lévy component is omitted. As in contrast to ModRIO, random walks are performed by using the Gaussian (or normal-) distribution. Robots using GaussRIO walk with a Brownian motion without any random large jumps. The search pattern is therefore more compact, as in Figure 2b), which might help to keep a higher connectivity within the swarm.

EXPERIMENT DESIGN AND RESULTS

Three experiments measuring time efficiency, swarm connectivity and noise tolerance were performed on each of the three techniques; StdRIO, ModRIO and GaussRIO. Each test was repeated five times to enable us to "see beyond noise". Five repetitions were chosen as this number has been found to offer a good trade-off between accuracy and processing time (Law & McComas, 1990; Hoad et al., 2007).

Prior to each test the robots were uniformly distributed throughout the arena. This means that there is a chance for some robots to start out nearby or even inside a shelter, which may cause lower shelter times. To avoid having tests with different initial advantages between the techniques, they all share the same initial robot distribution.

The number of robots is tested for $1 \leq N_A \leq 25$, because swarms of 26 or more saturated the search space and showed tendencies towards a hyper-scalable system for the chosen arena size of 10×10×10 meters. Other values for implementation specific variables are displayed in Table 1.

Table 1. Parameters for implementation specific variables.

Variable	Information	Value
N_A	Number of robots	1 to 25
C_0	Inertia weight	0.7
C_{max}	Roach parameter	1.43
P_s	Base probability	0.20
υ_{max}	Velocity constraint	4 (m/s)
A_1	Group parameter	0.49
A_2	...	0.63
A_3	...	0.65
t_{hunger}	Hunger interval	100

The following three sections will describe the different experiments in detail together with the achieved result.

ARGoS Simulator

The researchers were offered an exclusive version of the ARGoS 3.0, a swarm robotics simulator developed at the Artificial Intelligence research laboratory of the Université Libre de Bruxelles, IRIDIA (Pinciroli et al., 2012). It is an open source discrete-time 3D space simulator built on free C++ software libraries. While the main code is in C++, the environment i.e. the arena, physics engines, robots, sensors and other entities are setup using XML configuration files. The robot controller with the robot code was written in Lua scripting language instead of the default C++. Lua has a very simple application programming interface, which made it fast and easy to work with. Regardless if the controller is running simulated or real robot, it is allowed to access the robot on-board devices. The benefit of this is that a compiled code from simulation to real robots is transparent. This is one of the main reasons for choosing ARGoS as simulation platform for this research, besides providing high flexibility and computational efficiency making it possible to run on most computers. Another reason is that the researchers were in close contact with the creator of this simulation tool, which helped facilitate the process of getting started with the software.

ARGoS 3.0 allows simulation of three different types of robots; ground moving, climbing and flying (Figure 13). One can modify and create own robots, but in this research the Eye-bot model from IRIDIA was chosen.

The eye-bot was equipped with light sensors and a range and bearing system for communication purposes. The light sensors allow the robot to detect light sources. The robot has multiple light sensors, equally spread around its body. Each sensor reading returns an angle in radians and also a value of light intensity. The angle corresponds to where the sensor is positioned with respect to the front of the robot. This way the robot "understands" which direction is less illuminated.

The range and bearing system allows the robots to perform *localized communication*. That means that when a robot receives a message it can at the same time distinguish the position of the sender with respect to its own point of view. For this reason, the robots need to be in direct line of sight to commu-

Figure 13. The robots that have been modelled inside the ARGoS simulator are based on real robots.

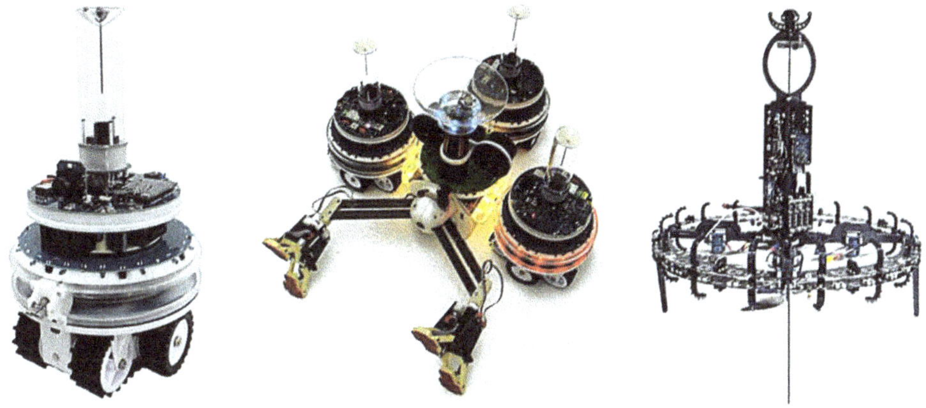

nicate. When robots connect, they can exchange data, vertical angles, horizontal angles and distance to the message source (Figure 14).

For StdRIO an additional positioning sensor is used for the robot to recognize its exact position.

Time Efficiency

In the first test three corner shelters were used, similar to the previous experiments done with real cockroaches. Each shelter could house the entire swarm; this was to prevent any robot from being denied entry to an approached shelter. When all robots had been initially placed the experiment commenced, and the time it took for the robot to reach a shelter was recorded (this will be referred to as "shelter time"). The test continued by increasing the number of robots by one, repeating the test, until 25 robots had been initiated. When performing tests with multiple robots they are likely to find shelters at different time. Some robots might even leave their shelters at several occasions by the time that another robot reaches one. Hence, when all of the robots have visited a shelter at least once, the experiment is stopped and the time is recorded. The shelter time is of great importance for this research since it provides knowledge on how long time beforehand the robots need to predict harsh weather conditions.

Figure 14. The package of information received from each robot. The robots can send a maximum of 10 bytes of data, but do only utilize one of these in the ModRIO and GaussRIO system. In addition to this the robot, via the range and bearing system, also receives the vertical angle, the horizontal angle and the range from itself to the source of the message.

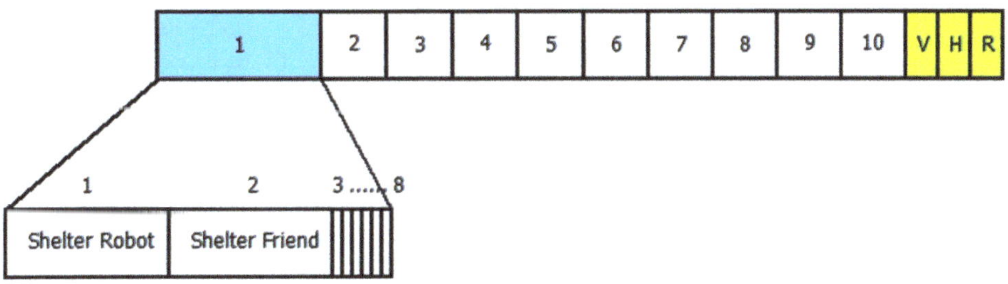

For the first test the trend line is represented by a third degree polynomial model. All three techniques indicate a performance growth with increasing number of robots. In minor groups GaussRIO does not show a distinguishing sign of improved performance but instead indicate a marginally slower shelter time for 15 to 16 agents in relation to its initial performance with fewer robots. Nevertheless, together with ModRIO, this technique can be considered the most constant one throughout the experiment, although they both show a slight decline in shelter time at higher populations. StdRIO on the other hand has the slowest shelter time for small groups, with nearly two times higher than that of GaussRIO for a single robot.

From Figure 18 in the first experiment we can establish the fact that there is no significant difference in performance between ModRIO and GaussRIO. This could mean that the effect of Lévy flight may be exaggerated or that an even larger search volume is needed to distinguish their dissimilarities. Nonetheless, the models predict an even shorter amount of shelter time as the swarm population increases, which perhaps indicate that the search space is being saturated with the relatively large number of robots. The fact that StdRIO showed the highest dependency on larger search groups and worst overall results, might be a sign that individual memory is overrated and may therefore not provide an advantage in all problem domains.

Figure 15. For StdRIO the trend line is much steeper with the highest values achieved with small groups of robots. The technique makes a remarkable recovery with larger swarms.

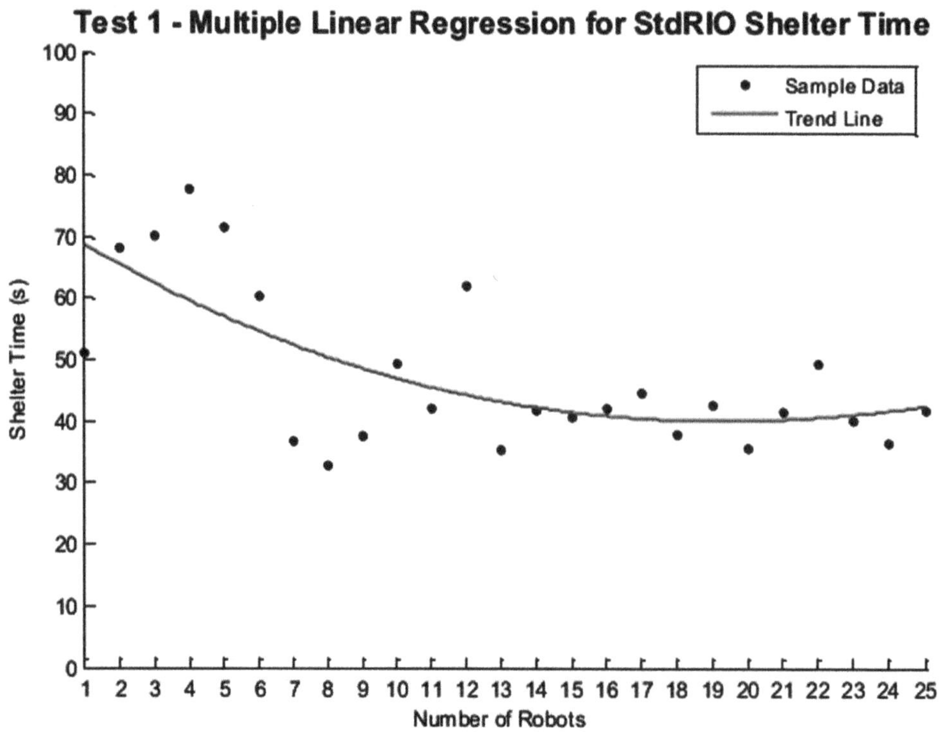

Figure 16. A third degree polynomial fit to the measured shelter times in the first experiment for the ModRIO method. The graph shows how the swarm becomes slightly more efficient with a larger number of robots.

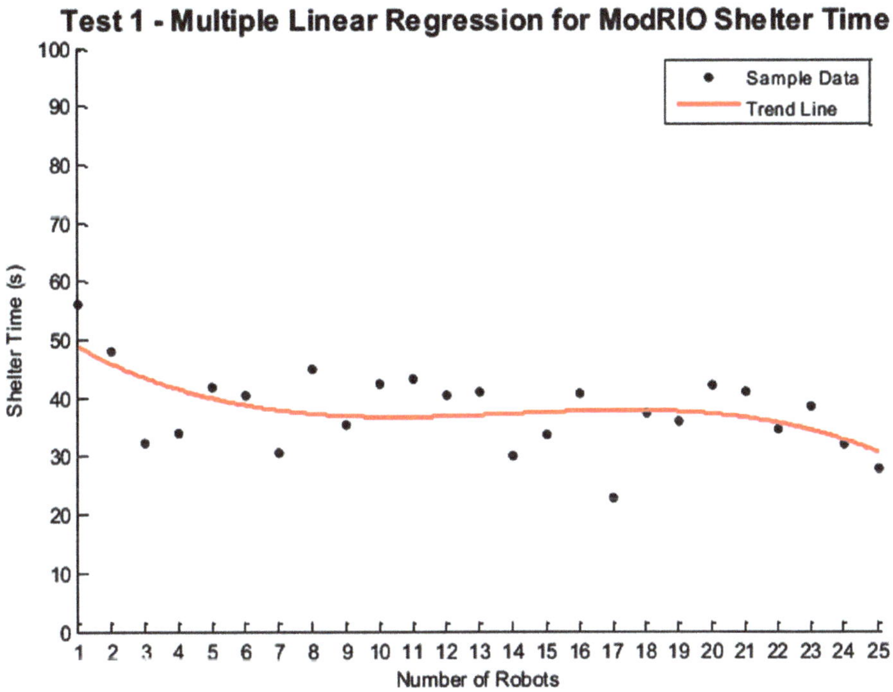

Figure 17. A fitted curve on the GaussRIO technique, showing slower shelter times for up to 16 robots, but more promising results for even larger groups.

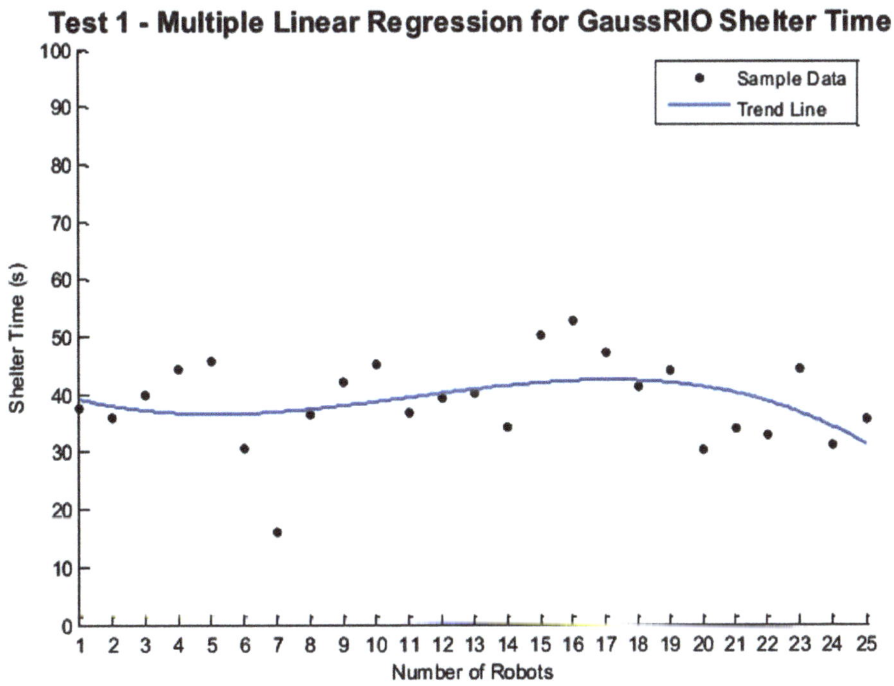

Figure 18. A comparison of the trend lines of the three techniques. StdRIO is the most dependent on larger quantities of robots as it finally catches up to the other two methods that show a fairly constant efficiency.

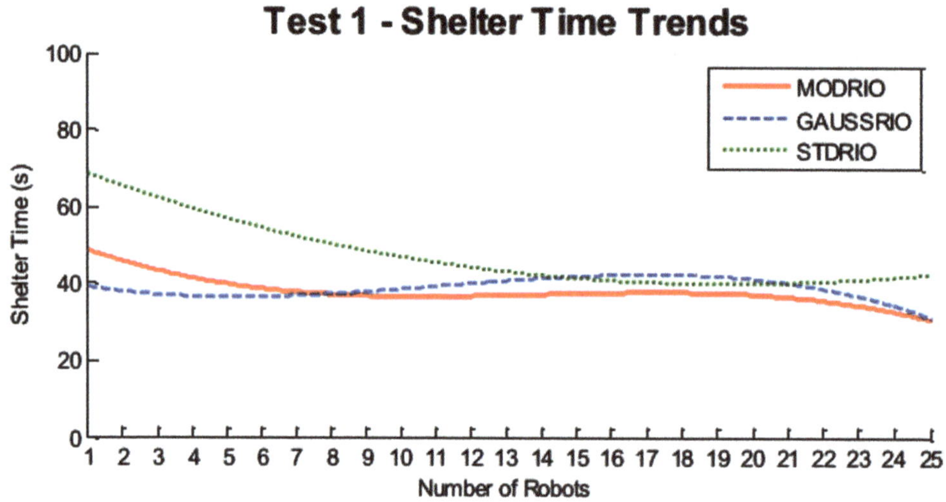

Shelter Connectivity

The second test used a fixed number of robots and instead varied the number of shelters, capacities and their position in the search space. Each shelter could house a number of robots so that the entire swarm population could be equally divided into the shelters. Thus for n shelters the capacity of each shelter is $1/n$ fractions, e.g. three shelters can each hold a third of the entire swarm population, while six shelters can house one sixth. This forces the swarm to split up so that we could monitor the collective behaviour of the group in a weaker state of connectivity. In cases where $1/n$ would imply fractions of robots, the quote was rounded upwards to the nearest integer.

The number of robots was chosen by the maximum number of shelters, with the aim of having one robot in each shelter in the final experiment. Empirical results showed that for a search volume of this size would become saturated with 10 shelters or more. The decision was therefore made to use no more than nine. The shelter distribution is shown in Figure 19. The shelters were chosen to be placed as far from each other as possible to increase the diffusive search patterns of the robots.

Because the robots are much more likely to leave empty shelters, the swarm will have a hard time to converge when shelters are too small. The leave shelter and find food behaviour was therefore stopped after a time corresponding to the highest shelter time of all techniques from the shelter time experiment, i.e. 166.9 seconds. A robot would after this time stay inside shelters it entered without leaving. Roaming robots would, once more, be given an equal amount of time (166.9 seconds) to find shelter before the simulation ended. After a completed simulation it was investigated how many robots failed to hide.

Both ModRIO and GaussRIO show almost identical behaviours in this experiment. The techniques are successful in keeping the majority of robots safe. Yet, an unexpected dip occurs at seven shelters, with capacities of two robots each, causing a robot loss of roughly 40% of the swarm population for both techniques. StdRIO shows the worst result with the lowest shelter population down to 33% of the entire swarm.

Figure 19. The different shelter configurations used in the second experiment.

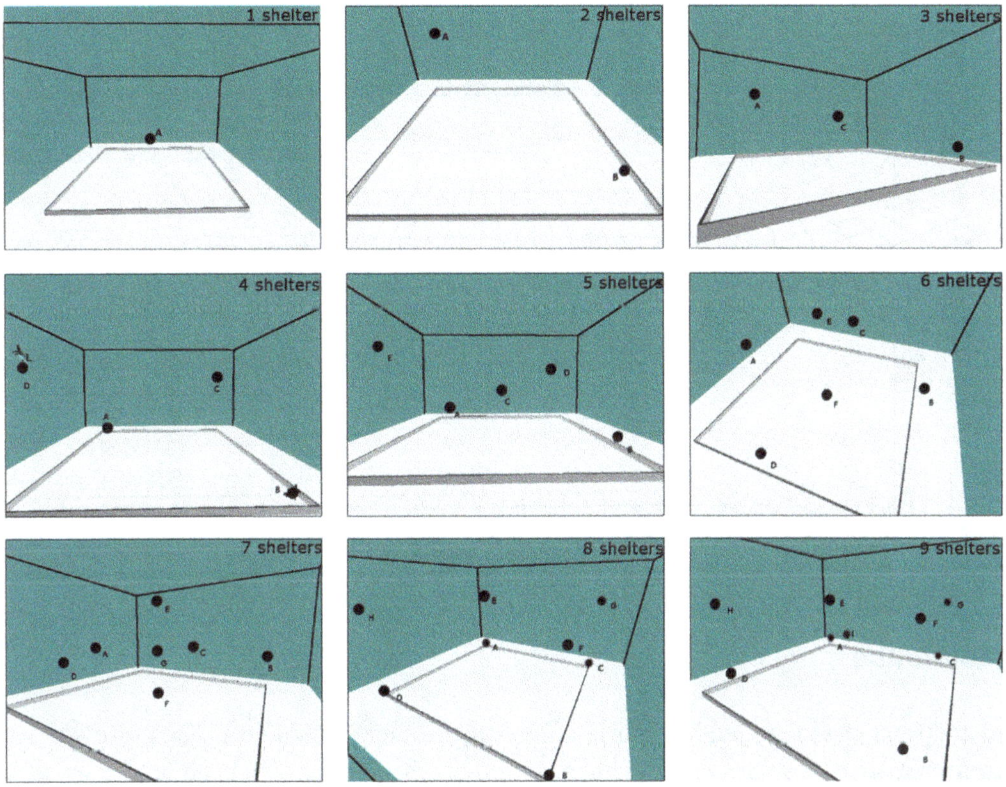

In the second experiment with results displayed in Figure 20, shelter memory seemed to cause more damage than good. The robots showed an ignorant behaviour where they always preferred the first shelter they visited, regardless how many friends it contained. In most trials this resulted in robots trying to enter a shelter which was already full, instead of finding a new one. When the time for shelter finding was up, the robots were still flying around some shelters waiting for other robots to leave so they could take their positions. This was never the case for ModRIO and GaussRIO, where there is no memory to remember what shelter was previously visited or not. Thus, those two techniques had a higher success rate with minimal loss of swarm agents. Still a rather interesting result showed that none of the techniques succeeded well on the seven shelter configuration, where ModRIO and GaussRIO both lost about 40% of their robots. Despite that the setup had the same capacity of two robots as the six shelter configuration, which on average sheltered the entire swarm population for both techniques. The only difference is the added shelter in the middle of the arena. There is a possibility that such a shelter disturbs searching robots when it is full. Since robots have a higher chance of discovering such shelters than those that are on the edge of the search volume, it is usually the one that gets filled first. The remaining robots that get dragged towards that shelter realize that it is full only when they reach it. When they leave the area they may at a later time be pulled back in again, this behaviour can repeat itself until it is too late for the robots to find the corner shelters. This theory would also explain why the nine, five and three shelter configuration also showed signs of robot loss for the two techniques.

Figure 20. ModRIO and GaussRIO have a similar pattern and are successful in most cases except for the seven shelter configuration. Yet again StdRIO stands out and displays the worst results for all nine shelter formations.

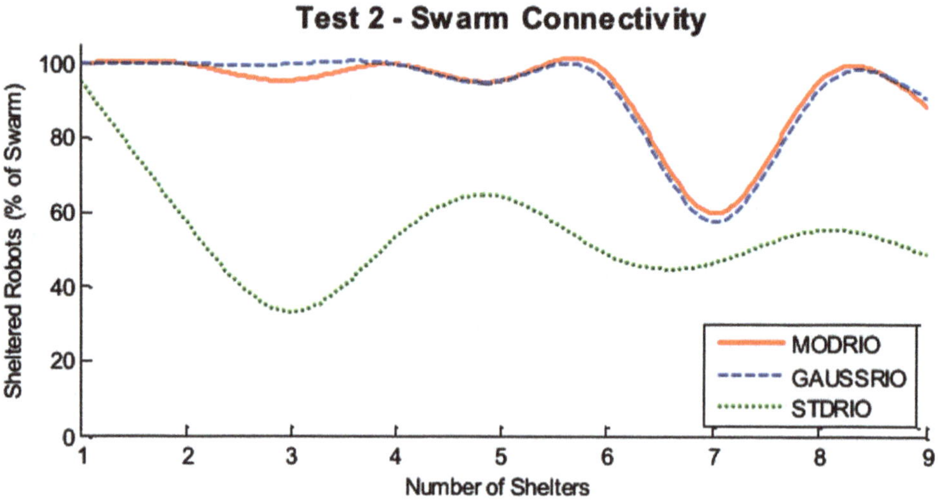

Noise Tolerance

The third and final test measured how noise affects the different techniques. The noise was generated from a uniform distribution and was implemented so that whenever a robot experienced noise, it would result in a complete loss of a data package. The arena was set up as in the shelter time experiment and 12 robots were used, which is approximately half of the maximum population of the first experiment. The noise rate started at 0% and was incremented by four percent to acquire the same step size as in the first experiment.

GaussRIO had the highest performance with some trivial oscillations showing a minor overall decrease in performance. ModRIO more than doubled its shelter time between 0 and 10%noise. The performance diminished with increased noise level and the shelter time was at one point four times higher than that of the noise free test. StdRIO was at first undisturbed by the noise but started to grow in an exponential manner for noise levels above 20%. With permanent communication loss this technique more or less quadrupled, while the other two techniques improved. A forth degree polynomial model is used to present the behaviour trends.

The final experiment surprisingly showed that ModRIO technique was the most vulnerable to noise, although StdRIO exponentially grew and took over at the higher noise levels (see Figure 25). The lack of Lévy flight might be a fairly reasonable argument to why GaussRIO would seem unaffected by noise while ModRIO is not. It was discovered during the experiments that the noise created confusion where robots "hesitated" when following their neighbours. This slow decision making process comes from when the robots lose the connection to a shelter friend because of the noise and starts to perform the blind random walk again. During this period ModRIO with its Lévy flight had time to jump outside the communication range, thus permanently losing its shelter friend. GaussRIO on the other hand is more prone to have a more compact search pattern and can therefore resume its path towards its shelter friend when the connection is regained after the noise. Figure 25 also shows that the results from ModRIO and

GaussRIO improve near a 100% loss of communication. It is possibly the effect of only performing blind random walks without getting pulled or hesitating because of surrounding neighbours.

Data Evaluation

Results from the experiments showed a varying shelter time, even by using the mean value of five consecutive trials for each step the values are very much scattered. Consequently it is difficult to observe a trend for such measurements, even though this is to be expected for experiments that are based on randomness. The curve fitting toolbox, *cftool*, in MATLAB was used to create and compare different trend lines. Evaluating goodness of fitted data can be considered a research topic in itself. In some cases the chosen model should have the reasonable statistics to be considered a "good fit". This involves checking statistics such as sum of square error (SSE), R-square and root mean square error (RMSE), etc. In other cases a graphical approach is the best way to go since it presents an easily observed predicted model. The latter is also the most common approach when presenting results. In this chapter however, both statistical and graphical methods were considered to determine the best fit.

CONCLUSION

All of the three proposed systems; StdRIO, ModRIO and GaussRIO are capable of finding shelters for the whole swarm under one minute in average, inside an $100m^3$ arena with three corner shelters. The results for ModRIO and GaussRIO also revealed tendencies of super-scalable systems for up to 25 robots in a swarm.

Additionally, ModRIO and GaussRIO both allow a swarm of robots to seek out shelters without getting trapped inside local optima and flat regions in a three dimensional search space.

Figure 21. Average data from the three distinct techniques indicating that GaussRIO is considerably faster.

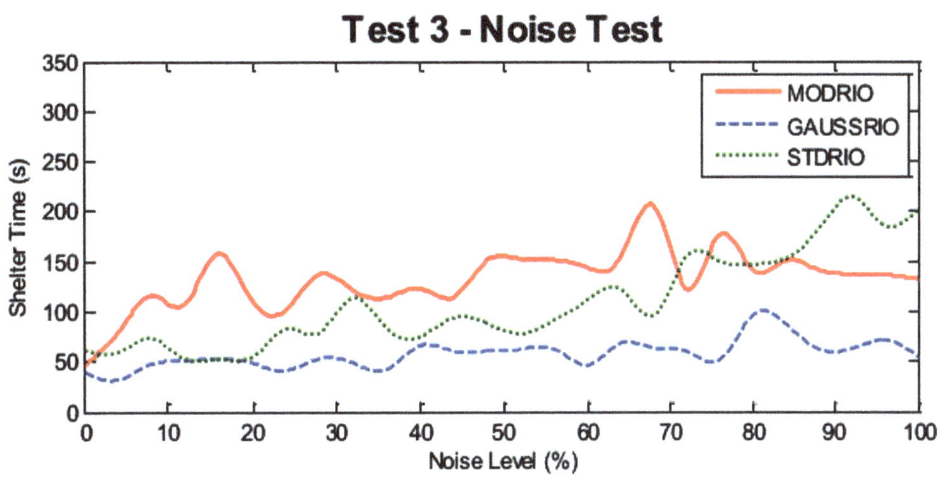

Figure 22. StdRIO seems to be following the same trend as GaussRIO for low level of noises, until the shelter time accelerates and eventually reaches values four times greater than that of a noise free test.

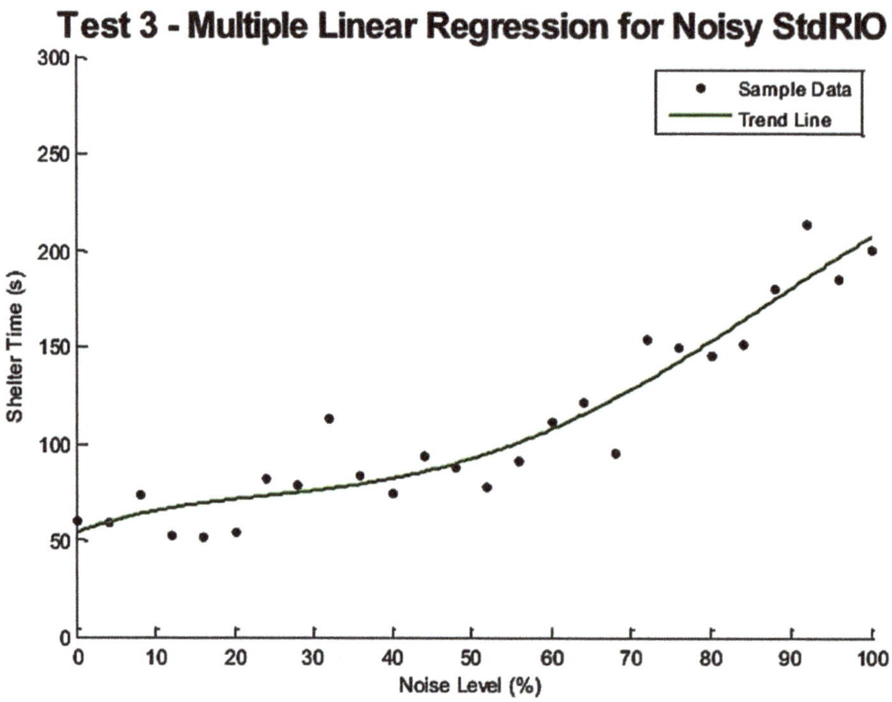

Figure 23. With introduced noise to the system ModRIO has a great leap in shelter time. Although having higher variations in the observed data, the technique follows a similar increasing trend as GaussRIO.

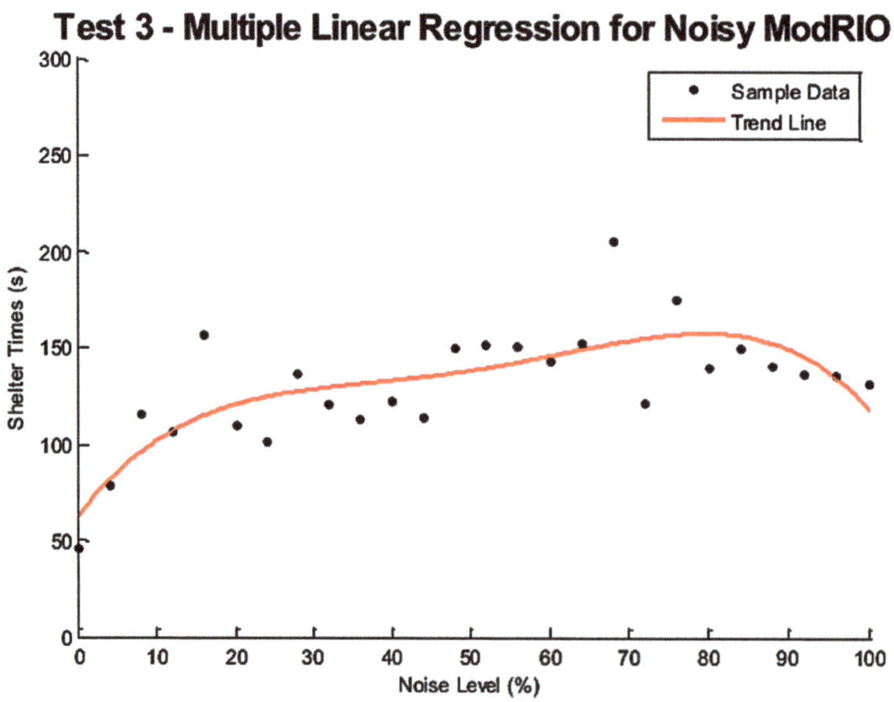

Figure 24. Even though there is an increase in shelter time, the noise seems to leave GaussRIO generally unperturbed.

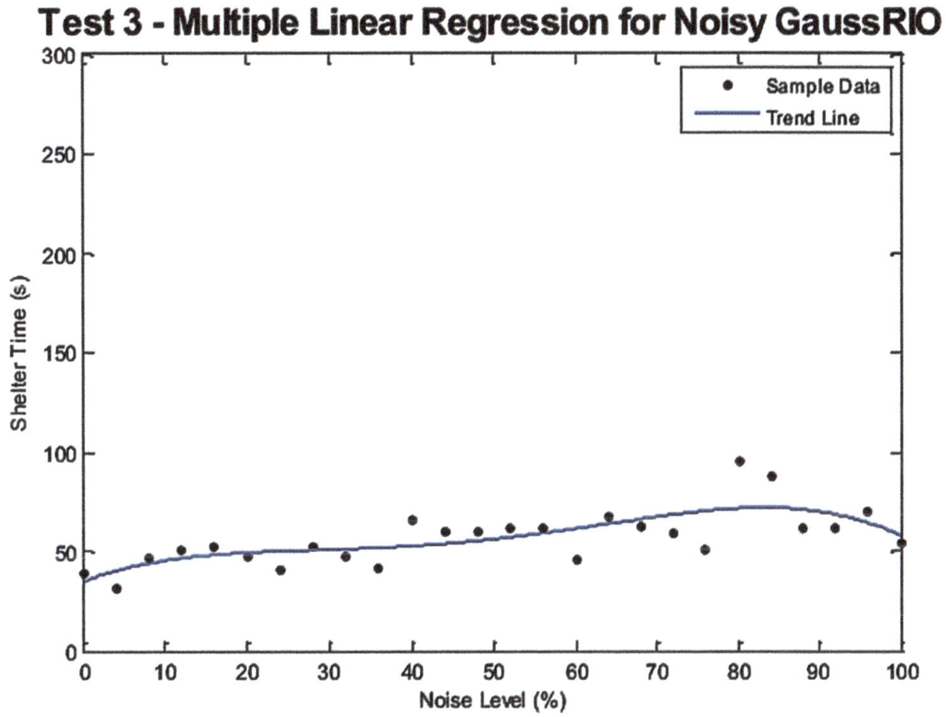

Figure 25. Comparison of the three trend lines showing the noise tolerance level for the techniques and which method is more sensitive than the others.

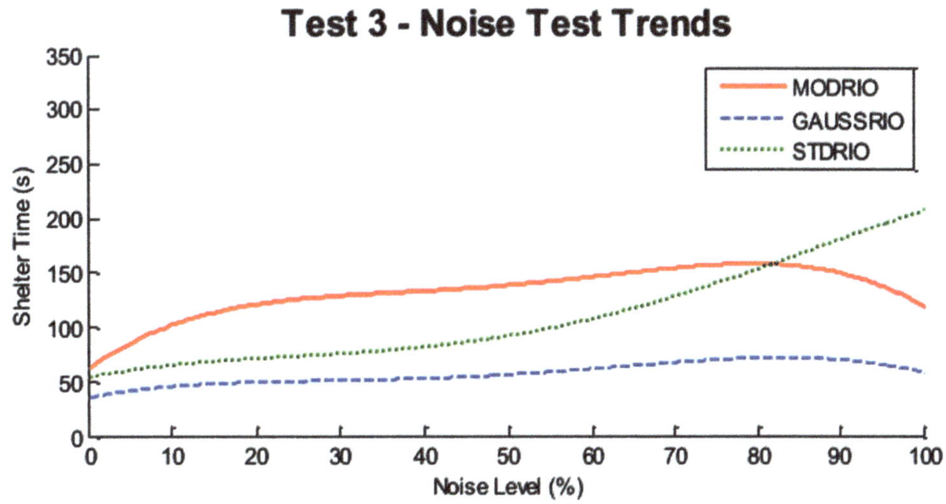

Because of the similar performance between ModRIO and GaussRIO perhaps the implementation of Lévy flight may have no particular influence on a system, or conceivably a larger search volume can prove otherwise. The one single difference that could be interpreted is the greater noise sensitivity of ModRIO, while GaussRIO seems rather unaffected to noise.

It was verified that StdRIO had the worst overall results from the experiments, which makes us question the necessity of individual memory in multi modal optimization.

This research area is yet to be explored and to this authors knowledge there is a lack of studies done on swarm intelligence for flying robots, particularly in the context of the plateau or flat-region problem. It can therefore be said that this research has a meaningful scientific contribution and can be considered as a suitable stepping stone for future development of autonomous flying robot swarms.

FUTURE RESEARCH

A series of extensions to this research can be undertaken to further investigate shelter seeking behaviours for flying robots. One suggestion is to evaluate the functionality of the proposed techniques across varying arena sizes with real life environments, where the population can be scaled up beyond 25 individuals. A different approach to full shelters may have to be developed to counter disrupted search patterns as was discovered in the connectivity experiment.

The technique could also be combined with collision avoidance techniques.

Presenting different qualities of shelters and investigating how this affects the collective decision-making would also be an interesting avenue for further research as the goal of the swarm is to find the best suited shelter. This does not necessary mean a darker shelter. Escape behaviour of cockroaches studied by Camhi & Tom (1978) supports the theory that cockroaches run away from wind sources, which would explain why a cockroach runs straight into the nozzle of the vacuum cleaner or moves just before being crushed by a shoe. The agricultural experiment station conducted a test where they found that wind is such a strong repellent that cockroaches even prefer to stay in well-lit areas than dark windy ones (Appel et al., 1997). This can be taken into consideration to better define a good hiding place. One must also take into account what is harmful to the quad copters themselves, e.g. moisture, dust or other aerosols.

On a final note, in this research the robots were unable to share their previous experiences with neighbours because they require memory to do so. Although the robots were made more selfish, there is no doubt that teamwork through connectivity and networking plays the largest role in a creating an intelligent swarm without specifically using smart individuals.

ACKNOWLEDGMENT

The authors would like to thank Sven Molin and The Swedish Foundation for International Cooperation in Research and Higher Education (STINT) under institutional grant IG2011.2025 for making this research collaboration possible. We would also like to express gratitude to Carlo Pinciroli from IRIDIA for giving us access to the latest version of ARGoS and for assisting us in the process of uncovering the inner workings of the simulator.

REFERENCES

Amé, J., Halloy, J., Rivault, C., Detrain, C., & Deneubourg, J. L. (2006). Collegial Decision Making Based on Social Amplification Leads to Optimal Group Formation. *Proceedings of the National Academy of Sciences of the United States of America, 103*(15), 5835–5840. doi:10.1073/pnas.0507877103 PMID:16581903

Appel, A., Oswalt, D., Smith, L., & Koen, J. (1997). Moving Air Repels Cockroaches. *Highlights of Agricultural Research, 44*(2). Retrieved from http://www.aaes.auburn.edu/comm/pubs/highlightsonline/summer97/cockroaches.htm

Bansal, J. C., Singh, P. K., Saraswat, M., Verma, A., Jadon, S. S., & Abraham, A. (2011). Inertia Weight strategies in Particle Swarm Optimization. *Nature and Biologically Inspired Computing (NaBIC)*, 633-640. doi:10.1109/NaBIC.2011.6089659

Beheshti, Z., & Shamsuddin, S. (2013). A Review of Population-Based Meta-Heuristics Algorithm. *International Journal of Advances in Soft Computing & Its Application, 5*(1), 2-35. Retrieved from http://home.ijasca.com/data/documents/IJASCA07_REVISED.pdf

Camhi, J. M., & Tom, W. (1978). The Escape Behavior of the Cockroach Periplaneta Americana. *Journal of Comparative Physiology. A, Neuroethology, Sensory, Neural, and Behavioral Physiology, 128*(3), 193–201. doi:10.1007/BF00656852

Campo, A. (2010). *On the Design of Self-Organized Decision Making in Robot Swarms* [Doctoral dissertation]. Université Libre de Bruxelles, Brussels, Belgium. Retrieved from https://iridia.ulb.ac.be/~mdorigo/HomePageDorigo/thesis/phd/CampoPhD.pdf

Carlisle, A., & Dozier, G. (2001). An off-the-shelf PSO. *Proceedings of the Particle Swarm Optimization Workshop.*

Daltorio, K. A., Tietz, B. R., Bender, J. A., Webster, V. A., Szczecinski, N. S., Branicky, M. S., ... Quinn, R. D. (2013). A model of exploration and goal-searching in the cockroach, Blaberus discoidalis. *Adaptive Behavior, 21*(5), 404–421. doi:10.1177/1059712313491615

Ganihar, D., Libersat, F., Wendler, G., & Camhi, J. M. (1994). Wind-evoked evasive responses in flying cockroaches. *Journal of Comparative Physiology. A, Neuroethology, Sensory, Neural, and Behavioral Physiology, 175*(1), 49–65. doi:10.1007/BF00217436 PMID:8083847

Garnier, S., Jost, C., Jeanson, R., Gautrais, J., Asadpour, M., Caprari, G., & Theraulaz, G. (2005). Collective decision-making by a group of cockroach-like robots. *Proceedings of the 2005 IEEE Swarm Intelligence Symposium SIS '05* (pp. 233-240). 10.1109/SIS.2005.1501627

Gazi, V., & Passino, K. M. (2011). *Swarm stability and optimization* (2011 ed.). New York, NY: Springer. doi:10.1007/978-3-642-18041-5

Green, D. G. (2014). *Of ants and men: The unexpected side effects of complexity in society*. Berlin, Germany: Springer-verlag Berlin And Heidelberg Gmbh & Co. Kg. doi:10.1007/978-3-642-55230-4

Greiner, H. (2013, June 18). Flying Robots. *MIT Technology Review*. Retrieved from http://www.technologyreview.com/view/516061/flying-robots/

Halloy, J., Sempo, G., Caprari, G., Rivault, C., Asadpour, M., Tâche, F., ... Deneubourg, J. L. (2007). Social Integration of Robots into Groups of Cockroaches to Control Self-Organized Choices. *Science, 318*(5853), 1155–1158. doi:10.1126cience.1144259 PMID:18006751

Havens, T. C., Spain, C. J., Salmon, N. G., & Keller, J. M. (2008). Roach Infestation Optimization. *IEEE Swarm Intelligence Symposium SIS 2008*, 1-7. 10.1109/SIS.2008.4668317

Hoad, K., Robinson, S., & Davies, R. (2007, Winter). Automating DES Output Analysis: How Many Replications To Run. *Proceedings of the Simulation Conference* (pp. 505-512). 10.1109/WSC.2007.4419641

Bjerknes, J. D., & Winfield, A. F. T. (2013). On Fault Tolerance and Scalability of Swarm Robotic Systems. In A. Marinoli, F. Mondada, N. Correll, G. Mermoud, M. Egerstedt, M. Ani Hsieh, ... K. Støy (Eds.), *Distributed autonomous robotic systems* (pp. 431–444)., doi:10.1007/978-3-642-32723-0_31

Jeanson, R., Rivault, C., Deneubourg, J., Blanco, S., Fournier, R., Jost, C., & Theraulaz, G. (2005). Self-organized aggregation in cockroaches. *Animal Behaviour, 69*(1), 169–180. doi:10.1016/j.anbehav.2004.02.009

Kennedy, J. N., & Eberhart, R. C. (1995). Particle Swarm Optimization. *Proceedings of the IEEE International Conference on Neural Networks '95* (Vol. 4, pp. 1942-1948). doi:10.1109/ICNN.1995.488968

Kernbach, S. (2013). *Handbook of collective robotics: Fundamentals and challenges*. Boca Raton, FL: Pan Stanford. doi:10.1201/b14908

Klavins, E. (2004). Communication Complexity of Multi-robot Systems. *Algorithmic Foundations of Robotics V, 7*, 275–291. doi:10.1007/978-3-540-45058-0_17

Kumar, V., & Michael, N. (2012). Opportunities and challenges with autonomous micro aerial vehicles. *The International Journal of Robotics Research, 31*(11), 1279–1291. doi:10.1177/0278364912455954

Law, A. M., & McComas, M. G. (1991). Secrets of Successful Simulation Studies. *Proceedings of the Simulation Conference Winter 91* (pp. 21-27). 10.1109/WSC.1991.185587

Lin, J. H., Cho, C. W., Yang, C. H., & Tsai, H. L. (2012). A Chaotic Levy Flight Bat Algorithm for Parameter Estimation in Nonlinear Dynamic Biological Systems. *Journal of Computer and Information Technology*, 2(2), 56–63. Retrieved from http://www.academypublish.org/papers/pdf/2632.pdf

Liu, C., Wu, H., Yang, G., & Wei, Z. (2010). Path Planning of Flying Robot for Powerline Inspection Based on Improved Particle Swarm Optimization. *Proceeding of the International Conference on Intelligent System Design and Engineering Application (ISDEA)* (Vol. 2, pp. 48-52). 10.1109/ISDEA.2010.347

Pinciroli, C., Trianni, V., O'Grady, R., Pini, G., Brutschy, A., Brambilla, M., & Dorigo, M. (2012). Swarm Intelligence. *ARGoS: A Modular, Parallel, Multi-Engine Simulator for Multi-Robot Systems* (pp. 271–295). Berlin, Germany: Springer.

Påhlsson, P., Strand, L., Powel, S., Stenari, D., Helgesson, M., Olsson, C. S., . . . Guisti, T. (2011). Marvel The Martian airborne research vehicle. *Proceedings of the 20th ESA Symposium on European Rocket and Balloon Programmes and Related Research*. Retrieved from http://spaceflight.esa.int/pac-symposium_archives/files/papers/s5_9pahlss.pdf

Small wonders: What ants can teach us. (2011, July 24). *CBS News*. Retrieved from http://www.cbsnews.com/news/small-wonders-what-ants-can-teach-us/

Definition of Swarm intelligence. (n. d.). *Dictionary.com*. Retrieved from http://dictionary.reference.com/browse/swarm%20intelligence

UAV Design [Blog post]. (2013, May 7). Retrieved from http://uavdesignguide.com/

Viswanathan, G. M., Buldyrev, S. V., & Havlin, S., daLuz, M. G. E., Raposo, E. P., & Stanley, H. E. (1999). Optimizing the success of random searches. *Nature*, *401*(6756), 911–914. doi:10.1038/44831 PMID:10553906

Yang, X., & Deb, S. (2010). Eagle Strategy Using Lévy Walk and Firefly Algorithms for Stochastic Optimization. *Proceedings of the Nature Inspired Cooperative Strategies for Optimization (NICSO 2010)*, 101-111. doi:10.1007/978-3-642-12538-6_9

KEY TERMS AND DEFINITIONS

Centralized System: A master-slave based system where all information and commands are passed through the central unit (master). There is no sporadic sharing or decision-making amongst the local nodes without the central power being involved.

Decentralized System: An individual decision-making authority that guides the system components. Everything is divided into colonies, each with one decision-making unit, and one single authority in the central core. Thus the central core controls the lower ranks that in turn control their respective local nodes. The power is decentralized but the central unit still has control over the entire system.

Distributed System: An assemblage of nodes or computers with equal authority that interact and share their computational resources and activities. From a user perspective one would see it as a single system because the nodes together act as if they were one large computer. The architecture of a distributed system is such that the nodes are autonomous and could in principal work independently.

Heterogeneous Swarms of Robots: A swarm of unique robots with divergent capabilities and divergent roles.

Homogenous Swarms of Robots: A unity of robots all having identical structure and capabilities.

Metaheuristic Optimization: An optimization with the use of an algorithm based on stochastic components, e.g. nature inspired algorithms.

Quad Copter: A rotorcraft with four propelled rotors that are mounted symmetrically on a cross frame. By controlling the rotation of the rotors, one can manoeuvre the rotorcraft.

Swarm Intelligence: Swarm Intelligence (SI) is a decentralized or a distributed system where multiple units using very simple components can make quick and smart decisions without the help of an external controller. Although the units are equally "smart", the overall intelligence of the group is higher than the intelligence of the single unit.

Unmanned Aerial System: Nowadays industry has adopted the term Unmanned Aerial systems (UAS) rather than using the previously preferred term unmanned aerial vehicle (UAV), which is an aircraft that does not carry any pilot, but is either autonomous, monitored, or controlled by a pilot from ground or from another vehicle. UAS not only refers to the UAV platform itself, but also to the system as a whole, i.e. the ground-based controller and the communication system between them.

This research was previously published in the Handbook of Research on Design, Control, and Modeling of Swarm Robotics edited by Ying Tan, pages 687-717, copyright year 2016 by Information Science Reference (an imprint of IGI Global).

Section 2

Control, Design, and Simulations

Chapter 4
Cross-Layer Scheme for Meeting QoS Requirements of Flying Ad-Hoc Networks:
QoS Requirements of Flying Ad-Hoc Networks

Bilal Muhammad Khan
National University of Sciences and Technology Islamabad, Pakistan

Rabia Bilal
Usman Institute of Technology, Pakistan

ABSTRACT

Recently, Flying Ad-hoc Networks (FANETs), enabling ad-hoc networking between highly mobile Unmanned Aerial Vehicles (UAVs), are gaining importance in several military, commercial and civilian applications. The sensitivity of these missions requires precise and prompt data delivery. Thus, the most important communication requirements that need to be addressed while designing FANETs are of high reliability and low latency. Considering these demands, this chapter focusses on mobility models, MAC protocols and routing protocols.

INTRODUCTION

Background

With the world turning into a global village due to technological advancements, automation in all aspects of life is gaining utmost importance. Wireless technologies have resulted in addressing the ever increasing demands of portable and flexible communication. Wireless ad-hoc networks which allow communication between devices without the need of any central infrastructure are gaining significance with monitoring and surveillance applications. A relatively new research area of ad-hoc networks is Fly-

DOI: 10.4018/978-1-5225-8365-3.ch004

ing Ad-hoc Networks (FANETs), governing the autonomous movement of Unmanned Aerial Vehicles (UAVs) (Bekmezci, Sahingoz, & Temel, 2013).

In the present day world, Unmanned Aerial Vehicles (UAVs) are considered a potential substitute of manned airplanes for several military, civilian and commercial operations. These aircrafts are used to conduct autonomous missions in remote areas in which the involvement of human personnel is considered dangerous and unsafe. Due to technological revolution in electronic and networking systems, UAVs may not only provide cost benefits but also help in efficient and adaptable mission completion for sensitive applications such as traffic monitoring, remote sensing, disaster monitoring, search operations, border surveillance and relaying networks.

The concept of single UAV systems is not new with their history associated with World Wars. Such systems comprise of an infrastructure which communicates with the UAV. This requires each UAV to be equipped with complex hardware systems to maintain its communication with the ground station, thus raising the disadvantages of high cost, reliability issues and link breakages. This led to the development of multi-UAV systems in which a network is formed between multiple UAVs which are made to communicate and collaborate so as to enhance the capability of single UAV systems. There are two distinct approaches for multi-UAV systems. The first is to form an infrastructure-based network between UAVs such that the communication between different UAVs in the network is regulated through a ground base station or satellite (Frew & Brown, 2008). This network model raises several design challenges as each UAV requires dedicated and complex hardware installment. For missions that may require higher number of UAVs, infrastructure-based multi-UAV systems are not feasible and reliable. Moreover, vigorous environmental conditions may cause link breakages between UAV and infrastructure resulting in data loss and increased delays. The second approach is to form an ad-hoc network between them. Such a network between multiple UAVs is termed as a FANET. All UAVs in the network are allowed to carry UAV to UAV communication and only a group of UAVs interact with the ground station. This eliminates the deployment of complex hardware in each UAV. Moreover, even if one of the UAVs turns down, there is no link breakage to the ground station due to ad-hoc network between UAVs.

Ad-hoc nature of FANETs allows their deployment in far-flung mission areas (Martinez-de et al., 2012). This helps in disaster management in places with impaired communication infrastructure or where prompt network establishment is required for receiving precise data from the affected region. Unlike (Mobile Ad-hoc Networks) MANETs and Vehicular Ad-hoc Networks (VANETs), FANETs have some critical requirements which need to be met. High mobility, low node density, low latency, large node distances and frequent topology change are some of the distinguishing characteristics of FANETs.

Motivation and Need

FANETs is an emerging research area and most of the research being done is in military and defense sector and hence not open source. The research has been conducted with the aim of sharing the findings with the research community so that affordable and efficient network models can be used for civilian applications to achieve desirable outcomes.

The amount of work done over the higher layers of FANETs is not enough. Moreover, a significant portion of research in FANETs focuses on the optimization and diversified use of Commercial Off-The-Shelf (COTS) equipment which adds to communication cost. Not many studies exist on Medium Access

Control (MAC) and routing protocols' performance in this domain. There is a lot of room to work and make good progress in the focused arena. Hence, this work can act as a stimulator for further progress in this research domain.

FANETs are not only expected to boost performance in battlefield and military applications but they also possess a great inclination towards social applications. This research work focusses on investigating the technological aspects of FANETs that must be considered for upgrading the quality of life and environment around us.

OVERVIEW OF FANETS

FANETs is a relatively new and potential study area and holds a lot of room for further research and development. This chapter discusses the social impact of FANETs along with a detailed overview of communication requirements and challenges faced by the research community.

Social Significance of FANETs

The usage of FANETs for several military, civilian and commercial applications is expected to deliver favorable results in terms of reliable and delay bounded data delivery. Traffic monitoring, remote sensing, border surveillance, disaster management, and relaying networks are some areas where FANETs can be used to uplift the quality of life as shown in Figure 1.

FANETs as Monitoring Networks

The use of UAVs for surveillance and monitoring applications is practically being implemented since long. However, with the advent of FANETs, the concept is expected to be revolutionized in terms of reliable and delay bounded data delivery.

Figure 1. FANETs for social applications

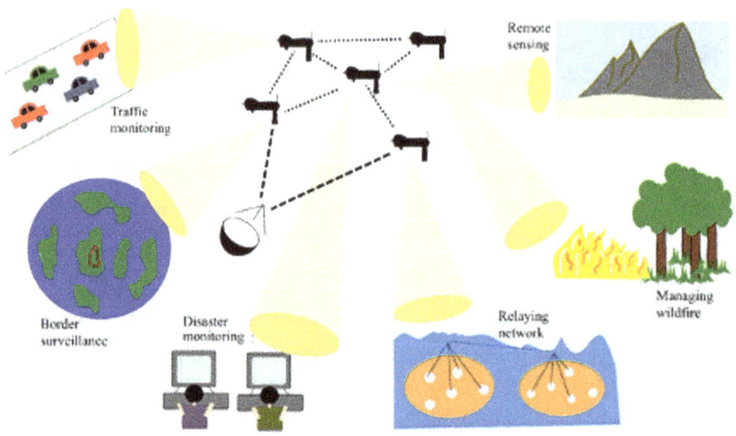

Reconnaissance Missions

In surveillance applications such as border patrol systems (Geng, Zhang, Wang, Fuh, & Teo, 2013) and urban surveillance (Semsch, Jakob, Pavlicek, Pechoucek, & Michal, 2009), UAVs are playing an important role in minimizing human intervention. Aerial reconnaissance missions range from information collection on battle fields to traffic monitoring and law enforcement activities. Monitoring tasks may include taking images of the objects and sites of interest spread over wide areas (Lim, Park, Ryoo, Choi, & Cho, 2010). For instance, in border policing swarm of UAVs can not only detect unplanned human disturbance including weapons and drugs but also illegal border crossing. The collected information can be immediately processed and transmitted to intelligence control rooms. However, the sensitivity of the data needs high accuracy for prompt action. All such surveillance missions exhibit complicated and ad-hoc nature and are intolerant to false alarms.

Traffic Monitoring

Roadway traffic monitoring is also a potential application where UAVs can replace intensive labor and complicated observing infrastructures (Coifman, McCord, Mishalani, Iswalt, & Ji, 2006). The deployment of detectors or human personnel throughout the roadway network is indeed a costly and time sensitive solution. UAVs which can monitor and report incidents or traffic management data are an economically and socially feasible option as they do not require moving on roads. Moreover, due to their high speeds, urgent event specific data can be transferred to boost up the rescue or destroy missions (George & Sousa, 2010).

Agricultural Management

Agriculture production management requires crop plant health to be monitored. Although manned aircrafts are being used in this sector but the emerging concept of small autonomous UAV swarms is believed to be more beneficial as they can perform field operations over smaller fields with more precision. This means that UAVs can focus specific sites and application of fertilizers can be performed accordingly (Huang, Thomson, Hoffmann, Lan, & Fritz, 2013).

By employing FANETs for the above discussed monitoring applications, coordination can be established amidst UAVs to achieve mission objectives cooperatively. FANETs can not only increase the coverage area but also scalability and survivability of missions (Bekmezci, Sahingoz, & Temel,2013). The type of communication carried between UAVs in such mission scenarios may include control data, video and image data and Voice over IP data (Pojda, Wolff, Sbeiti, & Wietfeld, 2011) depending on nature of the tasks assigned. Also, the scenarios and mobility patterns of the UAVs differ depending on the application and coverage requirements.

FANETs as Relaying Networks

Autonomously operated UAVs are also serving as a relaying network for MANETs and Wireless Sensor Networks (WSN) (Heimfarth & De Araujo, 2014). A relaying network is used to maintain communication links between remote source and destination which cannot communicate directly due to obstacles. To efficiently and securely transmit information collected by ground sensors to distant control centers and to increase communication range of ground relaying nodes, UAVs have been used as airborne communication relays (Jawhar, Mohamed, Al-Jaroodi, & Sheng Zhang, 2013; Cetin & Zagli, 2011). Areas such as forests and marshlands, which are inaccessible for ground vehicles due to their remote placements, can be prevented from disasters by WSN and UAV networks. Also, lifetime of the remote sensor networks can be increased by using UAVs as mobile relays since maintenance of such networks is a daunting task. Wildfire is a concerning problem all over the world. UAVs cannot only help in preventing such incidents by transmitting the sensed information related to location of the fire-prone areas but in post-fire tasks as well (Barrado *et al.*, 2010). Such event driven applications require that the information should be available to fire managers urgently so such applications tend to meet certain delay bounds.

FANETs are expected to enhance the performance of UAV relaying networks as the use of multiple UAVs can further increase the communication range. Moreover, in comparison to ground relaying networks swarms of UAVs can relay data with minimal latencies due to high mobility and robust connectivity. They move high up in the sky and are not met with terrain constraints. Also, there is clear Line of Sight (LoS) between ground nodes and UAVs which further ensures reliable delivery. The energy of sensor nodes in distant areas can be preserved by dedicating a group of nodes to directly communicate with the UAV network. This in turn minimizes the number of hops required to relay the information which in the absence of FANETs require multiple hops and ground relaying nodes. In this manner scientists can conduct quality research in dangerous remote spots, which are considered unsafe for technology personnel such as volcano-prone areas, without visiting them.

Figure 2. A block diagram of FANET node

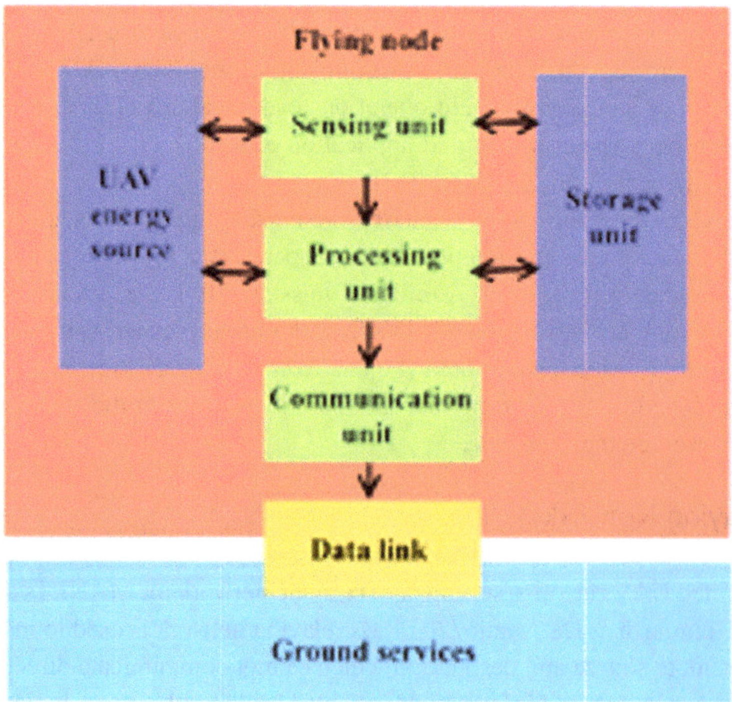

Distinguishing Features of FANETs

FANETs belong to the family of MANETs and VANETs. Figure 2 depicts block diagram of a FANET node. However, technical requirements of FANETs differ from the latter two with respect to node speed, mobility pattern, node density and energy consumption. The dissimilarities are discussed below:

- In FANETs, nodes are highly mobile with a typical UAV speed of 30-460 km/hr (Bekmezci, Sahingoz, & Temel, 2013; Huba & Shenoy,2012). Such high mobility factor accounts for challenging communication design.
- UAVs in a FANET are deployed on critical missions. A predetermined path is set to accomplish such operations. Since they fly in formations, special mobility models and path planning strategies are required for FANETs rather than commonly used Random Way Point (RWP) models in MANETs (Bouachir, Abrassart, Garcia & Larrieu, 2014).
- The UAVs comprising a FANET are scattered in the sky and they are much fewer in number forming a UAV swarm unlike large number of mobile devices or vehicles which form ad-hoc networks on ground (Bekmezci, Sahingoz & Temel, 2013).
- FANETs do not suffer from energy consumption problems because the power resource of the UAVs fulfills the energy requirement of FANET communication hardware and the constraints of battery powered devices are not met.

Communication Challenges

FANETs suffer challenges such as high mobility, frequent topology change, minimal delay and high reliability requirement. In order to carry efficient and vigorous communication between UAVs these challenges have to be encountered.

High Mobility

UAVs in FANETs demonstrate much rapid motion than MANETs and cannot be catered by conventional methods like mobile IP (Bekmezci, Sahingoz & Temel, 2013). Moreover, the quality of wireless links fluctuates due to high speed of UAVs which is addressed by MAC layer. Link stability and transmission range gets varied due to signal strength uncertainties. To study how link fluctuations caused by high mobility may be detrimental in FANET applications, consider a routing protocol which keeps a check on its neighbors by sending topology control messages before genuine data transmission. If the link quality goes down after the transmission of control packets, the link becomes void. Hence, the routing decisions will not be precise leading to performance degradation. MAC layer addresses this issue by maintaining Quality of Service (QoS) for one hop transmission (Cai, Yu, Li, Zhou, & Lamont, 2012) whereas routing layer ensures reliable end-to-end delivery (Pojda, Wolff, Sbeiti, & Wietfeld, 2011). One potential solution to encounter high mobility of UAVs has been addressed by predicting the location and movement of UAVs through heuristic techniques (Newaz, Pratama, & Chong, 2013). On the favorable side, when FANETs are used as relaying networks, the high mobility factor gives desired outcomes in establishment of delay tolerant links between geographically distant infrastructures.

Rapid Topology Change

Frequent topology changes result from high mobility which has to be considered in FANET design. Other factors which cause redefinition of routes include environmental obstacles such as mountains, failure of existing UAVs in the network, injection of new UAVs, possible mission updates or operational requirements. Due to weather conditions and geographical uncertainties, applications such as remote sensing become victim of bad links between UAVs (Bekmezci, Sahingoz, & Temel,2013). All such factors responsible for topological changes cause increased routing overhead, delays and low successful packet delivery rates. The topology change rate can be neglected by enlarging the transmission range of the UAVs i.e. employing a MAC layer protocol which supports large transmission range. The trade-off for adopting this solution will be complexity and energy consumption but UAVs have enough hardware support to cater such issues.

Low Latency Requirement

Search and destroy operations and disaster monitoring require minimal latency as the information needs to be transmitted at a very high rate. However, latency requirement of a FANET is application dependent. A crude way for acquiring minimal latency is that hop counts are decreased by long transmission ranges resulting in improved latency results. Moreover, the concept of priority schemes can also be used in FANETs (Shetty, Sudit, & Nagi, 2008). Priority based routing protocols can be used to attain different QoS for various message types. Congestion control protocols are vital for performing efficient collision control and coordination among UAVs (Bodanese, de Araujo, Steup, Raffo, & Becker, 2014). An appropriate choice of protocol suite can address the demand of delay bounds.

High Reliability

Affordability and accessibility of UAVs are determined by their reliability factor. FANETs are used in sensitive military and monitoring applications which require guaranteed data delivery (Coifman, McCord, Mishalani, Iswalt, & Ji, 2006; Barrado *et al.*, 2010). In FANETs, reliability is achieved by forming an ad-hoc network between UAVs so that if a UAV experiences a failed link it can still communicate to the infrastructure via other UAVs. In the absence of an ad-hoc set-up amid UAVs, each UAV needs to be connected to the infrastructure. Under such scenario if a UAV fails, important information related to the mission could be lost. Routing protocols based on data aggregation can serve the demand of highly accurate data (Huba & Shenoy, 2012). In such protocols data loss due to link failures is addressed by allowing the intended destination to request certain data from other UAVs in the network, thus ensuring the receipt of fully reliable information considering military and monitoring centric applications.

Communication Types in FANETS

UAV communication deals with exchange of data between UAVs and between UAVs and infrastructure i.e. ground station or satellite. Based on the intended destination, UAV communication in FANETs is classified into two main types as shown in Figure 3.

UAV to UAV

Inter-UAV communication carried in an ad-hoc manner gives rise to FANETs. Information is exchanged between UAVs without a central controller using different routing algorithms. These UAVs are in vicinity of each other and allow multi-hop communication of command and control data between them e.g. about target object in search and destroy operation. Since the FANET topology changes due to removal or addition of new UAVs, maintaining efficient UAV to UAV communication leads to several research directions. Robust routing protocols that can guarantee packets' delivery from source UAV to destination UAV are being explored (Alshabtat & Dong, 2011; Lin, Sung, Li, & Yang, 2012). Moreover, reducing packet latency between UAVs, due to rapid scenario change, is also an important research area. Although the same task could be performed using infrastructure based multi UAV systems but UAV to UAV data links in FANETs can reliably convey the information even if a UAV cannot form communication link with the infrastructure (Bekmezci, Sahingoz, & Temel, 2013).

UAV to Infrastructure

In FANETs, only a subgroup of UAVs communicates with the infrastructure. This type of communication conveys information about intended tasks over a wider scale and requires exchange of heavier data payloads. Few UAVs amongst the group send updates to the infrastructure e.g. on-site images in traffic monitoring. Extended communication range is required for the exchange of messages between UAV and satellite or ground station.

FANET Design Considerations

FANETs are characterized by rapid changes and activities during network operation. The design should be highly scalable with adaptive path planning and robust communication protocols. A brief insight of these technical requirements is summarized below:

Topologies

In order to maintain coordination and collaboration, peer-to-peer connections are formed between UAVs in a FANET scenario. The tasks can be completed using single cluster or multicluster formations. Single cluster networks can best serve to accomplish homogenous and small scale missions. However, the need for multicluster networks arises when certain UAVs have to perform multiple missions covering wide terrains. In such case, UAVs group together into small clusters with ad-hoc network within each cluster. The cluster head of each cluster is not only designated to carry downlink communication but to also communicate with cluster heads of other clusters. The clusters may be in same or different mission areas and serve to enhance the scalability of the network as shown in Figure 4. Such models are used in scenarios which involve large mission areas as search tasks can be performed more efficiently and collaboratively. For multicluster scenario, selection of cluster head is a serious design issue when one of the cluster heads goes down. Hence, FANETs should be designed for highly dynamic scenarios.

Figure 3. UAV-to-UAV and UAV-to-infrastructure communication scenario

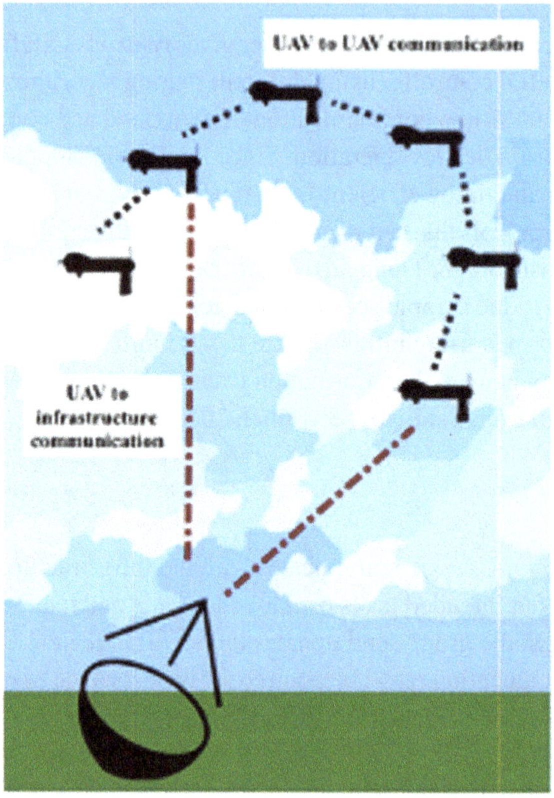

Figure 4. Clustered FANET scenario

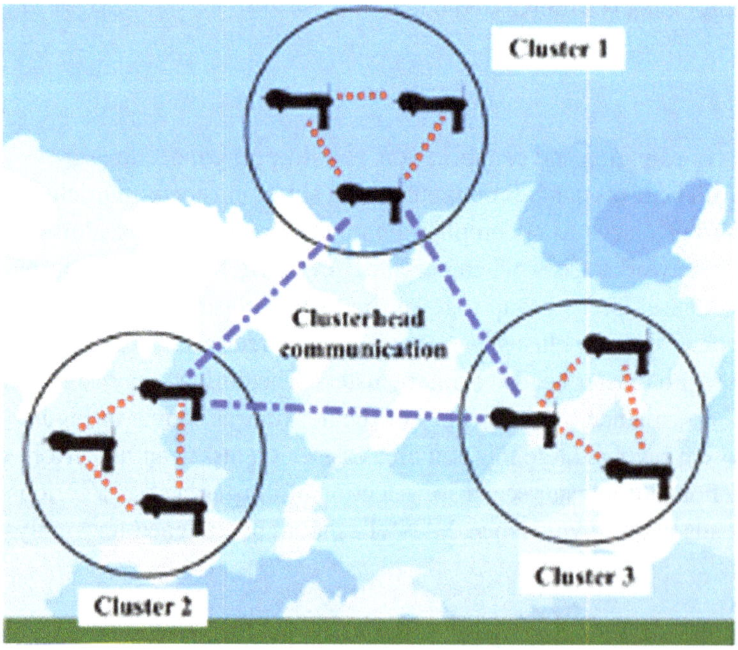

Mobility Models

Efficient communication in FANETs is also dependent on mobility models that capture trails and speed deviations of the UAVs. The selection of mobility model is application dependent. Some tasks require UAVs to move in a planned way on a predetermined path and do not exhibit random movement. Hence, random mobility models cannot be used to simulate FANETs in such scenarios. Reference Point Group Mobility Model (RPGM) can best represent motion of UAVs in performing autonomous military operations without centralized control. This model has a logical center node which governs the movement of the other nodes in its group and monitors their speed, location and direction. Many other models such as Manhattan Grid Mobility (MGM) model which can be used to emulate map based approach in which the movement of UAVs takes into account geographic restrictions (Geng, Zhang, Wang, Fuh, & Teo, 2013; Semsch, Jakob, Pavlicek, Pechoucek, & Michal, 2009) and Random Way Point (RWP) model can be used for patrolling applications where UAVs are allowed to adopt flexible trajectories. Gauss Markov mobility model can serve to model FANETs as relaying networks since the movement of UAVs is dependent on previous speed and directions (Jawhar, Mohamed, Al-Jaroodi, & Sheng Zhang, 2013; Cetin & Zagli, 2011).

Communication Protocols

Due to its unique characteristics, FANETs face communication design challenges. Being a subset of MANETs and VANETs and relatively a new research domain, existing MANET and VANET communication protocols are being tested for FANET scenarios. However, the characteristics and requirements of FANETs strongly differ from those of MANETs and FANETs. Hence, designing of dedicated communication protocols for MAC, network and transport layers still remains an open area of research in FANETs as insignificant amount of work has been done over the higher layers. Figure 5 describes the existing layered architecture for FANETs. To understand the function of MAC and network layer in this regard, widely used MAC protocol 802.11 and newly proposed 802.15.4 has been discussed below along with the essentials of routing protocols.

MAC Layer

MAC layer handles coordinated and scheduled transmission between several contending nodes. The important issues that FANET MAC layer should address are of variable link quality, packet delays and optimal channel utilization during real-time network operation. High bit error rates in FANETs make wireless communication links unreliable.

IEEE 802.11

For the MAC layer protocol, IEEE 802.11 has been used along with the optimization of commercial off the shelf hardware components such as directional and omnidirectional antennas. Due to its extensive availability and support for infrastructure-based as well as infrastructure-less modes, 802.11 is being used in on-going FANET research studies. Full duplex radio circuits and multipacket reception techniques have been used to obtain improved Chanel State Information (CSI) for the UAVs. Due to the availability

Figure 5. Existing layered architecture of FANETs

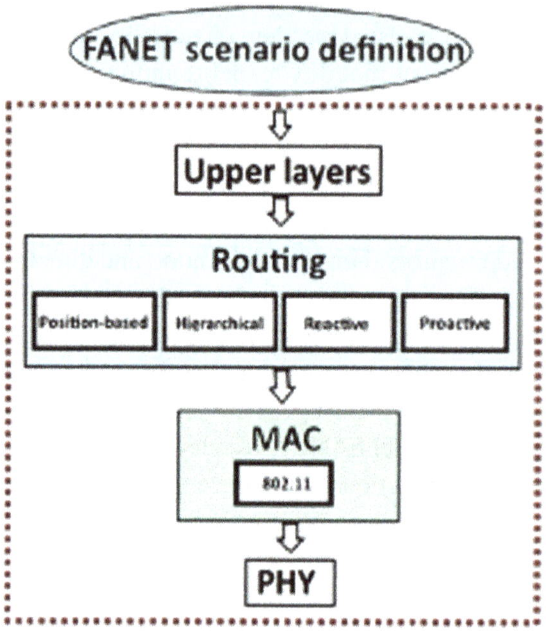

of up-to-date information about channel state, these schemes have exhibited improved results for latency and throughput even if the channel state information is not absolutely accurate.

For bandwidth hungry applications such as hunt and destroy operations and border patrolling, IEEE 802.11 can offer maximum advantages because it supports bandwidth up to 11Mbps and a typical range of 150m which can effectively facilitate multimedia images and videos.

IEEE 802.15.4

IEEE 802.15.4 could be another potential candidate for MAC layer protocol. It provides low power consumption, cost and complexity. This protocol was basically designed for devices with low energy requirement to be used in Wireless Personal Area Networks (WPANs). It can operate at a raw data rate of 250 kbps using 2.4 GHz ISM frequency band. Less bandwidth hungry applications such as monitoring and control, wildfire management, remote sensing and relaying networks require control information to be exchanged within the network. Such applications do not necessarily require a high data rate protocol such as 802.11. UAV to UAV communication can be realized efficiently by deploying a low data rate protocol such as IEEE 802.15.4. IEEE 802.15.4 has not been exploited so far for FANETs. For both single cluster and multi-cluster scenarios 802.15.4 can be used within the cluster so as to achieve better results in terms of reduced complexity and bandwidth minimization.

In comparison to 802.11, two important modes supported by 802.15.4 are beacon enabled or synchronous mode and non-beacon enabled or asynchronous mode. In beacon enabled mode, beacons are generated by a dedicated device known as the Personal Area Network (PAN) coordinator to the nodes who wish to transmit the data. The nodes then transmit their data to the coordinator using slotted Carrier Sense Multiple Access/Collision Avoidance (CSMA/CA) scheme. The slotted CSMA/CA uses superframe structure to split the time between beacons into 16 slots. These slots are divided into Contention Access Period (CAP) and Contention Free Period (CFP). CAP provides contention based medium access whereas CFP have Guaranteed Time Slots (GTS) for data transmission. Superframe order is related to the superframe duration i.e. the active period of the node during CAP and beacon order is used to define beacon interval i.e. the time between two beacons. For multicluster and hierarchical FANETs, synchronous mode can provide the desired benefits as it offers dedicated cyclic slots for heavy traffic.

In non-beacon enabled mode, the node transmits its data to the coordinator using unslotted CSMA/CA. For single cluster or small FANETs, asynchronous mode can be used as it offers low control overhead.

IEEE 802.15.4 can be expected to provide a winning situation for in-network communication between UAVs by modifying MAC layer and determining optimal values of backoff exponential, beacon order (BO), superframe order (SO) and CAP. By minimizing the backoff exponential value, backoff delay of the nodes can be shortened and increasing contention free period length in the superframe structure of 802.15.4 may take advantage in transmitting data leading to improved throughput performance. 100% duty cycle can be achieved by adjusting SO and BO values as energy resource constrain does not occur in FANETs due to dedicated UAV hardware.

To meet certain reliability bounds in FANET applications, controlled access MAC layer protocols can prove to be very promising. TDMA scheme allows nodes to transmit in dedicated time slots. In disaster management scenarios where QoS criteria need to be met, collision free transmissions can be done exploiting TDMA. In multicluster scenario, UAVs may communicate their bandwidth requirements to the cluster head which in turn assign timeslots accordingly.

Network Layer

Dynamic nature of FANETs causes sudden changes in the network topology and thus makes routing among UAVs a daunting task. Considering UAV to UAV communication, routing protocol plays an important role in reliable end-to-end data delivery and makes routing an engaging research prospect in FANET domain. Various reactive, proactive, position-based and hierarchical routing protocols have been implemented to meet the challenges of FANETs. The routing algorithm should be such that if a link cannot be established with one of the UAVs in a FANET, the network can still route messages through other UAVs. Various types of routing strategies that can be exploited in FANETs are mentioned below:

Proactive Routing

Such routing class of protocols tends to save the entire network topology in form of routing tables which are updated on periodic or event driven basis. In FANET studies, proactive protocols have resulted in minimized routing overheads but have also raised concerns as stale routing table cannot account for high mobility (Alshabtat & Dong, 2011).

Reactive Routing

If there is no prior maintenance of routes within the network and a route discovery process is initiated only when requested, the routing protocol used falls in the category of reactive routing. Reactive protocols have proved to address the demand of frequent and unstable topological changes that may occur due to weather conditions, structural obstacles or non-functioning UAVs. However, repetitive path finding results in exhaustive search (Pojda, Wolff, Sbeiti, & Wietfeld, 2011).

Geographic Routing

Geographic routing protocols route information based on the position information of the UAVs (Lin, Sung, Li & Yang, 2012). Benefits in terms of reduced routing overhead, computation time and energy resource are obtained. Since there is clear LoS between transmitter and receiver FANET nodes, hence geographic routing protocols can make use of exact information obtained by Global Positioning System (GPS). However, due to unique challenges of FANETs, such as exceptionally high mobility, determination of accurate positioning systems still remains an open area of research.

Hierarchical Routing

Hierarchical routing protocols run in multicluster scenarios by forming UAVs into groups. By performing efficient routing between cluster heads, hierarchical routing can increase robustness of FANETs along with improved results for large geographic area. Here the issue lies with cluster formation for which FANET dedicated algorithms have been proposed.

Some of the conventional protocols from the above categories like Directional Optimized Link State Routing Protocol (DOLSR) have met the requirements of latency by the use of directional antennas to reduce the number of multipoint relay nodes (Alshabtat & Dong,2011) . A dedicated geographic routing solution named Geographic Position Mobility Oriented Routing (GPMOR) has been proposed that not only determines location but also predicts movement of the UAVs in a FANET (Lin, Sung, Li & Yang ,2012). This approach proved to be beneficial for improved packet delivery ratio results. Dynamic Source Routing (DSR) and time slotted variant of Ad-hoc On-demand Distance Vector (AODV) have also been tested for FANETs (Forsmann, Hiromoto, & Svoboda,2007).

PERFORMANCE ANALYSIS OF MOBILITY MODELS FOR FANETS

Introduction

Efficient routing in FANETs is also dependent on mobility models that capture trails and speed deviations of the UAVs. Since the destination, speed and direction of nodes is not solely random, hence mobility model of MANETs such as Random Way Point (RWP) cannot be wholly used for predicting the behavior of routing protocols in a FANET scenario. However, semi random movement may arise as a result of environment changes leading to redefining path plans. Temporal-based models which generate correlated movement of nodes can be used for critical flight missions where the movement of UAVs should be systematic. Mobility models with spatial dependency can be used to generate group movements as of swarm of UAVs which move in collaboration with each other for dedicated missions. Another class of mobility models is geographic restriction based mobility models in which movement of nodes is subjected to the environment. This type of mobility pattern can be used to generate predefined pathways for UAVs.

In this chapter four available mobility models have been evaluated which include Reference Point Group Mobility Model (RPGM), Gauss Markov (GM) mobility model, Random Way Point (RWP) mobility model and Manhattan Grid Mobility (MGM) model selected from families of spatial dependent, temporal dependent, random based and geographic restriction based mobility models, respectively. The unique presentation of this chapter involves categorization of suitable mobility models for different applications in UAV domain. The effect of dynamic FANET environment, characterized by high mobility and generated by these models, is evaluated for different routing protocols: DSR, AODV and OLSR.

Related Work

Some studies have used existing MANET mobility models while others have proposed specific models for UAV ad-hoc networks. In (Wan, Namuduri, Zhou, & Fu, 2013), smooth turn mobility model has been proposed for Airborne Networks (AN). The model accounts for spatial and temporal dependency of airborne vehicles in highly random AN networks. The trajectory of vehicles is predicted by using the correlated information of speed and acceleration which is straight and smooth at turns.

In (Kuiper & Nadjm-Tehrani, 2006) two mobility models for group surveying scenario of UAVs have been proposed. The first model is based on memoryless movement of UAVs whereas the second model uses the idea of pheromones to provide dependent and coordinated movement among UAVs. The random model provides left, right and straight movements according to fixed probabilities. The study reveals that the models provide a trade-off between maximum coverage and maximum connectivity.

In (Bouachir, Abrassart, Garcia, & Larrieu, 2014), authors propose Paparazzi Mobility Model (PPRZM) based on Paparazzi system for the UAVs. The model supports five movements: Stay-At, Way-Point, Eight, Oval and Scan. Results show that PPRZM gives more realistic results in terms of End-to-End delay as compared to Random Way Point (RWP) Mobility model. Semi-Random Circular Movement (SRCM) mobility model has been proposed for UAV MANETs in (W. Wang, Guan, B. Wang & Y Wang, 2010). The UAV movement is around a fixed point with a random radius.

Gauss Markov mobility model has been used in (Lin, Sung, Li & Yang, 2012) along with the proposed Geographic Position Mobility Oriented Routing (GPMOR). The authors have utilized the best next hop discovery property of Gauss Markov Mobility model to make their routing protocol more robust towards route failures. Similarly, an optimized 3-D version of Gauss Markov mobility model has been used in (Broyles, Jabbar, & Sterbenz, 2010) for highly dynamic AN. In (Kuiper, 2008), the authors have tested their proposed protocol location aware routing for delay-tolerant networks (LAROD) using Random Way Point mobility for reconnaissance tasks.

Mobility Models

A brief discussion on the mobility models and the movement traces used in this study are given below.

Reference Point Group Mobility (RPGM) Model

Group mobility models can be used to simulate group of UAVs in performing autonomous military operations without centralized control. RPGM model falls in the category of mobility models exhibiting spatial dependency. Nodes are divided into groups and the movement of nodes within one group is dependent on each other. Every group has a group center that moves on a predefined path and defines the movement trend of the entire group. The direction and speed of every node in the group depends on its own reference point and a random deviating vector from group center (Camp, Boleng, & Davies, 2002). Figure 6 shows movement trace of RPGM.

Gauss Markov (GM) Mobility Model

In Gauss Markov mobility model, at the start a specific speed and direction is defined for each node (Camp, Boleng & Davies, 2002). This is a memory based model with speed and direction of a node at time instant T depending on speed and direction at time instant T-1. The values are updated after constant time intervals. Hence, Gauss Markov mobility model exhibits temporal dependency. The degree of randomness in the movement can be obtained through a parameter alpha (0<alpha<1). For a memoryless Gauss Markov mobility model alpha=0 and for full memory based model alpha=1.

Random Way Point (RWP) Mobility Model

Random Way Point is a commonly used model of MANETs. Initially, there is random deployment of nodes in the simulation area. The nodes remain stationary for a certain time termed as pause time and then start moving towards randomly chosen destinations. The chosen speed is uniformly distributed between [Vmin, Vmax] (Camp, Boleng, & Davies, 2002). In some applications like patrolling UAVs may be allowed to adopt flexible trajectories. Such scenarios can be modeled using Random Way Point mobility model. Unlike MANETs, UAVs cannot make sharp turns that occur in RWP. So, this random based model cannot provide realistic results for FANETs.

Manhattan Grid Mobility (MGM) Model

Manhattan Grid mobility model can be used to emulate map based approach in which the movement of UAVs takes into account geographic restrictions. The nodes move on pre-defined horizontal and vertical

Figure 6. Movement trace of two nodes under RPGM mobility model in 130 x 130 m area

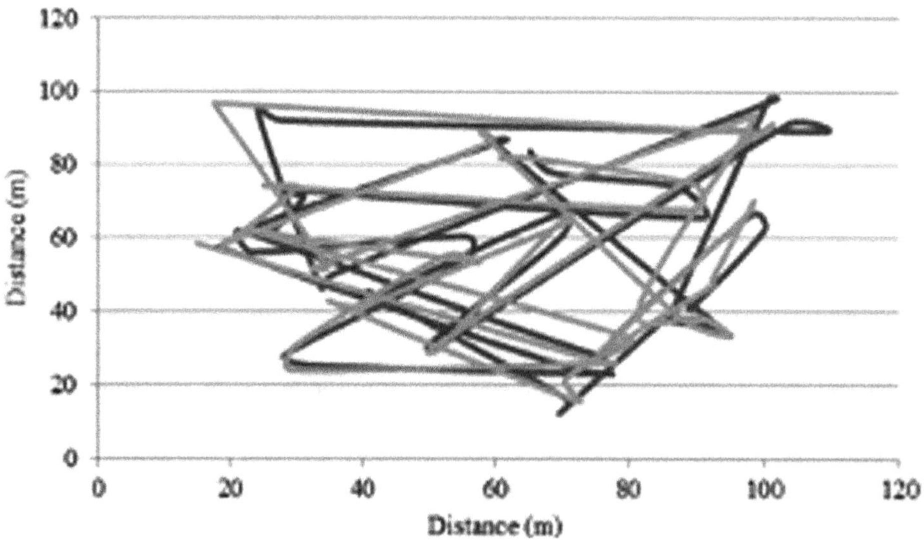

Figure 7. Movement trace of single node under GM mobility model in 130 x 130 m area

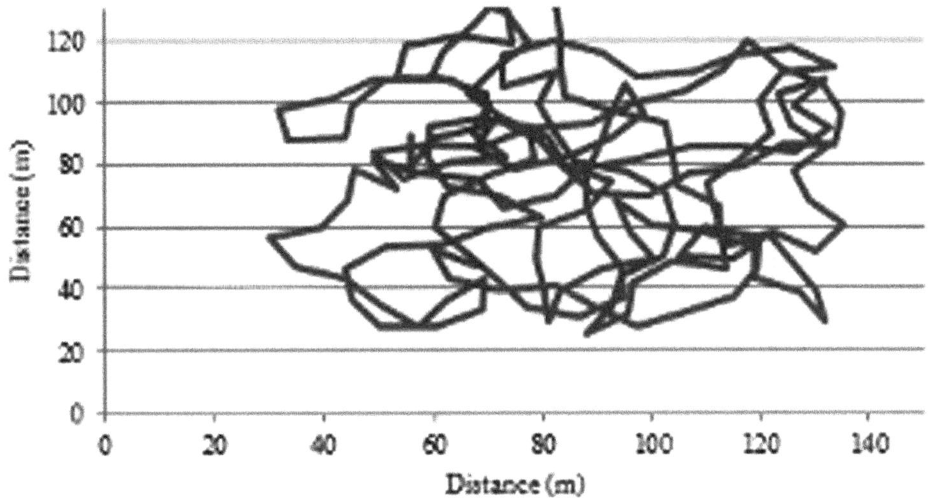

Figure 8. Movement trace of single node under RWP mobility model in 130 x 130 m area

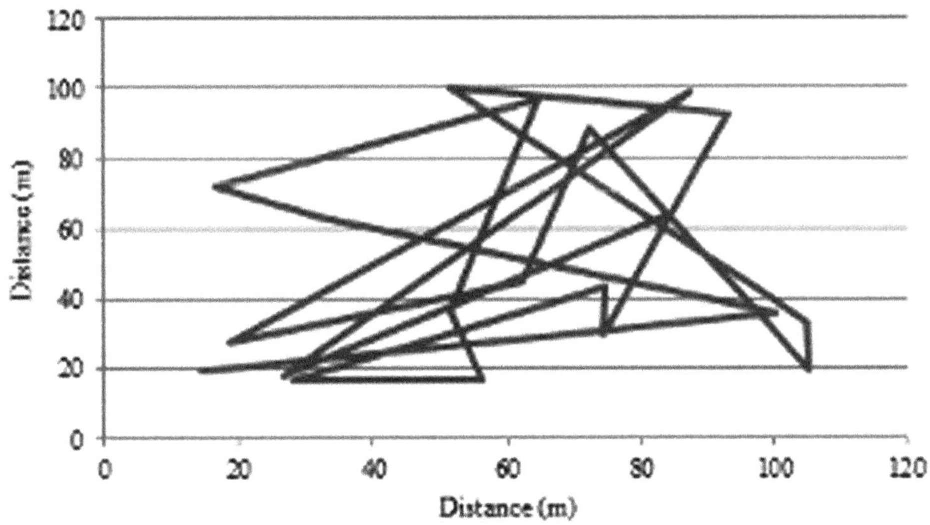

grids. The nodes are allowed to move in north, south, east or west direction. Straight movement of nodes on the defined grids occurs with probability 0.5 and the nodes take turns on corners with probability 0.25. The speed of a node in a lane at a certain time interval is dependent not only on the previous time interval but also on the nodes moving in the same lane.

Based on comprehensive study of FANET applications, mobility models have been categorized for various scenarios in Table 1.

Figure 9. Movement trace of single node under MGM mobility model in 130 x 130 m area

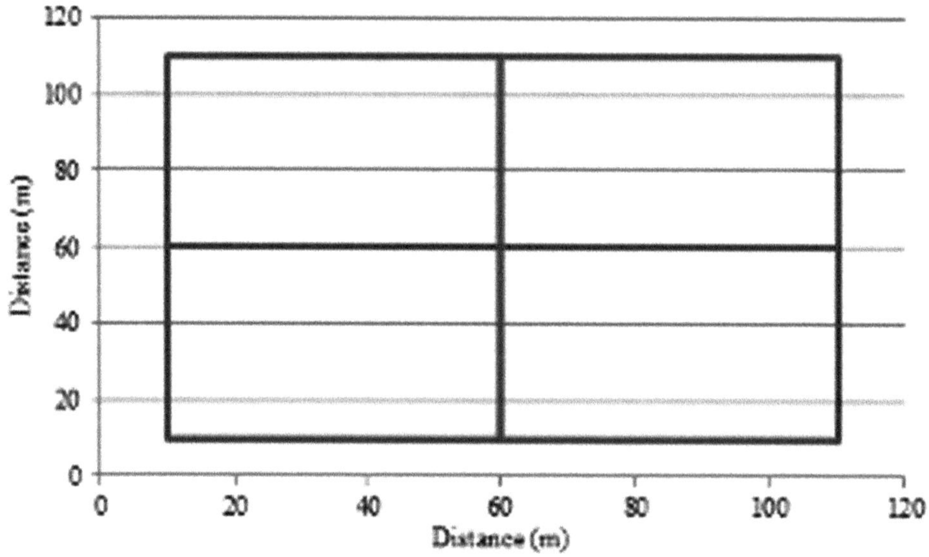

Table 1. Mobility models feasibility for FANET application scenarios

Mobility Model	Proposed Application Scenario
Manhattan Grid Mobility (MGM) model	Complex Urban Environments (Kothari, Postlethwaite & Gu, 2013)
Random Way Point (RWP) mobility model	Patrol systems (Girard, Howell & Hedrick, 2014)
Reference Point Group Mobility (RPGM) model	Disaster management, Search and destroy operations (Quaritsch et al.,2010; George & Sousa, 2010)
Gauss Markov	Multi-tier Airborne MANETs (Broyles, Jabbar, & Sterbenz, 2010)

Simulation Methodology

The simulations were performed on ns2 which is a discrete network simulator. To generate movement of nodes for different mobility models, BonnMotion utility was used which creates, analyzes and exports mobility scenarios to different network simulators (Aschenbruck, Ernst, Gerhards-Padilla & Schwamborn, 2010). Figure 10 shows the network topology used in for simulations. The network topology consists of 5 nodes where center node acts as a sink and a relaying node and it assigns timeslots such that two nodes contend for the medium in that duration and transmit their data to node 0 while the remaining two are in sleep mode. The effect of collision has been introduced for acquiring more comprehensive results. The nodes move with average speeds ranging from 10m/s to 100m/s (Bekmezci, Sahingoz, & Temel, 2013). The pause time was set to 5s to generate highly dynamic scenario with frequent topology changes. Constant Bit Rate (CBR) traffic model was used with packet size of 100 bytes transmitted at a constant rate of 50kbps. In FANETs a line of sight is maintained between UAVs (Bekmezci, Sahingoz, & Temel, 2013). Hence, the radio propagation model used is free space model as it works on the assumption of a single clear line of sight between transmitter and receiver. The simulation parameters given in table 2 have been selected in reference to (Alshabtat & Dong, 2011; Lin, Sung, Li, & Yang, 2012).

Figure 10. Network topology

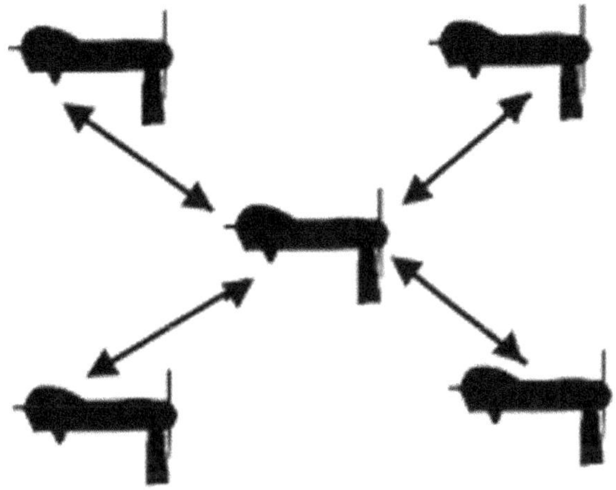

Table 2. Simulation parameters

Parameter	Value
Simulation tool	ns-2.35
Simulation time	200
MAC Protocol	802.11
Simulation Area	200 x 200
Routing Protocol	AODV, DSR, OLSR
Traffic	CBR
Number of nodes	5
Transmission Range	75m
CBR data rate	50 kbps
Speed	10,50,100,200,300,400,500 (m/s)
Channel capacity	11 Mbps
Mobility models	Reference Point Group Mobility Model (RPGM) Gauss Markov mobility model (GM) Random Way Point mobility model (RWP) Manhattan Grid Mobility model (MGM)

BonnMotion generates node movement patterns according to the mobility models specified by taking certain parameters as input such as maximum minimum speed, number of nodes, speed standard deviations, group change probabilities, pause time and simulation area and duration. The traces of node movements generated for different mobility models have been presented in the previous section.

Performance Metrics

The performance evaluation parameters used for FANET topology are described below:

Packet Delivery Ratio (PDR): It gives the ratio of data packets received at the destination successfully to the data packets generated by the source. This parameter is used to determine network reliability.

$$PDR = \frac{\sum_{1}^{N} D_{packets}}{\sum_{1}^{N} S_{packets}} \tag{1}$$

where:

$D_{packets}$ = Data packets received at the destination UAV
$S_{packets}$ = Data packets generated by the source UAV
N = Total number of packets

- **Average End to End (E2E) Delay:** This metric tells about the average time it takes for data packets to reach the destinations successfully.

$$Average\ End-to-End\ Delay = \sum_{1}^{N} \left(T_t + T_{rt} + T_b + T_p\right) \tag{2}$$

where:

T_t = Transmission time
T_{rt} = Retransmission time
T_b = Buffering time
T_p = Processing time
N = Total number of packets

- **Routing Overhead:** Routing overhead determines the total number of routing packets that took part in the transmission.

$$Routing\ overhead = \sum_{1}^{N} R_{packets} \tag{3}$$

where:

$R_{packets}$ = Routing packets
N = Total number of packets

Results and Analysis

Packet Delivery Ratio

Figures 11-14 depict the performance of different mobility models in terms of PDR for AODV, DSR and DSDV. It can be observed that for all scenarios, DSR maintains high PDR. The use of data link acknowledgements makes DSR robust to link failures. Moreover, since routing information is overheard by intermediate nodes, they learn multiple routes to a certain destination and include them in their caches. In case one link does not work from the cache, the alternate route can be used. AODV and OLSR show a decreasing trend in delivery ratio with the increase in speed. Since OLSR is table driven protocol, the routes in the table become outdated very quickly and the rediscovery of the broken link takes time. AODV due to its on-demand nature reacts more rapidly towards route failures.

It can be observed from Figure 11 that RPGM provides higher PDR for all routing protocols. This is because the movement of all nodes in RPGM is governed by defining a maximum distance from group center. Thus, the network is in a fully connected state most of the time. Also, as the speed increases, PDR does not drop below 90% because when UAVs are moving as a swarm and are in the radio range of each other, the effect of high mobility nullifies whereas in other mobility models the increase in node mobility causes decrease in PDR because of the uncoordinated node movements. Routing protocols play an important role by finding alternate valid routes. Thus, suitable mobility models along with appropriate routing protocol can address the issue of link breakages due to high mobility and can help in maintaining an acceptable PDR depending on the application for which FANET nodes are deployed.

Figure 11. PDR vs. speed for RPGM

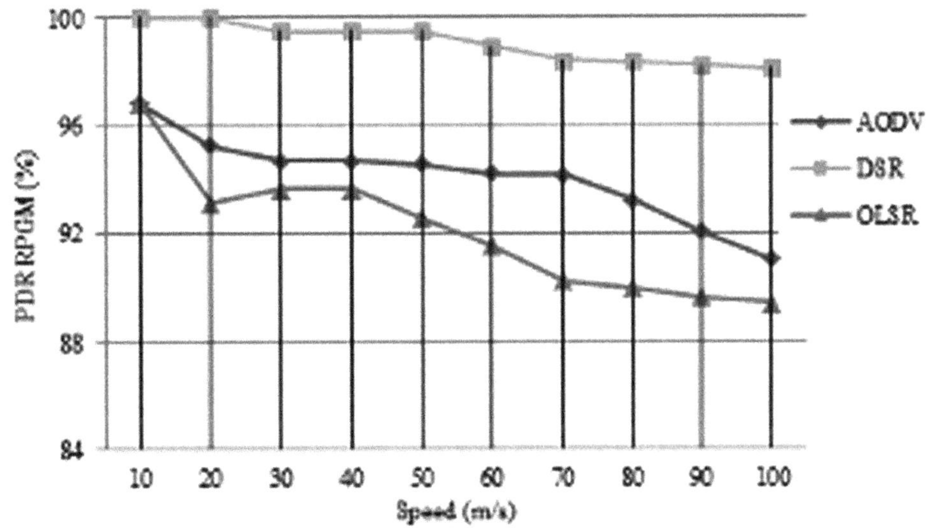

Figure 12. PDR vs. speed for RWP

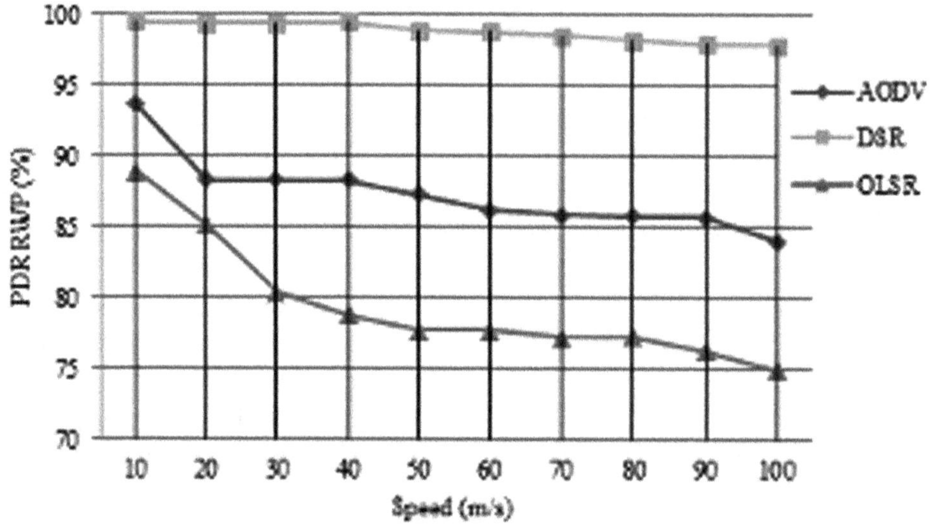

Figure 13. PDR vs. speed for GM

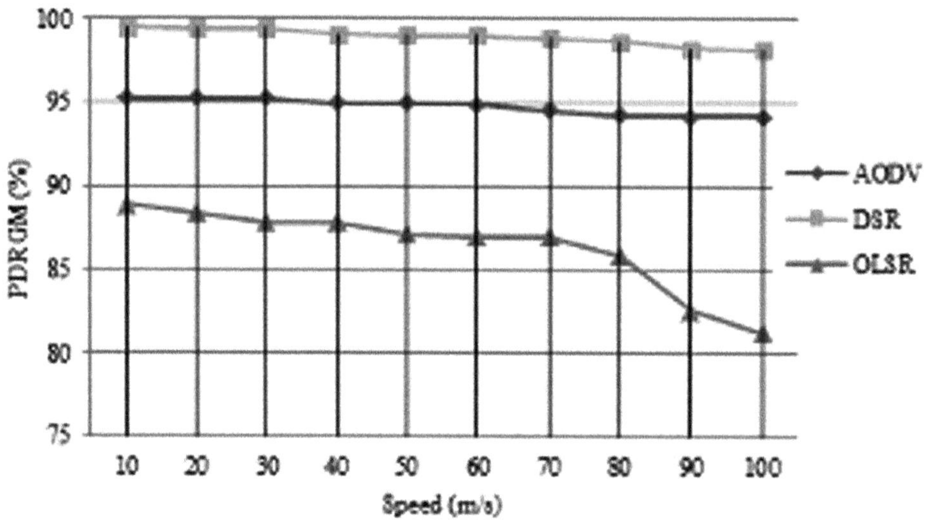

Average End-to-End Delay

Figures 15-18 show average end to end delay of AODV, DSR and OLSR for mobility models. It must be noted that the delays are effective delays of data packets that successfully arrive at the destination. As shown DSR exhibits higher delays followed by AODV and OLSR. The route caching mechanism of DSR increases the end to end delay of the network. Due to high mobility the cache contains more percentage of stale routes. If the mobile source node fails to find route to the destination it starts discovery for

Figure 14. PDR vs. speed for MGM

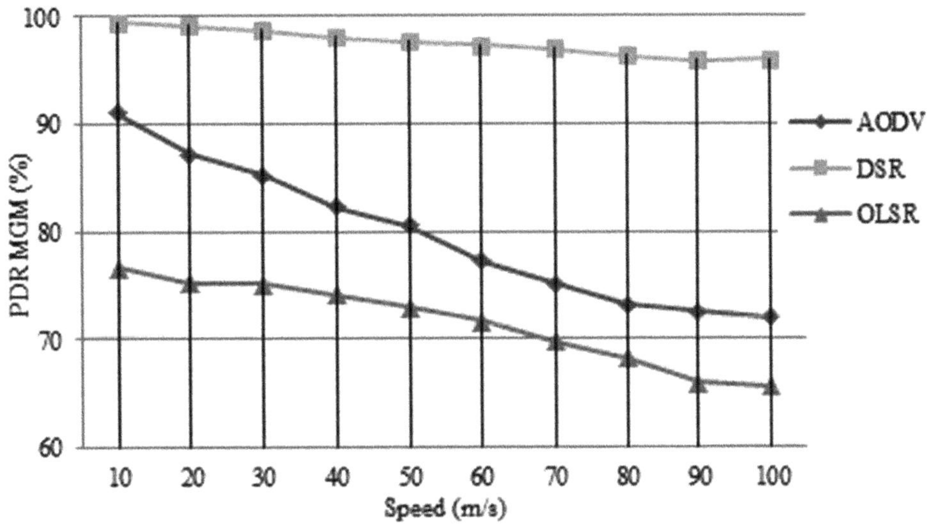

Figure 15. Average E2E delay vs. speed for RPGM

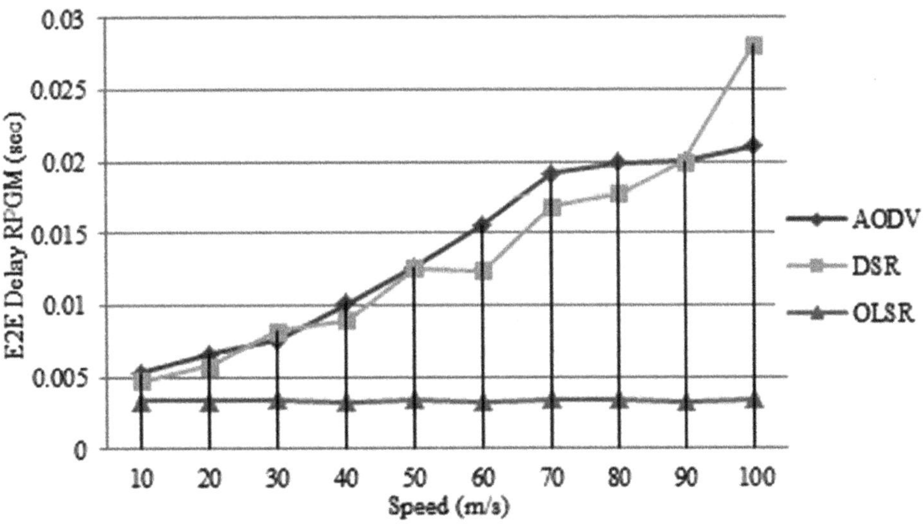

new route. The entire process increases the packet reception time. AODV on the other hand omits route cache discovery process and shows less average delay. OLSR has routes available all the time and can provide them when required. Due to low PDR of OLSR, fewer data packets are able to reach intended destinations within the simulation time and give low average end-to-end delay of data packets.

Figure 16. Average E2E delay vs. speed for RWP

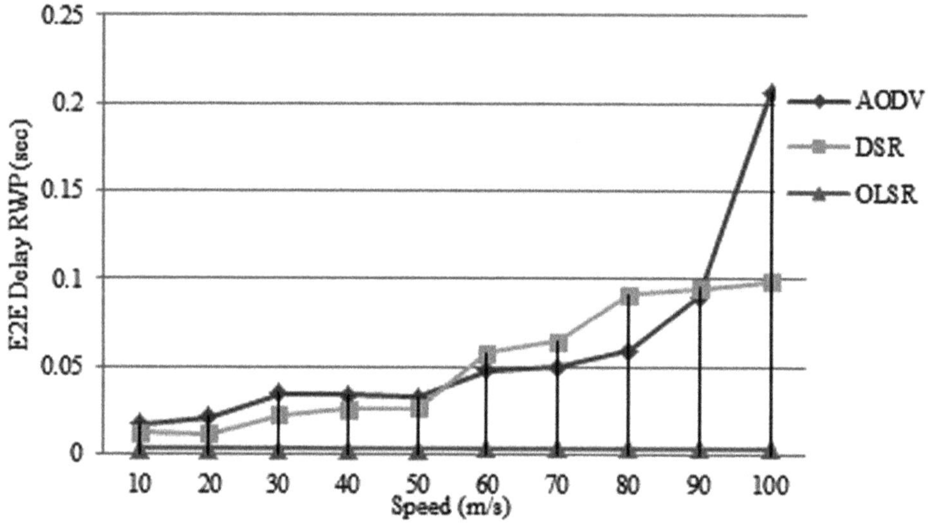

Figure 17. Average E2E delay vs. speed for GM

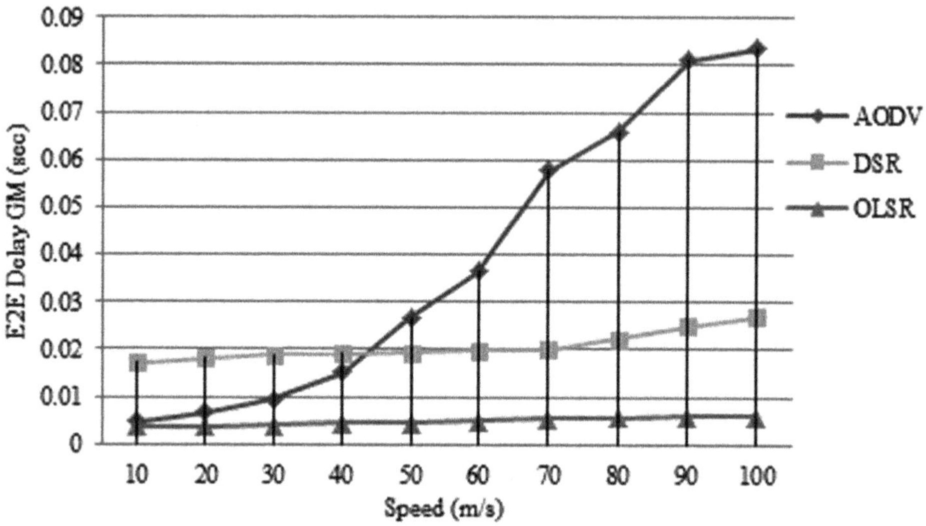

Routing Overhead

Figures 19-22 show control or routing overhead for AODV, DSR and OLSR. DSR demonstrates lowest routing overhead than AODV and OLSR. Since DSR also takes advantage of source route caches hence the number of route request packets generated is less than that of AODV. However, with the increase in speed, routing overhead of DSR also increases because frequent link failures make route cache mechanism less effective. OLSR depicts extremely large routing overhead due to the transmission of periodic HELLO and Topology Control (TC) messages. Unlike the used topology, dense and large networks can

Figure 18. Average E2E delay vs. speed for MGM

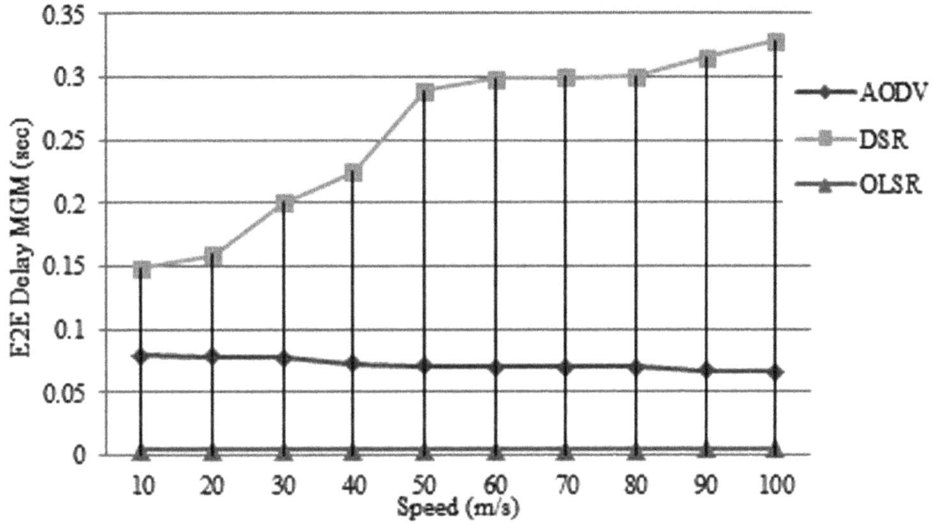

Figure 19. Routing overhead vs. speed for RPGM

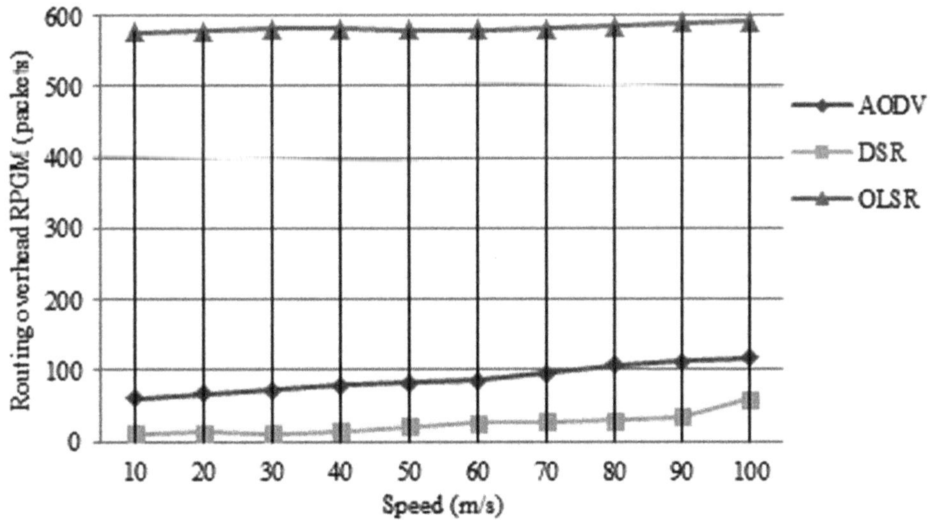

Figure 20. Routing overhead vs. speed for RWP

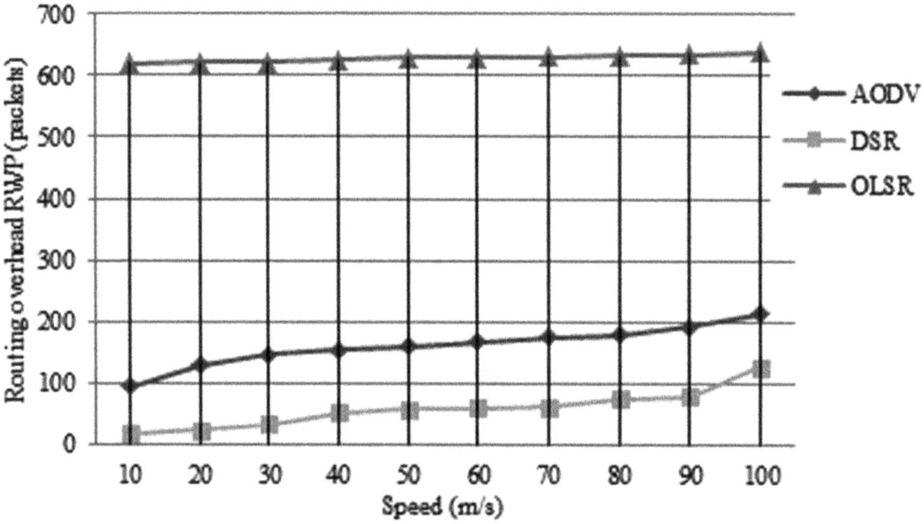

Figure 21. Routing overhead vs. speed for GM

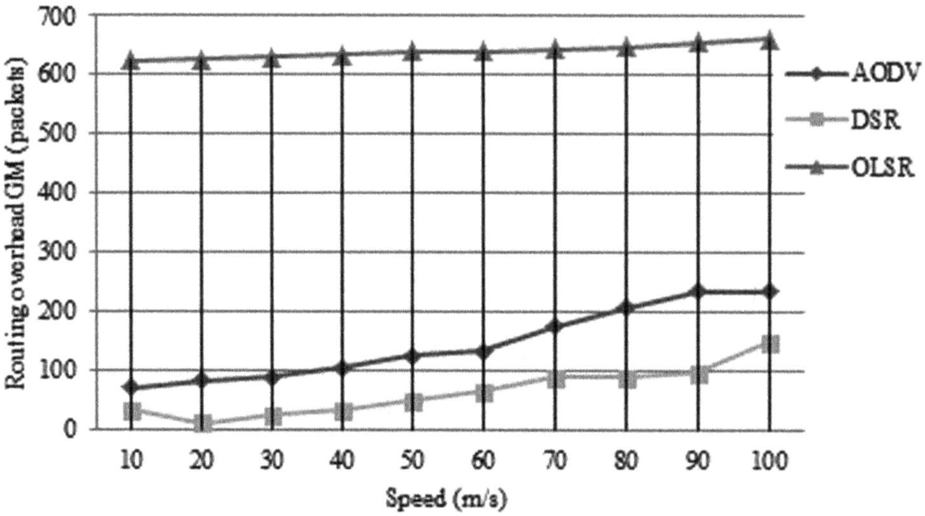

Figure 22. Routing overhead vs. speed for MGM

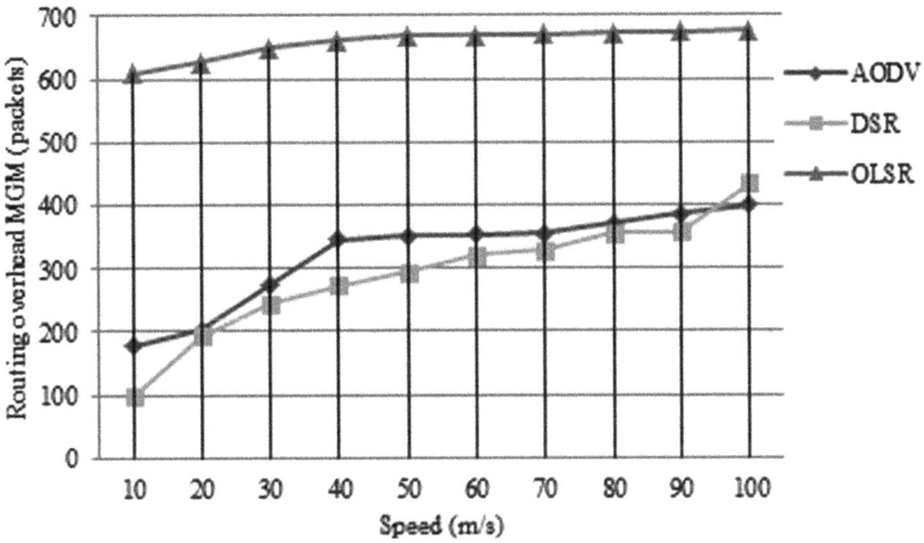

make the most of OLSR due to multi point relay selection mechanism which guarantees improved scalability in the sharing of topology information.

CONCLUSION

FANETs are used for critical and sensitive applications such as reconnaissance and patrolling and hence simulation analysis needs to be performed under realistic environments. Accurate mobility models are essential to foretell communication problems during protocols evaluation that may arise in real-world scenarios.

There is not a standard method to validate or select a mobility model for FANETs because different applications may require different movement patterns. Therefore, a detailed performance evaluation of mobility models has been done for the first time and different mobility models for various applications have been suggested. DSR, AODV and OLSR routing protocols have been evaluated using these models. DSR gave better results for highly dynamic environments generated by Reference Point Group Mobility, Gauss Markov, Random Way Point and Manhattan Grid mobility models.

REFERENCES

Alshabtat, A., & Dong, L. (2011). Low Latency Routing Algorithm for Unmanned Aerial Vehicles Ad-Hoc Networks. *Iranian Journal of Electrical and Computer Engineering*, *6*(1), 48–54.

Aschenbruck, N., Ernst, R., Gerhards-Padilla, E., & Schwamborn, M. (2010). BonnMotion - a Mobility Scenario Generation and Analysis Tool. In *3rd International ICST Conference on Simulation Tools and Techniques (SIMUTools)*, Brussels, Belgium. 10.4108/ICST.SIMUTOOLS2010.8684

Barrado, C., Messeguer, R., Lopez, J., Pastor, E., Santamaria, E., & Royo, P. (2010). Wildfire monitoring using a mixed air-ground mobile network. *IEEE Pervasive Computing / IEEE Computer Society [and] IEEE Communications Society*, *9*(4), 24–32. doi:10.1109/MPRV.2010.54

Bekmezci, İ., Sahingoz, O., & Temel, Ş. (2013). Flying Ad-Hoc Networks (FANETs): A survey. *Ad Hoc Networks*, *11*(3), 1254–1270. doi:10.1016/j.adhoc.2012.12.004

Bodanese, J., de Araujo, G., Steup, C., Raffo, G., & Becker, L. (2014). Wireless Communication Infrastructure for a Short-Range Unmanned Aerial. In *28th International Conference on Advanced Information Networking and Applications Workshops (WAINA)*, Victoria, BC. 10.1109/WAINA.2014.154

Bouachir, O., Abrassart, A., Garcia, F., & Larrieu, N. (2014). A Mobility Model for UAV Ad-hoc Network. In *International Conference on Unmanned Aircraft Systems*, Orlando, FL. 10.1109/ICUAS.2014.6842277

Broyles, D., Jabbar, A., & Sterbenz, J. (2010). *Design and analysis of a 3-D Gauss-Markov mobility model for highly-dynamic airborne networks.* In International Telemetering Conference (ITC), San Diego, CA.

Cai, Y., Yu, F., Li, J., Zhou, Y., & Lamont, L. (2012). MAC performance improvement in UAV ad-hoc networks with full-duplex radios and multi-packet reception capability. In *IEEE International Conference on Communications (ICC)*, Ottawa, Canada. 10.1109/ICC.2012.6364116

Camp, T., Boleng, J., & Davies, V. (2002). A survey of mobility models for ad hoc network research. *Wireless Communication and Mobile Computing*, *2*(5), 483–502. doi:10.1002/wcm.72

Cetin, O., & Zagli, I. (2011). Continuous Airborne Communication Relay Approach Using Unmanned Aerial Vehicles. *Journal of Intelligent & Robotic Systems*, *65*(1-4), 549–562. doi:10.100710846-011-9556-6

Coifman, B., McCord, M., Mishalani, R., Iswalt, M., & Ji, Y. (2006). Roadway traffic monitoring from an unmanned aerial vehicle. *IEEE Proceedings on Intelligent Transport Systems.*

Forsmann, J., Hiromoto, R., & Svoboda, J. (2007). A time-slotted on-demand routing protocol for mobile ad hoc unmanned vehicle systems. In *SPIE Defense and Security Symposium, Unmanned Systems Technology*, Orlando, FL. 10.1117/12.719355

Frew, E., & Brown, T. (2008). Networking Issues for Small Unmanned Aircraft Systems. *Journal of Intelligent & Robotic Systems*, *54*(1-3), 21–37. doi:10.100710846-008-9253-2

Geng, L., Zhang, Y., Wang, J., Fuh, J., & Teo, S. (2013). Mission planning of autonomous UAVs for urban surveillance with evolutionary algorithms. In *10th IEEE International Conference on Control and Automation (ICCA)*, Hangzhou, China. 10.1109/ICCA.2013.6564992

George, J., & Sousa, J. (2010). Search Strategies for Multiple UAV Search and Destroy Missions. *Journal of Intelligent & Robotic Systems*, *61*(1-4), 355–367. doi:10.100710846-010-9486-8

Girard, A., Howell, A., & Hedrick, J. (2014). Border patrol and surveillance missions using multiple unmanned air vehicles. In *43rd IEEE Conference on Decision and Control* (CDC), Nassau.

Heimfarth, T., & De Araujo, J. (2014). *Using Unmanned Aerial Vehicle to Connect Disjoint Segments of Wireless Sensor Network.* In IEEE 28th International Conference on Advanced Information Networking and Applications (AINA), Victoria, BC.

Huang, Y., Thomson, S., Hoffmann, W., Lan, Y., & Fritz, B. (2013). Development and prospect of unmanned aerial vehicle technologies for agricultural production management. *International Journal of Agricultural and Biological Engineering*, *6*(3), 1–10.

Huba, W., & Shenoy, N. (2012). Airborne surveillance networks with directional antennas. In *8th International Conference on Networking and Services (ICNS)*.

Jawhar, I., Mohamed, N., Al-Jaroodi, J., & Sheng Zhang, J. (2013). Data communication in linear wireless sensor networks using unmanned aerial vehicles. In *International Conference on Unmanned Aircraft Systems (ICUAS)*, Atlanta, GA. 10.1109/ICUAS.2013.6564725

Kothari, M., Postlethwaite, I., & Gu, D. (2013). UAV Path Following in Windy Urban Environments. *Journal of Intelligent & Robotic Systems*, *74*(3-4), 1013–1028. doi:10.100710846-013-9873-z

Kuiper, E. (2008). *Mobility and Routing in a Delay-tolerant Network of Unmanned Aerial Vehicles* (PhD Dissertation). Linköping University.

Kuiper, E., & Nadjm-Tehrani, S. (2006). Mobility Models for UAV Group Reconnaissance Applications. In *International Conference on Wireless and Mobile Communications (ICWMC)*, Bucharest, Romania. 10.1109/ICWMC.2006.63

Lim, C., Park, S., Ryoo, C., Choi, K., & Cho, J. (2010). *A path planning algorithm for Surveillance UAVs with timing mission constrains.* In International Conference on Control Automation and Systems (ICCAS), Gyeonggi.

Lin, L., Sung, Q., Li, J., & Yang, F. (2012). A novel geographic position mobility oriented routing strategy for UAVs. *Journal of Computer Information Systems*, *8*(2), 709–716.

Martinez-de Dios, J., Lferd, K., de San Bernabé, A., Núñez, G., Torres-González, A., & Ollero, A. (2012). Cooperation Between UAS and Wireless Sensor Networks for Efficient Data Collection in Large Environments. *Journal of Intelligent & Robotic Systems*. doi:10.100710846-012-9733-2

Newaz, A., Pratama, F., & Chong, N. (2013). Exploration Priority Based Heuristic Approach to UAV path planning. In RO-MAN IEEE, Gyeongju.

Pojda, J., Wolff, A., Sbeiti, M., & Wietfeld, C. (2011). Performance analysis of mesh routing protocols for UAV swarming applications. In *8th International Symposium on Wireless Communication Systems (ISWCS)*, Aachen. 10.1109/ISWCS.2011.6125375

Quaritsch, M., Kruggl, K., Wischounig-Strucl, D., Bhattacharya, S., Shah, M., & Rinner, B. (2010). Networked UAVs as aerial sensor network for disaster management applications. *Elektrotechnik und Informationstechnik, 127*(3), 56-63.

Semsch, E., Jakob, M., & Pavlicek, D. (2009). *Autonomous UAV Surveillance in Complex Urban Environments.* In IEEE/WIC/ACM International Joint Conferences on Web Intelligence and Intelligent Agent Technologies (WI-IAT), Milan, Italy.

Shetty, V., Sudit, M., & Nagi, R. (2008). Priority-based assignment and routing of a fleet of unmanned combat aerial vehicles. *Computers & Operations Research*, *35*(6), 1813–1828. doi:10.1016/j.cor.2006.09.013

Temel, S., & Bekmezci, I. (2013). *On the performance of Flying Ad-hoc Networks (FANETs) utilizing near space high altitude platforms* (HAPs). In *6th International Conference on Recent Advances in Space Technologies (RAST)*, Istanbul, Turkey. 10.1109/RAST.2013.6581252

Wan, Y., Namuduri, K., Zhou, Y., & Fu, S. (2013). A Smooth-Turn Mobility Model for Airborne Networks. *IEEE Transactions on Vehicular Technology*, *62*(7), 3359–3370. doi:10.1109/TVT.2013.2251686

Wang, W., Guan, X., Wang, B., & Wang, Y. (2010). A novel mobility model based on semi-random circular movement in mobile ad hoc networks. *Information Sciences*, *180*(3), 399–413. doi:10.1016/j.ins.2009.10.001

This research was previously published in the Handbook of Research on Recent Developments in Intelligent Communication Application edited by Siddhartha Bhattacharyya, Nibaran Das, Debotosh Bhattacharjee, and Anirban Mukherjee, pages 255-281, copyright year 2017 by Information Science Reference (an imprint of IGI Global).

Chapter 5
A Perceptual Computing Based Gesture Controlled Quadcopter for Visual Tracking and Transportation

Kumar Yelamarthi
Central Michigan University, USA

Raghudeep Kannavara
Intel Corporation, USA

Sanjay Boddhu
Finch Computing, USA

ABSTRACT

One of the fundamental challenges faced by an inexperienced user in portable unmanned aerial vehicle (UAV) such as quadcopters is flight control, often leading to crashes. Addressing this challenge, and leveraging upon the technological advancement in perceptual computing and computer vision, this research presents a modular system that allows for hand gesture based flight control of UAV, alongside a transport mechanism for portable objects. In addition to ascertain smooth flight control by avoiding obstacles in navigation path, real-time video feedback is relayed from the UAV to user, thus allowing him/her to take appropriate actions. This paper presents the design implementation by discussing the various sub-systems involved, inter system communication, and field tests to ascertain operation. As presented from testing results, the proposed system provides efficient communication between the subsystems for smooth flight control, while allowing for safe transport of portable objects.

DOI: 10.4018/978-1-5225-8365-3.ch005

1. INTRODUCTION

In the recent years, there has been tremendous research in the development of miniature unmanned aerial vehicles with a broad range of applications for both military and civilians (Achtelik, 2009, Moranduzzo, 2014). In situations where a given task might be either too complex or monotonous, it may be desirable to use these miniature UAVs. The advantages provided by these miniature UAVs led to the implementation in several applications such as law enforcement, security patrols in private property, agricultural surveillance, communications relay, aerial mapping, severe weather telemetry, search and rescue, disaster relief, industrial surveillance, remote sensing, and historical site surveillance to name a few. Of the diverse UAV's proposed over the past few years, quadcopters have received a board attention due to the several advantages it provides such as ease of implementation, relatively inexpensive to manufacture, great maneuverability, ability to carry small payload for variety of applications. While these quadcopters have a broad range of applications in aerial surveillance and monitoring, they suffer from the fundamental challenge of user interface and maneuverability.

One other recent technological advancement that has garnered interest and gaining popularity in the recent times is perceptual computing (Mendel & Wu, 2010). It is an application that provides the computer an ability to recognize what is going around it, and assists people to make subjective judgments accordingly using human gestures. This human gesture recognition is classified into two types: near-field and far-field, and required different depth cameras. When combined with effective computer vision software such as Intel Perceptual Computing SDK (Intel, 2015), some applications of these depth camera include close-range depth tracking to recognize position of each of user's hands, fingers; facial analysis for face detection and recognition through smiles, blinks etc., augmented reality to combine real-time images from camera and close-range tracking, and speech recognition. When efficiently implemented, this perceptual computing has the ability to simplify tasks that are relatively complicated for general public. Human gestures such as hand waiving and swiping are primitive communication methods, and when combined with perceptual computing leads to applications in diverse fields such as medicine distribution, surveillance of construction sites, assistive devices for object retrieval, aerial transportation etc.,

Infusing the advancements in both aerial quadcopters and perceptual computing, this paper presents the development and evaluation of a gesture controlled quadcopter system with advantages such as, improving usability to minimize user training for flight, modular system for broad applicability to diverse hardware, providing the ability for real-time video feedback for position accuracy and maneuverability. This paper is organized as follows. Section 2 presents the previous work in perceptual computing. Section 3 presents the design implementation of proposed system, and respective sub-system operation. Section 4 presents the preliminary results to prove the system feasibility. Section 5 presents details of extended distributed deployment testing, and conclusions are presented in Section 6.

2. PREVIOUS WORK

Perceptual computing is a new technology that is in its early stages of development. There are some consumer products that utilize this technology. The most prevalent and well known of these devices is Microsoft's Kinect sensor for the Xbox 360 (Kinect, 2015). This device is used as a game controller and interprets the user's large motion gestures, i.e. arm swinging and bending of the limbs, and is also used for localization of objects (Yelamarthi, DeJong, & Laubhan, 2014). This project is intended for use in a setting that requires more discrete gestures. The Microsoft Kinect is not designed to perceive gestures as small as the ones the user will input to the system. The Creative Labs Senz3D camera is fully capable of providing the desired perceptual computing for interpreting the discrete gestures (Creative, 2014).

In terms of item retrieval and delivery, aerial transportation systems using quad-copters are gaining popularity in today's tech world. Transportation systems like the Matternet are constrained by concepts such as aerial transportation and poor controllability (D'Andrea, 2014). Systems with fully integrated gesture control that are currently in use today are scarce. One such system consisted of an obstacle avoiding wheelchair that used the user's head gestures for control (Jia, 2007), as developed for elderly and disabled humans.

Another project used a Microsoft Kinect sensor which is capable of color and depth imaging. While color images may be useful in some applications, the sensor's depth perception capability is what has caught on with researchers. The depth sensing allows the sensor to detect the relative distances to objects, or rather obstacles. The user can easily interpret the depth from the black and white image that is generated as traversable areas will be darker than objects that would be considered an obstacle. Their system incorporates transmitting data over Wi-Fi to a server to be stored and processed. Proper functionality of the robot will be determined from the data sent to the server. Similarly, one other recently published research included (Yelamarthi, DeJong, & Laubhan, 2014) using the Microsoft Kinect sensor for space perception to identify objects in front of a blind user, and providing haptic feedback to allow him/her navigate indoor environments.

3. DESIGN IMPLEMENTATION

The system architecture of the proposed system is shown in Figure 1, and is comprised of two modules. The first module is a quadcopter system with both the quadcopter and the respective transceiver for flight control. The second module is an add-on hardware that serves as the brain of the quadcopter. The first component in the add-on module is a Creative Sensz3 depth and gesture recognition camera (Creative, 2014) designed to localize objects in short range, and track their respective movements to identify human gestures such as thumb up/down, hand swipe up/down, high five, hand swipe left/right etc., The second component is a portable computer used to process data from different sensors and receivers in the system. The third component is a microcontroller whose function is to generate signals for the quadcopter transceiver based on gestures recognized from the depth camera. The fourth component is a camera mounted on quadcopter to capture video of its surroundings, which is processed using onboard microcontroller, and relayed to the user through Wi-Fi transceivers positioned on the quadcopter and near the computer. The different gestures and their respective commands for flight control in the proposed system are shown in Figure 2.

The first step in system operation is capturing the human gestures using a depth camera, and is done so by using a Creative Sensz3D camera as in Figure 3. A custom computer program interprets data from the sensor to identify and localize hands, and its respective movements for gesture recognition. All the diverse gestures presented in Figure 2 have designated identification. In order to transfer this information to the onboard microcontroller, a custom program was written in .net language with the primary purpose of relaying this information through a serial port, which is further received for appropriate signal generation as in Figure 2 by the microcontroller. The communication and information flow and its respective modes in the proposed system are shown in Figure 4. To control the operation of the quad-copter, an

Figure 1. System architecture of proposed gesture controlled quadcopter

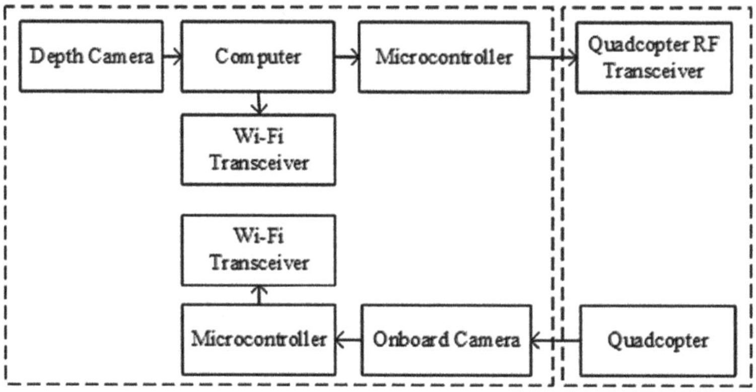

Arduino Uno microcontroller was implemented. The Arduino Uno is fully capable of accomplishing all the tasks that are required for the project. The Arduino loops through a cycle which sets the pulse widths for the six channel Pulse Position Modulated (PPM) signal. Each time the loop runs, the Arduino will check the serial stream for new data. If there is new data from the stream, the Arduino will change the pulse lengths of the corresponding channel of the PPM Signal. If there is no data in the stream, the default lengths of the six channels will be reinstated therefore the quad copter should hover and stay at the same location it is at. The Arduino will send a PPM signal to the quad-copter via the transceiver hardwired to the microcontroller. It will also be capable of interfacing with an XBee transceiver module in case any other communication is required between the quad-copter and the desktop control station. The quadcopter chosen for the test platform is a Cheerson CX-20 system shown in Figure 5. Having an onboard microcontroller made controlling the flight of the machine much simpler than if algorithms had to be developed from scratch. The CX-20 has built in hover mode which complements the discrete gesture controls that are to be used. For battery life, two 3000mAh batteries were chosen in order to provide a target flight time of about 15 minute.

In order to maintain the light weight theme of the project a custom device had to be designed. A passive mechanism as presented in Figure 5 is optimal for working within current energy constraints as well. This method was chosen for ease of implementation and proof of concept. An Arduino Yun microcontroller coupled with a Logitech webcam enables Wi-Fi video streaming onboard the quadcopter. The video feed will be accessed via a web browser by the person who is inputting gestures. Safety is always one of the most important aspects of projects and feedback is necessary for the operational standard. The camera is mounted on a servo motor so that the pilot has a variation of viewing angles during flight. It is expected that changing the view will be especially useful when the quad-copter is in hover mode and during landing.

4. EVALUATION RESULTS

Preliminary testing data has been collected to verify the system's ability to complete the design specifications. The categories where testing was performed include delay time for communication, and gesture recognition sampling rate. The gesture recognition sampling rate is an important parameter for the

Figure 2. Hand gestures and respective commands for quadcopter

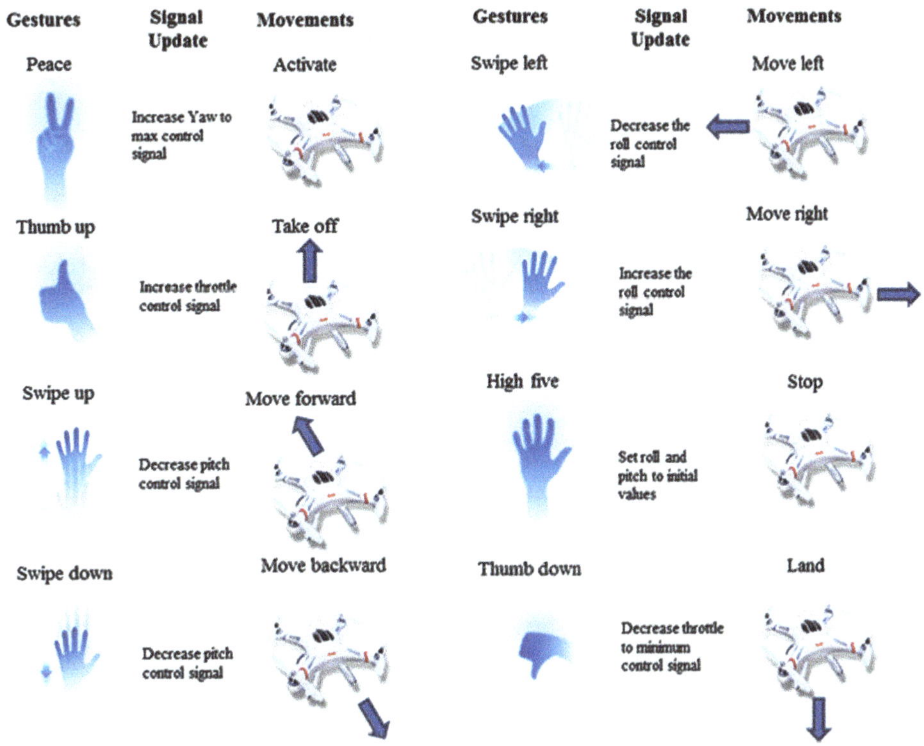

Figure 3. Senz3D depth camera used in proposed system

functionality of the system. In order for the design to be reactive and agile the camera must have a high sampling rate. The gesture recognition sampling rate was determined using a method where a standard time period was set and the amount of gestures that the camera could fully recognize was measured. The standard time period was chosen to be a minute. The testing was done in a series of trials each with its own gesture variations. The first trial began by having the input to the camera as the thumbs up gesture. The amount of gesture samples the camera perceived in this trial was 1765 samples. The second trial was performed using each of the different signals together. The amount of gesture samples decreased significantly to 512 samples. The last trial was performed with only one of the gestures that take more time to recognize which is the circle. The circle hand gesture is perceived when the user waves their

Figure 4. Information flow among sub-systems in proposed quadcopter system

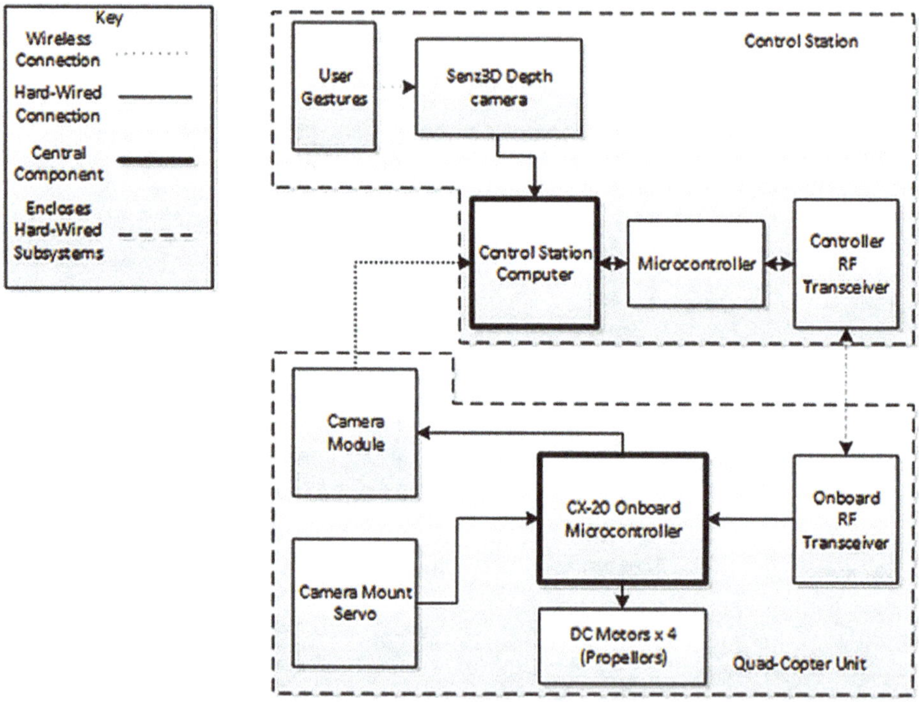

Figure 5. Proposed Gesture Controlled Quadcopter with portable transportation box

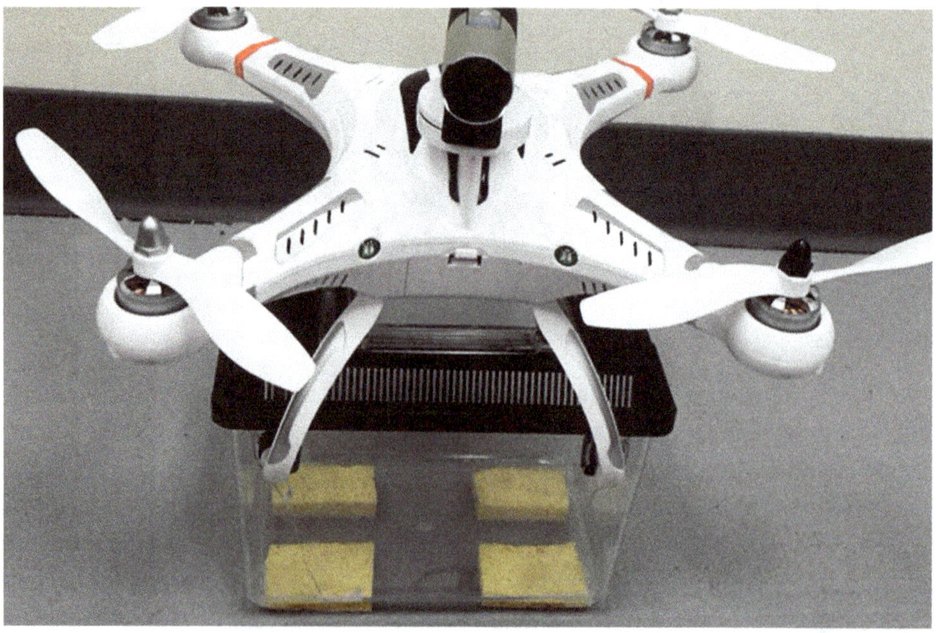

hand in a circular motion with a diameter of about 10cm. The number of gestures perceived with the circle gesture was the lowest at 96.

As the gesture recognition code is executed, the process takes time to be completed from reading the gesture until the designator for the code is sent to the Arduino microcontroller to transmit via radio frequency. The data in Table 1 contains the tested delay time from the point the user inputs the gesture to the point where the designator is sent to the microcontroller. The columns show the minimum, maximum and average delay for 10 trials for each gesture.

Further, extended testing was performed on the system with three different paths as presented in Figure 6. Test 1 was performed to ascertain quadcopter operation to travel in a straight line for a certain distance and land. During testing, it was found that the quadcopter experienced slight unintended rotation during take-off, resulting in moving forward at an angle. Repeated trials of this test has shown that the quadcopter travelled to the destination with an average position error of 1.8m at 25°. Test-2 was designed to find the accuracy by which the quadcopter could travel in multiple directions, where it was required to take-off, travel forward by a distance of 10m, travel to the right side by 10 m, repeat the same, and land. Repeated trials of this test has shown that the quadcopter travelled to the destination with an average position error of 1.2m at 41°. Test-3 was designed to find if accuracy of the quadcopter would improve for repetitive actions, where it was required to take-off, travel in a square pattern and land. Repeated trials of this test has shown that the quadcopter travelled to the destination with an average position error of 0.8m at 45°.

Table 1. Gesture recognition time

Hand Gesture	Minimum (sec)	Maximum (sec)	Average (sec)
Thumbs Up	0.785	1.124	0.959
Thumbs Down	0.641	1.276	0.940
Peace	0.542	1.313	0.896
Big Five	0.701	1.018	0.896
Double Thumbs Up	0.335	0.967	0.609
Double Thumbs Down	0.238	0.516	0.369
Double Peace	0.766	1.241	0.897
Double Big Five	0.758	1.455	1.026
Wave	0.814	1.567	1.071
Swipe Left	0.392	1.832	0.878
Swipe Right	0.722	2.030	1.282
Swipe Up	0.326	0.672	0.529
Swipe Down	0.609	2.095	1.288
Circle	0.449	0.986	0.705

Figure 6. Desired and actual path of quadcopter during test flights

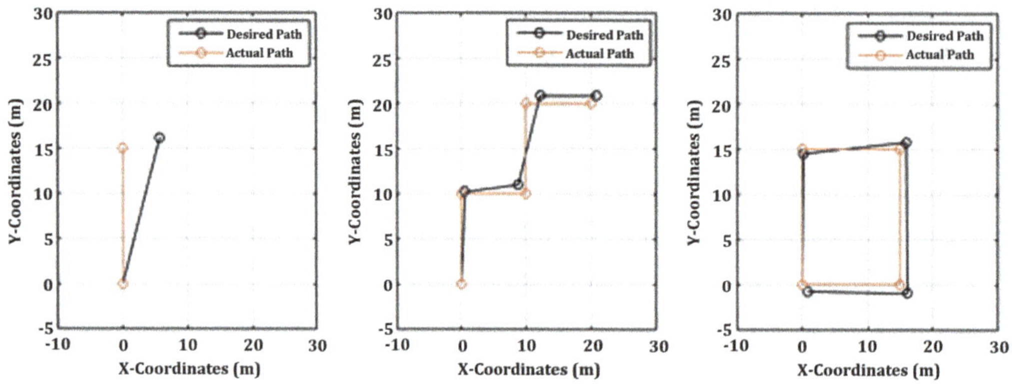

5. DISTRIBUTED DEPLOYMENT TESTING

The use of the interconnected convergent networks is proving to be useful for the real time situation assessment in monitoring and surveillance scenarios. These interconnected networks allow for the integration of multisensory data to introduce ease in the process of resource allocation and human-machine interaction, beyond limits of geospatial and temporal constraints. The developed Perceptual Computing based Quadcopter Control system (PCQCS) introduced in this paper, because of its modularity and interoperability in the design, could be easily be deployed with existing ubiquitous network architectures. Further, the developed PCQCS, would cater to scenarios like emergency management and critical surveillance missions (for law and order, military, etc...), where the logistics are constantly changing and are unpredictable (especially, when traditionally trained drone operator availability is low). The PCQCS addresses the issues with limited availability of trained resources, in the emergency mission scenarios, but they would be issues concerned with limited availability of network and central command control. To study the PCQCS's viability to emergency mission scenarios, the authors have integrated the PCQCS into one of the existing sensor fusion architectures, called Persistent Surveillance Storage Architecture (PSSA) (Boddhu, Williams, & Yue, 2015), which was proven to be effective in real-world time-critical situational awareness and human-centric sensing applications, where the delay in fusing the actuator, controller and sensor data, is unacceptable, and a central command control is an absolute requirement (Boddhu, 2013). From a logical perspective, the PSSA consists of four customizable software layers: Ingestion Services, Application Services, Data Services, and Dissemination Services as shown in Figure 7. In addition, the PSSA includes a common Core Services layer which is responsible for providing the infrastructure required to support the integration of the components in the customizable software layers. Although the Figure 7, only shows four types of sensor sources, the PSSA is designed to support the storage and exploitation of data from any sensor source by "plugging" in new ingestion services. Similarly, the PSSA is designed to support the distribution of sensor data and related exploitation data to any visualization platform by "plugging" in standards based and custom built dissemination services. More details about PSSA can be found in (Boddhu, Williams, & Yue, 2015).

In brief, the PSSA can be envisioned as an event-based system, where the events are modelled as standalone sensor data packets, tagged with timestamp, location information, event type and unique identifier. As shown in Figure 8, these PSSA events are generated by components called publishers and consumable by subscribers, with apt permissions. The initial test performed with single could instances of the PSSA, as network backbone, with multiple PCQCS was proven to be seamless and successful and the overall performance was same as the ones provided in the evaluation results section, for single PCQCS. Here, the quadcopter(s) of PCQCS, are modelled and integrated into PSSA, as a multi-model sensor system, with combination of publisher (its camera subsystem publishing video event data to operator in the command center) and subscriber (its low-level control subsystem subscribing to the control event data from the operator in the command center) in it. The other subsystem of PCQCS, involving human operators, kinetic systems and visualization (displays) are also modelled as multi-model sensor system, with publisher and subscriber modules, completing the distributed view of the proposed PCQCS in Figure 1. Further, an extended test to simulate a scenario of Quadcopters, going far beyond not only the line-of-sight, but across multiple PSSA cloud instances, called PSSA sky architecture (Boddhu, Williams, & Yue, 2015), was tested and verified.

6. CONCLUSION

Once the design was built and functional it ended up performing quite well. Communication speed within the system was excitingly smooth. The delay from the final video feed subsystem performed well with a peak delay of around one second, with almost negligible flight control delay. While the complex movements, such as the waving, swiping and circle gestures, took longer to register, the system still performed well. The technology of perceptual computing has proven to be useful in many different applications and

Figure 7. Logical view of the Persistent Sensor Storage Architecture

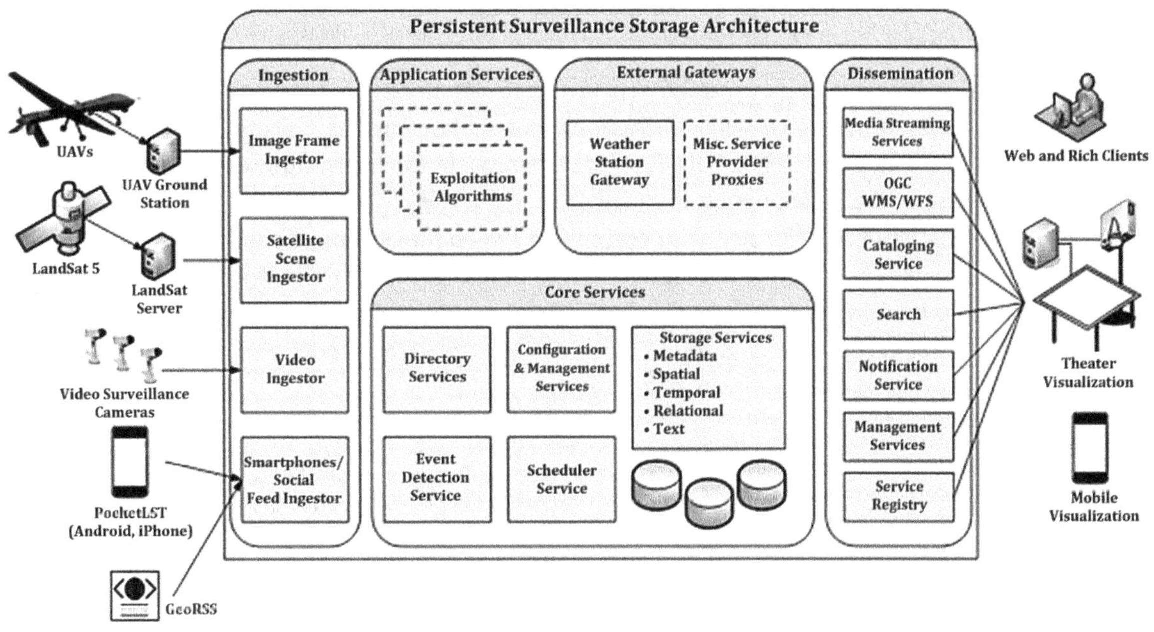

Figure 8. Key components of the Persistent Sensor Storage Architecture

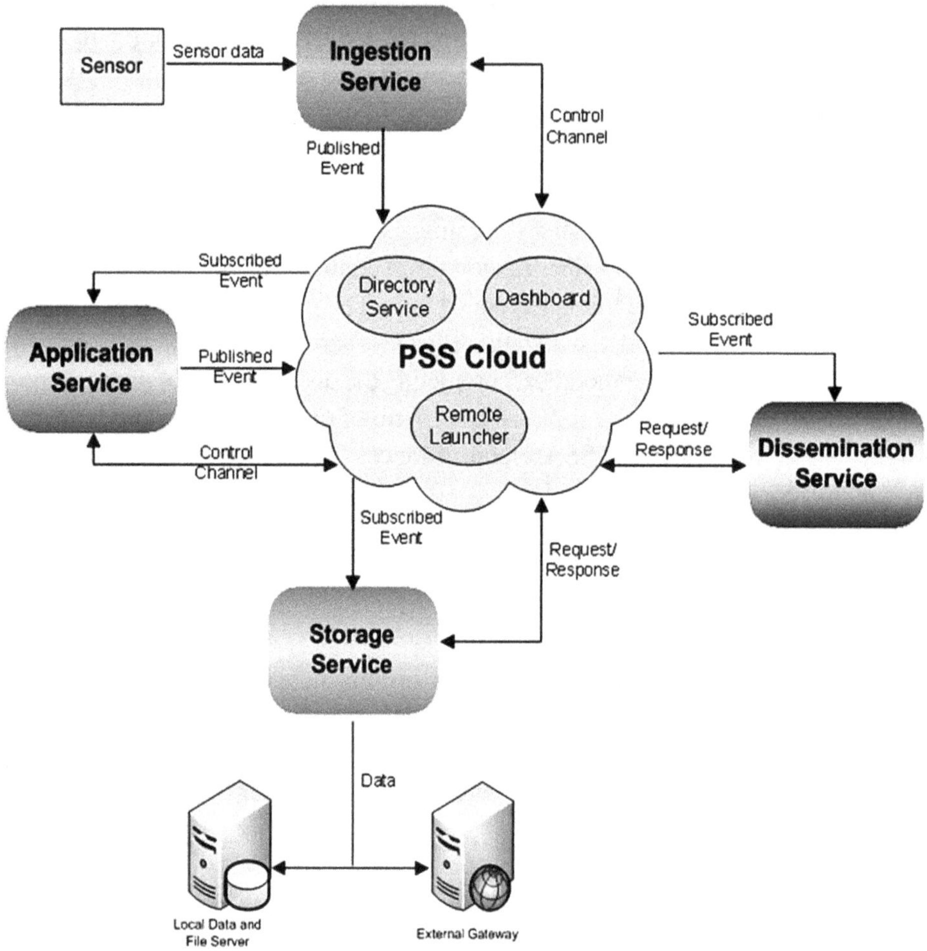

the application of quadcopter flight control for surveillance and monitoring was no different. The gesture controlled quadcopter system was successfully implemented to transport objects from one location to another with little human interference outside of the gestures. In the near future we plan to increase the accuracy and timing of the gesture control a well as increase the payload of the copter. Increasing the accuracy and timing will be performed with optimizing the coding structure used throughout the flow of information. The gesture control is currently a viable replacement for other existing controllers. Currently it seems only useful for applications that don't require extremely high mobility due to the lack of current continuity in the gesture recognition.

REFERENCES

Achtelik, M., Tianguang, Z., Kuhnlenz, K., & Buss, M. (2009). Visual tracking and control of a quadcopter using a stereo camera system and inertial sensors. *Proceedings of the International Conference on Mechatronics and Automation* (pp.2863-2869). 10.1109/ICMA.2009.5246421

Boddhu, S. K., McCartney, M., Ceccopieri, O., & Williams, R. L. (2013). A collaborative smartphone sensing platform for detecting and tracking hostile drones. In *SPIE Defense* (p. 874211–874211). Security, and Sensing. doi:10.1117/12.2014530

Boddhu, S. K., Williams, R. L., & Yue, X. G. (2015). A sensor fusion architecture for human-centric sensing applications. *Journal of. Multidisciplinary Engineering. Science Technology*, *2*, 1608–1615.

Creative Senx3D Depth Camera, Creative Labs. (n. d.). Retrieved from http://us.creative.com/p/web-cameras/creative-senz3d

D'Andrea, R. (2014). Guest Editorial Can Drones Deliver? *IEEE Transactions on Automation Science and Engineering*, *11*(3), 647–648. doi:10.1109/TASE.2014.2326952

Intel, Perceptual Computing SDK. (n. d.). Retrieved from https://software.intel.com/en-us/articles/intel-perceptual-computing-sdk-installer

Jia, P., Hu, H., Lu, T., & Yuan, K. (2007). Head gesture recognition for hands-free control of an intelligent wheelchair, *In Industrial Robot. International Journal (Toronto, Ont.)*, *34*(1), 60–68.

Kinect for Xbox 360. (n. d.). Microsoft Corporation. Retrieved from http://www.xbox.com/en-US/xbox-360/accessories/kinect

Mendel, J., & Wu, D. (2010). *Perceptual Computing: Aiding People in Making Subjective Judgments*. Wiley-IEEE Press. doi:10.1002/9780470599655

Moranduzzo, T., & Melgani, F. (2014). Automatic Car Counting Method for Unmanned Aerial Vehicle Images. *IEEE Transactions on Geoscience and Remote Sensing*, *52*(3), 1635–1647. doi:10.1109/TGRS.2013.2253108

Moranduzzo, T., & Melgani, F. (2014). Monitoring structural damages in big industrial plants with UAV images. *Proceedings of the IEEE International Geoscience and Remote Sensing Symposium* (pp.4950-4953) 10.1109/IGARSS.2014.6947606

Yelamarthi, K., DeJong, B. P., & Laubhan, K. (2014). A Kinect based vibrotactile feedback system to assist the visually impaired. *Proceedings of the IEEE International Midwest Symposium on Circuits and Systems* (pp.635-638) 10.1109/MWSCAS.2014.6908495

This research was previously published in the International Journal of Monitoring and Surveillance Technologies Research (IJMSTR), 3(2); edited by Nikolaos Bourbakis and Konstantina S. Nikita, pages 57-67, copyright year 2015 by IGI Publishing (an imprint of IGI Global).

Chapter 6
Multimodal Human Aerobotic Interaction

Ayodeji Opeyemi Abioye
University of Southampton, UK

Stephen D Prior
University of Southampton, UK

Glyn T Thomas
University of Southampton, UK

Peter Saddington
Tekever Ltd., UK

Sarvapali D Ramchurn
University of Southampton, UK

ABSTRACT

This chapter discusses HCI interfaces used in controlling aerial robotic systems (otherwise known as aerobots). The autonomy control level of aerobot is also discussed. However, due to the limitations of existing models, a novel classification model of autonomy, specifically designed for multirotor aerial robots, called the navigation control autonomy (nCA) model is also developed. Unlike the existing models such as the AFRL and ONR, this model is presented in tiers and has a two-dimensional pyramidal structure. This model is able to identify the control void existing beyond tier-one autonomy components modes and to map the upper and lower limits of control interfaces. Two solutions are suggested for dealing with the existing control void and the limitations of the RC joystick controller –the multimodal HHI-like interface and the unimodal BCI interface. In addition to these, some human factors based performance measurement is recommended, and the plans for further works presented.

DOI: 10.4018/978-1-5225-8365-3.ch006

INTRODUCTION

This chapter discusses the interfaces used in the control of small multirotor aerial robotic systems. In order to aid the discussion of these interfaces, a novel classification model for autonomy called the 'navigational control autonomy (nCA)' model, was presented and discussed intensively. This sets the background for discussing the RC joystick-controller control range and the control void existing beyond the tier-one nCA model. The limitations of the RC joystick controller were presented in the result section of this chapter. This chapter suggests two solutions that could address this control void revealed by the nCA model, as well overcome the limitation of the RC joystick controller. The first method suggests the development of an interface that emulates multimodal human-to-human communication methods to address higher-level interaction problems. The second method extends the first solution into a beyond normal human interaction of mind control via brain-computer interfaces. The limitations of these suggestions are also presented under the discussion section. It then concludes by suggesting some further research works resulting from this chapter's investigation.

This section briefly discusses the background of the unmanned aerial vehicle and introduces the term 'aerobot', which is used to refer to aerial robotic systems in this chapter's discussion. Some aerial robot applications are also presented in this section.

BACKGROUND

A particular class of unmanned aerial vehicles (UAVs) gaining wide popularity with applications cutting across diverse fields, is the small unmanned multirotor aircraft system. The most common application of this system is cost-effective image or data acquisition from remote locations, high altitudes, hazardous environments, or positions that are difficult or more expensive for a human to reach. Another rising application is drone delivery of goods, medical, or military supplies. However, in this research, a particular application of interest for these small multirotor aircraft systems is in aerial robotics. This research focuses on control for these aerial robots. Control for delivery applications or data/image acquisition can be achieved via a combination of manual or automatic control. However, for aerial robotic application, some higher-level control methods may be required. This chapter explores a few relevant control concepts.

Aerobot

The term aerobot was derived from the following two words 'AERial rOBOT', as a way of referring to small unmanned multi-rotor aircraft robotic systems (Abioye, Prior, Thomas, & Saddington, 2016). The term, as adopted in this chapter, is used to give these multi-rotor aircraft systems (often seen as toys) an elevated status of a proper robotic system. Some researchers who seem to share this position are Kim et al. (2016), Muscio et al. (2016), and Verma (2016). Kim et al. (2016) demonstrated the concept of a vision-guided flying robotic arm (aerial robotic manipulator). Muscio et al. (2016) developed a control model for aerial robotic manipulators. The authors particularly identified the industrial applications of these multi-rotor aircraft as Unmanned Aerial Vehicle-Manipulator (UAVM), when equipped with grippers or multi-joint robotic arm manipulator. Verma (2016) also presented the idea of a flying robotic projector system that can be used to convert any plain surface as a display screen. Clearly, these are multi-rotor UAV systems acting as robots. Therefore, aerobots can be considered as unmanned multi-rotor aircraft

Table 1. Civilian applications of aerial robots

S/No.	Category	Applications
1	Aerial Inspection	• Rail Network Inspection • Offshore oil drilling maintenance inspection
2	Disaster Relief	• Search and rescue operations • Healthcare supply delivery
3	Emergency Services	• Law enforcement applications such as portable aerial surveillance • Firefighting applications
4	Entertainment Applications	• Commercial Newsgathering • Filmmaking • Theatrical entertainment (Murphy et al., 2011) • Drone racing
5	Environmental Application	• Wildlife conservation • Habitat exploration • Flood monitoring • Open water monitoring
6	Field Application	• Aerial Survey & Mapping • Agricultural applications
7	Industrial	• Aerial robotic manipulators - flying robotic arms in manufacturing • Warehouse quick light-goods transportation • Indoor inspection
8	Photography	• Selfies • Sport Videography • Mountaineering expedition
9	Transportation	• Drone delivery of goods e.g. Amazon & DHL delivery trials • Drone rides transporting humans

systems with actuators, able to perform tasks analogous to fixed industrial robot manipulators. They are robot manipulators with an innate ability to hover or fly.

The applications of aerial robots are vast. Aerobots are being creatively adopted in many practical application scenarios such as working in human hazardous or radioactive environment, aiding search and rescue missions, performing military ISR (intelligence, surveillance, and reconnaissance) missions, undertaking logistics and transportation (DHL and Amazon examples), conducting space exploration missions (potential Mars rover attached aerobot missions), and videography/photography. Some civilian applications of aerial robots have been presented in Table 1, although some of these applications may be limited by current technological constraints.

However, compared to their industrial robotic manipulator counterparts, the precision and accuracy currently achieved by aerobots is relatively poor, hence limiting their direct practical applications in industrial setups. This is probably due to their inherent aerodynamic complexity of flying in three-dimensional space, the continuously changing references, dynamic environmental conditions, dynamic environmental objects, its un-location to a fixed origin, and a slow sensor response. In order to address some of these navigational limitations and improve the aerial robot's overall performance, some advancement in navigational autonomy may be required.

Autonomy Control Levels

By Autonomous, it is meant the ability to be self-governing/operating within given constraints or rule sets. (Payne, 2007)

The meaning of autonomy in robotics slightly differs from the political, philosophical, or biological meanings. This difference has been responsible for several contentions between experts on how this term is used. The argument as to whether robotic systems performing a task without external control commands are really doing so solely through their own ability to make decisions or through a decision-making method that has been pre-programmed into them. Especially when such systems, may be considered, to have been designed and developed to perform such specific functions and not a random system existing in nature. The designed artificial autonomous system process could be broken down into bits, understood, and hence predictable by the designer. Therefore, there seems to be a thin layer requiring further clarification as to whether:

- **Proponents:** A robotic system, performing a particular task without being told how to do such task, is simply autonomous or
- **Opponents:** Just executing a pre-defined program on how to make the right decisions that complete the task.

In order words, a clarification needs to be made as to whether executing a program, not of the task, but of how to decide on key elements that affects accomplishing the task, is autonomous or automatic. Clearly, executing a program of the task would be automatic as described by Clough (2002). In order to avoid being drawn into this debate, what is meant by autonomy in this chapter is defined in simple, clear, and unambiguous terms. Clough (2002) made a distinction between automatic and autonomous systems, "Automatic means that a system will do exactly as programmed, it has no choice. Autonomous means that a system has a choice to make free of outside influence…" According to Pfeifer and Scheier (1999), "autonomy means independence of control. This characterization implies that autonomy is a property of the relation between two agents, in the case of robotics, of the relations between the designer and the autonomous robot. Self-sufficiency, situated-ness, learning or development, and evolution increase an agent's degree of autonomy."

In this chapter, autonomy is conceptualised as being self-operating in a dynamic environment, within a given sets of rules or constraint, without relying on external control. This could be with the aid of a computer program working with sensors to guide decision-making, or some complicated artificial intelligence real-time learning and adapting algorithm. This could be a decision method that the operator may or may not understand, may or may not be able to predict, but is optimal enough to safely complete the assigned task to an acceptable standard. Autonomy, in this case, is void of continuous real-time control interaction during task execution but rather sparse real-time command, or just the initial command instruction may be given, in some cases.

Autonomy is an abstract concept difficult to measure empirically; hence, it is rather described qualitatively by analogically developed subjective hierarchical levels. The autonomy of an aerial robot can be considered in two context – navigation and payload. Usually, the payload autonomy level is expected to match the navigation autonomy as observed in Clough (2002) classification, which was adopted by the US Air Force Research Lab (AFRL) as the AFRL's ACL (Autonomous Control Levels). It was also used as a standard classification in other US military and government agencies, and by their contractors (Sholes, 2007). Clough (2002) identified eleven autonomy control levels:

1. Remotely Piloted Vehicle,
2. Execute Preplanned Mission,
3. Changeable Mission,
4. Robust Response to Real Time Faults/Events,
5. Fault/Event Adaptive Vehicle,
6. Real Time Multi-Vehicle Coordination,
7. Real Time Multi-Vehicle Cooperation,
8. Battlespace Knowledge,
9. Battlespace Cognizance,
10. Battlespace Swarm Cognizance,
11. Fully Autonomous.

The ACL incorporated a model of the human OODA (Observe, Orient, Decide, and Act) loop, a loop that seemed popular with the military (Clough, 2002). Sholes (2007) developed a simulation technique for evaluating multiple autonomy algorithms based on Clough AFRL (2002)'s ACL model. In a separate study by the Office of Naval Research (Horrigan & ONR's Committee, 2000) classified autonomy into the following six levels enumerated below:

1. Human Operated,
2. Human Assisted,
3. Human Delegated,
4. Human Supervised,
5. Mixed Initiatives,
6. Fully Autonomous.

Hill, Cayzer, and Wilkinson (2007) presented the modified PACT (Pilot Authority and Control of Task) autonomy level system, originally developed by the UK Defence Evaluation Research Agency (DERA), as a better classification of autonomy compared to the Clough's ACL. Firstly, they considered Clough's descriptive autonomy useful for vehicle level classification but inadequate for classification at functional levels. Secondly, Clough's ACL imposed an artificial ceiling on the autonomy level that can be achieved by an individual vehicle by introducing swarm behavioural requirements within it ACL. The PACT classification is as follows:

Level 0: Full pilot with no computer autonomy.
Level 1: Pilot assisted by computer only when requested.
Level 2: Pilot assisted by computer without pilot necessarily requesting it (computer advice continuously).
Level 3: Pilot backed up by computer.
Level 4a: Computer backed up by pilot – computer acts when authorized.
Level 4b: Computer backed up by pilot – computer acts unless pilot disapproves.
Level 5a: Computer monitored by pilot – computer informs human of actions.

Level 5b: Computer monitored by pilot – computer acts completely autonomously and need not inform the human of its decisions.

As may be observed, the PACT levels seemed similar to Horrigan and ONR Committee (2000) classification, which also closely aligns with the civilian aerial robot navigational control autonomy model presented in this chapter. In addition to the limitations already offered by Hill, Cayzer, and Wilkinson (2007), Clough's ACL was structured after military missions, therefore, its military-specifics context makes it difficult to adopt it in many emerging civilian applications. In addition, Clough's ACL was modelled as a generalised classification method for all types of military robotic systems – aerial, ground, and underwater.

Since autonomy, as a concept in robotics is a subjective non-quantitative (not measurable) analogical (but comparable) descriptor, perhaps its layers could be considered in fuzzy terms as overlapping strata expressed as a range of percentile rather than as discrete levels. Therefore, the autonomy levels presented in Figure 1, which is used as the basis of discussing the human aerobotic interfaces, is considered as a continuous, overlapping layer rather than quantized or discrete layers. For this reason, the presentation of the autonomy levels in Figure 1 is not presented in levels or vertical columns suggesting equal linear step size increments in autonomy. Sholes (2007) simulation plot and prediction of the growth of autonomy seems to supports this hypothesis of a non-linear step-size increase or change from one successive layer to another. Moreover, the concept of autonomy is a borrowed property from living organisms existing in nature and the ecological system, which is rarely linear.

Figure 1 presents a two-dimensional, expandable pyramidal model of Navigational Control Autonomy (nCA) levels for multi-rotor aerial robotic systems. In this model, autonomy is divided into three tiers: lower, intermediate, and upper tiers. Within each tier, autonomy level increases from left to right, and then across tiers from lower to intermediate to the uppermost tier. For this reason, it is termed the 'inverted lightning model of autonomy' as shown in Figure 2. This two-dimensional pyramidal model of autonomy is expandable rightwards and upwards to accommodate for new or missing autonomy level classes.

Unlike the ONR and Clough's Levels of Autonomy, the nCA model does not take into account the payload autonomy. However, payload autonomy is expected to match the aerial robot's navigation control level. The separation of navigational autonomy from payload autonomy simplifies the interaction relationship between an operator and an aerial robot (as considered in this chapter). Additionally, because autonomy classification may vary slightly across specific domains, the classification method used in this chapter have been developed specifically for multirotor aerial robotic systems' application in the civilian domain; as the AFRL's and ONR's classification were patterned towards military specific applications. Perhaps, with the rapidly evolving civilian applications of the aerobot, a less military objective-based classification may be worth considering.

Tier-One Control Levels

The nCA model described in Figure 1, has six components distributed in three tiers. The first tier consists of the following modes:

1. Rate control,
2. Attitude stabilisation control,
3. Altitude, position, and heading hold/assist Control.

Figure 1. Multirotor aerial robot navigation control autonomy (nCA) levels

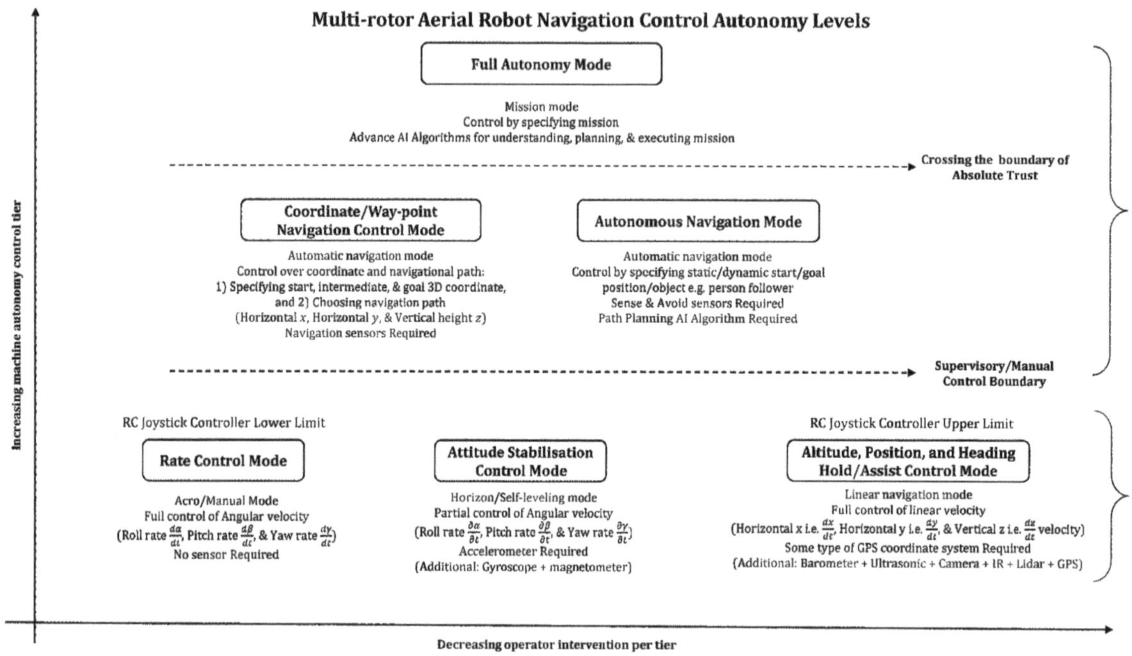

Rate control mode is the minimal level at which a human operator can control a small multirotor aircraft. In this mode, no sensor is needed. Full control of the three angular velocities (Roll rate $d\alpha/dt$, Pitch rate $d\beta/dt$, & Yaw rate $d\gamma/dt$) are given to the operator. This is usually performed with the aid of the RC joystick controller. A high level of skill is required to operate the multi-rotor aircraft at this level. This level is also called the Acro (short for Acrobatic) or Manual mode. RC hobbyist often prefer this mode when performing stunt manoeuvres for entertainment purposes. Flying at this level usually requires many training hours. It could take up to 600 or more hours of training to become an expert (from being a novice who has never flown before).

The Attitude stabilisation control mode is the next level after the rate control mode. In this mode, the operator is assisted with attitude stabilisation. The operator only has partial control over the three angular velocities (Roll rate $\partial\alpha/\partial t$, Pitch rate $\partial\beta/\partial t$, & Yaw rate $\partial\gamma/\partial t$). This is because the multi-rotor aircrafts flight controller, in this case, has a stabilization loop that continuously balances the aerobot horizontally, especially when the operator stops feeding new rate control data. The stabilization loop in this mode requires, at least, an accelerometer sensor. Additional or optional sensors may include the gyroscope and magnetometer. Data from a gyroscope could be collected and fused with the accelerometer data for a more accurate attitude stabilization control, because the accelerometer is prone to noise and the gyroscope to drift. A magnetometer could also be integrated to minimize yaw drift.

The next level after the attitude stabilisation, is the altitude, position, and heading hold/assist control mode. In addition to the attitude stabilisation from the previous mode, the operator is also assisted with the altitude, position, and heading direction. In this mode, the operator no longer has control over three angular velocity parameters. Control is performed by manipulating three linear velocities (Horizontal x i.e. dx/dt, Horizontal y i.e. dy/dt, & Vertical z i.e. dz/dt velocity). In this mode, more sensors are required, such as GPS, Barometer, Ultrasonic, Camera, Infrared, Lidar, etc. for localization within the system specified coordinate system.

Tier-Two Control Levels

The second tier of the nCA model consists of the following two modes:

1. Coordinate/waypoint navigation control mode,
2. Autonomous navigational mode.

The lowest level of the second tier is the coordinate/waypoint navigation control mode. This may also be called the automatic navigation mode. In this mode, low-level controls are completely abstracted from the operator. Navigation is specified by the user before execution. The start point, intermediate points, and goal point are pre-fed into the aerobot. Once execution starts, the multirotor flight controller is responsible for all low-level control required to achieve the goal point from the start point to the endpoint via the route specified by the operator. Execution would proceed as plan except the operator decides to abort the mission. In this mode, navigational sensors for location, continuously tracking flight route, are required. The system may or may not have a sense and avoid system and is, therefore, the responsibility of the operator to ensure that the specified route is safe enough for the aerobot to operate. The aerobot at this control level is expected to be able to perform some self-diagnostics check on itself, and relay issues requiring operator advice or intervention and may be robust enough to execute certain fundamental safety measures such as abort mission and return to launch.

The higher level of the second tier of the nCA model is the autonomous navigation mode. In this mode, the operator only needs to specify the goal position or object, which could be static or dynamic, as in the case of a person-following UAV. This mode requires: 1) sense and avoid (SAA) sensors and 2) a path planning AI algorithm.

Figure 2. The inverted lightning model of increasing autonomy

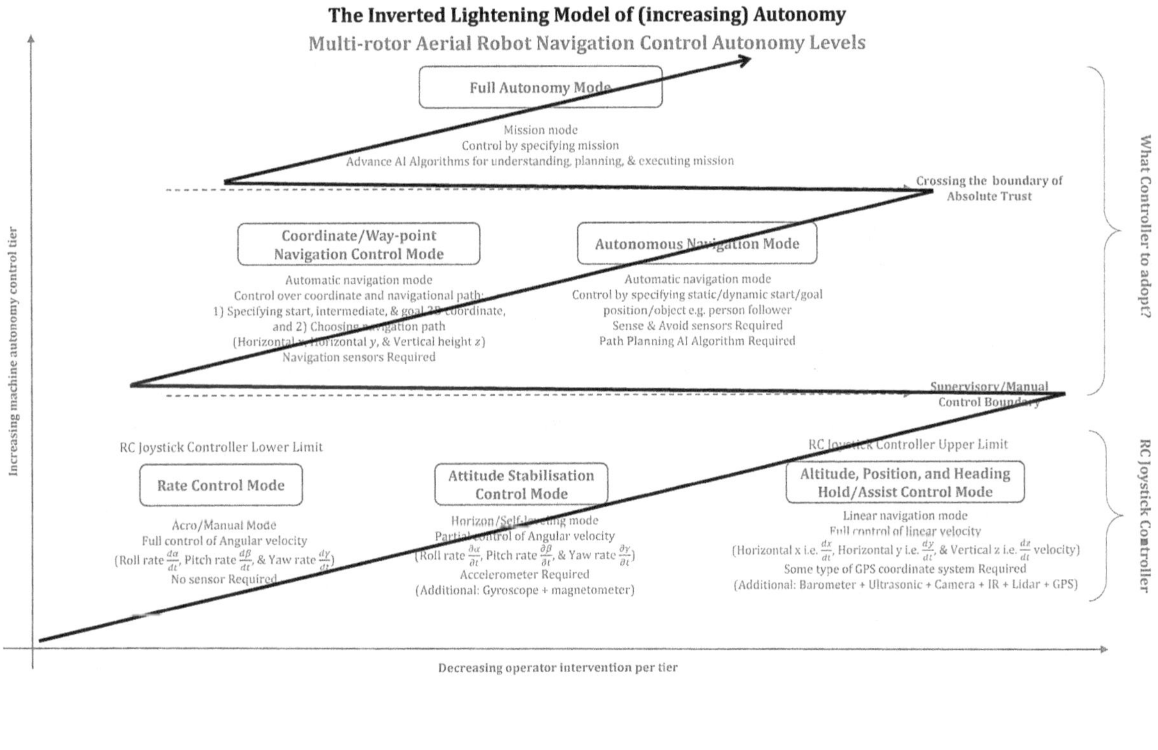

Tier-Three Control Level

The third tier consists of only one control mode – the full autonomy mode. At this point, the operator interaction with the aerobot is very sparse. The operator only needs to specify a mission and the aerobot takes care of the rest – planning and execution. In this mode, the aerobot must be well equipped with advanced AI algorithms, for understanding the prevailing scenarios or context within which mission has been given, planning the mission, and executing the mission.

Between tier one and tier two is the supervisory/manual control boundary. In order words, most of the operation performed in tier one is manual operator control, whereas tier two is mostly supervisory. The boundary separating tier-two from tier-three is the boundary of 'absolute trust.' For the aerobot autonomy to progress beyond this point, the operator must be willing to trust the aerobot's decisions absolutely.

AFRL vs. nCA vs. ONR Autonomy Control Levels

This section relates how the nCA level model maps to the ONR's and Clough's AFRL ACL models. In 2000, a research committee for the Office of Naval Research (ONR), proposed six robotic autonomy level (Horrigan & ONR's Committee, 2000). In 2002, the Airforce Research Laboratory (AFRL) UAV program motivated Clough's research into developing a measurement metrics for autonomous vehicles (Clough, 2002). This was later adopted by the AFRL as a metric method of measuring their autonomous vehicles. Since then, software has been developed by various vendors and researchers based on this measurement metrics for used by government and military agencies (Sholes, 2007). These two widely recognised standard metrics for autonomy control levels were developed particularly for the military domain. The nCA model was developed to serve a similar purpose in the evolving civilian application domain of autonomous vehicles. These three ACL models are compared side-by-side, with similar levels mapped to each other, as shown in Figure 3.

Figure 3. Mapping relationship between AFRL, nCA, and ONR autonomy levels

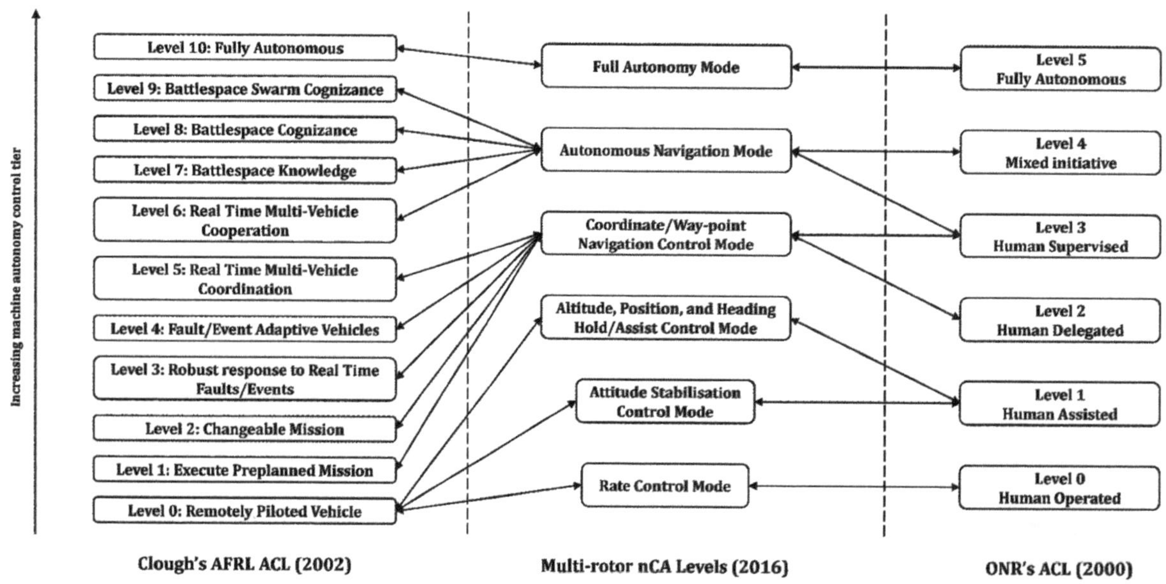

The AFRL's eleven-level model maps to the nCA six-level model as shown in Figure 3. AFRL's level zero (Remotely piloted vehicle) maps to the three tier one component of the nCA model. AFRL's level 1-5 maps to the first component of the second tier of the nCA model. AFRL's level 6-9 maps to the second component of the second tier of the nCA model. A one to one mapping is observed in the highest column of both hierarchies, as AFRL's level 10 (fully autonomous) maps to the nCA third tiers only component, full autonomy mode.

The ONR's six-level model maps to the nCA six-level model as shown in Figure 3. Both the bottom and top most level of both model have a one to one mapping. The ONR's level 1 maps to the second and third components of the first tier of the nCA model. The ONR's level 2 maps to the first component of the second tier of the nCA model. The ONR's level 3 maps to both components of the second ticr of the nCA model. The ONR's level 4 maps to the second component of the second tier of the nCA model.

The mapping relationships of the nCA model with the AFRL and ONR models describes how robotic systems classified in any of these scales could be re-classified under the nCA scale. It also shows how robust and fully representative the nCA scale is, of previous models and domain application, despite being designed for the emerging multi-rotor aerial systems in the civilian domain.

The Interface Problem

The Autonomy level of an aerial robot affects the range of selectable control communication interfaces that can be used for communicating control signals to it. Higher-level systems often require little or no operator intervention whereas low-level systems often require much operator control intervention. Different control interfaces offer varying level of intervention. For example, the RC controller enables the

Figure 4. Aerial robot control interfaces

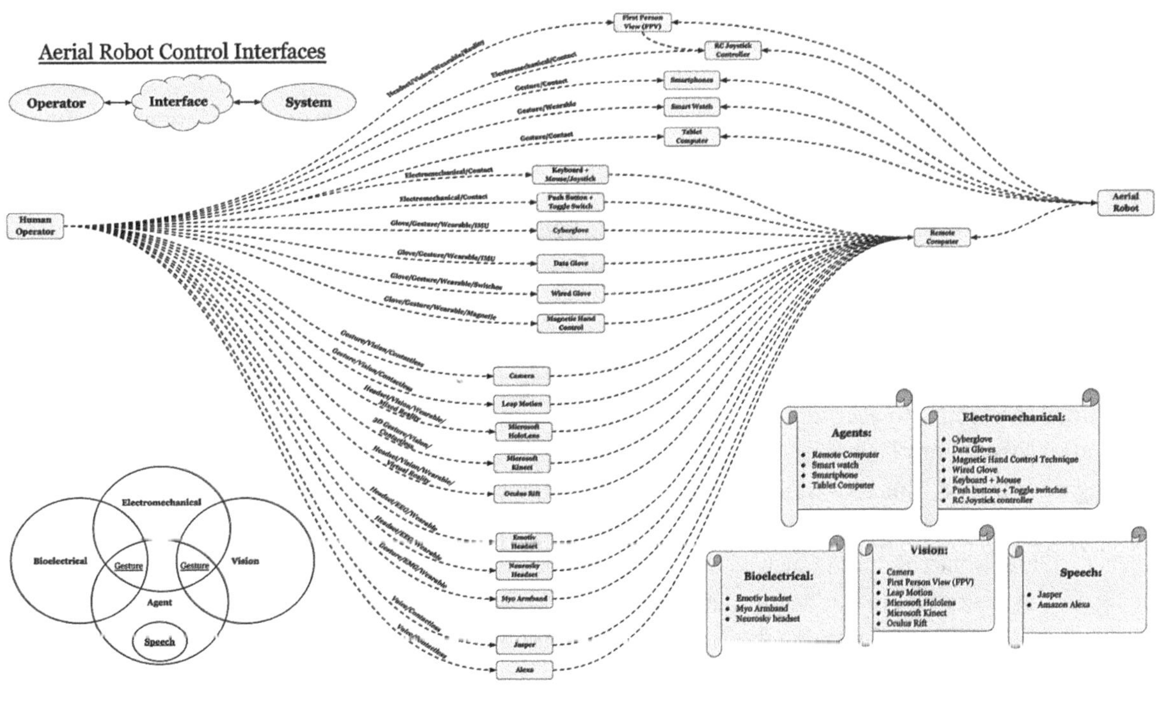

provision of much intervention and hence, it is suitable for low-level interaction; however, speech and gesture interfaces, though more convenient, offers less flexibility at such low-level. Figure 1 shows the application limits of the RC joystick controller. The RC joystick controller is a very suitable controller for the first tier of the nCA model. It has a dynamic range, with the rate control mode being its lower limit and the altitude-position-heading assist mode being the upper limit.

Although the RC joystick controller is widely accepted as a standard controller for the tier-one nCA model range, it is not without its limitations. These includes orientation un-intuitiveness (when the multirotor's front faces the operator), long training hours (estimated 600 hours to develop expert skill level), difficulty to learn for persons with neurological motor movement disorder (such as ataxia, dystonia, essential tremor, Parkinson's disease, and atypical Parkinsonism), and much cognitive effort requirement (depending on autonomy control level).

Therefore, if the RC joystick controller is a suitable interface for low-level (tier-one nCA levels) aerobotic interactions, but inadequate for higher-level interactions, then what controller should be adopted for higher-level (tier-two and tier-three nCA levels) interactions? This question forms the basis for investigating several human aerobotic interaction interfaces presented in the following sections.

RESEARCH INVESTIGATION: HUMAN AEROBOTIC INTERACTION

Human aerobotic interaction can be considered as an association of three component: an operator, an interface, and a system, as shown in Figure 4. The operator, in this case, is the human and the system is the aerobot. Therefore, this research investigation is a problem of 1) how to get the human operator, to communicate their control intentions, to the aerobot; and 2) how to get the aerobot, to provide real-time feedback, to the human operator, so that the operator can make a more informed control decision. In this section, unimodal and multimodal human aerobotic interaction interfaces would be discussed.

Unimodal Human Aerobotic Interfaces

As shown in Figure 4, the unimodal aerial robots interfaces discussed in this section would be grouped into five categories: agents, electromechanical, bioelectrical, vision, and speech.

Agents

According to Nwana, (1996) an agent can be referred to as a software or hardware component that is capable of acting (in an exact manner) on behalf of its user, in order to accomplish a specific task. Agents, in the context of this chapter, refers to intermediate intelligent or smart systems capable of 1) being pre-instructed or programmed by the operator, 2) processing such high-level programmes into low-level control instructions that can be 3) relayed to an aerial robot, with the aim of providing continuous guidance in order to accomplish the operator's pre-defined task or objective. The most commonly used aerial robot control agent as described in this section is a remote computer system, which could be running a Windows, Linux, or Mac operating system. Other popular agents include the tablet computers (such as Apple's iPad Pro and Samsung's Galaxy Tab S), Smartphones (such as the iPhone 6s and Sony Xperia L), and Smart watches (such as the Apple Watch 2 and Samsung Gear S2). The control level of the agents, as described in this section, is limited to the first component of tier-two in the nCA model.

As shown in Figure 4, besides the remote computer being an independent agent, it also acts as an intermediate processing/relay terminal for many other control interfaces (as discussed in later sections).

The remote computer can be pre-programmed for waypoint navigation (nCA tier-two, first component). It can accept electromechanical input and provide visual feedback of the aerial robot collected data and health status.

The tablet computer is often powerful enough to operate complex programs like the remote computer. It can also provide waypoint navigation, electromechanical contact/switch, touch screen, and accelerometer-based gesture interaction. In addition to this, it can also provide on-screen visual and haptic feedback of telemetry data, status, and other information.

The smartphone is a smaller version of the tablet computer. It can perform similar functions as the tablet computer although it has a smaller screen for visualising telemetry data, status, and other information.

The smart watch is a computing attempt to miniaturise the smartphone. Control using the smart watch could be relayed through the smartphone as demonstrated by Mark Ven (Reuters, 2015). Gesture control is achieved by collecting and processing the smart watch's IMU sensor data. Processing could be relayed through a smartphone, tablet computer, or a remote computer. Although, at the time of writing this chapter, the smartwatch is not capable of processing these data and hence directly controlling the aerial robot, this is a feat soon to be achieved by the collective efforts of the research communities at various research institutions and laboratories. However, unlike the other agents, the smart watch has the additional advantage of being a wearable technology.

Electromechanical

Electromechanical interfaces describe a class of devices that utilises mechanically actuated switches in controlling the flow or current in an electrical circuit. This regulation of the flow of current is used to signal control intentions to a computer, machine, or robotic system. The most popular electromechanical control device for an aerial robotic system is the RC joystick controller. Other electromechanical devices includes the keyboard, mouse, push buttons, toggle switches, Cyber glove, data glove, wired glove, Shift thumb drone controller (Lavars, 2016), IMU HandMagic device (Calella, Ortega, Rishe, Bernal, & Barreto, 2016), and magnetic hand gesture device (Yinghong et al., 2011) among several others. Most electromechanical controller application varies from nCA tier-one levels to the first component of the nCA tier-two.

The RC (Radio Controlled) joystick controller consists of two joysticks, a few toggle switches, push buttons, and variable potentiometers as shown in Figure 5 (HobbyKing & Turnigy, 2015). A joystick can be conceptualised as a two-dimensional potentiometer (continuous variable resistor) that generates an analogue voltage divider circuit in each dimension. With each dimensional axis being mapped to a control (e.g. thrust) with a continuous range of intensity/values bounded by two extremes levels. The RC joystick controller, unlike most other electromechanical devices, does not require additional or intermediate processing by a remote computer. It is capable of interacting directly with the aerobot's flight controller. Additionally, it may be augmented with an FPV (First Person View) headset, a smartphone, or a tablet computer to provide visual image data from the aerobot's camera. Telemetry data could also be feedback to the controller via the smartphone, tablet, or a small screen, installed on the controller. It can also be augmented to provide haptic feedback of flight turbulence.

Cyber glove and Data glove are examples of IMU-based, glove-like, hand-wearable, computer input and control device. An accelerometer sensor is used to measure hand gesture orientation. A gyroscope sensor may be used to measure the rate at which a gesture is formed, hence increasing the vocabulary of possible gesture vocabulary. To further increase the gesture vocabulary, some gesture gloves may have

Figure 5. 10-channel Turnigy TGY-i10 (mode 2) telemetry RC joystick controller

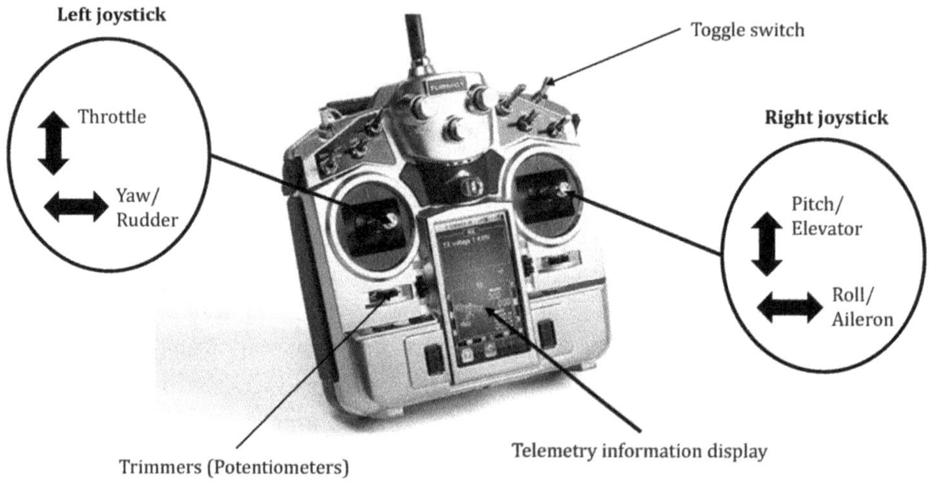

switches that can be triggered by specific hand gestures. These gesture gloves require processing on a remote computer. They could be wirelessly connected to the computer using wireless technologies like Bluetooth or simple connected via wires. Unlike the IMU-based Cyber-glove and Data-glove, Wired-glove is based on finger/hand motions triggering switches embedded along the wearable glove structure (Renitto & Thomas, 2014). Another gesture glove of interest is the Magnetic-Hand gesture technique that is based on small magnets installed on the fingernails, generating a magnetic field that is tracked by a magnetic sensor bracelet installed on the wrist (Yinghong et al., 2011). HandMagic is an IMU based wristband for sensing human hand gesture (Calella et al., 2016). Although it is similar to the IMU based gesture gloves, it is being worn on the wrist, hence does not capture finger motions. Similar to some RC joystick controllers, some wearable gesture devices are also capable of providing haptic feedback.

Other electromechanical devices such as keyboard, mouse, pushbuttons, toggle switches, and joysticks are also used in combination with a remote station for communicating operator intentions to the aerial robots. They could be used as inputs for either direct or pre-programed computer control. Waypoint mapping is an example of pre-programmed control.

Vision

Vision based control interfaces utilise the techniques of computer vision such as image capture/acquisition, filtering, segmentation, feature extraction, and classification, in identifying operator's control intentions. Image acquisition is usually performed with the aid of a regular camera, thermal camera, infrared camera, or some other specialised imaging system. Vision systems often require remote processing. Due to their low rate of feeding control signal which is in turn due to slow remote processing and communication medium lag, the vision control techniques are not suitable for high-rate low-level control function (first and second components of tier-one nCA model). Some devices in this category used for human aerobotic interaction includes regular standalone cameras, web cameras, stereo cameras, Leap Motion device, Microsoft Kinect. It may also include virtual, mixed, and augmented reality headsets such as the Oculus Rift and Microsoft HoloLens amongst others.

The following are examples of research were cameras and computer vision techniques are used. Soto-Gerrero and Ramrez-Torres (2013) demonstrated visual gesture interaction on a Parrot AR drone via an android-based mobile phone platform. Milanova and Sirakov (2008) and Qing et al. (2008) independently investigated systems to recognise expressions of joy, distress, surprise, interest, frustration, anger, disgust, fear, and a neutral expression among several human observable facial expressions. This could be potentially useful in emotional gesture interactions with aerobots.

Bioelectrical

Bioelectrical control interfaces rely on electrical nerve impulses generated by muscular or neurological human activities. These electrical impulses are usually detected with the aid of some biomedical instruments and probes. The detected impulses may be recorded for immediate analyses on a remote computer, where it is then cross-referenced with a previously collected signal database already mapped to the user's control intention (vocabulary of the bioelectrical signal), for identification of the user's control intention. The identified control intention is then relayed by the remote computer to the aerobot for appropriate execution. Digital Signal Processors (DSPs) are often used to provide high-speed processing to reduce processing time lag and ensure near real-time response to users control input.

The Emotiv and Neurosky headsets are two examples of electroencephalography (EEG) measuring devices that are being used by researchers to investigate brain-computer interaction for robotic systems such as aerial robots (Emotiv, 2014; Hammacher Schlemmer, 2015). Lafleur et al. (2013) demonstrated BCI (telepathic mind) control of a quadcopter in three-dimensional space, using an EEG cap. The MYO armband is another example of a bioelectrical device that is used to capture electromyography (EMG) signal. The MYO device is worn on the arm and is able to capture, consistently, a few unique arm gestures, which are then used in robotic control applications such as the aerobot navigation (Cacace, Finzi, & Lippiello, 2016; Nagar & Xu, 2015; Thalmic Labs, 2015).

Speech

Speech interfaces provide a method of communicating control intentions via voice commands. A microphone may be used to detect (and convert to an electrical signal) the sound wave generated by the operator's voice commands. The operator's speech command may be identified by querying a database of the aerobot's speech command vocabulary with the recorded speech signal. A successful match results in the matched control intention signal being sent by the speech processing system to the aerobot control system for control execution.

Harris and Barber (2014) investigated verbal speech interaction with a mobile robotic system. They converted the microphone-captured audio to text using the CMU Pocket Sphinx audio speech recognition (ASR) toolbox. With this method, a text command vocabulary can be defined instead of a speech command vocabulary; the transcript of the voice command can be compared with the predefined database of command, and the interpreted control intention is passed to the aerobot for execution. Recent advancement in speech technology is evident in a plethora of devices using the following voice assistant: Apple's Siri, Microsoft's Cortana, Google's Now, Amazon's Alexa, and the Jasper's ASR framework.

Multimodal Human Aerobotic Interfaces

According to Oviatt (2002), "Multimodal system process two or more combined user input modes – such as speech, pen, gaze, manual gestures, and body movements – in a coordinated manner…" A multimodal interface could be a combination of any two or more compatible, complementary, unimodal interfaces. Multimodal interaction offers the flexibility of switching sequentially between communication modes without breaking the interaction or communicating via parallel modes simultaneously (Turk, 2014). There are a number of reasons, why multimodal interfaces are being investigated and developed. One may be an attempt to overcome the limitations of the constituting unimodal interfaces. However, there is also the complication associated with fusing multiple unimodal interfaces together. In this research's investigation, the multimodal combination of interests is one that emulates the human multimodal interaction method, as would be discussed subsequent sections.

Interface Summary

Table 2 presents a summary of the merits and demerits of the unimodal and multimodal interfaces discussed in this section. It also highlights a few researchers and stakeholders involved in the development or investigation of such interfaces.

RESULTS

The Joystick-Type Controller Dilemma

Typically, the interaction between humans and aerobots fundamentally consists of a radio-controlled (RC) joystick transmitter controller, augmented with toggle switches, push buttons, and variable potentiometer, for improved control functionality as shown in Figure 5. The RC controller is plagued with a number of limitations. Firstly, learning to control an aerobot via the RC joystick controller requires an operator with a good neurological motor response. Although most UAV operators eventually adapt to the operation of this joystick-type controller and may become very skilful, getting used to it as a compound impulse response. Secondly, it is not a natural human interaction modality. Human interaction with one another occurs at a higher level of intelligence or autonomy. Thirdly, there may be side effects to briskly flicking both the left and right thumb, placed on the left and right joystick, in order to maintain horizontal level balance, within fractions of a second. Fourthly, there is the problem of too many controls. Besides operating the left and right joysticks, the operator might be required to operate switches to perform some other functions such as payload drop or snapshot from an onboard camera, at some specified time, altitude, distance, space, and location. Simultaneously executing this accurately may be difficult. Fifthly, there is the problem of un-intuitiveness of the controller, especially when the human operator and aerobot's forward orientation is out of alignment at 180 degrees. Sixthly, learning to fly the aerobot via the RC joystick controller often requires several training hours especially in the previously described un-intuitive orientation state. Seventhly, due to the control overload on a small unmanned multi-rotor aircraft's operator, the situational awareness of the operator is reduced; making it impossible to simultaneously control multiple aircraft (Cavett, Coker, Jimenez, & Yaacoubi, 2007).

Table 2. Aerobotic interface comparison

S/No	Interface Category	Interfaces	Merits	Demerits	Stakeholders/Researchers
1	Electromechanical	• RC Joystick Controller • Keyboard + Mouse + Joysticks • Push buttons + Toggle switches • Wired Glove • Magnetic Hand Gesture Technique • Data Glove • Cyber Glove • Shift drone thumb controller • Google Project Soli	• Fast response • Very reliable • Easier to implement	• Long hours dedicated to training • Not natural - not a normal human interaction method • Not intuitive - especially when re-oriented towards operator • Demands high cognitive workload	• Futaba; Turnigy; DJI; Hobbyking; FrSky; Spektrum; • Raytheon • 3D Connection • VDCI • Ma et al. (2010) • Lavars (2016) • Google (2014)
2	Vision	• Camera • First Person View (FPV) • Leap Motion • Microsoft HoloLens • Microsoft Kinect • Oculus Rift	• Natural • Easy to learn	• Difficult to implement • Requires higher processing power • Occlusion	• Gupta and Ma (2001) • Huang and Chung (2004) • Nimble VR (2012) • Blanz (2007) • Leap Motion Inc. (2010)
3	Bioelectrical	• MYO armband • Emotiv headset • Neurosky headset	• Assistive technology • Gesture interaction • Stealth interaction	• Not natural due to wearable device requirement • High cognitive concentration workload • Security concerns - snooping EEG signals to potentially steal certain information like login pins, credit card number, etc.	• Thalmic Labs (2013) • NeuroSky (2011) • Emotiv (2014) • Martinovic et al. (2012)
4	Speech	• Jasper • CMU Sphinx • Julius • Amazon Alexa	• Natural • Easy to learn	• Difficult to implement • Not immune to prevailing environmental noise	• Bastianelli et al. (2015) • Lee et al. (2003) • Mu-Chun & Ming-Tsang (2001) • Amazon; Google; • Microsoft; Apple;
5	Agent	• Remote Computer • Tablet Computer • Smartphone • Smartwatch	• Easy to learn • Touch-screen gestures • Orientation-tilt feature • Waypoint programming	• Relative precision requirement over a small screen area • Severe fatigue on user's wrist	Huhn and Haewon (2014)
6	Multimodal	HHI-like interaction (Speech + gesture)	• Natural • Easy to learn • Robust - complementing interfaces overcomes unimodal limitations	Difficult to implement	• Harris and Barber (2014) • Oviatt (2003) • Shah and Breazeal (2010) • Lackey et al. (2011) • Reeves et al. (2004) • Bolt (1980)

Finally, the level of control offered by the RC joystick controller is limited to tier-one of the nCA model. Interaction at a higher level of autonomy may require the adoption of some other interfaces.

DISCUSSION: SOLUTIONS AND RECOMMENDATIONS

In order to address the joystick-type controller dilemma, described in the result section, two solutions are suggested. The first solution is the development of an interface that emulates typical human-human multimodal speech and visual interaction. The second solution is the development of unimodal mind control (telepathic interaction) via brain-computer interfaces. In addition to discussing these two solutions, this section also discusses the need to measure the performance of such developed systems, from a human factors point of view. The limitation of each technique is discussed in each subsection.

HHI-Like Aerobotic Control Interface

Human-human interaction (HHI) is a multimodal combination of speech and non-verbal expressions such as gestures and gazes (Green, Chen, Billinnghurst, & Chase, 2007; Sharma, Pavlovic, & Huang, 1998). The natural human-to-human interaction often occurs within the context of a very small geographical area, hence the term 'near-interaction' (as used in describing Figure 6). The idea of near-interaction is used to indicate the limit of up to a few hundred metres. If the human is too far away, it would be difficult to be heard; and difficult to be seen when much farther away. Therefore in HHI, it is difficult, if not impossible, to interact via speech, gaze, gesture, and other non-verbal methods, beyond a small geographical location radius. This range limitation could also affect HHI-like human aerobotic interaction, as it becomes difficult for the onboard camera to identify the human, or for the microphone to detect the human voice amid several other environmental noise sources. At such long distances, other long distance communication technologies may need to be adopted, as with HHI interaction. Figure 6

Figure 6. Near interaction: human with aerobot and human with human
Abioye et al., 2016.

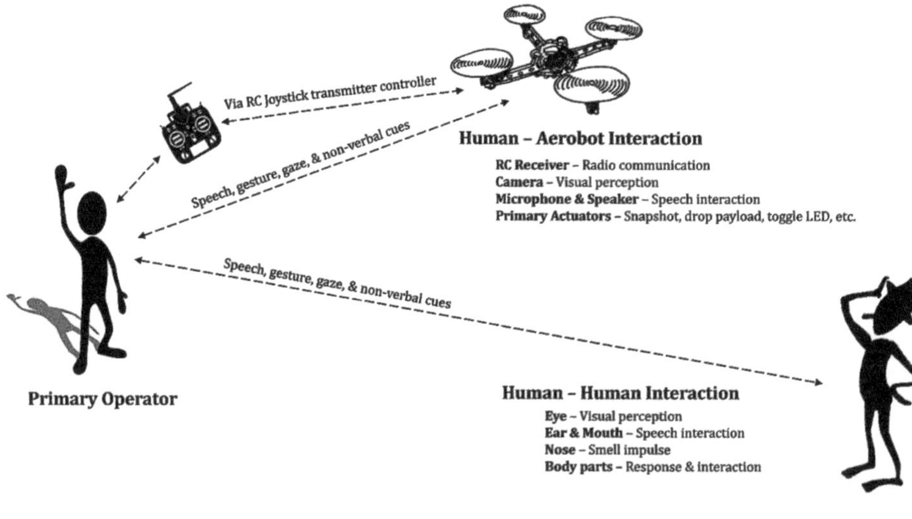

describes near human interaction between three actors. The first is an interaction between a human operator and an aerial robot. The second is an interaction between the human operator and another human. The human operator could potentially interact with the aerobot via RC joystick controller as shown or it could interact via normal human HHI-like methods. The second human in the figure, though not the primary operator, may also be able to initiate an interaction with the aerobot.

Generally, robots have been designed to reduce human workload, risk, cost, and human fatigue-driven errors. Fong and Nourbakhsh (2000) suggested that in order to achieve these objectives, it is crucial to make the human-robot interaction effective, efficient, and natural; which further suggests the use of multiple modalities of contact, dialogue, and gestures. According to Green et al. (2007), "It is clear that people use speech, gesture, gaze and non-verbal cues to communicate in the clearest possible fashion." Therefore, it may be necessary to develop systems that could adopt some of these natural human-human interaction techniques. The essence of human-robot collaboration is probably the fact that robots can complement the human effort by optimising problem-solving techniques, performing tasks faster, and in many cases with greater dexterity.

In Figure 6, a hand gesture interaction was initiated which could suggest a human-human or human- aerobot interaction. Some notable human-aerobot hand gesture interaction research includes the application of the data glove such as the AcceleGlove (Hernandez Jose, Kyriakopoulos, & Lindeman, 2002) and Magic glove (Chaomin, Yue, Krishnan, & Paulik, 2012), Magnetic hand tracking (Ma et al., 2010), Apple Watch (Reuters, 2015), and the MYO armband device (Nagar & Xu, 2015; Thalmic Labs, 2013). Gaze or eye tracking is a technique that seeks to determine the direction the eye is focused; perhaps, via the relative motion of the eye with respect to the head's position (Renitto & Thomas, 2014). Electrooculography (EOG) techniques could be used to approach this from a biomedical point of view.

Because the development of an HHI-like aerobotic interaction requires a multimodal approach (Sharma et al., 1998), there may be some technical complications associated with integrating the individual HCI interfaces – a control fusion problem. What happens when two unimodal interfaces receive contradictory commands? What happens when two contradictory actions result from individual unimodal interface interpretation of a particular human control interaction? What should be the conflict resolution model for such cases? Are the unimodal technologies sufficiently developed to meet the basic requirement of such multimodal interactions? These are just a few of many technical limitations that need to be addressed via further research investigation.

Brain Control Interfaces: Telepathic Interaction

One unimodal solution that could address the limitations of the RC joystick controller, unlocking higher control levels in the nCA model is telepathic mind control via brain-computer interfaces (BCI). The advancement in the development of brain-computer interfaces can be attributed to the understanding of the human neurology and the advancement in biomedical signal processing. These interfaces now find applications in various robotic systems. Telepathy is not necessarily a human-human interaction modality. Unlike the regular human interaction methods of speech and visual gesture, which have a "feel it and it happens" or "feel it as it happens" sensation, telepathic interaction emphasises a "think it and it happens" interaction. In natural gesture interaction, machines are taught to interact via natural human methods. However, in mind control interaction, the machines are being equipped to interact with humans in a way that humans do not normally interact with each other, a way that is beyond the normal human-human interaction capability. Humans skillful in the art of manipulating other humans via induction, deception, and psychological means such as the psychic, medium, or paranormal extrasensory perception or the sixth sense, may argue otherwise.

A challenge that the BCI system faces, even though it is still under development, is security. The mere thoughts of the human vulnerability to telepathic credit card hack elicits a fearful response, and a natural reluctance to the adoption of such technology (as a defensive strategy), until the security of such systems can be guaranteed (Martinovic et al., 2012). This means security is put at the forefront of this development rather than taking a back seat, as is often the case with new technologies. While this may suggest the possibility of absolute transparency in interaction, the breach of privacy and human right could be called to question.

Human Factors: Performance Measurement

A major consideration in the development of any system is how to measure the system's performance. According to (Lindquist, 1985), the effort-to-learn and the effort-to-use contribute to the usability of human-computer interaction (HCI) interfaces. The similarity of the developed interface to other known interfaces, the knowledge of similar interfaces, and the complexity of the developed interface, contributes to the effort-to-learn and usability of the developed system. Another performance metric of interest is probably a comparison of the interface's functionality with alternative existing interfaces. In this case, a comparison of the interface precision, accuracy, efficiency and effectiveness could be carried out. A performance benchmark could be drawn around the most generally accepted HCI interface such as the joystick-type controller for new aerobotic control interfaces.

Generally, a good interface design improves the overall user experience, makes interaction enjoyable, intuitive, easy to learn, and is functionally effective with an accepted level of precision and accuracy. As implied from (Chao, 2009), a good HCI is not about the human adapting to the limitation of the computer (as used to be the case), but about the computer adapting to the human's natural tendencies and expectations.

CONCLUSION

The discussion in this chapter focused on human aerobotic interaction interfaces. These are interfaces used by humans to interact with aerial robotic systems, which were introduced as 'aerobots', a term used extensively in this chapter. A number of these interfaces were presented. Most of the interfaces presented were unimodal, and a few were multimodal. In order to develop the premises for the discussion of the control interaction interfaces, the concept of autonomy in robotic systems was intensively explored. Because of the limitations of the existing autonomy models (such as the AFRL and ONR), in aerobot navigational control autonomy classification, a new autonomy classification model called the navigational control autonomy (nCA) model, was proposed and developed. It was compared with the other existing models. The nCA model differs from the other models in eliminating step-size increment, supporting fuzzy classification between levels, presenting levels in a pyramidal order, and appropriately mapping the limits of aerobotic control interfaces (such as the RC joystick controller). The nCA model was used as the basis for discussing the unimodal human aerobotic control interfaces, in this chapter. The nCA model identified a control void beyond the tier-one components of the nCA model. An attempt was made to address this by suggesting two solutions, a multimodal HHI-like interface and a unimodal

BCI interface. The limitations of each of these suggestions were also discussed. In addition to these, some human factors based performance measurement was recommended.

FURTHER WORKS

The need for intuitive control interaction interfaces for aerobots beyond tier-one components of the nCA autonomy model opens up an opportunity to explore smart novel interaction techniques. The next phase of this work would be exploring the practical application/implication of a multimodal HHI-like interface that combines speech with visual gestures in human aerobotic interaction. A working practical prototype would be developed to demonstrate the feasibility of this approach and to investigate the practical implications of this method. This would enable further investigation into some of the technical limitations presented in the discussion subsection of the HHI-like interface. Another potential solution to that may be investigated further would be the telepathic interaction via BCI interfaces.

REFERENCES

Abioye, A. O., Prior, S. D., Thomas, G. T., & Saddington, P. (2016). The Multimodal Edge of Human Aerobotic Interaction. In K. Blashki & Y. Xiao (Eds.), *International Conferences Interfaces and Human Computer Interaction* (pp. 243–248). Madeira, Portugal: IADIS Press.

Bastianelli, E., Nardi, D., Aiello, L. C., Giacomelli, F., & Manes, N. (2015). Speaky for robots: The development of vocal interfaces for robotic applications. *Applied Intelligence*. doi:10.100710489-015-0695-5

Blanz, V. (2007). A learning-based high-level human computer interface for face modeling and animation. *Proceedings of the 20th International Joint Conference on Artificial Intelligence, IJCAI 2007 - Workshop on Artificial Intelligence for Human Computing, LNAI* (*Vol. 4451*, pp. 296–315). Hyderabad, India: Springer Verlag. 10.1007/978-3-540-72348-6_15

Bolt, R.a. (1980). Put-that-there: Voice and Gesture at the Graphics Interface. *Proceedings of the 7th Annual Conference on Computer Graphics and Interactive Techniques - SIGGRAPH '80* (pp. 262–270). 10.1145/800250.807503

Cacace, J., Finzi, A., & Lippiello, V. (2016). Multimodal Interaction with Multiple Co-located Drones in Search and Rescue Missions. CoRR, abs/1605.0

Calella, J. C., Ortega, F. R., Rishe, N., Bernal, J. F., & Barreto, A. (2016). HandMagic: Towards User Interaction with Inertial Measuring Units. Proceedings of the *IEEE Sensors 2016,* Orlando, FL, USA. IEEE. doi:10.1109/ICSENS.2016.7808524

Cavett, D., Coker, M., Jimenez, R., & Yaacoubi, B. (2007). Human-computer interface for control of unmanned aerial vehicles. *Proceedings of the 2007 IEEE Systems and Information Engineering Design Symposium, SIEDS,* Charlottesville, VA, USA. IEEE. 10.1109/SIEDS.2007.4374014

Chao, G. (2009, March 8-10). Human-computer interaction: Process and principles of human-computer interface design. In *2009 International Conference on Computer and Automation Engineering ICCAE '09*, Bangkok, Thailand (pp. 230–233). IEEE. 10.1109/ICCAE.2009.23

Chaomin, L., Yue, C., Krishnan, M., & Paulik, M. (2012). The magic glove: A gesture-based remote controller for intelligent mobile robots. *Proceedings of the SPIE - The International Society for Optical Engineering*. doi:10.1117/12.912186

Clough, B. T. (2002). Metrics, schmetrics! How the heck do you determine a UAV's autonomy anyway? *Proceedings of the 2002 Performance Metrics for Intelligent Systems Workshop,* Gaithersburg, MD, USA (pp. 313–319).

Emotiv. (2014). Wearables for your brain. Retrieved from https://emotiv.com/

Fong, T., & Nourbakhsh, I. (2000). Interaction challenges in human-robot space exploration. *Proceedings of the Fourth International Conference and Exposition on Robotics for Challenging Situations and Environments* (pp. 340–346). http://doi.org/10.1145/1052438.1052462

Google. (2014). Google Project Soli: Your Hands Are the Only Interface You Will Need. Retrieved from https://atap.google.com/soli/

Green, S., Chen, X., Billinnghurst, M., & Chase, J. G. (2007). Human Robot Collaboration: An Augmented Reality Approach a Literature Review and Analysis. *Mechatronics*, *5*(1), 1–10. doi:10.1115/DETC2007-34227

Gupta, L., & Ma, S. (2001). Gesture-based interaction and communication: Automated classification of hand gesture contours. *IEEE Transactions on Systems, Man and Cybernetics. Part C, Applications and Reviews*, *31*(1), 114–120. doi:10.1109/5326.923274

Hammacher Schlemmer. (2015). The Mind Controlled UFO. Retrieved from http://www.hammacher.com/Product/Default.aspx?sku=84249&promo=Toys-Games-Remote-Control-Toys&catid=247

Harris, J., & Barber, D. (2014). Speech and Gesture Interfaces for Squad Level Human Robot Teaming. In R. E. Karlsen, D. W. Gage, C. M. Shoemaker, & G. R. Gerhart (Eds.), *Unmanned Systems Technology Xvi,* SPIE (Vol. 9084). doi:10.1117/12.2052961

Hernandez Jose, L., Kyriakopoulos, N., & Lindeman, R. (2002). The AcceleGlove a Hole-Hand Input Device for Virtual Reality. In *ACM SIGGRAPH Conference Abstracts and Applications*, 259.

Hill, A. F., Cayzer, F., & Wilkinson, P. R. (2007). Effective Operator Engagement with Variable Autonomy. *Proceedings of the 2nd SEAS DTC Technical Conference* (p. 7).

HobbyKing & Turnigy. (2015). Turnigy TGY-i10 10ch 2.4GHz Digital Proportional RC System with Telemetry (Mode 2). Retrieved from http://www.hobbyking.com/hobbyking/store/__58455__Turnigy_TGY_i10_10ch_2_4GHz_Digital_Proportional_RC_System_with_Telemetry_Mode_2_.html

Horrigan, F. A.ONR's Committee. (2000). *Review of ONR's Uninhabited Combat Air Vehicles Program.* Washington, D.C.: National Academy press.

Huang, C.-L., & Chung, C.-Y. (2004). A real-time model-based human motion tracking and analysis for human-computer interface systems. *EURASIP Journal on Applied Signal Processing*, *2004*(11), 1648–1662. doi:10.1155/S1110865704401206

Huhn, K., & Haewon, S. (2014). Evaluation of the safety and usability of touch gestures in operating in-vehicle information systems with visual occlusion. *Applied Ergonomics*, *45*(3), 789–798. doi:10.1016/j.apergo.2013.10.013 PMID:24231034

Kim, S., Seo, H., Choi, S., & Kim, H. J. (2016). Vision-guided aerial manipulation using a multirotor with a robotic arm. *IEEE/ASME Transactions on Mechatronics*, *4435*(c), 1–1. doi:10.1109/TMECH.2016.2523602

Lackey, S., Barber, D., Reinerman, L., Badler, N. I., & Hudson, I. (2011). Defining Next-Generation Multi-Modal Communication in Human Robot Interaction. *Proceedings of the Human Factors and Ergonomics Society Annual Meeting*, *55*(1), 461–464. http://doi.org/10.1177/1071181311551095

Lafleur, K., Cassady, K., Doud, A., Shades, K., Rogin, E., & He, B. (2013). Quadcopter control in three-dimensional space using a noninvasive motor imagery-based brain-computer interface. *Journal of Neural Engineering*, *10*(4), 046003. doi:10.1088/1741-2560/10/4/046003 PMID:23735712

Lavars, N. (2016). Thumbsteered drone leaves you with a free hand. Retrieved from http://newatlas.com/shift-drone-thumb/46188/

Leap Motion Inc. (2010). Leap Motion Controller. Retrieved from https://www.leapmotion.com/

Lee, T., Meng, H., Lo, W. K., & Ching, P. C. (2003). The State of the Art in Human-computer Speech-based Interface Technologies. *HKIE Transactions Hong Kong Institution of Engineers*, *10*(4), 50–61.

Lindquist, T. E. (1985). Assessing the usability of human-computer interfaces. *IEEE Software*, *2*(1), 74–82. doi:10.1109/MS.1985.230052

Ma, Y., Mao, Z.-H., Jia, W., Li, C., Yang, J., & Sun, M. (2010, May 9-12). Magnetic hand tracking for human-computer interface. *Proceedings of the 14th Biennial IEEE Conference on Electromagnetic Field Computation CEFC '10*, Chicago, IL, USA. IEEE. Doi:10.1109/CEFC.2010.5481499

Martinovic, I., Davies, D., Frank, M., Perito, D., Ros, T., & Song, D. (2012). On the Feasibility of Side-Channel Attacks with Brain-Computer Interfaces. *Usenixorg*, 1–16.

Milanova, M., & Sirakov, N. (2008, December 16-19). Recognition of emotional states in natural human-computer interaction. *Proceedings of the 8th IEEE International Symposium on Signal Processing and Information Technology ISSPIT '08* (pp. 186–191). IEEE. 10.1109/ISSPIT.2008.4775663

Mu-Chun, S., & Ming-Tsang, C. (2001). Voice-controlled human-computer interface for the disabled. *Computing & Control Engineering Journal*, *12*(5), 225–230. doi:10.1049/cce:20010504

Murphy, R., Shell, D., Guerin, A., Duncan, B., Fine, B., Pratt, K., & Zourntos, T. (2011). A Midsummer Nights Dream (with flying robots). *Autonomous Robots*, *30*(2), 143–156. doi:10.100710514-010-9210-3

Muscio, G., Pierri, F., Trujillo, M. A., Cataldi, E., Giglio, G., & Antonelli, G., … Ollero, A. (2016). Experiments on coordinated motion of aerial robotic manipulators. *Proceedings of the IEEE International Conference on Robotics and Automation (ICRA),* Stockholm, Sweden (pp. 1224–1229). 10.1109/ICRA.2016.7487252

Nagar, A., & Xu, Z. (2015). Gesture control by wrist surface electromyography. *Proceedings of the 2015 IEEE International Conference on Pervasive Computing and Communication Workshops* (pp. 556–561). doi:10.1109/percomw.2015.7134098

NeuroSky. (2011). Brainwaves. Not Thoughts. Retrieved from http://neurosky.com/biosensors/eeg-sensor/

Nimble VR. (2012). Nimble VR is joining Oculus. Retrieved from http://nimblevr.com/index.html

Nwana, H. S. (1996). Software Agents : An Overview. *The Knowledge Engineering Review, 11*(3), 205–244. doi:10.1017/S026988890000789X

Oviatt, S. (2002). Breaking the robustness barrier: Recent progress on the design of robust multimodal systems. *Advances in Computers, 56*(C), 305–341. doi:10.1016/S0065-2458(02)80009-2

Oviatt, S. (2003). Multimodal interfaces. In J. A. Jacko & A. Sears (Eds.), The Human-Computer Interaction Handbook: Fundamentals, Evolving Technologies, and Emerging Applications (pp. 286–304). London: Lawrence Erlbaum Associates, Incorporated.

Payne, K. (2007). *Autonomy - Proposal for a UK Route Map.* United Kingdom: BAE Systems.

Pfeifer, R., & Scheier, C. (1999). Embodied Cognitive Science: Basic Concepts. In R. Pfeifer & C. Scheier (Eds.), *Understanding Intelligence* (pp. 81–138). MIT Press.

Qing, C., Cordea, M. D., Petriu, E. M., Whalen, T. E., Rudas, I. J., & Varkonyi-Koczy, A. (2008). Hand-Gesture and Facial-Expression Human-Computer Interfaces for Intelligent Space Applications. *Proceedings of the IEEE International Workshop on Medical Measurements and Applications MeMeA '08* (pp. 1–6). doi:10.1109/MEMEA.2008.4542987

Reeves, L. M., Martin, J.-C., McTear, M., Raman, T., Stanney, K. M., Su, H., ... Kraal, B. (2004). Guidelines for multimodal user interface design. *Communications of the ACM, 47*(1), 57. doi:10.1145/962081.962106

Renitto, J. E., & Thomas, N. K. (2014). A Survey on Gesture Recognition Technology. *International Journal for Technological Research in Engineering, 1*(10), 1058–1060.

Reuters. (2015). Using The Force? No, it's an Apple Watch flying this drone. Retrieved from http://www.reuters.com/article/us-apple-watch-drone-idUSKBN0UE14Q20160101

Shah, J., & Breazeal, C. (2010). An Empirical Analysis of Team Coordination Behaviors and Action Planning With Application to Human-Robot Teaming. *Human Factors: The Journal of the Human Factors and Ergonomics Society, 52*(2), 234–245. doi:10.1177/0018720809350882 PMID:20942253

Sharma, R., Pavlovic, V. I., & Huang, T. S. (1998). Toward multimodal human-computer interface. *Proceedings of the IEEE, 86*(5 pt 1), 853–869. doi:10.1109/5.664275

Sholes, E. (2007). Evolution of a UAV autonomy classification taxonomy. *Proceedings of the IEEE Aerospace Conference Proceedings.* 10.1109/AERO.2007.352738

Soto-Gerrero, D., & Ramrez-Torres, J. G. (2013). A human-machine interface with unmanned aerial vehicles. *Proceedings of the 2013 10th International Conference Electrical Engineering, Computing Science and Automatic Control,* Mexico City, Mexico (Vol. 37, pp. 307–312). IEEE. http://doi.org/10.1109/ICEEE.2013.6676045

Thalmic Labs. (2013). Homepage - Gesture Control Has Arrived. Retrieved from https://www.myo.com/

Thalmic Labs. (2015). Myo - Real Life Applications of the Myo Armband. Retrieved from https://www.youtube.com/watch?v=te1RBQQlHz4

Turk, M. (2014). Multimodal interaction: A review. *Pattern Recognition Letters, 36*(1), 189–195. doi:10.1016/j.patrec.2013.07.003

Verma, P. (2016). Flying User Interface. *Adjunct Proceedings of the 29th Annual Symposium on User Interface Software and Technology UIST '16,* Tokyo, Japan (pp. 203–204). ACM. 10.1145/2984751.2984770

Yinghong, M., Zhi-Hong, M., Wenyan, J., Chengliu, L., Jiawei, Y., & Mingui, S. (2011). Magnetic Hand Tracking for Human-Computer Interface. *IEEE Transactions on* Magnetics, *47*(5), 970–973. doi:10.1109/TMAG.2010.2076401

KEY TERMS AND DEFINITIONS

Aerobot: A term derived from and referring to Aerial Robot.

Control: To operate a machine, a vehicle, or some other robotic systems such as an unmanned aerial vehicle, often with the aid of some device such as the RC joystick controller.

Gesture: Precise human body parts movement intended to communicate instructions to computers or robotic systems.

HCI: Abbreviation for Human-Computer Interaction.

Interface: A device used to operate a machine or robotic system e.g. joystick, keyboard, microphone.

Multimodal: A combination of two or more control interfaces, which could be, operated simultaneously or sequentially e.g. speech and gesture, keyboard and mouse.

nCA: Abbreviation for Navigational Control Autonomy, a concept introduced in this chapter.

Unimodal: A single control interfaces used to operate a machine, vehicle, or other robotic systems e.g. RC joystick controller.

This research was previously published in Smart Technology Applications in Business Environments edited by Tomayess Issa, Piet Kommers, Theodora Issa, Pedro Isaías, and Touma B. Issa, pages 39-62, copyright year 2017 by Business Science Reference (an imprint of IGI Global).

Section 3
Decision Support Systems

Chapter 7
Intelligent Fighter Pilot Support for Distributed Unmanned and Manned Decision Making

Jens Alfredson
Saab Aeronautics, Sweden

Ulrika Ohlander
Saab Aeronautics, Sweden

ABSTRACT

This chapter highlights important aspects of an intelligent fighter pilot support for distributed unmanned and manned decision making. First the background is described including current trends within the domain, and characteristics of a decision support system are discussed. After that a scenario and example situations are presented. The chapter also includes reflections of an intelligent fighter pilot support for distributed unmanned and manned decision making from the joint cognitive systems view, regarding human interoperability, and function allocation.

INTRODUCTION

This chapter aims at, identifying and systematically describe, system characteristics and contextual constraints of an intelligent fighter pilot support system for distributed unmanned and manned decision making. Specifically an analysis based on a literature review of state-of-the-art is presented as well as domain specific lessons learned from systems design within this field. Also, the chapter contributes more generally to guide and motivate students and researchers to perform similar contributions in adjacent domains of intelligent applications for heterogeneous system modelling and design.

In the technology-intensive domain of fighter aircraft it is important to design for human factors by regarding what is special about that specific context (Alfredson & Andersson, 2013), and to successfully apply cognitive design principles (Alfredson, Holmberg, Andersson, & Wikforss, 2011). For instance, human-centred automation guidelines could be applied to the fighter aircraft domain (Helldin, Falkman, Alfredson, & Holmberg, 2011). Also, design principles for adaptive automation and aiding have been

DOI: 10.4018/978-1-5225-8365-3.ch007

provided by Steinhauser, Pavlas and Hancock (2009). Already today and more so in future contexts, a fighter pilot has to interact with intelligent applications for heterogeneous systems where transparency is important (Helldin, 2014).

BACKGROUND

In the early days of aviation keeping the aircraft in the air was hard enough. The aviators were fully occupied by piloting. However, after some years of progress within the domain the aircraft could be better controlled in the air and there was also time to regard other activities. The pilots found time not only to aviate but also to navigate. In the military domain pilots could communicate what they had observed on the ground after they had landed, which could provide important reconnaissance information. Later, the history of military aviation is full of various types of aircraft performing very different missions in numerous scenarios. Fighter pilots of today make use of a flight control system or an "autopilot" or other functions to aid the piloting of the aircraft. Also, modern fighters are equipped with advanced sensor suits, high tech weapon systems, electronic warfare systems and many other subsystems that a modern fighter pilot has to manage. If you were ever given the opportunity to look at the instrumentation of a modern fighter aircraft performing a tactical mission you would probably see that instrumentation and displays to a great deal is used for tactical considerations and not only for flight instrumentation. The role of a fighter pilot has been transformed over time; from pilot to tactical decision maker.

This long term trend has led to current situations where a fighter pilot has to manage several tactical subsystems simultaneously and, at the same time, assess on going parallel tactical situations on the ground and/or in the air and make fast and important decisions to provide influence of the situations. Many situations are very applicable to naturalistic decision making as it were characterised by Klein, Orasanu, Calderwood, and Zsambok (1993). Situations are often complex, uncertain and dynamic characterised by high stakes, potentially risking both own and others life under extreme time pressure, calling for naturalistic decision making. Also, demanding situations may appear suddenly when performing almost any military mission, either it is an air-to-air mission, an air-to-surface mission or a reconnaissance mission. Also civil aviation can at times also be very demanding and dynamic at times, even though the military component of foes influencing the situation adds an extra need for specific decision making and corresponding decision support.

Today, and even more so in future aircraft systems, the pilot has to regard and interact with more and more information, if this trend will keep developing in the same direction and pace. This trend increases the need to support decision making. At the same time, there is a trend towards increasing abilities to support decision making. For instance, the computational power of modern avionics has increased substantially and the human-machine interaction technology has been improved, providing fighter aircraft engineers with new means of supporting fighter pilot interaction during demanding situations. There has been a long term trend towards increasing communication abilities between pilots not only including oral communication but also various means of data communication, allowing new means of communication between pilots as well as with command and control functions and more. Also, the cognitive ability for technical agents that the fighter pilot is in contact with, either direct by own manipulation and control, or indirect through another human, is increasing. Examples of technical agents that could influence the situation for a fighter pilot is a decision support system in the own aircraft as well as equivalent systems on other platforms, autonomous or highly automated unmanned aircraft or command and control decision support systems, as well as intelligent support systems for planning and evaluating the own mission. Also, some technical agents possess ability to support each other with data, including situation assessments, status of own system capabilities etcetera.

The further this trend is going the greater the abilities become to distribute decision making. One cognitive agent, human or technical, could very well pre-process information that another agent is making a decision based on, that yet another agent is executing etcetera. There are also tactical needs to distribute abilities between various cognitive agents. For instance, it might be more acceptable to expose unmanned systems to high risk settings at the same time as a human agent is assigned to judge the situation supported be decision support technology, all being distributed in the total system by location and/or in time. At the same time a parallel trend is emerging towards more and more competent unmanned aircraft, as the technology for unmanned systems is matured, possessing a wide spectrum of abilities. Hence, new opportunities to generate intelligent fighter pilot support for distributed unmanned and manned decision making emerge.

Within various civil sectors different trends that emphasize user subjective impression such as Kansaei Engineering (Nagamachi, 1989; 2011; Schütte, 2005) or user experience from the user-centred design tradition (Norman, Miller & Henderson, 1995; Hassenzahl & Tractinsky, 2006) have influenced design. Even though there have been variations in the actual design for military systems there has, for a long time, been an emphasis on performance as the utter importance for military systems. Even though concepts as mental workload and situation awareness has been used a lot to guide the design of military systems, their link to performance has been central and has been scrutinized for the fighter aircraft domain (Castor 2009; Svensson, Angelborg-Thanderz, & Sjöberg, 1993; Svensson, Angelborg-Thanderz, Sjöberg, & Olsson, 1997; Svensson & Wilson, 2002).

Also, it is of special importance that the performance of the total system is high rather than only the performance of individual subsystems, not to risk sub optimization. This is one of the reasons for why the joint cognitive system perspective (Hollnagel & Woods, 1983; 2005; Woods & Hollnagel, 2006) is a powerful base for analysis. However, for pragmatic reasons the system boundaries set for the analysis have to map the system level of the design process it is intended to support. Even though it is very important for a future user of a system that it supports the work that will be conducted in the applied context, not all these conditions are known when the design is performed and not all conditions could be influenced by the design decisions. Not knowing all design conditions and their relevance when conducting the design is a problem that has to be handled during the design and is sometimes referred to as the envisioned world problem (Dekker, 1996; Dekker & Woods, 1999). The fact that conditions other than the ones considered in the design process affect end users, can be regarded in the design process by assumptions and analysis, but cannot be controlled by system design alone. Conditions concerning the human, the methods used in the applied context, organizational factors and technology developed by other design organizations are all examples of factors of importance for the performance but with little or no influence from the system design process. To regard this broader scope, however, the design could be put in a wider context, such as the NATO human view presented by Handley and Smillie (2008).

Apart from striving to optimize the system design for users of the system currently being designed, the developing organisation may also have long term technology development goals. There has, for instance, been a long term trend towards highly automated systems and autonomy where the design organisation may have ambitions and visions. One way to handle this is to develop a strategy that aims towards that all products and product versions should be usable as well as to cope with the long term trend towards partial or fully autonomous systems. The strategy could include an automation strategy that could also assist in designing which functions to automate and which not to automate. To optimize the system design for users of the system the strategy could regard which activities humans are better at, which activities technical systems performs better, and which activities are best performed together by humans and technical systems, by thoughtful function allocation. Also, this function allocation does not have to be static, but designed to be dynamic. If the automation strategy involves guidance on how the dynamics of the function allocation should be designed it is more likely that the system design is more useful in aiding the user achieving high performance.

THE DECISION SUPPORT SYSTEM

Modern fighter aircraft are equipped with a range of sensors in order to give the pilot a view of the tactical environment and to support the launching of weapons. The information from the sensors is collected by the decision support system (DSS) which identifies and assesses the potential targets and threats and provides the analysed information to the pilot. The DSS controls the sensors and decides in what direction they should search and for what kinds of objects. The DSS helps the pilot to make decisions about actions, for example to dispense countermeasures and provide recommendations for delivery of weapons. The collected picture of the current situation the pilot receives from compiled information is often referred to as situation awareness (SA). It is crucial that the pilot can establish and maintain a high degree of SA in stressful dynamic situations with short time to take decisions.

SA and mental workload are widely used concepts among developers and system designers of human-machine interaction (HMI). Endsley (1988) defines SA as "The perception of the elements in the environment within a volume of time and space, the comprehension of their meaning, and the projection of their status in the near future". This construct is commonly used to explain why things went wrong and lack of SA is concluded as the cause of many accidents. Dekker and Hollnagel (2004) raise concerns about this use and label the constructs as "Folk Models". They argue that the concept is intuitively understood but not defined in a way that can be scientifically proven. Rather, one label is substituted for another, instead of trying to decompose into measurable parts. They propose that operator performance should be used as measure to validate design instead. Parasuraman, Sheridan and Wickens (2008) respond that these constructs do indeed have a scientific base and argue that the SA construct is useful in cognitive engineering. They argue that SA represents a continuous diagnosis of the state of a dynamic world. There is a "truth" and the user´s SA tells something about how much of this truth that is known to him, which has also been argued by Alfredson (2007). The term SA is widely used when formulating requirements for the development of interfaces and to describe the desirable performance of systems, such as DSS. In the case of fighter aircraft and DSS, SA could be regarded from two perspectives. The first is the picture of the situation the pilot gets from the sensors through the DSS, assuming that the DSS collects all the information that needs to be collected. In this case the pilot´s SA is depending on how good the DSS is to convey the information and this largely depends on the display interface. But it is not certain that the sensors have picked up everything there is to know. An object can be present but hidden and suddenly appear and surprise the pilot, by for example stealth technology. In this case it is rather the technical system which have faulted, not the pilot. Can the pilot be blamed for missing something the sensor system did not pick up?

The DSS is aiding the pilot in the tactical tasks of the mission; it is not connected to the actual flying of the aircraft. The complete SA should therefore be considered to include both awareness of the status of the aircraft (fuel etc.), the navigation and flying of the aircraft, knowing where the team members are and what they are doing and knowledge of the potential targets and threats in the environment. DSS supports the pilot in a part of his complete task. The discussed system is a hypothetical DSS in a fighter aircraft; it is simplified, and modelled for the purpose of this study.

The sensors on the aircraft are typically radar, electronic warfare, and infrared based systems that either actively (by sending out and receiving signals) or passively (by picking up signals) detect objects in the environment. This is performed at a distance greater than the human eye can see; therefore this kind of scenario is labelled beyond visual range (BVR). Here we will not get into the sensor technology, only presume that information about objects is obtained from the sensors and that the sensors need to be controlled and optimized in order to deliver the best possible measurements. Also, measurements made by other aircraft in the same tactical unit and the command and control (C2) will enter the DSS via data link.

The information that enters the DSS from the sensors typically consists of measurements of distances and bearings to various objects. The DSS fuses the information from the different sources to establish a fused object to make further analysis on. Then the DSS makes an estimate of what kind of aircraft it is most likely to be. Objects can for instance be classified as fighter, attack or other (transport, surveillance aircraft). The DSS also tries to establish the identity of the aircraft such as for example friendly, neutral, unknown or hostile.

The information is typically displayed to the pilot as a symbol on a digital moving map. The graphics on the map should optimally contain all the information the pilot needs.

In Figure 1 the flow of data and commands between the pilot and the sensors via the DSS is pictured.

The top functionalities for this system would be:

Figure 1. Overview of the analysed system

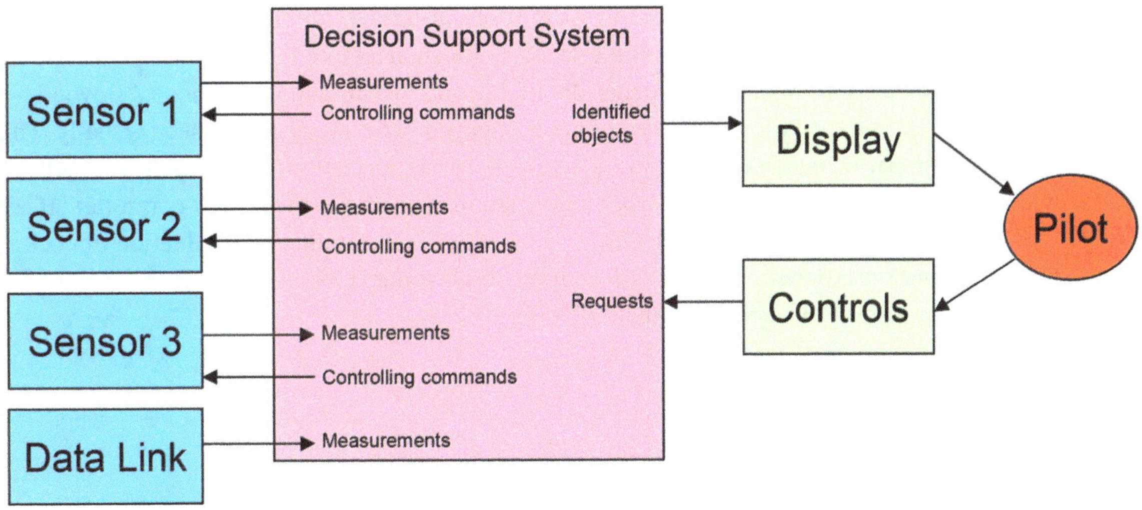

- Fuse data from different sensor sources
- Classify objects, (fighter, attack or other)
- Identify objects, (friendly, hostile, unknown)
- Assess situation, prioritize from most hostile object
- Control sensors for further measurements

It can be noted here that both the pilot and the DSS make assessments of the situation and both the pilot and the DSS can select objects for closer monitoring. But, as a rule, the pilot should always have the last word and the system adapt to his requests. Also, it should be noted that the term DSS usually refers to a system which helps in sorting out alternative ways of acting and aid the user in decision making. The DSS in this context should do a little more than that. It optimizes the use of the sensors and controls them according to pilot command in detail in order to extract as much information from the environment as possible. Aiming at high level interaction between the DSS and the pilot, a higher level of automation could influence the pilot to think more higher-level thoughts that has been shown in a commercial aviation context by Casner and Schooler (2014).

SCENARIO AND EXAMPLE SITUATIONS

Ollero and Maza (2007) listed the following civil application areas for unmanned aerial vehicles (UAV): "Aerial photography and cinematography", "Aerial mapping", "Meteorology", Environmental monitoring", "Agriculture and forestry", "Inspection", "Law enforcement and security applications", Disasters and crisis management", "Fire fighting", "Traffic surveillance", "Communications", and "Civil engineering", but also concluded that currently the main UAV applications are within the defence domain. Also in the future there will be defence related applications for unmanned aircraft. Even though the future scenarios are not known in detail, they will probably be related to the scenarios of today, for manned and unmanned aircraft.

There are a lot of possible scenarios were an intelligent fighter pilot support for distributed unmanned and manned decision making could be used. Below we are presenting an example scenario and discussing how it could be modified to represent one possible scenario for such an intelligent fighter pilot support. An example scenario could be an air-to-air scenario with beyond visual range conditions. "Beyond Visual Range (BVR) means a scenario where the enemy is engaged before they can be seen visually. This is possible due to the performance of the sensors available to the formation, either their own sensors (primarily the aircraft radars), and the sensors and information available to the fighter controller" (Castor, 2009, p. 21). Helander and Skinnars (2000a) used cognitive walkthrough for evaluation of cockpit design. The following breakdown of subtasks for an air-to-air combat beyond visual range has been identified and presented by Helander & Skinnars (2000b):

1. Waiting in aircraft ready to take off
   1.1 Sitting in aircraft waiting
   1.2 Start within one minute in specified direction
2. Receive order from Command-Control Center
3. Identify connection with Command-Control Center
   3.1 Receive uuencoded information on Tactical Display

3.2 Receive auditory information
3.3 Group Commander verifies mission
4. Take off from base
5. Follow directional commands given by Command-Control Center
6. Follow geographical command given by Command-Control Center
7. Determine target
   7.1 Commander coordinates mission
   7.2 Commander communicates mission
   7.3 Prioritize goals for own delivery of weapons
8. Achieve advantageous position
   8.1 Determine advantageous position
   8.2 Navigate to advantageous position
   8.3 Verify advantageous position
9. Prepare missile for target
   9.1 Verify target position
   9.2 Use radar information from outside source
   9.3 Use own radar to measure naval targets
10. Deliver missile
    10.1 Select advantageous time for delivery
    10.2 Deliver missiles
11. Supervise situation
    11.1 Verify results from weapons delivery
12. Finish mission
    Modify plans for return
13. Navigate to base assigned by Command-Control Center
    13.1 Check navigation points (position, radar or visual)
    13.2 Communicate with Traffic Control
    13.3 Select approach mode for landing
    13.4 Land aircraft
    13.5 Report mission
14. Select strategy to avoid attack against own aircraft
    14.1 Divert maneuver and drop weapons
    14.2 Let the aircraft drop and swing towards the ground
    14.3 Escape
    14.4 Disturb through
    14.5 Close range Air-to-Air, see section below
    14.6 Attack bomber aircraft using IR missiles, see Air-to-Surface below (pp. 16-17)

A complementary view containing some relevant technical and tactical parameters of air combat is suggested by Johansson (1999):

Jamming, electronic countermeasures and radar warning receiver

- Range of own jamming versus enemy radar
- Radar warning receiver range and accuracy

Radar
- Radar modes (e.g. Track While Scan, Continuous Wave)
- Aircraft radar range and angle of coverage

Weapon characteristics
- Mean speed
- Maximum range
- Missile seeker opening distance

Aircraft behaviour and tactics
- Absolute and relative altitude of aircraft in engagement
- Absolute and relative speed of aircraft in engagement
- Thrust
- Geometry, i.e. the relative position and ranges of the aircraft within the twoship or fourship to the enemy, and to other friendly aircraft, e.g. attack aircraft that are being escorted
- Aggressive or defensive stance and risk taking
- Intentions and Rules of Engagement for both sides
- Active versus passive use of radar

Numerical superiority
- Number of own aircraft
- Number of enemy aircraft

Command and Control
- Radar coverage for air surveillance radars (i.e. other friendly radars on the ground or in the air)
- Fighter controllers' threats classification capabilities

Loadout on aircraft
- Fuel
- Weapons loadout (quoted from Castor, 2009, p. 24).

There are various ways of performing BVR-combat, using various tactics and strategies and various systems and subsystems. However, as expressed by Castor (2009): "Regardless of engagement scenario the pure tactical goal would be to shoot down as many as possible of the enemy aircraft, while not getting shot down yourself" (p. 21). When new systems and subsystems are developed, often also new tactics are evolved in parallel. The use of manned and unmanned aircraft together is one example of that. Ollero and Maza (2007) has described challenges with multiple heterogeneous unmanned aerial vehicles, and more specifically Alfredson, Lundqvist, Molander, and Nordlund (2010) has described challenges related to the collaboration between a manned fighter aircraft and an unmanned aerial reconnaissance vehicle. Future fighter pilots have to interact, direct or indirect, with manned as well as unmanned fellow air vehicles in various scenarios and situations. How BVR-combat will then be performed is highly dependent on the capabilities of the future system and one speculation is that the potentials of distributed decisions will be explored further and that several users and technical systems will act together. The following situations, illustrated in Figure 2 and Figure 3, are examples that can be considered for analysis. They represent situations that are both tactically important and could also help illustrate the performance of the total system. The first case includes a team where one or several members are not using their radar in order not to emit energy and reveal themselves. One team member is relaying on the information he receives from the rest of the team via data link. It is crucial in this situation that the other team members are aware of this fact and realize that they are contributing with important information for this team member, for the situation depicted in Figure 2.

Another example is when a team is performing reconnaissance, see Figure 3. The total area of interest for the team is marked on the map and it should also be noted that it is a volume in space not just a flat area. How does each team member know what area to search next and how can the team best keep track of where the others recently have been searching?

THE JOINT COGNITIVE SYSTEMS VIEW

One way to describe users and technical systems is to regard them as a joint cognitive system – JCS (Hollnagel & Woods, 1983; 2005; Woods & Hollnagel, 2006). Agents in such a JCS could very well be interdependent to fulfil the tasks of the JCS that could act as one unit, although distributed in space. This chapter explores how the JCS perspective could be applied to an intelligent fighter pilot support for distributed unmanned and manned decision making. Later, what to automate is discussed. Such an analysis will benefit from distinguishing between various types of automation suited for specific tasks with known user cognitive demands (Taylor, Reinerman-Jones, Szalma, Mouloua, & Hancock, 2013). Similar to a study from the automotive domain on context adaptable driver information (Davidsson & Alm, 2014), this chapter will describe what do whom need and want when.

Often researchers that are applying the JCS approach argue that traditionally the focus in the field of Human Factors engineering has been to the interface between the user and the machine/computer (HMI) and therefore the purpose of the total system is at risk to be lost. Too much energy is spent on designing the interface instead of trying to optimize the output from the system. The JCS approach shifts the attention towards the functionality and performance. The JCS is not defined by what it is but rather by what it does.

Figure 2. Example situation. Team measurement (intersecting).

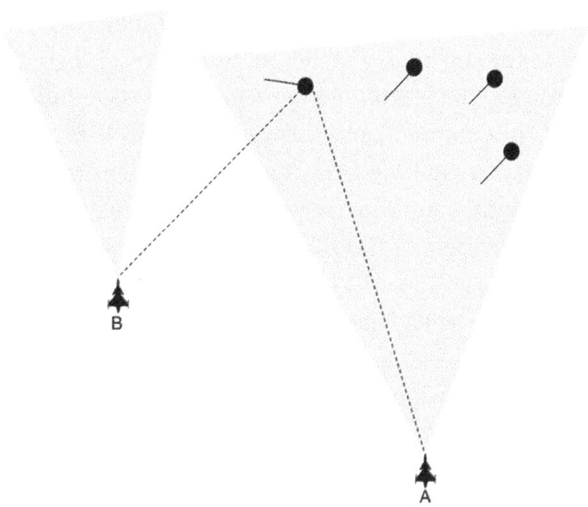

Figure 3. Example situation. Team reconnaissance.

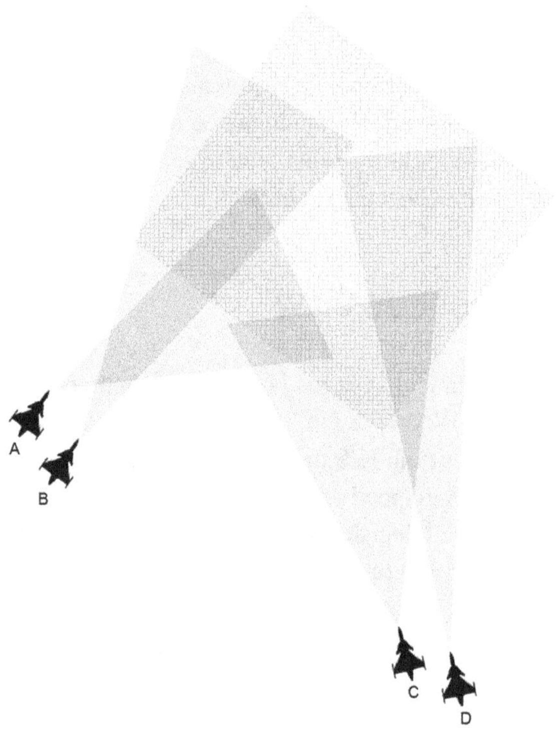

A JCS can be defined in different ways depending on the analysis but at least it consists of a human user (who is regarded as a cognitive system) and an artefact. A cognitive system is defined by Hollnagel and Woods (2005) as "a system that can modify its behaviour on the basis of experience so as to achieve specific anti-entropic ends" (p.22). The term anti-entropic here refers to the systems ability to resist the increase in entropy, i.e. the amount of disorder. An artefact is defined as something made for a specific purpose.

What parts of the system that should be included in the JCS analysis can be different depending on the context but typically the things that can be controlled should be included. An example is when analysing traffic the driver and the car are obviously included. But the roads can also be embraced since they are possible to control and change. The weather in the other hand is excluded. Even though the weather does influence traffic it can not be controlled and changed. The JCS in this case consists of the pilot, the aircraft and the DSS together. In this view the DSS is an artefact that mediates the relationship between the sensors and the pilot. The interaction is considered to be with the world through the artefact rather than with the artefact itself. The borders of the JCS can also be expanded to include several cooperating aircraft depending on the scope of the analysis. For example, the analysis can be done for a team of aircraft some of which can be unmanned and the JCS in this case would consist of several agents.

The artefact can be categorized depending on how transparent it appears to the user and how the user perceives it. In the transparency dimension there are two extremes, hermeneutic or transparent. In the case the artefact is transparent it is experienced by the user as a part of him rather than as a part of the application. It is easy for the user to build and maintain a mental model. The artefact can either serve as an extension of the user´s body and mind and therefore the relationship is considered to be embodied.

In this case the artefact is transparent and it is experienced by the user as a part of him rather than as a part of the application. The other dimension the artefact can be analysed in is as an interpreter and in this role it does all the communication between the user and the process. The artefact is opaque or non-transparent and is experienced more as being part of the process rather than the operator. In the case with the DSS the pilot cannot have any experience of what is going on than through the artefact since the scenario is taking place at a distance out of his sight and a lot of data fusion and identification is done by the system. This is considered to be an extreme case when the operator is dependent of the artefact, and the relation is labelled hermeneutic and means that the operator must go through the artefact to get to the world (Hollnagel and Woods, 2005). Another dimension of viewing the artefact is as either a tool or prosthesis. A tool is an artefact that amplifies the user´s cognitive functions while prosthesis rather takes over functions from the user. The DSS is a complex system and it will substitute the pilot in the control of the sensors and alleviate in the fusion of data, therefore it will be regarded as prosthesis. The transparency-exchangeability relation as proposed by Hollnagel and Woods (2005) distinguishes between two dimensions:

- Transparency: Hermeneutic vs. Embodiment
- Exchangeability: Tool vs. Prosthesis

The DSS is situated towards hermeneutic and prosthesis, that could be described as low transparency and high exchangeability. The combination of hermeneutic and prosthesis relation makes it challenging to develop a system that the user will understand and perform well with. It is probably advisable to try to design and construct the system in order for it to be perceived by the user as more of a tool with more transparency even if it is "actually" a hermeneutic prosthesis as in the case with the DSS. Our goal of the design would be to try to move towards tool and embodiment, striving for high transparency and low exchangeability.

The hermeneutic relation implies that is not possible for the pilot to have any experience of what is going on besides than through the DSS. However, one way for pilots to go "beside" the DSS would be to look at raw sensor data and try to make the assessments by themselves. This means that they would want to have access to the direct measurements from each sensor as well as the fused objects. The designers will then have to consider if this shall be provided. The risk being that the system gets even more complicated with many low level functions.

HUMAN INTEROPERABILITY

To regard the contribution from the design process in a wider context other aspects than technical has to be included in the analysis. Often new technology that is to be developed has very high expectations to meet increasing operational needs of the future that could not always be met. From studies of the command and control domain Persson (2014) has shown that several expected advantages of future technologies have not been met. However, relating technology to a wider context including human aspects when developing the technology may both adjust the expectations of the technology to more realistic levels as well as increasing the actual benefits from the technology by develop it considering for instance human operability, A network model for human interoperability proposed by Handley (2014) including the human view (Handley & Smillie, 2008) is applying the eight aspects: *Concept, Constraints, Functions, Roles,*

Human network, Training, Metrics, and *Human dynamics*. The human view "...provides an integrated set of products that can be used to inform and influence system design, especially for network-enabled systems" (Handley, 2014, p. 351). One of the main benefits for this kind of analysis is that it potentially aids human interoperability into systems design and support collaboration. Below, these aspects are discussed in the light of intelligent fighter support for distributed unmanned and manned decision making.

The human view product *Concept* is about the human in relation to operational demands and system components. This is a very high-level representation, and for an intelligent fighter support for distributed unmanned and manned decision making it is essential to have an idea of human contributions to decisions in this context. Figure 4, below, provides a depicted overview of cognitive agents, amongst some are supposed to be human. Cognitive agents in this context could for instance be an unmanned aircraft or any human interacting with that aircraft or a decision support system either on a manned aircraft, on the ground or distributed between several aircraft etcetera. Operational demands in this kind of setting are very much determined by the concept of operation in applied scenarios. As discussed above there are a variety of potential scenarios to regard for this kind of systems, amongst a BVR-combat is only one. However, it is assumed that the role of the human is important for total system performance and that several humans have to cooperate together with unmanned systems and intelligent subsystems for mission success. The abilities of system components are highly dependent on the technologies used in this systems and subsystems. However, some general abilities could be defined such as the ability for a technical system to be used as means of communication between human agents, or the ability for a technical system to aid a distributed decision process together with other technical systems and/or humans, or to autonomous execute tasks etcetera. Later in this chapter the ability and potential benefit of distributing tasks through function allocation will be further discussed.

The human view product *Constraints* is about the capabilities and limitations of the human in the system. Sub views presented by Handley and Smillie (2008) are: Manpower Projections, Career Progression, Establishment Inventory, Health Hazards, Human Characteristics, and Personnel Policy. Several of these categories contain aspects that could not at all, or only indirectly, be controlled by design decisions in a technical design process. However, it is important to regard these aspects also in the design process by making assumptions and analysis of potential future constraints of relevance for the technical systems under development. The human characteristics including physical characteristics of a pilot performing a mission is very relevant, if not crucial to regard in the design process. In fighter aircraft of today mental workload demands are occasionally very high and information overload could affect the performance negatively. A context where more information is available and decisions are designed to be distributed between several cognitive agents could very well add to mental demands if this is not supported by technology, training and organisational measures. Aspects to regard to design for human factors in development of fighter aircraft has been addressed by Alfredson and Andersson (2013) and what is special about a fighter pilot compared to many other users is that the pilot community is in some sense homogeneous, due to for instance selection and training. Apart from mental demands fighter pilots also have to cope with high G-loads, visually demanding environments and more, which also limits the potential design of a fighter aircraft.

The human view product *Functions* is about functions decomposed into tasks that have been allocated to the human, including a justification for why a human is performing the task and not a technical system. Later in this chapter function allocation is going to be discussed more separately. Apart from the actual function allocation this human view product includes the development of interface design guidelines dependent on the task requirements. Guidelines for human-centered automation have been studied for

Figure 4. The figure depicts an overview of cognitive agents in a joint cognitive system that supplies an intelligent pilot support for distributed unmanned and manned decision making.

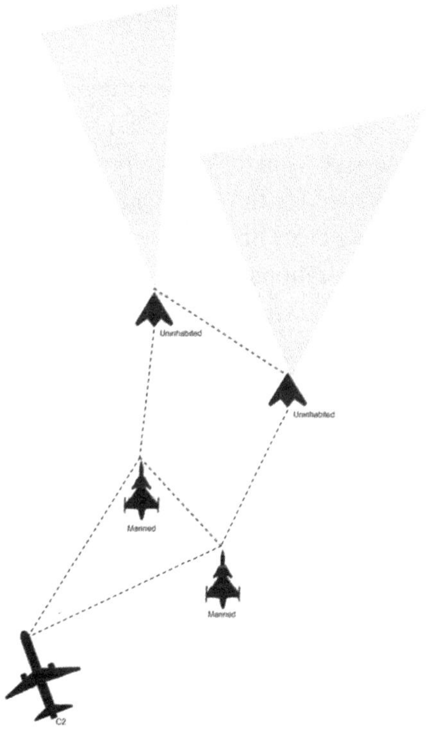

the fighter aircraft domain (Helldin, 2014; Helldin & Erlandsson, 2012; Helldin, Falkman, Alfredson, & Holmberg, 2011). It is likely that specific guidelines have to be defined or current guidelines have to be modified. For instance, guidelines probably have to be adjusted concerning principles of interaction for dynamic function allocation between various cognitive agents, as the competences of these agents evolve.

The human view product *Roles* is about a job function defining specific behaviour, and includes: Responsibility, Authority, Competencies, and Multiplicity (Handley & Smillie, 2008). Several of these categories contain aspects that could not at all, or only indirectly, be controlled by design decisions in a technical design process. However, it is important to regard these aspects also in the design process by making assumptions and analysis of potential future constraints of relevance for the technical systems under development. It is important to map the roles to the function allocation, and also to decide if roles are to be regarded as static or if they could change over time.

The human view product *Human Network* is about the human-to-human communication patterns, including team aspects, but also technology for distributed cognition, shared awareness etcetera. The human view product human network also includes type of interaction, such as coordination, supervision etcetera. Although much of the human-to-human communication only use technology for communication purposes there might also be indirect influence from technology to human-to-human communication. For instance reported Lyons and Stokes (2012) indications of that human-human reliance could become biased towards lower behavioural reliance on human aids when confronted to conflicting information from automation. For intelligent fighter support for distributed unmanned and manned decision making it is likely that humans will have to collaborate to achieve mission objectives. Also, there will be someone formally in command. However, apart from the formal structure there might be informal structures of a human network influencing the mission outcome and even formal and/or informal networks between human and non-human cognitive agents, possible expanding the scope of this human view product, if not handled in another human view product.

The human view product *Training* is about how aspects concerning training will affect the human. Most training aspects are partly or only indirectly controlled by design decisions in a technical design process. It is important to regard these aspects in the design process by making assumptions and analysis of potential future constraints of relevance for the technical systems under development. For instance, functionality for embedded training has a clear link between the design during the development process and the use of the future product in its operational use. Especially for intelligent fighter pilot support for distributed unmanned and manned decision making the implications to train for missions with mixed manned and unmanned components has to be regarded.

The human view product *Metrics* is about human factors metrics and human related performance criteria, target values etcetera. Involving human factors assessment, this product is broad and assessments may be effected by concerns in any of the other human view products. Recently, Berggren, Nählinder, and Svensson (2014) described important aspects of assessing command and control effectiveness. Regarding intelligent pilot support for distributed unmanned and manned decision making it is important not only to regard human factors assessment at the level of human system interaction for all humans involved, but also include assessment of behaviour and performance of the total system, including a joint cognitive system approach.

Last, but not least, the human view product *Human Dynamics* is about change over time and may concern any of the other products. Handley and Smillie (2008) provides the following examples of elements that the human view product Human Dynamics may include: states and state changes, conditions, time units, and performance measures. Scenarios belong to this human view product and for intelligent fighter pilot support for distributed unmanned and manned decision making it is important to regard not only the operational scenarios, such as the BVR-combat case described above, but also various use cases and examples of how a flexible system may perform in different ways in given operational scenarios.

FUNCTION ALLOCATION

Function allocation design challenges has to be regarded for an intelligent fighter pilot support for distributed unmanned and manned decision making, to achieve a high performing total system. There is a long research tradition in using knowledge of what humans are good at and machines are good at for achieving better design. For instance Fitts (1951) suggested principles to allocate functions between man and machine. Even though substitution-based function allocation methods have been criticised (Hollnagel, 1999; Dekker & Woods, 2002) it is still contributing to design in various domains. Also, dynamic function allocation described as levels of automation (Sheridan & Verplank, 1978; Taylor, 1997) is still contributing to design in various domains. However, Defense Science Board (2012), suggests to "abandon the debate over definitions of levels of autonomy and embrace a three-facet (cognitive echelon, mission timelines, human-machine system trade spaces) autonomous systems framework…" (p. 2).

Five key requirements for effective function allocation proposed by Feigh and Pritchett (2014), in a model of human-automation function allocation (Pritchett, Kim, & Feigh, 2014) are described below, that is: 1) "*Each agent must be allocated functions that it is capable of performing*", 2) "*Each agent must be capable of performing its collective set of functions*", 3) "*The function allocation must be realizable with reasonable teamwork*", 4) "*The function allocation must support the dynamics of the work*", and 5) "*The function allocation should be the result of deliberate design decisions*". Each of these aspects could be applied to the context of an intelligent fighter pilot support for distributed unmanned and manned decision making. By such an approach all cognitive agents, human and technical systems are analysed together which could be perceived as unsophisticated for any human that perceives human abilities and behaviour to be very different from those of technical systems. The analysis will for instance risk describing human characteristics in technical terms and thereby reduce humans into information processing agents. Also the opposite critique could be expressed, risking anthropomorphism when a technical system and its behaviour is analysed together with human agents. However, first, this analysis and the requirements below is not intended to mirror every aspect of a system in use, but is used to guide design decisions. Secondly, the reason for applying the requirements is that there are differences between the agents and if there are significant differences it is more challenging to develop methods and requirements that could be applied to all these various agents. However, it is due to the fact that there are significant differences between the agents that methods and requirements are needed to support the design process.

The first requirement proposed by Feigh and Pritchett (2014); *Each agent must be allocated functions that it is capable of performing*, is about each particular function relative to the capabilities of the agent that has been allocated the function. In the tradition of Fitts (1951) lists has described what men are better at and what machines are better at, in general. Specifically for an intelligent fighter pilot support for distributed unmanned and manned decision making an estimation of abilities of agents in that context has to be performed. In this context it is not only of interest which agent is better at performing a task, but also the risk/cost of performing a task has to be regarded since many missions may include substantial risks. For instance, if the performance of a task is associated to an increased risk of being shot down, there would be an incentive to allocate such a task to an unmanned rather than a manned aircraft as far as possible. Also, from either an economical or military strategically point of view it might be that some assets, either a flying platform itself or its subsystems or carried loads, may be very valuable and thereby the risk level of that entity should be minimized in comparison to other entities.

The second requirement; *Each agent must be capable of performing its collective set of functions*, is about understanding the agents limited resources. Even though one agent might be the best agent to perform the task according to its characteristics (e.g. according to a list of what men are better at and what machines are better at) it might not be prioritised or might not even be possible for that agent to perform the task due to the collective set of functions. For instance, workload might constrain the performance of a task, either the task might risk being performed badly or not performed at all due to high workload. For a technical system it might be limitations in processing capability or lack of memory or any other technical constraint, whilst for a human agent it might be an overload of some mental capability. Specifically for the fighter domain the visual demands of a fighter pilot has been very high at times. Team spatial working memory has been shown to affect team performance in networked supervisory control of unmanned air vehicles (McKendrick, Shaw, de Visser, Saqer, Kidwell, & Parasuraman, 2014). Specifically for an intelligent fighter pilot support for distributed unmanned and manned decision making it is also important to regard the spatial positioning of the aircraft relative each other before allocating the tasks, since some tasks may be better performed at one position rather than another. Also, in some military missions each agent's ability to perform a collective set of functions might vary a lot, due to for instance radar jamming or decreased functionality due to enemy fire etcetera.

The third requirement; *The function allocation must be realizable with reasonable teamwork*, is about teamwork functions that coordinates the work across the agents. In this sense human agents as well as technical systems are team members that have to communicate, interact and coordinate their actions. Some tasks need teamwork to be performed, whilst other tasks may be better performed by teamwork although they might have been possible to perform to some degree by a single agent. Specifically for an intelligent fighter pilot support for distributed unmanned and manned decision making it is important to consider the spatial positioning of the aircraft also for teamwork considerations. For instance, the distance between two aircraft that both use a passive sensor, such as an infra-red-search-and-track (IRST) sensor, is very important for determining the position of an object of interest. Also, the relative positioning of aircraft might be of importance for tactical reasons such as establishing potential threats at an enemy over a longer period of time etcetera.

The fourth requirement; *The function allocation must support the dynamics of the work*, is about the on going response to the work environment for each agent as well as for the total system. When functions are interdependent this could be more or less obvious and sometimes it is not obvious at all, but the function allocation should support the agents to cope with the dynamics of the situation. An alternative approach that could complement the striving of allocating functions to when they are needed is to gain resilience by increasing a team's ability to perform during unexpected conditions. Specifically for an intelligent fighter pilot support for distributed unmanned and manned decision making it is important to regard fast and drastic changes in the work setting, to cope with events such as aircraft that has been shut down or the sudden appearance of stealth aircraft on the battle scene etcetera. Also, the rules of engagement (ROE) that defines what to do during various circumstances has to be regarded at the same time as the function allocation supports the dynamics of the work.

The fifth and last of the requirements proposed by Feigh and Pritchett (2014); *The function allocation should be the result of deliberate design decisions*, is about integrating function allocation into the design process. In general it is important to make deliberate design decisions concerning function allocation early in the design as these decisions often influence the choice of technology to be used and very much influence criteria for human-machine interaction. Specifically for an intelligent fighter pilot support for distributed unmanned and manned decision making it is important to consider the very long life time of aircraft as well as the long cycles for development of new products compared to many other domains. Also, it is important not only to make deliberate design decisions on the function allocation of a specific aircraft, manned or unmanned, but also on the total system level of several cognitive agents collaborating using several manned and unmanned aircraft.

DISCUSSION

The purpose of the analysed DSS is to perform tasks for the pilot such as object search and identification. Both system complexity and time constraints would make it very difficult if not impossible for the pilot to both manually control the sensors and conclude from the data what kind of object is being measured. It would be very difficult for the pilot to go into the details and set up the sensors in the most optimal way. Not long ago the radar was to a large extent controlled by the pilot on a fairly low level and the raw data was plotted on a screen. In this setting the pilot probably felt that he was in control of the process and understood what the system was doing. When he had trouble locating and identifying an object he tried a different mode or perhaps changed some parameters. With the DSS the aim is to achieve a more capable system working at a faster pace with several channels of information. The challenge is to keep the pilot feeling in control and knowing what is going on in the system and still achieve powerful and fast automated functions by the DSS.

The division of decisions between the DSS and the pilot could include a strategy such as that the DSS handles for instance rule based decisions or suggestions, which can be predicted and programmed in the system (automated). The pilot could take the more difficult decisions, such as if to launch a weapon. He is able to adapt to the changing environment which is difficult for the DSS to manage. The type of decisions can serve as a guide to what functions should belong to the pilot and the DSS respectively.

The main purpose of the DSS is to measure objects with such good quality that weapons can be launched at hostile objects. Identification and classification of the objects is a key capability of the JCS. The performance of the DSS in the identification and classification of the objects is crucial. If the pilot experiences that the performance is bad there is a risk that he will lose trust in the DSS and perhaps try to make the assessment himself without the aid of the DSS. It is probably not possible to have total transparency in how the DSS controls and takes the decisions. Sometimes sophisticated algorithms are used and they can for example rely on Bayesian probability which is perhaps not totally intuitive. But in the case of uncertainty a measure to indicate this to the pilot could be provided.

CSE and the JCS view stress the need to design systems starting from the functionality and the purpose of the JCS. In this view it becomes clear why the system is being built and the requirements on the total system including the user can be identified. However, at some point during the development the need to make an interface between the artefact and the user still arises. Guidelines for what to automate and design style guides helps in the development.

CONCLUSION

To conclude, some long term trends have been identified for the domain. The role of a modern fighter pilot has transformed from pilot to tactical decision maker. Already today, and even more so tomorrow, the fighter pilot has an increasing need to be provided with decision support, and fortunately there are also an on-going parallel trend supporting this by technical progress. There are also increasing opportunities for distributing decision making, even between manned and unmanned aircraft.

Just like many other military applications it is important for an intelligent fighter pilot support for distributed unmanned and manned decision making to prioritize performance, and not only the performance of an individual aircraft and its subsystems but also the total system engaged in a mission, that could involve manned as well as unmanned aircraft in a JCS. It is important to have an automation strategy to achieve this at the same time as long term technology development goals for the design organization could be reached. The domain is characterized by long life cycles and long development times for aircraft compared to many other domains.

The automation strategy could specifically strive to help the DSS to be perceived as highly transparent and not very exchangeable, as a design goal. It is likely that specific guidelines has to be developed concerning principles of interaction for dynamic function allocation and it is important to map roles to the function allocation, regardless if the they are constant over time or not, taking into account both formal and informal structures among active cognitive agents. Concerning a DSS it is a challenge to keep the pilot feeling in control and knowing what is going on in the system and it is probably not possible to possess total transparency in how the DSS functions, but in the case of uncertainty a measure to indicate this to the pilot could be provided. As a part of an automation strategy guidelines could be provided to aid the development. For instance, the human pilot's ability to adapt to the changing environment could very well be explored further together with a DSS.

Specifically for intelligent fighter pilot support for distributed unmanned and manned decision making, concerning function allocation, it is important to not only regard which agent is best at performing a task but also to regard the risk/cost of performing a task in this kind of potentially hazardous context. Also, the risk of decreased functionality such as jammed radar or a shot down aircraft has to be regarded, as well as other drastic changes and dynamics of a mission. Another important factor to specially regard is the spatial positioning of aircraft which impact both the performance of an individual aircraft as well as team work between aircraft for certain tasks during a mission.

ACKNOWLEDGMENT

This research has been supported by the Swedish Governmental Agency for Innovation Systems (Vinnova) through the National Aviation Engineering Research Program.

REFERENCES

Alfredson, J. (2007). *Differences in situational awareness and how to manage them in development of complex systems* [Doctoral dissertation No. 1132]. Linköping, Sweden: Linköping University.

Alfredson, J., & Andersson, R. (2013). Designing for Human Factors in the Technology-Intensive Domain of Fighter Aircraft. *International Journal of Aviation Technology, Engineering and Management, 1*(2), 1–16. doi:10.4018/ijatem.2013070101

Alfredson, J., Holmberg, J., Andersson, R., & Wikforss, M. (2011). Applied cognitive ergonomics design principles for fighter aircraft. In D. Harris (Ed.), *Proceedings of the 9th International Conference on Engineering Psychology and Cognitive Ergonomics (EPCE 2011)* (pp. 473–483). Springer-Verlag. 10.1007/978-3-642-21741-8_50

Alfredson, J., Lundqvist, A., Molander, S., & Nordlund, P.-J. (2010). Decision support for the Gripen aircraft and beyond. *Proceedings of the 27th International Congress of the Aeronautical Sciences ICAS 2010.*

Berggren, P., Nählinder, S., & Svensson (2014). *Assessing command and control effectiveness: Dealing with a changing world.* Farnham, England: Ashgate.

Casner, S. M., & Schooler, J. W. (2014). Thoughts in flight: Automation use and pilots' task-related and task-unrelated thought. *Human Factors, 56*(3), 433–442. doi:10.1177/0018720813501550 PMID:24930166

Castor, M. (2009). *The use of structural equation modeling to describe the effect of operator functional state on air-to-air engagement outcomes* [Doctoral dissertation No. 1251]. Linköping, Sweden: Linköping University

Davidsson, S., & Alm, H. (2014). *Context adaptable driver information – Or, what do whom need and want when?* Applied Cognitive Ergonomics.

The role of autonomy in DoD systems. (2012). Defense Science Board Washington DC: Undersecretary of Defense.

Dekker, S. W. A. (1996). Cognitive complexity in management by exception: Deriving early human factors requirements for an envisioned air traffic management world. In D. Harris (Ed.), Engineering psychology and cognitive ergonomics, Volume I: Transportation systems (pp. 201-210). Aldershot, England: Ashgate.

Dekker, S. W. A., & Woods, D. D. (1999). Extracting data from the future: Assessment and certification of envisioned systems. In S. Dekker & E. Hollnagel (Eds.), *Coping with computers in the cockpit* (pp. 131–143). Aldershot, England: Ashgate.

Dekker, S. W. A., & Woods, D. D. (2002). MABA-MABA or Abracadabra: Progress on human automation cooperation. *Cognition Technology and Work*, *4*(4), 240–244. doi:10.1007101110200022

Endsley, M. R. (1988). Situation Awareness global assessment technique (SAGAT). *Proceedings of the IEEE National Aerospace and Electronics Conference* (pp. 789-795). New York: IEEE.

Feigh, K. M., & Pritchett, A. R. (2014). Requirements for effective function allocation: A critical review. *Journal of Cognitive Engineering and Decision Making*, *8*(1), 23–32. doi:10.1177/1555343413490945

Fitts, P. M. (1951). Human engineering for an effective air navigation and traffic control system. Ohio State University Research Foundation Report.

Handley, H., & Smillie, R. (2008). Architecture framework human view: The NATO approach. *Systems Engineering*, *11*(2), 156–164. doi:10.1002ys.20093

Handley, H. A. H. (2014). A network model for human interoperability. *Human Factors*, *56*(2), 349–360. doi:10.1177/0018720813493640 PMID:24689253

Hassenzahl, M., & Tractinsky, N. (2006). User experience – a research agenda. *Behaviour & Information Technology*, *25*(2), 91–97. doi:10.1080/01449290500330331

Helander, M., & Skinnars, Ö. (2000a). Use of cognitive walkthrough for evaluation of cockpit design. *In Proceedings of the 44th Annual Meeting of the Human Factors and Ergonomics Society* (pp.616-619). Santa Monica, CA: Human Factors and Ergonomics Society. 10.1177/154193120004400619

Helander, M., & Skinnars, Ö. (2000b). *Use of cognitive walkthrough for evaluation of cockpit design (HFA Paper No.2000-01)*. Linköping, Sweden: Swedish Centre for Human Factors in Aviation.

Helldin, T. (2014). *Transparency for Future Semi-Automated Systems* [Doctoral dissertation]. Örebro, Sweden: Örebro University.

Helldin, T., Falkman, G., Alfredson, J., & Holmberg, J. (2011). The Applicability of Human-Centred Automation Guidelines in the Fighter Aircraft Domain. *Proceedings of the 29th Annual European Conference on Cognitive Ergonomics* (pp. 67-74). Universitätsdruckerei Rostock, Germany 10.1145/2074712.2074727

Hollnagel, E. (1999). From function allocation to function congruence. In S. Dekker & E. Hollnagel (Eds.), *Coping with computers in the cockpit* (pp. 29–53). Aldershot, England: Ashgate.

Hollnagel, E., & Woods, D. D. (1983). Cognitive systems engineering: New wine in new bottles. *International Journal of Man-Machine Studies, 18*(6), 583–600. doi:10.1016/S0020-7373(83)80034-0

Hollnagel, E., & Woods, D. D. (2005). *Joint cognitive systems: Foundations of cognitive systems engineering*. Boca Raton, FL: Taylor & Francis. doi:10.1201/9781420038194

Johansson, C. (1999). Modellering av luftstridsavdömningar [Air Combat Resolution Modeling]. Linköping, Sweden: Linköping Institute of Technology.

Klein, G., Orasanu, J., Calderwood, R., & Zsambok, C. (Eds.). (1993). *Decision making in action: Models and methods*. Norwood, NJ: Ablex.

Lyons, J. B., & Stokes, C. K. (2012). Human-human Reliance in the context of automation. *Human Factors, 54*(1), 112–121. doi:10.1177/0018720811427034 PMID:22409106

McKendrick, R., Shaw, T., de Visser, E., Saqer, H., Kidwell, B., & Parasuraman, R. (2014). Team performance in networked supervisory control of unmanned air vehicles: Effects of automation, working memory, and communication content. *Human Factors, 56*(3), 463–475. doi:10.1177/0018720813496269 PMID:24930169

Nagamachi, M. (1989). *Kansei Engineering*. Tokyo: Kaibundo.

Nagamachi, M. (2011). *Kansei/Affective Engineering*. Boca Raton, FL: CRC Press.

Norman, D., Miller, J., & Henderson, A. (1995). *What you see, some of what's in the future, and how we go about doing it: HI at Apple Computer. Proceedings of CHI* (p. 155). New York: ACM Press.

Ollero, A., & Maza, I. (2007). *Multiple hetrogenous unmanned aerial vehicles. Springer Tracts on Advanced Robotics*. NY: Springer. doi:10.1007/978-3-540-73958-6

Parasuraman, R., Sheridan, T., & Wickens, C. (2008). Situation Awareness, Mental Workload and Trust in Automation: Viable, Empirically Supported Cognitive Engineering Constructs. *Journal of Cognitive Engineering and Decision Making, 2*(2), 140–160. doi:10.1518/155534308X284417

Persson, M. (2014). *Future technology support of command and control: Assessing the impact of assumed future technologies on cooperative command and control* [Doctoral dissertation]. Uppsala, Sweden: Uppsala University.

Pritchett, A. R., Kim, S. Y., & Feigh, K. M. (2014). Modeling human-automation function allocation. *Journal of Cognititve Engineering and Decision Making, 8*(1), 33–51. doi:10.1177/1555343413490944

Schütte, S. (2005). *Engineering emotional values in product design – Kansei engineering in development* [Doktoral dissertation No. 951]. Linköping, Sweden: Linköping University.

Sheridan, T. B., & Verplank, W. (1978). Human and Computer Control of Undersea Teleoperators. Cambridge, MA: Man-Machine Systems Laboratory, Department of Mechanical Engineering, MIT.

Steinhauser, N. B., Pavlas, D., & Hancock, P. A. (2009). Design principles for adaptive automation and aiding. *Ergonomics in Design*, *17*(2), 6–10. doi:10.1518/106480409X435943

Svensson, E., Angelborg-Thanderz, M., & Sjöberg, L. (1993). Mission challenge, mental workload and performance in military aviation. *Aviation, Space, and Environmental Medicine*, *64*(11), 985–991. PMID:8280046

Svensson, E., Angelborg-Thanderz, M., Sjöberg, L., & Olsson, S. (1997). Information complexity: Mental workload and performance in combat aircraft. *Ergonomics*, *40*(3), 362–380. doi:10.1080/001401397188206 PMID:11536799

Svensson, E., & Wilson, G. F. (2002). Psychological and psychophysiological models of pilot performance for systems development and mission evaluation. *The International Journal of Aviation Psychology*, *12*(1), 95–110. doi:10.1207/S15327108IJAP1201_8

Taylor, G. S., Reinerman-Jones, L. E., Szalma, J. L., Mouloua, M., & Hancock, P. A. (2013). What to automate: Addressing the multidimensionality of cognitive resources through systems design. *Journal of Cognitive Engineering and Decision Making*, *7*(4), 311–329. doi:10.1177/1555343413495396

Taylor, R. M. (1997). Human-Electronic Crew Teamwork: Cognitive Requirements for Compatibility and Control with Dynamic Function Allocation. In M. J. Smith, G. Salvendy, & R. J. Koubek (Eds.), *Design of Computing Systems, 21B* (pp. 247–250). Amsterdam: Elselvier.

Woods, D. D., & Hollnagel, E. (2006). *Joint cognitive systems: Patterns in cognitive systems engineering*. Boca Raton, FL: Taylor & Francis. doi:10.1201/9781420005684

This research was previously published in Intelligent Applications for Heterogeneous System Modeling and Design edited by Kandarpa Kumar Sarma, Manash Pratim Sarma, and Mousmita Sarma, pages 1-22, copyright year 2015 by Information Science Reference (an imprint of IGI Global).

Chapter 8
Applications of Decision-Support Systems in Sociotechnical Systems

Tetiana Shmelova
National Aviation University, Ukraine

Yuliya Sikirda
National Aviation University, Ukraine

ABSTRACT

In this chapter, the authors present conceptual models of decision support systems (DSS) and Expert systems (ES) for Human-Operator (H-O) of Air Navigation System (ANS), such as air traffic controller (ATC) in flight emergencies, Unmanned Aerial Vehicles (UAV) operator, Safety Management System (SMS), etc. The authors made an analysis of the International civil aviation organization (ICAO) documents on risk assessment and the impact of the social environment on the aviation system. Automated system of pre-flight information preparation with intelligent module for support the decision making (DM) about aircraft departure are presented and program realization of systems are shown. The authors obtained algorithm of determining the optimal aerodrome for the forced landing of aircraft is provided. Expert systems of aviation enterprise's estimation were developed. Inhomogeneous factors of internal and external management environment of aviation enterprise were generalized using set-theoretical approach. This gave possibilities to define that the level of safety of aviation activity has the greatest influence among factors of internal environment and global economic situation – among factors of external environment. The authors demonstrate some interesting applications of DM in Socio-Technical Systems (STS). The DSS for H-O of ANS in the emergency situational were developed. Examples of DSS: DSS of ATC in the emergency situational; the flight dispatcher's DSS; DSS of UAV's operator, etc. In addition, the chapter presents some cases of DSS developed by the authors and colleagues at National Aviation University, Ukraine.

DOI: 10.4018/978-1-5225-8365-3.ch008

DECISION SUPPORT SYSTEMS IN AIR NAVIGATION SYSTEM

Support of the safe functioning of ANS is one of the most important scientific and technical problems. Statistical data show that human errors account for up to 80% of all aviation accidents (Cross-Cultural Factors in Aviation Safety, ICAO, 2004). According to ICAO's data (State of Global Aviation Safety, ICAO, 2013), comparing with 2011, in 2012 the number of incidents in the world decreased on 21%, in 2013 – on 13%; number of fatalities in 2012 decreased on 10%, in 2013 – on 55%. Consequently, the least number of deaths was fixed in 2012-2013, starting from 2004. As a result of decreased number of incidents, and increased number of departures, the frequency of incidents in the world in 2012 was reduced to 3,2 events per million departures and in 2013 – to 2,8 events per million departures. This is the lowest value since the ICAO started to monitor the frequency of incidents in the world. Aviation systems cannot be wholly free from dangerous factors and connected with them risks, while, the elimination of aviation events and serious incidents continues to be the final goal of human activity in the sphere of aviation safety. Neither human activity nor systems created by it guarantee a total absence of operating errors and their consequence (Safety Management Manual ICAO, 2013). In such a way, safety is a dynamic characteristic of aviation with the help of which risk factors for flight safety should steadily decrease. It is important to note that adoption of efficiency indices of ensuring flight safety is frequently influenced by internal and international standards and also by cultural features (Manual on Global Performance of the ANS, ICAO, 2009). While risk factors for flight safety and operating errors are under control, such opened and dynamic system as civil aviation may be controlled providing the necessary balance between flight performance and safety requirements for passengers and their baggage.

Latest demands of international aviation organizations (ICAO) directed on implementation of integrated approach for the improvement of aviation safety. One of the ways to increase safety by timely is support of pilot in emergency situations (Report of the Twelfth Air Navigation Conference, 2012). Approach based on the characteristics (the performance-based approach – PBA), the principles of informed DM to achieve the desired/required results and to use facts and data for DM (Manual on Global Performance of the Air Navigation System, 2009) ICAO constantly develops and improves proactive, based on the risks evaluation, methods, directed on the farther decrease in number of aviation events in the world. Also ICAO encourages aviation communities to recognize the importance of adherence of the single global approach for safety improvement and monitoring (State of Global Aviation Safety, 2013).

So called aeronautical DM is DM in a unique environment – aviation. It is a systematic approach to the mental process used by pilots to consistently determine the best course of action in response to a given set of circumstances. It is what a pilot intends to do, on the basic of latest information he has (Pilot's Handbook of Aeronautical Knowledge, 2008).

Decision Support Systems (DSS) may be defined as «interactive computer systems intended to support different types of activity during the decision making (DM) including poorly-structured and unstructured problems» (Tarasov, Gerasimov, 2007; Gerasimov, Lokazyuk, 2007). Decision support systems contain common sets of components. These components include data related components, algorithm related components, user interface and display related components. The data related components are made of modules that ingest data, format data, store data, transfer data and archive data. The algorithm related components utilize diverse methods such as atmospheric models, rule based algorithms, fuzzy logic algorithms, statistical algorithms, algorithm of DM under uncertainty and risk, data mining algorithms. Development of the automated systems in aviation is associated with anthropocentric approach and on the principles of human-oriented automation, required new approaches to solving the problems of man-machine interface (Human Factors in CNS/ATM Systems (1994). One of the effective means of flight safety improvement is the inclusion of DSS to the automated systems of air traffic control (ATC) (Kharchenko, Shmelova & Sikirda, 2012; Shmelov, Sikirda.&Belyaev, 2014). The problems of DSS's informational provision have been considered in the works of scientists (Tarasov, Gerasimov, 2007;. Gerasimov, Lokazyuk, 2007; Sirodzha, 2002). The authors have obtained models of the DM by Air Navigation System's H-O and of the development of flight situations that are used as a part of an air traffic controllers DSS in flight emergencies to assist in DM for choosing of the optimal alternative of action with minimal damage in the conditions of incomplete and uncertainty information and as a part of a flight dispatcher's DSS for assistance in DM regarding the AC departure (Kharchenko, Shmelova & Sikirda,2012; Shmelov, Sikirda.&Belyaev, 2014).

Provision of informational support of DM by aircraft operator is possible in the conditions of data collection, processing and displaying of automation within a concept of perspective global CNS / ATM system (Communication, Navigation, Surveillance / Air Traffic Management), developed ICAO (Global Air Navigation Plan, ICAO, 2007). A modern approach, founded on the characteristics (performance-based approach – PBA) (Manual on Global Performance of the ANS, 2009), based on the next three principles: the main accent on desired/necessary results: DM, oriented on desired/necessary results; using facts and data while DM. Herein the principle "using facts and data while decision making" admits that tasks shall comply with the widely known in Western management criteria SMART, that correspond to abbreviation of five English words: specific, measurable, achievable, relevant and time bound. So called aeronautical DM is DM in a unique environment – aviation. It is a systematic approach to the mental process used by pilots to consistently determine the best course of action in response to a given set of circumstances. It is what a pilot intends to do, on the basic of latest information he has. The DSS may be defined as interactive computer systems intended to support different types of activity during the DM. Nowadays a new approach for flight safety provision is forming in global practice. The ICAO's term «Safety Management System» integrates operations and technical systems with the management of financial and human resources to ensure aviation safety or the safety of the public (Safety Management Manual, 2009).

Given statement implies binding into one, in a single system different objects and subjects of aviation activity. The subjects of aviation activity are designers and manufacturers of aviation equipment, airline operators, and organizations engaged in aircraft maintenance and repair, and also handling companies. The air traffic services providers, aviation educational institutions, research and design aviation organizations, aviation authorities are included to them. Each of subjects of aviation activity is used to provide manufacturing processes and functioning of civil aviation the appropriate objects: aircrafts, their components and equipment, industrial equipment, ground vehicles, aerodrome and on-route equipment (radio- and electro-technical facilities), engineering technical buildings, other movable and immovable assets, etc. And it is a very complex task to join the named elements of aviation. Since the period of «organizational era» (1990s – present) flight safety has been considered systematically, taking into account organizational, human and technical factors. Also at that date, a notion «aviation incident because of organizational reasons» appeared in aviation, taking into account impact of organizational culture and politics on the control systems efficiency for flight safety risk factors. The «Swiss Cheese Model» by James T. Reason shows that significant role in the incidents causation belongs to organizational and management factors (Safety Management Manual, 2009).

Building an effective SMS of aviation activity requires an integrated research of environment in which aviation enterprises operate. Many research related to the determination of influence of factors of internal and external management environment of aviation enterprise on the efficiency of its business. Estimating of the level of influence of factors of aviation enterprise's management environment on the efficiency of SMS is a perspective task. The structural analysis of management environment of aviation enterprise was carried out for DSS ANS. And, as follows from the analysis, inhomogeneous factors which influence on the aviation activity were classified and formalized. Inhomogeneous factors of internal and external management environment of aviation enterprise were systematically generalized using set-theoretical approach. The influence of factors of internal and external management environment on the aviation enterprise's aviation activity was determined (Shmelova, Sikirda, Assaul & Stasiuk, 2015). It is considered that aviation is the most fail-safe type of transfer. In as little as century, aviation, in the sphere of flight safety, rose through the ranks from unstable system to the first «ultra-safe» system in the history of transport, it means system in which the number of catastrophic failures, in the sphere of safety, make up less than one per one million of production cycles (Normal operations safety survey, 2008). The authors present investigation of management environment of aviation enterprise in ANS, determined the influence of factors of internal and external management environment of aviation enterprise on aviation activity. Safety management of aviation enterprise is directed on the formation and provision the achieving its goals rationally using present resources (labour, material, financial, informational, etc.).

Aviation enterprise is under continuous pressure from different branches of internal and external environment and use as ES in aviation. Aviation enterprise's management environment is a set of circumstances and factors inside and around that affect DM process (Figure 1). The ES for H-O of ANS were developed. Internal environment of aviation enterprise is a complex of components connected with each other by means of certain structures within it. The main variables of internal environment of aviation enterprise management:

- **Enterprise Goals:** Specific final state or desired result, which must be achieved (safety, regularity and economic efficiency of aviation activity; while in the context of aviation, achievement of safety means to achieve state in which the possibility of harm to persons or of property damage is reduced to, and maintained at or below, an acceptable level);
- **Enterprise Tasks:** Predictable work, series of works or part of it, which must be done with the help of previously defined method at predetermined time (detection of negative factors that influence on safety, regularity and economic efficiency of aviation activity; management of risk factors);
- **Aviation Enterprise Structures:** Logical relationships, the interaction of organizational forms of technical processes, inputs processing, constructed in such a way that it is possible to achieve the goals of aviation enterprise in the most effectively way (functional interaction diagrams);
- **Technologies:** Conversion process in the system, which consists of programs testing and implementation of operations over resources in order to transform them into the desired product (established sequence of operational development, production and sale of air, navigation, airport, maintenance, handling, etc. services);
- **People of Aviation Enterprise (Central Factor):** Aviation personnel with their abilities, aptitudes, needs, expectations, perceptions, attitude and values (crew members, personnel of technical maintenance, etc).

Figure 1. Environment of the aviation enterprise's management

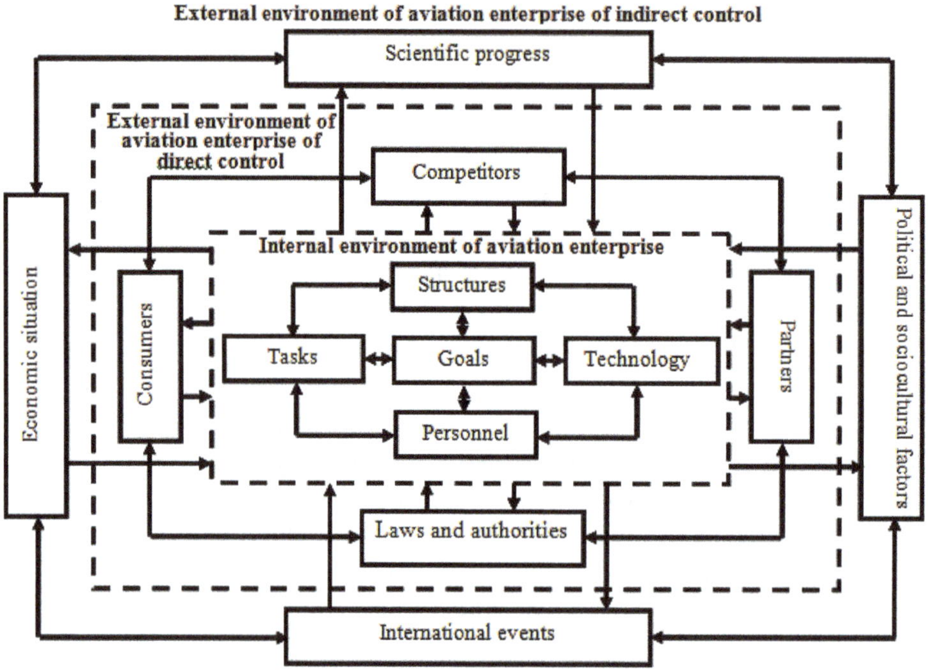

Internal variables as usual are called sociotechnical subsystems because they have social component (people) and technical components (other internal variables). External environment of aviation enterprise management is a complex of elements which are not part of aviation enterprise but have some influence on it. The main characteristics of external environment of aviation enterprise are relation of its factors, complexity, mobility and uncertainty. External environment of aviation enterprise management is divided into two types:

1. **External Environment of Direct Influence:** An environment that includes elements that directly affect the operations of aviation enterprise and feel the direct impact of its operations. They are:
   a. **Consumers of Air (Passengers, Corporate Clients, VIP-Clients, etc.) and Other Services in Aviation:** Natural or legal persons who use services for personal needs or for needs of their organization;
   b. **Competitors:** Producers of similar air, navigation, airport, maintenance, handling, etc. services on the market;
   c. **Partners:** Legal or natural persons who are co-owners of the aviation enterprise or temporary partners according to any transaction, agreement (air navigation service providers, airports, handling and cleaning companies, catering firms, hotels, etc.);
   d. **Laws and Authorities:** Legal and normative acts of bodies of higher legislative and executive powers or their special bodies that oversee the compliance with state requirements (international and local regulations;

2. **External Environment of Indirect Influence:** A complex of factors that do not influence directly, don't an immediate impact on the operations of aviation enterprise, but will eventually be shown to them. They are:
   a. **Economic Situation:** Tendency and level of inflation, interest rate, level of unemployment;
   b. **Progress in Science and Technology:** Correspondence between techniques and present requirements/standards;
   c. **Political Factors:** Tendency in the sphere of legislation, court cases, etc.;
   d. **Sociocultural Factors:** Set of attitudes, values, norms, beliefs, and behavior, etc;
   e. **International Events:** The processes which take place outside the country and influence on the central rate, business struggle, etc.

Any aviation enterprise is operating in the environment and functioning in the case when the environment gives it that opportunity. There should be an idea as of internal so of external environment in order to determine the level of safety of certain aviation enterprise, its potential and development trend and also its place in the environment. The study of the internal and external environment allows aviation enterprise to reduce the negative impact of the elements on the level of safety, and increase opportunities to improve safety. Decomposition of factors which influence on safety was done in order to provide structural analysis of management environment of aviation enterprise (Figure 2).

Content and formal description of parameters is given in table 1.The structural analysis of management environment of aviation enterprise was carried out in order of significance decreasing of its factors: firstly, the factors of internal management environment were analysed, secondly – factors of external management environment of direct influence and, ultimately, factors of external management environment of indirect influence. In such a way, from the point of systematic approach the factors of internal and external management environment of aviation enterprise which influence on its aviation activity were determined.

Generalization of inhomogeneous factors of internal and external management environment of aviation enterprise was carried out using set-theoretical approach. This gave possibilities to consider structural hierarchy, heterogeneity, dynamic instability of factors and to define conditions for their assessment:

$$\overline{F} = \overline{F}_{ie} \cup \overline{F}_{eedi} \cup \overline{F}_{eeii}, \tag{1}$$

where $\overline{F}_{ie} = \left\{\overline{G}, \overline{Ta}, \overline{S}, \overline{Te}, \overline{Pe}\right\}$ – is set of factors of internal management environment of aviation enterprise (goals $\overline{G}$, tasks $\overline{Ta}$, structures $\overline{S}$, technologies $\overline{Te}$, personnel $\overline{Pe}$); $\overline{F}_{eedi} = \left\{\overline{C}, \overline{Co}, \overline{Pa}, \overline{L}\right\}$ – is set of factors of external management environment of aviation enterprise of direct influence (consumers $\overline{C}$, competitors $\overline{Co}$, partners $\overline{Pa}$, laws and authorities $\overline{L}$); $\overline{F}_{eeii} = \left\{\overline{ES}, \overline{SP}, \overline{PF}, \overline{SF}, \overline{IE}\right\}$ – is set of factors of external management environment of aviation enterprise of indirect influence (economic situation $\overline{ES}$, scientific and technical progress $\overline{SP}$, political factors $\overline{PF}$, sociocultural factors $\overline{SF}$, international events $\overline{IE}$.

Figure 3 shows graphical interpretation of method of inhomogeneous factors generalization of internal and external management environment using set-theoretical approach.

Figure 2. Decomposition of factors of environment of the aviation enterprise's management

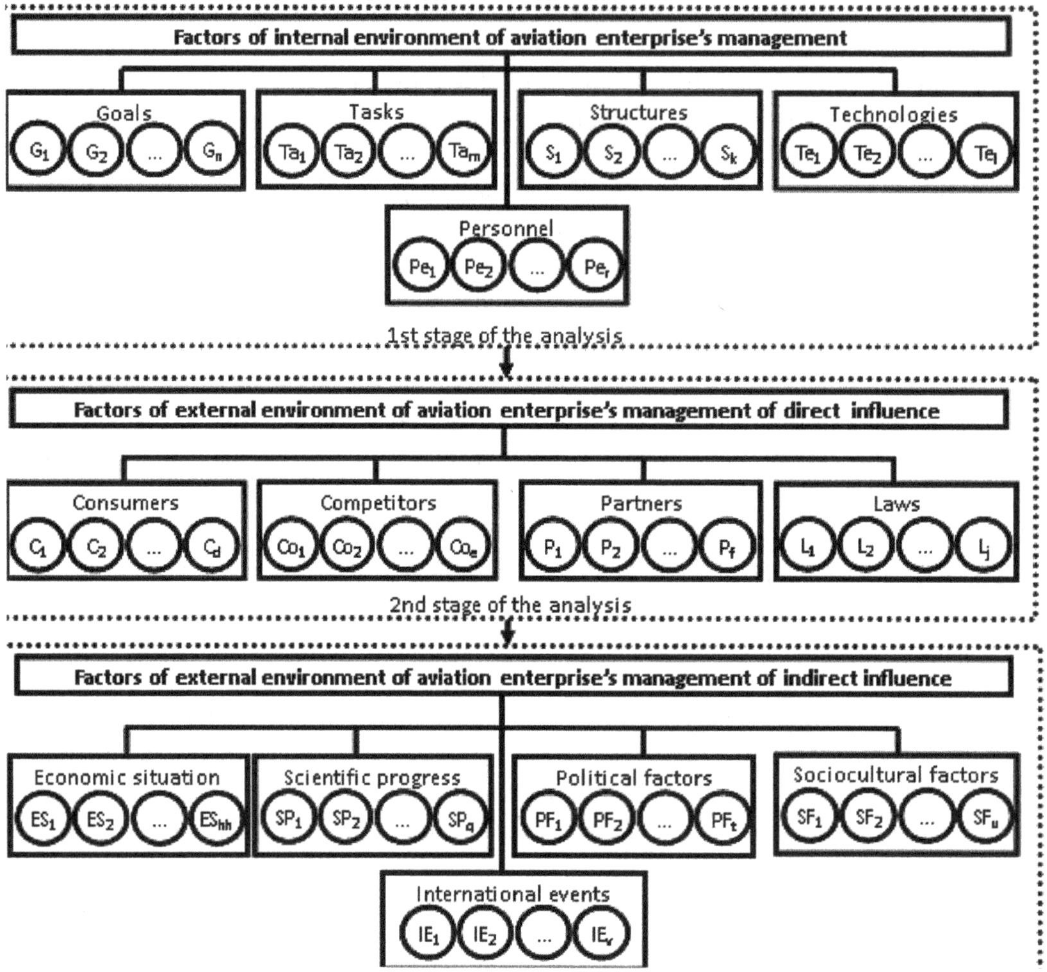

Method of expert estimation was used in order to define the level of influence of factors of internal and external management environment of aviation enterprise on aviation activity. Respondents from among pilots and air traffic controllers of different age categories, with different professional experience filled in the offered questionnaires and defined individual preferences concerning influence of factors of internal and external management environment of the aviation enterprise on aviation activity. Priority influence of factors on aviation activity was established by comparison of weight coefficients. As an example, Figures 4 and 5 show the influence of goals and international events of internal and external management environment correspondingly on aviation activity. Graphical presentations of weight coefficients on Figures 4 and 5 are corresponding systems of preferences of experts. The enterprises was analysed for using experts from aviation enterprises. There were defined, that the greatest influence has:

- The level of safety of aviation activity (factor of internal management environment of aviation enterprise);
- The global economic situation (factor of external management environment of aviation enterprise).

Table 1. Decomposition of factors of external management environment of aviation enterprise

Factors	Parameters	Coding
Economic situation $\overline{ES} = \{ES_1, ES_2, ..., ES_h\}$	Cost of living	SE1
	Minimum salary	SE2
	Minimum pension	SE3
	Average earnings	SE4
	Inflation rate	SE5
	Bank interest rate	SE6
	Unemployment rate	SE7
	State of investment activities in transport domain	SE8
	Development of tourism	SE9
Scientific and technical progress $\overline{SP} = \{SP_1, SP_2, ..., SP_q\}$	Level of moral depreciation of aircraft	ST1
	Level of moral depreciation of special techniques	ST2
	Level of moral depreciation of buildings and constructions	ST3
	Progressiveness of technological schemes on development and realization of aviation services	ST4
	Progressiveness of technological schemes from ground, technical, airport, air navigation services	ST5
Political factors $\overline{PF} = \{PF_1, PF_2, ..., PF_t\}$	Stability of political regime	PF1
	Level corruption	PF2
	Crime rate	PF3
	Level of state regulation in aviation sphere	PF4
Sociocultural factors $\overline{SF} = \{SF_1, SF_2, ..., SF_u\}$	Social structure of society	SF1
	National and household traditions of population	SF2
	Popularity of air traffic	SF3
International events $\overline{IE} = \{IE_1, IE_2, ..., IE_v\}$	Global economic situation	IE1
	Strategic-military situation	IE2
	Influence of certain states	IE3
	Nature effect	IE4
	State of raw material and natural resources	IE5

Factors of internal and external management environment of aviation enterprise which influence on aviation activity were determined from the point of systematic approach. Aviation enterprise is a complex of basic elements such as goals, tasks, structures, technologies, personnel, and also environment – political, economic, cultural, market which interacts and to which it has to adapt. Internal environment of aviation enterprise is a source of life force and includes the potential which give possibilities to exist and survive in a defined time period but also it can be a source of problems. External environment is a source of resources needed for maintenance of its internal potential at the necessary level in order to achieve aviation enterprise's goals. Such level of accuracy of tasks determination may be achieved only using the way of consistent and structural description of inhomogeneous components of aviation - avia-

Figure 3. Graphical interpretation of method of inhomogeneous factors generalization of internal and external management environment of aviation enterprise

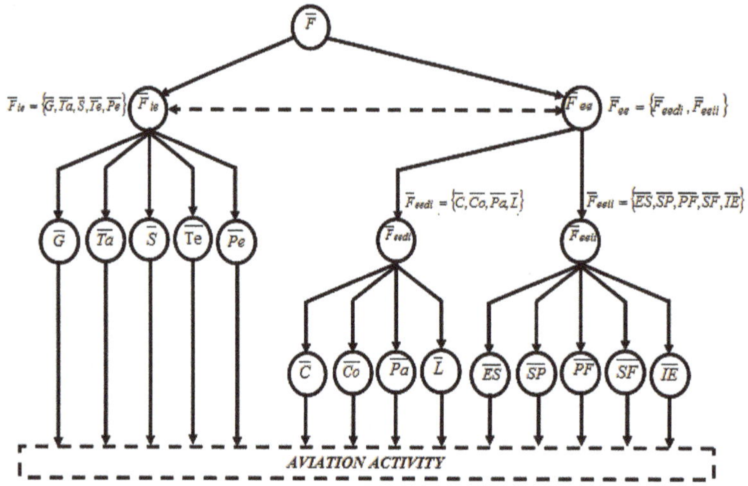

Figure 4. Influence of factors of internal management environment of aviation enterprise

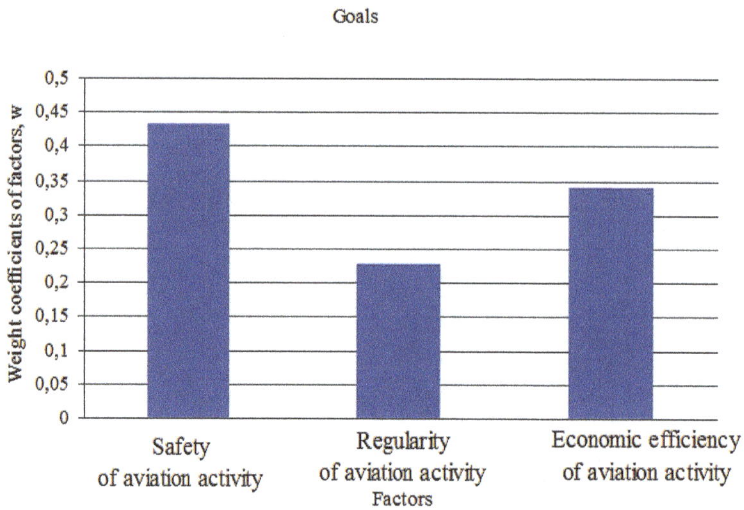

tion enterprises, aviation personnel, aviation infrastructure, techniques, rules and information that is used for creating of conditions and applying of aerospace by aircraft users.

The aviation system is a complex system that requires investigation of the human contribution to safety and an understanding of how human performance may be affected by its multiple and interrelated components. The HF remains a major cause of aviation accidents. That is why current management system needs to improve. Human factors (HF)s and its effects on safety performance continue to evolve and develop. Reduce of HF risks in aviation safety it's one of important tasks on todays. The theory of the HF is gradually developed, tested and institutionalize at the end of the previous century, and now ICAO presents the safety case for cultural interfaces and systematic approach in aviation safety with reference to main establishment of conceptual of safety models. Now, SMS is continuation of develop-

Figure 5. Influence of factors of external management environment of indirect influence of aviation enterprise

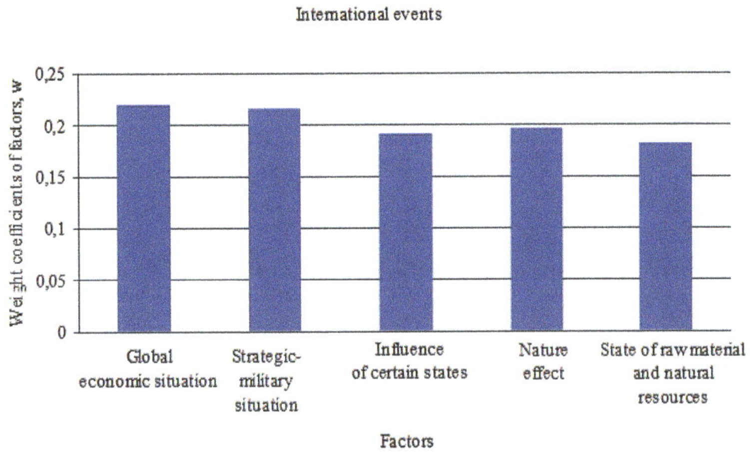

ment of HF models (Safety Management Manual, ICAO, 2013). The SMS is a systematic and explicit approach defining the activities by which safety management is undertaken by an organization in order to achieve acceptable or tolerable safety. The objective of a SMS is to provide a structured management approach to control safety risks in operations. Effective safety management must take into account the organization's specific structures and processes related to safety of operations.

Safety management provides the means by which organizations can control the processes that could lead to hazardous events, in order to ensure that the risk of harm or damage is limited to an acceptable level. Much of this activity focuses on hazards as they are identified through such processes and activities as the investigation of safety occurrences, incident reporting systems and safety oversight programmes. Safety assessments provide another proactive mechanism for identifying potential hazards and finding ways to control the risks associated with them (Safety Management Manual, ICAO. 2013). There are steps of the safety assessment process in SMS:

Step 1: Development (or procurement) of a complete description of the system to be evaluated and of the environment in which the system is to be operated;
Step 2: Identification of hazards;
Step 3: Estimation of the severity of the consequences of a hazard occurring;
Step 4: Estimation of the likelihood of a hazard occurring;
Step 5: Evaluation of risk;
Step 6: Mitigation of risk;
Step 7: Development of safety assessment documentation.

There are many DSSs must be developed for safety assessment for each situation and stage of flight of aircraft (AC). Provision of informational support of DM by aircraft operator is possible in the conditions of data collection, processing and displaying of automation within a concept of perspective global CNS / ATM system (Communication, Navigation, Surveillance / Air Traffic Management), developed by International Civil Aviation Organization (Global Air Navigation Plan, 2007). The "system", as defined for the purpose of safety assessment, will always be a sub-component of some larger system, such as ANS. For example, even if the assessment encompasses all services provided within an aerodrome, this can be considered a sub-component of a larger regional system, which in turn is a subcomponent of the global aviation system. A detailed description of the system should include (Safety Management Manual, 2013):

- The purpose of the system;
- How the system will be used;
- The system's functions;
- The system's boundaries and the external interfaces; and
- The environment in which the system will operate.

The safety impact of a potential loss or degradation of the system will be determined, in part, by the characteristics of the operational environment in which the system will be integrated. The description of the environment should therefore include any factors that could have a significant effect on safety. These factors will vary from one case to another. They could include, for example, traffic characteristics, airport infrastructure and weather-related factors. Representation of the ANS in the form of a STS first makes possible to take into account the influence of social, cultural environment of people who DM. Modern approaches to control some factors (psycho-physiological, behavioural, ergonomic, professional, etc.) do not take into account the functional state of a H-O under conditions of dynamic changes of external and internal factors. The ambient conditions determine the reaction of H-O and this reaction changes the environmental conditions accordingly. One of possible approaches to solve these problems may be through formalization and mathematical description based on a system analysis H-O's actions as a complex STS. Taking into account the influence on DM by H-O the professional factors (knowledge, habits, skills, experience) as well as the factors of non-professional nature (individual-psychological, psycho-physiological and socio-psychological) enables to predict the H-O's actions and flight situation development with the aid of the informational, analytic and diagnostics complex for research operator behaviour in extreme situation and to develop DSS of ANS operators for each situation and each stage of AC's flight (Kharchenko, Shmelova & Sikirda, 2012, 2016).

So, DSSs contain common sets of components. These components include data related components, algorithm related components, user interface and display related components. The data related components are made of modules that ingest data, format data, store data, transfer data and archive data. The algorithm related components utilize diverse methods such as atmospheric models, rule based algorithms, fuzzy logic algorithms, statistical algorithms, algorithm of DM under uncertainty and risk, data mining algorithms. Development of the automated systems in aviation is associated with anthropocentric approach and on the principles of human-oriented automation, required new approaches to solving the problems of man-machine interface. One of the effective means of flight safety improvement is the inclusion of DSS to the automated systems of air traffic control (ATC) (Shmelova & Sikirda, 2014). The authors have obtained models of the DM by Air Navigation System's human-operator (H-O) and of the development of flight situations (Kharchenko, Shmelova & Sikirda, 2012, 2016) that are used as a part of an ATC's DSS in flight emergencies to assist in DM for choosing of the optimal alternative of action with minimal damage in the conditions of incomplete and uncertainty information and as a part of a flight dispatcher's DSS for assistance in DM regarding the AC departure. So, DSSs is a computerized system designed to help a user make decisions. The DM is seen as a process of selecting the optimal alternative action or permissible subset of the set of possible alternatives. The solution can be interpreted as the result of a particular choice of alternative actions. The alternative decisions are possible scenario during the DM.

So, DM problems and people, who make them, recently deserved more attention. This is due to the increasing of dynamism of the environment, increasing interdependence of many decisions, the high rate of scientific and technological progress. Managers who make decisions are faced with difficult choices, with need to consider the set of alternatives. To evaluate the options the knowledge of experts, sophisticated analytical calculations, scientific research, information technologies are used. Questions of decision support at all stages of the process (goal identification, development and DM, organization and implementation of control) are becoming more relevant (Figure 6).

Figure 6. Scheme of the preparation, making and implementation of decisions

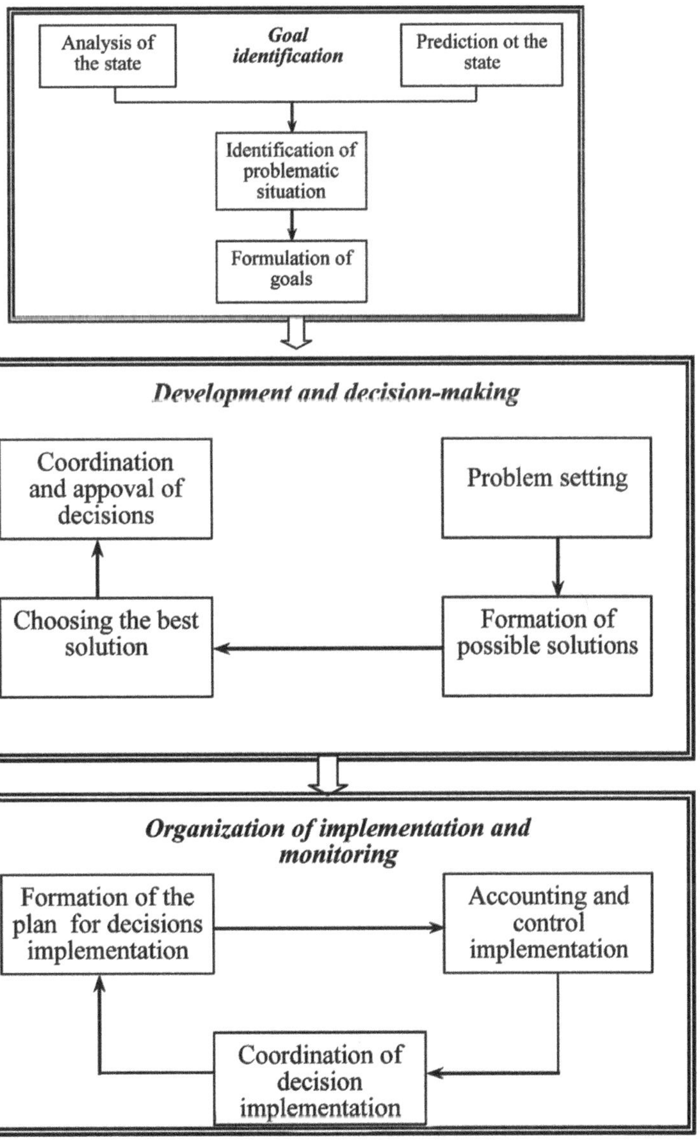

In fact, the problem consists in automation of creative work of the responsible employees groups of organizational management - managers of all levels and those who make decisions under actual conditions of their work. Three fundamental components of DSS architecture are:

- The database (or knowledge base) - information structure that reflects the status and relationship of objects analysed;
- The model base - a set of mathematical, logical, linguistic and other models used for comparative analysis of multi-alternative decision;
- The user interface.

Problems of DM in organizational management are most unique and unusual, but in their situational basis they have the following common features:

- Unique situation of choice;
- Difficult nature to assess the alternatives under consideration;
- Lack of certainty of the actions consequences (after action uncertainty);
- Availability of diverse set of factors that must be considered during decision making;
- The presence of a person or group of persons responsible for DM.

Computer support of technician's activity does not require sophisticated methodological framework and is implemented within the framework of conventional information systems. As for executives and professionals, it is important to determine the nature and characteristics of preparing and decisions making based on the level of organizational management in order to create effective software of DSS as systems which use formal rules and object management model together with the database manager and personal experience to develop and validate variants of management decisions (appeared at the beginning. 70's).

Systems of this kind do not provide the information DM, but participate in it. The main components of DSS are presented in Figure 7.

The main methods of analytical modelling to support DM are:

1. **Factor Analysis ("Cause- Consequence"):** Change the values of some variables (factors) and their relationships (formulae) for the investigation of values changes of other, dependent, and variable.
2. **Optimization Analysis:** Getting the best value of the objective function taking into account the restrictions imposed by the selection of variables.

Figure 7. The main components of DSS

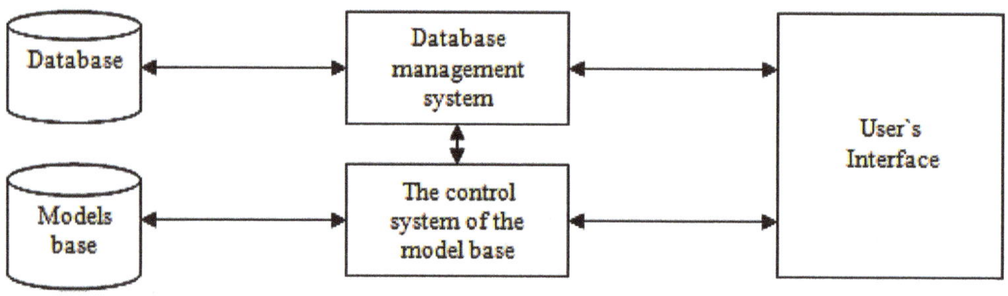

3. **Correlation and Regression Analysis:** Determining the type of relations between the dependent variables and factors.
4. **Trend Analysis:** Predicting the dynamics of the control object by building trends - segments of numerical series;
5. Control system models base;
6. **The User's Interface:** The way of user interaction with the system.

Building a DSS should ensure implementation of its important properties:

1. **Interactivity of DSS:** Means that the system responds to various types of action that one intends to influence the computational process, especially in interactive mode. Human and system exchange information at a rate that is comparable with the speed of human information processing.
2. **Integrity of DSS:** Is complex of system component for data management and the communication with users during decision support.
3. **Power of DSS:** Means the ability of a system to respond to the most significant question.
4. **Availability of DSS:** The ability to provide answers to the user requests in the right form and at the right time.
5. **Flexibility of DSS:** Describes the possibility of adapting to needs and situations changing.
6. **Reliability of DSS:** Means the ability of the system to perform the required functions for a given long period.
7. **Robustness of DSS:** This is the system's ability to recover in case of error situations, both external and internal origin. For example, there are mistakes of the input data or malfunctioning hardware in a robust system.

The DSS for H-O of ANS in the emergency situational were developed. Examples of DSS: DSS of ATC in the emergency situational; the flight dispatcher's DSS; DSS of UAV's operator, etc.

Decision Support Systems of an Air Traffic Controller in Flight Emergencies

Research has shown, the choice of the optimal variant of the flight completion in emergencies, which requires AC's forced landing, requires the operator to analyse the significant amount of diverse information. For a comprehensive account of the factors that affect the DM by an air traffic controller, has been built to adaptive DSS, which allows take into account the dynamic characteristics of the state of the control object (AC) and of the external environment (ATC zone). The main tasks of a ATC DSS in flight emergencies are:

- Data collection about the state of the control object (AC), external environment (ATC zone) and the individual qualities of the person, who makes the decision;
- Creation of action strategy in flight emergency (continue flight to the destination (alternate) aerodrome or perform a forced landing);
- Modelling of DM by H-O in flight emergency;
- Prediction of the flight situation development during DM by H-O in flight emergency;
- Construction of the AC's field of approachability in case of necessity of forced landing;

- Evaluating the effectiveness of alternative decisions in flight emergency and formation of recommendations for determination of the optimal variant of flight completion.

The general concept of adaptive system for informational support of the DM by air traffic controller in flight emergency, which requires AC's forced landing, is presented in the form of a set of subsystems (Figure 8).

Structure of ATCs DSS in flight emergencies is shown in Figure 9. Functions of an air traffic controllers DSS in flight emergencies are expanded through information-analytical diagnostic complex for research of patterns of H-O's activity. By using complex, air traffic controller can predict the flight situation development, based on the operational monitoring of the pilot's emotional state and determining the effect of social environment on him.

The specialized program complex «Prompt» (Kharchenko, Shmelova & Sikirda, 2012, 2016) has been created, in order to provide informational support to an air navigation system's operator. This was done to enable him to make the decision concerning the selection of the optimum flight completion strategy in flight emergencies, which require a forced landing, in a timely manner. The program complex was created with the aid of the visual system for designing programs Delphi 5, which is based on the special version of the programming language Pascal – Object Pascal and supports the main principles of the object-oriented programming. With the aid of the program «Prompt» the subsystems forming the strategies of actions, predicting the development of a situation, defining the characteristics of the alternative variants of the flight completion, evaluating the efficiency of the potential alternatives and building the optimum decision were realized.

The program "Prompt" enables handling of two non-standard situations – an engine failure and fire on board of an aircraft (Shmelova, Kharchenko, Sikirda, Gerasimenko, 2011). When the system receives a message about the unusual situation origination, the additional panel for providing the recommendations concerning the possibility of the flight continuation or necessity to execute a forced landing appears on the operator's monitor. When a type of situation – engine failure – is to be handled the number of the engines which have failed is to be input; when the type – fire – is to be handled, the condition of the aircraft is to be input and the proper advice appears on the monitor (Figure 10).

Figure 8. A conceptual model of an air traffic controllers DSS in flight emergencies, which is adaptive to changes in the state of the controlled object and of the external environment

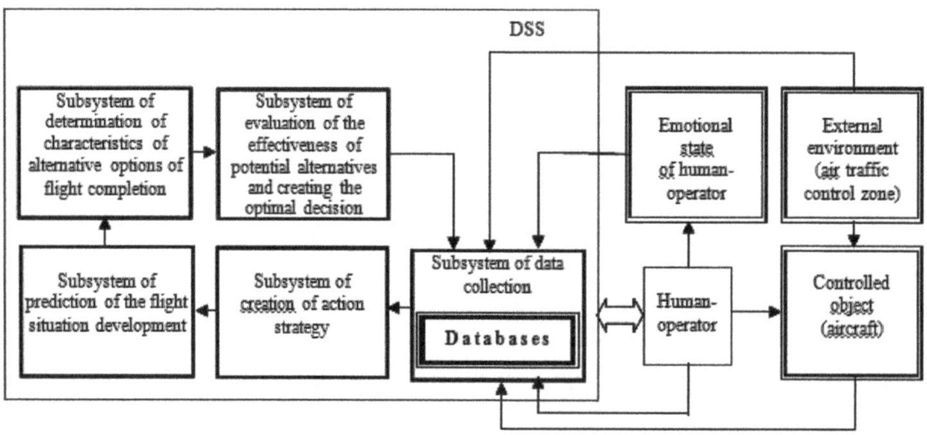

Figure 9. The structure of DSS of an air traffic controller in flight emergency, which requires AC's forced landing: AD – aerodromes, L – locality, LG – landing grounds; E – emotional state of H-O

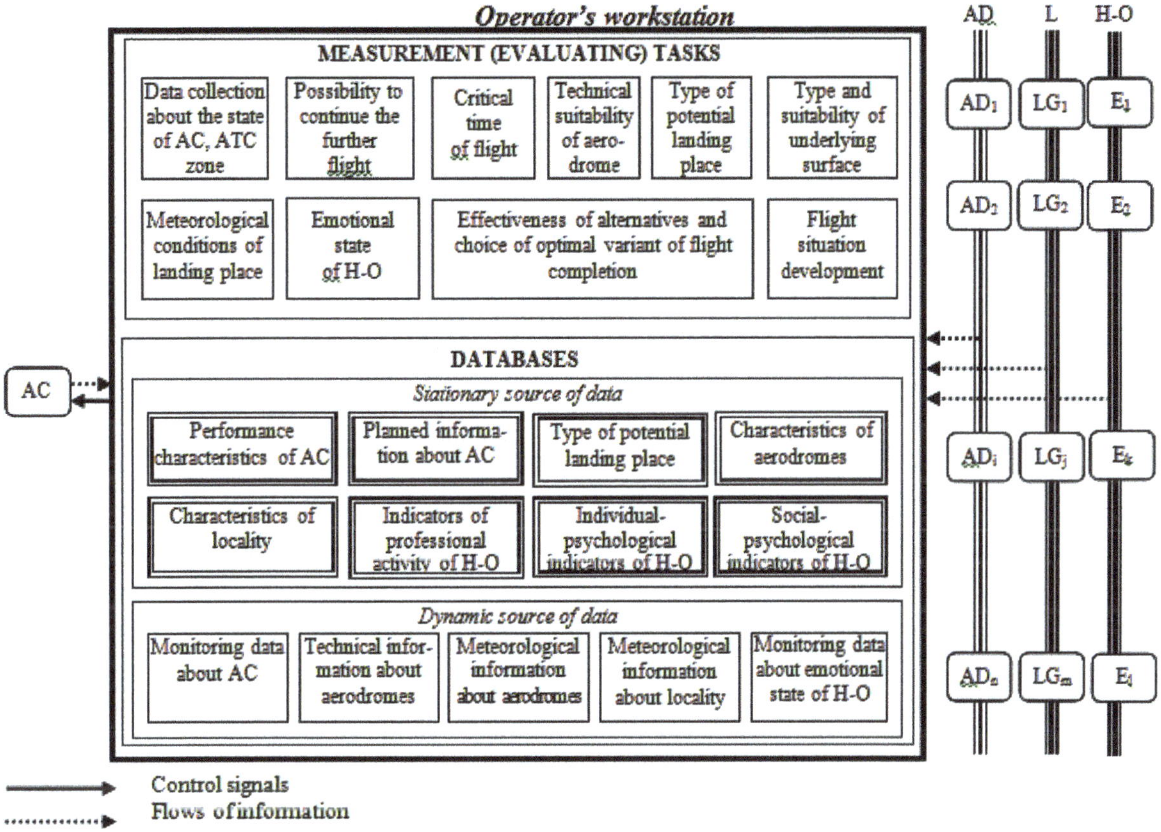

Figure 10. Realization of the subsystem of forming the strategy of action

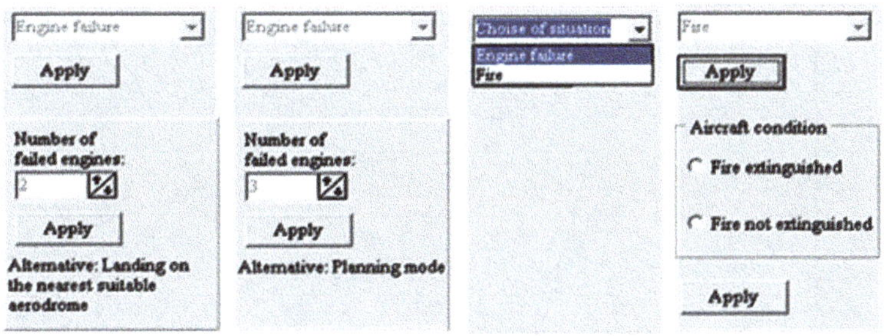

When the recommendation concerning the necessity to accomplish a forced landing received, the field of approachability for the aircraft is built on the monitor, also values of the potential loss required for selection of a definite flight completion alternative as well as the coordinates of the potential landing sites (azimuth, range) are presented on the monitor. The program envisages the formation of the field of approachability for two extreme cases – the regime of planning and regime of an immediate descent (Figures 11, 12).

Figure 11. The field of approachability and defined potential loss for a few variants of the flight completion in the event of the complete failure of the aircraft engines

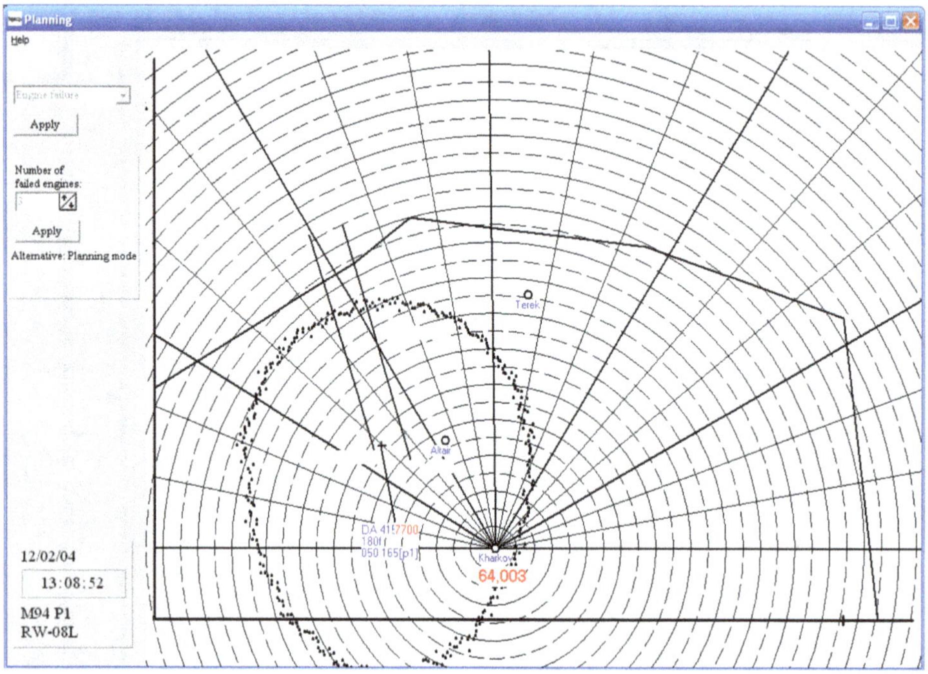

Figure 12. The field of approachability and defined potential loss for a few variants of the flight completion in case of the fire on board the aircraft which was not extinguished

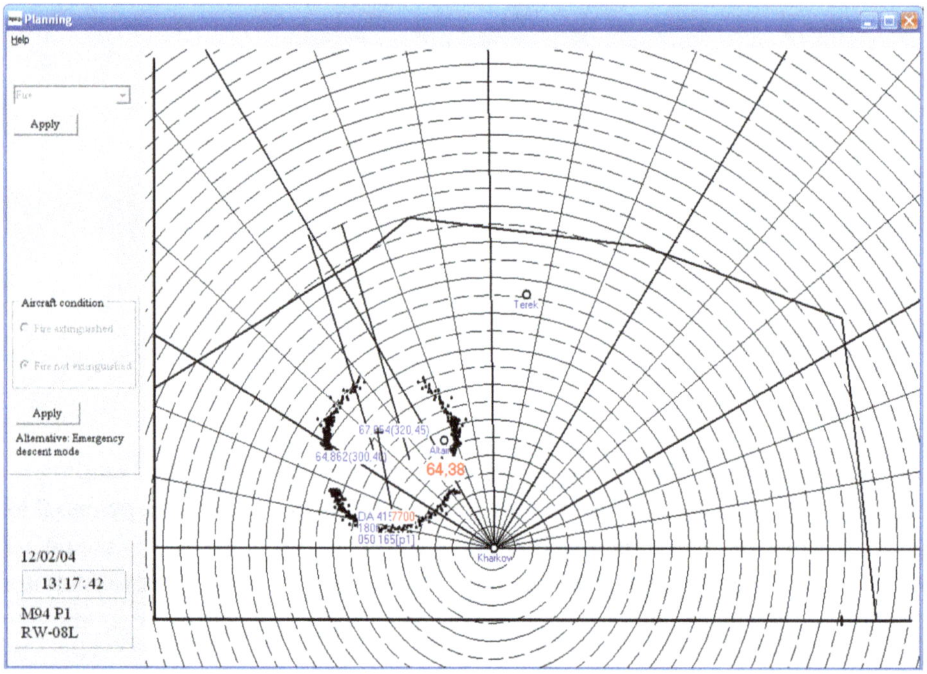

The program complex «Prompt» allows:

- To give the recommendations concerning the possibility of the further flight continuation or the necessity to accomplish a forced landing of the aircraft with the aid of the interface suitable for the user;
- To define the field of approachability of the aircraft in case when the necessity to make a forced landing arises;
- To form the evaluation of the alternative variants of the flight completion and define the optimum variant using the potential-loss minimization criterion.

High-quality pre-flight preparation of an aircraft crew is an important stage of the flight. During pre-flight preparation the pilot must handle a large amount of diverse information and make decision about (not) departure. In addition, the pre-flight preparation may take place under time pressure, so the solution of the pre-flight preparation tasks is advisable to entrust to the automated system of pre-flight information preparation (AS PIP) with intelligent module for support of the decision making about aircraft departure (Figures 13, 14).

The authors developed computer programs for DSS (Shmelova & other, 2011-2015) of flight pilot, UAVs operator, etc. The result of pre-flight preparation is decision making by pilot regarding the aircraft departure / not departure. The main factors influencing the decision making are take-off weight of AC and meteorological conditions at the aerodrome of departure, destination (alternate) aerodrome (Shmelova, Kharchenko,Artemenko, Otryazhiy, 2011).

Based on the requirements that are set out in regulatory documents, module for support of the decision making about aircraft departure entrusted with the following main tasks:

- Determination of compliance of actual take-off weight of AC to permitted;
- Determination of compliance of actual meteorological conditions at the aerodrome of departure to permitted;
- Determination of the presence of hazardous weather phenomena enroute;
- Determination of compliance of actual meteorological conditions at the destination (alternate) aerodrome to permitted.

Figure 13. Conceptual model of AS PIP

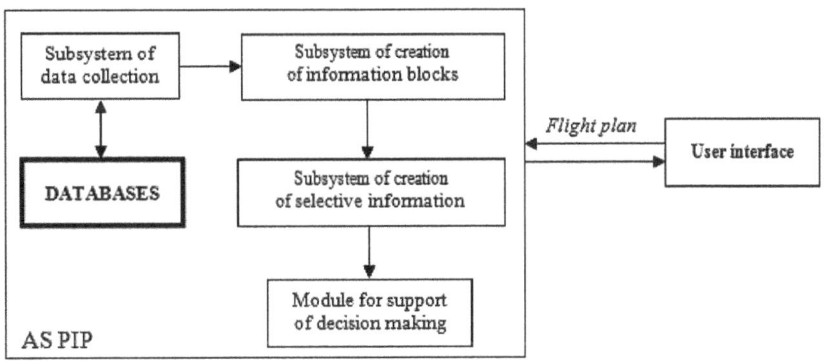

These data are input to the model, which generates recommendations for decision making regarding the aircraft departure. If any of the factors does not comply with the requirements, departure is prohibited. Compliance or non-compliance conditions for decision making about departure are represented in polar coordinates as a diagram "Spider-CWI". Another problem solved by the AC crew during pre-flight preparation is choice of the alternate aerodromes. According to the regulations, alternate aerodromes are chosen based on the following factors:

- Minimum of the First Pilot;
- Meteorological conditions at alternate aerodromes;
- Amount of fuel on board;
- Distance to the alternate aerodrome.

In flight emergencies for ensuring the flight safety and the cost-effectiveness of the flight, finding an optimal alternative for passengers, cargo and crew, in a flight dispatcher's DSS is used multifactorial model of choice of landing aerodrome, which takes into account more additional parameters: danger of the situation, type of the flight (regular, primary), technical characteristics of the AC, technical specifications of aerodromes (for example, condition of runways, navigational aids, lighting system), air navigation and airport charges, etc.

Decision Support Systems of Unmanned Aerial Vehicle's Operator

Remotely piloted aircraft systems (RPAS) are a new component of the aviation system, one which ICAO, States and the aerospace industry are working to understand, define and ultimately integrate. These systems are based on cutting-edge developments in aerospace technologies, offering advancements which may open new and improved civil/commercial applications as well as improvements to the safety and efficiency of all civil aviation. Unmanned aircraft has several advantages, namely low operating cost, good concealment and flexibility, simplicity and availability of technology compared to manned aircraft and Unmanned Aerial Vehicles (UAV) can be used in cases where the usage of manned aircraft is impractical, expensive or risky (Reg Austin., 2010). The goal of ICAO in addressing unmanned aviation is to provide the fundamental international regulatory framework through Standards and Recommended Practices (SARPs), with supporting Procedures for Air Navigation Services and guidance material.

Figure 14. The structure of intelligent module for support the decision making about aircraft departure

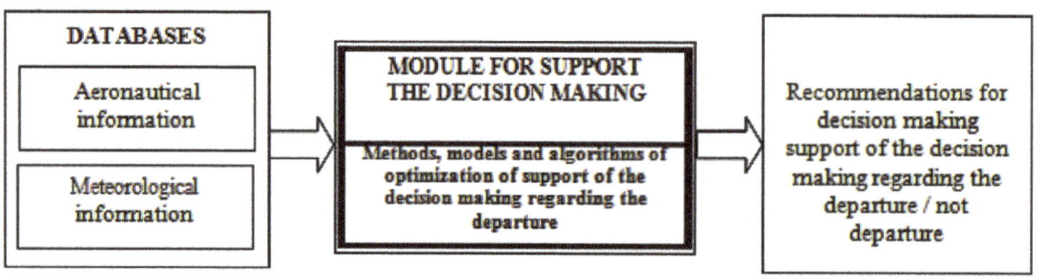

The main advantage of using UAVs is tasks that involve risk to humans and efficiency in solving economic problems. Obviously, UAVs are effective in monitoring forest fires, search and rescue operations in the processing of agricultural crops, relay communications and the movement of goods. In this sense, the usage of UAVs is more appropriate: to relay communications in those places - where the antenna coverage cannot be set because of difficult terrain, agriculture, with aerial photography, moving cargo, for military purposes. Emergency situations may occur when flying both in manual, and in the autonomous management. For operations carried out "manually", plays an important role the HF and a significant part of emergency arises due to wrong actions of the operator.

Let we have some UAV that performed different tasks purposes. Air traffic controller using technological procedures "ASSIST" (Acknowledge, Separate, Silence, Inform, Support, Time) decides in emergency situations of flight. At a certain stage of flight is probable extraordinary or emergency situations (for example: loss of control, engine failure, etc.), where it is some risk to lost UAVs. Taking into account the high cost of UAVs it is proposed to build an algorithm of UAV's operator actions using module «ASSSIST» (Acknowledge, Separate, Synergetic ((Coordinated, Cooperation, Consolidation)) Silence, Inform, Support, Time) for each type of UAV. Module «ASSSIST» includes in Distributed Decision support system (DDSS) and has models of the Decision Making (DM) by H-O under Certainty, Risk and Uncertainty (Shmelova, 2016).

The chapter analyses the problem of organization of integrated flying remotely piloted aircrafts. The structure of the distributed DSS remotely piloted aircraft (RPA) solutions. Useful of the database of local DSS operators of RPAS is NoSQL database, For rational aggregating requests and increase the processing speed as a database as NoSQL solution. For planning and flight control UAV developed a Distributed Decision Support System (DDSS), which represents a complex system with complex interactions geographically distributed local DDSS operators of RPA. During the flight UAV's (RPA's) may be controlled by RPAS. Remotely piloted aircraft controlled with remote piloting station (RPS) with the management and control line (C2). Together with other components such as the starter equipment and equipment for the return, if it is used, RPA, RPS and the line C2 constitute RPAS (Manual of Remoted Piloted Aircraft Systems (RPAS), ICAO, 2015).

At any given time t_i k-RPA can be controlled from only one j-th RPS, if necessary, at time t_{i+1} to be transmitted to the control $(j + 1)$th RPS for using DDSS (Figure 15). This transfer flight control of the j-th RPS to $(j + 1)$ -th RPS to be safe and effective, which is provided through the local DSS operators RPAS. According to the recommendations of the ICAO guidelines (Manual of Remoted Piloted Aircraft Systems (RPAS), ICAO, 2015) task system can perform one or more nodes (local DSS operators RPAS). With the formation of the database addresses issues related to the inclusion of RPA the existing regulatory framework of civil air navigation system; description and classification of UAVs and related components; rules of flight, such as instrument flight rules (IFR) and Visual Flight Rules (VFR) flights in the visual line of sight (VLOS) and beyond line of sight (BVLOS).

From the DDSS organization depends on its effectiveness. In determining the structure of the system are the following objectives:

- Study of problems that solves DSS;
- Selection algorithm implementation problems;
- Determine the levels and components in the system;
- Distribution of tasks to nodes (levels);
- Definition of technical means complex systems;

Figure 15. The structure of distributed RPS Mission Control UAVs

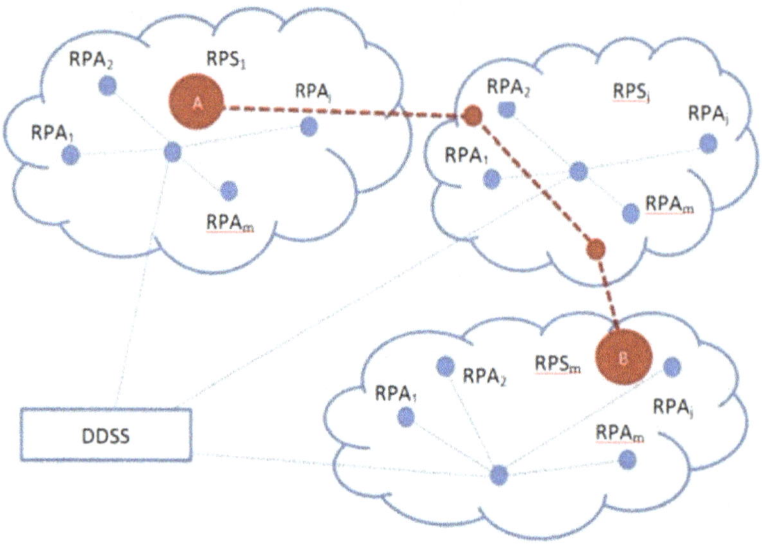

- Relevant content and management database.

Indeed, UAV piloted by remote piloting station (RPS) using line management and control (C2). For piloting UAVs use RPS, which is both portable and station is located indoors or outdoors, be stationary or mobile (mounted on the vehicle / ship / aircraft). To coordinate interaction and exchange of information between remoted pilots developed database of local RPS NoSQL. During developing a database of local RPS, UAV users, it was made UAS components analysis, UAV, RPS, C2, and so on. Information block local DSS operators RPAS within a distributed DSS drawn up in accordance with the recommendations of ICAO (Manual of Remoted Piloted Aircraft Systems (RPAS), ICAO, 2015) Instructions for assembly applications for private use under the planning of flights contains the following components:

- Characteristics of RPAS, RPA, RPS, communications, navigation, surveillance (conduct of the lines and control (C2), GNSS, ADS-B, etc.);
- Certification RPAS, RPA, RPS;
- RPAS obligations of the operator;
- Safety management;
- DAA;
- Connection with the ATM;
- Procedures manual landing planes, airports;
- RPA operating procedures, rules of operation, etc.

Unlike relational model, which stores logic business core application in different physical table in order to normalize, NoSQL storage with these entities operate as an integral of objects. So instead of creating a new table for each separate entity, we simply create objects of entities defined for these attributes. Pre-flight planning should include consideration to alternate aerodromes / recovery sites, as appropriate, in the event of the emergency or meteorological-related contingency. Before selecting an alternate recovery / landing, location the remote pilot should consider the adequacy fuel / energy, reserves, reliability of C2 links with the RPA, ATC communication capability as necessary and meteorological conditions at the alternate. For using known criteria of decision making under uncertainty are finding optimal landing aerodrome in FE. Using this approach, the task, will ignore the specific objects and their attributes and will work with individual flows with a holistic entities. This allows you to use a standardized approach to consolidating all information flows in the subject area, and to identify for each use prioritize and rules RPAS and flights RPA.

FUTURE RESEARCH DIRECTIONS

Further research should be directed to the solution of practical problems of actions UAV's operator in case of emergencies, software creation. Models of FE development and of DM by UAV's in FE will allow predicting the H-O's actions with the aid of the informational-analytic and diagnostics complex for research H-O behaviour in extreme situation. To develop new DSS / ES of ANS's H-O (pilots, ATCs, UAV, SMS – operators) in flight emergencies and in other situations too. To investigate applied tasks of the DM in STS by an H-O in aviation systems, Chemical production, Energy, military industry, etc. Further research should be directed to the solution of problem in prerequisites for a emergency situation and preventing catastrophic situations too. DSS of an air traffic controller in flight emergencies gives to the aviation operator possibility to quantify evaluation of alternatives of the flight situation development and to operatively choose an action strategy with minimal potential damage in the conditions of incomplete and uncertainty information. DSS of a flight dispatcher provides fast search of the information and support of the aviation operator during DM regarding the possibility of departure or regarding the optimal aerodrome in case of forced landing of AC. It is reasonable to use an air traffic controllers DSS during the real ATC in flight emergencies, a flight dispatcher's DSS – in the automated system of pre-flight information preparation.

What's Next?

On next steps we are planning consider other expert systems and decision support systems for STS complex system, such as hydraulic engineering, chemical and military industries, gas and oil pipelines, nuclear power plants and transport etc. It is necessary to analyze the factors influencing the decision-making of operators in these systems in order to predict the development of the technoogenic catastrophe and prevent it using expert systems and decision support systems. It is reasonable to use DSS during the real ATC in flight emergencies, a flight dispatcher's DSS – in the automated system of pre-flight information preparation.

CONCLUSION

Conceptions of the ATCs DSS in flight emergencies and of a flight dispatcher's DSS have been presented, program realization of systems has been shown. DSS of an air traffic controller in flight emergencies gives to the aviation operator possibility to quantify evaluation of alternatives of the flight situation development and to operatively choose an action strategy with minimal potential damage in the conditions of incomplete and uncertainty information. DSS of a flight dispatcher provides fast search of the information and support of the aviation operator during DM regarding the possibility of departure or regarding the optimal aerodrome in case of forced landing of AC.

Conceptual models of DSS for air traffic controller in flight emergencies and automated system of pre-flight information preparation with intelligent module for support the DM about aircraft departure are presented and program realization of systems are shown. Algorithm of determining the optimal aerodrome for the forced landing of aircraft is provided. Global operational air traffic management concept that reflects the ICAO vision of a single coherent and based on global fashion mutually ANS systems should ensure collective consolidation to management of traffic. Consolidation of information flow plays an important role in air traffic management. In today's rapid development of aviation to the person receives a large amount of information from a variety of control systems. To make the right decision information which comes to air traffic controllers should be structured and consolidated information flows. There is a need to establish a system that will structure the information to collect and consolidate a disparate information flows and the output file turnkey solutions for air traffic management. To implement this system plays an important role choose Database. Analysis of existing solutions shows the need to use NoSQL database. These systems are capable of providing the required speed. And thanks to a special approach to data aggregation becomes possible standardization of methods of consolidation flows.

REFERENCES

Austin, R. (2010). *Unmanned Aircraft Systems: UAVs Design, Development and Deployment*. John Wiley & Sons Ltd. doi:10.1002/9780470664797

Federal Aviation Administration. (2008). Aeronautical Decision Making. In Pilot's Handbook of Aeronautical Knowledge. FAA-H-8083-25. 17.1–17.32. Author.

Gerasimov, B., & Lokazyuk, V. (2007). *Intelligent Decision Support Systems*. Kyiv: European University Press.

International Civil Aviation Organization. (1994). *Human Factors in CNS/ATM Systems. Introduces the Concept of Human-Centered Automation: Human Factors Digest Nº 11. Cir. ICAO 249-AN/149*. Montreal, Canada: Author.

International Civil Aviation Organization. (2004). *Cross-Cultural Factors in Aviation Safety: Human Factors Digest Nº 16. Circ. ICAO 302-AN/175*. Montreal, Canada: Author.

International Civil Aviation Organization. (2005). *Global Air Navigation Plan. Doc. ICAO 9750-AN 963*. Montreal, Canada: Author.

International Civil Aviation Organization. (2008). Normal operations safety survey (NOSS). Doc. ICAO 9910-AN 473. Montreal: Author.

International Civil Aviation Organization. (2009). *Manual on Global Performance of the Air Navigation System. Doc. ICAO 9883*. Montreal, Canada: Author.

International Civil Aviation Organization. (2012). *Report of the Twelfth Air Navigation Conference.* AN-Conf/12-WP/162. Montreal: Author.

International Civil Aviation Organization. (2013). *State of Global Aviation Safety*. Montreal, Canada: Author.

International Civil Aviation Organization. (2013). Safety Management Manual (SMM) (3rd ed.). Doc. ICAO 9859-AN 474. Montreal: Author.

International Civil Aviation Organization. (2015). *Manual of Remoted Piloted Aircraft Systems (RPAS). Doc. ICAO 10019/AN 507*. Montreal, Canada: Author.

Kharchenko, V., Shmelova, T., Bondarev, D., & Stratiy, A. (2017). Modelling of Decision Making of Unmanned Aerial Vehicle's Operator in Emergency Situations. *Proceedings of the National Aviation University*, *1*, 20–28.

Kharchenko, V., Shmelova, T., & Sikirda, Y. (2012). *Decision-Making of Operator in Air Navigation System*. Kirovograd: KFA of NAU.

Kharchenko, V., Shmelova, T., & Sikirda, Y. (2016). *Decision-Making in Socio-Technical Systems.* Kyiv: NAU.

Shmelova, T., & Bondarev, D. (2016). Graph Theory Applying for Quantitative Estimation of UAV's Group Flight into Aerial Photography. *Electronics and Control Systems*, *4*(46), 128–133.

Shmelova, T., Bondarev, D., & Znakovska, Y. (2016). Modeling of the Decision Making by UAV's Operator in Emergency Situations. In *4th International Conference on Methods and Systems of Navigation and Motion Control (MSNMC-2016)*. Kyiv: NAU. 10.1109/MSNMC.2016.7783099

Shmelova, T., Kharchenko, V., Artemenko, O., & Otryazhiy, V. (2011). *Computer Program «Choosing the Pre-Flight Information and Decision Making about Departure for Automated System of Pre-Flight Information Preparation (AS PIP)»: certificate of registration of copyright in the product Nº 40062*. Academic Press.

Shmelova, T., Kharchenko, V., Sikirda, Y., & Gerasimenko, O. (2011). *Computer Program «Optimizing the Choice of Alternative Variant of Aircraft Flight Completion in Flight Emergencies Prompt»: certificate of registration of copyright in the product Nº 3787*. Academic Press.

Shmelova, T., Sikirda, Y., Assaul, A., & Stasiuk, O. (2015). Structural Analysis of Management Environment of Aviation Enterprise from the Point of Systematic Approach. *Proceedings of the National Aviation University*, *2*, 3–12.

Shmelova, T., Sikirda, Y., & Belyaev, Y. (2014). Informational Support of Air Navigation System's Human-Operator. *Scientific Works of the National University of Food Technologies*, *20*(4), 7–18.

Shmelova, T., & Stratij, A. (2015). *Computer Program "Aggregation of heterogeneous information flows": certificate of registration of copyright in the product Nº 59466*. Academic Press.

Shmelova, T., Yakunina, I., & Moiseenko, V. (2014). *Computer Program "Network analysis of a special case in flight": certificate of registration of copyright in the product Nº 55587*. Academic Press.

Shmelova, T., Yakunina, I., & Yakunin, R. (2013). *Computer Program "TestExtraordinaryIncident": certificate of registration of copyright in the product Nº 50826*. Academic Press.

Shmelova, T., Yakunina, I., Znakovskaya, E., Ivleva, G. (2014). *Computer Program "Decision making under uncertainty": certificate of registration of copyright in the product Nº 56819*. Academic Press.

Shmelova, T., & Znakovskaya, E. (2015). *Computer Program "Decision Tree": certificate of registration of copyright in the product Nº 62163*. Academic Press.

Sirodzha, I. (2002). *Quantum Models and Methods of Artificial Intelligence for Decision Making and Management*. Kyiv: Naukova Dumka.

Tarasov, V., & Gerasimov, B. (2007). *Intelligent Decision Support Systems: Theory, Synthesis, Efficiency*. Kyiv: International Academy of Computer Science and Systems.

KEY TERMS AND DEFINITIONS

AC: Aircraft.

Alternative: A possible scenario during the decision-making.

ANS: Air Navigation System.

Artificial Intelligence Systems - AIS: Systems that perform the inherent human intellectual activities associated with the perception and processing of knowledge, reasoning and relevant communication.

AS PIP: Automated system of pre-flight information preparation.

ATC: Air Traffic Control.

ATCO: Air Traffic Control's Operator.

Base Models: A set of mathematical, logical, linguistic and other models used for comparative analysis of multi-alternative decision.

CNS/ATM: Communication, Navigation, Surveillance / Air Traffic Management.

Database: An information structure that reflects the status and relationship of objects being analyzed.

Decision Making: The process of selecting the optimal alternative action or permissible subset of the set of possible alternatives.

Decision Support System (DSS): A computerized system designed to help a user make decisions.

DM: Decision-making.

EJM: Expert Judgment Method.

ES: Emergency situation.

FE: Flight Emergencies.

Frame: Structure to describe stereotypical situation prevailing characteristics of the situation (slots) and values (placeholders slots).

HF: Human factor.

H-O: Human-operator.

ICAO: International civil aviation organization.
RPA: remotely piloted aircraft.
Solution: The result of a particular choice of alternative actions.
UAV: Unmanned Aerial Vehicles.
User Interface: The way of user interaction with the system.

This research was previously published in Socio-Technical Decision Support in Air Navigation Systems edited by Tetiana Shmelova, Yuliya Sikirda, Nina Rizun, Abdel-Badeeh M. Salem, and Yury N. Kovalyov, pages 182-214, copyright year 2018 by Engineering Science Reference (an imprint of IGI Global).

Section 4

Detection, Imaging, and Sensor Technologies

Chapter 9
Skin Detection With Small Unmanned Aerial Systems by Integration of Area Scan Multispectral Imagers and Factors Affecting Their Design and Operation

Stephen R. Sweetnich
Air Force Institute of Technology, USA

David R. Jacques
Air Force Institute of Technology, USA

ABSTRACT

Dismount skin detection from an aerial platform has posed challenges compared to ground-based platforms. A small, area scanning multispectral imager was constructed and tested on a Small Unmanned Aerial System (SUAS). Computer vision registration, stereo camera calibration, and geolocation from autopilot telemetry were utilized to design a dismount detection platform. The test expedient prototype was 2kg and exhibited skin detection performance similar to a larger line scan hyperspectral imager (HSI). Outdoor tests with a line scan HSI and the prototype resulted in an average 5.112% difference in Receiver Operating Characteristic (ROC) Area Under Curve (AUC). This research indicated that SUAS-based Spectral Imagers are capable tools in dismount detection protocols.

DOI: 10.4018/978-1-5225-8365-3.ch009

INTRODUCTION

Integrating Multispectral Imaging (MSI) on a Small Unmanned Aerial System (SUAS) may present a valuable tool for dismount detection. SUAS offer versatile platforms for Military, police, or Search and Rescue (SAR) operators. The U.S. Geological Survey, the U.S. Air Force, and others have demonstrated the utility of SUAS in natural disasters such as the Fukushima Daiichi meltdown to take photographs at altitudes that were hazardous to human pilots (U.S. Geological Survey, 2011). SUAS can extend our capacity to perform missions in dangerous environments with limited personnel due to their autonomous operating capability (Aeryon Labs Inc., 2013). Sensors such as MSI and hyperspectral imaging (HSI) can expand the capabilities of SUAS. Currently, skin detection is achieved with large and resource intensive sensor assets. HSI can remotely identify materials such as skin but it has proven difficult to utilize the data quickly and efficiently. In SAR, border patrol, or ISR operations, the ability to identify skin can save lives. If rescue crews were able to pinpoint survivors on the ground by differentiating rubble and skin, they could provide immediate support at their location. Situational awareness of the mission area may be improved by identification of material and dismounts. SUAS are quickly deployable, versatile sensor platforms, which can provide invaluable information in many domains. This research utilizes results from aerial testing of an unmanned platform to propose a spectral imaging system that can detect dismounts from an SUAS. Also, factors that pertain to accurate dismount detection with an SUAS are explored.

SCOPE

Reasearch Objective

The main focus of this research was to measure the feasibility of a skin detection multispectral sensor on an SUAS for Search and Rescue operations. To measure the utility of a system, subjective and objective metrics were used. By using standardized metrics, the detection system was compared against a wide range of detection systems. There were four main investigative questions:

- Is MSI effective for dismount detection on an SUAS?
- What type of MSI system is needed for an aerial platform?
- Which camera parameters are most critical for aerial dismount detection?
- What metrics are best to compare dismount detection ability?

Assumptions and Limitations

The system for skin detection was limited to the sensors available at AFIT, including the MSI imager, visible spectrum (VIS) imager, and bandpass filters. The SUAS used for the skin detection was limited to the airframes available at the AFIT Autonomy and Navigation Technology (ANT) Center. For the time available for flight test, a specific loiter pattern was utilized, which was optimized to the camera parameters (e.g., resolution, focal length). Due to the cost and size of a digital video acquisition and transmission system, onboard storage of imagery was performed in flight test. Weather conditions (sunlight, rain, wind …) were unavoidable due to limited test range time, and they affected flight duration, autopilot waypoint tracking, and sensor parameters. The research assumed that each dismount had skin which was visible from the air. A flat earth model was used for processing flight data.

Materials and Equipment

Multiple components were utilized for the SUAS, including an airframe, communications, video, and ground control subsystems. Each of these components was selected to meet the requirements specified in the literature (Beisley, 2012; Liggins, Hall, & Llinas, 2008), and experimentation. The major airframe components were COTS hardware that included a modified electric Sig Rascal 110 RC aircraft, a programmable autopilot, and sufficient batteries for test flights. The communications components included 900 MHz wireless telemetry modems, and 5.8 GHz wireless video transmitters. The video subsystem components included a filter wheel, bandpass filters, a programmable stepper motor driver and microcontroller, along with separate batteries to power the sensor components. The GCS included video and telemetry receivers, mission planning software, MATLAB and ENVI for image processing software, and frame grabbing hardware (Excelis Inc, 2013; The Mathworks Inc., 2014). The entire system was flight tested at Camp Atterbury, IN with U.S. Army, U.S. Air Force, and Co-Operative Engineering Services Inc (CESI).

BACKGROUND

SUAS and Aerial Imagery

Small UAS have begun to utilize inexpensive and widely available sensors to conduct ISR, and their utility has been proven (Beard & McLain, 2012; Gundlach, 2012). Classification of objects in a scenario can improve the situational awareness for multiple stakeholders. One operational scenario was explored by (Merino, Fernando, Martinez-de Dios, Ferruz, & Ollero, 2006) who focused on geolocation of forest fires from an unmanned helicopter. This research identified many factors to consider with platform (airship vs. helicopter) onboard processing time, off board processing, communications, and software required for fire detection. However, the research did not consider the specific electromagnetic wavelengths to use for feature extraction or application to dismount detections. Nevertheless, this research demonstrated the capability of SUAS to perform object identification in hazardous environments.

Image Geometry and Projection

Imagery acquired from an aerial platform requires consideration of aircraft attitude and position. The impact of attitude and position of an SUAS is found from many sensors onboard the aircraft. The Inertial Measurement Unit (IMU) provides information on body frame position an attitude, which is used for both vehicle control calculations and conversion between body frame and inertial (earth) frame coordinates. A body fixed camera maintains a relatively simple geometric relationship to the body frame as compared to a gimbaled camera. Each coordinate frame is related by a rotation matrix, and they can be defined intuitively using the Euler angle system. (Beard & McLain, 2012)

Since the IMU data is given in the body frame, one can then relate to any frame of reference with a transformation matrix. The coordinate transformation from the body frame to the inertial frame, is given by Equation, where θ, ϕ, ψ are used to define the orientation angles, and $\theta_{sensor}, \phi_{sensor}, \psi_{sensor}$ define the camera orientation relative to the body frame. The transformation from the body to camera frame is given by Equation, assuming that the camera is at (or near) the CG (Beard & McLain, 2012). The final aimpoint vector, $\vec{A}$, can be obtained by applying the rotation matrix given by Equation . Thus, any sensor aimpoint can be determined approximately from the autopilot telemetry. It is approximate, because there are inherent errors and drift associated with autopilot sensors; this makes the data noisy (Beard & McLain, 2012). The sensor footprint from an SUAS will change over time because the aircraft is moving at each point in time. Figure 1 shows the camera frame of reference with respect to the inertial reference frame and the sensor aimpoint.

$$R_b^i = \begin{bmatrix} c_\theta c_\psi & s_\phi s_\theta c_\psi - c_\phi s_\psi & c_\phi s_\theta c_\psi + s_\phi s_\psi \\ c_\theta s_\psi & s_\phi s_\theta s_\psi + c_\phi c_\psi & c_\phi s_\theta s_\psi - s_\phi c_\psi \\ -s_\theta & s_\phi c_\theta & c_\phi c_\theta \end{bmatrix}$$

$$R_b^c = \begin{bmatrix} -s_{\theta_{sensor}} & c_{\theta_{sensor}} c_{\psi_{sensor}} & c_{\theta_{sensor}} s_{\psi_{sensor}} \\ 0 & -s_{\psi_{sensor}} & c_{\psi_{sensor}} \\ c_{\theta_{sensor}} & s_{\theta_{sensor}} c_{\psi_{sensor}} & s_{\theta_{sensor}} s_{\psi_{sensor}} \end{bmatrix}$$

$$R_c^i = R_b^i \left[R_c^b \right]^T = R_b^i R_c^b$$

Figure 1. The coordinate system for a camera equipped airframe

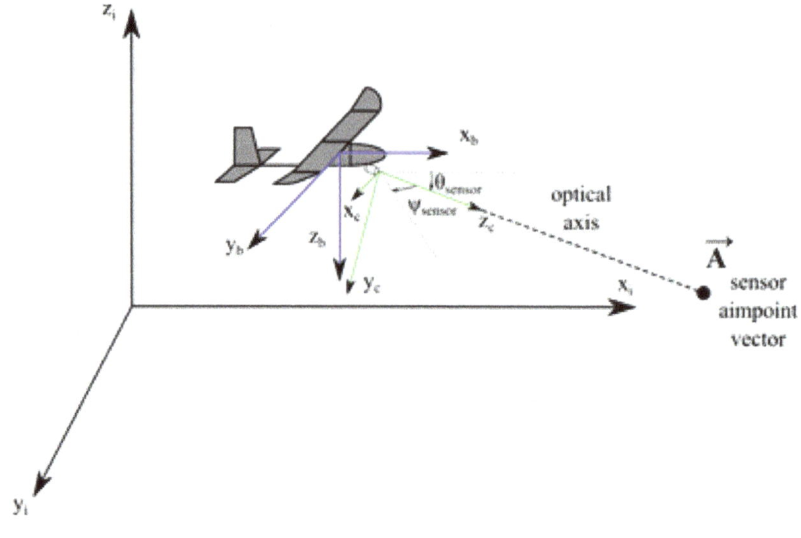

Multi-Spectral/Hyperspectral Sensors

Imagers suitable for skin detection were of the main focus to this research; however, previous research indicated a few difficulties with using VIS/RGB for skin detection. Benefits to VIS imagers included their small form factor, complexity, and cost. VIS colorspace algorithms from previous work attained skin detection true positives averaging 87% and false positives averaging 20% using the Compaq database (Vezhnevets, Sazonov, & Andreeva, 2003). Work proposed by Gomez used a rules based threshold for RGB values and attained false positives as low as 8.0% (Gomez & Morales, 2002). However, Assigning RGB threshold values for the multitude of skin types is challenging due to different pigmentation (Kakumanu, Makrogiannis, & Bourbakis, 2007).

These methods fail to respond to changes in skin type, this may be due to light scattering from melanin, melanin, hemoglobin, filamentous proteins (Lister, Wright, & Chappell, 2012). Nunez studied skin in the multispectral regime and discovered the relationship between the absorption and reflection properties of skin with infrared light in the VIS and SWIR spectrums. This work attained consistent true positives of 95% with false positives of 0.6%, surpassing the performance of all RGB colorspace methods. Furthermore, this method was invariant of skin type due to the universal reflection properties of skin melanosomes, blood content, oxygenation, dermal thickness, collagen, and subcutaneous reflections in the VIS and SWIR domains (Nunez, 2009). The method from Nunez, is indifferent to skin type and provides a robust detector for all types of skin pigments; however, it requires an imager sensitive to the infrared electromagnetic spectrum such as an MSI.

Multispectral imagers (MSI) operate at a few broadband wavelengths while hyperspectral imagers (HSI) operate in a few hundred narrow wavelengths on the electromagnetic spectrum from 0.4-2.5μm (Rao & Arora, 2004). These imagers create a data cube from spectral measurements. Many inexpensive focal plane sensors are responsive up to the near infrared (NIR), but expensive components such as diffraction gratings or prisms are currently needed to segment light at different wavelengths across the sensor plane.

There are three main types of HSI scanning methods: point scanning, line scanning or push broom, and area (focal plane) scanning. A point scanning imager is used to capture one single pixel of data for multiple wavelengths. For example, the ASD FieldSpec 3 can be used to capture highly accurate reflectance data across the spectrum (350-2500nm) but is limited to close range optical fiber or contact probe for data collection (ASD Inc., 2012). The push broom HSI imager utilizes a dispersion device such as a prism or optical fibers to distribute different wavelengths of a single line of pixels. An example of a line scan imager is the SpecTIR AisaDUAL built from two SpecIM HSI (SpecIM, 2013). A focal plane detector, snapshot, or area scanner, keeps the sensor behind a bandpass filter and objective optic (Grahn, Geladi, & Burger, 2007; Hagan & Kudenov, 2013). In this type of imager, the sensor receives light from a wavelength until the filter is changed. An area scan imager was constructed in research by Peskosky (2010), but used one entire SWIR camera for each wavelength in a single fore optic or monocular system. The camera built by Peskosky would be size prohibitive on an SUAS due to focal length requirement and spacing from additional sensors on the multi-channel system. Other area scan techniques perform band separation with a patterned filter array in front of a sensor; however, these types of imagers are not widely available (Eichenholz & Dougherty, 2009). Illustrations of each HSI mechanism are shown in Figure 2 and are adapted from (Grahn et al., 2007).

Figure 2. A few types of HSI Imagers and example configurations are shown. Top, portrays a point scan imager. Middle, depicts a line scan imager. Bottom, depitcts an area scanning, or snapshot imager arrangement

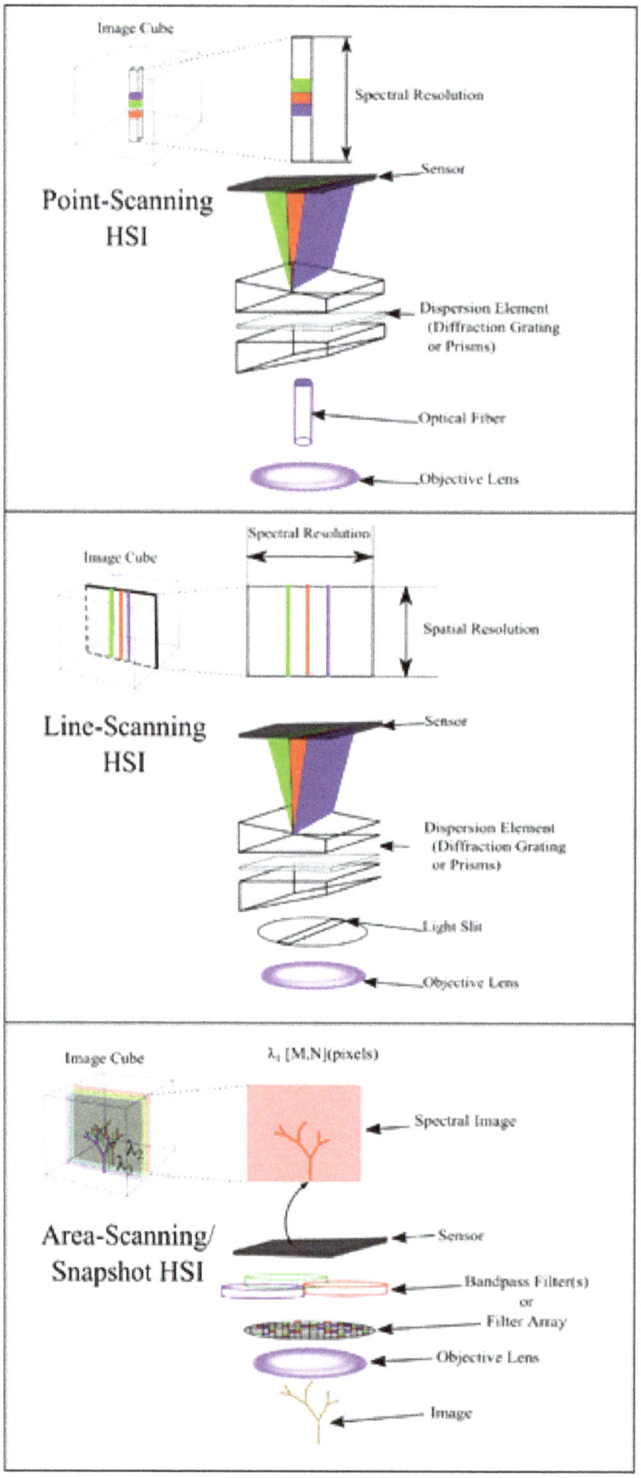

Sun Angle

The incident angle of light can impact the detection of an object. Koch (2011) denotes the objects' material properties as an independent variable of the apparent reflectance. Skin also has unique reflective properties that make detection with HSI difficult. The relationship between light angle and material property can be modeled by a Bidirectional Reflectance Distribution Function (BRDF), and the reader is referred to the literature for more research on this subject (Comninos, 2011; Koch, 2011). Different angles of reflected light may result in inconsistencies in the detector's rule based threshold. Models for the BDRF have been derived with empirical and physically-based methods; the models have been used to enhance the realism of 3-D computer graphics (Comninos, 2011). Figure 3 depicts the sun angle and incidence angle of reflection on a dismount with non-orthogonal illumination where α is the sun zenith, θ_i is the angle of incidence, and θ_d is the viewing angle from the dismount surface normal. For any light based sensor in changing environmental conditions, an accurate model of illumination is needed to make robust detections.

Detection Algorithms

For design of an aerial dismount detection platform, multiple skin detection strategies were considered. Comparison of skin detection techniques provided insight on the feasibility of VIS or RGB imagers for aerial detection. VIS detection techniques provide a sensible alternative to MSI because they minimize costly hardware or preprocessing (Kakumanu et al., 2007). Kakumanu et al. found that reasonable accuracy could be obtained by Gaussian Mixture Models and Neural Networks with the VIS bands in different color spaces; however, threshold values for different skin types, lighting, and environments posed a challenge for these techniques (2007).

Figure 3. Sun angle affects the measured skin reflectance on a dismount from a sensor

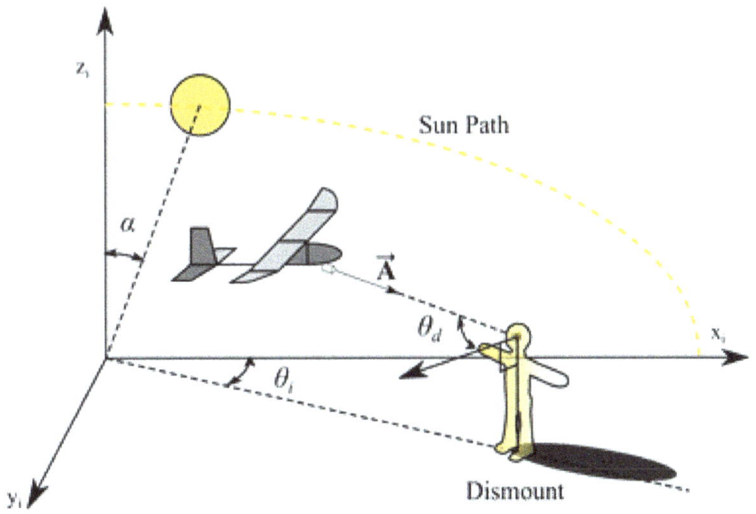

Work by Nunez, and then Petrosky, and Beisley, explored the benefits of skin detection with SWIR bands. Nunez found significantly higher accuracy from normal RGB histogram detection by adding SWIR wavelengths (Nunez, 2009). The principle behind SWIR detection methods is that skin reflects water similarly to vegetation. The author adapted agricultural analysis methods such as the normalized difference vegetation index (NDVI) to detect skin. Altering the SWIR wavelengths allowed for accurate skin detection, regardless of skin pigmentation. After an image is collected, another crucial step is converting the raw pixel values to reflectance. Reflectance is a unitless measure of light which is unique for different types of materials; spectral libraries can define objects by their spectral signature or fingerprint (NASA, 2011). More information on radiometric conversions for HSI can be found in the literature (Beisley, 2012; NASA, 2011; SpecTIR, 2012). The importance of converting aerial imagery to reflectance, is to normalize the data despite environmental reflections, sun angle, or weather conditions. Highly accurate spectral measurements can be obtained with a point scanning HSI such as a spectroradiometer, and signatures were obtained for several skin types as truth data.

There were several detection algorithms utilized for this research:

- Normalized Difference Skin Index (NDSI): This computational combination of visible spectrum (VIS) and MSI imagery is used to classify objects. The algorithm was effectively validated with prior research from (Beisley, 2012; Koch, 2011; Nunez, 2009; Peskosky, 2010). The NDSI was adapted from the NDVI, which was used classically to perform vegetation analysis for aerial agriculture exploitation (Nunez, 2009). Equation is the Normalized Difference Skin Index (NDSI) algorithm that is applied to the MSI data, where ρ is the reflectance at a specific wavelength. The normalized green red index (NDGRI), which utilizes VIS bands to minimize vegetation detections, is given by Equation . The combined detection algorithm is given by Equation . Finally, a threshold is placed on $d_{NDSI}(x)$.

$$NDSI = \frac{\rho(1080nm) - \rho(1580nm)}{\rho(1080nm) + \rho(1580nm)}$$

$$NDGRI = \frac{\rho(540nm) - \rho(640nm)}{\rho(540nm) + \rho(640nm)}$$

$$d_{NDSI}(x) = NDSI_x - \begin{Bmatrix} NDGRI_x > 0 \to 1 \\ NDGRI_x < 0 \to 0 \end{Bmatrix}$$

- Matched Filter (MF): Matched Filters have been commonly used for detections for their robustness and simplicity (Beisley, 2013). The matched filter can provide rapid means of detection compared with other detection algorithms (Excelis, 2013). The matched filter searches for a spectral signature of skin, using a spectral signature as *a priori* knowledge. The score for a MF is defined by (5) where x is a pixel vector of all spectral bands, s^T is the spectral signature vector (transposed) of the desired object such as skin, and $\|x\|$ is the l_2 Norm of the vector x. A high score indicates a high detection probability.

$$MF(x) = \frac{s^T x}{\|s\|\|x\|}$$

Many comparable algorithms exist for spectral classification, such as Adaptive Cosine Estimators, Constrained Energy Minimization, Mahalanobis Distance, Spectral Information Divergence, Spectral Angle Mapper, Support Vector Machines (Excelis Inc, 2013; Manolakis, Marden, & Shaw, 2003). However, due to similarities in ROC performance and scope abatement, these detections were left for future research.

Metrics for Performance

Several different performance metrics were utilized in this research:

- **ROC Analysis:** Receiver Operating Characteristics (ROCs) are commonly used to quantify performance of a detection system. ROC curves display classification true positive rate and false positive rate over a varying detector threshold. For ROC analysis, truth labels were generated from the imagery. Next, the detection output or score, was thresholded and compared to the truth data. For the NDSI Algorithm, a single score method was adapted to generate a ROC (Beisley, 2012). Other detection methods could be compared by an objective metric such as area under the curve (AUC), where higher AUC denotes better performance.
- **NIIRS and GIQE:** A general metric for judging pixel density requirements is the National Image Interpretability Rating Scale (NIIRS). The NIIRS scale stratifies sensor Image Quality (IQ) from (1-9) with 9 being the highest level of quality; however, it is largely subjective and only quantifies details down to 5cm (Gundlach, 2012; Kerkes J. P. & Hsu, 2004). An objective metric was developed to predict NIIRS ratings based on empirically measured scales and is called the General Image Quality Equation (GIQE) (Gundlach, 2012).
- **Cumulative Probability of Blur Detection (CPBD):** The blur metric measures the image distortion due to motion blur, focus, or other aberrations. It utilizes a digital filter to detect edges in an image, and then applies a threshold window for edges that appear blurred. Another technique to calculate blur may apply a digital filter for recreating blur and then compare the filtered image to the original image. The principal behind the metric is if an image is already blurred, then the difference between the original and distorted will be less significant. The benefits of the blur metric are that it requires no reference image, in comparison to the Mean Square Error, Structural Similarity Index, or Peak Signal to Noise metrics. The algorithm for the CPBD metric is described in detail in the literature (Narvekar & Karam, 2011).

EXPERIMENTAL DESIGN AND CONSTRUCTION

SUAS MSI Video Subsystem

The SUAS MSI utilized a short wave infrared (SWIR) and VIS imager. For skin detection, a filter wheel was placed in front of the SWIR to change bands, and filters with wavelengths historically unique to the reflectance of human skin were utilized (Beisley, 2012; Nunez, 2009). A dual SWIR system was developed by Peskosky (2010); however, it relied on heavy optic table components and a wired connection to high performance computers. Only a single SWIR camera was required for the proposed MSI, which minimized weight and data transmission requirements. Lens focal length was selected based on an iterative analysis of lens size vs pixel density from test data at expected altitudes and flight conditions. Figure 4 shows the relationship between pixel size and sensor angle with the selected lenses. The SWIR was outfitted with a 35mm lens and the VIS was outfitted with a 12mm focal length lens to maintain similar image sizes given their sensor format. Additional processing algorithms were required to spatially register frames from both imagers. Stereo camera calibration software from was used to determine the epipolar geometry and the projective transformation required (Bouguet, 2013)

The camera mount was designed to add minimal weight to the airframe, and was body fixed. The baseplate and aircraft mounting assemblies were machined from aluminum and allowed for 30 - 45 degree sensor angle adjustments on the airframe, which was a Sig Rascal 110. The camera and filter wheel components were mounted with vibration damping material in compression. The baseplate was affixed to the airframe under the CG and the APM autopilot. The mount was first constructed as 3-D solid models to determine placement, weight balance, and where additional reinforcement was required on the airframe.

To capture images over multiple wavelengths, a filter wheel configuration was chosen based on cost and availability. The filter wheel alternated two different bandpass filter types to accommodate the NDSI algorithm. It was driven by a microcontroller with Hall Effect sensors to sense filter changes. The filter wheel housing was constructed with a 3-D printer from lightweight composites to minimize weight while keeping tight tolerances. Alignment of the filters was critical for eliminating aberrations. Figure 5 shows the detailed model of the MSI assembly. What this analysis depicts, is that sensor pixel density positively correlates with working distance; however, there is a compromise with frame overlap.

Ground Scanning Imager

A ground based imager was used to compare performance of the prototype MSI. This larger imager was a prototype line scan HSI based off the AisaDUAL with the ability to collect 363 bands from 400-2500nm. It used two sensors, SWIR and VNIR, and each had a lens with focal lengths of 18.5 and 22.5 mm respectively.

Test and Evaluation

The algorithm for processing video was unique to the SUAS MSI that was constructed. Since the filter wheel was spinning while the aircraft was moving, the images had to be overlaid to create an image cube. The process was as follows: A frame was captured with the filter in the first position, λ_1. Next, the filter wheel rotates to the second position, λ_2, and another frame is captured. It was necessary to achieve proper overlap between the first frame, F_1, and the second frame, F_2, for $i = 1,2 \ldots n$. To spatially align pixels in F_1 and F_2, registration was done using SIFT and RANSAC for point feature extraction and homography transformation. Figure 6 illustrates the footprint of frames F_1 to F_n for two wavelengths, λ_1, λ_2. For the d_{NDSI} skin detection algorithm, only four bands were required for the image cube. The VIS camera's Bayer patterning allowed it collect 2 images simultaneously, but the SWIR imager

Figure 4. Comparison of different sensor angles using test flight telemetry at 32m AGL. For a filter change rate of 0.5 Hz, the corresponding frame overlap %, working distance in meters, and pixel size for SWIR and VIS sensors in mm is displayed

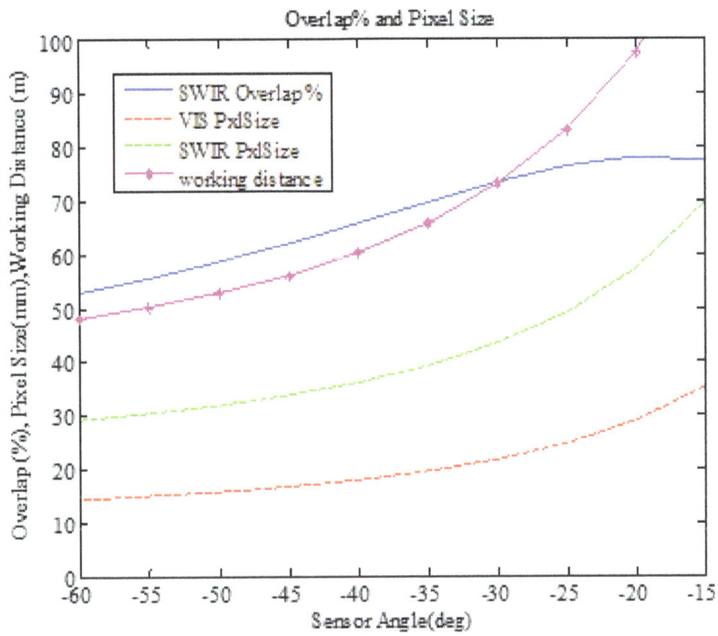

Figure 5. MSI Sensor Assembly with Aircraft Mounting

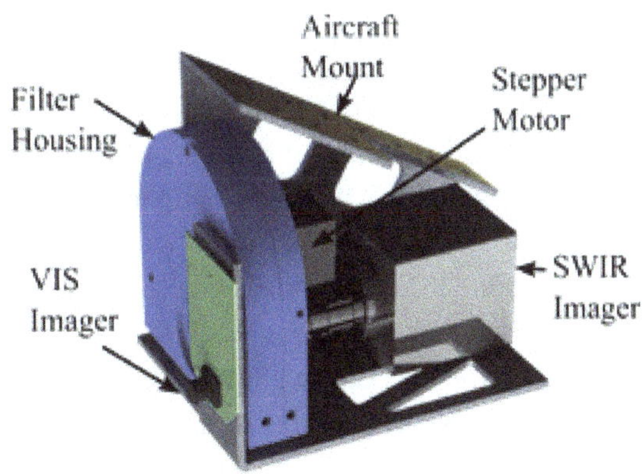

was most readily configured to collect one band per frame. Since the VIS camera had no filter wheel, the frames were capture at a faster rate.

Image registration is a technique that has been used by satellites makers to digitize large maps. For aerial surveillance, a snapshot imager may not register images to create a map due to moving objects or dismounts. Registration takes pixels from each image and stiches them together. Line scan imagers that collect a single column of pixels have almost no overlap. This restricts their usage to still images. Snapshot imagers have the ability to capture real time footage, and more ideal for aerial skin detection (Hagan).

Figure 6. Frame capture sequence for spectral image construction. Area scan type imager captures a frame of each wavelength at specified time intervals

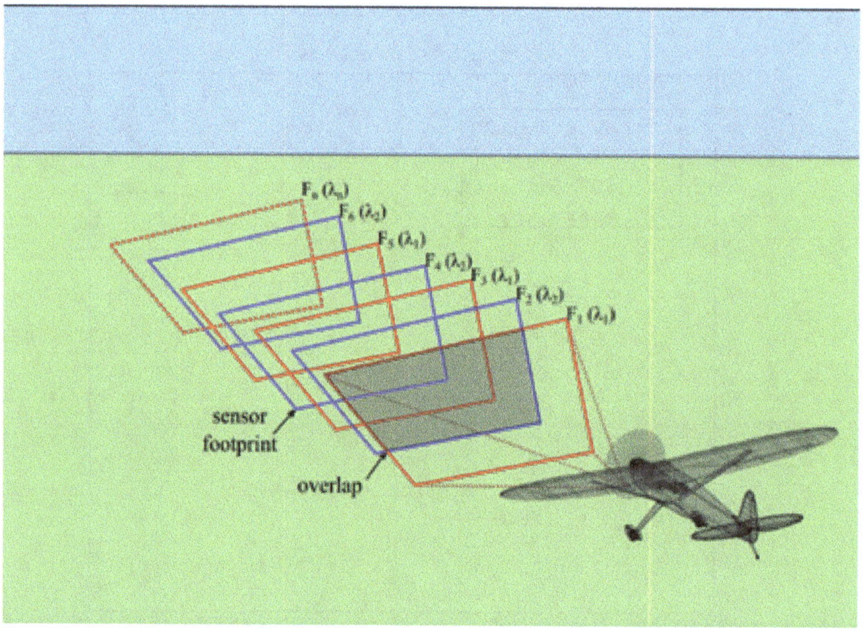

Ground Imager Comparison

The SpecTIR HSI uses a digital frame grabber to store 12 bit information for each band (SpecIM, 2013). This improves upon the 8 bit compression from the ADC used by the SUAS MSI. The SpecTIR benefits from a simple registration alignment and is closer to a parallel stereo arrangement. Figure 7 shows detection results using d_{NDSI} from the SpecTIR MSI. Figure 8 shows detection results from the SUAS MSI. The same bands were used for both imagers for a relevant comparison.

The ground test dismount detection scenario compared the ground based HSI and SUAS MSI. The NIIRS score was largely dependent upon GSD, and similar values were obtained for each imager. The two sensors were compared over the 4 bands which were available to both imagers: 532, 640, 1050, and 1550nm. When detecting skin however, the HSI consistently obtained greater AUC values compared to the SUAS MSI. This may be attributed to registration errors, optical transmission, aberrations, digital resolution, or exposure settings. The CPBD score was near 1 for both, which meant that blur was minimal, which made sense for stationary imagers and dismounts. Statistical significance was found between NIIRS/GIQE, and AUC. Table 1 presents the results of the sensor comparison.

The dual camera system as researched by Peskosky used the same SWIR imagers, and used similar wavelengths as the SUAS MSI. This larger imager had no reliance on frame overlap or moving filters; however, preliminary tests revealed the single camera system performed similarly in accuracy on ground based dismounts.

Figure 7. Left, SpecTIR detection from dNDSI. Right, ROC Curves for two skin detection algorithms generated with truth map. Inlay left, VIS image of scene. The matched filter only using 4 out of 363 available bands produced. Radiometric corrections were done using Flat Field Correction. These detections produced Area Under the Curve (AUC) values of: AUC_{dNDSI} *=0.9694,* AUC_{MF}*=0.9478*

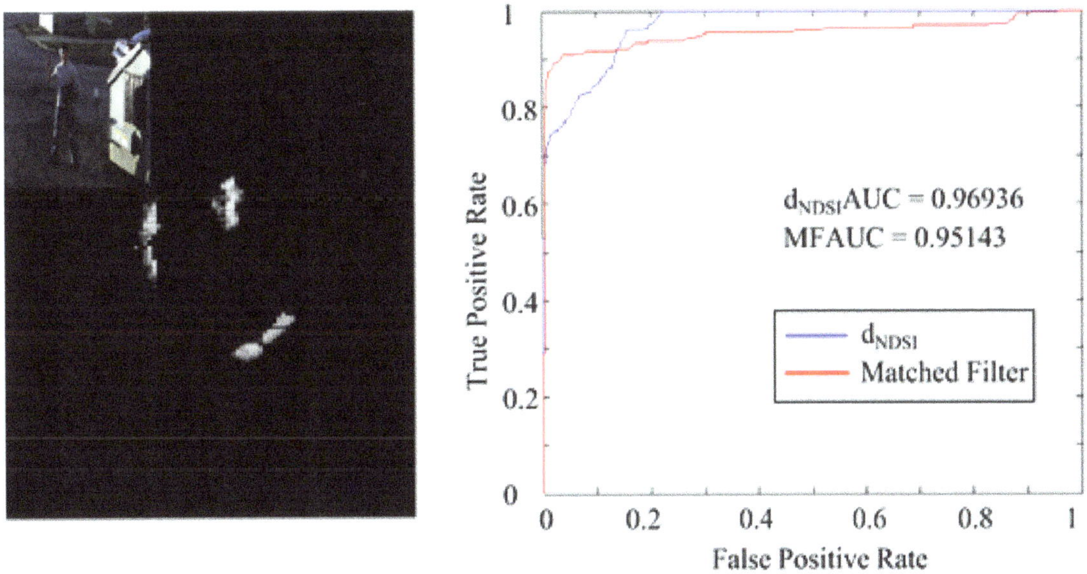

Figure 8. Left, SUAS MSI detection from dNDSI. Right, ROC Curves for two skin detection algorithms generated with truth map. Inlay left, VIS image of scene. The matched filter using 4 available bands. Radiometric corrections were done using Flat Field Correction. These detections produced Area Under the Curve (AUC) values of: AUC_{dNDSI}*=0.8912,* AUC_{MF}*=0.9052*

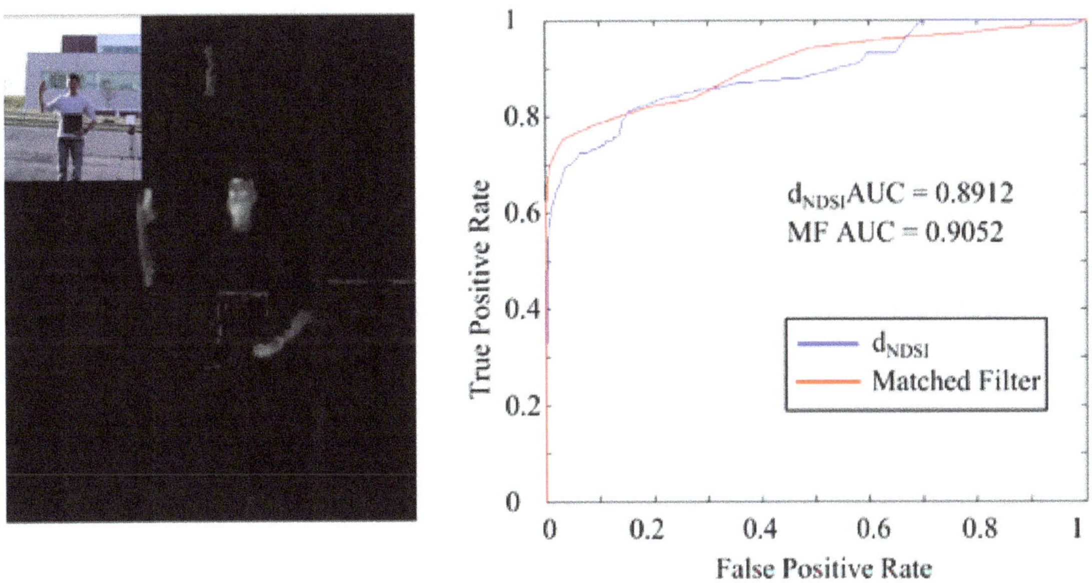

Table 1. Comparison of sensors

Parameter	Ground HSI	SUAS MSI
Specifications		
Focal Length	18.5-22.5 mm	35mm
FOV(vertical)	32°	26°
Spectral Bands	363	4
Design	Line Scanning	Filter Wheel
Range (5.0 cm pixel)	50 m	71 m
Frame Collection Rate	60 Hz	0.5 Hz
Digital Resolution	12-bit	8-bit
Mass	>15kg	1 kg
Performance		
NIIRS / GIQE (60 m)	8.15	8.70 (PD = 6.53%)[1]
CPBD	0.983	0.999 (PD = 1.75%)
d_{NDSI} AUC[2]	0.974	0.912 (PD = 6.55%)*
MF AUC[3]	0.972	0.937 (PD = 3.67%)*

Flight Test

Flight test was conducted at Camp Atterbury, IN, in the fall of 2013. Altitudes for data collection ranged from 30 to 56m, and airspeed ranged between 10 and 25 m/s. With the 30-45 degree camera mount, slant distance ranged from 42-112m. Sensor slant ranges were approximately at these altitudes. For skin detection tests. Dismounts were arranged 10m on an airport runway and ramp; they wore clothing that exposed at least 50 pixels of skin at the given slant range and IR camera resolution. Wind and wind gusts made groundspeed less predictable and ranged from 5 to 15 m/s. A combination of both manually controlled (RC piloting) and autopilot controlled test points were utilized. Cloud cover ranged from scattered clouds, to overcast with 500m AGL ceilings. These conditions allowed the system to be characterized under different wind and lighting. Higher wind speeds made patterns and camera angles less predictable. With different lighting, proper radiometric correction became important to utilize the raw image data.

Testing the imager in flight required consideration of many factors unique to a moving aerial platform. Proper exposure and gain settings were necessary to limit saturation. Synchronization between frame acquisition and autopilot telemetry was required for geolocation. Wind speed and direction impacted roll, pitch, and yaw rates; these factors were directly proportional to motion blur. Motor vibration also impacted image quality and increased distortion. Figure 9 displays a rectangular pattern from flight testing. The effects of wind are evident from the actual ground path, sensor aimpoint, and footprint or field of view. Flight test time was valuable and approximately 7 hours of aerial imagery was recorded. Ample tests were performed to validate the system, however, new system configurations would warrant additional flight tests.

Figure 9. Flight test data with ground path, sensor aimpoint, and sensor image plane projective transforms or footprint

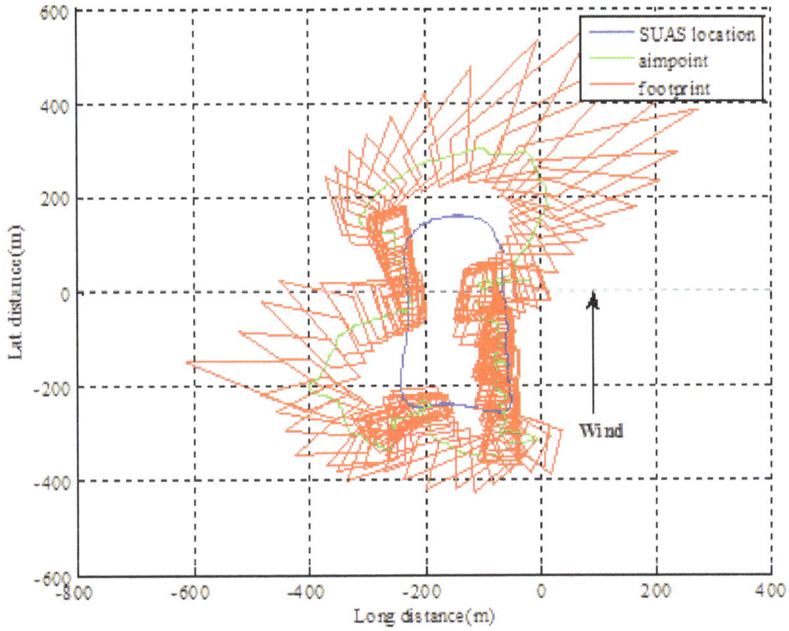

Registration and Geolocation

Compared to line scanning imager types, an area scan can generally provide more spatial resolution and viewing area on a target. Geolocation is further aided by overlap between image frames. A road scene was observed with two wavelengths from the SUAS MSI and is displayed in Figure 10; multiple frames from the SWIR were registered with SIFT and RANSAC to create a panorama. Despite camera movement, overlap between successive frames allowed band-to-band registration and then geolocation. Figure 11 displays geolocation in the VIS bands using autopilot telemetry for position and aimpoint. The registered frames are a result of projective transformation from autopilot position. This method for registration is significant due to the potential accuracy obtained with proper autopilot and frame synchronization. Registered image maps can be obtained by combining feature matching and autopilot telemetry; this may enhance dismount location accuracy.

CONCLUSION

Summary

The literature and results both show that Spectral imaging with a SUAS is difficult due to cost, computation, communications, and environmental forces on small sized airframes. Nevertheless, the SUAS MSI exhibited dismount detection performance comparable to a larger, more capable sensor with only a 5.112% difference for ROC AUC values. Skin detection may be a valuable asset for military, police,

Figure 10. Left, 1050nm video frame, after projective transformations from SIFT and RANSAC applied. Right, 1550nm video frames

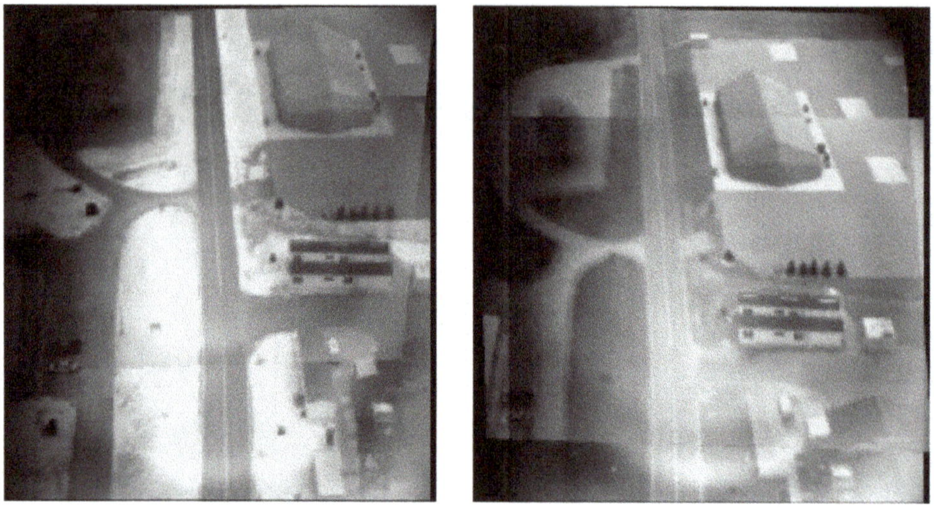

Figure 11. Projective transformation of a video sequence from autopilot telemetry

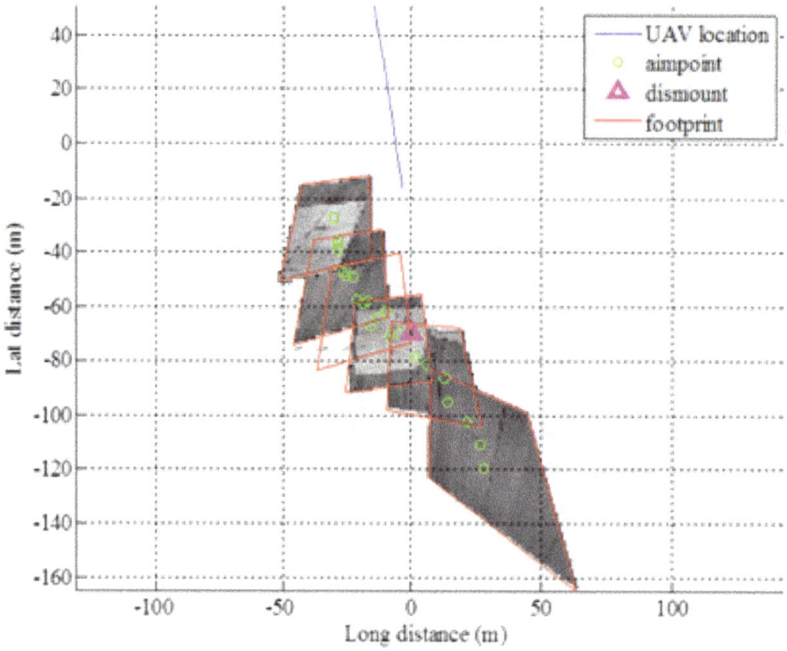

or first responders. Small mobile platforms such as SUAS can provide lightweight, expedient, low cost dismount detection with further development. Refinements to the system can certainly improve its ability to capture, transmit, process data, and yield valuable results for users. With all these issues considered, the SUAS MSI performed well given the time and cost constraints on its development.

Proposed Modifications

The proposed solution had a few issues that hindered performance, but could have been avoided. The first was using analog output from the SWIR to capture imagery; upgrading the system to a digital frame grabber would increase data quality. The second modification addresses communications for video data. Analog transmission did not retain the same level of image quality as the video stored on board because static and noise. Video on the ground station was only usable when the SUAS was line of sight. Digital may enable higher resolution data transmission. The third modification addresses lens selection. The refractive lens material affected the quality of imagery from both SWIR bands. Due to the separation in wavelength between 1050nm and 1550nm, a different lens material or a reflective lens would eliminate longitudinal chromatic aberrations. The forth modification addresses the filtering system. The current equipment operated with a filter speed and frame trigger at 30Hz; higher frame rate would effectively allow more precise geolocation and overlap between frames. 3-D spatial transformation from both attitude and vision processing would have also been helpful for accurate registration, and this was difficult due to synchronization between autopilot telemetry and video. Leveraging both highly accurate autopilot capabilities and fast feature detection capabilities of computer vision with fusion provides opportunity for higher probability of dismount detection.

Area Scan Imagers

Aerial platforms can specifically benefit from area scan type imagers as used in this and other research. Specifically, real time, single channel area scanning imagers present the largest potential for aerial object detection platforms. Single focal plane, single channel operation eliminates parallax corrections and spatial transformations that are required for multiple focal plane, multi-channel systems (Geelen, Tack, & Lambrechts, 2014; Kriesel et al., 2012). Implementation of an MSI with a single channel, and single focal plane can reduce size and weight requirements for an airframe. Geelen and others (2014), enabled a small form factor camera to capture multispectral imagery with a monolithically integrated bandpass filter. Using such an array to capture imagery may decrease computation requirements by eliminating band-to-band registration.

Future Research

This research set the groundwork for aerial skin detection with SUAS; however, further research can improve the technology readiness level of a detection system. Applications of autonomous decision making and waypoint routing may decrease operator workload. Applying new video technologies to SUAS such as micro-scale HSI that utilize tunable or patterned arrayed filters may allow real time scene dismount detection; efficient detection and radiometric calibration techniques may streamline reflectance conversions and improve classification performance in dynamic lighting environments; pursuing cost effective digital video recording and transmission can increase first responder situational awareness. Onboard SUAS detection processing can with embedded microcontrollers can eliminate bandwidth requirements and decrease ground control reliance. These technologies would increase the capability and efficiency of the SUAS MSI for future applications.

ACKNOWLEDGMENT

This work would not have been possible without the Air Force Institute of Technology ANT Center and Electrical Engineering department. The ANT Center provided SUAS flight test equipment and expertise. Lt Col Jeffrey Clark with the AFIT EE department provided hardware and software which were essential. The Air Force Research Lab Human Effectiveness and Sensors directorates also provided facilities and equipment for testing and construction. Finally, the students that worked with the author to construct and prepare for the tests were instrumental in this research.

REFERENCES

Aeryon Labs Inc. (2013). Aeryon scout SUAS supports critical florida wildfire research. Retrieved from http://www.aeryon.com/news/latest-news/pressreleases/541-uaf-scout-research.html#sthash.PlRsviGy.dpuf

Beard, R. W., & McLain, T. W. (2012). *Small unmanned aircraft:Theory and practice* (1st ed.). Princeton, Oxford: Princeton University Press. doi:10.1515/9781400840601

Beisley, A. P. (2012). *Spectral detection of human skin in VIS-SWIR hyperspectral imagery without radiometric calibration* (Masters Thesis). (AFIT/GE/ENG/12-03).

Bouguet, J. (2013). Camera calibration toolbox for matlab. California Institute of Technology: Comninos, P. (2011). Mathematical and computer programming techniques for computer graphics. U.S: Springer. doi:ISBN-13: 978-1-8233-902-9

Eichenholz, J. M., & Dougherty, J. (2009). Ultracompact fully integrated MegaPixel MultiSpectral imager. *SPIE 7218: Integrated Optics: Devices, Materials, and Technologies XIII,* San Jose, CA. (721814) doi:10.1117/12.809527

Excelis Inc. (2013). ENVI (5.0th ed.)

Geelen, B., Tack, N., & Lambrechts, A. (2014). A compact snapshot multispectral imager with a monolithically integrated per-pixel filter mosaic. *Proc.SPIE, 8974*, 89740-89740-8. doi:10.1117/12.2037607

Gomez, G., & Morales, E. F. (2002). Automatic feature construction and a simple rule induction algorithm for skin detection. Paper presented at the *In Proc. of the ICML Workshop on Machine Learning in Computer Vision*, 31-38.

Grahn, H. F., Geladi, P., & Burger, J. E. (2007). Multivariate images, hyperspectral imaging: Background and equipment. In H. F. G. Grahn P. (Ed.), Techniques and applications of hyperspectral image analysis (1st ed., pp. 1-16). West Sussex, England: John Wiley and Sons. doi:ISBN: 978-0-470-01086-0

Gundlach, J. (2012). Designing unmanned aircraft systems: A comprehensive approach American Institute of Aeronautics and Astronautics. doi:ISBN: 978-1-60086-843-6

Hagan, N., & Kudenov, M. W. (2013). Review of snapshot spectral imaging technologies. *Optical Engineering (Redondo Beach, Calif.), 52*(9). doi:10.1117/1.OE.52.9.090901

Inc, A. S. D. (2012). FieldSpec 4 user manual [ASD Document 600979] (B ed.) ASD Inc.

Kakumanu, P., Makrogiannis, S., & Bourbakis, N. (2007). A survey of skin-color modeling and detection methods. *Pattern Recognition*, *40*(3), 1106–1122. doi:10.1016/j.patcog.2006.06.010

Kerkes, J. P., & Hsu, S. M. (2004). Spectral quality metrics for terrain classification. *Imaging Spectrometry X*, *5546*, 382–389. doi:10.1117/12.560025

Koch, B. M. (2011). *A multispectral bidirectional reflectance distribution function study of human skin for improved dismount detection* (Masters Thesis). (AFIT/GE/ENG/11-22).

Kriesel, J., Scriven, G., Gat, N., Nagaraj, S., Willson, P., & Swaminathan, V. (2012). Snapshot hyperspectral fovea vision system (HyperVideo). *Proceedings of the Society for Photo-Instrumentation Engineers*, *8390*, 83900T–83900T-6. doi:10.1117/12.918643

Liggins, M., Hall, D., & Llinas, J. (2008). In M.E.L., D.L.H. (Eds.), Handbook of multisensor data fusion: Theory and practice, second edition. Boca Raton, FL: CRC Press.

Lister, T., Wright, P. A., & Chappell, P. H. (2012). Optical properties of human skin. *Journal of Biomedical Optics*, *17*(9), 0909011. doi:10.1117/1.JBO.17.9.090901 PMID:23085902

Manolakis, D., Marden, D., & Shaw, G. A. (2003). Hyperspectral image processing for automatic target detection applications. *Lincoln Laboratory Journal*, *14*(1), 79–116.

Merino, L., Fernando, C., Martinez-de Dios, J. R., Ferruz, J., & Ollero, A. (2006). A cooperative perception system for multiple UAVs: Application to automatic detection of forest fires. *Journal of Field Robotics*, *23*(3-4), 165–184. doi:10.1002/rob.20108

Narvekar, N. D., & Karam, L. J. (2011). A no reference perceptual image sharpness metric based on a cumulative probability of blur detection. *IEE Transactions on Image Processing, 20*(9), 2678-2683. doi:978-1-4244-4370-3/09

NASA. (2011). Landsat 7: Science data user Handbook**
**. Retrieved from http://landsathandbook.gsfc.nasa.gov/data_prod/prog_sect11_3.html

Nunez, A. S. (2009). *A physical model of human skin and its application for search and rescue* (Doctoral Dissertation). (AFIT/DEO/ENG/09-14).

Peskosky, K. (2010). *Design of a monocular multi-spectral skin detection, melanin estimation, and false-alarm suppression system* (Masters Thesis). (AFIT/GE/ENG/10-24).

Rao, R., & Arora, M. K. (2004). Image processing overview. In K. Pramod, & M. K. Arora (Eds.), Advanced image processing techniques for remotely sensed hyperspectral data (pp. 51-82) Springer. doi:ISBN: 978-3-662-05605-9

Spec, I. M. (2013). Airborne imager datasheets. Retrieved from http://www.specim.fi/index.php/products/airborne/

SpecTIR. (2012). ProSpecTIR instrument calibration and data processing overview. Reno, NV: SpecTIR.

The Mathworks Inc. (2014). *MATLAB release 2014b. Natick, Massachusetts, US: U.S. Geological Survey. (2011). United states geological survey unmanned aircraft systems roadmap 2010 - 2025. (No. Version 1.0 rev. A)*. Denver, CO: Opustek.

Vezhnevets, V., Sazonov, V., & Andreeva, A. (2003).
A survey on pixel-based skin color detection techniques Graphicon, Moscow, Russia. 85-92.

ENDNOTES

1 PD: Percent (%) Difference, * indicates statistical significance (p-value ≤ 0.05)

2 Averaged over a sample size of n=3 images, SUAS MSI: $\sigma_{dNDSI} = 0.0200$, Ground: $\sigma_{dNDSI} = 0.0143$

3 Averaged over a sample size of n=3 images, SUAS MSI: $\sigma_{MF} = 0.0208$, Ground: $\sigma_{MF} = 0.0245$

This research was previously published in the International Journal of Monitoring and Surveillance Technologies Research (IJMSTR), 2(3); edited by Nikolaos Bourbakis and Konstantina S. Nikita, pages 67-84, copyright year 2014 by IGI Publishing (an imprint of IGI Global).

Chapter 10
Compute-Efficient Geo-Localization of Targets From UAV Videos:
Real-Time Processing in Unknown Territory

Deendayal Kushwaha
Defence R&D Organisation, India

Sridhar Janagam
Defence R&D Organisation, India

Neeta Trivedi
Defence R&D Organisation, India

ABSTRACT

Unmanned Air Vehicles (UAVs) have crucial roles to play in traditional warfare, asymmetric conflicts, and also civilian applications such as search and rescue operations. Though satellites provide extensive coverage and capabilities crucial to many remote sensing tasks, UAVs have distinct edge over satellites in dynamic situations due to shorter revisit times and desired area/time coverage. The course, speed and altitude of a UAV can be dynamically altered, details of an activity of interest monitored by loitering over the area as desired. A fundamental requirement in most UAV operations is to find geo-coordinates of an object in the captured image. Most small, low-cost UAVs use low-cost, less accurate sensors. Matching with pre-registered images may not be possible in areas with low details or in emergency situations where terrain may have undergone severe sudden changes. In these situations that demand near real-time results and wider coverage, it is often enough to provide approximate results as long as bounds on accuracies can be established. Even when image registration is possible, it can benefit from these bounds to reduce search space thereby saving execution time. The prime contributions of this paper are computation of location of target anywhere in the image even at larger slant ranges, optimized algorithm to compute terrain elevation at target point, and use of visual simulation tool to validate the model. Analysis from simulation and results from real UAV flights are presented.

DOI: 10.4018/978-1-5225-8365-3.ch010

INTRODUCTION

Unmanned aerial vehicles (UAVs) are increasingly being recognized as capability multipliers in both military and civil settings. They play crucial role in Intelligence, Surveillance, Target Acquisition, and Reconnaissance missions of the military and are also of great assistance in disaster recovery missions. They carry daylight TV and/or infrared sensors, usually in a stabilized steerable turret or gimbal. Though satellites provide extensive coverage and capabilities crucial to many remote sensing tasks, UAVs have distinct edge over satellites in dynamic situations due to shorter revisit times and desired area/time coverage. The course, speed and altitude of a UAV can be dynamically altered, details of an activity of interest monitored by loitering over the area as desired. Most real-life missions require geo-coordinates of objects on the video captured through UAV. Military may want it to eliminate a potential threat; disaster recovery team may want it to plan the recovery. The results must pass the spatio-temporal accuracy tests; precise efforts must be delivered to a specific location in short time, so as to avoid attacking a target that has moved, minimizing collateral damage, or sending rescue team to a place where it is no longer needed.

Computing the geo-location involves performing series of transformations (Barber, Redding, McLain, Beard, & Taylor, 2006) across reference frames using position and attitude data of the UAV, and the azimuth and elevation angles of the turret. However, all sensors come with inherent inaccuracies. Also, flat or declining military budgets and requirement to keep the UAVs more affordable for civilian applications motivates using low-cost sensors, which obviously are more inaccurate.

Geographical Information Systems (GIS) that store huge amounts of geo-spatial data provide some means for compensating for the sensor errors. The image captured through the UAV can be registered with a pre-registered image to enhance accuracy (Conte, Hempel, Rudol, Lundström, Duranti, Wzorek & Doherty, 2008; Conte, Rudol, Wzorek, Petitti, Iocchi, & Doherty, 2008; Sim, Jeong, Lee, Parkm, & Kim, 1999) . However, in places where there are not many details on the terrain, or on sea, this method is of little help. Further, in disasters like flood, earthquake, tsunami etc., terrain may have undergone substantial changes in the areas of interest and hence registration may fail.

Given these constraints, it is required to have a comprehensive base geometrical model for calculating geo-location from sensor data, prove its correctness under zero errors and prove bounds on accuracies under different error conditions. The accuracy bounds help the users to plan for the worst-case response. The bounds are also helpful in reducing the search space for image registration where possible. Since registration is a highly computation-intensive task, reducing search space is a very attractive option.

This paper discusses a model for obtaining geo-location for a target in real time from UAV videos, taking into account the digital elevation data as well. The novelty of the approach lies in the usage of the combination all parameters and terrain elevation data to calculate target coordinates at the center as well as off the center location of the image. Note that under non-zero, especially under high roll and pitch conditions, the image may represent an arbitrary quadrilateral and off-center calculations may not be trivial. The prime contributions are computations for location of off-center targets, optimized algorithm to compute intersection of the sensor Line of Sight with terrain at target location and use of a visual simulation tool to provide accuracy bounds. Results from real flights of a tactical UAV with specific sensor inaccuracies are also presented.

PRIOR WORK

Significant work has been done in the field of target geo-location computation. Bearing-only target tracking has been studied extensively; Ponda, Kolacinskiyand and Frazzoliz (2009) suggest taking multiple measurements from different angles. However, they discuss, non-orthogonal measurements result in higher combined errors, and at longer ranges the propagation of attitude error as well as propagation error is large. With many surveillance UAVs flying at high altitudes, orthogonal measurements requires large separation between measuring positions. Also, Mallick (2007) states that most inaccuracies in geo-location calculation are due to inaccuracies in estimating the Euler angles of the aircraft. Normalization of these estimation errors requires large number of measurements, which may not be realistic in many real-life applications.

Taking advantage of the fact that sensor noise is random one study suggests (Conte, Hempel, Rudol, Lundström, Duranti, Wzorek & Doherty, 2008) improvements by way of offline (most often computation-intensive) post-processing. Post-processing is not suitable for real-time applications.

Many authors (Conte, Hempel, Rudol, Lundstrom, Duranti, Wzorek & Doherty, 2008; Conte, Rudol, Wzorek, Petitti, Iocchi & Doherty, 2008; Sim, Jeong, Lee, Parkm Kim, Lee, & Kim, 1999) propose use of reference imagery to enhance geo-location accuracy. As discussed earlier, this may not be possible in many settings.

Most existing work on transformation-based modeling assumes flat terrain model. This produces erroneous results in presence of terrain height variations. To cater to these height variations, some authors (Conte, Hempel, Rudol, Lundstrom, Duranti, Wzorek & Doherty, 2008) propose use of range finding techniques. This involves additional hardware amounting to additional weight and volume on the aircraft, and requires use of active sensing, not always desirable.

Terrain height has been considered by Kumar, Sawhney, Samarasekera et al. (2001). They mention the use of terrain height data to find the intersection of the image ray with the terrain profile. The authors suggest improving the precision of geo-coordinates by aligning current video frames to previously obtained calibrated reference images and terrain height data. As discussed earlier, in the scenario considered in this paper, we assume that reference imagery is not available.

Lyle, Long, Niessner, Ross, Geiger, Sinsley and Horn (2008) propose precise marking on the ground at known locations and calculating target position with respect to them. While this method gives excellent accuracy, this luxury is not available in most cases considered in this paper.

Barber, Redding, McLain, Beard and Taylor (2006) proposes flying in a circle around the target to cancel constant bias errors. In many military applications, it may amount to airspace violation. Certain disaster situations also may not allow crossing some hazardous boundaries. The authors assume a flat terrain model.

Saeedi, Lowe and Lawrence (2003) use a trinocular set of stereo images to extract range from point feature correspondences. Chroust and Vincze (2003) take a similar approach using a bincoular set of stereo images. Stereo imaging requires an increased hardware set compared to monocular imaging systems. Addition of this hardware is undesirable, and in some cases impossible. Also, use of stereo requires a certain minimum distance between the two sensors and precise mounting arrangement. This is clearly a problem in smaller vehicles.

Xie and Li (1997) calculates geographical coordinates of a target by computing correlation between real time images. Duan (2002) and Sim, Park, Kim, Lee, and Kim (2002) present geo-location approach using sequential aerial images where range information is extracted by stereo modeling of two successive image frames. The most serious shortcoming of this approach is the inability to work without ground control points.

Different methods of target localization are discussed by Madison, DeBitetto, Olean, and Peebles (2008). The authors discuss how aircraft sensors contribute to localization errors and suggest use of LoS filtering or Structure from Motion. During simulations at ADE, it has been found that these methods work best when measurements are taken from many different directions, and when at least some of the aircraft sensor parameters are absolutely accurate. If they are not, the algorithms sometimes fail to converge.

The method proposed by Johnston (2006) is similar to the one used in this paper but we additionally take into account the asymmetry in the image projection, propose method to deal with terrain height variations, and to validate the model in a visual simulation environment.

Almost all the existing work assumes constant pixel-to-meter ratio at each point in the image, an assumption readily violated for non-zero roll/ pitch and gimbal elevations other than 90°. Ponda, Kolacinskiy and Frazzoliz (2009) suggested that the pixel location of the target can be transformed into two bearing angles. This model can handle the resultant asymmetry. This paper derives necessary math for this and uses it for calculations.

To the best of our knowledge, this is the first work reporting the geometric model in its entirety and computation efficiency in dealing with terrain height data, and the first attempt at validating the model using visual simulation tool to establish accuracy bounds.

METHODOLOGY

The fundamental concept behind geo-location calculation is transformation of coordinates from pixel frame to camera frame, camera frame to gimbal frame, gimbal frame to aircraft frame and finally aircraft frame to world coordinate frame. The UAV has six Degrees of Freedom (DoF), three translational and three rotational. The gimbal has two DoF, one in azimuth and the other in elevation.

Targets away from the center location present challenges due to the fact that perspective look due to rotational movement of camera results in non-uniform pixel-to-meter ratio throughout the image. For surveillance UAVs flying at high altitudes and at a standoff distance to avoid violation of airspace or other constraints, the effect of this non-uniformity gets magnified. For this reason, the angles from the camera center line to the pixel location where target is present in the image must be considered. Also, since image-based measurements only provide bearing information, terrain elevation data must be used along with the aircraft altitude for calculating range.

The model proposed by Barber, Redding, McLain, Beard and Taylor (2006) is an excellent reference for geometry of the measurement process for the target calculation model under flat terrain approximation and constant pixel-to-meter ratio.

The inertial frame to the vehicle frame transformation is essentially a translation. Vehicle frame to the body frame transformation consists of rotations based on the Euler angles and body frame to the gimbal frame transformations amount to rotations by the azimuth and elevation angles. If the center of gravity (CG) of the aircraft and rotation center of the gimbal do not coincide, an additional translation is required.

Handling Asymmetry

It can be readily seen (Figure 1) that under the rotational transformations of the aircraft, the assumption of constant pixel-to-meter ratio does not hold, since these transformations cause the image plane to be not in parallel with the terrain below and hence the projection of the image plane on the terrain is asymmetric in every direction. To handle this asymmetry, we use the technique proposed by Ponda, Kolacinskiyand & Frazzoliz (2009), where the pixel location of the target is transformed into two bearing angles (ψ_{pix}, θ_{pix}) from the camera pointing axis to a vector that passes through the target and through the camera focal point (Figure 2), and given as shown in Box 1, where (x_{pix},y_{pix}) is the pixel location of target point on the image, X_{pix} and Y_{pix} are half of the image width and height in pixels respectively and FOV is the diagonal field of view of the camera in radians. Aspect ratio of 4:3 is assumed.

Consider the Line-of-Sight (LoS) originating at the focal point of the camera and ending at the target point on ground, having length *r* meters. The projection of this LoS on all the three axes of world coordinate system is given by Equation 3:

$$\begin{bmatrix} x_t \\ y_t \\ -h_t \end{bmatrix} = r * (R_a^w * R_g^a * R_c^g * R_{pix}^c)_2 \tag{3}$$

where x_t, y_t are distances of the target point from the aircraft location in X and Y directions, h_t is the altitude of the ground target relative to the aircraft, and the subscript *2* refers to the 2nd column of the matrix. We must take 2nd column because Ψ_{pix}, θ_{pix} are measured from the Y axis. $R_a^w = R_\psi * R_\theta * R_\varphi$ is the transformation from aircraft to world coordinates, $R_g^a = R_\beta * R_\alpha$ is transformation from gimbal to aircraft coordinates, $R_c^g = I$ is the transformation from camera to gimbal coordinates (identity because camera is fixed to the gimbal and both have the same origin), and $R_{pix}^c = R_{\psi_{pix}} * R_{\theta_{pix}}$ is the transformation from LoS to camera frame. All rotations are referenced to left-handed coordinate frames. Further, ψ, θ and ϕ represent heading, pitch and roll respectively of the aircraft, and α and β represent elevation and azimuth respectively of the gimbal.

Box 1.

$$\psi_{pix} = \tan^{-1}\frac{P'O'}{OO'} = \tan^{-1}\left(\frac{x_{pix} * 0.8 * \tan\left(\frac{FOV}{2}\right)}{X_{pix}}\right) \tag{1}$$

$$\theta_{pix} = \tan^{-1}\frac{PP'}{OP'} = \tan^{-1}\left(\frac{y_{pix} * 0.6 * \tan\left(\frac{FOV}{2}\right) * \cos(\tan^{-1}(x_{pix} * 0.8 * \tan\left(\frac{FOV}{2}\right) / X_{pix}))}{Y_{pix}}\right) \tag{2}$$

Figure 1. Asymmetric pixel-to-meter ratio for non-zero roll, pitch

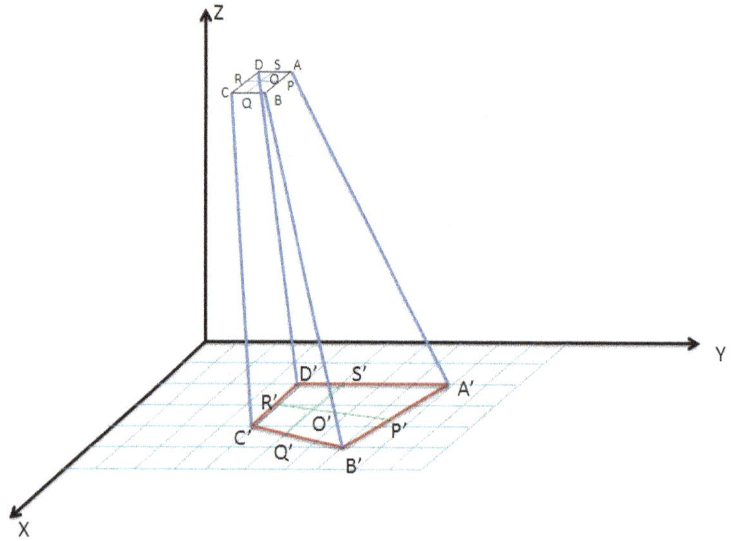

Figure 2. Angles formed at camera center by off-center pixel

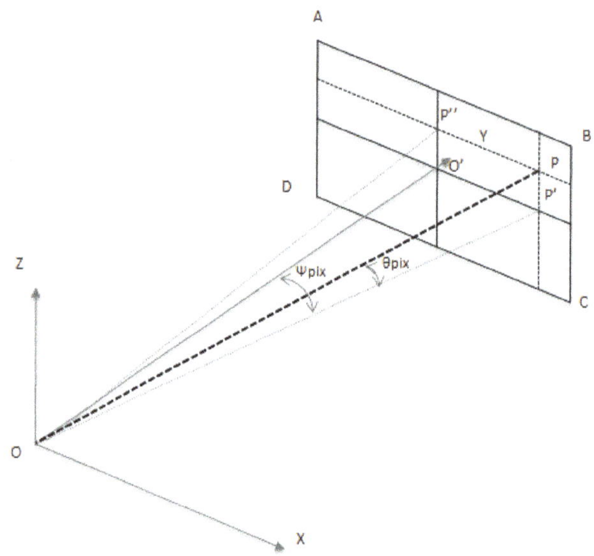

Finally, the geo-referenced location of the target point (X_{tgt}, Y_{tgt}), given aircraft location (X_{ac}, Y_{ac}) is given by:

$$X_{tgt} = X_{ac} + x_t; Y_{tgt} = Y_{ac} + y_t \quad (4)$$

Equation 3 requires altitude of the aircraft relative to the target; however, aircraft sensors only provide altitude above mean sea level (AMSL). This requires knowing the terrain height at target location AMSL.

This cannot be 'read' from terrain data since we do not know target location yet. When range finder is available, range is used along with bearing to directly determine (x_t, y_t). The streams of work assuming flat terrain model use average ground level height. In the following sections we propose computationally efficient method for finding the target height without the use of range finder.

Effect of Terrain Height

Flat terrain assumption results in incorrect calculations (Figure 3), and the error will depend on the height of the target above (or below) the average ground level. The point where the camera's LoS intersects for the first time with the height profile is the correct target point. Consider this LoS in 3D world coordinates. When moving away from the camera, the intersection point is the first (X, Y) where terrain height is equal to or more than the Z on LoS. To find this point, we can move in steps from the camera focal point along the LoS, checking the heights at each point. However, there is a lot of redundant computation here.

From digital elevation data, we know the minimum (H_{min}) and maximum (H_{max})terrain height in the region. The intersection point must lie above H_{min} and below H_{max}. We make use of this fact to reduce the search space. The geo-location calculation steps therefore get updated as follows:

1. Execute Equation 4 with $h_t = (h - H_{max})$ to get (X_1, Y_1, Z_1);
2. Repeat with $h_t = (h - H_{min})$ to get (X_2, Y_2, Z_2).

The points (X_1, Y_1, Z_1) and (X_2, Y_2, Z_2) are shown in Figure 3 for a side view and in Figure 4 for top view. The blocks in Figure 4 correspond to the resolution of the terrain height data. We used DT2 data format with resolution of 30m. Bilinear interpolation can be used for better accuracy. This resolution is a matter of tradeoff between accuracy and execution time.

To further save execution time, the minimum and maximum heights in each DT2 file were extracted in offline mode and stored in the unused fields in the DT2 header.

Finding the target point (X, Y, Z) in Figure 4 now amounts to finding the (X, Y) blocks between these two points starting from (X_1, Y_1) and checking if the terrain height at that block is equal to or more than the LoS height at that block.

This problem of finding the blocks between two points is exactly the same as that of finding intermediate pixels to be illuminated for drawing line between two pixels on a raster display. In computer graphics, Bresenhem's line drawing algorithm (Harrington, 1987) is proven to be computationally efficient, one of the reasons being that it works on integer parameters. We use this algorithm to find the intermediate blocks as shown in Algorithm 1 and Figure 5.

EVALUATION, VERIFICATION, AND VALIDATION

For verification, typical known inputs were used. Manual calculations using Equation 3 provide expected result whereas the value output by the program is actual result. Results were found to be correct to 1m.

For validation, a visual simulation tool 'Vega Prime' from Presagis, was used. The tool allows an aircraft with 6 DoF to be placed at a desired location in the 3D world coordinate system, allows a 'gimbal' with 2 DoF and cameras with different Fields-of-View (FoV) to be attached to the aircraft, and allows

Figure 3. Effect of terrain height variations on target calculation

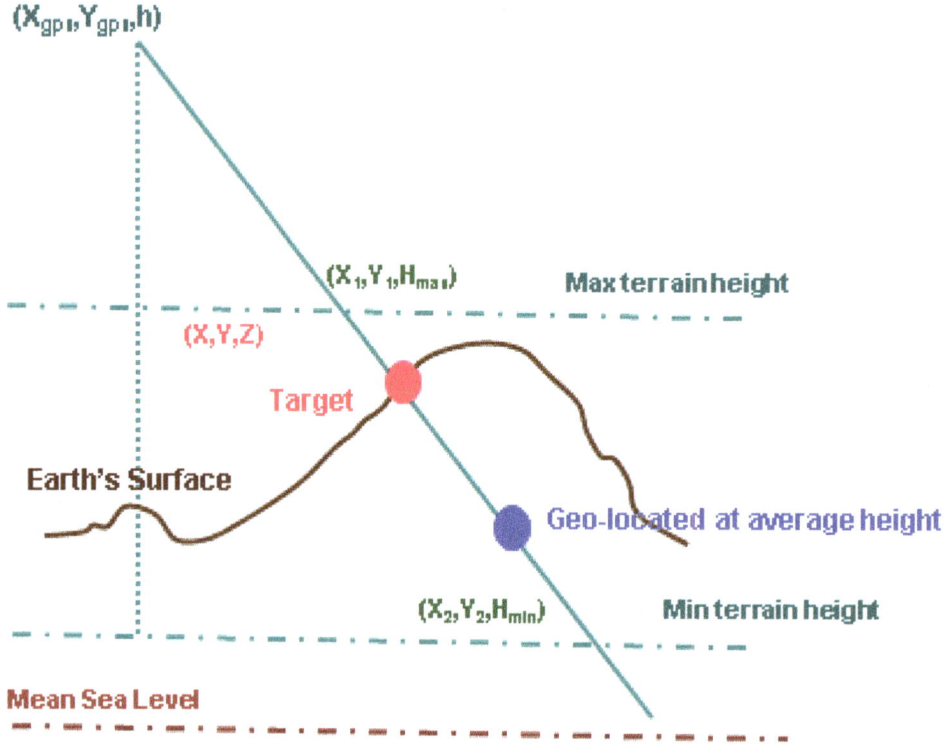

Figure 4. Top view of the LoS intersecting with the terrain at the max and min heights

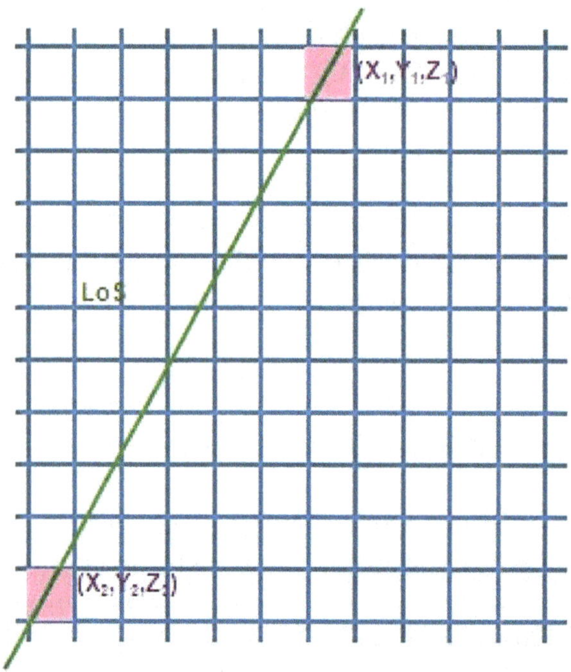

Algorithm 1.

```
Start = (X1, Y1, Hmax)
End = (X2, Y2, Hmin)
Mid = Start
WHILE (Mid(x,y) is not equal to End(x,y))
    Calculate Mid(z) using similar triangle
    principle (Figure 5)
    IF (Mid(z) ≤ terrainHeight at Mid(x,y))
        Declare Mid(x,y) as intersection
        point
        Break from the loop
    ENDIF
    Mid(x,y) = Next point using
    Bresenhem's algorithm
ENDWHILE
```

Figure 5. Calculation of height at intermediate points

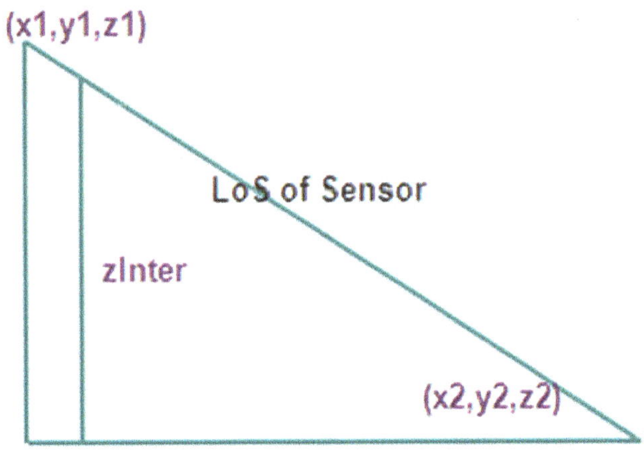

assignment of desired values to the various aircraft and gimbal parameters. Once positioned, aircraft can 'capture' image of the terrain.

To be able to select a definite known location, special-looking terrain area of 27*27km^2 was created using the Presagis terrain creation tool 'Creator'. Real terrain elevation data was used. Box-type objects were placed on this terrain at every 500 meters in X and Y directions. The boxes were placed atop a 'pole' so that the location of object when seen from UAV can be pin-pointed. Color coding was used to avoid mistakes in identifying the object. This is important because UAV camera is like a 'soda straw'; it is very difficult to get the bigger picture from a scene. Color coding along with object placement pattern provides 'surveyed' geo-location. Figure 6 shows sample images captured after terrain was visualized taking into consideration elevation data, the aircraft assigned a position and an orientation in the 3D space, and the gimbal assigned specific values for azimuth and elevation. The terrain color specifies the sector on terrain, and the block color specifies the location of the block relative to the sector(s). Together they provide the true latitude-longitude positions.

These images are fed to the target calculation model and results checked. The procedure is repeated with different values of the aircraft and gimbal parameters.

For any captured image, there is a minimum pixel-to-meter ratio. The model cannot give accuracies better than this figure. Second, the selection of object is limited by the human ability to click on an exact desired pixel.

Inaccurate sensor data can be modeled as noise in the parameters. Taking typical figures from known sensors, zero mean Gaussian noise model was used with known variances as given in Table 1.

Simulation Results

Table 2 lists percentage of cases for various error limits. Over 1,000 test cases were generated for error-free data and over 100,000 test cases (100 random samples for each error-free case) were used for noisy data.

Figure 6. Image frames captured; the blocks are objects placed at 500m distance, color codes used for object location

Table 1. Sensor inaccuracies considered

Parameter	Noise Variance
Roll	0.5°
Pitch	0.5°
Heading	1.0°
Altitude	25m
Gimbal Azimuth	0.25°
Gimbal Elevation	0.25°

Table 2. Errors in simulated scenarios

RMS Error	No Noise	Noisy Data
Error ≤ 50m	98%	95%
50 < Error ≤ 100m	1.99%	4%
Error > 100M	0.01%	1%

Results From Real UAV Flight

The model was used with a tactical UAV designed and developed by ADE. The UAV operates at altitudes of 2-3km, can fly for over 4 hours up to 160km from the ground control station. It carries a gimbaled payload assembly that can house a daylight TV and an infrared camera simultaneously. Its primary sensors are GPS, baro altimeter, magnetic heading sensor, vertical gyro and rate gyros. The UAV can fly in autonomous and manual modes. Flights data is received on ground through a jam resistant RF link. The geo-location model was used to compute locations of targets captured from this UAV. About 15 targets were surveyed in advance, and compared with locations calculated from the UAV video and flight data.

In our UAV, gimbal Azimuth and Elevation readings are accurate to 0.1°, altitude to 50m, and roll and pitch to 1°. Heading is better for stable flight, with error up to 2°.

Given the above accuracy figures, simulated results tally with the actual flight experiments, where most errors are contained within 150m RMS. This figure was acceptable to our users. The results, summarized in Table 3, include cases where flight was straight and level, and also during gentle maneuvers. No sensor calibration was performed.

IMPROVISATION OVER EXISTING METHODS

As discussed earlier, methods that use registration with pre-registered images (Conte, Hempel, Rudol, Lundström, Duranti, Wzorek & Doherty, 2008; Conte, Rudol, Wzorek, Petitti, Iocchi, & Doherty, 2008; Sim, Jeong, Lee, Parkm, & Kim, 1999) could fail in many dynamic situations and in cases where land does not have enough distinct visual features. Since the approach in this paper is different, a direct comparison cannot be obtained.

Methods based on triangulation converge only when at least some of the flight parameters are accurate. The authors have verified this for many real-life datasets. Most small, low-cost UAVs use low-cost sensors, which do not provide necessary accuracy. A direct comparison cannot be made in this case, either.

Table 3. Errors in real flight experiments

RMS Error	Results	Remarks
Error ≤ 50m	10%	15 targets captured about 200 times in 3 flights
50 < Error ≤ 100m	62%	
100 < Error ≤ 150m	22%	
Error > 150m	6%	

As long as terrain elevation data is accounted for in computing the geolocation, and the resolution is comparable, results from this algorithm and others that use geometric transformations for geolocation computation will match. However, the proposed algorithm achieves the target with significantly lower computation time, an analysis of which is provided in the Tables 4 and 5.

On most processors, floating point operations consume many more cycles compared to integer operations. For example, a MPC7450 microprocessor takes 5 cycles for floating point multiplication, 5 for floating point addition, 5 for floating point comparison, 1 for integer addition/ comparison. On this processor, the proposed implementation offers 4:1 speedup.

To analyze the improvement on real data set, UAV flight data was studied for flights in an area where $H_{max} = 1100$m, $H_{min} = 400$m. The targets were captured at $h = 3000$m, $\tau = 30^0$. For this setup, on an average, 80% reduction in search area and 4 times improvement in the remaining 20% search is obtained. As a result, real-time operations are possible using modest computing power, which makes it possible to design low-power (and hence less heat dissipating), lightweight, cost-effective solutions for real-time target localization.

CONCLUSION

Real-time operations, flight path restrictions, and other practical situations impose limitations on the methods for calculating geo-location. In such cases, approximations are acceptable provided error bounds can be established. We have implemented one such model in computationally efficient manner taking terrain height into consideration. It provides best possible error bounds for given sensor accuracies, and was validated with the help of a visual simulation tool. For noise-free inputs, the model performs correctly within the limits of the pixel-to-meter ratio and the human ability to click at a particular pixel. For noisy data, error bounds are established for given sensor accuracies. To the best of our knowledge, this is the first time such a validation has been reported.

Table 4. Number of iterations to compute terrain intersection

Method	Worst Case[1]	Best Case
Brute Force	$(h - H_{min})/\tan(\tau)$	$(h - H_{max})/\tan(\tau)$
Reduced Search Space	$(H_{max} - H_{min})/\tan(\tau)$	1

Note: τ is look angle, h is the height of the UAV and $h >> H_{max}$.

Table 5. Number of operations per iteration

Operation	Naïve Algorithm	Bresenhem's Line Drawing Algorithm
Division	0	0
Multiplication	1 Floating Point	0
Addition/ Subtraction	2 Floating Point	3 Integer
Comparison	1 Integer	2 Integer

The results from actual UAV trials have been reported. Sensors can be selected depending on requirements of accuracy. Pre-flight sensor calibration improves accuracy.

In disaster situations, previously registered images may not match with the changed terrain. In such cases, this approximation can be used with geographical information systems (GIS) for a quick recovery planning. For military operations at sea or in places with few details on the terrain, this approximation with bounded errors allows planning with the best possible precision under the conditions.

Presently we are working on using the reduced search space for quicker image registration and use of auxiliary models as well as sensor calibration to enhance accuracies.

ACKNOWLEDGMENT

The authors thank Director, ADE for his permission to publish this paper. The support from Group Director, Mr. VS Chandrashekar, Sc ‘G’, ADE, is duly acknowledged. Thanks are also due to Ms. Rashmi S, ADE, and the team of CMC engineers for their immense support towards validation. The work of Dr. Jharna Majumdar, ex-Scientist, ADE towards early development is also acknowledged.

REFERENCES

Barber, D. B., Redding, J. D., McLain, T. W., Beard, R. W., & Taylor, C. N. (2006). Vision-based target geo-location using a fixed-wing miniature air vehicle. *Journal of Intelligent & Robotic Systems*, *47*(4), 361–382. doi:10.100710846-006-9088-7

Boid, D., Brown, A., Nylund, M., & Sulllivian, D. (2001). GI-Eye II GPS/inertial system for target geo-location and image geo-referencing. *NAVSYS Corparation*, White Paper.

Chroust, S. G., & Vincze, M. (2003). Fusion of vision and inertial data for motion and structure estimation. *Journal of Robotic Systems*, *21*(2), 73–83. doi:10.1002/rob.10129

Conte, G., Hempel, M., Rudol, P., Lundström, D., Duranti, S., Wzorek, M., & Doherty, P. (2008). *High accuracy ground target geo-location using autonomous micro aerial vehicle platforms*. Honolulu, Hawaii: AIAA Guidance, Navigation, and Control Conf and Exhibit. doi:10.2514/6.2008-6668

Conte, G., Rudol, P., Wzorek, M., Petitti, P., Iocchi, L., & Doherty, P. (2008). Use of geo-referenced images with unmanned aerial systems, Proc. Intl. Conf. simulation, modeling and programming for autonomous robots, Italy.

Duan, L. F. (2002), Research on target locating methods based on image sequences, University of Science and Technology of China Press., pp. 31-89

Harrington, S. (1987). *Computer graphics - a programming approach*. McGraw-Hill Publications.

Jiaxiang, Y., Deyun, X., & Ludong, J. (2009). New geo-location approach based on camera coordinates and common points on multiple images. *Defence Science Journal*, *59*(1), 43–48. doi:10.14429/dsj.59.1483

Johnston, M. G. (2006), Ground object geo-location using UAV video camera, Proc. 25th *IEEE/AIAA Digital Avionics Systems Conference*, pp. 1-7, 10.1109/DASC.2006.313770

Kumar, R., Sawhney, H. S., Samarasekera, S., Pope, A., Hsu, S., Tao, H., & ... (2001). Aerial video surveillance and exploitation[[doi:]. *Proceedings of the IEEE, 89*(10), 1518–1539. doi:10.1109/5.959344

Lyle, J., Long, N., Niessner, A. F., Ross, A., Geiger, B. R., Sinsley, G. L., & Horn, J. F. (2008). *Vision-based target geolocation and optimal surveillance on an unmanned aerial vehicle*. Honolulu, Hawaii: AIAA Guidance, Navigation, and Control Conf.

Madison, R., DeBitetto, P., Olean, A. R., & Peebles, M. (2008). Target geolocation from small unmanned aircraft system, Proc. *IEEE Conf Aerospace Systems*, pp. 1-19, 10.1109/AERO.2008.4526560

Mallick, M. (2007). Geolocation using video sensor measurements, Proc. *10th Intl Conf Information Fusion*, Canada.

Ponda, S. S., Kolacinskiyand, R. M., & Frazzoliz, E. (2009). Trajectory optimization for target localization using small unmanned aerial vehicles, *Proc. AIAA Guidance, Navigation, and Control Conf*, Chicago, Illinois. 10.2514/6.2009-6015

Presagis Content creation: www.presagis.com/products/ content_creation/creator/ last accessed 01-09-2011

Presagis Visualization. www.presagis.com/products/ms /visualization/ vegaprime/ last accessed 01-09-2011

Saeedi, P., Lowe, D. G., & Lawrence, P. D. (2003). 3D localization and tracking in unknown environments, Proc. IEEE Conf. *Robotics and Automation,* Vol. 1, pp. 1297–1303. doi:

Sim, D. G., Jeong, S. Y., Lee, D. H., Parkm, R. H., Kim, R. C., Lee, S. U., & Kim, I. C. (1999). Hybrid estimation of navigation parameters from aerial image sequence[[doi:]. *IEEE Transactions on Image Processing*, *8*(3), 429–435. doi:10.1109/83.748898 PMID:18262886

Sim, D. G., Park, R. H., Kim, R. C., Lee, S. U., & Kim, I. C. (2002). Integrated position estimation using aerial image sequences[[doi:]. *IEEE Transactions on Pattern Analysis and Machine Intelligence, 24*(1), 1–18. doi:10.1109/34.982881

Xie, L. L., & Li, Y. R. (1997). A new approach to improve target location accuracy for remotely piloted vehicles (RPV) with digital terrain maps. *J. Northwestern Polytech. Univ., 16*(1), 61–64.

ENDNOTE

[1] The best case is when target is at terrain height=H_{max} and the worst case is when it is at H_{min}.

This research was previously published in the International Journal of Applied Geospatial Research (IJAGR), 5(3); edited by Donald Patrick Albert and Samuel Adu-Prah, pages 36-48, copyright year 2014 by IGI Publishing (an imprint of IGI Global).

Chapter 11
Alternative Methods for Developing and Assessing the Accuracy of UAV-Derived DEMs

Dion J. Wiseman
Brandon University, Canada

Jurjen van der Sluijs
University of Lethbridge, Canada

ABSTRACT

Digital terrain models are invaluable datasets that are frequently used for visualizing, modeling, and analyzing Earth surface processes. Accurate models covering local scale landscape features are often very expensive and have poor temporal resolution. This research investigates the utility of UAV acquired imagery for generating high resolution terrain models and provides a detailed accuracy assessment according to recommended protocols. High resolution UAV imagery was acquired over a localized dune complex in southwestern Manitoba, Canada and two alternative workflows were evaluated for extracting point clouds. UAV-derived data points were then compared to reference data sets acquired using mapping grade GPS receivers and a total station. Results indicated that the UAV imagery was capable of producing dense point clouds and high resolution terrain models with mean errors as low as -0.15 m and RMSE values of 0.42 m depending on the resolution of the image dataset and workflow employed.

INTRODUCTION

Terrains models are three-dimensional representations of a part of Earth's surface (Li et al., 2004). These include Digital Elevation Models (DEM) and Digital Terrain Models (DTM), which are both bare Earth surfaces; the later including topographic elements such as slope, aspect, curvature, and gradient; and Digital Surface Models (DSM) that portray all surface features such as vegetation and built structures (Li et al., 2004). All three are typically raster data models that convey a visual impression of the landscape that is easier to interpret and understand than two-dimensional maps since they more accurately represent our perception of Earth.

DOI: 10.4018/978-1-5225-8365-3.ch011

Terrain models are valuable tools for Earth surface interpretation and analysis and are used in a variety of disciplines. They are important data sets for geomorphological research because surficial features can be visualized, processes operating on landforms modeled, and rates of surface change predicted (Cavalli et al., 2008; Frankl et al., 2010). They are also frequently used in hydrological modeling to delineate watersheds, natural streams, and sinks (Hopkinson et al., 2009; Murphy et al., 2008) and have also been used extensively for land use planning and site selection and suitability analyses, including the prediction of fish densities using stream channel gradients (Wissmar et al., 2010) and for the identification of critical habitat for endangered species according to prescribed surface characteristics (Moran & Button, 2011). In addition, they are often used to perform line of sight and viewshed analyses to determine optimum locations for structures and facilities such as communication towers and wind turbines (Akella et al., 2010; Möller, 2006) and are frequently used for erosion modeling, calculating the loss of glacial ice, or performing cut-and-fill analyses to determine the volume of earth, rock, or water located within a defined area of interest (Chen et al., 2006; Keutterling & Thomas, 2006; Thoma et al., 2005).

Terrain models are commonly developed using three-dimensional coordinates derived from one of three sources: ground surveys, remote sensing techniques, or existing hardcopy or digital topographic maps. The resulting spot elevations or point clouds are then processed to produce a terrain model using an interpolation algorithm that produces a raster model at a user-defined spatial resolution. Ground surveys traditionally employed plane table surveys and theodolites to acquire the horizontal and vertical position of control points across an area of interest (Wolf, 2002). However, modern surveying techniques typically utilize either a total station, combining a theodolite and electronic distance measuring device (EDM), or Global Positioning System (GPS) receiver (Wolf, 2002). While total stations are capable of collecting coordinates to within millimetres, they are generally labour intensive and, thus, only used when working on small, localized study areas a few hectares or acres in size (Rayburg et al., 2009). Conversely, survey or mapping grade GPS receivers are capable of attaining decimetre level accuracy and covering considerably larger areas with significant increases in efficiency (Baptista et al., 2011). The ability to accurately record three-dimensional coordinates at almost any user defined sampling interval enables ground surveys to produce high resolution terrain models with minimal to moderate investment in equipment, however, labour costs are typically quite high.

Remote sensing techniques include the use of active or passive sensors to acquire elevation data either directly (e.g. LIDAR) or through image processing techniques based on the principles of photogrammetry and aerial triangulation to derive three dimensional coordinates at positions visible within at least two adjacent, overlapping images (Wolf & DeWitt, 2000). Because both active and passive sensors typically collect data at a significant distance from Earth's surface, the geographic extent covered is much larger than ground surveys. As a result, sensors aboard IKONOS, Quickbird, SPOT-5, and ASTER have often been used to acquire imagery to generate regional or global terrain models, albeit with a poorer spatial resolution than ground surveys (ASTER GDEM Validation Team, 2011; Reinartz et al., 2006; Toutin, 2004b; Toutin, 2008). In comparison, active and passive sensors carried aboard aircraft obtain data of higher spatial resolution, but for much smaller (i.e. local to regional) geographic extents (St-Onge et al., 2008; Williams & Jenkins, 2009).

Existing hardcopy topographic maps display elevation through the use of contour lines which can be digitized or scanned and processed to obtain terrain surface data (Hopkinson et al., 2009; Jurado-Expósito et al., 2008). Although a wide variety of maps can be used, the resulting terrain model should only be generated at a resolution commensurate with the scale of the original map since its accuracy is entirely dependent on the accuracy of the source map.

As a general rule of thumb, the cost of terrain models increases when higher accuracies, larger geographic extents, or denser sampling intervals are required (Table 1). The cost of terrain models ranges from data freely available online (e.g. low resolution global DEMs produced through space shuttle and satellite missions) to data costing tens of thousands of dollars in equipment and labour costs (e.g. ground surveys, high resolution airborne imagery, LiDAR). Vertical accuracies range from millimetres to tens of metres; however, the most accurate of these systems (*e.g.* total stations, DGPS, and LIDAR) are appropriate only at local to regional extents and have moderate to poor temporal resolution as a result of cost, time, and labour required to complete such surveys.

As a result of these limitations, the need for a cost effective alternative for the development of high resolution DEMs has been identified (e.g. Eissenbeiß, 2009; Hugenholtz et al., 2013). A promising alternative has been the application of Unmanned Aerial Vehicles (UAVs) equipped with relatively inexpensive, consumer grade digital cameras. Such systems are capable of acquiring high resolution imagery of localized areas with relatively little preparation time and very quick turnaround time (e.g. hours). Several software applications, employing common photogrammetric techniques, are available for the extraction of three-dimensional coordinates (i.e. point clouds) from the resulting imagery. For example, UAV systems have been successfully used for developing three-dimensional models of archaeological sites (Chiabrando et al., 2011; Eisenbeiß, 2009; Gomez-Lahoz & Gonzalez-Aguilera, 2009), the hydrology of permafrost and arctic environments (Scheritz et al., 2008; Turner et al., 2012), and the location and form of highways, buildings, rockslides, and forest stands (Haarbrink & Eisenbeiß, 2008; Wallace et al., 2012).

Although numerous studies have demonstrated the ability of UAVs to derive DEMs, few have conducted accuracy assessments conforming to either the Federal Geographic Data Committee (FGDC) (1998) or American Society for Photogrammetry and Remote Sensing (ASPRS) protocols, or others protocols proposed by Li (1991), Aguilar et al. (2007), or Hohle & Hohle (2009). These protocols recommend: 1) an assessment against reference data sets at least three times more accurate than derived data points (e.g. total station, dGPS, or LIDAR), 2) comparison against a minimum of 20 and recommended 30 well distributed reference points, 3) that the distribution of errors be tested for normality using standardized tests of skewness, and 4) that the 95% confidence limits of the vertical accuracy be reported. In some cases the number of reference points has been fewer than 30 (Grenzdorffer et al., 2008; Rosnell and Honkavaara, 2012), while in other studies (Eisenbeiß, 2009; Fonstad et al., 2013; Hugenholtz et al., 2013; Turner et al., 2012) large numbers of reference points have been used, however, the normality of the distribution of errors has not been tested nor have the confidence limits been reported. As suggested by Aguilar et al. (2007), Congalton and Green (2009); and Höhle and Höhle (2009) it is likely that 95% confidence intervals are not commonly reported because of the low number of reference points frequently used, and that the root mean square error (RMSE) is often converted to an established confidence level assuming a mean error of zero and a normal distribution of elevation residuals, even though this is often not the case for vertical positioning errors. Furthermore, reference data are typically compared to interpolated products derived from point clouds, rather than conducting a direct comparison of derived data points to reference data points, in spite of the recognition that interpolation algorithms introduce systematic errors (Wise, 2000) and affect the magnitude of residuals (Poon et al., 2005).

Table 1. Comparative accuracies and cost of DEM approaches[1]

Approach	Reported Accuracy		Extent	Cost ($)	References
	Lowest	Highest			
Topographic Maps	6 m (mean)	3 m (mean)	Regional to continental	Generally free	Hopkinson et al. (2009) Rayburg et al. (2009). Walker & Willgoose (2006)
Total Station	5 mm (mean)	2 mm (mean)	Site	1,000's	Lambrou & Pantazis (2010)
dGPS Surveys	0.37 m (RMSE)	0.22 m (RMSE)	Site to local	1,000's	Baptista et al. (2011) Schloderer et al. (2011)
Satellite Photogrammetry					
GeoEye Const.	4.78 m (RMSE)	1.2 m (RMSE)	Regional	1,000's	Poon et al. (2005) Toutin (2004a) Zhang & Gruen (2004)
DigitalGlobe Const.	6.70 m (RMSE)	1.3 m (RMSE)	Regional	1,000's	Toutin (2004b)
ASTER	31.74 m (RMSE)	6.1 m (RMSE)	Regional	Free (for USA) to 1,000's	Eckert et al. (2005) San & Suzen (2005) Toutin (2008)
ASTER GDEM	17 m (95%)	8.2 m (RMSE)	Global	Free	ASTER GDEM Validation Team (2011) Li et al. (2012)
SPOT-5	10 m (RMSE)	2.2 (RMSE)	Regional	1,000's	Reinartz et al. (2006) Zhang & Gruen (2004) Toutin (2006)
Aerial Photography	1.69 m (99%)	0.17 m (mean)	Local to regional	10,000's	Scholten & Wewel (2000) St-Onge et al. (2008) Westaway et al. (2003)
RADAR					
Satellite / Shuttle	78 m (RMSE)	3 m (mean)	Global	Free	Chen et al. (2007) Li et al. (2012) Rodriguez et al. (2006) Toutin (2010)
Airborne	10.67 m (RMSE)	1.5 m (RMSE)	Regional	10,000's	Andersen et al. (2008) Hodgson et al. (2003) Mikhail et al. (2001)
LiDAR	0.93 m (RMSE)	0.15 m (RMSE)	Local to regional	10,000's	Csatho et al. (2008) Hodgson et al. (2003) Hopkinson et al. (2009) Jensen (2007) Rayburg et al. (2009)

[1]This table portrays the general range of achieved accuracies, irrespective of the source of ground reference data.

Therefore, the purpose of this research was to investigate the application of an UAV for the development of high resolution DEMs of localized areas, and to provide a detailed accuracy assessment, including 95% confidence intervals using reference data sets containing a large number of points. This study also aims to assess whether the interpolation of UAV-derived point clouds introduces systematic errors and increases the magnitude of errors by evaluating residuals measured through a direct point to point (i.e. reference point to derived point) and point to raster (i.e. reference point to interpolated raster) approach.

METHODS

Study Area

The study area was located in south-west Manitoba, Canada, measured 336 by 235 metres or 7.9 hectares, and was specifically selected so that it included both morphologically simple and complex features as well as bare and vegetated surfaces (Figure 1). The region is part of the Assiniboine Delta and consists of distal glaciofluvial sediments deposited as outwash fans in glacial Lake Agassiz by meltwater turbidity currents (Matile and Keller, 2004). These late Plesitocene sediments have been reworked by Holocene aeolian processes such that the area is now characterized by parabolic sand dunes and localized blowouts with local relief up to 10 m. Vegetation at the study site was predominantly sparse mixed short grass prairie including creeping juniper (*Juniperus horizontalis*), prairie sage (*Artemisia ludoviciana*), and needle-and-thread grass (*Stipa comata*).

Reference Datasets

Accurate reference datasets were necessary to evaluate the accuracy of derived terrain models. The American Society for Photogrammetry and Remote Sensing (ASPRS) recommends that the independent source of reference data should be at least three times more accurate than the instrumentation being used

Figure 1. Location of study area within province of Manitoba (a), relative to the city of Brandon (b), and view across study area looking east (c)

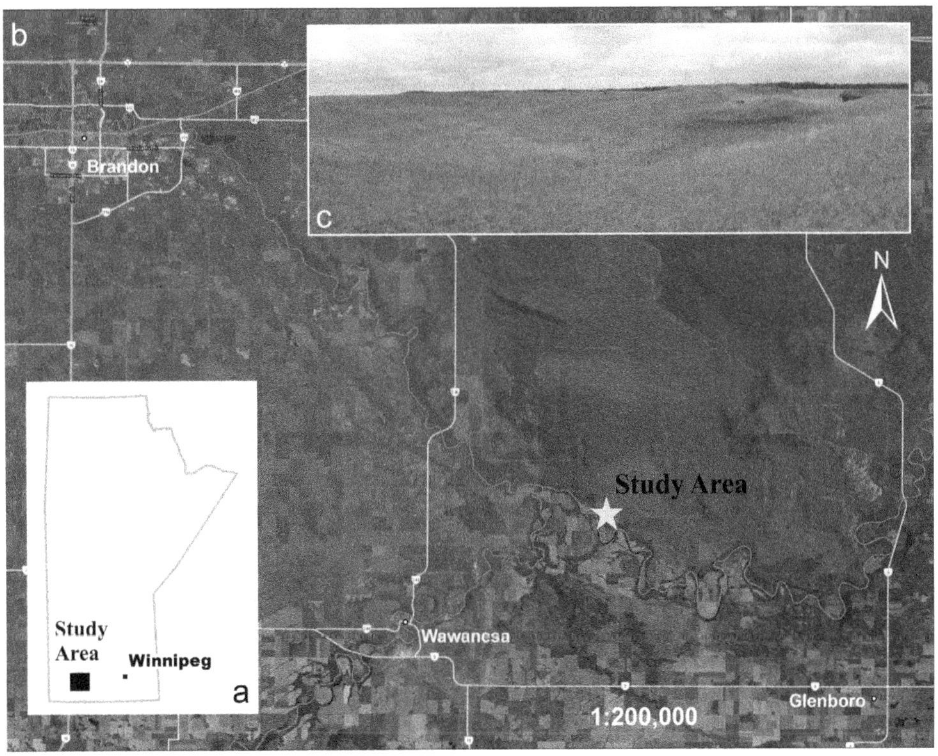

to develop the terrain models being tested, whenever possible (American Society for Photogrammetry and Remote Sensing, 2004). For this research project both dGPS and total station surveys were conducted to produce reference datasets. These techniques are capable of measuring horizontal and vertical coordinates to decimeter and centimeter level accuracies respectively, and have been used extensively to produce high resolution DEMs of localized study areas with high degrees of accuracy (Lambrou & Pantazis, 2010; Rayburg et al., 2009).

dGPS Survey

A Trimble Pro XRS dGPS rover and base station were used to acquire a reference dataset of the entire study area in June, 2011. The survey consisted of 32 transects walked in a north-south direction, with an average spacing between transects of 10 – 20 m and intervals between positions of one second (Figure 2a). In addition, nine strategically positioned transects were conducted along ridge and valley features and other lines of topographic inflection to capture the complex dune features of the landscape. In total, 12,667 control points were recorded in 5.5 hours of fieldwork.

The dGPS data were downloaded and differentially corrected using Trimble Pathfinder Office 2.7 software. The resulting shapefile was imported in ESRI ArcScene for data quality analysis, however, no inconsistencies were found. The Ordinary Kriging method in ArcGIS 10 Geostatistical Wizard was used to interpolate the 12,667 sample points. To suppress the generation of surficial artifacts as a result

Figure 2. Location of dGPS survey transects across (a) the entire study area and (b) the total station points within the Sandpit site

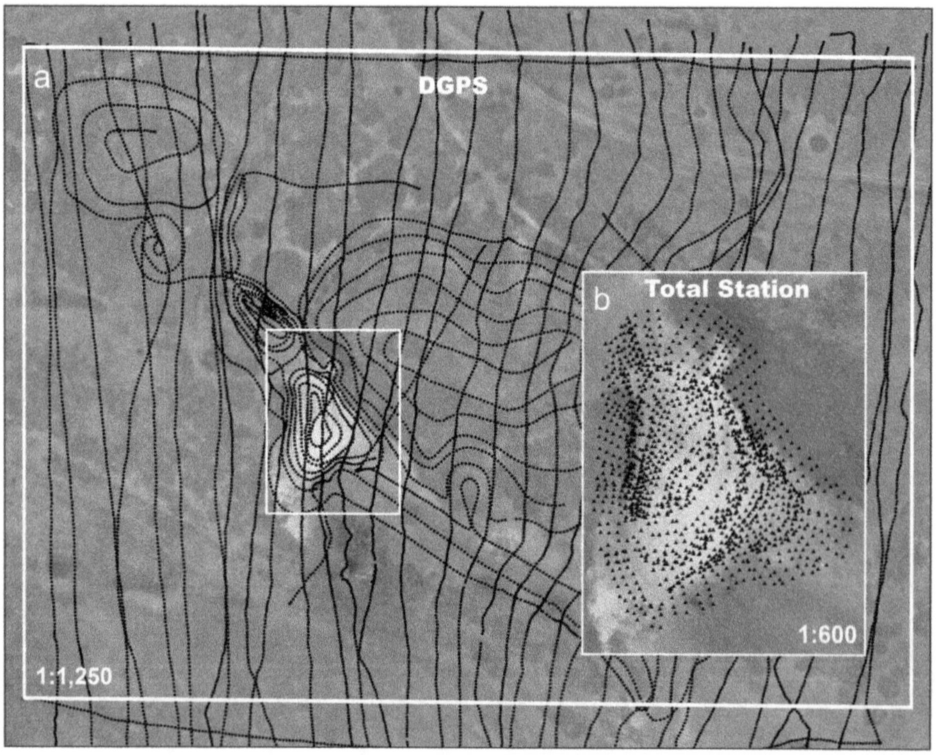

of the north-south transects, a custom semivariogram was created to determine the optimal size of the neighbourhood and number of sample points from which the interpolation algorithm would estimate the height of each output grid cell. The DEM was generated at both 20 cm and 40 cm spatial resolutions so that the reference DEMs and UAV-derived DEMs would be of the same resolution. As reported by Westaway et al. (2003), the highest attainable DEM resolution derived from an image source is typically five times greater than the ground resolution cell size of the imagery, which in this case was 4 and 8 cm, resulting in UAV-derived DEM resolutions of 20 cm and 40 cm.

Total Station Survey

Because of the high labour costs associated with dense total station surveys, a smaller study site measuring approximately 44 m by 56 m (0.17 ha) was selected from within the larger study area. This smaller area, referred to as the "Sandpit", was surveyed at a sampling interval of approximately 30 cm with a Nikon Nivo 5M total station set up at two vantage points, providing both east and west views of the site (Figure 2b). To guarantee data consistency during the survey, nine previously established benchmarks located outside the survey area were regularly measured with the total station and checked for significant positional differences. The total station is used is capable of measuring positions to within 3 mm (+/- 2 ppm). In total 1,422 control points were measured in 11.5 hours of fieldwork within the Sandpit study site.

The total station survey data were downloaded in text format (.csv) using Trimble Data Transfer software, imported into ArcGIS 10, and converted to a shapefile. The dataset was visually checked for data inconsistencies in ArcScene and a total of six faulty measurements were filtered out. Because the sampling interval was consistent throughout the survey, no custom semivariogram was necessary to interpolate the dataset. The remaining 1,416 control points were used to interpolate a DEM using the default Kriging method with a twelve sample point variable radius neighbourhood. Similar to the dGPS reference dataset, the resulting DEM was generated at both 20 cm and 40 cm spatial resolutions.

Acquisition of UAV Imagery

For this study, a UAV system called CropCam, manufactured by Micropilot, Inc. of Stony Mountain, Manitoba, Canada was used to acquire high resolution imagery of the study area. CropCam is an autonomous UAV based on a model sailplane airframe equipped with a Micropilot MP2028^{g} autopilot and Trimble GPS module. The UAV weighs 2.7 kg, has a 2.4 m wingspan, and is powered by four lithium-polymer batteries running an Axi brushless electric motor. In addition to autonomous flight capabilities, CropCam features a radio telemetry system which allows the ground crew to manually operate the UAV using either a 5 channel RC transmitter, or Ground Control Station (GCS) featuring a 2.4 GHz radio modem and HORIZON^{mp} software. The UAV was outfitted with a Pentax Optio A40 digital camera which features a 1/1.7-inch CCD sensor with 12 megapixels (4000 x 3000 pixels), a f/2.8 lens, 7.9 mm focal length, and camera shake reduction mechanism. Images are saved in JPEG format at an 8 to one compression ratio. This provides near lossless image quality and is well under the compression ratio of 15:1 recommended by Lam and Yuan (2001) for deriving terrain models from JPEG imagery.

Two aerial surveys were designed using the UAV flight planning software Lentsika (CePED Ltd.). Both flights were planned with 65 percent side- and end-lap and the study area was buffered by 80 m to ensure stereoscopic overlap and complete coverage of the study area. The flight lines were oriented in an east-west direction to minimize the number of legs necessary to cover the rectangular study area. The flights were conducted at an altitude of 160 m and 320 m, resulting in imagery with a spatial resolution of 4 cm and 8 cm respectively. Prior to image acquisition, 17 large white pails were placed at regular intervals throughout the study area and the total station was used to collect accurate UTM coordinates for each pail. These pails were also used as ground control points to independently georeference the imagery.

The UAV was deployed on June 28, 2011 at 15:26 and 16:50 Central time under sunny conditions. The Sun was 48.5° and 35.4° above the horizon, respectively, satisfying the recommendations of Jensen (2000) of ideal time of day for aerial imagery acquisition. A moderately calm wind (≈ 20 km/h) was present from the SSE, resulting in a slight crabbing of the airplane. The first UAV flight, at 320 m altitude, acquired 25 colour images in a 12 minute flight. The second UAV flight, at 160 m altitude, acquired 81 colour images in a 16 minute flight.

Generating UAV-Derived DEMs

The UAV imagery was processed using two software applications; a commercially available photogrammetric application called LISA (Geo-Software GmbH), as well as an online freeware application called Photosynth (Microsoft Inc.). Both the 4 cm and 8 cm resolution imagery was processed for the entire study area and Sandpit study site.

The LISA Workflow

LISA is a digital photogrammetric workstation capable of processing satellite imagery, traditional aerial photography, and UAV imagery, whereby its FOTO module features camera definition and calibration, interior and exterior orientation, a bundle block adjustment and an area-based matching stereo correlation for digital surface reconstruction (Wundram & Löffler, 2007). LISA provides a photogrammetric workflow similar to PCI Geomatics or Leica Photogrammetry Suite, however, at a much lower commercial cost. Before any of the imagery were processed, LISA required a camera calibration to model the radial distortion of the consumer grade lens. Seventy-seven target marks were measured on a calibration plate to model the distortions and create a camera calibration file.

After the camera calibration, the 4 cm resolution UAV imagery was imported and converted to LISA's native .IMA image format. A ground control point file was generated in the control point editor, displaying the UTM coordinates of the 17 white pails. The white pails were manually identified in each of 30 adjacent pairs of overlapping images to determine the absolute orientation for each stereo model (*i.e.* image pair). Tie points were manually measured in each stereo-model to strengthen the relative orientation between stereo models.

After the ground control points and additional tie points were measured, image strips in an east-west orientation were defined and subsequently generated. The measurement of tie points between image strips solved the relative orientation of each image strip to the neighbouring strips. A bundle block adjustment was then used to resolve the position and orientation of each pixel within the images, match objects in adjacent images in order to determine their parallactic angles, and derive the elevation at specific point locations.

The cross correlation algorithm employed by LISA is a statistical comparison that analyzes the correlation coefficient of pixel values from sub arrays (*i.e.* rectangular blocks of pixels) identified by a roving window in both the left and right image. To ensure that pixels were accurately matched, the highest correlation coefficient found in the roving window was compared against a correlation threshold of 0.7. If the correlation coefficient equaled or exceeded the correlation threshold, the pixel match was considered successful and was used in the overall parallax determination. A roving window (*i.e.* neighbourhood) of 13 x 13 pixels was advised as optimal given that the study area contained spectrally homogenous groupings of pixels (pers. comm. Wilfried Linder, 2011).

To reduce the number of reference data points generated, stereoscopic measurements were made at a 0.3 m sampling interval. Due to file size limitations, the resulting vector point file was split in half and exported in text file format (.csv) then imported into ArcGIS 10, where the two files were merged and converted to shapefile format. The default Kriging method was used to obtain a continuous elevation surface at 20 cm spatial resolution using a variable radius neighbourhood of 12 points to derive a DEM. Data values beyond the extent of the study area were included in the interpolation to minimize erroneous edge effects. The same workflow was applied to process the 8 cm resolution UAV imagery resulting in a 40 cm resolution DEM.

Microsoft Photosynth

Both the 4 cm and 8 cm resolution imagery were also processed using Microsoft Photosynth to investigate the capability of computer vision techniques to derive DEMs from UAV imagery. Photosynth was chosen mainly because the program is fully automated and free of charge, thus, allowing a comparison in performance and cost with more traditional digital photogrammetric applications such as LISA. The described Microsoft Photosynth workflows were originally developed by Craig (2010) and Willis (2010).

Using computer vision techniques, Photosynth is capable of analyzing a sequence of images and determining their three-dimensional position relative to each other. Photosynth is fully automated, so does not allow the parameters to be changed for its matching algorithm, and the relative orientation of the imagery is determined without the manual selection of tie points or camera calibration. Instead, the images are matched using proprietary computer vision algorithms and the resulting mosaic and point cloud are visualized on the online Microsoft Photosynth user profile.

The freeware application SynthExport (Hausner, 2010) was used to download the point cloud in .ply format to store the three dimensional data. Meshlab (ISTI CNR, 2010), an open-source software application for editing point clouds, was used to filter out erroneous points generated by Photosynth, and the resulting point cloud was saved as a tab-delineated text file. After removing the .PLY header information in Microsoft Notepad, the point cloud was loaded in Menci Scanview (Menci Software, 2011) to measure the local coordinates of the 17 ground control points. Because Photosynth assigns an arbitrary coordinate system to store the relative position of each point, the absolute coordinates of the white pails previously measured were used to convert the local coordinates of the Photosynth point cloud to UTM coordinates using the open source CoordTrans function available in the Java Graticule 3D (JAG3D) program using a 9 parameter, least-square, Gauß-Helmert transformation for rotation, scaling, and translation (Losler, 2011).

The transformed point cloud was copied to a Microsoft Excel spreadsheet, saved in .csv file format, then imported into ArcGIS 10 and converted to a point shapefile. Similar to the LISA dataset, the default Kriging interpolation function with a variable radius neighbourhood of 12 points was used to generate DEMs at both 20 cm and 40 cm spatial resolution for the 4 cm and 8 cm resolution imagery.

Data Analysis

The accuracies of the UAV-derived point clouds and resulting DEMs were evaluated through three separate quantitative analyses.

Direct Comparison Without Interpolation

Because the interpolation of point data producing a DEM may cause the unintended propagation of errors (e.g. Westaway et al., 2003), the first analysis relied on the direct comparison of measured reference data values (*i.e.* dGPS and total station coordinates) against UAV-derived data values (*i.e.* LISA and Photosynth point clouds). For the Sandpit study site the accuracy of the LISA and Photosynth derived datasets were assessed against the total station reference dataset. In addition, the portion of the dGPS reference dataset collected for the entire study area that overlapped with the Sandpit site was evaluated against the more accurate total station reference dataset to verify that the dGPS dataset was suitable to be used as the reference dataset for the entire study area. Points in the derived datasets were spatially linked to the nearest point in the reference datasets and only those points that were within a 20 cm horizontal distance of the reference points and on slopes of less than 45 degrees were selected to ensure that any elevation change that occurred would be less than the vertical precision of the DEM. The total number of paired observations for the Sandpit study site and entire study area, respectively, were 1411 and 10,309 for the LISA dataset, 247 and 1,353 for the 4 cm Photosynth dataset, and 52 and 317 for the 8 cm Photosynth dataset.

Commonly, the accuracy of a terrain model is measured using the root mean square error (RMSE) and mean error (ME). RMSE measures the absolute fit of the model to a more accurate dataset, whereas the ME measures the systematic under- or over-estimation of the elevation values (Höhle and Höhle, 2009). Using the paired derived and reference data values, the mean error, root mean squared error, and maximum interval of error at the 95% probability level were calculated for each dataset. The mean error is given by Höhle and Höhle (2009):

$$\mu = \frac{1}{n}\sum_{i=1}^{n}\Delta hi \quad (1)$$

where "Δh" is the difference between the reference data value and the derived data value at location "*i*", and "*n*" is the number of check points.

Höhle and Höhle (2009) provides the RMSE formula:

$$RMSE = \sqrt{\frac{1}{n}\sum_{i=1}^{n}\Delta hi^2} \quad (2)$$

where "Δh" is the difference between the reference data value and the derived data value at location "*i*", and "*n*" is the number of check points.

The maximum interval of error at the 95% probability level is given by Congalton and Green (2009), and uses the mean error of the sample (ì), the standard deviation estimator of elevation errors in the dataset (S_v), and appropriate z-value (±1.96), in the following formula:

$$\textit{Estimated interval of error at } 95\%\,\textit{probability} = \mu \pm 1.96\left(\mathrm{Sv}\right) \tag{3}$$

With respect to mapping standards set by American Society for Photogrammetry and Remote Sensing (2004), a measure of skewness was used before the maximum interval of error at the 95% probability is calculated. Equation 3 was applied in situations where skewness ≤ ±0.5, while the nonparametric 95th percentile method was used if the (skewness ≥ ±0.5). The test of skewness ($ð^3$) uses the number of reference points (n), mean error of the sample (ì), the standard deviation estimator of elevation errors (S_v), as well as the errors of each check point (x_i), to determine if the elevation errors are normally distributed.

$$ð^3 = \frac{\sum (xi - \mu)^3}{n\,\mathrm{Sv}^3} \tag{4}$$

Comparison of UAV-Derived DEMs to Reference Points

To reduce the possibility of error propagation, the second analysis compared the accuracy of the UAV-derived DEMs directly to the reference data points, rather than an interpolated reference DEM. For the Sandpit study site the interpolated LISA and Photosynth DEMs were compared to the total station dataset, while the dGPS dataset was used to evaluate the accuracy of the UAV-derived DEMs of the entire study area. At each measured point in the reference dataset the elevation of the UAV-derived DEM was extracted using the "Extract Values to Points" function in ArcGIS10. The accuracy of the derived DEMs was determined through the calculation of the mean error (Equation 1), root mean squared error (Equation 2), and the maximum interval of error at the 95% probability level (Equation 3), or nonparametric 95th percentile method (Equation 4).

RESULTS

Evaluation of dGPS Reference Dataset

The average Positional Dilution of Precision (PDOP) of dGPS points collected was $\mu = 2.10$ ($\sigma = 0.41$). The estimated horizontal precision ranged from 0.01 to 0.381 m ($\mu = 0.024$ m, $\sigma = 0.012$ m) and vertical precision ranged from 0.019 to 0.641 m ($\mu = 0.050$ m, $\sigma = 0.030$ m).

To verify the suitability of the dGPS survey as a reference dataset, elevation differences between the dGPS positions and total station coordinates for the Sandpit study site were evaluated. The accuracy assessment indicated that the mean difference between dGPS and total station positions was -0.10 m, with an RMSE of 0.15 m. The relatively large errors are most likely due to poorer satellite geometry and sky view within the bowl of the Sandpit area, as indicated by higher PDOP values recorded ($\mu = 2.60$) and lower vertical precision ($\mu = 0.08$ m, $\sigma = 0.02$ m). Nevertheless, relative to the expected RMSE of the UAV-derived DEMs, the dGPS dataset was deemed to be a suitable alternative reference dataset for the entire study area.

UAV-Derived Point Clouds

LISA failed to process the 8 cm resolution imagery due to reduced stereoscopic overlap and image sharpness caused by the UAV crabbing into the wind, excessive image motion, and incorrect focus. These factors resulted in poor contrast between distinct features and their surroundings and textural dissimilarities between the same features in different photos to the extent that the stereo-correlation image matching algorithm failed to calculate a DEM (pers. comm. Wilfried Linder, 2011). However, Photosynth's structure-from-motion algorithm, which is invariant to image scaling, rotation, and illumination, successfully derived a total of 46,231 data points from the same imagery, of which 20,062 were located within the study area.

From the 4 cm spatial resolution imagery, LISA derived 1,355,286 data points, of which 878,848 were located within the study area (11.1 points m^{-2}). In comparison, Photosynth derived a total of 197,816 data points from the same imagery, with 95,776 located within the study area (1.2 points m^{-2}). Within the Sandpit extent LISA generated 19,162 data points from the 4 cm resolution imagery, while Photosynth derived 2,708 and 468 points from the 4 cm and 8 cm spatial resolution imagery, respectively.

Direct Comparison Without Interpolation

Direct comparison without interpolation indicated that the points derived from the 4 cm resolution imagery using LISA were of considerably higher accuracy than those derived from either the 4 or 8 cm imagery using Photosynth (Figure 3). In comparison with the 1,416 total station reference data points collected at the Sandpit study site, LISA generated a point cloud with a mean error of -0.31 m and a RMSE of 0.46 m. With reference to the 10,350 dGPS reference data points collected over the entire study area, LISA generated a point cloud with a mean error of -0.15 m and a RMSE of 0.42 m. The Photosynth 8 cm imagery dataset produced the second best results, with a mean error of 1.11 and 0.76 m and a RMSE of 1.18 and 1.37 m, for the Sandpit site and entire study area, respectively. The Photosynth 4 cm imagery dataset produced the worst results, with a mean error of 2.16 and 0.81 m and a RMSE of 2.17 and 1.74 m, for the Sandpit site and entire study area, respectively.

The skewness of elevation differences between derived and reference datasets indicated that some exceeded the $\pm$ 0.5 threshold. Consequently, the nonparametric 95th percentile method was used to obtain the maximum interval of error for these datasets. For the two normally distributed datasets Equation 3 was used to obtain the maximum interval of error at the 95% probability for these datasets. Figure 4 indicates that for the Sandpit site the maximum interval of error is similar for the LISA and Photosynth datasets (1.46 m vs. 1.13 m) even though there are differences in mean error. At the study area extent the results suggest that the LISA dataset has a much smaller interval than the two Photosynth datasets. The similarity of both the parametric ("^{p}") and non-parametric ("^{np}") statistic for the LISA 4 cm dataset indicates that the nature of the statistical measure has little influence on the calculated interval of maximum error. Whereas 95% of the Photosynth errors fall within a range size of 4.22 and 5.97 m (8

Figure 3. Direct comparison of total station and dGPS elevation compared to UAV-derived point clouds

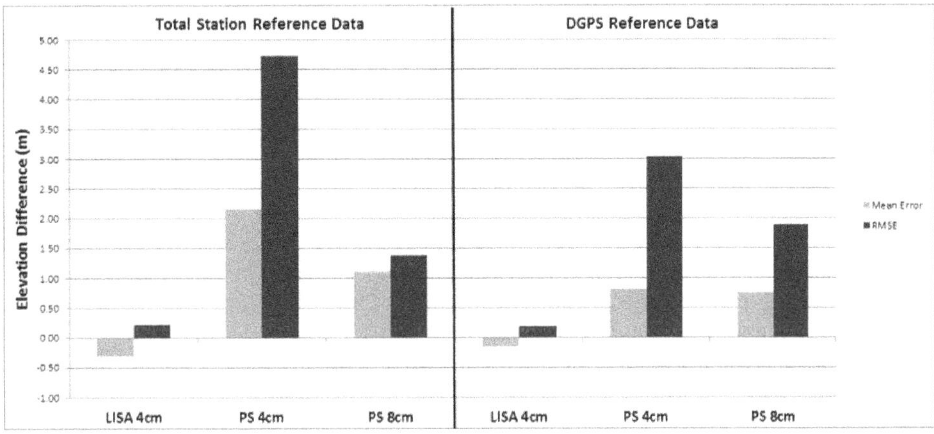

cm dataset and 4 cm dataset, respectively), 95% of the LISA errors fall within a range size of 1.55 and 1.79 m, depending on the statistical test used.

Comparison of Derived DEMs to Reference Data Values

The UAV-derived DEMs of the entire study area are shown in Figures 5a and b for the LISA 4cm and the Photosynth 4 cm datasets, respectively.

Although all three models have captured the bowl and ridge characteristics of the surveyed topography, the comparison of derived DEMs to reference data values indicated that the interpolated DEM derived from the 4 cm imagery using LISA was of higher accuracy than the DEMs generated using the Photosynth workflow (Figure 6). In comparison with the total station reference values at the Sandpit site, LISA generated a DEM with a mean error of -0.31 m and a RMSE of 0.45 m. With reference to the dGPS reference dataset for the entire study area, LISA generated a DEM with a mean error of -0.15 m and a RMSE of 0.42 m. The Photosynth 8 cm imagery dataset produced the second best results, with a mean error of 1.16 and 0.63 m and a RMSE of 1.20 and 1.31 m for the Sandpit site and entire study

Figure 4. Maximum interval of error (95%) of UAV-derived DEM compared to reference data points

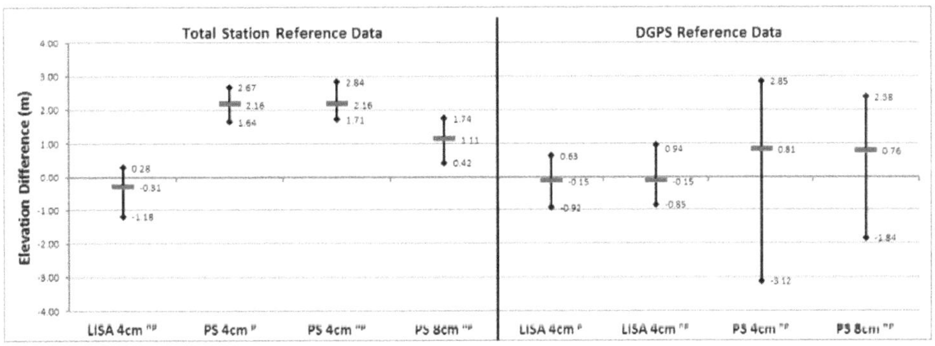

Figure 5. UAV-derived DEM from LISA 4cm dataset (a) and Photosynth 4 cm dataset (b)

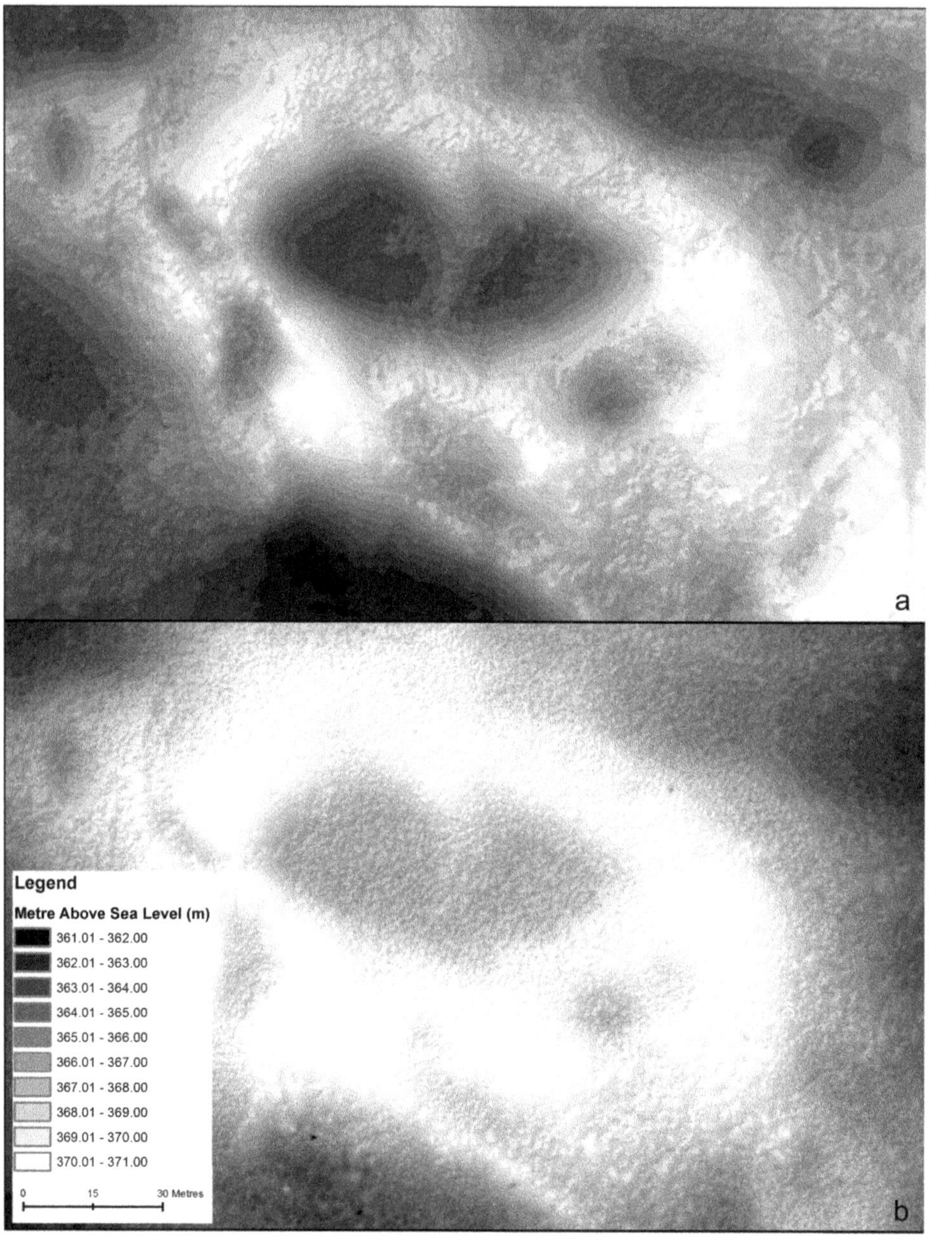

area, respectively. The Photosynth 4 cm imagery produced the worst results, with a mean error of 2.15 and 0.88 m and a RMSE of 2.17 and 1.71 m, for the Sandpit site and entire study area, respectively.

Figure 7 indicates that for the Sandpit site the maximum interval of error is similar for the LISA and Photosynth datasets due to similarities in standard deviation. At the study area extent the results suggest that the LISA dataset has a much smaller interval than the two Photosynth datasets. The similarity of both the parametric ("p") and non-parametric ("np") statistic of the LISA 4 cm dataset again indicates that the nature of the statistical measure (*i.e.* parametric versus non-parametric) has little influence on the calculated interval of maximum error.

Figure 6. Comparison of total station and dGPS Elevation compared to the interpolated UAV-derived point clouds

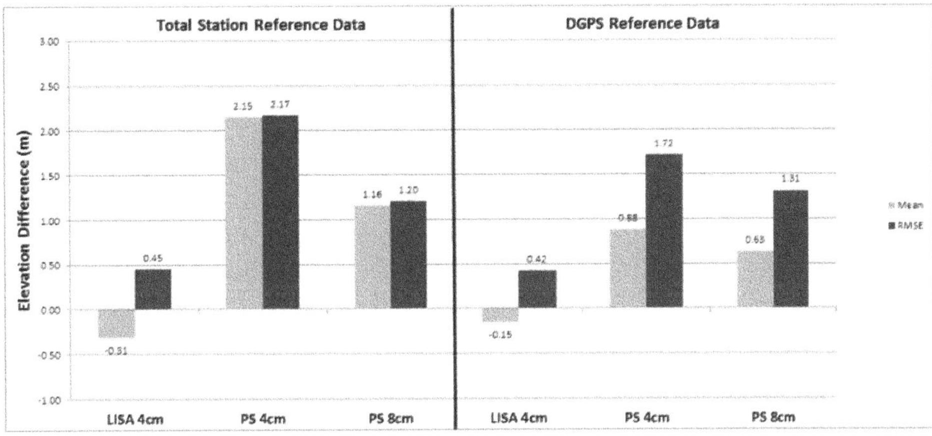

Figure 7. Maximum interval of error (95%) of UAV-derived DEMs compared to reference data points

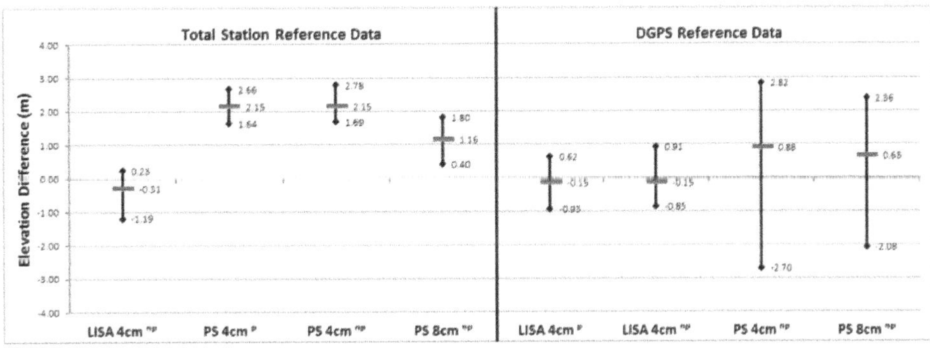

DISCUSSION

The majority of the datasets exceeded the skewness threshold, thereby highlighting the need for non-parametric accuracy evaluations. In addition to the exceeded skewness thresholds, the mapped errors also revealed that the two Photosynth datasets show a definite convex pattern with increasing negative errors towards the boundaries of the DEMs similar to what has been reported by Rosnell and Honkavaara (2012), while the errors produced by LISA appeared to be randomly distributed. The elevation errors of the Photosynth datasets are skewed due to the convex nature of the point clouds, and because Photosynth is fully autonomous this convex shape could not be improved by changing image matching parameters. The removal of the radial lens distortion before the Photosynth process could potentially alleviate this problem; however, this remains to be investigated.

As reported in the literature and other DEM protocols, the RMSE accuracy alone does not provide a DEM user with a full description of how well each cell in the DEM represents the true elevation. The RMSE values achieved in this study (i.e. LISA = 0.45 m, Photosynth = 1.20 m) fall within the range of what has been achieved in other recent studies (Rosnell and Honkavaara, 2012; Turner et al., 2012;

Hugenholtz et al., 2013). Although the RMSE measures differ considerably between the two software applications, the range of the 95% confidence intervals are similar for the smaller Sandpit study site because the strongest outliers in the Photosynth DEMs occurred outside this area near the edges of the larger study area. Consequently, for the entire study area the 95% confidence interval of Photosynth is considerably larger than LISA. This suggests that both software applications are capable of generating DEMs of similar accuracy after the systematic error (i.e. the mean error) is corrected, and the strongest outliers removed. It is evident that reporting the 95% confidence intervals provides the end user with additional useful information not reported by the mean error and RMSE alone.

With respect to the influence of the interpolation of point clouds, the similarities between Figure 3 and 6, as well as between Figure 4 and 7 indicate that the interpolation of the point clouds did not introduce systematic errors or increase the magnitude of residuals as previously reported in other DEM studies (e.g. Poon et al., 2005; Wise, 2000). This may be in part because of the relatively high density of points used for the Kriging interpolation (i.e. 1.2 – 11 points m-2), thereby minimizing the impact of the interpolation process (Chaplot et al., 2006; Li et al., 2004). To assess whether this observation holds true for smaller ground reference sample sizes (e.g. the more common n = 20, 30, 100) and point cloud densities, further investigations with regards to bootstrapping the datasets are required.

The reason why the 8 cm imagery produced a better result than the 4 cm imagery in the Photosynth workflow could not be determined. The 4 cm imagery was of higher quality in terms of spatial resolution, stereoscopic overlap, image sharpness, and contrast. The base-to-height ratio remained constant between the 4 and 8 cm imagery and, thus, the parallactic angles between images did not change. The poorer accuracy of the 4 cm imagery may be caused by the amplification of artifacts generated by the Photosynth matching algorithm, as well as the polynomial function applied in the transformation of the point cloud (e.g. greater number of points increases the noise level). Given the many unknown processing components of the Photosynth image matching technique (e.g. point cloud generation and the establishment of an internal coordinate scheme) it was not possible to rigorously check the influence of a wide range of image matching parameters that could have influenced the DEM accuracy.

CONCLUSION

The purpose of this research was to evaluate the application of an UAV for the development of high resolution DEMs and expand upon previous studies by providing a detailed accuracy assessment including 95% confidence intervals. Two workflows, describing the application of both a commercially available and free softcopy photogrammetric application were used to test the influence of the image processing workflow upon DEM accuracy. The accuracy of both workflows was determined by comparing the UAV derived DEMs against reference datasets collected using total station and dGPS surveys, whereby particular attention was paid to the influence of point cloud interpolation and the introduction of systematic errors by conducting both a direct, point to point comparison without interpolation, and a UAV-derived DEM to reference point comparison. The results indicated that the UAV imagery was capable of producing dense point clouds and high resolution terrain models with RMSE values ranging between 0.42 m and 1.72 m, depending on the image dataset used. Whereas the mean error and RMSE differed considerably between the two applications, the 95% confidence intervals suggest that DEMs derived for small areas of interest can be of equivalent accuracy when post-processed and corrected for the mean error. This study highlights the importance of 95% confidence intervals as an essential metric in accuracy assessments of UAV-derived DEMs, as generally recommended by the scientific literature and international terrain model protocols. The results also suggest that the Kriging algorithm used did not decrease the accuracy of the resulting DEM, because the density of data points was relatively high. The influence of both the number of ground reference points and point cloud density on the skewness of the DEM as well as the DEM accuracy (e.g. the mean, RMSE, and 95% confidence intervals) remains to be investigated.

REFERENCES

Aguilar, F. J., Aguera, F., & Aguilar, M. A. (2007). A Theoretical Approach to Modeling the Accuracy Assessment of Digital Elevation Models. *Photogrammetric Engineering and Remote Sensing*, *73*(12), 1367–1379. doi:10.14358/PERS.73.12.1367

Akella, M. R., Delmelle, E., Batta, R., Rogerson, P., & Blatt, A. (2010). Adaptive Cell Tower Location Using Geostatistics. *Geographical Analysis*, *42*(3), 227–244. doi:10.1111/j.1538-4632.2010.00791.x

American Society for Photogrammetry and Remote Sensing. (2004). ASPRS Guidelines - Vertical Accuracy Reporting for Lidar Data. In M. Flood (Ed.), *Vol. American Society* (p. 20). American Society for Photogrammetry and Remote Sensing.

Andersen, H. E., McGaughey, R. J., & Reutebuch, S. E. (2008). Assessing the influence of flight parameters, interferometric processing, slope and canopy density on the accuracy of X band IFSAR derived forest canopy height models. *International Journal of Remote Sensing*, *29*(5), 1495–1510. doi:10.1080/01431160701736430

ASTER GDEM Validation Team. (2011). In D. Meyer (Ed.), *ASTER Global Digital Elevation Model Version 2 – Summary of Validation Results* (p. 26).

Baptista, P., Cunha, T. R., Matias, A., Gama, C., Bernardes, C., & Ferreira, O. (2011). New land-based method for surveying sandy shores and extracting DEMs: The INSHORE system. *Environmental Monitoring and Assessment*, *182*(1-4), 243–257. doi:10.100710661-011-1873-5 PMID:21301958

Cavalli, M., Tarolli, P., Marchi, L., & Dalla Fontana, G. (2008). The effectiveness of airborne LiDAR data in the recognition of channel-bed morphology. *Catena*, *73*(3), 249–260. doi:10.1016/j.catena.2007.11.001

Chaplot, V., Darboux, F., Bourennane, H., Leguédois, S., Silvera, N., & Phachomphon, K. (2006). Accuracy of interpolation techniques for the derivation of digital elevation models in relation to landform types and data density. *Geomorphology*, *77*(1-2), 126–141. doi:10.1016/j.geomorph.2005.12.010

Chen, R.-F., Chang, K.-J., Angelier, J., Chan, Y.-C., Deffontaines, B., Lee, C.-T., & Lin, M.-L. (2006). Topographical changes revealed by high-resolution airborne LiDAR data: The 1999 Tsaoling landslide induced by the Chi–Chi earthquake. *Engineering Geology*, *88*(3-4), 160–172. doi:10.1016/j.enggeo.2006.09.008

Chen, Y., Shi, P., Li, J., Deng, L., Hu, D., & Fan, Y. (2007). DEM accuracy comparison between different models from different stereo pairs. *International Journal of Remote Sensing*, *28*(19), 4217–4224. doi:10.1080/01431160600606841

Chiabrando, F., Nex, F., Piatti, D., & Rinaudo, F. (2011). UAV and RPV systems for photogrammetric surveys in archaelogical areas: Two tests in the Piedmont region (Italy). *Journal of Archaeological Science*, *38*(3), 697–710. doi:10.1016/j.jas.2010.10.022

Congalton, R. G., & Green, K. (2009). *Assessing the Accuracy of Remotely Sensed Data: Principles and Practises* (2nd ed.). Boca Raton, Florida, U.S.A.: CRC/Taylor & Francis.

Craig, N. (2010). Structure from Motion Point Clouds to Real World Coordinates, from http://www.personal.psu.edu/nmc15/blogs/anthspace/2010/02/structure-from-motion-point-clouds-to-real-world-coordinates.html

Csatho, B., Schenk, T., Kyle, P., Wilson, T., & Krabill, W. B. (2008). Airborne laser swath mapping of the summit of Erebus volcano, Antarctica: Applications to geological mapping of a volcano. *Journal of Volcanology and Geothermal Research*, *177*(3), 531–548. doi:10.1016/j.jvolgeores.2008.08.016

Eckert, S., Kellenberger, T., & Itten, K. (2005). Accuracy assessment of automatically derived digital elevation models from aster data in mountainous terrain. *International Journal of Remote Sensing*, *26*(9), 1943–1957. doi:10.1080/0143116042000298306

Eisenbeiß, H. (2009). *UAV Photogrammetry. (Doctor of Sciences)* (p. 18515). Zurich, Switzerland: ETH Zurich.

Federal Geographic Data Committee. (1998). Geospatial Positioning Accuracy Standards. Part 3: National Standard for Spatial Data Accuracy.

Fonstad, M. A., Dietrich, J. T., Courville, B. C., Jensen, J. L., & Carbonneau, P. E. (2013). Topographic structure from motion: A new development in photogrammetric measurement. *Earth Surface Processes and Landforms*, *38*(4), 421–430. doi:10.1002/esp.3366

Frankl, A., Nyssen, J., Calvet, M., & Heyse, I. (2010). Use of Digital Elevation Models to understand and map glacial landforms — The case of the Canigou Massif (Eastern Pyrenees, France). *Geomorphology*, *115*(1-2), 78–89. doi:10.1016/j.geomorph.2009.09.032

Gomez-Lahoz, J., & Gonzalez-Aguilera, D. (2009). Recovering traditions in the digital era: The use of blimps for modelling the archaeological cultural heritage. *Journal of Archaeological Science*, *36*(1), 100–109. doi:10.1016/j.jas.2008.07.013

Grenzdorffer, G. J., Angel, A., & Teichert, B. (2008). *The Photogrammetric Potential of Low-cost UAVs in Forestry and Agriculture.* Paper presented at the XXIst International Society of Photogrammetry and Remote Sensing Congress, Beijing, China.

Haarbrink, R. B., & Eisenbeiß, H. (2008). *Accurate DSM Porduction from Unmanned helicopter Systems.* Paper presented at the XXI International Society of Photogrammetry and Remote Sensing Congress - Technical Commission I, Beijing, China.

Hausner, C. (2010). SynthExport, from http://synthexport.codeplex.com/

Hodgson, M. E., Jensen, J. R., Schmidt, L., Schill, S., & Davis, B. (2003). An evaluation of LIDAR- and IFSAR-derived digital elevation models in leaf-on conditions with USGS Level 1 and Level 2 DEMs. *Remote Sensing of Environment*, *84*(2), 295–308. doi:10.1016/S0034-4257(02)00114-1

Höhle, J., & Höhle, M. (2009). Accuracy assessment of digital elevation models by means of robust statistical methods. *ISPRS Journal of Photogrammetry and Remote Sensing*, *64*(4), 398–406. doi:10.1016/j.isprsjprs.2009.02.003

Hopkinson, C., Hayashi, M., & Peddle, D. (2009). Comparing alpine watershed attributes from LiDAR, Photogrammetric, and Contour-based Digital Elevation Models. *Hydrological Processes*, *23*(3), 451–463. doi:10.1002/hyp.7155

Hugenholtz, C. H., Whitehead, K., Brown, O. W., Barchyn, T. E., Moorman, B. J., LeClair, A., & ... (2013). Geomorphological mapping with a small unmanned aircraft system (sUAS): Feature detection and accuracy assessment of a photogrammetrically-derived digital terrain model. *Geomorphology*, *194*, 16–24. doi:10.1016/j.geomorph.2013.03.023

ISTI CNR. (2010). MeshLab, from http://meshlab.sourceforge.net/

Jensen, J. R. (2000). *Remote sensing of the environment: an earth resource perspective*. Upper Saddle River, N.J.: Prentice Hall.

Jensen, J. R. (2007). *Remote sensing of the environment: an earth resource perspective*. Upper Saddle River, NJ: Pearson Prentice Hall.

Jurado-Expósito, M., López-Granados, F., Peña-Barragán, J. M., & García-Torres, L. (2008). A digital elevation model to aid geostatistical mapping of weeds in sunflower crops. *Agronomy for Sustainable Development*, *29*(2), 391–400. doi:10.1051/agro:2008045

Keutterling, A., & Thomas, A. (2006). Monitoring glacier elevation and volume changes with digital photogrammetry and GIS at Gepatschferner glacier, Austria. *International Journal of Remote Sensing*, *27*(19), 4371–4380. doi:10.1080/01431160600851819

Lam, K., Li, Z., & Yuan, X. (2001). Effects of JPEG Compression on the Accuaracy of Digital Terrain Models Automatically Derived from Digital Aerial Images. *The Photogrammetric Record*, *17*(98), 331–342. doi:10.1111/0031-868X.00187

Lambrou, E., & Pantazis, G. (2010). Evaluation of the Credibility of Reflectorless Distance Measurement. *Journal of Surveying Engineering*, *136*(4), 165–171. doi:10.1061/(ASCE)SU.1943-5428.0000029

Li, P., Shi, C., Li, Z., Muller, J. P., Drummond, J., Li, X., & ... (2012). Evaluation of ASTER GDEM using GPS benchmarks and SRTM in China. *International Journal of Remote Sensing*, *34*(5), 1744–1771. doi:10.1080/01431161.2012.726752

Li, Z. (1991). Effects of Check Points on the Reliability of DTM Accuracy Estimates Obtained from Experimental Tests. *Photogrammetric Engineering and Remote Sensing*, *57*(10), 1333–1340.

Li, Z., Zhu, Q., & Gold, C. (2004). *Digital Terrain Modeling - Principles and Methodology*. CRC Press. doi:10.1201/9780203357132

Losler, M. (2011). JAG3D - CoordTrans - Koordinatentransformationprogram, from http://derletztekick.com/software/koordinatentransformation

Matile, G. L. D., & Keller, G. R. (2004). Surficial geology of the Brandon map sheet (NTS 62G), Manitoba, scale 1:250,000 *Surficial Geology Compilation Map Series*. Winnipeg, Manitoba: Manitoba Industry, Economic Development and Mines, Manitoba Geological Survey.

Menci Software. (2011). ScanView, from http://www.menci.com/component/content/article/30-other-apps/264-scanview

Mikhail, E. M., Bethel, J. S., & McGlone, J. C. (2001). *Introduction to modern photogrammetry*. New York, USA: Wiley.

Möller, B. (2006). Changing wind-power landscapes: Regional assessment of visual impact on land use and population in Northern Jutland, Denmark. *Applied Energy*, *83*(5), 477–494. doi:10.1016/j.apenergy.2005.04.004

Moran, K., & Button, C. E. (2011). A GIS model for identifying eastern spadefoot toad (Scaphiopus holbrookii) habitat in eastern Connecticut. *Applied Geography (Sevenoaks, England)*, *31*(3), 980–989. doi:10.1016/j.apgeog.2011.01.004

Murphy, P. N. C., Ogilvie, J., Meng, F.-R., & Arp, P. (2008). Stream network modelling using lidar and photogrammetric digital elevation models: A comparison and field verification. *Hydrological Processes*, *22*(12), 1747–1754. doi:10.1002/hyp.6770

Poon, J., Fraser, C. S., Chunsun, Z., Li, Z., & Gruen, A. (2005). Quality Assessment of Digital Surface Models Generated From IKONOS Imagery. *The Photogrammetric Record*, *20*(110), 162–171. doi:10.1111/j.1477-9730.2005.00312.x

Rayburg, S., Thoms, M., & Neave, M. (2009). A comparison of digital elevation models generated from different data sources. *Geomorphology*, *106*(3-4), 261–270. doi:10.1016/j.geomorph.2008.11.007

Reinartz, P., Müller, R., Lehner, M., & Schroeder, M. (2006). Accuracy analysis for DSM and orthoimages derived from SPOT HRS stereo data using direct georeferencing. *ISPRS Journal of Photogrammetry and Remote Sensing*, *60*(3), 160–169. doi:10.1016/j.isprsjprs.2005.12.003

Rodriguez, E., Morris, C. S., & Belz, J. E. (2006). A Global Assessment of the SRTM Performance. *Photogrammetric Engineering and Remote Sensing*, *72*(3), 249–260. doi:10.14358/PERS.72.3.249

Rosnell, T., & Honkavaara, E. (2012). Point cloud generation from aerial image data acquired by a quadrocopter type micro unmanned aerial vehicle and a digital still camera. *Sensors (Basel)*, *12*(1), 453–480. doi:10.3390120100453 PMID:22368479

San, B. T., & Suzen, M. L. (2005). Digital elevation model (DEM) generation and accuracy assessment from ASTER stereo data. *International Journal of Remote Sensing*, *26*(22), 5013–5027. doi:10.1080/01431160500177620

Scheritz, M., Dietrich, R., Scheller, S., Schneider, W., & Boike, J. (2008). *Digital Elevation Model of Polygonal Patterned Ground on Samoylov Island, Siberia, Using Small-Format Aerial Photography.* Paper presented at the Ninth International Conference on Permafrost, University of Fairbanks, USA.

Schloderer, G., Bingham, M., Awange, J. L., & Fleming, K. M. (2011). Application of GNSS-RTK derived topographical maps for rapid environmental monitoring: A case study of Jack Finnery Lake (Perth, Australia). *Environmental Monitoring and Assessment*, *180*(1-4), 147–161. doi:10.100710661-010-1778-8 PMID:21136293

Scholten, F., & Wewel, F. (2000). *Digital 3D-Data Acquisition with the High Resolution Stereo Camera-Airborne (HRSC-A)*. Paper presented at the XXXIII International Symposium on Photogrammetry and Remote Sensing, Amsterdam, the Netherlands.

St-Onge, B., Vega, C., Fournier, R. A., & Hu, Y. (2008). Mapping canopy height using a combination of digital stereo-photogrammetry and lidar. *International Journal of Remote Sensing*, *29*(11), 3343–3364. doi:10.1080/01431160701469040

Thoma, D. P., Gupta, S. C., Bauer, M. E., & Kirchoff, C. E. (2005). Airborne laser scanning for riverbank erosion assessment. *Remote Sensing of Environment*, *95*(4), 493–501. doi:10.1016/j.rse.2005.01.012

Toutin, T. (2004a). Comparison of Stereo-Extracted DTM from Different High-Resolution Sensors: SPOT-5, EROS-A, IKONOS-II, and QuickBird. *IEEE Transactions on Geoscience and Remote Sensing*, *42*(10), 2121–2129. doi:10.1109/TGRS.2004.834641

Toutin, T. (2004b). DSM generation and evaluation from QuickBird stereo imagery with 3D physical modelling. *International Journal of Remote Sensing*, *25*(22), 5181–5192. doi:10.1080/014311604100 01726030

Toutin, T. (2006). Generation of DSMs from SPOT-5 in-track HRS and across-track HRG stereo data using spatiotriangulation and autocalibration. *ISPRS Journal of Photogrammetry and Remote Sensing*, *60*(3), 170–181. doi:10.1016/j.isprsjprs.2006.02.003

Toutin, T. (2008). ASTER DEMs for geomatic and geoscientific applications: A review. *International Journal of Remote Sensing*, *29*(7), 1855–1875. doi:10.1080/01431160701408477

Toutin, T. (2010). Impact of Radarsat-2 SAR Ultrafine-Mode Parameters on Stereo-Radargrammetric DEMs. *Geoscience and Remote Sensing. IEEE Transactions on*, *48*(10), 3816–3823.

Turner, D., Lucieer, A., & Watson, C. (2012). An Automated Technique for Generating Georectified Mosaics from Ultra-High Resolution Unmanned Aerial Vehicle (UAV) Imagery, Based on Structure from Motion (SfM) Point Clouds. *Remote Sensing*, *4*(12), 1392–1410. doi:10.3390/rs4051392

Walker, J. P., & Willgoose, G. R. (2006). A Comparative Study of Australian Cartometric and Photogrammetric Digital Elevation Model Accuracy. *Photogrammetric Engineering and Remote Sensing*, *72*(7), 771–779. doi:10.14358/PERS.72.7.771

Wallace, L., Lucieer, A., Watson, C., & Turner, D. (2012). Development of a UAV-LiDAR System with Application to Forest Inventory. *Remote Sensing*, *4*(12), 1519–1543. doi:10.3390/rs4061519

Westaway, R. M., Lane, S. N., & Hicks, D. M. (2003). Remote survey of large-scale braided, gravel-bed rivers using digital photogrammetry and image analysis. *International Journal of Remote Sensing*, *24*(4), 795–815. doi:10.1080/01431160110113070

Williams, M. L., & Jenkins, L. G. (2009). GeoSAR and BDInSAR: Combining "X" with "P" for Tropical Forest "C. *Photogrammetric Engineering and Remote Sensing*, *75*(7), 738–743.

Willis, M. (2010). How to Create a Digital Elevation Model from Photosynth point clouds, from http://palentier.blogspot.ca/2010/12/how-to-create-digital-elevation-model.html

Wise, S. (2000). Assessing the quality for hydrological applications of digital elevation models derived from contours. *Hydrological Processes*, *14*(11-12), 1909–1929. doi:10.1002/1099-1085(20000815/30)14:11/12<1909::AID-HYP45>3.0.CO;2-6

Wissmar, R. C., Timm, R. K., & Bryant, M. D. (2010). Radar-Derived Digital Elevation Models and Field-Surveyed Variables to Predict Distributions of Juvenile Coho Salmon and Dolly Varden in Remote Streams of Alaska. *Transactions of the American Fisheries Society*, *139*(1), 288–302. doi:10.1577/T08-204.1

Wolf, P. R. (2002). Surveying and Mapping: History, Current Status, and Future Projections. *Journal of Surveying Engineering*, *128*(3), 79–107. doi:10.1061/(ASCE)0733-9453(2002)128:3(79)

Wolf, P. R., & DeWitt, B. A. (2000). *Elements of photogrammetry: with applications in GIS*. Boston, MA: McGraw-Hill.

Wundram, D., & Löffler, J. (2007). High-resolution spatial analysis of mountain landscapes using a low-altitude remote sensing approach. *International Journal of Remote Sensing*, *29*(4), 961–974. doi:10.1080/01431160701352113

Zhang, L., & Gruen, A. (2004). *Automatic DSM Generation from Linear Array Imagery Data*. Paper presented at the XXth International Symposium on Photogrammetry and Remote Sensing Congress, Istanbul, Turkey.

This research was previously published in the International Journal of Applied Geospatial Research (IJAGR), 6(3); edited by Donald Patrick Albert and Samuel Adu-Prah, pages 58-77, copyright year 2015 by IGI Publishing (an imprint of IGI Global).

Chapter 12
A Practical UAV Remote Sensing Methodology to Generate Multispectral Orthophotos for Vineyards:
Estimation of Spectral Reflectance Using Compact Digital Cameras

Adam J. Mathews
Oklahoma State University, USA

ABSTRACT

This paper explores the use of compact digital cameras to remotely estimate spectral reflectance based on unmanned aerial vehicle imagery. Two digital cameras, one unaltered and one altered, were used to collect four bands of spectral information (blue, green, red, and near-infrared [NIR]). The altered camera had its internal hot mirror removed to allow the sensor to be additionally sensitive to NIR. Through on-ground experimentation with spectral targets and a spectroradiometer, the sensitivity and abilities of the cameras were observed. This information along with on-site collected spectral data were used to aid in converting aerial imagery digital numbers to estimates of scaled surface reflectance using the empirical line method. The resulting images were used to create spectrally-consistent orthophotomosaics of a vineyard study site. Individual bands were subsequently validated with in situ spectroradiometer data. Results show that red and NIR bands exhibited the best fit (R^2: 0.78 for red; 0.57 for NIR).

DOI: 10.4018/978-1-5225-8365-3.ch012

1. INTRODUCTION

1.1. Use of Compact Digital Cameras in Remote Sensing

The low cost and high availability of compact (point-and-shoot) digital cameras has led to usages beyond that of recreational or professional capture of natural color photography (Dean, Warner, & McGraw, 2000). Digital camera ease of operation, speed of image capture, efficient processing, and lightweight design lend utility to aerial image capture (King, 1995). Coupled with an unmanned aerial vehicle (UAV) to create an unmanned aerial system (UAS), digital cameras can inexpensively provide very high spatial and temporal resolution data for research in soils (Levin, Ben-Dor, & Singer, 2005), tree inventory and biomass (Dean et al., 2000), land management (Rango, Laliberte, Herrick, Winters, Havstad, Steele, & Browning, 2009), and agriculture (Hunt, Hively, Daughtry, & McCarty, 2008; Lebourgeois, Beguye, Labbe, Mallavan, Prevot, & Roux, 2008; Lelong, Burger, Jubelin, Roux, Labbe, & Baret, 2008; Ritchie, Sullivan, Perry, Hook, & Bednarz, 2008) including viticulture (Smit, Sithole, & Strever, 2010; Turner, Lucieer, & Watson, 2011). See Everaerts (2008) and Watts, Ambrosia, and Hinkley (2012) for informative reviews on UAV and UAS technology. Camera system design for UAV-based remote sensing varies by the nature of the images to be captured; as evidenced by studies that alter digital cameras for their specific needs, while others employ cameras with off-the-shelf configurations. With or without modification, digital cameras remotely record spectral information of a target of interest (Dean et al., 2000; Levin et al., 2005; Ritchie et al., 2008). Levin et al. (2005) utilized an unaltered digital camera (UDC) to record the spectral properties of soils with three spectral bands of blue, green, and red. Most digital cameras, however, have the ability to sense wavelengths beyond the visible spectrum with minor alteration (Cheng & Rahimzadeh, 2005).

A typical digital camera has the following vital components: a sensor (either a charge-coupled device [CCD] or a complementary metal oxide semiconductor [CMOS]) made up of an array of sensor elements or sensels (each of which later become picture elements or pixels in the captured image), a Bayer filter (or other color filter), a lens, and a hot mirror. This study focuses on the CCD sensor, which senses light and converts this spectral information into digital numbers (DNs), typically with 8-bit radiometric resolution (0-255). The Bayer filter is responsible for splitting the incoming visible wavelengths into separate bands so brightness for each can be recorded (i.e., blue, green, and red; stored in reverse order and referred to as RGB). The lens de-magnifies the scene to properly represent physical objects and their geometric relationships within a captured image. The hot mirror is an internal spectral filter that limits detector sensitivity to visible wavelengths (i.e., the sensor will not record energy in the near-infrared [NIR] region). In most cases the hot mirror wavelength cut-off is somewhere around 670-690 nm (Dean et al., 2000; Ritchie et al., 2008), whereas most CCDs have a spectral range up to 900 nm (Dare, 2008; Lelong et al., 2008). Therefore, the hot mirror forces the CCD to sense only the desired portions of the spectrum to more easily replicate the visible light range for natural color photography. For studies that require NIR sensitivity, the hot mirror can be removed and replaced with clear glass to allow the CCD to sense NIR wavelengths (Cheng & Rahimzadeh, 2005). In the case of vegetation studies for instance, NIR sensitivity is important to be able to monitor differentials in crop health, thus the hot mirror must be replaced (Dare, 2008; Ritchie et al., 2008). System designs must be creative to integrate the NIR band usually by use of two cameras. Ritchie et al. (2008), for example, employed an UDC combined with an altered digital camera (ADC) to gather four spectral bands for analysis (UDC—blue, green, red; ADC—NIR).

Image format is another important image acquisition consideration when using digital cameras as remote sensors. Image format is important because compression of images alters the way DNs are computed and stored (Dean et al., 2000). RAW and TIFF image formats are presumed to be better because they do not alter the image (Cheng & Rahimzadeh, 2005; Lebourgeois et al., 2008). RAW and TIFF images are made up of unprocessed, uncompressed pixel data as captured by the CCD. The downsides to such measurements are increased file size and capture time (Lelong et al., 2008). An alternative to RAW and TIFF formats is JPEG, which has proven adequate for scientific analyses (Hunt, Cavigelli, Daughtry, McMurtrey, & Walthall, 2005; Lelong et al., 2008; Levin et al., 2005). Compared to other formats, JPEG offers the convenience of small file size, quick capture ability, and easier processing.

1.2. Improving the Quality of Collected Images

Regardless of the type of sensor employed, radiometric and geometric effects must be addressed in the processing of aerial imagery to properly create image products for future analysis. Radiometric corrections aim to remove inconsistencies produced by (1) atmospheric effects like scattering and/or absorption of incoming or reflected energy between the sensor and the surface, and (2) topographic effects that may create shadowing and other unequal reflectance over space within the collected images (Jensen, 2005). Geometric corrections address internal and external errors caused by the sensor and aircraft (Jensen, 2005). Internal errors are predictable and therefore easier to correct like distortion created by the camera lens. External errors are flight-dependent and include altitude changes as well as sensor viewing geometries dictated by aircraft attitude. External errors are accentuated in UAV image capture due to low flying heights and reduced control of in-flight motion (Lelong et al., 2008). Low flying height also results in small image footprints, which requires mosaicking of images to create the desired ground coverage (Laliberte, Goforth, Steele, & Rango, 2011; Turner, Lucieer, & Watson, 2012).

1.2.1. Radiometric Corrections

"Radiometric calibration of a sensor involves determining the relationship between image brightness as measured in digital image units [DNs] and actual radiance or reflectance of the target" (King, 1995, p. 258). This is normally completed by capturing known spectral targets during flight and establishing a correlative relationship with in situ spectroradiometer measurements (King, 1995; Smith & Milton, 1999). Conversion to spectral reflectance allows for a standardized method by which quantitative comparisons can be made over time with accuracy and confidence (Smith & Milton, 1999). One method of conversion is the empirical line method (Jensen, 2005; Smith & Milton, 1999). This method assumes that targets of differing reflectance are captured by the sensor that can be compared to actual target reflectance as measured by a spectroradiometer. At the very least one spectral target with very low reflectance (dark) and one with very high reflectance (light) should be included in the scene. Spectral targets with reflectances along the spectrum between the light and dark objects are encouraged to be included,, not forgetting that similar variation must exist in NIR reflectance (if NIR calibration is necessary). The empirical line method essentially correlates brightness measured with the CCD to reflectance measured by a spectroradiometer. Linear regression is then used to create a slope line equation to predict reflectance of non-target pixels (Karpouzli & Malthus, 2003; Smith & Milton, 1999). Each band of imagery requires a different prediction equation "which attempt to remove both illumination and atmospheric effects" that can then be "applied to the remotely sensed data to produce images in units of [relative] reflectance" (Smith & Milton, 1999, p. 2654).

In the case of using digital cameras as remote sensors, the empirical line method has proven practical and successful in a number of instances (Levin et al., 2005; Ritchie et al., 2008). Levin et al. (2005) utilized repeating targets (plastic chips) of black, gray, white, blue, green, red, and also a known 100 percent reflectance target, whereas Ritchie et al. (2008) used a ColorChecker chart with 24 standardized colored squares as reflectance targets. These targets were designed for small-scale calibration of digital camera images taken on-the-ground. The difficulty with this method in the case of digital cameras for quantitative research relates to the properties of the digital camera sensor that are, for the most part, unknown and unpublished by the manufacturer, specifically the wavelength interval captured by each band. Lelong et al. (2008) were fortunate to know the properties of the sensor they employed: a Canon EOS-350D with blue (420-500 nm), green (490-580 nm), red (570-640 nm), and NIR (720-850 nm) bands. Levin et al. (2005), however, did not know this information and were therefore forced to select specific wavelengths that were assumed to best represent and correlate with DNs from each band collected by the CCD: Olympus Camedia C-920 with blue (460 nm), green (510 nm), and red (640 nm) bands. Jensen, Apan, Young, and Zeller (2007) provided an example of a typical CCD sensitivity that showed blue centered on 450 nm, green on 550 nm, and red on 650 nm, with each band exhibiting a wavelength interval of around 80-120 nm.

After this calibration, large-scale targets can be captured within aerial imagery such as tarpaulins or other specially designed spectral targets to aid in conversion to reflectance of field captured imagery. These targets must cover substantial area on-the-ground depending on the spatial resolution of the imagery (Hunt et al., 2008). With small-footprint UAV imagery, Laliberte et al. (2011) noted difficulty capturing targets within all images. In converting UAV image DNs to reflectance using spectral targets, Laliberte et al. (2011) found estimates to be more accurate when the conversion was applied to the entire mosaic rather than to single images (prior to mosaicking).

Prior to conversion to reflectance, correction for vignetting within each image may be necessary. Aerial imagery collected by digital cameras often displays a radial pattern of brightness falloff near the image edges that can vary based on exposure (Kelcey & Lucieer, 2012). Vignetting corrections range from producing and applying anti-vignetting filters (Lelong et al., 2008) to the complete removal (masking) of edge pixels (De Biasio, Arnold, Leitner, McGuinnigle, & Meester, 2010), although leaving images unaltered and not correcting for vignetting is not uncommon (Turner et al., 2012). Other radiometric issues including general brightness differences between collected images can be addressed with color balancing and/or histogram equalization (Niethammer, Rothmund, Schwaderer, Zeman, & Joswig, 2011), although such brightness adjustment is not always a part of mosaicking methodologies (Rango et al., 2009; Turner et al., 2011; Turner et al., 2012).

1.2.2. Geometric Corrections

Internal geometric errors like radial and tangential distortions created by digital camera lenses are less predictable and stable than traditional aerial photograph sensors, but are still able to be estimated and corrected (Turner et al., 2012). Kelcey and Lucieer (2012) for instance used the free software Agisoft Lens (Agisoft LLC, St. Petersburg, Russia) to calculate lens distortion coefficients and apply a correction to their UAV images, although the literature indicates this correction is often ignored. Other sources of error such as external geometric errors are consistently present with UAV collected data because of the difficulty of maintaining consistent flying height, roll, pitch, and yaw (Lelong et al. 2008; Turner et al., 2012), which leads to image mosaics with differing spatial resolution and viewing angles. Selectivity in regard to which images to include in the mosaic becomes important (i.e., nadir-only at or near the same flying height; Turner et al., 2011; Turner et al., 2012).

Proper georectification of UAV collected images is traditionally based on a highly accurate digital terrain model (DTM) of the imaged area. UAV-based methodologies have benefitted from recent advances in Structure from Motion (SfM), which much like traditional photogrammetry utilizes image overlap from multiple perspectives of the collected images to identify keypoints (Snavely, Seitz, & Szeliski, 2008) and generate an accurate point cloud model of the surface (Kaminsky, Snavely, Seitz, & Szeliski, 2009). Once georeferenced, SfM point clouds can be filtered and interpolated to create highly accurate, site-specific digital terrain models (DTMs; Mathews & Jensen, 2013) and/or digital surface models (DSMs; Dandois & Ellis, 2010) upon which image mosaics are georectified (Turner et al., 2012). Adjustment of imagery to the DTM accounts for any topographic effects at the image site (Baboo & Devi, 2011).

1.3. Objective

The objective of this paper was to design a practical and inexpensive data collection and processing methodology to generate orthophotomosiacs with multiple bands each with pixel values in units of reflectance. This was completed using only compact digital cameras and a small UAV, which is contrary to similar studies that utilize specialized, more expensive sensors and UAVs (Baluja, Diago, Balda, Zorer, Meggio, Morales, & Targuila, 2012; Bellvert, Zarco-Tejada, Girona, & Fereres, 2013; Primicerio et al., 2012; Turner et al., 2011). Ultimately, the system would have the ability to collect very high spatial resolution aerial imagery (less than 10 cm) in four spectral bands (blue, green, red, and NIR), each of which could be converted to estimates of reflectance using the empirical line method. To address the objective, this paper is presented in two parts: (1) on-ground camera testing to determine the sensitivity and capability of the digital camera sensors and (2) application of the camera system in the field to create spectrally-consistent orthophotos of a study vineyard site in the Texas Hill Country. A per-band validation was implemented to gauge the quality of each of the four collected bands for future analyses. A per-band validation is contrary to similar research that evaluates orthophoto quality using NDVI or other vegetation index values (Hunt et al., 2005; Primicerio, Di Gennaro, Fiorillo, Genesio, Lugato, Matese, & Vaccari, 2012). This UAV and digital camera system design was aimed at identifying spatial heterogeneity in grapevine canopy vigor across vineyards, however, it could be used for a number of applications that require very high spatial resolution imagery with visible and NIR bands (precision agriculture and crop management, small area land cover change, etc.).

2. MATERIALS AND METHODS

2.1. Digital Cameras

This study utilized two Canon PowerShot A480 compact digital cameras (Canon U.S.A., Inc., Lake Success, NY, USA), one UDC and one ADC. The A480s have a resolution of 10 megapixels (3648 by 2736 pixels) with a 1/2.3-inch type CCD sensor and cost approximately $150 USD each. The ADC was purchased with the hot mirror removed and replaced with clear glass as described by Cheng and

Rahimzadeh (2005). The same model camera (A480) was utilized for both the UDC and ADC to consistently replicate potential geometric (focal length, megapixels) and spectral (sensor, Bayer filter) distortions between cameras.

Following Levin et al. (2005) and Ritchie et al. (2008), all UDC and ADC images were captured on manual setting with sunlight white balance and no image adjustment. In addition, the UDC ISO level was fixed at 80. The ADC ISO level was fixed at 400 due to the speed of the UAV carrying the camera, increasing the cameras' ability to effectively capture the scene. All images were captured in JPEG format following Levin et al. (2005). Although JPEG format undergoes compression which may slightly alter the saved data, JPEGs provide the most practical means to capture hundreds of aerial images while in-the-field due to its quick recording ability and smaller file size. This is contrary to RAW files, which in testing needed several more seconds per image to save and were around three times larger in file size (JPEG: 2-5 MB, RAW: 15 MB). The high speed capturing ability is necessary when using faster moving aircraft like planes to ensure proper image overlap.

Four spectral bands were captured: blue, green, red, and NIR. Like Ritchie et al. (2008), both cameras were utilized to obtain these bands with the UDC capturing the three visible bands and the ADC capturing three NIR bands. To block visible wavelengths from being sensed by the ADC, a NIR-transmitting filter was placed in front of the ADC lens similar to Dean et al. (2000). This filter permitted the ADC CCD to only sense wavelengths 750 nm and longer. As was discussed previously, presumably most digital camera sensors (CCDs) cannot sense wavelengths greater than 900 nm, which forced the ADC to collect three bands (still assigned as RGB) within this 750-900 nm range. Placing the NIR filter on the outside of the camera, instead of replacing the hot mirror with a NIR filter, was preferred so other spectral filters (i.e., red) could be implemented during future acquisitions.

2.2. On-Ground Image Collection and Camera Calibration

Ground-based calibration data were collected on two cloud-free days near noon: 23-November-2011 at 11:30 AM and 27-November-2011 at 12:45 PM. Data were captured at the Texas State University campus in San Marcos, Texas, USA (29°53′18″N, 97°56′32″W). On both occasions, the following datasets were collected: (1) 2-3 images from both cameras capturing within the field of view (FOV) a calibration panel of color targets, a known 100 percent Spectralon reflectance target, and two leaves for distinct NIR response, and (2) spectroradiometer readings of each of these objects. The calibration panel of spectral targets was crafted out of black, gray, white, purple, yellow, blue, green, and red paint swatches affixed to foamboard (near-Lambertian surfaces as suggested by Smith & Milton, 1999). The swatches were collected from a local hardware store and exhibited a flat color finish. Each color square was 5 cm by 5 cm. Two of each color swatch were included to account for variability in spectral response. Calibration panel design was similar to the set-up produced by Levin et al. (2005).

Images were taken perpendicular to the targets (nadir) at a distance of approximately one meter (see Figure 1a-b for sample images). An ASD FieldSpec Pro spectroradiometer (Analytical Spectral Devices, Inc., Boulder, CO, USA) collecting a value for every nanometer within a spectral range of 350-1050 nm was utilized to collect spectral signatures of the objects immediately following image capture. The ASD device consisted of a backpack unit and a fiber optic cable connected to a 25° field of view handheld optic (pistol-grip with integrated leveler for nadir positioning). Measurements were logged with a connected laptop computer. Throughout spectroradiometer data collection for on-ground and aerial-based calibration and validation, measurements were taken by a single operator standing on the far side of the sun to avoid casting shadows into the measurement area (McCoy, 2005). Data were collected as values of reflectance by regularly calibrating the ASD with the Spectralon reference panel. Measurements of spectral targets were taken at nadir approximately 10 cm above each target aimed at the center of each target. A minimum of two measurements were taken of each target and later averaged. DNs were obtained for each target using ENVI 4.8 (Exelis, Inc., Boulder, CO, USA) to define 19 regions of interest (ROI) as shown in Figure 1a. ROIs were drawn central to each target area to avoid adjacency effects (Dean et al., 2000; Levin et al., 2007). The mean DN value was taken from each ROI to represent each target. Single images from each sample were used to calculate mean DN after observing minimal variation in DNs across images within each sample.

Mean DNs were then compared to spectroradiometer-measured reflectance for each band. Both cameras used a Bayer filter to capture three bands of data (UDC—blue, green, red and ADC—NIR [called ADC-Blue], NIR [ADC-Green], NIR [ADC-Red]). The three NIR bands recorded DNs for energy reflected within a 750-900 nm interval. The optimal NIR band, to be used in future analysis, was chosen based on the linear relationship between DNs and spectroradiometer measurements. The band with the highest R^2 value was chosen as the best representation of the NIR wavelengths. The optimal wavelength interval for each band was determined by testing a series of wavelength intervals. Tested intervals for each band were centered on 450 nm (blue), 550 nm (green), 650 nm (red), and 850 nm (NIR) following the typical CCD sensitivity illustrated by Jensen et al. (2007). The following wavelength intervals were tested for the blue band from narrow to wide: 450 nm, 445-455 nm, 440-460 nm, 420-480 nm, and 400-500 nm. Similarly, the same intervals were used for the other bands (i.e. green: 550 nm, 545-555 nm, 540-560 nm, etc.; red: 650 nm, 645-655 nm, etc.), except for narrow intervals in the NIR band (850 nm, 845-455 nm, and 840-860 nm) because the wavelength interval range was more confidently known. These intervals

Figure 1. Sample images of spectral targets captured with the UDC (a) and the ADC (b) with the regions of interest (ROIs) shown (a) in different colors (see online publication for color)

determined the range at which spectral signatures captured by the spectroradiometer would be averaged and compared to mean ROI-specific DNs for each target.

The relationship between spectroradiometer-measured reflectance and camera recorded DNs was modeled using linear regression (R^2) and root mean square error (RMSE). This was completed for both samples to account for any variability that may exist with the digital camera and spectroradiometer measurements over time. For each band, the wavelength interval with the best combination of R^2 (high) and RMSE (low) for both samples was selected as the most representative wavelength interval. These wavelength intervals were then used with in situ spectroradiometer measurements to convert aerial image DNs to reflectance using the empirical line method.

2.3. Application of UAV-Based Image System to the Study Vineyard and Generation of Analysis-Ready Image Products

2.3.1. Study Site and Data Collection

The vineyard study site is located in the Texas Hill Country American Viticultural Area near Fredericksburg, Texas, USA (120 km west of Austin). This study focused image collection on two contiguous vineyard blocks consisting of 39 vine rows each with up 90 vertical trellis-trained 5-10 year-old Tempranillo (*Vitis vinifera*) vines covering approximately 1.9 ha (4.8 acres). Figure 2 shows the vineyard layout with the study vines outlined in red. The study blocks are separated with a dotted red line. The eastern block is older than the western block and provides desirable variation in canopy size and reflectance characteristics. At the time of data collection, the area beneath the vine canopies consisted of bare soil, while the inter-row space was covered in grass. The vineyard has a gentle slope from its highest point in the northwest corner to the lowest point in the southeast corner.

Aerial images of the study vineyard were captured using the previously discussed digital cameras mounted (at separate times) on a Hawkeye UAV (www.ElectricFlights.com, Kingsland, TX, USA; see Figure 3 a-b). The Hawkeye is a lithium battery-powered kitewing plane with a 1.5 m wingspan, single rear-facing propeller, and single nadir-facing camera mount. The Hawkeye was flown both autonomously with pre-programmed autopilot and manually. Ardupilot (code.google.com/p/ardupilot/) was used to allow for autopilot capability. This Arduino-based, open-source hardware and software combination based its flight path on pre-programmed, ordered waypoints (X, Y, and Z coordinates). When autopilot was switched on, Ardupilot controlled the UAV autonomously using the on-board GPS and the waypoint locations. When flown on autopilot, the Hawkeye is prone to becoming stagnant in gusty winds, particularly when flying into the wind. Manual flight control, therefore, was initiated during wind gusts to avoid unnecessary reduction in ground covered and battery power loss. Takeoff and landing were also manually controlled. To launch the UAV, the Hawkeye was thrown forward at shoulder height with the propeller running and manual control initiated. After gaining altitude under manual control, autopilot was enabled with a switch on the remote control. The autopilot was pre-programmed to follow the flight path provided in Figure 2. Flight paths were orientated north-to-south to keep the UAV flying directly into or with the winds, serving to steady the aircraft and capture higher quality images. Otherwise, the UAV was likely to sway and decrease the cameras ability to capture nadir images. Upon completion of aerial data collection manual flight control was reinitiated, the altitude of the UAV was gradually decreased, the UAV was guided to touchdown in the landing area and retrieved.

Figure 2. The study vineyard blocks separated with a dashed line and enclosed with a solid line. The UAV takeoff/landing area is enclosed with a rectangle, GCPs are shown as X's, and the UAV flight path for low-altitude aerial image capture is shown with the line with arrows (see online publication for color)

Prior to image capture, several spectral targets were placed along the northeastern edge of the vineyard to aid the empirical line method transformation. Six flat, colored foamboard spectral targets sized 0.75 m by 0.5 m (black, white, grey, blue, green, red) were placed on flat ground spaced about 2 m apart near the takeoff/landing area (see Figure 2). The targets were large enough to ensure that many pure pixels would represent each target within the captured images that could then be averaged and compared to measured reflectance. Other targets were placed in the field to help georeference the collected imagery. Five total ground control points (GCPs) were positioned following Mathews & Jensen (2013). Locations of the GCPs are shown with X's in Figure 2 at the corners and eastern edge of the study vineyard. The 0.6 m by 0.6 m GCP targets were crafted out of durable foamboard and painted with flat hues of red, and small concentric black and white circles at the center, similar to Aber, Marzoff, and Ries (2010). These highly visible targets were easy to find in aerial images and served to provide additional spectral targets to aid in SfM keypoint matching as well as conversion of DNs to reflectance. GCPs were located using a Trimble GeoXH GPS with an external Zephyr antenna that averaged 200 separate positions for each location (X,Y: NAD83 UTM Zone 14N; and Z: NADV88). Acquisition of GPS positions was limited to a maximum position dilution of precision (PDOP) of three. Following GPS data collection, differential correction was completed using the Trimble GPS Analyst Extension in ArcGIS (ESRI, Redlands, CA, USA).

Figure 3. The kitewing UAV (a) with mounted digital camera (b) and other hardware (see online publication for color)

UAV-based aerial images were captured under clear skies on 16-May-2012 at approximately 11:00 AM to minimize shadowing between the vine rows. At this time of the growing season the vines were flowering, which has served as a useful time to gauge relative vine health and eventual production (Hall, Lamb, Holzapfel, & Louis, 2011). Images were captured at a flying height of approximately 125 m and before noon due to lower wind speeds at that time (around 15 kph). Images were captured within two 15 minute time periods coinciding with the life of each lithium battery. Upon landing, the UAV battery and cameras were switched to continue capturing data. Images were automatically captured every second throughout flight using the Canon Hackers Development Kit (CHDK; chdk.wikia.com) intervalometer script preinstalled on the camera SD cards. During aerial image collection, camera settings remained the same as were used for on-ground testing.

Immediately following image capture, spectroradiometer data were collected at the vineyard site with the ASD FieldSpec Pro. A minimum of two measurements of each of the spectral targets, including the red portion of each GCP, were collected and averaged. A stratified sample of vine measurements was implemented for every fifth row starting with the first row (easternmost) and ending at row 30. Within rows, every tenth vine was sampled starting from the northern end of the vine rows (alternating starting vines from first to fifth vine; i.e. row 1-vine 1; row 1-vine 10; row 5-vine 5; row 5-vine 15; etc.). Vine canopy reflectance was sampled by aiming the spectroradiometer pistol-grip fiber optic sensor at a level nadir direction approximately 0.5 m above the vine canopy directly above the vine trunk. At this height above the canopy, the spot or target size of the ASD optic was around 0.23 m in diameter similar to vine canopy measurements taken by Rodríguez-Pérez, Riaño, Carlisle, Ustin, & Smart (2007). The sensor was then moved slightly to cover more vine canopy area and account for variation in spectral response. Multiple measurements were also captured to account for variation due to wind and potential canopy movement (McCoy, 2005). Two to four observations of each vine canopy were taken and averaged. The same procedure was used to sample each vine. All samples were taken with the ASD spectroradiometer by the same operator standing on the far side of the canopy (in relation to the sun) to not cast shadows into the vine canopy being sampled. In total, the spectral signatures of 61 vine canopies were collected. The sampled vine locations (vine trunks) were located using the previously discussed GPS unit and processing method with the number of averaged positions reduced from 200 to 30.

2.3.2. Data Processing

The data processing workflow from UAV image capture to export of the final orthophotos is provided with Figure 4. The same processing chain was completed separately for both the RGB and NIR images. After capturing aerial images of the vineyard site, a cursory manual image filtering was employed to remove any non-nadir images and other poor quality (blurry) images caused by aircraft motion. This image subset was used throughout the rest of the processing. Image histogram equalization was applied to the image subset to account for any brightness differences between individual images using ERDAS ImageEqualizer (Intergraph Corp., Madison, AL, USA). ImageEqualizer collects image statistics for all of the images and applies a normalization procedure prior to exporting new, radiometrically-adjusted images. Without such spectral adjustment, spatial heterogeneity in surface reflectance was visually apparent in the output mosaics (see Figure 5). Figure 5a displays increased brightness in the central-western portions of the study vineyard. Brightness variation is also apparent in the NIR imagery (Figure 5c) in the southernmost parts of the study vineyard. Histogram equalization served to correct such image-to-image brightness variation (Figure 5b,d). This normalization also served to minimize the effect of vignetting by brightening pixels near image edges.

Following image normalization, images were aligned using Agisoft PhotoScan (Agisoft LLC, St. Petersburg, Russia). PhotoScan provides a SfM-based automatic image alignment procedure that creates point clouds with arbitrary coordinates. Input images are corrected for internal geometric distortions in PhotoScan by automatic import of digital camera EXIF data on lens and focal characteristics (applies lens distortion coefficients). Across the input images, point clouds for RGB and NIR images were created. Following point cloud creation, manual editing was necessary to remove obvious outlier points or noise (usually well above or below the ground surface). SfM point clouds typically include noise that necessitates removal (Dandois & Ellis, 2013; Mathews & Jensen, 2013).

Following noise removal, the point clouds were georeferenced within PhotoScan by manual identification of the GCP targets within images and by inputting their respective UTM coordinates. PhotoScan then transformed and optimized the point cloud with arbitrary coordinates to real-world UTM coordinates (X, Y, Z). Within PhotoScan, geometry was then built to both (1) create the underlying DTM surface onto which orthophotos were rectified/generated, and (2) determine the spatial resolution (pixel size as well as pixel array layout and image coverage) of the output orthophoto by analyzing the input images. In PhotoScan, the DTM surface was derived by creating a mesh from a simplified/filtered version of the point cloud (like a TIN). The spatial resolution of the orthophotos was determined by the lowest altitude image within the image set. This image exhibited the highest spatial resolution that all of the other images were upsampled to.

At this stage, preliminary RGB and NIR orthophotos were exported from PhotoScan, which include all input images blended (mean DNs at edge overlap). Figure 4 specifies further manual editing at this point, specifically trial and error image masking or removal of undesirable areas within images (i.e., vignetting corners if still present, areas of distortion). This was necessary because the geometry of vineyards and similar row crops provide a difficult surface to automatically create orthophoto mosaics from due to the canopy being elevated above the ground surface on a trellis system. This elevated canopy combined with the wide FOV of the digital cameras created areas of high distortion within images. In images captured with an area array like a CCD, distortion naturally occurs in the pixels away from the principal point. This is further accentuated in UAV imagery due to low flying height and variability in aircraft attitude. Figure 6 shows a more traditional example of aerial image capture of vineyards where in this case one image captured the entire vineyard within its FOV (black camera at or above 250 m flying height). In this case, all of the vine rows would look to be near-planimetrically correct because they fall with a "usable" field of view (UFOV) showed with dark lines extending from the FOV to the ground surface (Figure 6), without the appearance of "leaning" vine rows away from the principal point. In the case of UAV-captured images at an altitude of 125 m (ideal locations shown with black cameras in Figure 6), these areas of distortion within images are increased due to the UFOV shrinking with reduc-

Figure 4. Data workflow from UAV image capture to final Orthophoto generation

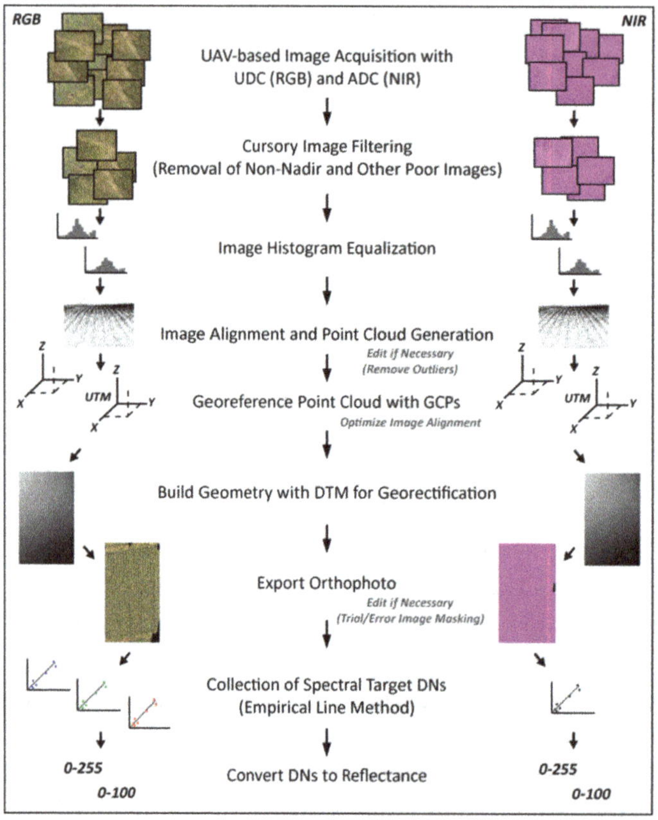

tion in altitude. In this example, instead of the whole vineyard being successfully imaged, only five rows of vines are accurately represented (multiple images needed with mosaic). This is further complicated by the more realistic picture of UAV camera geometry (gray cameras in Figure 6) where flying height is inconsistent and not every image is perfectly nadir-facing. A significant amount of manual masking (within PhotoScan) of unusable areas of images (outside the UFOV) within both the RGB and NIR image sets was necessary. In some extreme cases, entire images were completely excluded from production of the final orthophoto.

The final orthophotos (two mosaics) were exported from PhotoScan following the trial/error masking process: a RGB mosaic from the UDC images and a NIR mosaic from the ADC images. The ADC-Blue and ADC-Red bands were dropped from the NIR orthophoto since ADC-Green was determined to best gauge NIR. All three NIR bands were kept in the earlier stages of processing to aid SfM image alignment, which was more successful with the additional pixel information provided by multiple bands. Average DNs of spectral targets were collected from all four bands avoiding target edge pixels. Mean DNs were compared to measured reflectance to create linear regression models to transform each band's pixels from DNs (0-255) to percent reflectance (0-100). All six spectral targets were intended to be included in creation of all slope line equations as suggested by Smith & Milton (1999), but this was only completed for the NIR band. For the visible bands, two dark objects (black target; GCP red target for the blue and green bands, blue target for red band) and one light object (white target) were used due an observed tendency of the visible bands to over- and underestimate brightness of color targets. The non-black dark targets were chosen based on having at or near the same recorded brightness as the black target. Regression equations were created for each band and applied to every pixel in that band using the raster calculator in ArcGIS.

2.3.3. Output Validation

Using the output mosaics and spectroradiometer measurements, each output band was validated to compare measured reflectance to estimated reflectance. Validation ROIs were obtained for each vine location using 0.48 m diameter buffers of the GPS-located vine trunks. More narrow and broad buffers

Figure 5. Comparison of Image Mosaics (RGB: a, b; NIR: c, d) with (b, d) and without (a, c) Image Histogram Equalization (see online publication for color)

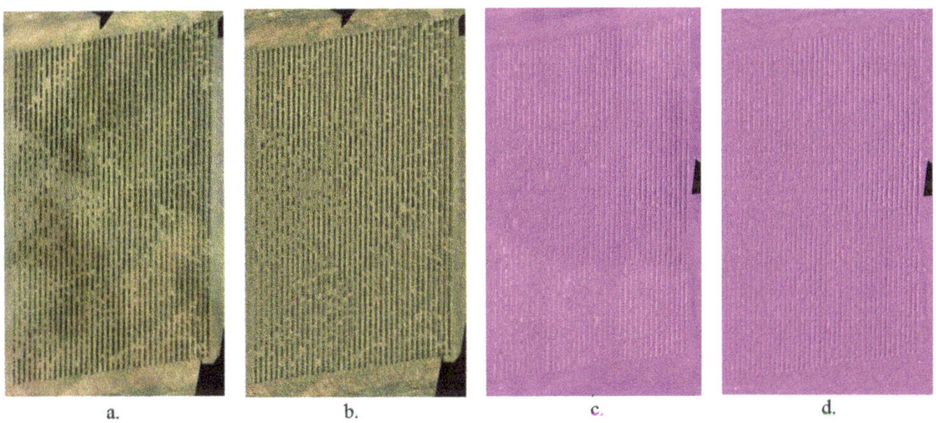

Figure 6. High- vs. Low-Altitude Aerial Image Capture of Vineyard Canopy (ideal in black and realistic in gray)

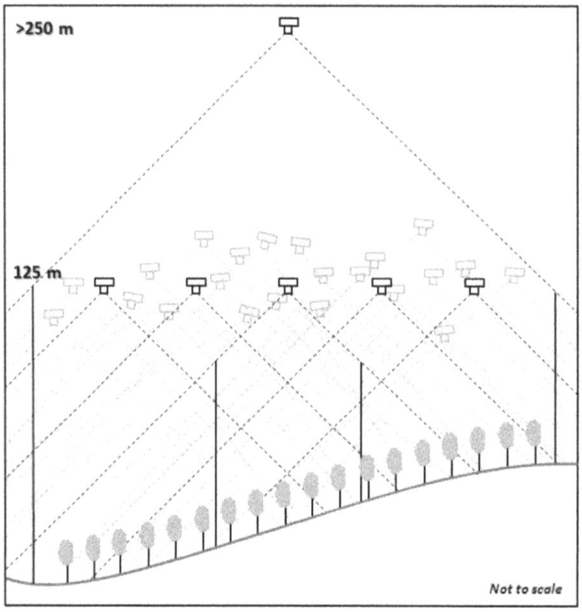

were tested (no buffer, 0.12 m, 0.24 m, and 1 m), but were considered less representative of the canopy size and spectroradiometer FOV (multiple measurements over the canopy). This 0.48 m buffer size was near the extent of the canopy width for most vines, although smaller canopies and canopies with sizable gaps bring in the possibility of spectral contamination from underlying soil. All pixels with centroids within the 0.48 m buffers were averaged and compared to spectroradiometer-measured reflectance. In addition to vine samples, reflectance values from on-ground spectral targets not included in the DN to reflectance conversion were also compared (including GCP targets and other color targets like blue, green, and red). Linear regression was used to compare each band's estimated and measured reflectances and determine the suitable bands for future analyses.

3. RESULTS

3.1. On-Ground Image Collection and Camera Calibration

Results from the wavelength interval analysis for the visible bands are provided in Table 1. The wavelength intervals determined to best represent the sensor for each band are displayed in bold. The blue band accounted for a significant amount of variation, especially within narrow wavelength intervals. The 440-460 nm interval was selected as the best fit between samples with R^2 and RMSE values of 0.937 and 8.250 for sample 1 and 0.923 and 9.628 for sample 2. Although, the interval of 445-455 nm and single wavelength of 450 nm had slightly higher R^2 values, these were considered to be less robust due to greater variation between samples 1 and 2. The green band DNs exhibited a strong relationship with spectroradiometer measurements contained with a relatively wide interval of 500-600 nm, yielding R^2

and RMSE values of 0.973 and 6.057 for sample 1 and 0.965 and 7.242 for sample 2. The red band R^2 values were the most consistent among the different wavelength intervals, with R^2 of 0.974 and 0.960. The 620-680 nm range was chosen based on the combination of high R^2 and low RMSE. Overall, the blue band was slightly less successful in correlating DNs with measured reflectance (R^2 less than 0.94) when compared to the green and red bands (both with R^2 greater than 0.96).

The ADC provided three NIR bands, of which the green band (regardless of wavelength interval) consistently provided DNs that highly correlate with measured reflectance (Table 2). The ADC-Green band was therefore used as the NIR band throughout this paper (hereafter referred to as just 'NIR'). Within this band, the widest wavelength interval had the highest R^2 values accounting for greater than 98% of the variation in measured reflectance of both samples. RMSE values for this wavelength interval also were desirably low (3.587 and 4.220).

Using the selected wavelength intervals, the per-band relationships between DNs and reflectance are provided in Figure 7(a-d). In each case, all of the 19 ROIs are plotted twice (once for each sample). For each band there was a noticeable offset between the origin and where the fit lines intersect the X-axes. This was present with all bands. For each band the dark object (black spectral target) registered mean DNs greater than zero, meaning that the digital cameras measured the dark objects as being more reflective than expected when compared to the spectroradiometer measured reflectance. The low RMSE and high correlation for the NIR band are graphically demonstrated with observations falling close to, or on the fit line. However, this is not the case with the visible bands, where less variability in spectro-radiometer values was accounted for by the regression models. Additionally, the spectral signatures of several of the calibration panel targets are shown in Figure 8 (the first blue, green, red, leaf, white, and

Table 1. Spectral calibration test results for UDC (visible bands)

UDC-Blue			**UDC-Green**			**UDC-Red**		
Interval (nm)	***Sample 1 (R^2; RMSE)***	***Sample 2 (R^2; RMSE)***	***Interval (nm)***	***Sample 1 (R^2; RMSE)***	***Sample 2 (R^2; RMSE)***	***Interval (nm)***	***Sample 1 (R^2; RMSE)***	***Sample 2 (R^2; RMSE)***
450	0.941; 8.028	0.925; 9.52	550	0.940; 9.730	0.958; 8.383	650	0.973; 6.683	0.963; 7.998
445-455	0.940; 8.083	0.927; 9.386	545-555	0.940; 9.691	0.958; 8.367	645-655	0.973; 6.681	0.963; 8.029
440-460	**0.937; 8.250**	**0.923; 9.628**	540-560	0.943; 9.482	0.959; 8.307	640-660	0.973; 6.710	0.961; 8.126
420-480	0.923; 9.111	0.906; 10.514	520-580	0.958; 7.986	0.963; 7.735	**620-680**	**0.974; 6.547**	**0.960; 8.189**
400-500	0.909; 9.436	0.894; 10.710	**500-600**	**0.973; 6.057**	**0.965; 7.242**	600-700	0.973; 6.478	0.958; 8.331

Table 2. Spectral calibration test results for ADC (NIR bands)

ADC-Blue			**ADC-Green**			**ADC-Red**		
Interval (nm)	***Sample 1 (R^2; RMSE)***	***Sample 2 (R^2; RMSE)***	***Interval (nm)***	***Sample 1 (R^2; RMSE)***	***Sample 2 (R^2; RMSE)***	***Interval (nm)***	***Sample 1 (R^2; RMSE)***	***Sample 2 (R^2; RMSE)***
840-860	0.939; 8.853	0.923; 10.010	840-860	0.988; 3.903	0.984; 4.526	840-860	0.946; 8.352	0.929; 9.600
820-880	0.937; 8.991	0.921; 10.125	820-880	0.988; 3.903	0.984; 4.526	820-880	0.946; 8.361	0.929; 9.600
800-900	0.933; 9.250	0.917; 10.341	**800-900**	**0.990; 3.587**	**0.986; 4.220**	800-900	0.945; 8.406	0.929; 9.625

Figure 7. Relationships between spectroradiometer measured reflectance (Y-axis) and digital camera measured brightness (X-axis) of spectral targets within blue (a), green (b), red (c), and NIR (d) bands

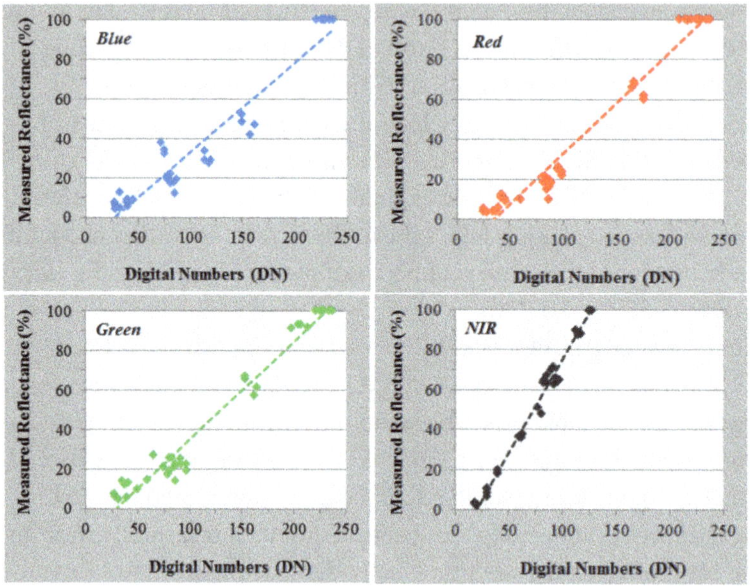

Figure 8. Spectral signatures of selected targets from sample 1 with camera spectral ranges and optimal band wavelength intervals

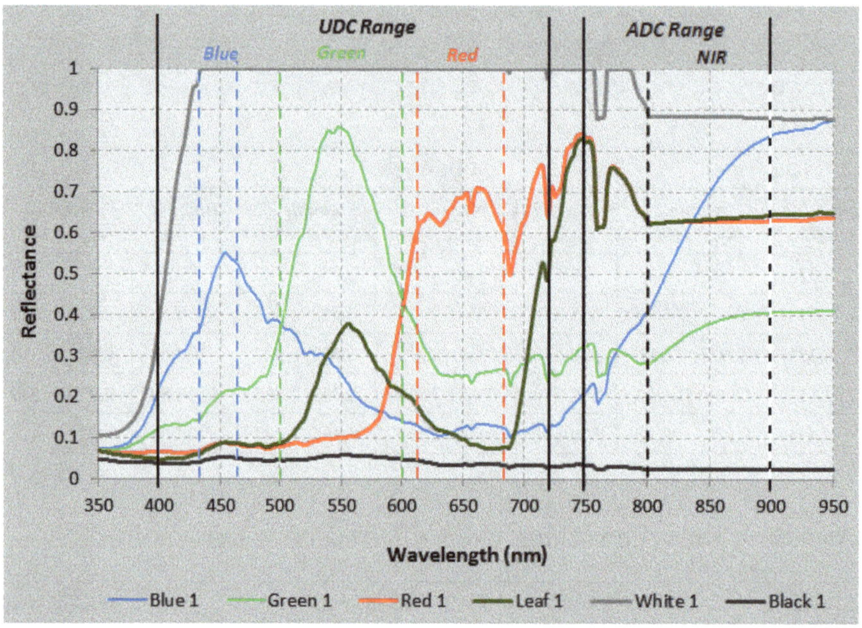

black targets). The spectral sensitivities of the UDC and ADC are provided, within which the selected wavelength intervals for each band are displayed.

3.2. Application of UAV-Based Image System to the Study Vineyard and Generation of Analysis-Ready Image Products

3.2.1. Orthophoto Generation and Per-Band Reflectance Conversion

Of the 150 total RGB images collected, 59 were selected as high-quality and nadir or near-nadir facing, which were used in model creation. In total, 25 of these images were used to export the final orthophoto mosaic. As for NIR, 45 of the 191 images captured were used to create the NIR model. Of this subset, only 13 images were needed to create the NIR orthophoto. Fewer images were needed for NIR because the NIR flight yielded higher quality images overall, mostly attributed to more consistent, slightly higher flying height (due to reduced winds during the NIR flight). This resulted in the NIR orthophoto having a lower spatial resolution. Both orthophoto mosaics yielded spatial resolutions of less than 4 cm. The RGB images had a resolution of 2.59 cm, and the NIR image had a resolution of 3.45 cm. Orthophotos were converted from DNs to reflectance using the equations for each band provided in Table 3. The light and dark targets used for the empirical line conversion yielded R^2 of 0.98 or greater for all bands (all models $p <= 0.085$). Each band of the converted, final orthophotos of the study vineyard is shown in Figure 9(a-h).

3.2.2. Per-Band Validation

The validation results including R^2, RMSE, scatterplots with trend lines and 1:1 lines are provided in Figure 10(a-d). For each band, the vine values are clustered, whereas the spectral target points are scattered due to their distinctness. Overall, the red (Figure 10c) and NIR (Figure 10d) bands exhibited the highest validation R^2 values of 0.78 and 0.57 respectively. Viewing the distribution of all points around the 1:1 line is representative of the relationship between measured and estimated. The NIR band in particular appears more robust than the other bands due to the tight clustering around the provided 1:1 relationship line. The vine values for the red (Figure 10c) band however are variable, shown along the bottom of the graph increasing the RMSE value for this band. This over- and sometimes underestimation of reflectance was also observed in the other visible bands of blue (Figure 10a) and green (Figure 10b). The blue and green bands yielded R^2 values of 0.35 and 0.02 respectively. More importantly though, the green vine samples are clustered similarly to the red and are even more variable. Vine reflectance was overestimated by the green band more so than the red. Vine values for the blue band were often underes-

Table 3. Derived reflectance conversion equations for each band

Band	Conversion Equation	R^2	RMSE	Spectral Targets Used
Blue	Reflectance = (0.7076 * DN) - 71.149	0.999	2.753	White, Black, Red GCP
Green	Reflectance = (0.6016 * DN) - 48.385	0.982	10.158	White, Black, Red GCP
Red	Reflectance = (0.6162 * DN) - 55.824	0.991	6.631	White, Black, Blue
NIR	Reflectance = (0.5124 * DN) - 11.675	0.992	3.420	All

Figure 9. Generated orthophoto bands of blue (a,e), green (b,f), red (c,g), and NIR (d,h) at the whole-vineyard (a-d) and partial-vineyard (e-h) scales

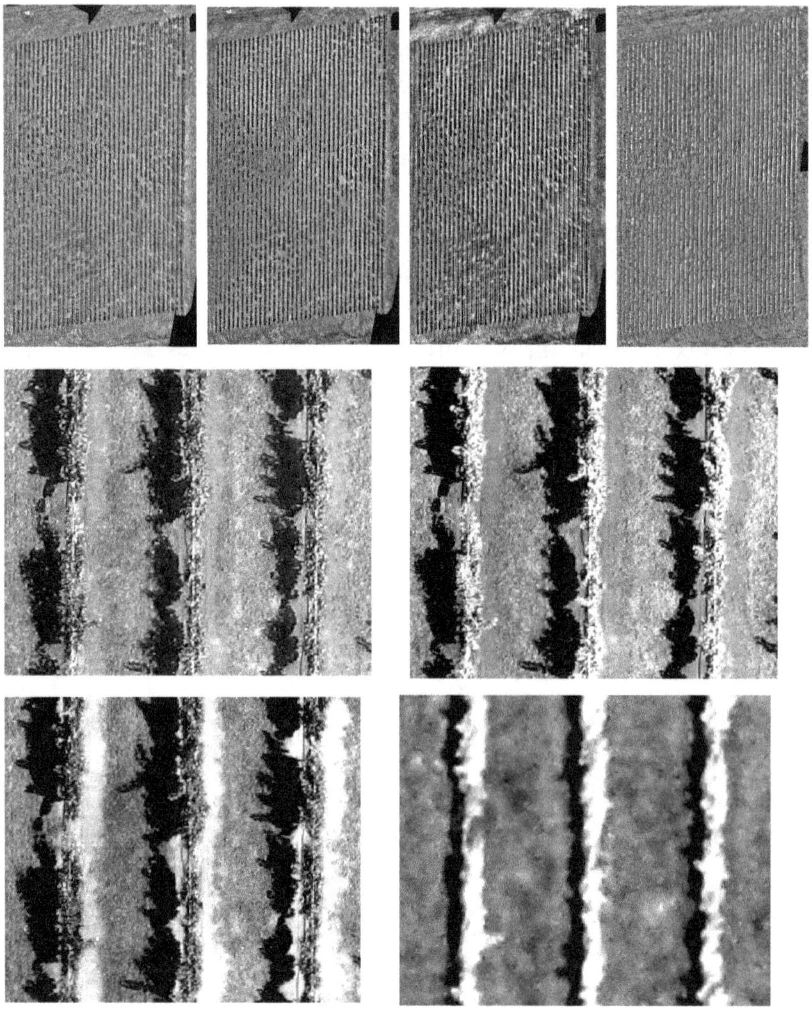

timated with values at, near, or less zero. Some values ended up slightly negative following conversion, leaving the remaining samples off the graph. The over- and underestimation of reflectance in the visible bands, particularly the blue and green bands, also occurred with the spectral targets (outliers shown in the upper-left and lower-right portions of Figures 10a and 10b).

4. DISCUSSION

4.1. Validation Results

The estimated reflectance results were weaker than desired for the visible bands, particularly with blue and green, where blue resulted in extremely low values and green resulted in extremely high, variable

Figure 10. Validation results comparing measured (Y-axis) and estimated (X-axis) reflectance of vines and spectral targets within each band

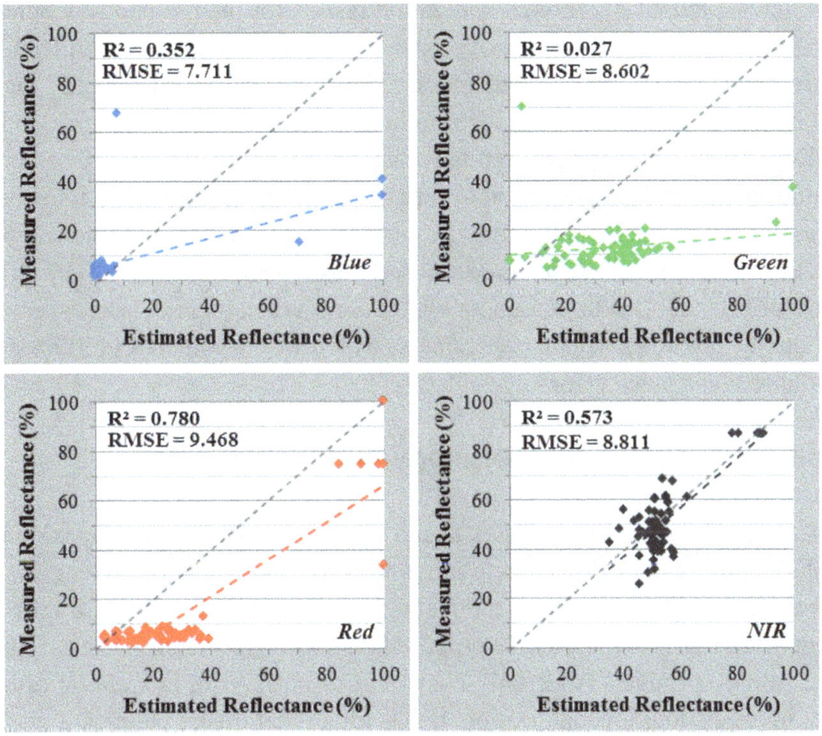

values. Shorter wavelengths are typically more difficult to measure due to atmospheric scattering, which may have played a slight role. At such a low flying height though, these effects should have been minimal. Similar to the other visible bands, the modeled red band did not account for much of the variability in the validation data. The red band, however, exhibited the best fit of the visible bands due to its accurate estimations of spectral target reflectance unlike the blue and green bands (see Figure 10a-d; spectral target reflectance values are found on or very near the fit line). The tendency of the visible band models to over- and underestimate brightness (DNs and subsequent reflectance) might be attributed to the CCD and Bayer filter's susceptibility to accentuate color as it is measured. Hunt et al. (2005) discussed similar findings where, in the majority of images captured, mean DNs of non-light (object) spectral targets were greater than values of 250. The sensors used in this study were designed for and marketed to a recreational audience that limits the amount of customization possible. Even using the custom settings outlined in the Materials and Methods section, digital cameras still automate much of the data collection process. The JPEG file format may have also contributed to this error even though similar studies insist otherwise (Lebourgeois et al., 2008).

In terms of future qualitative analyses, the produced orthophotos are highly accurate representations of the study vineyard that are very useful for surveillance mapping. As for quantitative analyses, the weak validation R^2 values for all bands does not support future work without a substantial amount of error. In cases where such error can be tolerated (which is very likely due to the vastly reduced cost of UAV-based image collection), the red and NIR bands could be employed due to their higher R^2 values. Of all bands, NIR appeared graphically (Figure 10d) to best estimate vine reflectance (clustered around the fit line) without the high variability that the visible bands exhibited (low RMSE). Similarly, Jensen et al. (2007) noted that digital camera-collected NIR information played the most significant role in predicting grain yields. Statistically though, validation results for the red band accounted for the most variation in measured reflectance. Overall, poor results may relate to the general difficulty of establishing a statistical relationship between ground-measured spectra and aerial imagery due to spatial and spectral resolution differences between the spectroradiometer and the digital cameras. This difficulty was discussed by Lelong et al. (2008) and led to relative analyses (within each image set and not image set to image set) with DNs instead of conversion to reflectance. In addition, due to its very heterogeneous nature with leaves of varying angles and sizes, canopy gaps, trellis hardware (wire and posts), and other background noise/spectral contamination (bare soil and grass), vine canopy is a very difficult surface to model in terms of reflectance. Due to this, using vine canopy as a validation is an especially challenging endeavor. Additionally, unknown sensor capabilities, which were estimated in this study (wavelength intervals), might have contributed error.

4.2. Practicality of Provided Methodology and Future Work

The proposed data workflow provides a practical and very inexpensive means by which to collect very high spatial resolution orthophotos. Processing time and effort, though, remains lengthy, specifically the effort needed to mask the input images to generate high quality image products. Future research would benefit from increased automation within the proposed data workflow in an effort to reduce overall time needed to produce image products (spectrally-consistent or not). Also, in creating orthophotomosaics with and without image histogram equalization, it was found that image alignment was less successful with images that were adjusted, necessitating more manual effort to remove noise within the point clouds. This is due to the slightly adjusted DNs that, when changed within all images, can reduce image-to-image matching. It would therefore be useful to create a methodology that could improve SfM results of spectrally-adjusted images. UAV image processing software is rapidly becoming more sophisticated though, which may soon provide the ability to match images and then adjust spectral properties within a single application.

Of the more successful bands of red and NIR, the NIR band was less likely to over- or underestimate spectral reflectance. This might be due to the ADC only being able to sense a narrow portion of the spectrum, which minimizes the degree to which the Bayer filter and CCD affect the DNs of the captured scene. In this way, future work may wish to use a spectral filter with the UDC (or ADC) to collect single bands within the visible range. For example, a green filter (blocking blue and red light) would force the sensor to collect less spectral information overall, which may increase accuracy of green information collected within three (RGB) bands. How the Bayer filter and CCD might record this spectral data is unknown. Future work may wish to create a time series of orthophotos to gain a fuller sense of the usefulness of digital camera-based spectral reflectance estimations.

In keeping with the practical nature of this research, band-to-band registration between cameras was not performed. Band-to-band registration between two cameras is impractical in many UAV applications because UAV (planes, quadcopters, blimps) payloads are extremely low (less than 400g in many cases where one digital camera weighs around 150-200 g with batteries), therefore necessitating separate flights with each camera. Even if the UAV has the payload capacity to carry two cameras, further problems can arise: mounting the cameras very near each other proves difficult with many do-it-yourself UAVs and capturing images simultaneously without a servo and the additional weight of the servo (with motor) further complicates system creation. If desired at a later time, bands can always be co-registered by way of resampling the rasters to a lower spatial resolution.

As for vineyard-specific future work, analysis of the generated image products is needed to further assess the usefulness of the data to viticulturists. Due to band-to-band registration being disregarded, future analysis with the red and NIR bands would likely need to take an object-oriented approach or resample the imagery to a coarser spatial resolution to properly register the bands to one another.

5. CONCLUSION

This study used two Canon PowerShot A480 digital cameras (UDC and ADC) to collect UAV aerial imagery within four spectral bands of blue, green, red, and NIR at a vineyard site in central Texas. A data workflow methodology was proposed to address radiometric and geometric corrections within the collected images, which using SfM image processing exports very high spatial resolution orthophotomosaics. The empirical line method was employed to transform DNs collected by the camera system to values of reflectance. Through a validation procedure it was found that the red and NIR bands were the most accurate at estimating reflectance (although with a significant amount of error), and therefore are better suited for future analyses. The blue and green bands were less accurate and were highly prone to over- and underestimate reflectance.

ACKNOWLEDGMENT

The author wishes to thank John Klier for help in collecting the aerial images at the vineyard site. Additional thanks are extended to Kumudan Grubh, Matthew Patton, and David Yelacic for help in collecting the on-ground spectroradiometer data in San Marcos. The author would like to further acknowledge Jennifer Jensen and Nathan Currit for providing helpful advice throughout the project. This work was supported by a doctoral research stipend provided by the Texas State University Graduate College.

REFERENCES

Aber, J. S., Marzoff, I., & Ries, J. B. (2010). *Small-Format Aerial Photography: Principles, Techniques, and Geosciences Applications*. Oxford, UK: Elsevier.

Baboo, S. S., & Devi, M. R. (2011). Geometric Correction in Recent High Resolution Satellite Imagery: A Case Study in Coimbatore, Tamil Nadu. *International Journal of Computers and Applications*, *14*(1), 32–37. doi:10.5120/1808-2324

Baluja, J., Diago, M. P., Balda, P., Zorer, R., Meggio, F., Morales, F., & Tardaguila, J. (2012). Assessment of vineyard water status variability by thermal and multispectral imagery using an unmanned aerial vehicle (UAV). *Irrigation Science*, *30*(6), 511–522. doi:10.100700271-012-0382-9

Bellvert, J., Zarco-Tejada, P. J., Girona, J., & Fereres, E. (2013). Mapping crop water stress index in a 'Pinot-noir' vineyard: Comparing ground measurements with thermal remote sensing imagery from an unmanned aerial vehicle. *Precision Agriculture*. doi:10.100711119-013-9334-5

Cheng, C., & Rahimzadeh, A. (2005). *Hacking Digital Cameras*. Indianapolis, Indiana: Wiley Publishing, Inc.

Dandois, J. P., & Ellis, E. C. (2010). Remote sensing of vegetation structure using computer vision. *Remote Sensing*, *2*(4), 1157–1176. doi:10.3390/rs2041157

Dandois, J. P., & Ellis, E. C. (2013). High spatial resolution three-dimensional mapping of vegetation spectral dynamics using computer vision. *Remote Sensing of Environment*, *136*, 259–276. doi:10.1016/j.rse.2013.04.005

Dare, P. M. (2008). Small Format Digital Sensors for Aerial Imaging Applications. *The International Archives of the Photogrammetry, Remote Sensing, and Spatial Information Sciences*, *XXXVII*(B1), 533–538.

De Biasio, M., Arnold, T., Leitner, R., McGuinnigle, G., & Meester, R. (2010). UAV-based environmental monitoring using multi-spectral imaging. *Proceedings of SPIE – The International Society for Optical Engineering, 7668*, 1-7.

Dean, C., Warner, T. A., & McGraw, J. B. (2000). Suitability of the DSC460c colour digital camera for quantitative remote sensing analysis of vegetation. *ISPRS Journal of Photogrammetry and Remote Sensing*, *55*(2), 105–118. doi:10.1016/S0924-2716(00)00011-3

Everaerts, J. (2008). The use of unmanned aerial vehicles (UAVs) for remote sensing and mapping. *The International Archives of the Photogrammetry, Remote Sensing, and Spatial Information Sciences*, *XXXVII*(B1), 1187–1191.

Hall, A., Lamb, D. W., Holzapfel, B. P., & Louis, J. P. (2011). Within-season temporal variation in correlations between vineyard canopy and winegrape composition and yield. *Precision Agriculture*, *12*(1), 103–117. doi:10.100711119-010-9159-4

Hunt, E. R. Jr, Cavigelli, M., Daughtry, C. S. T., McMurtrey, J. III, & Walthall, C. L. (2005). Evaluation of Digital Photography from Model Aircraft for Remote Sensing of Crop Biomass and Nitrogen Status. *Precision Agriculture*, *6*(4), 359–378. doi:10.100711119-005-2324-5

Hunt, E. R., Hively, W. D., Daughtry, C. S. T., & McCarty, G. W. (2008). Remote Sensing of Crop Leaf Area Index Using Unmanned Airborne Vehicles. In ASPRS Pecora 17, Denver, CO, November 18-20 (pp. 1-9).

Jensen, J. R. (2005). *Introductory Digital Image Processing: A Remote Sensing Perspective*. Upper Saddle River, NJ: Pearson Prentice Hall.

Jensen, T., Apan, A., Young, F., & Zeller, L. (2007). Detecting the attributes of a wheat crop using digital imagery acquired from a low-altitude platform. *Computers and Electronics in Agriculture*, *59*(1-2), 66–77. doi:10.1016/j.compag.2007.05.004

Kaminsky, R. S., Snavely, N., Seitz, S. T., & Szeliski, R. (2009). Alignment of 3D Point Clouds to Overhead Images. In *Proceedings of Second IEEE Workshop on Internet Vision*, Miami, FL, June 20-25 (pp. 63-70). 10.1109/CVPRW.2009.5204180

Karpouzli, E., & Malthus, T. (2003). The empirical line method for the atmospheric correction of IKONOS imagery. *International Journal of Remote Sensing*, *24*(5), 1143–1150. doi:10.1080/0143116021000026779

Kelcey, J., & Lucieer, A. (2012). Sensor Correction of a 6-Band Multispectral Imaging Sensor for UAV Remote Sensing. *Remote Sensing*, *4*(12), 1462–1493. doi:10.3390/rs4051462

King, D. J. (1995). Airborne Multispectral Digital Camera and Video Sensors: A Critical Review of System Designs and Applications. *Canadian Journal of Remote Sensing*, *21*(3), 245–273. doi:10.1080/07038992.1995.10874621

Laliberte, A., Goforth, M. A., Steele, C. M., & Rango, A. (2011). Multispectral Remote Sensing from Unmanned Aircraft: Image Processing Workflows and Applications for Rangeland Environments. *Remote Sensing*, *3*(11), 2529–2551. doi:10.3390/rs3112529

Lebourgeois, V., Beguye, A., Labbe, S., Mallavan, B., Prevot, L., & Roux, B. (2008). Can Commercial Digital Cameras Be Used as Multispectral Sensors? A Crop Monitoring Test. *Sensors (Basel, Switzerland)*, *8*(11), 7300–7322. doi:10.33908117300

Lelong, C. C. D., Burger, P., Jubelin, G., Roux, B., Labbe, S., & Baret, F. (2008). Assessment of Unmanned Aerial Vehicles Imagery for Quantitative Monitoring of Wheat Crop in Small Plots. *Sensors (Basel, Switzerland)*, *8*(5), 3557–3585. doi:10.33908053557

Levin, N., Ben-Dor, E., & Singer, A. (2005). A digital camera as a tool to measure colour indices and related properties of sandy soils in semi-arid environments. *International Journal of Remote Sensing*, *26*(24), 5475–5492. doi:10.1080/01431160500099444

Mathews, A. J., & Jensen, J. L. R. (2013). Visualizing and Quantifying Vineyard Canopy LAI Using an Unmanned Aerial Vehicle (UAV) Collected High Density Structure from Motion Point Cloud. *Remote Sensing*, *5*(5), 2164–2183. doi:10.3390/rs5052164

McCoy, R. M. (2005). *Field Methods in Remote Sensing*. New York, USA: The Guilford Press.

Niethammer, U., Rothmund, S., Schwaderer, U., Zeman, J., & Joswig, M. (2011). Open source image-processing tolls for low-cost UAV-based landslide investigations. *International Archives of the Photogrammetry, Remote Sensing and Spatial Information Sciences, XXXVIII-1/C22.*

Primicerio, J., Di Gennaro, S. F., Fiorillo, E., Genesio, L., Lugato, E., Matese, A., & Vaccari, F. P. (2012). A flexible unmanned aerial vehicle for precision agriculture. *Precision Agriculture*, *13*(4), 517–523. doi:10.100711119-012-9257-6

Rango, A., Laliberte, A., Herrick, J. E., Winters, C., Havstad, K., Steele, C., & Browning, D. (2009). Unmanned aerial vehicle-based remote sensing for rangeland assessment, monitoring, and management. *Journal of Applied Remote Sensing*, *3*(1), 033542. doi:10.1117/1.3216822

Ritchie, G. L., Sullivan, D. G., Perry, C. D., Hook, J. E., & Bednarz, C. W.G. L. Ritchie; D. G. Sullivan; C. D. Perry; J. E. Hook; C. W. Bednarz. (2008). Preparation of a Low-Cost Digital Camera System for Remote Sensing. *Applied Engineering in Agriculture*, *24*(6), 885–896. doi:10.13031/2013.25359

Rodríguez-Pérez, J. R., Riaño, D., Carlisle, E., Ustin, S., & Smart, D. R. (2007). Evaluation of Hyperspectral Reflectance Indexes to Detect Grapevine Water Statuts in Vineyards. *American Journal of Enology and Viticulture*, *58*(3), 302–317.

Smit, J. L., Sithole, G., & Strever, A. E. (2010). Vine Signal Extraction – an Application of Remote Sensing in Precision Viticulture. *South African Journal of Enology & Viticulture*, *31*(2), 65–74.

Smith, G. M., & Milton, E. J. (1999). The use of the empirical line method to calibrate remotely sensed data to reflectance. *International Journal of Remote Sensing*, *20*(13), 2653–2662. doi:10.1080/014311699211994

Snavely, N., Seitz, S. M., & Szeliski, R. (2008). Modeling the world from internet photo collections. *International Journal of Computer Vision*, *80*(2), 189–210. doi:10.100711263-007-0107-3

Turner, D., Lucieer, A., & Watson, C. (2011). Development of an Unmanned Aerial Vehicle (UAV) for Hyper Resolution Mapping Based Visible, Multispectral, and Thermal Imagery. In *Proceedings of 34th International Symposium of Remote Sensing Environment*, Sydney, Australia, April 10-15 (pp. 1-4).

Turner, D., Lucieer, A., & Watson, C. (2012). An Automated Technique for Generating Georectified Mosaics from Ultra-High Resolution Unmanned Aerial Vehicle (UAV) Imagery, Based on Structure from Motion (SfM) Point Clouds. *Remote Sensing*, *4*(5), 1392–1410. doi:10.3390/rs4051392

Watts, A. C., Ambrosia, V. G., & Hinkley, E. A. (2012). Unmanned Aerial Systems in Remote Sensing and Scientific Research: Classification and Considerations of Use. *Remote Sensing*, *4*(6), 1671–1692. doi:10.3390/rs4061671

This research was previously published in the International Journal of Applied Geospatial Research (IJAGR), 6(4); edited by Donald Patrick Albert and Samuel Adu-Prah, pages 65-87, copyright year 2015 by IGI Publishing (an imprint of IGI Global).

Chapter 13
The Role of Affective Computing for Improving Situation Awareness in Unmanned Aerial Vehicle Operations:
A US Perspective

Jonathan Bishop
Centre for Research into Online Communities and E-Learning Systems, Belgium

ABSTRACT

Unmanned aerial vehicles (UAVs), commonly known as drones, are a robotic form of military aircraft that are remotely operated by humans. Due to lack of situation awareness, such technology has led to the deaths of civilians through the inaccurate targeting of missile or gun attacks. This chapter presents the case for how a patented invention can be used to reduce civilian casualties through attaching an affect recognition sensor to a UAV that uses a database of strategies, tactics and commands to better instruct fighter pilots on how to respond while in combat so as to avoid misinterpreting civilians as combatants. The chapter discusses how this system, called VoisJet, can reduce many of the difficulties that come about for UAV pilots, including reducing cognitive load and opportunity for missing data. The chapter concludes that using UAVs fitted with VoisJet could allow for the reduction of the size of standing armies so that defence budgets are not overstretched outside of peacetime.

DOI: 10.4018/978-1-5225-8365-3.ch013

INTRODUCTION

The use of remote systems like drones, technically known as unmanned aerial vehicles (UAVs), is starting to move out of the first wave of public and media attention, and now needs to shift toward a more cohesive debate with productive and practical ends. Indeed, it has been argued that a problem with the design of existing UAVs driven by policy makers, who are not militarily qualified, has been illustrative of the claim that new technological practices that aim to bridge physical distance provide a more efficient moral distance and make it difficult for people to exercise moral responsibility (Coeckelbergh, 2013). This has been associated with Internet abuse such as trolling, where people will abuse others for their own enjoyment without considering the consequences (Bishop, 2013; Bishop, 2014; Hardaker, 2013; McCosker, 2013), but should be considered a flaw that needs adjusting in current drone systems where the actions of pilots would be dehumanised. The use of UAVs by the US Government provides greater freedom in those military personnel carrying out operations on their behalf, as use of foreign military personnel and the agreement of their government will be less necessary (Byman, 2013). A greater use of UAVs also means that there will be fewer casualties arising out of improvised explosive devices (IEDs) as air-based operations provider greater safety for military personnel in this regard (Reynolds, 2013). The problem however is evident that the way UAVs are currently designed does not allow for the efficiency of operations and the safety of civilians and other non-military targets.

Uses and Limitations of UAVs in Combat Operations

The NATO forces are becoming more dependent on UAVs, and their use has been driven by the US and such policy choices have consequences, regardless of whether their military leaders believe these are right or wrong (Kean et al., 2004, p.376). Any government in the world will want to increase its national security whilst limiting loss of life to those in its military and the civilians in the localities where its military operations are engaged. However, before trying to understand the technology of UAVs and how they can be improved, it is important to understand why there is a growing need for the US military to be allowed to use UAVs in military engagements in order to safeguard the democratic rights of Americans whom they must protect whoever is in power in the White House.

BACKGROUND

There is one thing more certain than death or taxes and that is that humans will always be in conflict with one another, regardless of which difference or transgression created that conflict. One of the longest lasting wars in history is that which is taking place in the Holy Land between those Jews that identify as Israelis and those Arabs that identify as Palestinians. During the first millennium BCE this part of the world, one might wish to call the Holy Land, was known as Israel in the Hebrew language and as Palestine in Roman Latin (i.e. they were the same place). It was only in the mid-21st century CE that there was a separation of the territories caused by meddling politicians who formed part of the United Nations. The people in the Holy Land around 900 BCE were known as the Israelites, and there were 12 Semitic tribes in total, forming part of a united kingdom. These tribes split around 820 BCE with the enthronement of the son of King Solomon, Rehoboam, who threatened "God's wrath" on the 10 tribes in the North, who have now become the Jewish Semitics (i.e. Jews) and the two tribes in the South whom supported him have now become Arabic Semitics (i.e. Arabs). These two Semitic races have been warring ever since the fall of Rehoboam and international interference has led to some of the greatest atrocities. This includes the massacre of the Jews by King Richard I in York in 1190 CE (Jones, 2000) and the same by the Hitler Government in Germany between 1933 and 1945 CE (Rees, 2005), right up to the bombing of the World Trade Center in New York. This act of war is according to those in the FBI a direct result of political support for the Netanyahu Government by the politicians in Washington (Drezner, 2011; Kean, Hamilton, & Fitzgerald, 2004; Kean et al., 2004, p.147). The bombing of three of its sister cities (i.e. London, Madrid, Paris), in all cases by Jihads (i.e. radical Arabs), is also believed to be because of the political leadership in the US Government. The Netanyahu Government have a policy of driving out Arabs from the Holy Land with the support of Zionists (i.e. radical Jews), which is something the Jihads take exception to. The Netanyahu Government is not the first to want to drive Arabs from the Holy land to create a single country liked existed around 900 BCE, as significant attempts were made in the early Middle Ages, starting with the People's Crusade of 1096 CE right up until the Ninth Crusade ending in 1172. The Netanyahu Government's plan of mass eradication of Arabs, is to end the historical existence of the Israelites (i.e. Jews, Arabs) to create a single state made up only of so-called Israelis (i.e. Jews), with the eradication of Palestinians (i.e. Arabs). This has meant that the Western world has become less safe in relation to those countries who backed and still support the post-World War II settlement to break up the Holy Land, leaving the military to find affordable solutions to what some see a costly mistake. Indeed, it has been suggested that the US Government reassess its policy in relation to the support of the Netanyahu Government by some of its own security officials (Kean et al., 2004; Kean & Hamilton, 2006). Since the election of Nobel Peace Prize winner Barrack Obama as President in 2008, there has been a shift in the way the US Government is approaching the three millennia long war in the Holy Land (Drezner, 2011). Barrack Obama and his successors, and the US military whom they exercise control over, have a huge mess to clean up from the previous politicians in the US Government. This may come in the form of negotiating a settlement between the Zionist and Jihadist forces that threaten world peace, whether the state-led attacks of the Netanyahu Government in the case of the former, or the gruella-led attacks of Hamas and Al-Qaeda in the case of the latter. Whilst Al-Qaeda could be considered a grave threat to the peace and safety of Western nations, comparable to Japan in the Second World War, in this context one might prefer to compare Hamas to the First and Second Continental Congresses that secured US independence even though they were viewed as terrorists at one point.

The rhetoric of successive US politicians in branding anti-Zionist forces as terrorists – such as for the attacks by Jihads on the World Trade Center – has given the military political clout to enter countries where Jihadist training camps exist to counter the actions of US politicians (Mobley, 2012). These anti-Zionists seek to make further attacks against America and the allies of the US Government, such as NATO countries, because of US politicians, making life difficult for the US military, whom are tasked with protecting America, whomever is in power, and whatever their policies happen to be. The sudden rise of media-savvy Jihads supported by Western citizens, like ISIS (or ISIL, Islamic State, Al-Qaeda in Iraq), may in part be down to the empathy of these Arabs and others to the Palestinians being terrorised in their homeland (i.e. the Holy Land). Since media attention in 2014 turned towards Israeli attacks on Gaza the membership of Jihadist organisations has significantly grown. This has resulted in the State of Israel, which occupies much of the Holy Land, using UAVs in their operations as well as sporadic missile attacks designed to cause maximum destruction. It is unclear whether it is possible to have a time where all Israelites – Arabs, Jews and Gentiles – will be able to share the Holy Land in the same way Europeans move freely in the European Union.

There have been significant demands on Western military forces to take forward ground operations to deal with ISIS in Iraq, but public opinion on the issue is fragile followed the failed 2003 liberation of Iraq. The ground-led operations carried out for successive US presidents by the US military, has led to a huge number of deaths of US and other NATO military personnel, such as through IEDs, as well as civilians in the countries targeted. The innovative use of unmanned aerial vehicles (UAVs) by the military for dog-fighting as well as currently targeted drops appears to be the only effective means to hold back those anti-Zionist forces seeking to attack NATO member states, while the politicians continue to make poor policy decisions that the military, and indeed non-governmental organisations, have been left to pick up the pieces of. The current design of UAVs has led to a number of innocent civilian deaths in countries harbouring Jihadists that are military targets for NATO forces. It is therefore necessary to re-design future UAVs to take account of the need to protect civilian lives. One way to achieve this might be to add affective and augmentive sensors and software to increase the amount of information available to UAV pilots, so as to increase their situation awareness.

AN EVALUATION OF AN INVENTION FOR REDUCING HUMAN CASUALTIES IN UAV-BASED MILITARY ENGAGEMENTS: INTRODUCING VOISJET

It is clear that unmanned aerial vehicles will play an important part in defence and other military operations for the foreseeable future, but their use will have to move away from being given direction by military commanders and for more judgement on the pilots' part to be utilised. The known civilian deaths such as weddings and other social engagements, as well as unknown ones, such as where a missile may have hit the wrong person, means that the technology for assisting pilots in identifying threats and avoiding causalities needs to be improved. This section sets out how a system, known as 'VoisJet,' could be used to assist UAV pilots in correctly targeting enemy personnel, without the need for direction from military commanders.

Figure 1. Overview of processes for VoisJet

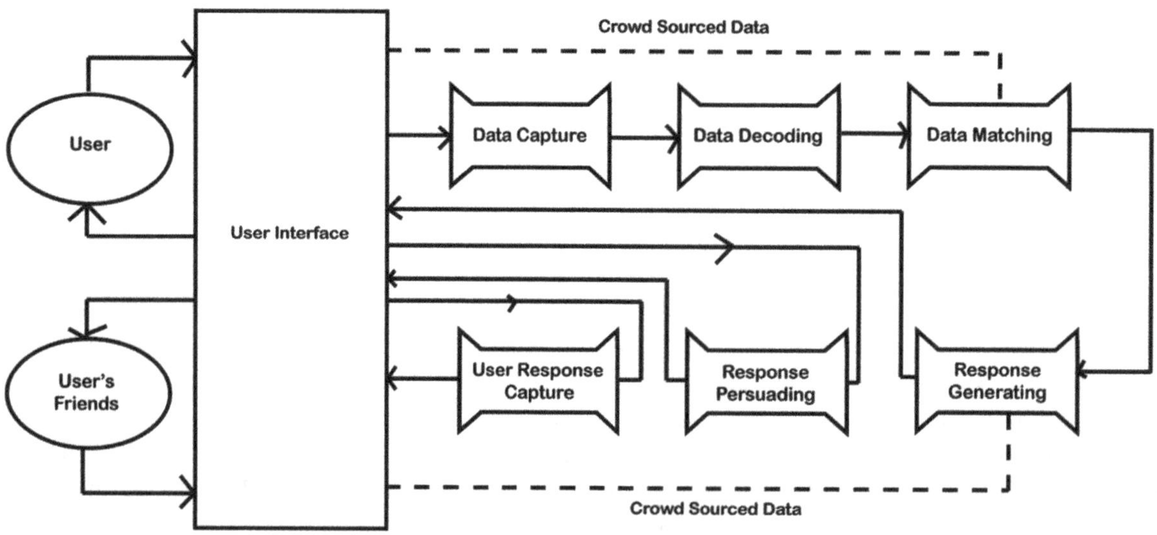

VoisJet operates by using sensors to capture information, followed by decoding that information into a readable format by its bridge between external inputs and internal processes. The type of sensors used can be extended and work together to achieve an overall picture of what is going on in the area at which the pilot of a UAV has it targeted at. There will now be discussed some of the existing algorithms that could be fed into VoisJet. Systems which may make useful inputs for VoisJet are Point Clouds, which using open source libraries can filter, estimate features, and remove outliers from distorted capture, which have already been tested to be effective with UAVs (McLeod et al., 2013).

Table 1, which is drawn from Endsley, Bolte, & Jones (2010) shows many of the problems with unmanned vehicles and their aerial versions that can impair situation awareness of the pilots. This next section will be looking at the way VoisJet can improve situation awareness and thus reduce human casualties.

Reducing Poor Sensory Data, Intermittent Data, and Time-Lags in UAVs

A primary challenge in developing effective virtual reality systems is designing the human interface hardware which allows rich sensory information to be presented to users in natural ways (Rosenberg & Stredney, 1996). It has been argued that autonomic systems can provide increased manageability and higher availability by executing in an autonomous execution environment and by having rich sensory information at their disposal (Durham, Milenkovic, & Cayton, 2006). Table 1 shows that rich sensory information (visual, auditory, haptic, etc.) about the in combat operation of a UAV is essential, especially relating to the state of the UAV in the operating environment (Endsley et al., 2010). Many UAV designs result in visual or cognitive sub-optimisation, which can be fatal (Endsley et al., 2010). Research has

Table 1. Limitations on situation awareness in current UVs and UAVs

Factor	Description
Poor sensory data, intermittent data, and time-lags	Rich sensory information (visual, auditory, haptic, etc.) about the in combat operation of a UV is essential, especially relating to the state of the UV in the operating environment. Many UVs result in visual or cognitive sub-optimisation, which can be fatal.
Difficulties in UV localisation	Disorientation in UVs is a common problem that seriously affects situation awareness. Being able to have a sense of localisation – where the operator is aware of what is going on in combat – is essential. Lack of situation awareness on the physical orientation and location of the vehicle can cause problems in any task that requires a special understanding of the area.
Demanding tasks in complex environments	UV operations often involve perceptually demanding tasks, especially visually, making it that more demanding. In the case of UAVs operators' visual environment is often impoverished, with object properties (e.g. surface texture and colour) as well as poor spatial and chromatic resolution of the VDU through disrupted temporal and optical flow.
Low-Level data overload and interface design	The interfaces for UVs often only provide low-level data regarding vehicle systems and parameters. Particularly with UAVs, the situation awareness of operators is poor, often because of lack of key information or poor salience.
Multitasking in UV operations	Control over UVs often require multi-tasking and systems to date have been poor at obtaining it. In terms of UAVs this has often been in the form of not being able to target accurately, leading to crashes.
Lack of support for multi-person operations	UVs and in particular UAVs may be operated based on the needs of other team members who may possess critical information on the mission context, which can add additional task loading. This includes communications with commanders at observing those on the ground.
Increased autonomy for unmanned vehicle operations	Where autonomy is added to a UV problems in acquisition and maintenance of situation awareness are multiplied when the human must "operate, monitor and supervise" those UV systems under their control.

established that speech input is significantly better than manual input in terms of task completion time, task accuracy, flight/navigation measures, and pilot ratings in terms of UAVs (Draper, Calhoun, Ruff, Williamson, & Barry, 2003). UAVs can provide regular yet intermittent data transmission which is essential for the success of defence operations (Yang, Chen, Ammar, & Lee, 2005). VoisJet thus has an important role in reducing poor sensory data, intermittent data and time-lags. As an advice-giving agent it can process all the sensory information that the pilot is normally burdened with and also that which is absent, meaning they can make decisions based on their experience, rather than have to make do with incomplete information. The problems with time-lag, where the information collected by the UAV and presented to the operator can be removed by VoisJet through the processing being done machine-side and not by the pilot. This would enable the operator to make quicker decisions with less risk of incomplete information. One might there want to think of VoisJet like a Cloud application. The data processing is done server-side (on the UAV) and the decision-making is done client-side (by the pilot/operator).

Resolving Difficulties in UAV Localisation

Manned aerial operations are complex on their own, but in such situations fighter pilots have an awareness of the situation making it easy for judgements to be made that take account of the immediate circumstances. Traditional UAV operations can be as limiting to a fighter pilot as a PC is to a novice user, who relies on a third-party to convey information for them to be situation aware. The aim therefore should be for UAVs to provide instant access to the information a fighter pilot needs to perform the mission being asked of them. Indeed, whenever a task necessitates both local spatial task performance and knowledge of the global environment simultaneously, the operator at present is often forced to trade off performance in one task to maintain performance in the other task, which can result in poor performance in both local and global tasks and disorientation (Yee, 2013). As can be seen in Table 1, disorientation in UAVs is a common problem that seriously affects situation awareness of operators. Being able to have a sense of localisation – where the operator is aware of what is going on in combat – is thus essential (Endsley et al., 2010). Lack of situation awareness on the physical orientation and location of the vehicle can cause problems in any task that requires a special understanding of the area (Endsley et al., 2010). Even in non-military operations it is clear that the greatest improvements in situation awareness could be achieved by adding a localisation device to the UAV (Chisholm, Cui, Lum, & Chen, 2013). Through VoisJet it will be possible for the terrain to be evaluated for threats or to ensure civilian life is not lost by mistake. For instance, reports of wedding parties being interrupted by UAV strikes firing in error can be avoided, as the situation awareness algorithms in VoisJet can let the pilot know whether someone is likely to be a civilian or combatant and the best course of action to take – such as in a hostage situation.

Improving Management of Demanding Tasks in Complex Environments

Unmanned Aerial Vehicles (UAVs) are developed to perform more demanding tasks in civic and military applications such as geological surveys, aerial photography for mapping, security and environmental monitoring (Ando, Oguchi, & Yamada, 2013). As can be seen from Table 1, UAV operations often involve perceptually demanding tasks, especially visually, making it that more demanding (Endsley et al., 2010). In the case of UAV operators, their visual environment is often impoverished, with object properties (e.g. surface texture and colour) as well as poor spatial and chromatic resolution of the VDU through disrupted temporal and optical flow (Endsley et al., 2010). With VoisJet, as fighter pilots would not need to make judgements on what something is, and would be given options about what courses of action they can take, then it could play greater emphasis on their judgement in terms of knowing what to do, as opposed to working out what is going on in such a way they lack the time to make good judgements.

Improving Low-Level Data Overload and Interface Design in UAVs

Table 1 shows that some of the interfaces for UVAs only provide low-level data regarding vehicle systems and parameters and the situation awareness of operators is often poor because of lack of key information or poor salience (Endsley et al., 2010). Low-level data for critical decision-making that is not readily available is necessary in UAVs, yet there are often demands for a single data stream that combines drone video feeds, cell phone intercepts, and targeting radar in UAVs (Gomez & Chimento, 2011). With VoisJet the most pertinent information would be available to the operator so that they have the appropriate number of options for the appropriate operating context they are in. The advancement in multi-touch technology might make it easier for pilots to access relevant information between various screens, as effortlessly as one might switch from one smartphone application to another.

Multitasking in UAV Operations

Control over UAVs often requires multi-tasking as Table 1 shows, yet to date systems have been poor at obtaining it and this often results in pilots not being able to target accurately, leading to crashes (Endsley et al., 2010). With VoisJet being based on a multi-agent system, of which one embodiment can cope with over 250,000 action states, means the need for multi-tasking at a decision-making level could be reduced, leaving pilots to focus on operating the UAV based on their tacit experience and knowledge. As VoisJet is based around providing information and suggestions by computing several different inputs simultaneously, then it might mean that operation-specific interfaces could be designed to display that information depending on the specific goals of the pilot and the mission, of which the strategies could be varied by moving from one screen to another.

Overcoming Lack of Support for Multi-Person Operations in UAVs

As Table 1 makes evident, UAVs may be operated based on the needs of other team members who may possess critical information on the mission context, which can add additional task loading, which can included communications with commanders at observing those on the ground (Endsley et al., 2010). The advantage of VoisJet is that it can use information from the actions of one person to advise another. This need not be as simple as letting a UAV operator know whether someone is a combatant or civilian, but can also advise them in relation to other human-centred systems, like pilots of other UAVs on the same mission. Known as 'blue-on-blue' attacks, poor situation awareness can lead to pilots targeting a colleague after mistakenly thinking they are an enemy combatant. Using VoisJet would make this less likely, as the user interface for the UAV could be adjusted to make sure that the pilot is aware when a colleague is in range. As VoisJet can take into account clothing as signifying whether someone is a threat, then it would make it easier for a UAV pilot to avoid targeting an ally if they wear a specific sign to indicate their allegiances. The risk is that enemies could also use this information to either deceive as to their identity or to target those signifying their identity.

Allowing for Increased Autonomy in UAV Operations

Table 1 shows that where autonomy is added to a UAV that problems in acquisition and maintenance of situation awareness are multiplied when the human must "operate, monitor and supervise" those UAV systems under their control (Endsley et al., 2010). VoisJet would remove a lot of the cognitive load from achieving full situation awareness by making many of the calculations needed by a pilot so they can focus on using the data as opposed to having to generate it first.

The strategic and tactical component of VoisJet is based on the operational demands of those leading military operations, and can be customised to each set of circumstances. It has been argued that current concepts of operation require multiple operators for every unmanned system, a ratio will be difficult to sustain if the projected increase in use is to be believed and if the wider use of UAVs is to be realized, then a more desirable manning concept must be identified (Hughes, 2012). This is where VoisJet comes into its own – it is advanced situation awareness capability of giving pilots a set of options relating to the situation they are in, where they can exercise the judgement needed in standard aerial operations.

IMPLICATIONS AND FUTURE RESEARCH DIRECTIONS

The advantage of an invention like VoisJet is that it will make military engagements using UAVs more effective and practical, meaning military operations have more missions completed with fewest causalities. This could therefore mean that defence budgets could be cut as the money needed to train UAV pilots would be significantly reduced, as it may only be necessary to have a university education with minimal specialist training to take part in a military campaign. This would mean that military personnel would only be engaged as and when necessary as they would be working in normal jobs the rest of the time. This would have the effect of reducing the size of standing armies as the skill and training currently needed would be eliminated. The need for university graduates trained in military hacking will increase, as cyber-warfare and cyber-terrorism become the norm. The brute-force based on-the-ground operations associated with wars to date will be all-but eliminated. Ground-operations will be for specific purposes – such as interception of opposition personnel, and even then the technology that forms part of VoisJet can be used to improve the effectiveness of those engagements. Future research will have to look at how a military reduced in size and budget can improve other aspects of national security, such as expanded departments of homeland security, where those military personnel who work abroad during wartime, can be part of national security services during peacetime.

DISCUSSION

Unmanned aerial vehicles (UAVs), known in the vernacular as 'drones' are a military aircraft that looks certain to stay as a cost effective way to proceed with air-based military operations without risk to human life in terms of its pilots. Human life is at risk from such pilots however, as existing systems have major issues in terms of maintaining situation awareness. This chapter therefore has evaluated a system – called VoisJet – which can capture action-data from one person to provide information and advice to another. This can mean that UAV operators have a clear idea of what is going on around them so that they do not have to make complex calculations that increase cognitive load. VoisJet can at present cope with at least 250,000 action states, which goes way beyond what is consciously possible with the human mind, especially when operating in high pressured military engagements. The chapter has proposed means for resolving difficulties in UAV localisation as well as reducing poor sensory data, intermittent data, and time-lags in UAVs. The chapter has shown how VoisJet can allow for increased autonomy in UAV operations as well as assist with overcoming lack of support for multi-person operations in UAVs. Furthermore it shows that multitasking in UAV operations can be improved with VoisJet taking over many calculations and also how low-Level data overload and interface design in UAVs can be improved, as can the management of demanding tasks in complex environments.

It is clear that with the advancements that can be made to UAVs through VoisJet, then public policy makers will have to reconsider how they fund national security operations. The costs in terms of training military personnel would be reduced by VoisJet-based UAVs, as they could be operated by university graduates with minimal training. Thus the size of a nation's standing army could be reduced by such graduates holding other job roles during peacetime, and the more skilled personnel can be part of national security bodies, such as the FBI and CIA in the USA or the NCA in the UK.

REFERENCES

Ando, M., Oguchi, K., & Yamada, M. (2013). Exponential stabilization of quad-rotor helicopter based on exact linearization and disturbance rejection with on-line estimation of disturbances. In *2013 Proceedings of the SICE Annual Conference (SICE),* (pp. 1180-1185). Japan.

Bishop, J. (2013). The effect of deindividuation of the internet troller on criminal procedure implementation: An interview with a hater. *International Journal of Cyber Criminology*, *7*(1), 28–48.

Bishop, J. (2014). Representations of 'trolls' in mass media communication: A review of media-texts and moral panics relating to 'internet trolling'. *International Journal of Web Based Communities*, *10*(1), 7–24. doi:10.1504/IJWBC.2014.058384

Byman, D. (2013). Why drones work: The case for Washington's weapon of choice. *Foreign Affairs*, *92*(4), 32–43.

Chisholm, R. A., Cui, J., Lum, S. K., & Chen, B. M. (2013). UAV LiDAR for below-canopy forest surveys. *Journal of Unmanned Vehicle Systems*, *1*(01), 61–68. doi:10.1139/juvs-2013-0017

Coeckelbergh, M. (2013). Drones, information technology, and distance: Mapping the moral epistemology of remote fighting. *Ethics and Information Technology*, *15*(2), 87–98. doi:10.100710676-013-9313-6

Draper, M., Calhoun, G., Ruff, H., Williamson, D., & Barry, T. (2003). Manual versus speech input for unmanned aerial vehicle control station operations. In *Proceedings of the Human Factors and Ergonomics Society Annual Meeting, 47*(1) 109-113. Denver, CO. 10.1177/154193120304700123

Drezner, D. W. (2011). Does Obama have a grand strategy: Why we need doctrines in uncertain times. *Foreign Affairs*, *90*, 57.

Durham, L. M., Milenkovic, M., & Cayton, P. (2006). Platform support for autonomic computing: A research vehicle. Paper presented at the *IEEE International Conference on* Autonomic Computing, 2006 (ICAC'06), pp. 293-294. 10.1201/9781420009354.ch16

Endsley, M. R., Bolte, B., & Jones, D. G. (2010). *Designing for situation awareness: An approach to user-centered design*. London, UK: Taylor & Francis Group.

Gomez, E. A., & Chimento, J. (2011). Information access challenges: Data fission needs of the field expert. In *Proceedings of the 16th ICCRTS,* Québec, QC. pp. 151-160.

Hardaker, C. (2013). "Uh.... not to be nitpicky, but... the past tense of drag is dragged, not drug": An overview of trolling strategies. *Journal of Language Aggression and Conflict*, *1*(1), 57–85. doi:10.1075/jlac.1.1.04har

Hughes, T. C. (2012). Sources of adaptive capacity during multi-unmanned aerial vehicle operations (Doctoral thesis). Ohio State University, USA.

Jones, S. R. (2000). The roots of antisemitism. In I. Davies (Ed.), *Teaching the Holocaust: Educational dimensions, principles and practice* (pp. 11–24). London, UK: Bloomsbury Publishing.

Kean, T. H., & Hamilton, L. H. (2006). *Without precedent: The inside story of the 9/11 commission*. New York, NY: Knopf.

Kean, T. H., Hamilton, L. H., Ben-Veniste, R., Kerrey, B., Fielding, F. F., Lehman, J. F., & ... . (2004). *The 9/11 commission report*. New York, NY: W. W. Norton & Company.

McCosker, A. (2013). Trolling as provocation: YouTube's agonistic publics. *Convergence (London)*. . doi:doi: 1354856513501413

McLeod, T., Samson, C., Labrie, M., Shehata, K., Mah, J., Lai, P., ... Elder, J. H. (2013). Using video acquired from an unmanned aerial vehicle (UAV) to measure fracture orientation in an open-pit mine. *GEOMATICA*, *67*(3), 173–180. doi:10.5623/cig2013-036

Mobley, B. W. (2012). *Terrorism and counterintelligence: How terrorist groups elude detection*. New York, NY: Columbia University Press.

Rees, L. (2005). *Auschwitz: The Nazis & the final solution*. New York, NY: Random House.

Reynolds, M. A. (2013). Return of the Maxim gun? Technology and empire in an age of austerity. *Global Discourse*, *3*(1), 120–128. doi:10.1080/23269995.2013.807603

Rosenberg, L. B., & Stredney, D. (1996). A haptic interface for virtual simulation of endoscopic surgery. *Studies in Health Technology and Informatics*, *29*, 371–387. PMID:10172846

Yang, J., Chen, Y., Ammar, M., & Lee, C. (2005). Ferry replacement protocols in sparse manet message ferrying systems. In *Proceedings of the* IEEE *Wireless Communications and Networking Conference* (vol. 4, pp. 2038-2044). Atlanta, GA.

Yee, A. S. (2013). An investigation of the use of real-time image mosaicing for facilitating global spatial awareness in visual search (Doctoral thesis). University of Toronto, Canada.

KEY TERMS AND DEFINITIONS

Action-Data: Information picked up by a sensor that describes what is occurring in the situation being monitored.

Autonomy: The extent to which a person can make decisions for themselves in any given operational environment.

Hamas: A group of Islamic militants who although not Jihadists are committed to using force in order to protect the Arab Semite territories from being occupied by Zionists. It can be compared to the First and Second Continental Congresses, which was a military force that existed prior to US independence and then became the US Department of Justice.

Jihadist: A person who believes that the Holy Land and other Middle East nations occupied since the Crusades should be based wholly on Islamic law and uses force rather than democratic means to achieve that.

Sensory Data: Action-data that is made available to a person for processing using their own senses and judgement.

Situation Awareness: Situation awareness refers to the extent to which someone is able to know what is going on in any given situation and the options available.

Unmanned Aerial Vehicle: An unmanned aerial vehicle (UAV) is an aircraft that is piloted by a remote operator in order to achieve a particular objective, such as to eliminate enemy combatants.

VoisJet: VoisJet is a method for assisting UAV pilots in making decisions while involved in military operations through reducing the cognitive load and other factors that can lead to errors of judgement.

Zionist: A person who believes that the Holy Land should only consist only of Jewish Semites, and use undemocratic means to exclude Arab Semites and other religious/ethic groups from that territory.

This research was previously published in the Handbook of Research on Synthesizing Human Emotion in Intelligent Systems and Robotics edited by Jordi Vallverdú, pages 404-414, copyright year 2015 by Information Science Reference (an imprint of IGI Global).

Chapter 14
Chemical Plume Tracing and Mapping via Swarm Robots

Wei Li
California State University, Bakersfield, USA

Yu Tian
Shenyang Institute of Automation, China

ABSTRACT

This chapter addresses the key issues of chemical plume mapping and tracing via swarm robots. First, the authors present the models of turbulent odor plumes with both non-buoyant and buoyant features, which can efficiently evaluate strategies for tracing plumes, identifying their sources in two or three-dimensions. Second, the authors use the Monte Carlo technique to optimize moth-inspired plume tracing via swarm robots under formation control, which includes a leader to perform plume tracing maneuvers and non-leaders to follow the leader during plume tracing missions. Third, the authors introduce a variety of robot-based plume tracers, including ground-based robots, autonomous underwater vehicles, or unmanned aerial vehicles. Finally, the authors prospect the further research in this area, e.g., applying swarm robots to detect oil or gas leak, or to investigate subsea chemical pollution and greenhouse gases.

INTRODUCTION

A potential application of a swarm robot system is to search for environmentally interesting phenomena, unexploded ordnance, undersea wreckage, and sources of hazardous chemicals or pollutants. Navigating the swarm robot system in response to real-time sensor information to find the plume, trace the plume towards its source, and identify the source location, is referred to as Chemical Plume Tracing (CPT). Factors that complicate CPT include the chemical source concentration being unknown, the advection distance of any detected chemical being unknown, significant filament intermittency and meander of a chemical plume developed in turbulent fluid flow environments, and the flow variation with both location and time.

DOI: 10.4018/978-1-5225-8365-3.ch014

Since last decades, there has been a growing interest to apply robot-based chemical plume tracers in environmental monitoring (Dunbabin & Marques, 2012; Oyekan & Hu, 2014) and in searching for sources of hazardous chemicals or pollutants (Cowen and Ward, 2002). Ishida et al. (1996) used an array of sensors to track the plume by estimating the three-dimensional direction toward the odor source. Russell (2001) included robotic implementation of algorithms that estimate statistics of the plume such as the plume centroid. Marques et al. (2002) performed plume-tracing tests using mobile robots in laboratory environments. Li et al. (2001) developed, evaluated and optimized both passive and active plume tracing strategies inspired by moth behavior. The moth-inspired plume tracing strategies were implemented on a REMUS underwater vehicle for the in-water test runs in November and April 2002 at the San Clemente Island of California and in June 2003 in Duck, North Carolina (Farrell et al., 2005; Li et al., 2006). The field experiments successfully demonstrated tracking of chemical plumes over 100 m and source identification on the order of tens of meters in the near shore, oceanic fluid flow environments, where plumes were developed under turbulence, tides and waves. The most recent CPT in-water test run via an autonomous underwater vehicle at Dalian Bay in China (Tian et al., 2014) also validated effectiveness of the moth-inspired CPT strategies.

Animal swarms are typical distributed systems with flexibility and autonomy for homing, foraging or mate-seeking. The swarm robotics inspired from nature is a combination of swarm intelligence and robotics showing a great potential in several aspects (Beni, 2005; Liu et al., 2010; Tan and Zheng, 2013; Brambilla et al, 2013). For example, a multi-robot system inspired by animal swarms suits to CPT missions very well. Hayes et al. (2002) used multiple robots to improve a Spiral Surge Algorithm in the field of swarm intelligence in order to find a plume and to trace the plume to its source location. Zarzhitsky et al. (2004) presented an approach to CPT based on the physics of fluid dynamics, upon source localization using the divergence theorem of vector calculus. Krishnanand and Ghose (2009) addressed the problem of multiple signal source localization via robotic swarms. Kang and Li (2012) expanded moth-inspired plume tracing via a single robot to multiple robots, which included a mechanism determining a leader vehicle to perform moth-inspired plume-tracing maneuvers and a formation algorithm controlling non-leaders to follow the leader during plume-tracing missions. Marjovi and Marques (2014) presented an analytical approach to the problem of odor plume finding by a swarm robot system.

The chapter starts with the models to achieve short time-scale structure of odor plumes. These computational plume models suit to evaluate algorithms for tracing chemical plumes and identifying the chemical sources via a single robot or multiple robots in two or three dimensions. The chapter discusses the Monte Carlo technique for optimizing moth-inspired CPT via multiple autonomous robots with a leader to perform moth-inspired plume-tracing maneuvers and non-leaders to follow the leader during CPT missions. The results demonstrate CPT performance achieved by swarm robots, which automatically switch their roles superior to the single robot. The chapter presents a variety of robot-based chemical plume tracers, including ground-based DaNI robots, autonomous underwater vehicles (AUVs), or unmanned aerial vehicles (UAVs). The chapter also prospects the further research, e.g., applying swarm-robot-based plume tracers into the oil and gas industry.

Background

Olfactory-based mechanisms have been hypothesized for a variety of biological swarm behaviors (Vickers, 2000; Zimmer and Butman, 2000): homing by Pacific salmon (Hassler and Scholz, 1983), foraging by Antarctic procellariiform seabirds (Nevitt, 2000), foraging by lobsters (Basil and Atema, 1994; Devine and Atema, 1982), foraging by blue crabs (Weissburg and Zimmer-Faust, 1994), and mate-seeking and foraging by insects (Cardé, 1996; Cardé and Mafra-Neto, 1996; Elkinton et al., 1987). Typically, olfactory-based mechanisms proposed for biological entities combine a large-scale orientation behavior based in part on olfaction with a multisensory local search in the vicinity of the source. Long-range search is documented in Antarctic procellariiform seabirds, possibly by olfactory landmark navigation, over 1000 km (Nevitt, 2000). Plume tracing by moths at ranges of 100-1000 m is described on p. 276 in Cardé and Mafra-Neto, (1996): Rau and Rau (1929) provided perhaps the most extensively documented case of long-distance mate finding by luring male silkworm moths released several kilometers from caged virgin females held on a second-story roof of their home in St. Lous, Missouri (Rau and Rau, 1929). C.H. Fernald also demonstrated evident location of distant females in the gypsy moth (Lymantria dispar) in New England: a few males released several kilometers away located females after several hours (Forbush and Fernald, 1896). Unfortunately, recapture of released males "cannot disentangle the contributions of ranging flight which might bring a male into contact with a pheromone plume and of pheromone-dictated maneuvers which close contact between pheromone sender and receiver" (Cardé and Mafra-Neto, 1996).

The CPT methods can be roughly classified into: anemotaxis (Belanger et al., 1996; Hayes et al., 2002), chemotaxis (Grasso et al., 1999), chemo-anemotaxis (Li et al., 2001; Grasso and Atema, 2002; Wu et al., 2013), fluxotaxis (Zarzhitsky et al., 2005), and infotaxis (Vergassola et al., 2006; Voges et al., 2014). The fact that various biological entities use olfactory-based search with high degrees of success has prompted research into the design of autonomous vehicles capable of olfactory based search and chemical plume tracing. Such autonomous vehicle capabilities have applicability in searching for environmentally interesting phenomena, hazardous chemicals, and pollutants. In addition to vehicle development, sensor development, and analysis of the fluid environment, design and optimization of the chemical plume tracing strategies are of critical importance to the solution of this problem.

DARPA/ONR sponsored an interdisciplinary team to investigate biologically inspired chemical plume tracing strategies under the CPT program (Cowen and Ward, 2002). Grasso et al. (2000) implemented biomimetic strategies on their robot lobster to analyze chemotaxis-based algorithms and to evaluate biomimetic strategies. Grasso and Atema (2002) employed a single sensor or two sensors detecting fluorescence to compare the plume-tracing performance of three plume-tracing strategy variations. Other studies (Liao and Cowen, 2002; Weissburg et al., 2002) proposed sensor array-based strategies and suggested that search strategies based on following the 'edge' of a plume as opposed to the centerline are robust. Li et al. (2001) proposed both the passive and active plume tracing strategies inspired by moth behavior. The moth-inspired plume tracing strategies were validated through the in-water test runs in November and April 2002 at the San Clemente Island of California and in June 2003 in Duck, North Carolina (Li et al., 2006; Farrell et al., 2005).

Chemical plume tracing and mapping via swarm robots should be able to improve the CPT performance in a large-scale search area. For example, the studies (Kang & Li, 2012) demonstrated that the tacking of a plume via a fleet of robots is superior to that via a single robot. In addition, a swarm robot system with the distributed structure is efficient and reliable to trace multiple chemical plumes and to build their maps in natural environments. For example, the in-water experiments of tracing four chemical plumes and identifying their source locations in an operation area of 367×1094 m (greater than 60 football fields) under the near-shore ocean environments, where the plumes were developed under turbulence, tides and waves, lasted several hours (Farrell et al., 2005). The performances of these CPT missions must be significantly improved if swarm robot-based plume tracers could be employed.

PLUME MODELS, ALGORITHM DESIGN, AND EVALUATION

In order to solve the problem of chemical plume mapping and tracing via swarm robots, the key issues including sensor development, analysis of the fluid environment, plume tracer design, must be addressed. This chapter discusses the computational models for simulating chemical plumes, the strategies for evaluating CPT algorithms, and the techniques for building robot-based plume tracers by integrating a variety of sensors.

Plume Models

Evaluation of CPT strategies requires hardware and software evaluation platforms. During the initial design stages, software evaluation is preferred, since such tools allow competing strategies to be evaluated under identical conditions for various environmental scenarios. The tradeoffs between field-testing, wind tunnel testing, and simulation testing are discussed in Belanger and Willis (1996). To calculate reliable performance statistics, numerous batches of such evaluations are required. Therefore, the computational efficiency of the software simulation tools is a key concern. To ensure that strategies that perform well in simulation will also perform well in hardware experiments, the simulation must contain the key features that complicate the CPT problem. The authors herein discuss a plume simulation implementation that addresses the tradeoffs between computational efficiency and inclusion of realistic complicating features.

Computational plume models can represent concentration by its mean or probability density function. For example, dispersion in a turbulent medium is frequently represented as a Gaussian distribution such as (Gifford, 1960; Gifford, 1968; Sutton, 1947; Sutton, 1953)

$$C(x,y,z)=\frac{Q}{2\pi S_y S_z \bar{u}}\exp\left(-\left(\frac{y^2}{2S_y^2}+\frac{z^2}{2S_z^2}\right)\right), \tag{1}$$

where $\bar{C}(x,y,z)$ represents the average concentration profile as a function of position. In this expression, and throughout this chapter, x, y, and z denote the downwind, crosswind, and vertical position coordinates relative to the odor source with x positive along the mean wind direction. S_y and S_z are the standard deviations of the time-averaged plume concentration in the crosswind and vertical directions. The application of Gaussian plume concentration models is well established and the results they produce are well understood. Probability density plume models, however, do not set out to provide information related to the short time-scale signature of the concentration, the instantaneous peak concentration, or the intermittent nature of the plume. Although the fine-scale characteristics of a plume are of little interest in many cases, they are of considerable importance in CPT studies.

To generate realistic simulated plumes for CPT studies as shown in Figure 1, the following three characteristic structural features are of primary concern. First, the sensed plume at a fixed location should have an intermittent internal structure that closely duplicates experimental fine time-scale observations. Second, the plume shape that results from a semi-continuous release of odor should be sinuous and time varying. Third, the plume shape and fluid flow history should be coherent.

Figure 1. The filament-based simulated plume exhibits significant puff intermittency and meander

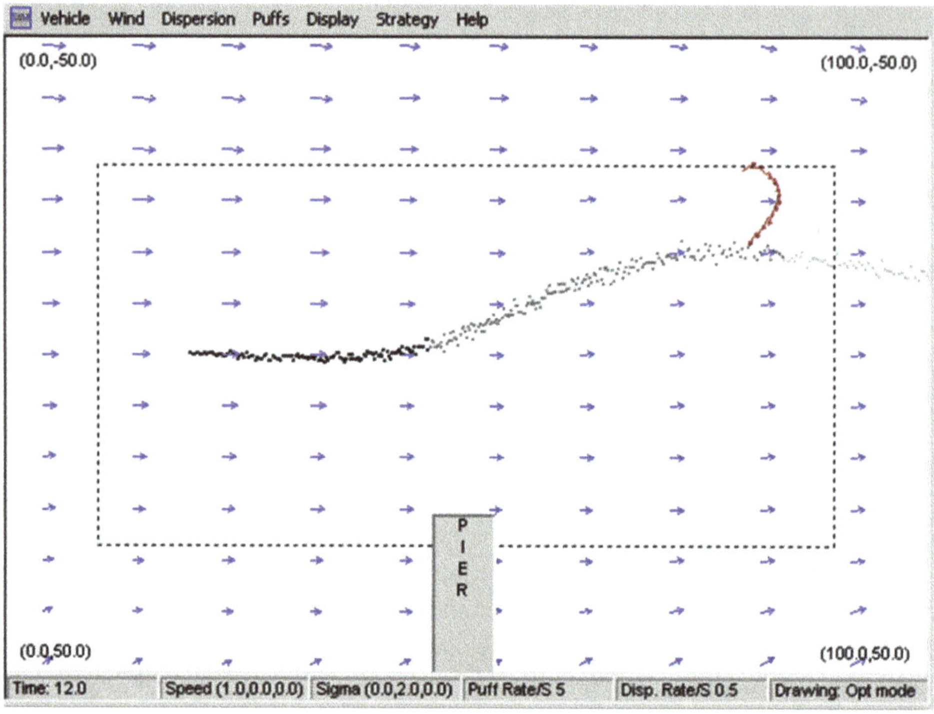

The first characteristic feature is achieved herein by representing the odor plume as a sequence of puffs (Belanger and Willis, 1996; Gifford, 1959). Puffs are released sequentially at the source location. Each puff is composed of *n* filaments. After release, the filament center location is determined by integration of the wind velocity: $\dot{\mathbf{p}}_i = \mathbf{v}(\mathbf{p}_i)$, where $\mathbf{p}_i$ and $\mathbf{v}(\mathbf{p}_i)$ are the 3 dimensional position of the *i*-th filament and wind velocity at the location of the *i*-th filament. To achieve the second characteristic feature, the model incorporates a dynamic model of the wind. This model allows the wind, which transports the filaments, to be a continuous but varying function of position and time. Changes in wind direction cause the plume to meander (form a snake-like path when viewed from above). As the wind velocity changes, the instantaneous wind direction at given positions within the plume's boundaries sometimes will not point either toward the odor's source nor be coincident with the direction of the plume's centerline.

A chemical released at a source location will be manipulated by turbulent and molecular diffusion while transported advectively by the wind. As described in Aylor et al. (1976), the odor dispersion process is dominated by turbulent dispersion. Turbulent dispersion involves a wide range of length (or eddy) scales. Eddies larger than the scale of a section of the plume (i.e., a puff) transport the puff as a whole, causing the ensemble of puffs to appear as a sinuous plume. Eddies smaller than the puff mix the components of the puff causing little puff motion or growth. Eddies on the order of the puff size cause significant growth/distortion of the puff and motion of individual filaments relative to the instantaneous plume centerline.

To develop a corresponding plume simulation model, the velocity vector is decomposed into three components: $\mathbf{v}_d$, $\mathbf{v}_m$, and $\mathbf{v}_a$. A distinct process will implement each component. The decomposition of the velocity spectrum is motivated by the numerical implementation, but can be interpreted theoretically in terms of the eddy scales described above. To clarify this correspondence, it is first necessary to state that the puffs, for simulation, will be further decomposed into parcels referred to herein as filaments. The effect of the smallest eddies (i.e., slow growth of the filaments) of the wind fluid flow process (modeled by $\mathbf{v}_d$) will be implemented as an increase in filament size and change in shape. The term $\mathbf{v}_a$ represents the portion of the wind process with characteristic length much larger than the filaments. This portion of the wind process transports each filament as a body; therefore, the term $\mathbf{v}_a$ represents advection. The advection portion of the velocity is represented as a continuous (in time and space), but temporally and spatially varying function, so that a sequence of filaments released at the source will result in a sinuous trail of filaments leaving the source. The term $\mathbf{v}_m$ represents the intermediate range of scales that transports (i.e., stirs) the filaments within the body of the plume.

Although the position of individual molecules cannot be conveniently simulated, the idea is useful to consider for the purpose of defining the simulation. Therefore, the position of a single molecule is given by $\dot{\mathbf{p}}_m = \mathbf{v}_d + \mathbf{v}_m + \mathbf{v}_a$. Because the intermediate and advective terms affect all molecules within a filament similarly, the model can be decomposed into two processes–the transport of the filament and the changing shape of the filament. The position of the i-th filament will be represented by $\mathbf{p}_i$. The changing position of the i-th filament is represented by $\dot{\boldsymbol{p}}_i = \mathbf{v}_{mi} + \mathbf{v}_a$, where the implementation of the $\mathbf{v}_{mi}$ and $\mathbf{v}_a$ processes is discussed in next paragraphs. Physically, the growing and changing shape of the filament is a very interesting process dominated by turbulent diffusion. However, numerical simulation of diffusion processes is a complex and time-consuming task. Therefore, at this level some compromise is required to achieve reasonable simulation times while still achieving the objectives of the simulation model. The approach herein is to use a template filament shape that can change size and orientation. The simulation results are not sensitive to the choice of template shape, as long as the change in shape is physically reasonable.

First is the model of $\mathbf{v}_a = \left(\overline{u}, \overline{v}, \overline{w}\right)$ suitable for simulation implementation. The advection model is implemented in two dimensions (i.e., x and y) with the assumption that for the area of interest $\mathbf{v}_a = (\overline{u}, \overline{v}, \overline{w})$ is independent of z. This assumption is motivated by the assumption that the plume tracing will occur at a nearly constant altitude within a few meters of the ocean floor or earth surface. Within the atmospheric surface layer, it is assumed that compared with turbulent forces, the following forces on an air parcel are negligible: Coriolis, geostrophic winds, and molecular viscosity. It is also assumed that time-averaged pressure terms can be neglected, so that the time-mean motion equations are

$$\begin{aligned}\frac{\partial \overline{u}}{\partial t} = & -\overline{u}\frac{\partial \overline{u}}{\partial x} - \overline{v}\frac{\partial \overline{u}}{\partial y} - \overline{w}\frac{\partial \overline{u}}{\partial z} \\ & -\frac{\partial \overline{u'u'}}{\partial x} - \frac{\partial \overline{u'v'}}{\partial y} - \frac{\partial \overline{u'w'}}{\partial z}\end{aligned} \tag{2}$$

$$\frac{\partial \bar{v}}{\partial t} = -\bar{u}\frac{\partial \bar{v}}{\partial x} - \bar{v}\frac{\partial \bar{v}}{\partial y} - \bar{w}\frac{\partial \bar{v}}{\partial z} - \frac{\partial \overline{v'u'}}{\partial x} - \frac{\partial \overline{v'v'}}{\partial y} - \frac{\partial \overline{v'w'}}{\partial z}, \tag{3}$$

with the continuity equation

$$0 = \frac{\partial \bar{u}}{\partial x} + \frac{\partial \bar{v}}{\partial y} + \frac{\partial \bar{w}}{\partial z}, \tag{4}$$

where $u(t) = \bar{u}(t) + u'(t)$. To conserve computational resources, the simplest form of *K*-closure method (Gifford, 1959; Pielke, 1994) was used to numerically solve these equations at a grid of points on the region of interest for a specific set of time-varying boundary conditions (Farrell et al., 2012). The term $\mathbf{v}_a$ derived in this section has the sole function of transporting the fine-scale (filaments) of the plume. Therefore, the grid point separation Δ is defined to be larger than the scale of the filaments. The advective wind velocity at locations between the grid points is determined by interpolation based on the advective wind at the adjacent grid points. This approach yields a continuous, spatially varying, and time varying model of the advection of the plume elements with a reasonable characterization of the physics. Therefore, the wind velocities at distinct but nearby points are correlated, but spatially varying. The above equations were solved numerically for boundary conditions generated by a mean flow plus a colored noise process. When a sequence of filaments is released at a fixed location, the advection terms $(\bar{u}, \bar{v}, \bar{w})$ generate a meandering trail of filaments.

Second is the model for $\mathbf{v}_m$. The model for $\mathbf{v}_m$ is the scale of the flow between Δ and the filament size. Assuming that Δ is specified judiciously, this scale of the flow distributes the filaments about the plume centerline. For computational reasons, this scale of the flow is not modeled through finite difference methods. Instead, the velocity of the *i*-th filament relative to the centerline $\mathbf{v}_{m_i}$ is modeled as a random process implemented in state-space notation (Luenberger, 1979) as

$$\dot{\omega} = A\omega + B\varphi \;\; \mathbf{v}_{m_i} = C\omega + D\varphi, \tag{5}$$

where $\omega \in \Re^n$; *A*, *B*, *C*, and *D* are appropriately sized matrices; and φ is a white noise process with power spectral density σ_φ^2. The transfer function from φ to $\mathbf{v}_{m_i}$ is

$$\frac{\mathbf{v}_{m_i}}{\varphi} = H(s) = C(sI - A)^{-1}B + D, \tag{6}$$

where *s* is the Laplace variable. Therefore, $\mathbf{v}_{m_i}$ is a colored noise process with spectral density defined by

$$PSD_{\varphi_m}\left(s\right) = H^*\left(s\right)H\left(s\right)\sigma_{\varphi}^2, \tag{7}$$

where $H^*\left(s\right)$ is the complex conjugate transpose of H.

When summed, $\mathbf{v}_{m_i}$ and $\mathbf{v}_a$ describe the velocity of the filament center as the filament is bodily transported over the field of interest. Given a large number of filaments, the overall instantaneous concentration at $\mathbf{x}=(x, y, z)$ due to all the filaments is just the sum of the concentrations at that location contributed by each filament:

$$C\left(\mathbf{x},t\right) = \sum_{i=1}^{N} C_i(\mathbf{x},t)\frac{\text{molecules}}{\text{cm}^3}, \tag{8}$$

where N is the number of filaments currently being simulated (Farrell et al., 2012).

Sutton et al. (2008) used object-oriented programming techniques to upgrade the above models by defining the advection fluid flow model class, plume model classes, sensor classes, plume tracing strategy classes, and vehicle model classes. The upgraded models allow efficiently evaluating strategies for tracing multiple plumes, identifying their sources, and even in three dimensions. The visual effect on plumes is improved by using Open Graphics Library (OpenGL).

Tian et al. (2013) expanded the two-dimensional models into the three-dimensional ones with buoyant stem, as shown in Figure 2. The three-dimensional plume models consider the vertical advection of each plume filament caused by the plume buoyancy and initial momentum except for the filament horizontal advection. The vertical component $\overline{w}_b$ for each plume filament at time t is calculated by (Anfossi, 1985; Anfossi et al., 1993):

$$\overline{w}_b\left(t\right) = \frac{H\left(\cdot,t+\Delta t\right) - H\left(\cdot,t\right)}{\Delta t}, \tag{9}$$

where $H(\bullet, t)$ is a model that describes the plume centerline rise with time, t is the time since the plume filament has been discharged from the source, and Δt is the simulation time step.

Evaluation of Moth-Inspired Plume Tracing Strategies

The authors use the plume model developed above to evaluate the moth-inspired CPT algorithms since the location of pheromone-emitting females by flying male moths is considered to be a remarkable case. The chemical plume tracing behaviors exhibited by moths could be summarized as follows. When a moth detects pheromone, it tries to maintain contact with the plume and to move upwind toward the source location, with the maneuver being a short sprint predominantly in the upwind direction. Repeated pheromone encounters result in the moth progressively approaching the chemical source. When a moth has not detected pheromone for a sufficiently long period of time, it ceases upwind movement and performs progressively widening crosswind excursions, termed "casting". In this case, the moth appears to be searching for pheromone near the position where it was last detected. This reacquiring behavior

Figure 2. The plume simulation model includes buoyant stem

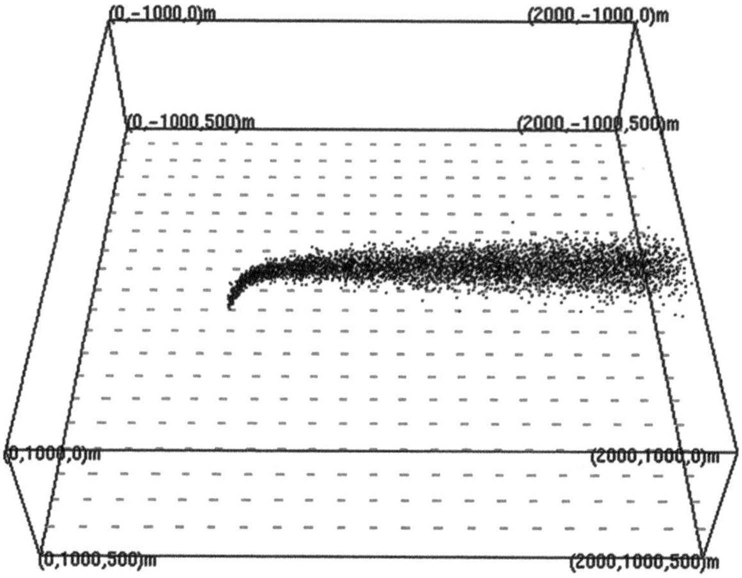

can continue for several seconds until either chemical is again detected or the moth behavior changes. If the moth fails to detect pheromone again, it may return to the Find-Plume behavior used initially for location of the plume.

Inspired by the moth plume-tracing behavior, Li et al. (2001) developed a behavior-based planning strategy for tracking of turbulent chemical plumes. The strategy takes the information from the chemical sensor and the flow sensor as input, and outputs the commanded robot heading ψ_c and speed υ_c. Considering the full spectrum of field behaviors for conducting chemical plume tracing missions in real-world environments, the strategy consists of four fundamental behaviors coordinated in a subsumption architecture: finding the plume (Find-Plume), maintaining the plume (Maintain-Plume), re-acquiring the plume (Reacquire-Plume), and identifying the source location (Declare-Source), as shown in Figure 3.

The inhibition approach is used to arbitrate between the commands from the various behaviors to determine the commanded output, as shown by the circles labeled with an '**S**' in Figure 3. The output of the subsumption module is connected to the top input of another subsumption module. If the top input to the subsumption module has any value, then the output of the subsumption module assumes that value; otherwise the output of the subsumption module takes the value of its left input. For the CPT strategy, the Declare-Source behavior has the highest priority. Once that behavior has declared the source location, it defines its commanded outputs to drive the vehicle to a home location. The Declare-source output commands block the commands from any other behaviors. The Maintain-Plume behavior includes Track-In and Track-Out activities. Track-In tries to make rapid progress toward the source while chemical is being detected. Track-Out manipulates the tracer heading relative to upflow in the time immediately following the loss of chemical detection to try to rapidly re-contact the plume. These activities have the second and third highest priorities. For these activities to be controlling the vehicle, the tracer will either be detecting or have recently detected the plume. The Find-Plume behavior is designed to dominantly implement a cross-flow search for the entire predefined operational area without any assumptions about the location of the plume source. The subsumption architecture also embeds Moving to Field for sending

Figure 3. Subsumption architecture for chemical plume tracing

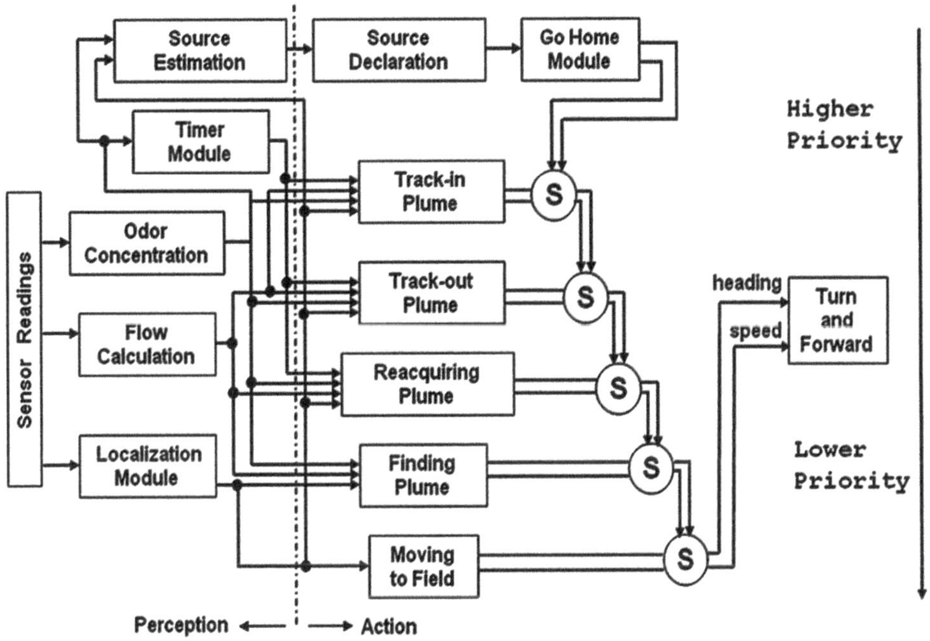

the plume tracer before staring a CPT mission and Go Home Module for driving the plume tracer home after completing the CPT mission.

Determining the optimal direction to search for a chemical plume in a time-varying flow is addressed theoretically in Dusenbery (1989) and Sabelis and Schippers (1984). The commanded heading for Find-Plume is defined as $\psi_c = f_{dir}(t, x, y) + \mathrm{sign}(\eta)\Delta\psi(t)$, which is an offset to the computed flow direction $f_{dir}(t, x, y)$ at time t and the tracer location (x, y). $\eta=0.5(Y_{max} + Y_{min}) - y$ and its sign will be ± 1, where $[X_{min}, X_{max}] \times [Y_{min}, Y_{max}]$ specifies the plume-tracing operational area. The variable $\Delta\psi(t)$ is an offset angle used for up-flow or down-flow search, and can only take on one of the two constant values $\Delta\psi(t)_{up}$ (for up-flow search) or $\Delta\psi(t)_{down}$ (for down-flow search). How to choose an initial direction to start the Find-Plume behavior is discussed in Li et al. (2006).

The Maintain-Plume and Reacquire-Plume are inspired by the behaviors hypothesized from observations of pheromone plume-tracing moths. The Maintain-Plume replicates the behavior of moths to track a plume and includes Track-In and Track-Out activities. Track-In tries to make rapid progress toward the source while chemical is being detected. Track-Out manipulates the tracer to rapidly re-contact the plume immediately following the loss of chemical detection, as shown in Figure 4. Track-In and Track-Out are described by:

$$\begin{cases} \psi_c = f_{\mathrm{dir}}\left(t, x, y\right) \\ \quad + 180^\circ + \Delta\psi(t)_{\mathrm{Track-In}} \qquad t \in T_{\mathrm{Track-In}} \\ v_c = v_{\mathrm{Track-In}} \end{cases} \tag{10}$$

$$\begin{cases} \psi_c = f_{\text{dir}}\left(t, x, y\right) \\ \quad + 180^\circ + \Delta\psi(t)_{\text{Track-Out}} & t \in T_{\text{Track-Out}}, \\ \upsilon_c = \upsilon_{\text{Track-Out}} \end{cases} \tag{11}$$

where $\Delta\psi(t)_{\text{Track-In}}$ and $\Delta\psi(t)_{\text{Track-Out}}$ are the offset angles for Track-In and Track-Out, $\upsilon_{\text{Track-In}}$ and $\upsilon_{\text{Track-Out}}$, and $T_{\text{Track-In}}$ and $T_{\text{Track-Out}}$ are commanded tracer speeds and durations for Track-In and Track-Out, respectively.

If Track-Out fails to detect chemical plumes within a given period of time, then the tracer switches to Reacquire-Plume behavior which maneuvers the tracer to search for the plume in the vicinity of the most recent chemical detected location, as a moth does with the casting behavior. A cloverleaf shaped trajectory was used to implement the Reacquire-Plume behavior to cast for the lost chemical plume. The commanded tracer heading is calculated using the line of sight guidance method: $\psi_c = \text{actan2}(y_i - y, x_i - x)$, where (x_i, y_i) is the sub-goal located on the cloverleaf, whose center is located on the most recent chemical detected location (x_{last}, y_{last}).

The Declare-Source behavior implements an algorithm SIZ_F to identify the plume source location, which is derived from the two moth-inspired behaviors: Maintain-Plume and Reacquire-Plume. Maintain-Plume is broken down into Track-In and Track-Out activities due to intermittency of a chemical plume transported in a fluid flow environment. A chemical tracer alternatively utilizes Maintain-Plume and Reacquire-Plume in making progress towards the source location in the up-flow direction. In a typical scenario of plume tracing, the plume tracer activates Track-In once it detects the chemical, e.g., the activities during ΔT_1 and ΔT_3 in Figure 4. It continues Track-Out when it loses contact with the chemical within λ seconds, e.g., the activity during ΔT_2 in Figure 4. After λ seconds, it switches to Reacquire-Plume for casting the plume, e.g., the activity after ΔT_4 in Figure 4.

Figure 4. Source identification algorithms are derived from moth-inspired plume tracing strategies. A plume tracer records (x_{last}, y_{last}) as a last chemical detection point (LCDP) if it cannot re-catch the plume within λ seconds during Track-Out activity

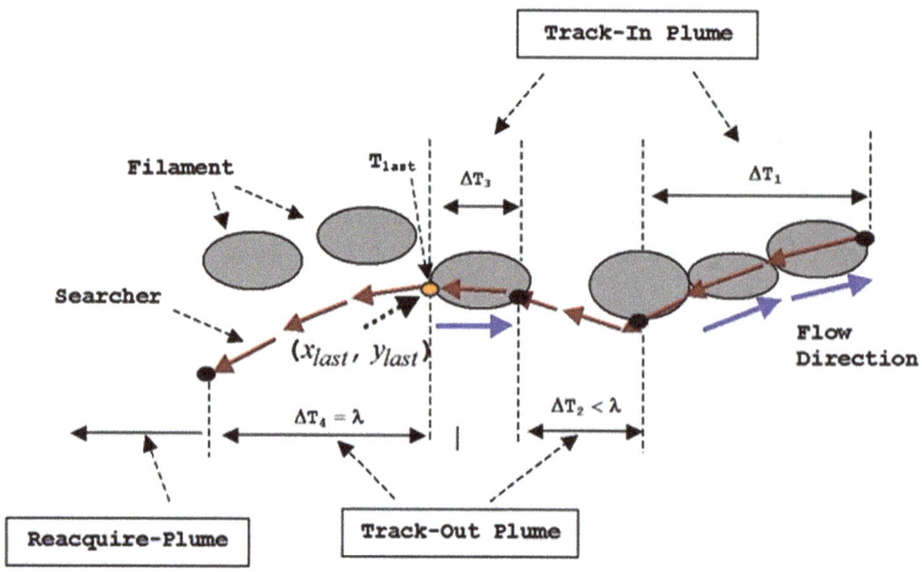

Figure 5. The plume tracer generates most of the LCDPs in the vicinity of the source location, when it overshoots the source and reacquires the lost plume

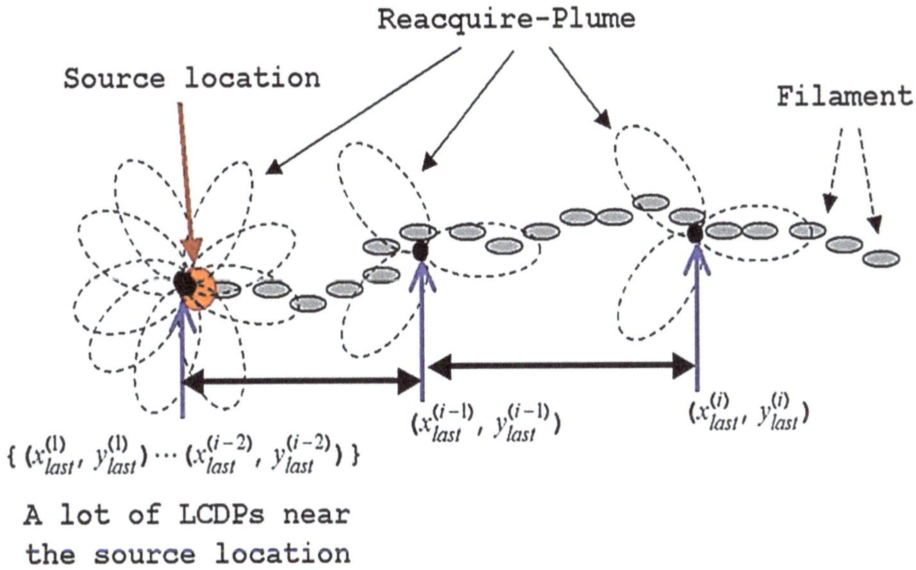

A chemical detection point at which the robot loses contact with the chemical plume for λ seconds is defined as a LCDP, e.g., point (x_{last}, y_{last}) at T_{last} in Figure 4. For identifying the source location, the coordinates of (x_{last}, y_{last}) are specified in a coordinate system with the origin defined by the center of an operation area. The Reacquire-Plume behavior is implemented by a Cloverleaf trajectory with the center (x_{last}, y_{last}), as shown in Figure 5. Note that one leaf is aligned with the down-flow direction for the plume tracer to rediscover the chemical when it has passed the source location. During a Reacquire-Plume activity, the plume tracer either detects the chemical or completes the Cloverleaf trajectory N_{re} times (N_{re} =2 or 3 for the in-water tests (Li et al., 2006)). The length of each leaf of the Cloverleaf trajectory is constrained to be larger than the robot turning radius. If N_{re} repetitions are completed without a chemical detection, the robot reverts to Find-Plume. Here, a LCDP node is defined by:

```
struct LCDP_Node
{
double T_last, x_last, y_last;
double conc, f_dir, f_mag;
double x_flow, y_flow;
};
```

where T_{last} is the time when the LCDP is detected, (x_{last}, y_{last}) are the coordinates of the plume tracer at T_{last}, *conc* is the chemical concentration at (x_{last}, y_{last}) and T_{last}, (*fdir, fmag*) are the flow direction and magnitude at (x_{last}, y_{last}) and T_{last}, and (x_{flow}, y_{flow}) are the coordinates in a new coordinate system defined according to the current flow direction. For convenience, (x_{last}, y_{last}) is also used to represent the LCDP in the following discussion.

The robot detects a new LCDP and inserts its node into the priority queue when it switches its behaviors from Maintain-Plume to Reacquire-Plume. The queue sorts the LCDP nodes in a new coordinate system, defined in order of the current up-flow direction, f_{dir}+180°. Its x axis is aligned with the f_{dir} direction, and its origin is located at (x_{last}, y_{last}). The algorithm maps each LCDP in the queue into the new coordinate system by:

$$\begin{bmatrix} x_{\text{flow}} \\ y_{\text{flow}} \end{bmatrix} = \begin{bmatrix} \cos(f_{\text{dir}} + 180^{\circ}) - \sin(f_{\text{dir}} + 180^{\circ}) \\ \sin(f_{\text{dir}} + 180^{\circ}) + \cos(f_{\text{dir}} + 180^{\circ}) \end{bmatrix} \begin{bmatrix} x_{\text{last}} \\ y_{\text{last}} \end{bmatrix} \tag{12}$$

The x_{flow} components determine the LCDP nodes' priorities according to the current up-flow direction. The smallest x_{flow} has the highest priority. The more LCDPs the priority queue accumulates, the more information about the source location the robot gathers. LCDPs are used to develop the source identification algorithms SIZ_F.

The SIZ_F algorithm maintains all LCDPs in the order of the current up-flow direction using the priority queue. SIZ_F holds a constant size, ε_F, and makes the source identification by the following iterative construct: First, SIZ_F calculates $\left(x_{\text{last}}^{(m)}, y_{\text{last}}^{(m)}\right)$ of all the LCDPs; Second, SIZ_F find the point, $p_{\max}$, with the largest distance to $\left(x_{\text{last}}^{(m)}, y_{\text{last}}^{(m)}\right)$:

$$D_{\max} = \max\left\{\sqrt{\left(x_{\text{last}}^{f(i)} - x_{\text{last}}^{(m)}\right)^2 + \left(y_{\text{last}}^{f(i)} - y_{\text{last}}^{(m)}\right)^2}\right\} \qquad \left(i = 1, 2, \cdots N_{\text{all}}\right) \tag{13}$$

from the priority queue, where a superscript f indicates that the LCDPs are sorted in the order of the most recent up-flow direction, and N_{all} is the total number of LCDPs detected during a CPT mission. If D_{max} is greater than, ε_F, SIZ_F removes the LCDP with p_{max} from the set of LCDPs. These calculations repeat until all remaining LCDPs are close enough to $\left(x_{\text{last}}^{(m)}, y_{\text{last}}^{(m)}\right)$, i.e., they all are located inside SIZ_F. If the number of the remaining LCDPs is greater than N_{min}, SIZ_F identifies its most up-flow LCDP as the plume source. The SIZ_F algorithm has three parameters: the SIZ_F size, ε_F, the initial value, N_{ini}, and the integer, N_{min}, which indicates the minimum number of LCDPs remaining inside SIZ_F for the source identification. The SIZ_F algorithm also has two the adjustable parameters ε_F and N_{min}. The parameter, N_{ini}, defined in the algorithms works as a filter to block some invalid LCPDs only when N_{dec} or N_{min} is very small.

For CPT missions via swarm robots, it is essential to re-define the Find-Plume, Maintain-Plume, Reacquire-Plume, and Declare-Source behaviors. The Find-Plume behavior controls the multiple vehicles to search for a plume in the entire operational area using cross-flow trajectories without any assumptions about the location of the chemical plume and its source. The Reacquire-Plume behavior controls the multiple vehicles to follow a cloverleaf-shaped trajectory or its variation to cast the lost plume. Obviously, the Find-Plume or the Reacquire-Plume maneuvers using multiple robots increase the chance of finding or casting the plume. The multiple vehicles system activates its Maintain-Plume behavior once one of the vehicles detects a chemical plume. During Maintain-Plume maneuvers, the leader vehicle utilizes the Track-In and Track-Out activities to move up-flow toward the source location, whereas the non-leader vehicles serve as guarders to follow the leader. For chemical source declaration via the multiple robots, the inputs to the SIZ_F algorithm are the LCDPs collected by all the robots, instead of a single one. The robots' roles are assigned at activating the Maintain-Plume behavior. The source identification behavior does not change their roles and only the leader robot declares the source location. Its algorithm is in detail described in Li (2010).

The proposed strategy for the plume tracing via swarm robots inherits the demonstrated advantage of the moth-inspired plume tracing. Being superior to the single robot system, the swarm robots system determines dynamically its leader during the course of CPT missions according to the robots' statuses and environment information so that each robot in the swarm robots system potentially acts as a leader. Through communication between the vehicles, the leader has to direct the non-leaders to find the plume, trace the plume, cast the plume, and declare the plume source location. Figure 6 shows the behavior switching diagram-based coordination mechanism for the swarm robots system, which is designed by adding a Formation-Keeping behavior to the diagram for a single robot system (Kang and Li, 2012).

The symbol *d* denotes the behavior switch that is on when chemical is detected, s denotes the behavior switch that turns on when source is declared, c denotes the behavior switch that is on when the robot's role is switched, $\bar{d}$ denotes the behavior switch that is on when chemical is not detected for some time interval, $\bar{s}$ denotes the behavior switch that is on when source is not declared, and the symbol $\bar{c}$ denotes the behavior switch that is on when the vehicle's role remains unchanged. Once the leader changes its role, it acts as a follower by switching its behavior to Formation-Keeping. For the followers, their fundamental behavior is Formation-Keeping. Each follower probably switches its role into that of a leader

Figure 6. Behavior switching diagram-based coordination mechanism for swarm robots to trace a chemical plume

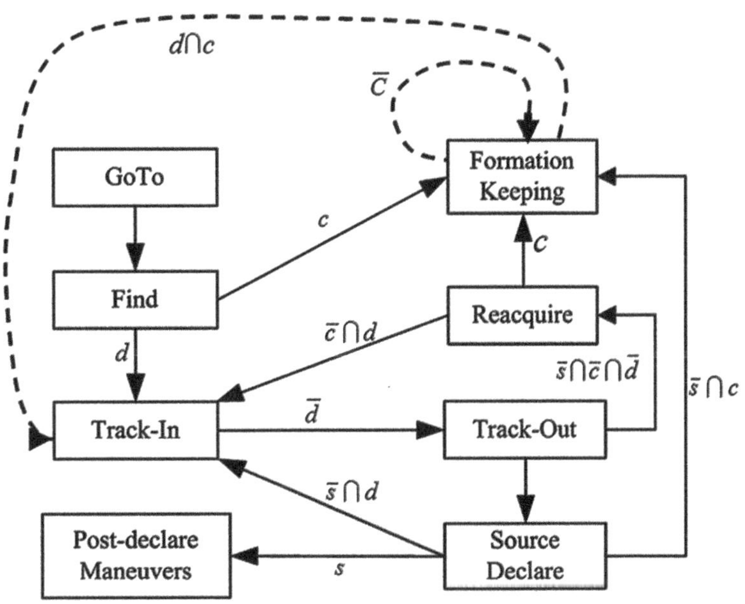

when it triggers a Track-In activity, as indicated by a dash-dot arrow labeled with *d∩c* in Figure 6, and at the same time the former leader becomes a follower.

The authors discuss a swarm robot system built by three robots, which perform CPT missions while maintaining a triangular formation. A formation approach maintains control of the multiple vehicles in two levels. It first determines a sub-target for the leader according to the current robots' status, and then generates trajectories to steer the robots toward the target in a formation shape. A leader is controlled by the moth-inspired plume tracing algorithms. In order to keep the swarm robots in a given formation, the kinematic equations of the robot need to be established for generating trajectories for followers relative to the leader (Kang & Li, 2012). Tables 1 and 2 list the results of the evaluation studies in three aspects: reliability, accuracy, and time cost. The results show that CPT via the multiple vehicles significantly reduces the time cost for tracing the plume and declaring the source location, and slightly improves the source location accuracy.

The Monte Carlo technique was used to evaluate the formation control strategy in the simulated fluid-advected environment discussed above. The boundary conditions of the simulation scenario were set as follows. The operational area is specified by [0, 100]×[-50, 50] in meters. The filament release rate is 5 filaments per second, the simulation time step is 0.01 s, and the mean fluid velocity is 1.0 m/s. The measured fluid flow is corrupted by additive noise that is a white normal random process. The plume source is located at (10, 0) in meters, which is unknown to the robots fleet. For the Monte Carlo simulation, the CPT missions need to be repeated a large amount of times under the same conditions. A variety of home locations were selected for the Monte Carlo simulation. Tables 1 and 2 use the home location defined as (110, 40) in meters outside the operational area because this distance between the home and plume source locations is long enough for evaluating the performances of the Find-Plume, Maintain-Plume, Reacquire-Plume, and Declare-Source behaviors. The operation time is limited to

Table 1. Time cost for CPT test runs on single vehicle vs multiple vehicles

Strategies (1000 Runs)	Average Time, s	Time 150-200 s	Time 200-250 s	Time 250-300 s	Time 300-350 s	Time 350-400 s	Time Greater Than 400 s
Single vehicle	229.8s	61/1000 (6.1%)	754/1000 (75.4%)	164/1000 (16.4%)	16/1000 (1.6%)	5/1000 (0.5%)	0/1000 (0.0%)
Multiple vehicles	195.1s	645/1000 (64.5%)	329/1000 (32.9%)	21/1000 (2.1%)	2/1000 (0.2%)	2/1000 (0.2%)	1/1000 (0.1%)

Table 2. Source identification accuracy for CPT test runs on single vehicle vs multiple vehicles

Strategies (1000 Runs)	Average Source Error, Meters	Estimated Source Location	Error Within 0.0-0.5 m	Error Within 0.5-1.0 m	Error Within 1.0-1.5 m	Error Within 1.5-2.0m	Error Greater Than 2.0m
Single vehicle	0.2195m	$\bar{X}$ =10.146m $\bar{Y}$ =0.0018m	878/1000 (87.8%)	87/1000 (8.7%)	28/1000 (2.8%)	6/1000 (0.6%)	1/1000 (0.1%)
Multiple vehicles	0.1782m	$\bar{X}$ =10.1066m $\bar{Y}$ =-0.0022m	921/1000 (92.1%)	61/1000 (6.1%)	14/1000 (1.4%)	2/1000 (0.2%)	2/1000 (0.2%)

T_{max}=1000 s. For each CPT mission, the robot fleet starts around the home location and returns to the home location when the robot fleet identifies the source location or reaches its maximum mission time T_{max}. The evaluation studies are to compare the CPT performances via a single robot with those via swarm robots. Both strategies run in the same simulated fluid-advected environment concurrently. The simulation studies keep changing the chemical plume in a fluid environment over time, and 1000 CPT test runs are performed for the evaluation studies. The robots speeds can be adjusted between 0.5 and 4.0 m/s. Figures 7-10 show a test run of the CPT mission, including the Find-Plume, Maintain-Plume, Reacquire-Plume, and Declare-Source behavior activities. The robot fleet starts its CPT mission around the home location. Robot 1 that is selected as the leader activated the Find-Plume behavior, while robots 2 and 3 follow the leader to perform a cross-flow search for the chemical plume in the operational area. Since robot 3 first detects the plume, it becomes the leader, as shown in Figure 7. Once robot 3 activates the Maintain-Plume behavior as the new leader and uses Track-In and Track-Out activities toward the chemical source location, while the other robots start to follow the new leader via communication. For example, robot 1 switches its role to a follower, so it moves to the left side of the leader to build another triangular formation and maintains this formation by following the leader, as shown in Figure 8. When robot 3 loses its contact with the chemical for a certain number of seconds, it activates Reacquire-Plume behavior to cast for the lost chemical, and robots 1 and 2 follow the leader on cloverleaf-shaped trajectories. As robot 2 first casts the lost plume during Reacquire-Plume activity, it switches its role to the leader and triggers a new Maintain-Plume activity. In the meantime, robot 3 turns its role to that of a follower and moves to the right side to build a new triangular formation. Robot 2 leads the robot fleet to reach the source location, as shown in Figure 9. Figure 10 demonstrates Declare-Source activities of the robot fleet to identify the source location. Some of CPT missions may fail due to the significant filament intermittency and meander of the plumes developed in fluid-advected environments.

CHEMICAL PLUME TRACERS

It is important to validate the moth-inspired CPT algorithms via robot-based plume tracers, so this section introduces how to design robot-based plume tracers, including ground-based autonomous robots (DaNI robots), aerial unmanned vehicles, and autonomous surface/underwater vehicles.

Figure 11 shows the ground-robot based chemical plume tracer, developed in the Robotics Lab at California State University, Bakersfield (CSUB). The used platform is a DaNI robot from National Instruments with an open architecture mobile robot. It comes preassembled and has two motors, encoders, and an ultra-sonic sensor. It uses the sbRIO-9632 embedded control and acquisition device integrating a 400 MHz industrial processor, a user-reconfigurable field-programmable gate array (a 2M gate Xilinx Spartan FPGA), and I/O on a single printed circuit board. It features 110 3.3V (5V tolerant/TTL compatible) digital I/O lines, 32 single-ended/16 differential 16-bit analog input channels at 250 kS/s, and four 16-bit analog output channels at 100 kS/s. It also has three connectors for expansion I/O using board-level NI C Series I/O modules. The sbRIO-9632 offers a -20 to 55 °C operating temperature range, and includes a 19 to 30 VDC power supply input range, 128 MB of DRAM for embedded operation, and 256 MB of nonvolatile memory for storing programs and data logging. It also features a built-in 10/100 Mbit/s Ethernet port you can use to conduct programmatic communication over the network and host built-in Web (HTTP) and file (FTP) servers. A user also can use the RS232 serial port to control peripheral devices. Its programming environment is LabVIEW developed by National Instruments (NI).

Figure 7. A test run of moth-inspired plume tracing via multiple robots is under formation control. Find-Plume activity: robot 1 leads robot 2 and 3 to find the plume, but robot 3 becomes the leader when it first detects the plume

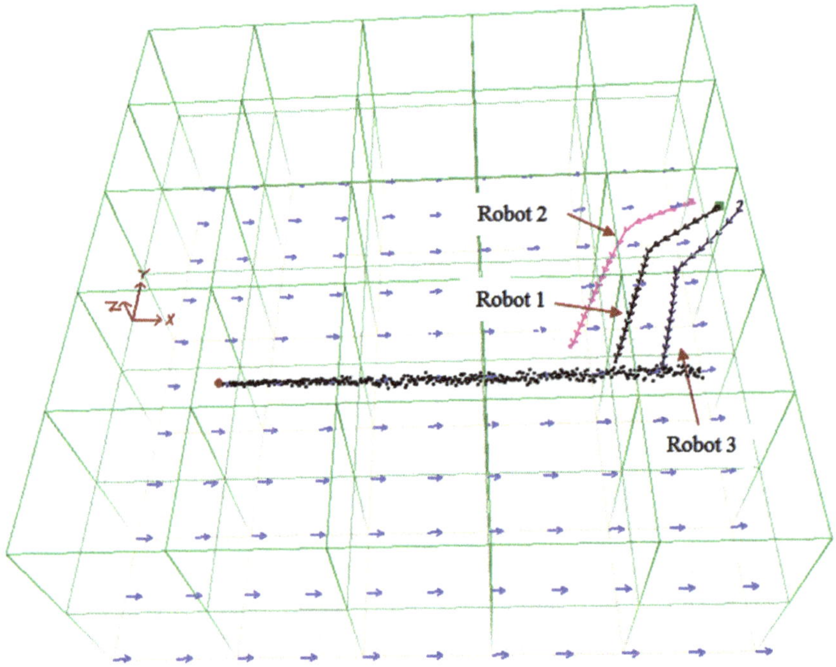

Figure 8. Maintain-Plume activity: robot 3 leads robots 1 and 2 to trace the plume toward to the source

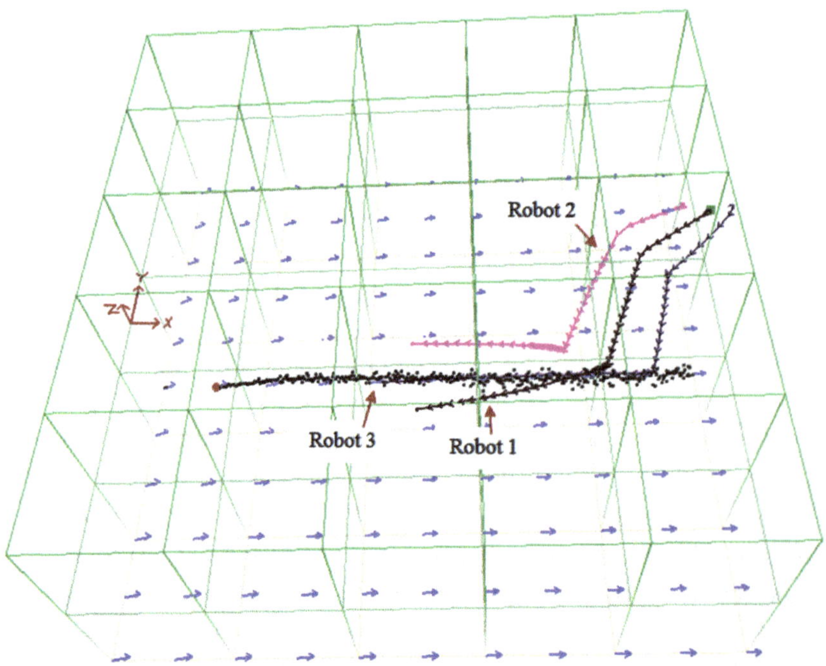

Figure 9. Reacquire-Plume activity: robot 2 first detects the plume during Reacquire-Plume and becomes the new leader, and robots 1 and 3 follow it toward to the source location

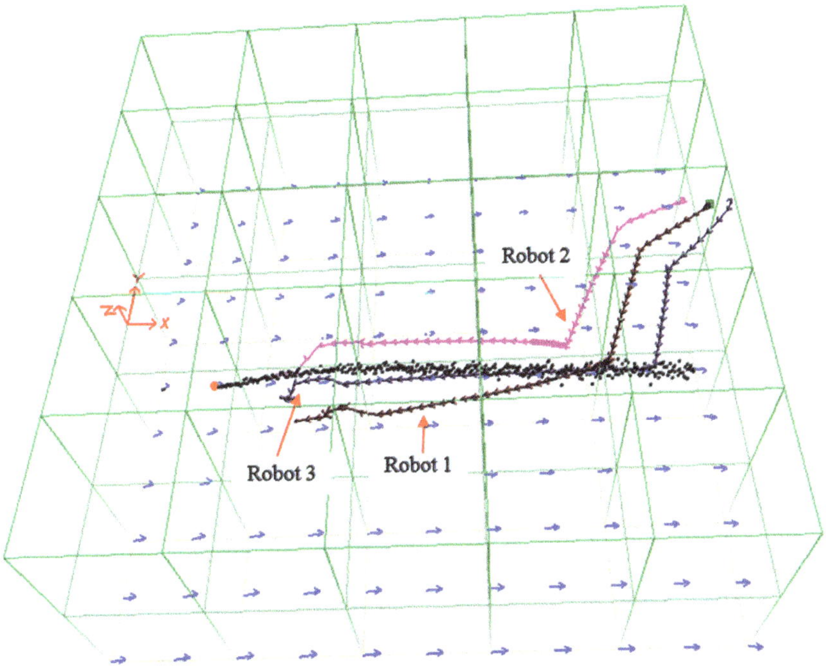

Figure 10. Declare-Source activity: all vehicles run around the source location to identify the chemical source by alternating the leader and followers' roles

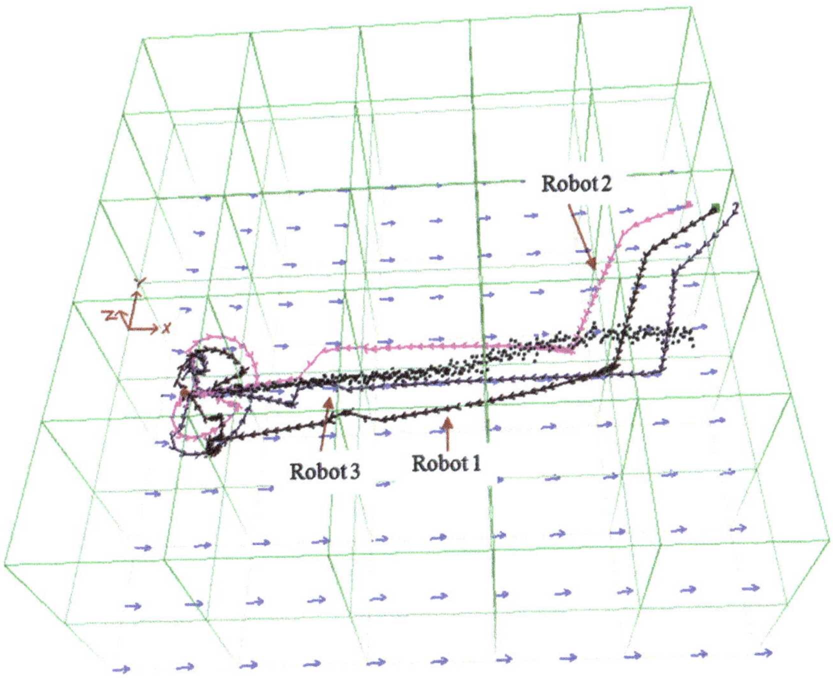

Figure 11. DaNI robot based chemical plume tracers with the three Figaro chemical sensors, developed in Robotics Lab, at CSUB

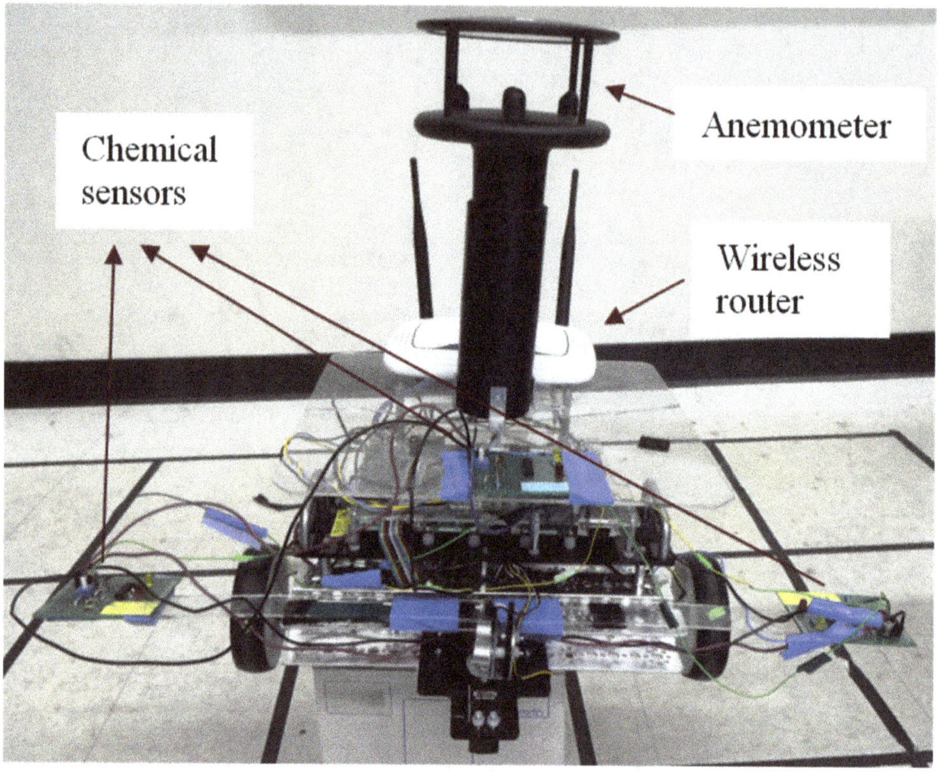

A DaNI robot is equipped with a Parallax ultrasonic sensor. It detects objects by emitting a short ultrasonic burst and then "listening" for the echo. Under the control of a host microcontroller (trigger pulse), the sensor emits a short 40 kHz (ultrasonic) burst. This burst travels through the air at about 1130 feet per second, hits an object, and then bounces back to the sensor. The PING sensor provides an output pulse to the host that terminates when the echo is detected. Meaning the width of this pulse corresponds to the distance to the target. The DC motors use 12V of power and offer 300 oz-in of torque and 152 RPM. The encoders use 5V of power and offer 100 cycles per revolution and 400 pulses per revolution.

The NI Robotics Starter Kit, or the DaNI robot, comes preassembled, but since it is open architecture it allows for different sensors to be added on. The first thing to add on was a wireless router to enable wireless communication with the robot. In order to configure the router, first the setup disc that comes with the router is run on a computer. The router is then connected to the computer and to the Internet, both using an Ethernet cable. The setup disc checks if the connections between the DaNI robot and the computer are established. For the setup, the name for the wireless network and the password are required. After that, it is ready to be plugged into the DaNI robot and connect to a computer for wireless communication. Next, an AXIS M1011 network camera is added to the DaNI robot to allow various visual projects and to view things from the robot's perspective. The camera was mounted on the robot and connects to the router using an Ethernet cable. The camera provides multiple, individually configurable video streams in H.264 as well as Motion JPEG and MPEG-4. It also offers video quality at 30 frames per second in VGA resolution. The NI Measurement and Automation software can check if the camera works or not and configure the camera format and image resolution.

The chemical sensor made by Figaro uses a metal oxide semiconductor layer on an alumina substrate of a sensing chip with an integrated heater. When the chemical is detectable, the sensor's conductivity increases depending on the chemical concentration in the air, where the humidifier that pumps alcohol as the chemical source. An electrical circuit board converts chemical concentrations into electrical output signals, as shown in Figure 12.

The Gill WindSonic wind sensor is a low-cost anemometer shown in Figure 13, which utilizes Gill's ultrasonic technology to provide wind speed and direction data via one serial or two analogue outputs. To confirm correct operation, outputs are transmitted together with an instrument status code. It features a robust, corrosion-free polycarbonate housing which makes it very lightweight. The WindSonic sensor has no moving parts, offering maintenance-free operation in a wide range of applications. The WindSonic sensor is a robust ultrasonic wind speed and direction sensor with aluminum alloy construction. The sensor is solid-state with no moving parts, using ultrasonic measurement technology to detect wind speed and direction at speeds up to 60m/s (134mph). The robust aluminum alloy housing is hard-anodized to ensure suitability in harsh marine environments, and the optional heating system allows operation down to -40°C. The WindSonic sensor provides a marine-standard NMEA 0183 output, with options for RS232, 422 and 485 outputs. A single 9-way connector and three mounting holes for attachment to a 1.75" pipe ensure installation is straightforward. The WindSonic sensor has a very low maintenance overhead.

Figure 14 shows the AUV-based plume tracer, developed by Shenyang Institute of Automation (SIA), Chinese Academy of Sciences. Six thrusters control the AUV. The AUV is equipped with multiple sensors, including a depth sensor, an electronic compass, a Cyclops-7 underwater fluorometer (http://www.turnerdesigns.com) in Figure 15 to detect the Rhodamine dye, and a Doppler velocity log (DVL), Workhorse Navigator with 1200 kHz (http://www.rdinstruments.com) shown in Figure 16, to measure the AUV' velocities relative to the sea bottom and the reference water layer. In order to run the field experiments, the two vertical thrusters are used to keep the AUV's depth, the two forward horizontal

Figure 12. The electrical circuit board for the Figaro chemical sensor

Figure 13. Gill WindSonic sensor is a low-cost anemometer, which provides wind speed and direction data via one serial or two analogue outputs

thrusters control the AUV's heading, and the two afterward horizontal thrusters control the AUV's speed. The vehicle's ideal speed is about 0.5 m/s. Using the two horizontal thrusters to control the heading of the AUV delivers the better maneuverability of the vehicle, e.g., the vehicle turning radius on the order of several meters is smaller than the REMUS AUV controlled by a main thruster and a rudder, so at low speed, the AUV is able to change its heading without much variation of its position.

Figure 17 shows an UAV-based plume tracer with the two chemical sensors developed in Robotics Lab at CSUB. The used platform is a quadcopter made by 3D Robotics Inc (http://3drobotics.com/). The quadcopter based on a APM 2.6 or PixHawk autopilot system includes four arms, body plates, a power distribution board, four 880 kV brushless motors with bullet connectors and electronic speed controllers (ESCs), one complete 10-inch propeller set (four individual 10x4.7 propellers), an I2C splitter module, a GPS with compass, and GPS mast.

APM 2.6 is a complete open source autopilot system and requires a GPS unit with an onboard compass or an external compass module for full autonomy, which allows the user to turn any fixed, rotary wing or multirotor vehicle into a fully autonomous vehicle; capable of performing programmed GPS missions with waypoints. Pixhawk is an advanced autopilot system designed by the PX4 open-hardware project and manufactured by 3D Robotics. It features advanced processor and sensor technology from ST Microelectronics® and a NuttX real-time operating system, delivering incredible performance, flexibility, and reliability for controlling any autonomous vehicle.

The two same chemical sensors used for the DaNI robot are adhered on an additional nylon plate to convert chemical (alcohol) concentrations into electrical output signals. One chemical sensor is fixed underneath the nylon plate and the other above the nylon plate to investigate aerodynamic effect on the chemical puffs when the propellers are spinning

Figure 14. The AUV-based plume tracer employed in the in-water tests is developed at Shenyang Institute of Automation, Chinese Academy of Sciences

Figure 15. Cyclops-7 fluorometer

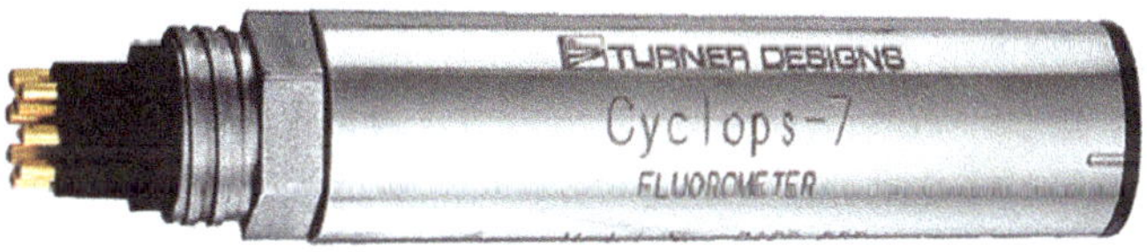

Figure 16. Workhorse Navigator 1200 kHz DVL

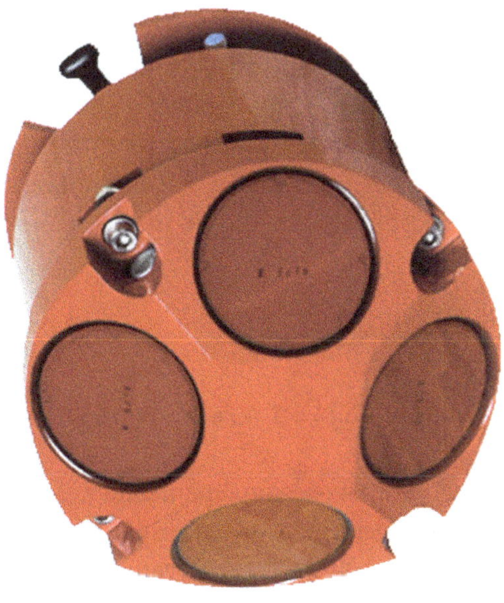

Figure 17. An UAV based chemical plume tracer with the two Figaro chemical sensors, developed in Robotics Lab, at CSUB

Except for a barometer sensor, an ultrasonic sensor, Matbotix XL-EZ4, is added to the quadcopter to better keep its altitude at a fixed level for CPT missions. The ultrasonic sensor has a max range of 7.65 meters and uses a narrow beam that reduces the chance of interference. A camera is mounted on the quadcopter to provide live feedback to base.

The UAV-based plume tracer uses a lithium polymer (LiPo) battery that allows the quadcopter to fly 20 and 40 minutes. The programming environment for quadcopter is MissionPlanner (http://copter.ardupilot.com/). One of programming approach is to use GPS and Google Maps to direct the quadcopter through waypoints. Figure 18 shows a path created by six waypoints, including the home location. Once the program executes, the plume tracer automatically flies to its home waypoint, follow the path, and finally return home. This technique is very good for building plume maps in a given area. MissionPlanner allows Python scripts to control the plume tracer by using basic commands, such as arming and disarming the motors, increasing or decreasing throttle, and controlling pitch or yaw orientations based on multiple sensor feedback, to trace a chemical plume in the three dimensions.

EXPERIMENTS OF CPT MISSIONS

This section presents experimental results on CPT missions via some of the plume tracers described in the previous section.

The first experiment reported here is moth-inspired plume tracing under turbulent airflow created in the Robotics lab environment. For the experiments, a Koala robot (http://www.k-team.com) is developed to the plume tracer by adding an AXIS M1011 network camera, a WindSonic anemometer, and two

Figure 18. A path that the UAV follows is created by six waypoints, including the home location

Figaro chemical sensor boards, and a WiBox dual port wireless device servers that enable to connect equipment to 802.11 b/g wireless networks vial serial or Ethernet. Its programming environment is upgraded from C to LabVIEW 2013.

The experimental environment includes a crane drop shape ultrasonic humidifier that pumps alcohol as a chemical source and a fan that generates turbulent airflow, as shown in Figure 19. The first step for the experiments was controlling the plume tracer to follow the airflow without the presence of chemical puffs. The plume tracer turned left or right according to given wind velocities in meters per second and wind directions in degrees to calibrate the controllers for the wheels speed on both sides of the robot. This behavior is responsible for zigzag movement, which is the remarkable plume-following maneuver of male moths to track pheromones released by a female moth.

For testing the moth-inspired plume tracing strategy, the plume tracer only uses one chemical sensor in conjunction with the wind sensor to track air-borne chemicals. A 3.8 m × 4.6 m arena area in the Robotics lab was utilized to conduct the experiments of CPT missions. A chemical source location and a starting position were chosen randomly. The maximum initial distance between the plume tracer and the chemical source was 4.5 m. For the experiments, the fan with fix heading produced turbulence-dominated airflow to develop chemical plumes with significant intermittency of chemical puffs; while the fan with varying heading produced turbulence-dominated airflow to develop chemical plumes not only significant intermittency of chemical puffs but also with significant plume meander. The fan varied its heading between -45 and +45 degree and the maximum wind velocity of 4.0 m/s was measured. The experiments show that 100% success rate was achieved using moth-inspired plume tracing algorithms under both the types of airflow conditions. The tracking of the chemical plumes with significant meander spent more time to localize the source locations. The experiments show that both Find-Plume and Require-Plume seem inactive when the tracking of chemical plumes in a small-scale operation area, such as in the lab

Figure 19. Moth-inspired plume tracing under turbulent airflow environments via a Koala robot based plume tracer. A fan with a fix heading or a varying heading generates turbulent airflow and a humidifier pumps alcohol as the chemical source

environment, because the plume tracer can easily find the chemical plume with roaming maneuvers and require the chemical plume just staying at the position where it loses contact with the chemical plume.

The second experiment reported here is tracking of chemical plumes developed in both diffusive and turbulent airflow environments using a Petri net based model without need for the wind anemometer, as shown in Figure 20. It is challenging to develop a generalized algorithm to effectively trace both types of chemical plumes due to the significant differences of their kinetic and dynamic properties. The idea is to utilize to model the change relationships of chemical concentrations acquired by the two Figaro sensors mounted on the both sides of a DaNI robot during a CPT mission. Because the relationships imply the effects of flow variation on chemical puffs, a flow sensor is eliminated. In order to express and maintain the knowledge of chemical concentration changes using the Petri net, it is essential to design a mapping algorithm for generating the Petri net from production rules. The Petri net model was implemented on the DaNI robot using LabVIEW. The chemical plume tracing experiments achieved 93.8% and 87.5% of source localization rates under both turbulent and diffusive airflow environments respectively (Yuan et al., 2014).

Most of the existing studies mainly focused on the CPT issue and validated their algorithms in a small operation area in range of centimeters to few meters or in a simulated environment, as discussed above, but the tracking of chemical plumes in the near-shore oceanic environments, where the plumes are developed under turbulence, tides and waves, is much more challenging. Figure 21 shows a Rhodamine dye plume developed from its source at the sea bottom that is approximately lowered 20 meters

Figure 20. Petri nets based plume tracing under diffusive airflow environments via a DaNI robot based plume tracer equipped with the two Figaro chemical sensors without the wind sensor

from the surface at the San Clemente Island of California in November 2002 and Figure 22 shows a Rhodamine dye plume with length over 350 meters in near-shore oceanic environments at the Dalian Bay of China in October 2010.

The factors that significantly challenge the CPT issues are: First, chemical puffs only distribute over a very narrow region in a large operation area under natural environments, so a plume tracer has to consider how to efficiently find the chemical plume by searching the entire operation area without any prior knowledge about the plume concentration distribution. Second, the chemical plumes developed in the near-shore oceanic environments, as displayed in Figure 22, exhibit local maxima with sharp spatial gradients, i.e., chemical concentrations at some locations where the plume looks like broken are very low, but at other locations with a large dye area colored by the deep red are very high, due to turbulence and beating of waves. Third, the lack of correlation between the fluid flow direction and the plume long axis often causes the plume tracer to lose contact with the plume because the instantaneous fluid flow direction within the plume with the significant meander (a snakelike path) is not always aligned with the plume's long axis shown in Figure 22, so the plume tracer has to re-contact the plume in a local area where the plume is lost. Next, the declaration that the source has been found is extremely complicated because the chemical source concentration being unknown, the advection distance of any detected chemical being unknown, and the flow variation with both location and time. Therefore, conducting CPT missions in near-shore ocean environments has to consider the full spectrum of field behavior that include Find-Plume, Maintain-Plume, Reacquire-Plume, and Declare-Source coordinated by the subsumption architecture shown in Figure 3.

The moth-inspired plume tracing algorithms were implemented on a REMUS AUV for the tracking of the Rhodamine dye plume, which was developed under turbulence, tides, and waves. The REMUS vehicle is equipped with a variety of sensors, including a fluorometer for detecting the Rhodamine dye

Figure 21. The Rhodamine dye plume was developed in the field experiments in November 2002 at the San Clemente Island of California

Figure 22. The Rhodamine dye plume developed in the field experiments in October 2010 at the Dalian Bay of China (A snapshot of the plume with length over 350 m)

plume, side-scan sonar, and conductivity, temperature, and depth sensors. The REMUS also carries sensors to provide information about the current operating state of the vehicle, including the vehicle latitude and longitude, depth and altitude, heading and speed, and water flow velocity.

The in-water tests conducted in November 2002 on San Clemente Island, California, demonstrated the first successful CPT via the REMUS-based plume tracer in near-shore ocean environments (Li et al. 2006). An operation area is approximately 300 m × 112 m. The long axis is rotated by 47 degree west of north is approximately parallel to the shoreline. Initial positions were chosen randomly and the source locations were unknown prior to the CPT missions. Figure 23 shows the mission MSN007r2 documented over distance of 411 meters from the first detection point to the identified source location. During this experiment, the CPT algorithm calculated the flow varied in magnitude between 10 and 15 cm/s and in direction between 110 and 147 degree.

The follow-up field tests were conducted in June 2003 in Duck, North Carolina (Farrell et al. 2005). The operation area for the MSN003 mission is 367 m × 1094 m (greater than 60 football fields). For this test run, the commanded speed was set as 2 m/s and the typical AUV altitude as 2.0 m. The actual altitude varies by plus or minus 0.7 m relative to the commanded altitude. To challenge the CPT algorithms, the first chemical detection to occur was as far as possible from the chemical source. Therefore, the source is located near the upflow edge of the operation area and the AUV starts the test run near the downflow edge of the operation area. This test run recorded the successful tracking of the Rhodamine dye plume with the longest distance over 976 meters between the first point of chemical detection and the identified source location, as shown in Figure 24. The declared source location is 13 m south and 2 m east of the ground truth source location.

The recent CPT missions were conducted in October 2010 at Dalian Bay, China, via the AUV-based plume tracer developed by Shenyang Institute of Automation, as shown in Figure 14. The vehicle's ideal speed is about 0.5m/s. Figure 25 shows the test result of an adaptive plume mapping mission, i.e., when the AUV left the plume for a given distance, it ended at the current track-line and turned to the next track-line to build the plume map. The blue lines on the AUV trajectory represent flow directions and magnitudes. The origin of the operation coordinate system was defined as the location where the AUV was launched, and the plume source was located near the origin of the coordinate system. The AUV performed the CPT missions from the up-flow to the down-flow direction. The AUV started its

Figure 23. The first successful CPT in near-shore ocean environments was conducted in November 2002 on San Clemente Island, California. The mission MSN007r2 documented over distance of 411 meters from the first detection point to the identified source location

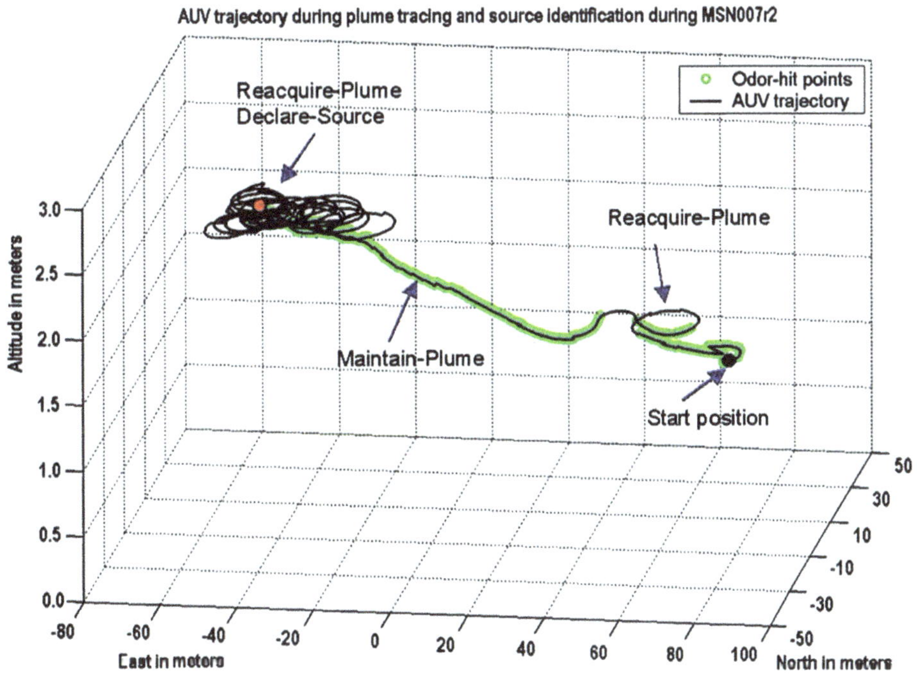

Figure 24. The MSN003 mission at Duck in June 2003 tracked a chemical plume with the longest distance over 975 meters between the first point of chemical detection and the identified source location

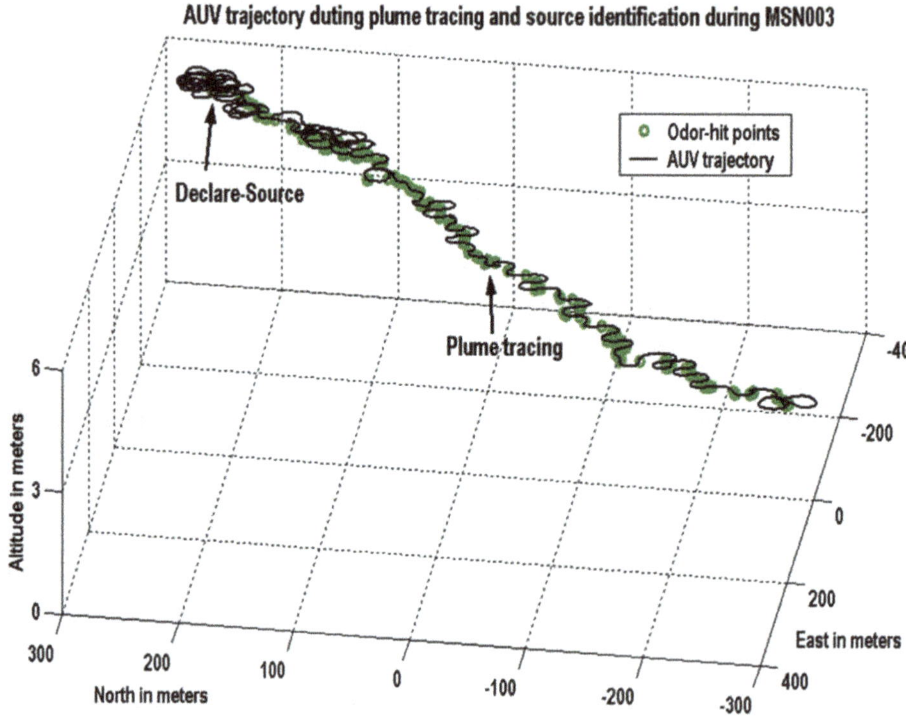

mission from (0, 0) m. It first navigated to (20, -20) m, and then started detecting the Rhodamine dye plume. This mapping mission lasted for 850 s and ended at position (62, 21) m. The space between two track-lines was set as 10 m. The mean depth and speed of the AUV were measured about 1.6 m and 0.45 m/s. Figure 25 shows that the plume widths were 15.5 m, 22.4 m, 29.7m, and 43.7m with the distances of 30.0 m, 40.0 m, 50.0 m, and 60.0 m away from the chemical source location, and the centerline direction of the plume were calculated as 27.9 deg, 11.9 deg, and 11.7 deg between the distances 30.0 m and 40.0 m, 40.0 m and 50.0 m, and 50.0 m and 60.0 m. Figure 26 displays the map in three dimensions. Figures 27-28 shows two CPT missions conducted in October 2010, including the Find-Plume, Track-In, and Track-Out activities during the course from the home positions to the source locations. The AUV-based plume tracer started at the home location by activating the Find Plume behavior to search for the Rhodamine plume in an operation area with 300 m × 200 m. The plume tracer switched its activity from Finding-Plume to Track-In. The Track-In activity navigated the AUV with an angle of 20 degrees relative to the up-flow direction to trace the Rhodamine plume. When the tracer did not detect the plume with a distance of 10 m, the tracer activated the Track-Out activity to across the plume. The data from the two missions shows that the plume tracer took over 1000 s to complete the CPT missions, the plume tracer tracked the plume below the sea surface about 1.7 m with an average speed of 0.434 m/s, and the plume tracer travelled a distance over 100.0 m between the home positions and the source locations.

Figure 25. A Rhodamine dye plume map was constructed in two dimensions (expressed in a defined local coordinate system, the dashed lines are the plume centerline and plume boundary, respectively)

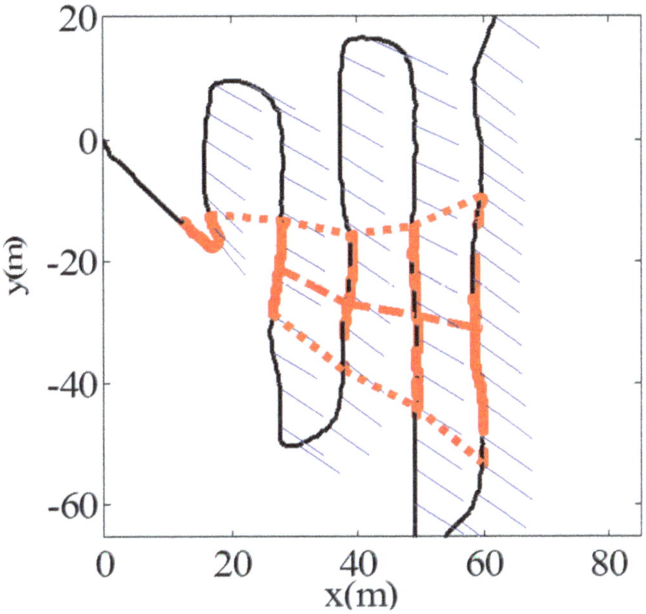

Figure 26. The map was displayed in three dimensions (red dots indicate points where the chemicals were detected)

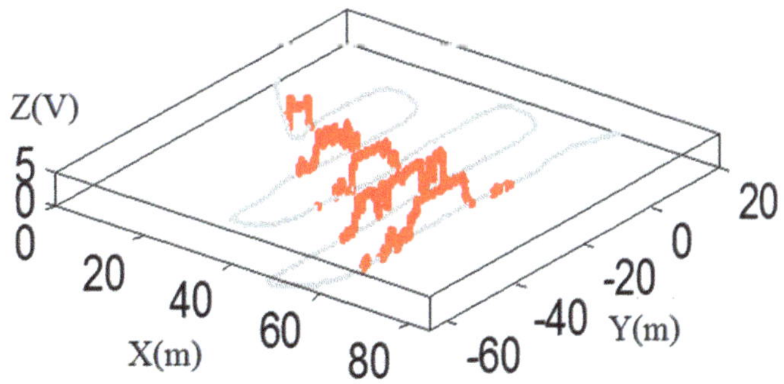

FUTURE RESEARCH DIRECTIONS

Until now, most of the existing studies on chemical plume tracing via swam robots are conducted in simulated environments (Hayes et al., 2002; Zarzhitsky et al., 2004; Kang and Li, 2012). In order to trace and map chemical plumes via swam robots in natural environments, the future research will focus on field tests of controlling the formation of swarm robots to trace chemical plumes and build the plume maps.

First, a fleet of grounded-based robots with the distributed structure developed in the Robotics lab, at CSUB, will be employed to validate the strategy for moth-inspired plume tracing via swarm robots under formation control (Kang and Li, 2012). Because each robot is equipped with multiple sensors, including a wireless router and the two chemical sensor boards, as shown in Figure 29, the plume tracers in the swarm

Figure 27. The first CPT mission tracked the Rhodamine dye plume over 100 m (expressed in a defined local coordinate system)

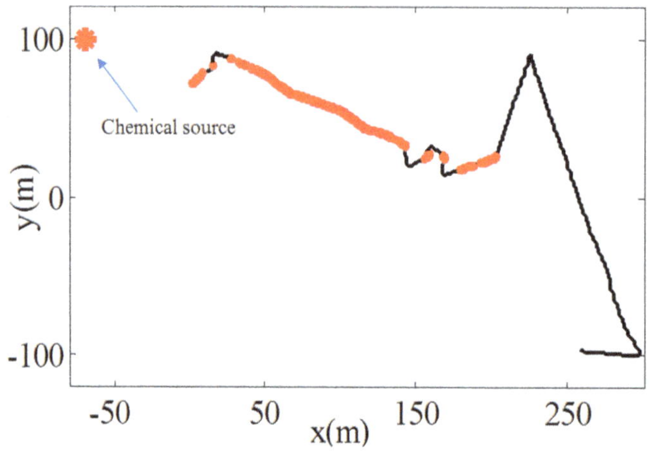

Figure 28. The second CPT mission tracked the Rhodamine dye plume over 100 m (expressed in a defined local coordinate system)

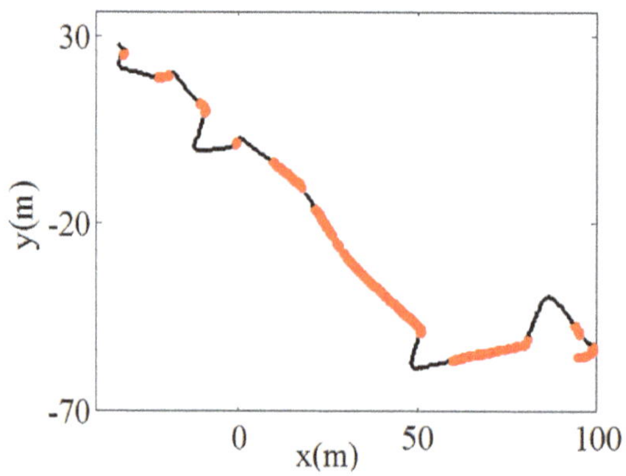

robot system are able to talk each other in LabVIEW Robotics projects. Each plume tracer performs its maneuver and shares the collected information with the others, such as, chemical concentrations, airflow orientations and magnitudes, and its current state. The swarm robots are able to dynamically form their shape to perform CPT missions according to the formation control mechanism. Especially, conducting CPT missions via the swarm robots under formation control in an open and large-scale operation area is on schedule. The scheduled field tests will investigate the impacts of swarm robots' formation shapes and numbers on chemical plume tracing and mapping performances.

Second, two AUV-based plume tracers developed by the Yiping Li's group at SIA will conduct further CPT missions in near-shore ocean environments (Zeng et al., 2014), as shown in Figure 30. The mechanical structure of the AUV differs from the structure of the DaNI robots as the DaNI robots are able to drive their wheels differentially to change their orientation with a near zero turning radius. In contrast, the AUV requires a large (5-10 m) turning radius to change orientation. Therefore, it is important to test the AUV kinematics, formation control errors, and reliability of the AUVs. Prior to CPT missions

Figure 29. A swarm robot system, built upon the DaNI robots, for chemical plume tracing and mapping is developed in the Robotics Lab, at CSUB

via swarm robots in near-shore ocean environments, the AUVs were controlled to follow the predefined paths under formation control in a lake near Shenyang city, as shown in Figure 31. Figure 32 plots the trajectories yielded by the two AUVs during a formation control experiment. Such an experiment is very important to calibrate the parameters for chemical plume tracing and mapping by formation control of the two AUVs, e.g., to determine the threshold of the chemical sensors by limiting surroundings noise.

Figure 30. Two AUV-based plume tracers are developed at SIA

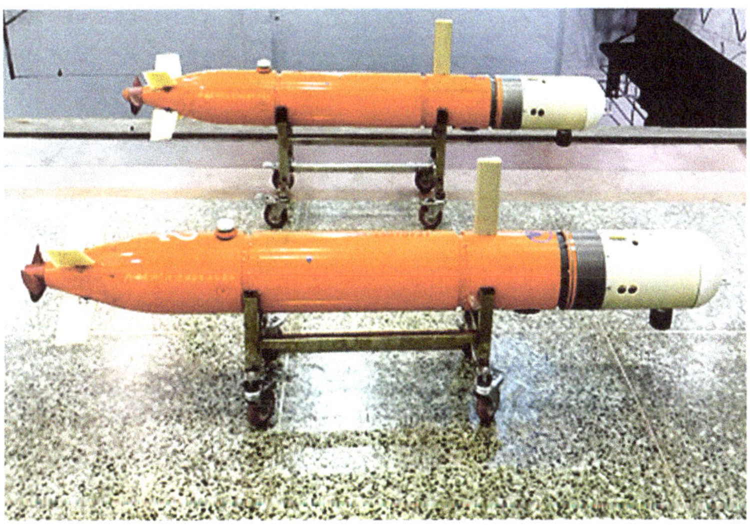

Figure 31. Coordinated path-following experiment on a formation is performed in a lake near Shenyang city

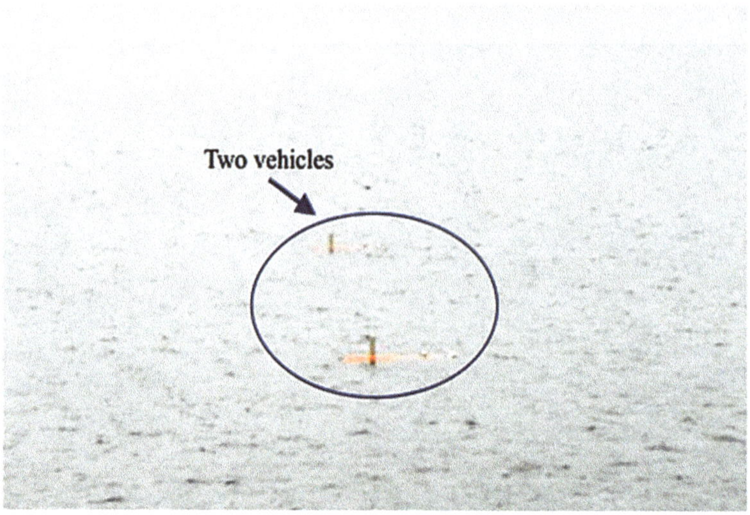

Figure 32. Trajectories of the two AUVs are recorded during the formation control experiment

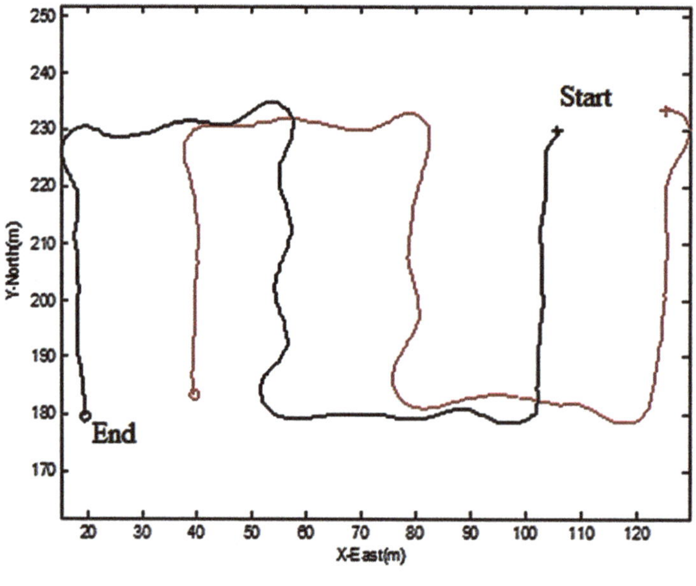

Over the decade, drones have become increasingly popular due to low cost, flexibility, and ability to carry additional sensors, which provides opportunities to develop relatively low operational cost data acquisition systems. The UAV-based based plume tracers, including quadcopters and hexcopters, are developed in Robotics Lab, at CSUB. In order to conduct field tests via the UAVs-based plume tracers, three key issues have to be solved. First is to implement the Find-Plume, Maintain-Plume, Reacquire-Plume, and Declare-Source behaviors of the moth-inspired algorithms by the three types of motions: roll, yaw, and pitch, as shown in Figures 33 and 34. Second is to investigate effects on chemical puff distributions of aerodynamics under spinning propellers. A data acquisition system, consisting of a chemical sensor board and a NI sbRIO-9636 embedded device, is developed for collecting chemical concentration data. The sbRIO-9636 has a 400 MHz processor, 512 MB nonvolatile storage, 256 MB DRAM for deterministic control and analysis, and a reconfigurable Xilinx Spartan-6 LX45 FPGA. The chemical data acquisition system is calibrated on the NI ELVIS II+ station under static airflow condition. Collecting chemical concentration data by the chemical sensors that are placed on the hexcopter arms is tested under spinning propellers. Third is to conduct the experiments on chemical plume tracing and mapping via the formation control of a fleet of UAVs. The purpose for developing the UAV-based swarm robots is to monitor air pollution and to identify pollution sources.

Figure 33. The moth-inspired behaviors are tested on the quadcopter

Figure 34. The moth-inspired behaviors are tested on the hexcopter

In general, chemical plume tracing and mapping via swarm robots in natural environments is very challenging because the chemical source concentration being unknown, the advection distance of any detected chemical being unknown, significant filament intermittency and meander of a chemical plume developed in turbulent fluid flow environments, and the flow variation with both location and time. In order to apply the swarm robots in plume tracing and mapping, the following issues have to be addressed: 1. Model chemical plumes, including turbulent chemical plumes or atmospheric plumes in 2 or 3 dimension; 2. Develop effective and practical algorithms for chemical plume tracing and mapping via the swarm robots and evaluate and optimize their performances; 3. Design chemical plume tracers, which are equipped with a variety of sensors; 4. Establish robot kinematics and dynamics and design effective controllers for the swarm robots to control the swarm robots in formation precisely; 5. Conduct field tests in natural environments need to validate the algorithms for the swarm robots and to test the reliability and flexibility of the chemical plume tracers.

CONCLUSION

This chapter discusses the key issues of chemical plume tracing and mapping via swarm robots. CPT in a natural environment is very challenging because chemical source concentration is unknown, the advection distance of any detected chemical is unknown, and flow directions and magnitudes vary with both location and time. Developing the plume model for evaluating CPT strategies via swarm robots is one of the most important steps toward to applications in the real world. First, the tracking of simulated plumes allows evaluating different CPT strategies under the uniform conditions. Second, generating simulated plumes with significant filament intermittency and meander challenges the CPT strategies. An alternative way to evaluate CPT strategies is to trace chemical plumes via swarm robots in a laboratory environment. Although the tracking of a chemical plume generated in the laboratory environment is not so challenging, this way still can be used to test the performance of chemical sensors and formation control mechanism for swarm robots.

Chemical plume tracing and mapping in the real world via swarm robots allows to monitor pollutions or to search for sources of hazardous chemicals or pollutants in large-scale natural environments. Potential applications of AUV-based plume tracers are to search for hydrothermal vents on mid-ocean ridges or to investigate subsea chemical pollution and greenhouse gases by building plume maps. Opportunities for oil and gas companies to use UAV-based plume tracers are vast: inspecting flare stacks, detecting and identifying corrosion or leaks in pipelines and other tanks. Applications of plume tracing and mapping via swarm robots into agriculture, e.g., to identify insect active areas, is appealing.

REFERENCES

Anfossi, D. (1985). Analysis of plume rise data from five TVA Steam Plants. *Journal of Climate and Applied Meteorology*, *24*(11), 1225–1236. doi:10.1175/1520-0450(1985)024<1225:AOPRDF>2.0.CO;2

Anfossi, D., Ferrero, E., Brusasca, G., Marzoratti, A., & Tinarelli, G. (1993). A simple way of computing buoyant plume rise in Lagrangian stochastic dispersion models. *Atmospheric Environment*, *27*(9), 1443–1451. doi:10.1016/0960-1686(93)90130-Q

Aylor, D. E., Parlange, J. Y., & Granett, J. (1976). Turbulent dispersion of disparlure in the forest and male gypsy moth response. *Environmental Entomology*, *5*(5), 1026–1032. doi:10.1093/ee/5.5.1026

Basil, J., & Atema, J. (1994). Lobster orientation in turbulent chemical plumes: Simultaneous measurements of tracking behavior and temporal chemical patterns. *The Biological Bulletin*, *187*, 272–273. PMID:7811821

Belanger, J. H., & Willis, M. A. (1996). Adaptive control of odor-guided locomotion: Behavioral flexibility as an antidote to environmental unpredictability. *Adaptive Behavior*, *4*(3-4), 217–253. doi:10.1177/105971239600400302

Beni, G. (2005). From swarm intelligence to swarm robotics. Swarm Robotics, Lecture Notes in Computer Science (Vol. 3342, pp. 1-90).

Brambilla, M., Ferrante, E., Birattari, M., & Dorigo, M. (2013). Swarm robotics: A review from the swarm engineering perspective. *Swarm Intelligence*, *7*(1), 1–41. doi:10.100711721-012-0075-2

Cardé, R. T. (1996). Odour plumes and odour-mediated flight in insects in olfaction in mosquito-host interactions. *Proceedings CIBA Found. Symp.* (pp. 54-70).

Cardé, R. T., & Mafra-Neto, A. (1996). Mechanisms of flight of male moths to pheromone. In R. T. Cardé & A. K. Minks (Eds.), *Insect Pheromone Research. New Directions* (pp. 275–290). New York: Chapman and Hall.

Cowen, E. A., & Ward, K. B. (2002). Chemical plume tracing. *Environmental Fluid Mechanics*, *2*(1/2), 1–7. doi:10.1023/A:1016247516041

Devine, D. V., & Atema, J. (1982). Function of chemoreceptor organs in spatial orientation of the lobster, homarus americanus: Differences and overlap. *The Biological Bulletin*, *163*(1), 144–153. doi:10.2307/1541504

Dunbabin, M., & Marques, L. (2012). Robotics for environmental monitoring: Significant advancements and applications. *IEEE Robotics & Automation Magazine*, *19*(1), 24–39. doi:10.1109/MRA.2011.2181683

Dusenbery, D. B. (1989). Optimal search direction for an animal flying or swimming in a wind or current. *Journal of Chemical Ecology*, *15*(11), 2511–2519. doi:10.1007/BF01014727 PMID:24271593

Elkinton, J. S., Schal, C., Onot, T., & Cardé, R. T. (1987). Pheromone puff trajectory and upwind flight of male gypsy moths in a forest. *Physiological Entomology*, *12*(4), 399–406. doi:10.1111/j.1365-3032.1987.tb00766.x

Farrell, J. A., Murlis, J., Long, X., Li, W., & Cardé, R. T. (2002). Filament-based atmospheric dispersion model to achieve short time-scale structure of odor plumes. *Environmental Fluid Mechanics*, *2*(1/2), 143–169. doi:10.1023/A:1016283702837

Farrell, J. A., Pang, S., & Li, W. (2005). Chemical plume tracing via an autonomous underwater vehicle. *IEEE Journal of Oceanic Engineering*, *30*(2), 428–442. doi:10.1109/JOE.2004.838066

Gifford, F. A. (1959). Statistical properties of a fluctuating plume dispersion model. In F. N. Frankiel & R. A. Shepard (Eds.), *Advances in Geophysics, Atmospheric Diffusion and Air Pollution* (Vol. 6, p. 117). New York: Academic Press. doi:10.1016/S0065-2687(08)60099-0

Gifford, F. A. (1960). Peak to average concentration ratios according to a fluctuating plume dispersion model. *International Journal of Air Pollution*, *3*, 253–260. PMID:13705193

Gifford, F. A. (1968). An outline of theories of diffusion in the lower layers of the atmosphere. In D. H. Slade (Ed.), *Meteorology and Atomic Energy* (pp. 65–116). US Atomic Energy Comm.

Grasso, F. W., & Atema, J. (2002). Integration of flow and chemical sensing for guidance of autonomous marine robots in turbulent flows. *J. Environ. Fluid Mech.*, *1*, 1–20.

Grasso, F. W., Consi, T. R., Mountain, C. D., & Atema, J. (2000). Biomimetic robot lobster performs chemo-orientation in turbulence using a pair of spatially separated sensors: Progress and challenges. *Robotics and Autonomous Systems*, *30*(1-2), 115–131. doi:10.1016/S0921-8890(99)00068-8

Hassler, A. D., & Scholz, A. T. (1983). Olfactory imprinting and homing in salmon. New York: Spring-Verlag, 13-40.

Hayes, A. T., Martinoli, A., & Goodman, R. M. (2002). Distributed odor source localization. *IEEE Sensors Journal*, *2*(3), 260–271. doi:10.1109/JSEN.2002.800682

Ishida, H., Kagawa, Y., Nakamoto, T., & Moriizumi, T. (1996). Chemical source localization in the clean room by an autonomous mobile sensing system. *Sensors and Actuators. B, Chemical*, *33*(1-3), 115–121. doi:10.1016/0925-4005(96)01907-7

Kang, X., & Li, W. (2012). Moth-inspired plume tracing via multiple autonomous vehicles under formation control. *Adaptive Behavior*, *20*(2), 131–142. doi:10.1177/1059712311433131

Krishnanand, K. N., & Ghose, D. (2009). A glowworm swarm optimization based multi-robot system for signal source localization. *Design and Control of Intelligent Robotic Systems Studies: Computational Intelligence*, *177*, 49–68. doi:10.1007/978-3-540-89933-4_3

Li, W. (2010). Identifying an odour source in fluid-advected environments, algorithms abstracted from moth-inspired plume tracing strategies. *Applied Bionics and Biomechanics*, *7*(1), 3–17. doi:10.1155/2010/287801

Li, W., Farrell, J. A., & Card, R. T. (2001). Tracking of fluid-advected odor plumes: Strategies inspired by insect orientation to pheromone. *Adaptive Behavior*, *9*(3-4), 143–170. doi:10.1177/10597123010093003

Li, W., Farrell, J. A., Pang, S., & Arrieta, R. M. (2006). Moth-inspired chemical plume tracing on an autonomous underwater vehicle. *IEEE Transactions on Robotics*, *22*(2), 292–307. doi:10.1109/TRO.2006.870627

Liao, Q., & Cowen, E. A. (2002). The information content of a scalar plume-A plume tracing perspective. *Environmental Fluid Mechanics*, *2*(1/2), 9–34. doi:10.1023/A:1016231701928

Liu, K., Tan, Y., & He, X. G. (2010). Lecture Notes in Computer Science: Vol. 6145. *An adaptive staged PSO based on particles' search capabilities. Advances in Swarm Intelligence* (pp. 52–59).

Luenberger, D. G. (1979). *Introduction to dynamic systems*. New York: John Wiley.

Marjovi, A., & Marques, L. (2014). Optimal swarm formation for odor plume finding. *IEEE Transactions on Cybernetics*, *44*(12), 2302–2315. doi:10.1109/TCYB.2014.2306291 PMID:25415939

Marques, L., Nunes, U., & de Almeida, A. T. (2002). Olfaction-based mobile robot navigation. *Thin Solid Films*, *418*(1), 51–58. doi:10.1016/S0040-6090(02)00593-X

Nevitt, G. A. (2000). Olfactory foraging by Antartic procellariiform seabirds: Life at high Reynolds number. *The Biological Bulletin*, *198*(2), 245–253. doi:10.2307/1542527 PMID:10786944

Oyekan, J., & Hu, H. S. (2014). Biologically-inspired behaviour based robotics for making invisible pollution visible: A survey. *Advanced Robotics*, *28*(5), 271–288. doi:10.1080/01691864.2013.871578

Pielke, R. A. (1994). *Mesoscale meteorological modeling*. San Diego: Academic Press.

Rau, P., & Rau, N. L. (1929). The sex attraction and rhythmic periodicity in the giant saturniid moths. *Trans. Acad. Sci. Saint Louis*, *26(2)*, 82–221.

Russell, R. A. (2001). Tracking chemical plumes in constrained environments. *Robotica*, *19*(04), 451–458. doi:10.1017/S0263574700003283

Sabelis, M. W., & Schippers, P. (1984). Variable wind directions and anemotactic strategies of searching for an odour plume. *Oecologia*, *63*(2), 225–228. doi:10.1007/BF00379881

Sutton, J., & Li, W. (2008). Development of CPT_M3D for multiple chemical plume tracing and source identification. *Proceedings of the 2008 Seventh International Conference on Machine Learning and Applications* (pp. 470-475). 10.1109/ICMLA.2008.64

Sutton, O. G. (1947). The problem of diffusion in the lower atmosphere. *Quarterly Journal of the Royal Meteorological Society*, *73*(317-318), 257–281. doi:10.1002/qj.49707331704

Sutton, O. G. (1953). *Micrometeorology*. New York: McGraw-Hill.

Tan, Y., & Zheng, Z. Y. (2013). Research advance in swarm robotics. *Defence Technology*, *9*(1), 18–39. doi:10.1016/j.dt.2013.03.001

Tian, Y., Li, W., & Zhang, A. (2013). A lagrangian particle random walk model for simulating a deep-sea hydrothermal plume with both buoyant and non-buoyant features. *China Ocean Engineering*, *27*(2), 215–230. doi:10.100713344-013-0019-7

Tian, Y., Li, W., Zhang, A., Yu, J., Zhang, Q., & Li, Y. (2014). From simulation to validation: Moth-inspired chemical plume tracing with an autonomous underwater vehicle. *Proceedings of IEEE/MTS Oceans '14*, Canada.

Vergassola, M., Villermaux, E., & Shraiman, B. I. (2006). 'Infotaxis' as a strategy for searching without gradients. *Nature*, *445*(7126), 406–409. doi:10.1038/nature05464 PMID:17251974

Vickers, N. J. (2000). Mechanisms of animal navigation in chemical plumes. *The Biological Bulletin*, *198*(2), 203–212. doi:10.2307/1542524 PMID:10786941

Voges, N., Chaffiol, A., Lucas, P., & Martinez, D. (2014). Reactive searching and infotaxis in odor source localization. *PLoS Computational Biology, 10*(10), e1003861. doi:10.1371/journal.pcbi.1003861 PMID:25330317

Weissburg, M. J., Dusenbery, D. B., Ishida, H., Janata, J., Keller, T., Roberts, P. J. W., & Webster, D. R. (2002). A multi-disciplinary study of spatial and temporal scales containing information in turbulent chemical plume tracing. *Environmental Fluid Mechanics, 2*(1/2), 65–94. doi:10.1023/A:1016223500111

Weissburg, M. J., & Zimmer-Faust, R. K. (1994). Chemical plumes and how blue crabs use them in finding prey. *The Journal of Experimental Biology, 197*, 149–175.

Wu, W. C., Chang, D. S., & Zhang, F. M. (2013). A bio-inspired robust 3D plume tracking strategy using mobile sensor networks. *Proceedings of the 52nd IEEE Conference on Decision and Control*, 4571-4578.

Yuan, J., Oswald, D., & Li, W. (2015). Autonomous tracking of chemical plumes developed in both diffusive and turbulent airflow environments using Petri nets. *Expert Systems with Applications, 42*(1), 527–538. doi:10.1016/j.eswa.2014.08.005

Zarzhitsky, D., & Spears, D. F., Spears, William. M., & Thayer, D. R. (2004). A fluid dynamics approach to multi-robot chemical plume tracing. *Proceedings of the Third International Joint Conference on Autonomous Agents and Multiagent Systems* (Vol. 3, pp. 1476-1477).

Zarzhitsky, D., Spears, D. F., & Spears, W. M. (2005). Swarms for chemical plume tracing. *Proceedings of 2005 IEEE Swarm Intelligence Symposium* (pp. 249-256).

Zeng, J., Li, S., Li, Y., Wang, X., & Yan, S. (2014). Performance of the portable autonomous observation system. *Proceedings of MTS/IEEE Oceans'14*, Tapei. 10.1109/OCEANS-TAIPEI.2014.6964292

Zimmer, R. K., & Butman, C. A. (2000). Chemical signaling processes in the marine environment. *The Biological Bulletin, 198*(2), 168–187. doi:10.2307/1542522 PMID:10786939

KEY TERMS AND DEFINITIONS

Animal Swarms: A large number of small animals, e.g., a group of ants, bees, or social wasps, establish formations exhibiting a collective behavior, when they are in motion.

Chemical Plume Tracing: Navigating a robot system in response to real-time sensor information to find the plume, trace the plume towards its source, and identify the source location.

Formation Control: Formation control is a coordinated control for a fleet of robots to follow a predefined trajectory while maintaining a desired spatial pattern.

Moth-Inspired Plume Tracing Strategy: A plume tracing strategy is inspired by the maneuvers of moths flying upwind along a pheromone plume toward the source location.

Olfactory-Based Navigation: Olfactory-based navigation uses the chemical odor information as navigation landmark to identify the odor source location.

Plume Model: Plume models simulate the propagation of a chemical plume in diffusive or turbulent-dominated fluid flow environments.

Robot-Based Plume Tracer: Robot systems, including ground-based autonomous robots, aerial unmanned vehicles, and autonomous surface/underwater vehicles, are able to trace chemical plumes and identify their source locations.

Swarm Robots: A robot system with the distributed architecture controls a team of robots to form a desired collective behavior from the interactions between the robots and interactions of robots with the environment through the internet communication.

This research was previously published in the Handbook of Research on Design, Control, and Modeling of Swarm Robotics edited by Ying Tan, pages 421-455, copyright year 2016 by Information Science Reference (an imprint of IGI Global).

Section 5
Government and Public Welfare Use

Chapter 15
A Centralized Autonomous System of Cooperation for UAVs- Monitoring and USVs- Cleaning

Salima Bella
University of Oran1 Ahmed Ben Bella, Algeria

Assia Belbachir
Polytechnic Institute of Advanced Sciences, France

Ghalem Belalem
University of Oran1 Ahmed Ben Bella, Algeria

ABSTRACT

This article proposes an approach which deals with the problem of monitoring ocean pollution and cleaning dirty zones using autonomous unmanned vehicles. The authors present a cooperative agent-based planning approach for heterogeneous unmanned vehicles with different roles. Unmanned aerial vehicles (UAVs) monitor multiple ocean regions and unmanned surface vehicles (USVs) tackle the cleaning of dirty zones. Due to the rapid deployment of these unmanned vehicles, and the increase of ocean pollution, it is convenient to use a fleet of unmanned vehicles. Thus, most of the existing studies deal with the monitoring of different zones, the detection of the polluted zones and then the cleaning of the zones. In order to optimize this process, the authors' solution aims to use one UAV and one USV to reduce the pollution level of the ocean zones while taking into account the problem of fault tolerance related to these vehicles.

DOI: 10.4018/978-1-5225-8365-3.ch015

1. INTRODUCTION

Water is the fundamental element in all socio-economic processes. The water quality, due to the illegal oil rejection, has become a major issue for public authorities, national and international instance and organizations. For example, IFREMER (Institut Français de Recherche pour l'Exploitation de la Mer) figured out more than 50 million tons of waste between 0 and 250m deep in the Bay of Biscay, 15000 tons of plastic bags which circulate between two waters and 50,000 tons of these bags rest at the bottom of the gulf (Ifremer, 2006).

At the same time, manufacturers in many sectors face a challenge between producing better and reducing the pollution. These sectors produce very diverse effluents which require at each time new investigations and the development of specific treatment processes. Therefore, this has stimulated and encouraged the improvement of existing de-pollution techniques and the development of new processes, satisfying and conforming with the increasingly restrictive international standards in such environments. For that purpose, there was a growing interest on several sectors to use unmanned vehicles at first in the military domain and later in civilian domains (Cox, Nagy, Skoog, Somers & Warner, 2004). With the increase in the number of these vehicles at lower cost, several applications have emerged such as the help to weather forecasts, evaluation of natural disasters such as the floods, the monitoring of industrial and maritime regions and the cleaning of the oceanic regions where soil and water pollution is dispersed.

Region monitoring is an important task for collecting information and physically detecting the phenomena. Generally, the task of surveillance handles the problem of image processing and updating for a given region (Jakob, Semsch, Pavlıcek, & Pechoucek, 2010). The task of cleaning of a zone represents a very big interest because of its significant not only environmental or economic, but also social impact. An effective ocean region monitoring is a precious action to detect the dirty zones. To cope with this situation, a big mobilization of human, material, energy or even software resources is necessary (Jakob et al., 2010). However, the posed challenge by the monitoring and cleaning problem is the reduction of the required time to monitor a region and analyze it to detect dirty zones by an autonomous unmanned aerial vehicle (UAV). Then make decisions to launch autonomous unmanned surface vehicle (USV) to clean these disasters by considering the reduced energy of each vehicle in order to reduce the execution time of the mission. Due to the decisional autonomy and the low cost of unmanned vehicles, they became a candidate solution in the deployment of such field: monitoring and cleaning dirty zones.

The cooperation of Unmanned Aerial Vehicle (UAV) and Unmanned Surface Vehicle (USV) attracts more and more researches, essentially for the complementary skills provided by each type of vehicle (e.g. payload, computing power, motion speeds, local vision, etc.). Thus, the deployment of UAV and USV in a cooperative way allows a wide range of applications.

In this regard, in order to improve the efficiency of these applications, some researchers have integrated a centralized decision. This decision includes a real or artificial agent that automatically plans or organizes unmanned vehicle activities, monitors the performance of their objectives, and monitors the state of the systems.

The main objective of this work is to propose a cooperative approach based on UAV and USV to carry out in a coordinated way a set of dedicated tasks: monitoring, detection and cleaning of dirty zones in an ocean zone.

In this paper, we propose a centralized architecture for UAVs and USVs. This paper focus on monitoring unknown ocean regions, detecting dirty zones and cleaning these zones using USVs. The geo-referenced images of the UAV are collected using the sensors that will be used in this work to create an exploration map of the region with known a priori path and to determine the positions of the dirty zones (the way points are predefined). The UAV sends the explored map and the data from the dirty zones to the general coordinator (represented by an artificial agent). After an analysis of these captured data, the USV uses its sensors and the provided map by the UAV, through the general coordinator, to navigate to the assigned dirty zone and clean it. The novelty in this work is that this proposed method of air-sea cooperation is extended by a fault-tolerance service for either UAVs or USVs by replacing failing vehicles by equivalent ones.

The paper is organized as follows: in Section 2, we present some related works; we describe, in Section 3, the proposed approach for the various semi-autonomous unmanned vehicles; we propose, in Section 4, a logical formalization of our proposal; and we illustrate, in Section 5, an example to simulate the operation of our approach. We conclude in Section 6 our work by synthesizing our approach and tracing directions of future works.

2. RELATED WORK

The field of heterogeneous unmanned and autonomous vehicles such as UAVs, UGVs (Unmanned Ground Vehicles) has received some attention while several reactive approaches and deliberative search methods have been proposed. The applications of such vehicles are diverse, they can be used for autonomous aerial surveillance of urban areas in a geometrical and complex environment using a group of UAVs as it is mentioned in the work of (Jakob et al., 2010). This work consists of finding the shortest path for the UAV that guarantees the coverage of all points of interest while taking into account occlusions. This work uses the "Zig-Zag algorithm" and a "Greedy algorithm" to allocate areas to drones in an optimal way.

UAVs can also collaborate with UGVs to accomplish a common mission such as inspection of the area, transport of objects, etc. The UAV provides an overview of the explored area to assist a human operator in supervising and guiding a UGV group (Harik, Guérin, Guinand, Brethé & Pelvillain, 2014). It is also used to locate and eliminate simulated mines (UneXploded Ordnance (UXO)) in an unknown environment where a UAV creates a map of the explored region to determine the UXOs positions to send towards a base station. The UGV uses its sensors and its extracted waypoints to visit each waypoint in order to eliminate UXOs using A* search algorithm to minimize the total traveled distance by the UGV (E. MacArthur, D. MacArthur & Crane, 2005).

In this last work (E. MacArthur et al., 2005), all the communications between the vehicles and the base station are carried out using the messages from JAUS (Joint Architecture for Unmanned Systems). This architecture manages the problem of interoperability between different unmanned systems such as the embedded system "Autonomous System cooperative (CAS)" proposed in (Hager, Kwon, Zarzhitsky, & Pack, 2011). The CAS includes a complete collection of material and software modules, designed for the real time coordination of the autonomous UAVs and UGVs for cooperative missions in an unknown middle. In the work of (Belbachir & Escareno, 2016), forestry monitoring is proposed, where the authors propose a distributed approach for the mission of localized forest fires in real time via a cooperative exploration strategy. The authors have integrated an autonomous decisional system adapted to UAVs using a task planning, combined with a drone controller to execute and control the generated path.

However, there are other works of monitoring (Martinic, 2014) (O'Young & Hubbard, 2007) (Finn & Wright, 2012), which try to solve the problem of the water pollution of seas and oceans (Valada et al., 2014) (Gal, 2011). The main causes of the pollution of these maritime spaces are the oil slicks caused by shipwrecks of oil tankers as well as the waste rejected by factories at the bottom of the ocean, etc. Thus, these ecological consequences have been catastrophic. An overall assessment of the mortality undergone by the marine species was attempted. For that purpose, we are trying to contribute to this problem by proposing a solution to monitor the ocean regions and clean dirty zones using a heterogeneous group of unmanned vehicles. We are proposing a centralized approach based on a cooperation between vehicles air-sea, and an exploration strategy to generate and plan the paths for UAVs and USVs, extended by a fault-tolerance service.

3. PROPOSED APPROACH

Seas and oceans are major elements in the hydrological equilibrium of the earth. Pollution of the water of the seas and oceans is one of the main pollution of the world with that of the atmospheric pollution. It is in this problem that we try to contribute to monitor the ocean regions and to clean dirty zones using UAV and USV. The objective of this work is to propose an approach based on cooperative air-sea vehicles: UAV and the USV.

3.1. Decisional Level of our System

In this part, we describe the decisional level and the different entities used in our centralized approach.

Figure 1 shows the decisional level of our deployment system. General coordinator, UAVs and USVs are monitoring each region and cleaning vehicles execute the operation of cleaning dirty zones. This cleaning vehicle is equipped with a device allowing the emission of light of LEDs type. This LEDs shows the level of the remaining energy.

Figure 2 shows the decision-making hierarchy in our proposed approach where the general coordinator has the highest decision, namely the launch of the tasks to initiate the monitoring, cleaning and recovery vehicles. The monitoring drone is loaded of a lower decision, because it is introduced by the coordinator and qualified as a semi-autonomous drone UAV. It has the role to oversee its cleaning vehicle. The

Figure 1. Deployment of our centralized system based on a Coordinator, UAVs and USVs

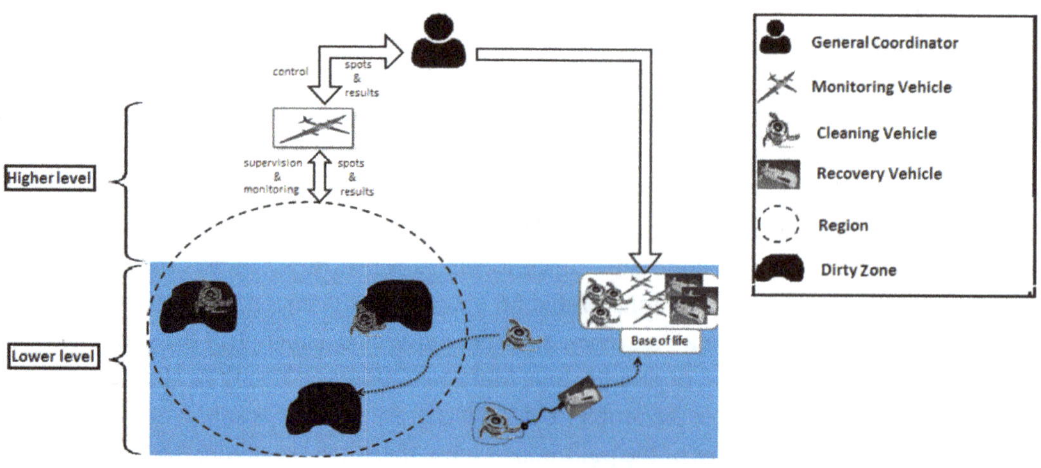

Figure 2. Decision-making hierarchy

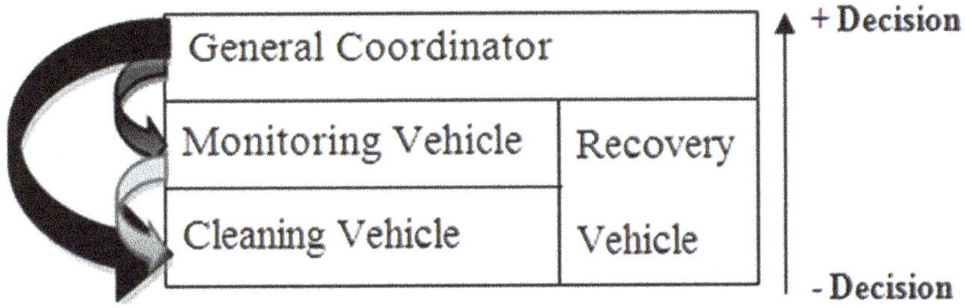

latter has for role to realize the action of cleaning of the zone which has been intended to emit a light according to the availability of its energy. This latter, is qualified like a semi-autonomous USV. The recovery vehicle is initiated by the general coordinator, aims to recover the failing vehicles according to a demand outside the base of life, it is qualified as a special unmanned vehicle.

3.2. Environmental Modeling

3.2.1. Description of the Environment

1. **Set of Tasks:** We describe the two high level tasks that are used by UAV and USV.
   a. **Monitoring Task:** This is the set of different monitoring tasks (t_m) of a given region.
   b. **Cleaning Task:** This is the set of different cleaning tasks (t_c) of a dirty zone.
2. **Set of Vehicles:** We define different used vehicles with their roles.
   a. *Monitoring vehicle UAV_{MR} (Supervisor Sup_{MR})* are homogeneous and identical UAVs. They are responsible of:
      i. Monitoring their regions, their dirty zones, their cleaning vehicles.
      ii. Supervising their cleaning vehicles (USV_{CZ}).
      iii. Launching and return the data and the results to the general coordinator ($General_{Crd}$).
   b. *Cleaning vehicle (USV_{CZ})* are homogeneous and identical UAVs. They are responsible of:
      i. Cleaning their dirty zones;
      ii. Launching and returning the collected data to their Sup_{MR} (UAV_{MR});
      iii. Coming-back to the base of life which represents a storage area of UAV_{MR}, USV_{CZ} and $Vehicle_{Rec}$.
   c. *Recovery vehicle* ($Vehicle_{Rec}$) These vehicles are special named recovery vehicle ($Vehicle_{Rec}$). In the sense where these vehicles make it possible to bring back the vehicles that are out of order towards the base of life.
3. **Descriptive Table of The Different Agents:** The following table (Table 1) shows different types of agents, their names and their roles.
4. **Set of Regions:** It is composed of two levels; an higher level (atmospheric space) where we find the UAV_{MR} and a lower level (maritime space) where we find the USV_{CZ}, base of life and dirty zones. We define our environment as follow:
   a. We divide our environment into a number of regions which need to be monitored.
   b. Using a single monitoring drone (UAV_{MR}) for each region.

Table 1. Descriptive table of various defined agents

Type of Agent	Agent Name	Description / Agent Roles
$General_{Crd}$	General Coordinator	- This agent is managing the base of life. - Launch each vehicle of its region with the different allocated tasks. - Responsible to collect the data related to regions and dirty zones from different vehicles.
$Vehicle_{Rec}$	Recovery vehicle (rescuer at sea)	Capable of recovering the faulted vehicles from the water surface to the base of life.
USV_{CZ} *(free)*	Cleaning vehicle available (free state)	It is a cleaning vehicle that has completed its task or is ready for another mission.
USV_{CZ} *(discharge)*	Cleaning vehicle in a discharged state	It is a cleaning vehicle, that has not yet completed its task and its energy is low. Which will not allow it to accomplish its cleaning task.
USV_{CZ} *(prepared)*	Cleaning vehicle is ready for the mission (prepared state)	It is a cleaning vehicle prepared in the base of life by the $General_{Crd}$ which will replace the USV_{CZ} *(discharge)*.
Sup_{MR} *(USV_{CZ}_discharge)*	Supervisor of USV_{CZ} *(discharge)*	It is the supervisor of the cleaning vehicle in the discharge state.
Sup_{MR} *(USV_{CZ}_free)*	Supervisor of USV_{CZ} *(free)*	It is the supervisor of cleaning vehicle in free state.

c. Using a cleaning vehicle for each dirty zone in each zone.

d. At each detected dirty zone by UAV_{MR}, only one cleaning vehicle is launched by $General_{Crd}$.

5. **Dirty Zones:** They represent dirty part where we find the water pollution, such as oil slicks, plastic waste, etc. In this work, we propose that the water color is the metric that shows that a region is dirty.
   a. We define the degree of dirt, where we assume four colors corresponding to the following intervals [0,25],] 25, 50],] 50, 75],] 75,100]. Each interval is represented respectively by the following color: the white (the water is clear where there is no or a little of dirt); the light brown (the dirt is low/weak); the dark brown (the dirt is average) and the black (the dirt is strong).
   b. Each zone is characterized by a list " $List_{zone}$ " which delimits its borders by "$Coordinate_{zone}$ ". The latter is composed by degree of dirt "$Degree_{cell}$ "; the position of the cell "$Position_{cell}$ " and it is attached to a zone by "$Position_{zone}$ ".
6. **Base of Life (Storage Area):** It is a zone where we store a fixed number of UAV, USV and $Vehicle_{Rec}$.

3.2.2. The Used Parameters Used for Our Proposal

The used parameters in this approach are represented by the following table (Table 2):

Table 2. The used parameters for our proposal

Parameters	Description
Threshold	It is a fixed threshold that allows classifying the zone coordinates according to the $Degree_{cell}$.
$List_{energy\_usv}$ $(Id_{USVCZ}, Cap_{energyCZ})$	This list contains the energy capacity of a USV during cleaning as well as its identifier.
$List_{zone}$ $(Degree_{cell}, Position_{cell}, Position_{zone})$	This list contains all the zone $Coordinate_{zone}$.
$List_{zoneS}$	It is the $List_{zone}$ sorted in relation to the degree of dirt.
$List_{threshold}$	This list contains the coordinates of the zone that are compared with the dirt threshold.
Nbr_{CZ}	A function that gives the number of selected USV_{CZ} by $General_{Crd}$.
$Parameters_{start\text{-}up(M)}$ $(Id_{region}, Position_{region}, Path_{region})$	A triplet of start parameters for each UAV_{MR} which represents the identifier, position, and path of a region.
$Parameters_{start\text{-}up(C)}$$(Id_{region}, Id_{zone}, Position_{zone}, Path_{zone}, Id_{UAVMR})$.	A triplet of start parameters for each USV_{CZ} which represents the identifier of a USV_{CZ}, identifier of a region, identifier of an zone thus its position and path.
$List_{characteristics}$ $(Id_{zone}, Id_{USVCZ}, Dur_{EC}, Cons_{EC}, Dur_{ED}, Cons_{ED})$	This list contains the USV_{CZ} characteristics: • Id_{zone}: identifier of an zone, • Id_{USVCZ}: identifier of a USV_{CZ}. • Dur_{EC}, $Cons_{EC}$: the duration and the energy consumption of cleaning, • Dur_{ED}, $Cons_{ED}$: the duration and the energy consumption of displacement.

3.3. Main Phases of Our Proposed Approach

Our approach is made up of a set of phases which are: assignment, preparation, recovery, monitoring, decision-making and cleaning phases.

3.3.1. Assignment Phase

The assignment phase is triggered by the general coordinator " $General_{Crd}$ ", it consists to initialize the vehicles monitoring according to the number of regions, associating a single drone for each region. The main objective of this phase is to allocate for each region one "UAV_{MR} ".

3.3.2. Preparation Phase

The preparation phase consists in preparing the methods of navigation, planning and the existence of vehicles in the base of life.

- **Navigation:** The navigation can be decomposed into two types: reactive navigation and global navigation, as described in (Chatila & Laumond, 1985) (Batalin, Sukhatme & Hattig, 2004). In our work, we used a global mapping navigation. The latter can be used either in metric (geometric), topological or occupation grid (Baba & Chatila, 2008). For the approach used, we have used grid-type occupation navigation. This type of navigation will allow us to reduce our search space where the vehicle discretizes the environment in the cells.

- **Path Planning:** Several path planning methods have been proposed as in (Chatila & Laumond, 1985) (Hachour, 2008) (Zhang, 2011) for example roadmap, cell decomposition, potential functions, etc. We advocate allocating a known path a priori for each vehicle so that it can reach its region or its dirty zone.
- **Trajectory Planning:** There are several strategies of explorations in the robotic domain. We can cite Greedy, Hungarian and K-mean Method (Kulich, Juchelka & Přeučil, 2015) (Juliá, Gil & Reinoso, 2012). In our work, we assume that:
    - A simple method where there are no solid (hard) obstacles that prevent the vehicle from navigating and planning its trajectory exists in its environment.
    - The vehicle has the ability to move in four directions (left, right, forward and back) in its environment to achieve its goal.
    - The step of path planning of drone is done periodically according to the data that it encounters.
- Checking the existence of vehicles in the base of life: this method starts by feeding the base of life, which consists of putting in this base of life (can be a boat, an island, ...) a number of vehicles that it is monitoring, cleaning and recovery. The following algorithm (Algorithm 1) illustrates the case where the monitoring vehicle will need to assign a cleaning vehicle to clean a dirty zone. It sends a request to the General Coordinator ($General_{Crd}$) and then the latter will check the existence of a USV_{CZ}. If the $General_{Crd}$ finds an available vehicle then it will activate a USV_{CZ} to reach the designated zone and associate a response to the query made by the Sup_{MR}; otherwise, it will request it to wait for a USV_{CZ} by registering this request in a waiting list.

3.3.3. Recovery Phase

We endowed our solution with a fault-tolerant service. This service is installed on a special vehicle called a repair vehicle ($Vehicle_{Rec}$). It allows to bring the failed vehicle towards the base of life (See "Algorithm 2").

3.3.4. Monitoring Phase

The monitoring phase consists mainly of two stages:

1. The launching phase of monitoring vehicles. In this step, the $General_{Crd}$ triggers the monitoring operation according to the following actions:
    - Assign for each UAV_{MR} a $Parameters_{start\text{-}up(M)}$ (Id_{region}, $Position_{region}$, $Path_{region}$) before starting.
    - UAV_{MR} is launched from the base of life, and it follows the path to reach its region.
    - Once UAV_{MR} (Sup_{MR}) arrives at its region, it follows the same steps:
        - Sup_{MR} captures the data of its environment from its initial position (its region).
        - Sup_{MR} discretizes its region in two levels: higher (atmospheric space) and lower (maritime space).
        - Sup_{MR} plans its displacement (trajectory) through the representation of its model so that it can fly under the good conditions and carry out the task of its region monitoring.
2. Step Collection and Analysis of the Coordinates Zone
    - Sup_{MR} periodically processes the data of its discretized region (Maritime space) to identify dirty zones using the algorithm (Algorithm 3).

Algorithm 1. Checking the existence of vehicles in the base of life

```
begin
while need a USV_CZ = true do
    General_Crd ← send the request by SUP_MR
    if General_Crd receives the request then
    search for USV_CZ in the base of life
    if USV_CZ = True then
    send the response (exists a USV_CZ) to SUP_MR
    else
    send the response (wait) to the SUP_MR
    end if
    General_Crd save the response
    SUP_MR receives the response
    end if
end while
end
```

Algorithm 2. Recovery phase

```
begin
if USV_CZ = state out of order then
General_Crd ← send the request by SUP_MR
Vehicle_Rec ← send the request by General_Crd
if Vehicle_Rec receives the request then
Vehicle_Rec returns the USV_CZ
end if
end if
end
```

- The Sup_{MR} sorts the $List_{zoneS}$ then saves it in $Liste_{Threshold}$, relative to a threshold, and then sends the latter to the $General_{Crd.}$
- We propose a diagram of activity shown in Figure 3, to describe the approach of Sup_{MR} and to detail the step of collecting and analyzing the zones.

3.3.5. Decision-Making Phase on the Number of Cleaning Vehicles

The $General_{Crd}$ analyzes the received data (the list of $List_{threshold}$). This phase is described by Algorithm4. The step of selecting the number of cleaning vehicles is described by Algorithm 5.

3.3.6. Trigger Phase of the Cleaning Process

After the execution of Algorithm 4, a decision is made to trigger the cleaning operation. The selected USV_{CZ} is assigned to the appropriate zone by the $General_{Crd}$ according to a known a priori path. But in this variant of our proposal, we considered that only one cleaning vehicle is assigned for one dirty zone, and each USV_{CZ} contains LEDs that allow to know the energy level (**Led**$_{green}$: Strong energy; **Led**$_{yellow}$: Average energy; **Led**$_{red}$: Low energy).

The cleaning process. This process consists of several steps:

Algorithm 3. Collect and analyze the zones

begin
while $\text{Collection}_{Period} \neq$ null **do**
Collection_{Data} ()
while $\text{Treatment}_{Data} \neq$ null **do**
for all Coordinate_{zone} captured **do**
save Coordinate_{zone} in List_{zone}
sort List_{zone} in List_{zoneS}
classify List_{zoneS} compared to threshold in $\text{List}_{threshold}$
end for
end while
end while
end

- $General_{Crd}$ sends the identifier of the selected USV_{CZ} to its supervisor (Sup_{MR}).
- Upon receipt of this message, Sup_{MR} sends its exploration map (discretized environment) to the $General_{Crd}$.
- USV_{CZ} will know the $Parameters_{start\text{-}up(C)}(Id_{region}, Id_{zone}, Position_{zone}, Path_{zone}, Id_{UAVMR})$ + the explored map by a message from the $General_{Crd}$..
- USV_{CZ} is launched on the water and its $\text{Led}_{\text{green}}$ is active.
- USV_{CZ} begins to swim by following the indicated path to reach its dirty zone.

Figure 4 describes briefly our used sequence diagram.

3.3.7. Cleaning Phase

- **Initialization:** The monitoring drone periodically monitors its own USV_{CZ} and saves its characteristics in a list called: $List_{characteristics}$.
- **Starting Cleaning:** USV_{CZ} plans its way from its supervisor's explored map so that it can move and at the same time clean its zone as long as the $\text{Led}_{\text{green}}$ is activated.
- **Termination of the Cleaning Operation:** This operation consists of identifying the termination of the cleaning operation and collecting the current characteristics ($List_{characteristics}$) of each USV_{CZ}.
  Case 1. The cleaning task is completed (Figure 5):
- When the USV_{CZ} finishes its cleaning task, it informs Sup_{MR} by sending a message.
- Upon reception of the message, Sup_{MR} informs the USV_{CZ} to return to the base of life.
- When the USV_{CZ} arrives at the base of life, it informs Sup_{MR} by sending a message.
- Upon reception of the message, Sup_{MR} stops saving the the characteristics of the USV_{CZ}.
- Sup_{MR} sends $List_{characteristics}$ of USV_{CZ} to the $General_{Crd}$ for registration.

Figure 3. Diagram of activity to collect and analyze the coordinates zone

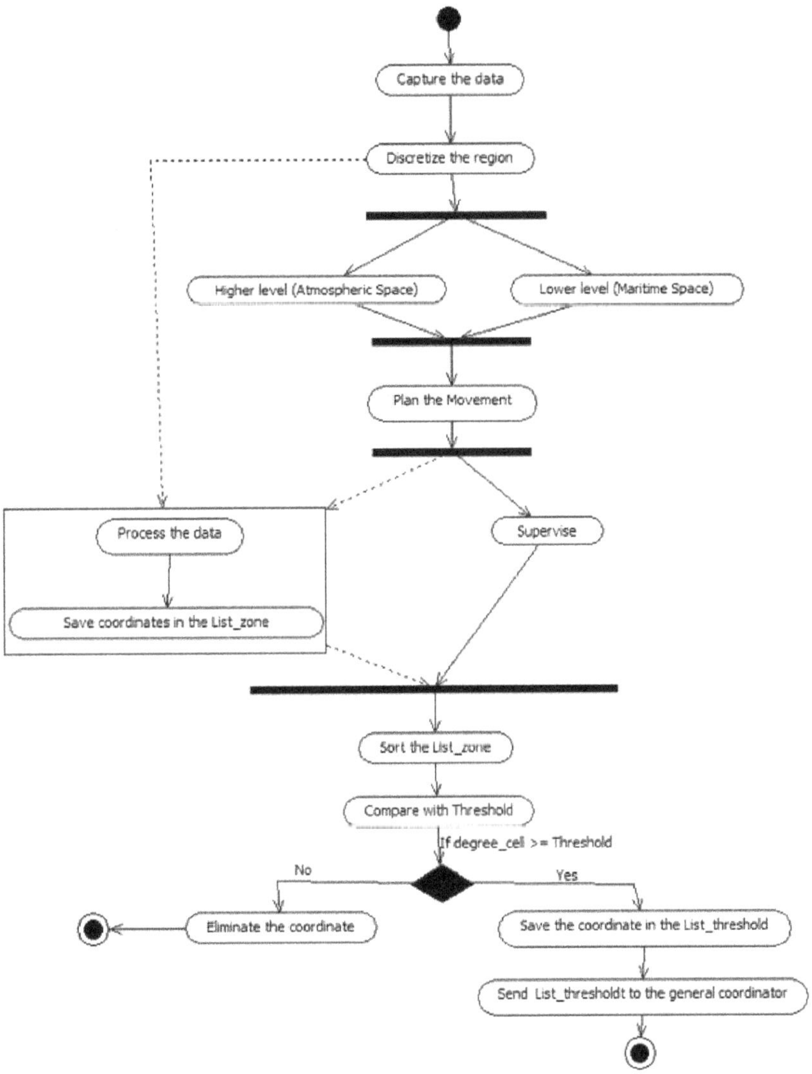

Algorithm 4. Decision on number of USV_{CZ}

1: **begin**
2: save $\text{Lists}_{threshold}$
3: calculate the average ($\text{Lists}_{threshold}$)
4: calculate the min ($\text{Lists}_{threshold}$)
5: calculate the number (USV_{CZ})
6: **end**

Algorithm 5. Calculate the number (USV_{CZ})

m : USV_{CZ} number in base of life
sum $\leftarrow$ 0
begin
for all average $\left(\text{List}_{threshold}\right)$ calculated **do**
$x \leftarrow \dfrac{\text{average}\left(\text{List}_{threshold}\right)}{\min\left(\text{List}_{threshold}\right)}$
save x in List_X
sum $\leftarrow$ sum + xy
end for
m $\leftarrow$ sum
$y = \dfrac{m}{x}$
for all x found in List_X **do**
 # calculate the number (USV_{CZ})
num $\leftarrow$ y*x
end for **end**

Figure 4. Sequence Diagram "Trigger the cleaning process for a USV_{CZ}"

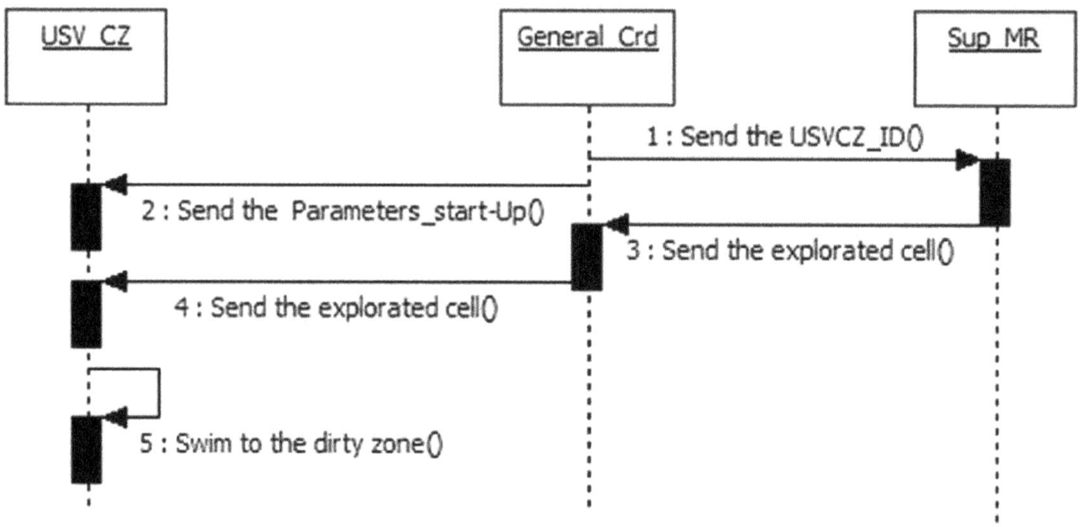

Case 2. The cleaning task is not completed and the yellow Led_{yellow} is activated: in this case, we have illustrated three situations.

- **The first situation is presented in Figure 6:**
  - During the monitoring of Sup_{MR}, the latter receives data periodically.

Figure 5. Diagram of activity for "Termination of the cleaning operation" (Case 1)

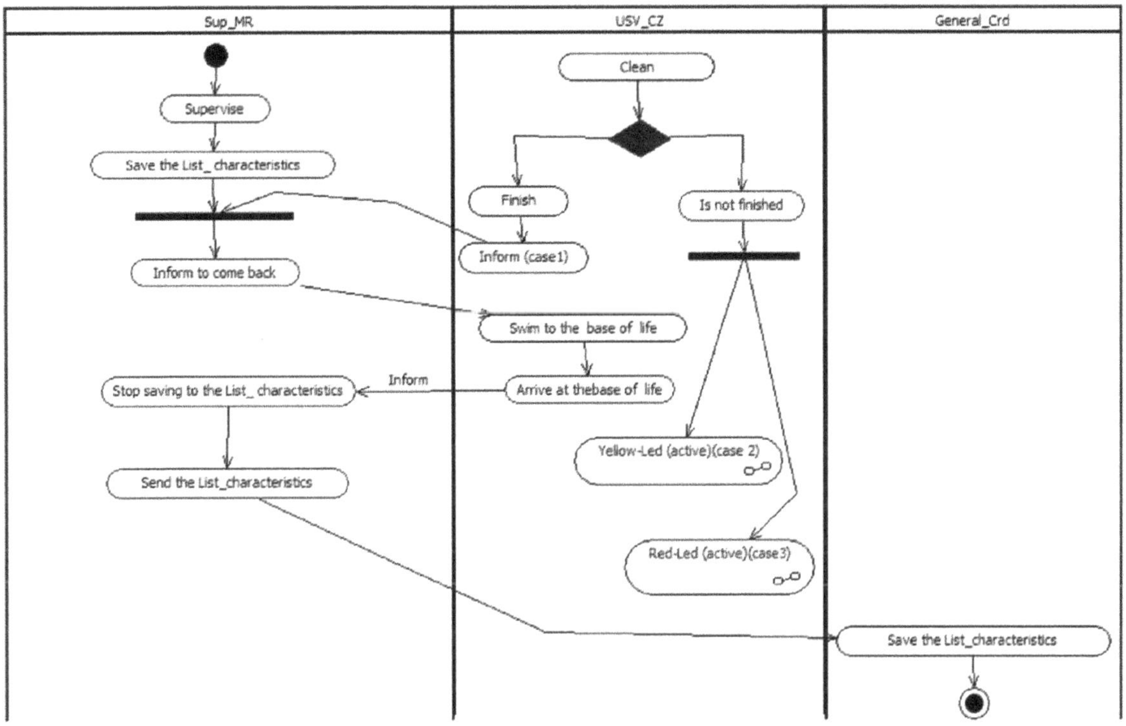

- At some point, the Led_{yellow} of the USV_{CZ} is activated so it sends a message to its Sup_{MR} to inform it that its energy is in an average level.
- Sup_{MR} saves information in $List_{energy\_usv}$ and processes it.
- After a deadline of wait, if Sup_{MR} does not receive another message (a message from another USV_{CZ} that has completed its task in another zone of the same region) before exceeding this deadline, then it passes to the second situation .
- Otherwise, Sup_{MR} processes the received message and then sends a message (inform about its new task with $Parameters_{start\text{-}up(C)}$ and the coordinate of the zone) to the USV_{CZ} (free) which is able to replace USV_{CZ} (unloaded*)* and at the same time inform the $General_{Crd}$.
- Sup_{MR} waits for the Led_{red} of USV_{CZ} (unloaded*)* activation message to launch the USV_{CZ} *(*free).

- **The second situation is shown in Figure 7:**
  - If there is no USV_{CZ} *(free)* in the same region, then Sup_{MR} *($USV_{CZ_}$discharge)* sends a message with the coordinates of the zone to the $General_{Crd}$.
  - $General_{Crd}$ broadcasts the received message to neighbors of Sup_{MR} *($USV_{CZ_}$discharge)* to select the USV_{CZ} (free) have completed their tasks among their USV_{CZ}.
  - If there is one or several USV_{CZ} (free), then each Sup_{MR} will send the message (the list of USV_{CZ} (free) + its energy capacity (LEDs) + $List_{characteristics}$) to the $General_{Crd}$.
  - $General_{Crd}$ returns the received message to Sup_{MR} *($USV_{CZ_}$discharge).*

Figure 6. Diagram of Activity of "Termination of the cleaning operation" Case2: the first situation)

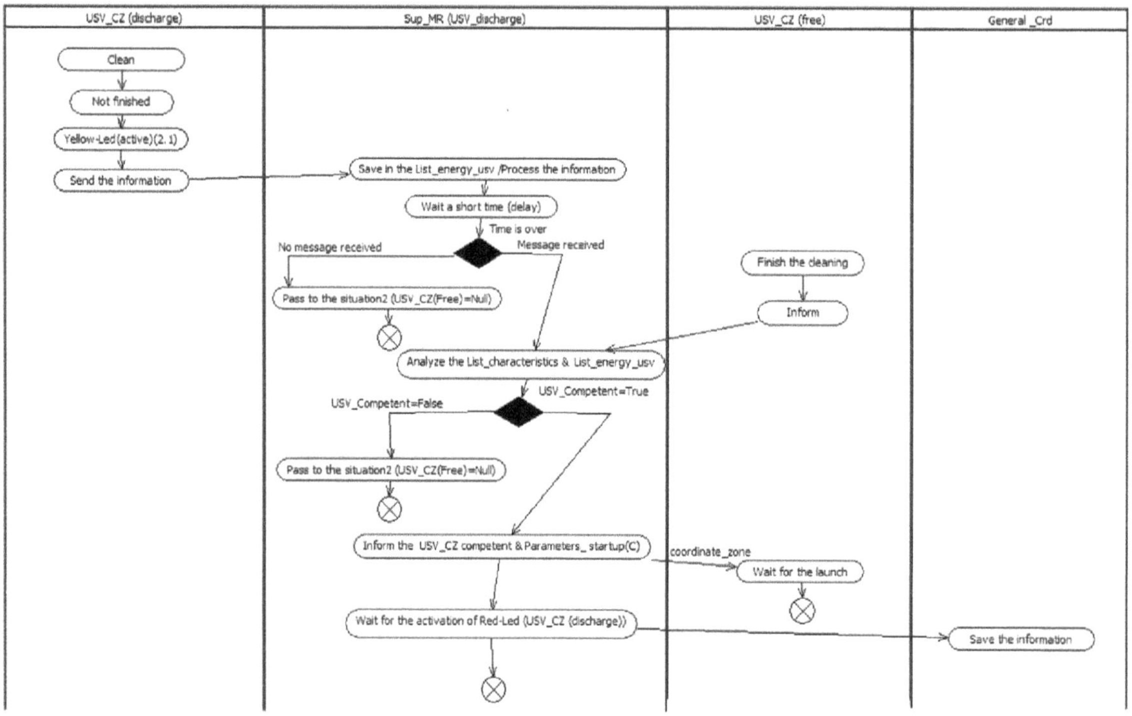

- $Sup_{MR}\ (USV_{CZ\_}discharge)$ will process the messages received from each Sup_{MR} then if there is not a USV_{CZ} *(free)* that is able to complete the task, then it passes to the third situation. Otherwise it chooses the USV_{CZ} (free) competent based on energy.
- $Sup_{MR}\ (USV_{CZ\_}discharge)$ sends a message (inform: USV_{CZ} *(free)* competent +the new task with its coordinate + $Parameters_{start\text{-}up(C)}$) to the $Sup_{MR}\ (USV_{CZ\_free})$ of the selected USV_{CZ} *(free)* by the intermediary $General_{Crd}$.
- $General_{Crd}$ disseminates this information to other Sup_{MR}.
- $Sup_{MR}\ (USV_{CZ\_}discharge)$ waits for the Led_{red} of USV_{CZ} *(discharge)* to be activated to notify the $General_{Crd}$.
- $Sup_{MR}\ (USV_{CZ\_}free)$ waits for the $General_{Crd}$ signal to inform its USV_{CZ} *(free)* to start.

- **The third situation is illustrated in Figure 8:**
  - If there is not one USV_{CZ} *(free)*, the other Sup_{MR} sends a message (USV_{CZ} *(free)* not available to the $Sup_{MR}\ (USV_{CZ\_}discharge)$ through the $General_{Crd.}$
  - $Sup_{MR}\ (USV_{CZ\_}discharge)$ sends another message (need for a USV_{CZ} with the coordinates of the zone and $Parameters_{start\text{-}up(C)}$) to the $General_{Crd}$.
  - $General_{Crd}$ processes the message and then chooses a USV_{CZ} according to the coordinates in the base of life.
  - Waits for the Led_{red} of the USV_{CZ} *(discharge)* to be activated to launch the signal to the $General_{Crd}$.
  - $General_{Crd}$ waits for the signal to throw the USV_{CZ} *(prepared)*.

Figure 7. Diagram of Activity of "Termination of the cleaning operation" (Case2: the second situation)

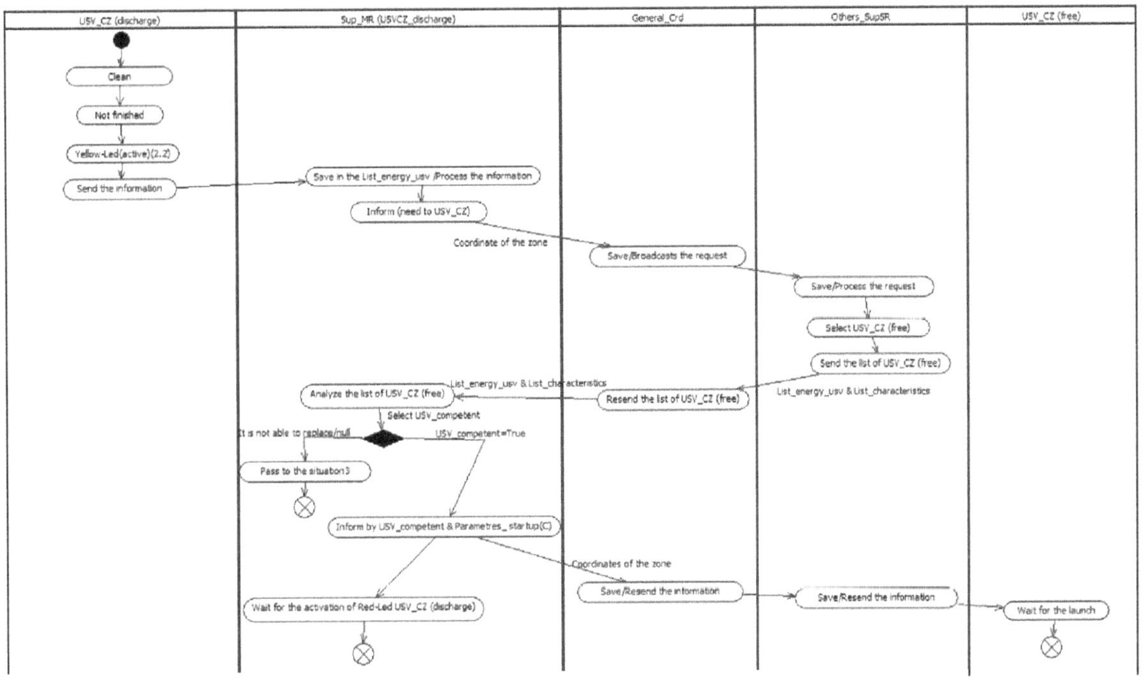

Figure 8. Diagram of Activity of "Termination of the cleaning operation" (Case2: the third situation)

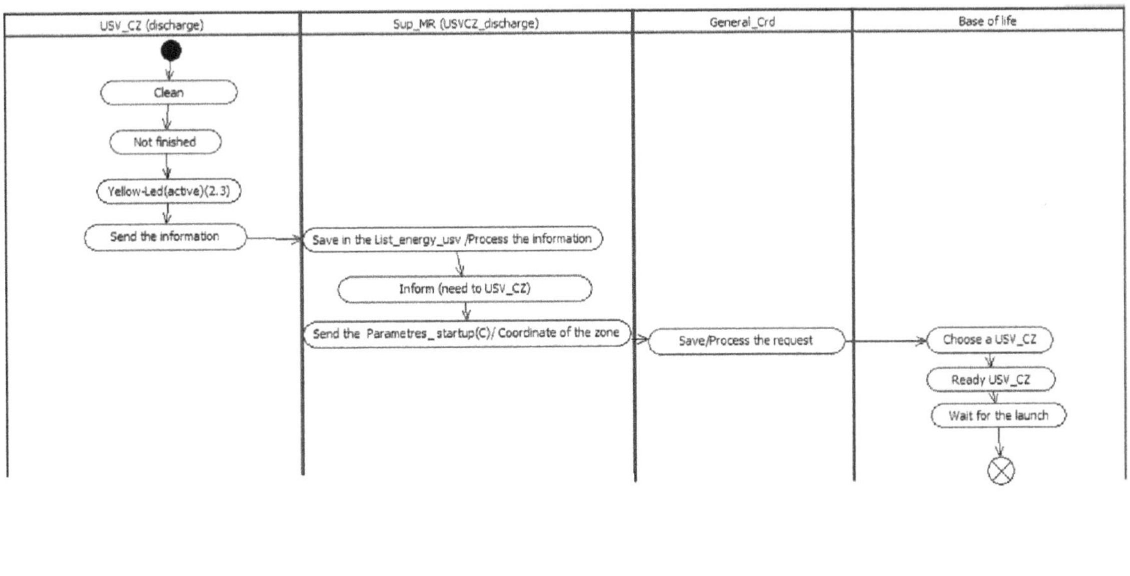

Case 3. The cleaning task is not completed and the Led$_{red}$ is activated: this case is realized for all the situations cited above (continued).

- When the energy of *USV$_{CZ}$(discharge)* is low so the Led$_{red}$ will be active and it sends this information to its supervisor (all situations).
- Upon receipt of this information, *Sup$_{MR}$(USV$_{CZ_}$discharge)* sends a message (inform about launch) towards:
    - Directly to *USV$_{CZ}$ (free)*: If that vehicle belongs to another zone in the same region (first situation).
    - *General$_{Crd}$* to another *Sup$_{MR}$(USV$_{CZ_}$free)*: To throw the *USV$_{CZ}$ (free)* to prepare it by the *Parameters$_{start\text{-}up(C)}$* so that it can calculate its trajectory and swim to its new dirty zone and clean it, and *General$_{Crd}$* broadcasts this information to the other *Sup$_{MR}$* (second situation).
    - *General$_{Crd}$* to *USV$_{CZ}$ (prepare)*: Launch the *USV$_{CZ}$ (prepared)* which is in the base of life in order to prepare it by the *Parameters$_{start\text{-}up(C)}$* and calculate its trajectory as well as swim towards its new dirty zone and cleaning (third situation).
- *Sup$_{MR}$(USV$_{CZ_}$discharge)* informs the *USV$_{CZ}$(discharge)* to return by the received message so that it can swim from its zone to the base of life (and to load) (all situations).
- *Sup$_{MR}$* sends the *List$_{characteristics}$* of the *USV$_{CZ}$ (discharge)* to *General$_{Crd}$* for the recording.
- *Sup$_{MR}$(USV$_{CZ_}$discharge)* re-launch a new *List$_{characteristics}$* for the *USV$_{CZ}$ (prepared)* or *USV$_{CZ}$ (free) (second and third situation).*

4. LOGICAL FORMALIZATION OF THE PROPOSAL

4.1. Conceptual Model for Planning

A conceptual model is a simple theoretical device for describing the main elements of a problem (Ghallab, Nau & Traverso, 2004). Most of the planning approaches described in (Ghallab et al., 2004) rely on a general model, which is common to other areas of computer science, the model of state-transition systems (also called *discrete-event systems)* (Ghallab, Nau & Traverso, 2004, 2016).

Formally, a state-transition system is a 4-tuple **Σ =(S, A, E, γ)**, where:

- $S=\{s1,s2\ ,\ldots\}$ is a finite or recursively enumerable set of states;
- $A=\{a1,a2\ ,\ldots\}$ is a finite or recursively enumerable set of actions;
- $E=\{e1,e2\ ,\ldots\}$ is a finite or recursively enumerable set of events; and
- $\gamma : S \times A \times E$ is a state-transition function.

A state-transition system may be represented by a directed graph whose nodes are the states in S. If $s' \in \gamma(s,u)$ where u is pair (a,e), $a \in A$ and $e \in E$, then the graph contains an arc from s to s' that is labeled with u. Each such arc is called a *state transition* (Ghallab et al., 2004).

Example 1: Figure 9 shows a state-transition system for a region involving two locations, a base of life (for example a boat) contains vehicles, an object of the crane type for picking up, puting down and releasing vehicles, a dirty zone, a monitoring vehicle. Here, the set of states is {s0,s1, s2, s3, s4}, the set of actions is {take, put, move1∧start-monitor, move2∧fin-monitor, discover, undiscover}, and there are no events. The arc (s2, s3) is labeled with the action " discover", the arc (s1, s0) with the action " stayinbase", and so forth.

The system in Figure 10 is similar to the system in Figure 9 but the difference is the use of a single cleaning vehicle to clean the dirty zone, There is no monitoring drone and the region is composed of three locations. Here, the set of states is {s0, s1,s2,s3, s4,s5}, the set of actions is{take, put, start-clean, fin-clean, move1, move2}, and there are no events. The arc (s0, s1) is labeled with the action "put ", the arc (s2, s3) with the action " start-clean", and so forth.

4.2. A Running Example "USV-Cleaning" and "UAV-Monitoring"

The planning procedures and techniques are illustrated on a simple yet nontrivial running example called the USV-Cleaning (or USV-C domain applied to cleaning vehicles) and another called UAV-Monitoring (or UAV-M domain applied to monitoring vehicles). The USV-C domain is a complement to the UAV-M domain.

The environment is a generalization of Example 1 (Figure 9) and (Figure 10) that presents regions. For the UAV-M domain, it represents an atmospheric space where monitoring vehicles are moving to monitor their regions. The USV-C domain represents a maritime space with several elements that can contain bases of life containing vehicles used, cranes to deposit cleaning vehicles and dirty zones.

An abstract version of these domains can be defined by giving seven finite sets of constant symbols:

Figure 9. A state-transition system for the simple domain UAV-Monitoring

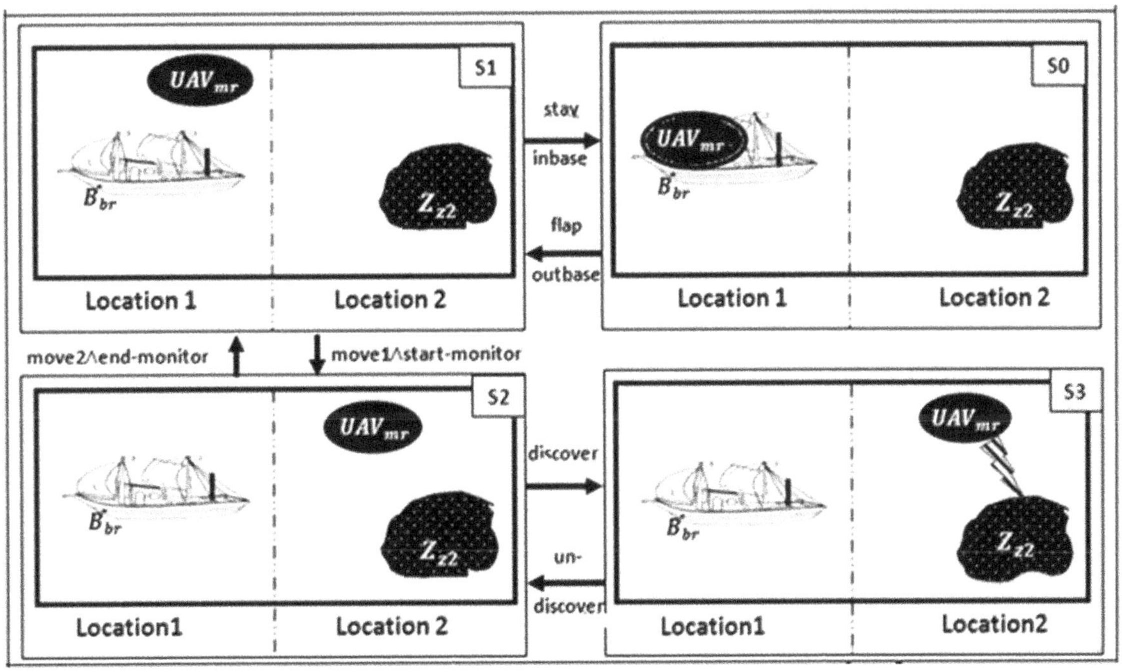

Figure 10. A state-transition system for the simple domain USV-Cleaning

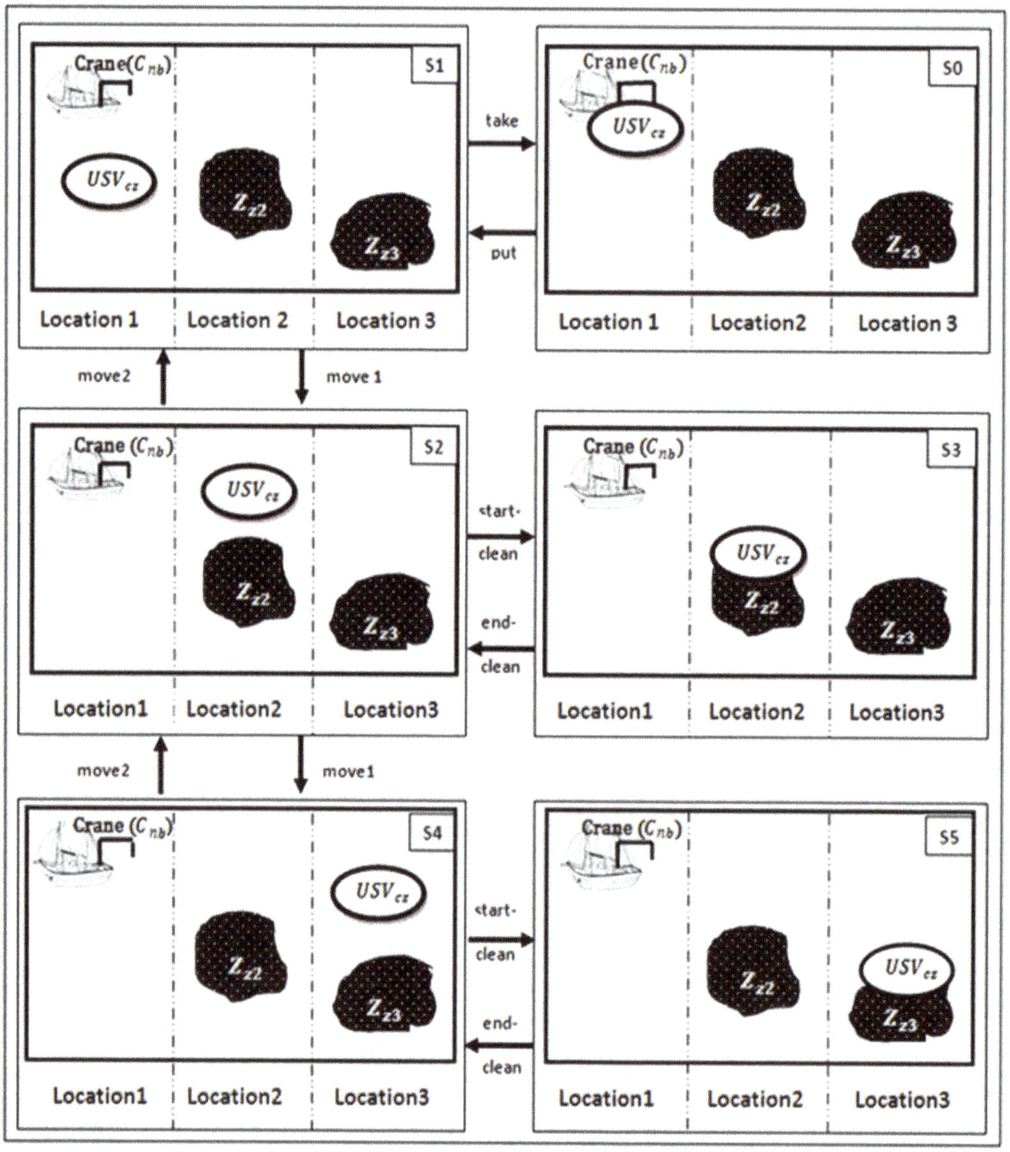

- ***A set of regions (R_r){ R_1, R_2, …}:*** a region *r (r=1,…R, R>0)* is a space of two levels (atmospheric and maritime) which may contain a base of life, one or more dirty zones, one or more vehicles and locations.
- ***A set of locations in each region (L_{lr}) {L_{1r}, L_{2r},…}:*** location *l* is part of a region r where *l=1,…L, L>0.*
- ***A set of base of life (B_{br}) {B_{1r}, B_{2r},…}:*** a base of life *b (b=1,…B, B≥0)* can be found in each region r.
- ***A set of dirty zones (Z_{zl}) {Z_{1l}, Z_{2l}…} :*** a dirty zone *z* is located in one or more locations *l* where *(z=1,…Z, Z>0).*

- ***A set of monitoring vehicle*** **(*UAV*$_{mr}$)** *{UAV$_{1r}$, UAV$_{2r}$...}:* Each monitoring vehicle *m (m=1,...M, M>0)* may be in a certain location of a region r. It may move to another adjacent location of the same region r.
- ***A set of cleaning vehicle*** **(*USV*$_{cz}$)***{USV$_{1z}$, USV$_{2z}$ ....}* **:** Each monitoring vehicle *c (c=1,...C, C>0)* may be in a certain zone *z* or location *l*, it can move to another adjacent zone or adjacent location either in the same region or in another adjacent region.
- **A set of cranes-type objects (*C*$_{nb}$)** *{C$_{1b}$, C$_{2b}$...}* **:** a crane-type object *n (n=1,...N, N≥0)* is found in every base of life *b*.

The current configuration of the two domains is denoted using instances of the following predicates, which represent relationships that change over time:

- ***at(NAME, E):*** vehicle *NAME={UAV$_{mr}$, USV$_{cz}$}* is currently at location *E={ L$_{lr,}$ R$_r$}*.
- ***start-clean (NAME, Z):*** vehicle *NAME.USV$_{cz}$* is ready for cleaning in the dirty zone *Z={Z$_{zl}$}*.
- ***end-clean(NAME, Z):*** vehicle *NAME.USV$_{cz}$* is finished the cleaning of the dirty zone Z.
- ***start-monitor(NAME, R) :*** vehicle *NAME.UAV$_{mr}$* is currently ready to start monitoring in the region *R={R$_r$}*.
- ***end-monitor(NAME, R):*** vehicle *NAME.UAV$_{mr}$* has now completed monitoring of its *R*.
- ***discover (NAME, Z):*** dirty zone Z is discovered by the drone *NAME.UAV$_{mr}$*.
- ***undiscover(NAME, Z):*** region does not contain a dirty zone Z or the dirty zone Z is not discovered by the drone *NAME.UAV$_{mr}$*.
- ***holding*(C, NAME)*:*** crane *C={C$_{nb}$}* is currently holding vehicle *NAME.USV$_{cz}$*.
- ***stayinbase(NAME, B):*** vehicle *NAME.UAV$_{mr}$* is currently in the base of life *B={B$_{br}$}*.
- ***flapoutbase(NAME, B):*** vehicle *NAME.UAV$_{mr}$* is outside its base *B*. It flies in the air.

We can enumerate the possible actions in the two domains **UAV-M** and **USV-C**:

- ***Move (NAME, E1, E2):*** a vehicle *NAME* moves from a location *E1={L$_{lr}$, R$_r$}* to some adjacent and unoccupied location *E2={L$_{lr}$', R$_r$'}*.
- ***Take (NAME, C, R):*** an empty crane *C* takes a *NAME.USV$_{cz}$* vehicle in the same region *R={R$_r$}*.
- ***Start-Clean (NAME, Z):*** a vehicle *NAME.USV$_{cz}$* begins cleaning a dirty zone *Z*.
- ***End-Clean (NAME, Z):*** a vehicle *NAME.USV$_{cz}$* has finishes cleaning a dirty zone *Z*.
- ***Start-Monitor (NAME, R):*** a vehicle *NAME.UAV$_{mr}$* begins the monitoring of a region *R*.
- ***End-Monitor (NAME, R):*** a vehicle *NAME.UAV$_{mr}$* finishes the monitoring of a region *R*.
- ***Discover (NAME, R, Z):*** a dirty zone Z is discovered by the vehicle *NAME.UAV$_{mr}$* in its *R*.
- ***Undiscover (NAME, R, Z):*** a dirty zone Z is not discovered by the vehicle *NAME.UAV$_{mr}$* in its region *R*.
- ***StayinBase (NAME, B):*** a vehicle *UAV$_{mr}$* remains in the base of life *B*.
- ***FlapoutBase (NAME, B)*****:** a vehicle *NAME.UAV$_{mr}$* flies outside base *B*.
  <u>**NB:**</u> The action **"Move"** is executed in parallel with the action **"Start-Monitor"** and **"End-Monitor"**.

4.3. Representations for Classical Planning

There are three different ways to represent classical planning problems (Ghallab et al., 2004):

- **In A Set-Theoretic Representation:** Each state of the world is a set of propositions, and each action is a syntactic expression specifying which propositions belong to the state in order for the action to be applicable and which propositions the action will add or remove in order to make a new state of the world.
- **In A Classical Representation:** This representation generalizes the set-theoretic representation using notation derived from first-order logic instead of propositions. States are represented as sets of logical atoms that are true or false within some interpretation. Actions are represented by planning operators that change the truth values of these atoms.
- **In A State-Variable Representation:** The main motivation here is to rely on functions instead of flexible relations. each state is represented by a tuple of values of n state variables $\{x_1,..., x_n\}$, and each action is represented by a partial function that maps this tuple into some other tuple of values of n state variables. This approach is especially useful for representing domains in which a state is a set of attributes that range over finite domains and whose values change over time.

In this work, we focus on the first representation (set-theoretic) to apply its planning on our proposition.

4.3.1. Set-Theoretic Representation

Example 2 illustrates a set-theoretic representation of the sets of the two domains (**USV-C** and **UAV-M**) which are described in Example 1.

Example 2

1. Let $L=\{p_1,...,p_n\}$ be a finite set of proposition symbols:
    a. **For the UAV-M domain:** L= {inair, inregion, inbase, at1, at2}, where:
    inair means that the UAV_{mr} drone is in the air;
    inregion means that UAV_{mr} drone is found in its region;
    inbase means that the UAV_{mr} drone remains on the base of life;
    at1 means that the UAV_{mr} drone is at location 1; and
    at2 means that the UAV_{mr} drone is at location 2.
    b. **For the USV-C domain:** L= {underwater, onzone, holding, at1, at2}, where:
    underwater means that the USV_{cz} vehicle is on the water;
    onzone means that the vehicle USV_{cz} is found in its dirty zone;
    holding means that the grue is holding the USV_{cz} vehicle ;
    at1 means that the vehicle USV_{cz} is at location 1; and
    at2 means that the vehicle USV_{cz} is at location 2.
2. Each state S is a subset of L,
    a. **For the UAV-M domain:** $S=\{s_0, ..., s_5\}$, where:
    s_0 = {inair, at1}; s_1 = {inbase, at2};
    s_2 = {inair, at2}; s_3 = {inbase, atl}; and

s_4 = {inregion, at1}; s_5 = {inregion, at2}.

b. **For the USV-C domain:** S={s_0, …, s_5}, where:
s_0 = {onwater, at2}; s_1 = {holding, at2};
s_2 = {onwater, atl}; s_3 = {holding, atl}; and
s_4 = {onzone, at1}; s_5 = {onzone, at2}.

3. Each action a ∈ A is a triple of subsets of L, which we will write as a= (precond (a), effects⁻(a), effects⁺(a)). The set precond (a) is called the *preconditions* of a, and the sets effects⁺(a) et effects⁻(a) are called the *effects* of a.
   a. **For the UAV-M domain:** A= {take, put, start-monitor, fin-monitor, move1, move2, discover, undiscover}, where:
   take=({inair}, {inair}, {onbase});
   put=({onbase}, {onbase}, {inair});
   start-monitor=({inregion}, (inregion}, {inair, at2});
   fin-monitor=({inair, at2}, {inair},{onbase, at1});
   discover =({inair,at1}, {inair}, {inair,at2});
   undiscover =({onbase}, {onbase}, {inair});
   move1=({at1}, {at1}, {at2}); and
   move2=({atl},{atl}, {at2}).
   b. **For the USV-C domain**: A= {take, put, start-clean, fin-clean, move1, move2}, where:
   take=({onwater}, {onwater}, {holding});
   put=({holding}, {holding}, {onwater});
   start-clean=({holding, atl}, (holding}, {onzone, at2});
   fin-clean=({onzone, at2}, {onzone},{holding, at1});
   move1=({at1}, {at1}, {at2}); and
   move2=({atl},{atl}, {at2}).

4.4. A General Example "UAV FTM and USV FTC"

We have illustrated an example of simple non-trivial execution called "UAVs FTM (For the monitoring) & USVs FTC (for the cleaning) " to combine the previous two "UAV-M " and " USV-C " domains.

The environment is a combination of the example (Figure 9) and (Figure 10) in a single example which presents regions. Each region is made up of an atmospheric space where monitoring drones are located, and a maritime space with several elements, namely: bases of life, crane-type objects, cleaning vehicles, dirty zones.

An abstract version of this domain is defined as the abstract version of two " UAV-M" and USV-C " domains, but there is one more constant symbol that is:

- ***A set of general coordinator (General$_g$) {Coord$_1$, Coord$_2$, ...}***: a coordinator general *g (g=1,...G, G>0)* is an agent responsible for the different elements and tasks of the domain.

The domain is represented as in Section "A Running Example" "USV-Cleaning" and "UAV-Monitoring" but in addition there are instances of the following predicates, which represent relationships that change over time:

- ***inform (NAME, Coord):*** *NAME.* UAV_{mr} vehicle informs the *Coord* $_=${*General*$_g$} of the presence of a dirty zone and prepare a *NAME.* USV_{cz} vehicle.
- ***notinformed (NAME, Coord):*** *NAME.* UAV_{mr} vehicle does not inform the *Coord* of the presence of a dirty zone and not to prepare a *NAME.* USV_{cz} vehicle.
- ***prepare(Coord, NAME):*** Coordinator *Coord* prepares *NAME.* USV_{cz} vehicle by the information of *NAME.* $UAV_{mr.}$
- ***notprepared (Coord, NAME):*** Coordinator *Coord* does not prepare the *NAME.* USV_{cz} vehicle and receives no information.

There are possible actions in the two UAV-M and USV-C domains by adding four other actions for

UAVs FTM & USVs FTC domains:

- ***Inform (NAME, Coord, Z}:*** the *NAME.* UAV_{mr} vehicle informs *Coord* about the presence of a dirty zone Z.
- ***Notinformed (NAME, Coord, Z}:*** the *NAME.* UAV_{mr} does not inform its *Coord* by the presence of a dirty zone Z.
- ***Prepare:*** the *Coord* prepares the *NAME.* USV_{cz} vehicle to the presence of the action "*Inform*".
- ***Notprepared:*** the *Coord* did not prepare the *NAME.* USV_{cz} vehicle to the presence of the action "*Notinformed*".

NB: The action "*Prepare*" executes following the action "*Inform*" and the same case for "*Notprepared*" and "*Notinformed*".

Example 3: Figure 11 shows a state transition system for a region involving two locations. It also composed by a base of life, a crane type object, a monitoring vehicle, a cleaning vehicle, a dirty zone. Here, the set of states is{s0, s1, s2, s3, s4, s5, s6, s7}, the set of actions is {take, put, move1∧start-monitor, move2∧fin-monitor, discover, undiscover, inform→ prepare, notinformed→ notprepared, move1, move2}, and there are no events. The arc (s0, s1) is labeled with the action " flapoutbase", the arc (s5,s4) with the action " take", and so forth.

5. EXAMPLE FOR SIMULATION

We propose an illustrative example to simulate the functioning of our proposal. To highlight the contributions of our approach, we focus on the following metrics: The total energy consumption of the selected USV_{CZ}, average of energy consumption of cleaning by USV_{CZ} and number of cleaned cells by the selected USV_{CZ}. In order to study the behavior of our proposal and to analyze its obtained results on simulation, we compare them to two approaches: direct displacement to another zone (DDZ) and indirect displacement to another zone (IDZ) (Passage through the base of life). A series of simulations were realized according to several parameters. Before starting the experiments, we will describe a virtual environment and its discretization.

Figure 11. A state-transition system for the simple domain UAV FTM & USVs FTC

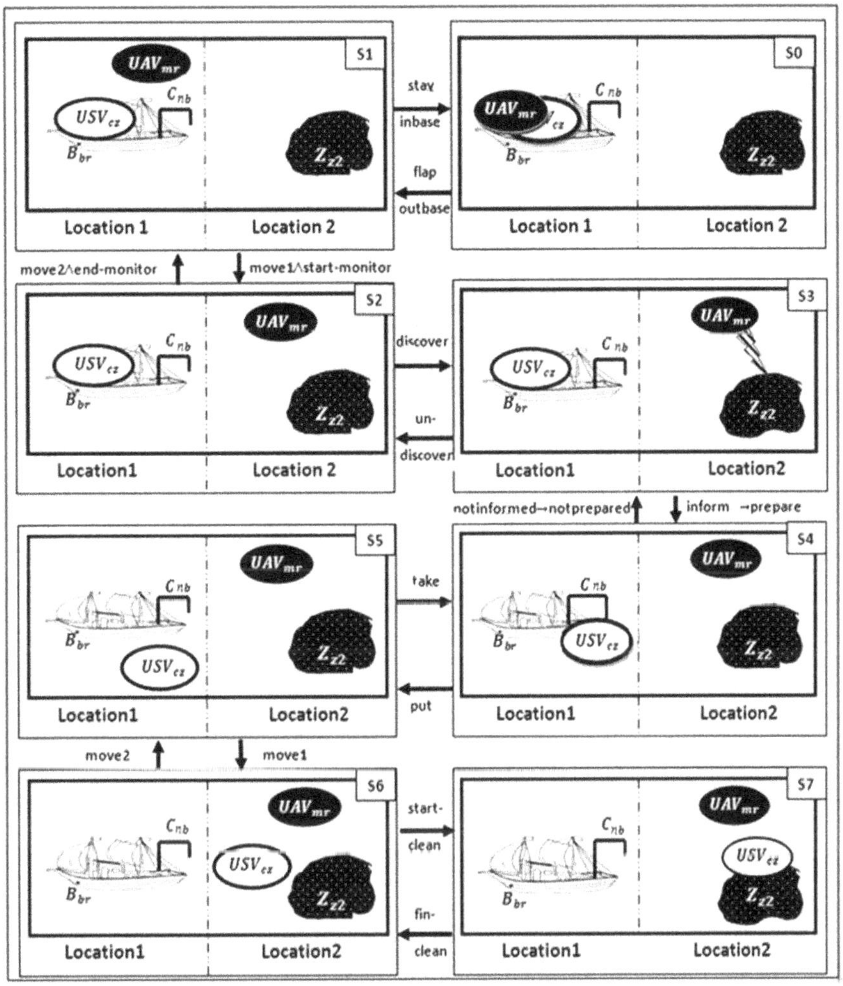

5.1. Virtual Environment

We have taken a simple example of a region "region1" that includes two locations "Location1 and Location2" locations. There is a dirty zone "zone1" in "Location1", a second dirty zone "zone2" and a base of life in "Location2". The following Figure 12 shows this virtual environment:

5.2. Discretization of the Environment (Region1)

After the injection of the degrees of dirt randomly (see Figure 12), and the launching the discretization step on this environment, we obtained the following matrix (see Table 3). The blue and brown cells represents respectively zone 1, zone 2.

For our environment, we assumed the following assumptions:

Figure 12. Example of a simplistic virtual environment

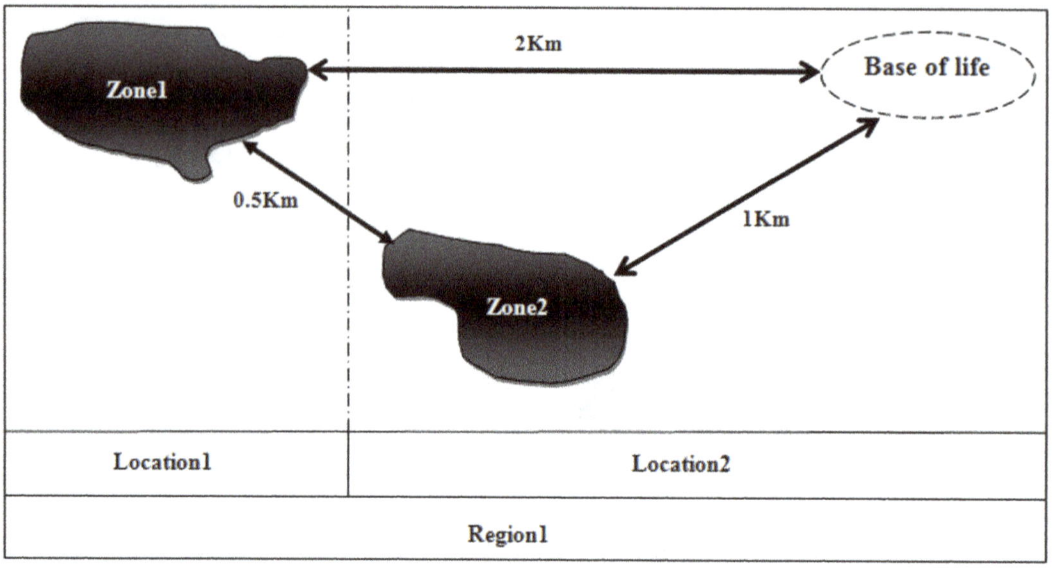

Table 3. Matrix of discretization of the environment

C_{ij}	1	2	3	4	5	6	7	8	9	10	11	12	13	14	15	16
1	12	10	18	13	19	14	9	11	3	12	16	12	2	9	2	3
2	16	11	22	20	17	9	8	5	2	4	18	16	8	1	5	7
3	19	9	31	13	18	41	25	22	4	9	100	100	90	83	68	11
4	12	30	29	6	7	19	9	19	23	14	93	83	70	87	68	13
5	9	80	55	39	63	85	25	23	16	10	43	37	38	55	40	14
6	8	53	48	63	74	74	63	89	12	9	76	69	80	57	74	15
7	10	70	91	98	75	59	82	100	11	10	3	6	17	41	43	17
8	16	2	12	17	9	8	18	23	15	16	11	19	9	4	7	22
9	20	14	18	14	9	5	11	22	19	20	12	18	17	3	21	9

- The threshold proposed has the role of classifying the list of coordinates of the dirty zone by the operation of comparison between its value (equal to 24%) and the $Degree_{cell}$ of each coordinate.
- The number of proposed regions is equal to 10 identical regions that have the same surface
- The number of USV_{CZ} for each zone is equal to 5 according to the algorithm proposed above (Algorithm5) but since our proposition is based on a USV_{CZ} then one assigns a USV_{CZ} to each dirty zone.
- Solid (hard) obstacles are unavailable that can prevent the vehicle from navigating and plan its trajectory in this environment.
- The energy of displacement between the cells is negligible.
- Total Energy Consumption (TEC) = Cleaning Energy Consumption (CEC) + Displacement Energy Consumption (DEC) to the zone or base of life.
- It is assumed that the needed energy to clean a black cell (too salty) is 5%, for an average dirt cell is 3% and for a small dirt cell is 1%.

5.3. Total Energy Consumption of Selected USV_{CZ}

To analyze the total energy consumption of selected USV_{CZ}. The obtained results from the simulations are summarized numerically in Table 4 and schematized by Figure 13.

This simulation (see Figure 13) allows us to estimate the behavior of the USV_{CZ} numbers in relation to the total energy consumption for both approaches. The selected USV_{CZ} is the vehicle that has consumed less energy than the other USV_{CZ} in zone 1; so, it is chosen to complete the cleaning of zone 2 (the case of a USV_{CZ}*(free)*). We notice that our approach (the blue curve) is below the curve of the IDZ approach (red curve). So our proposal is better than the IDZ approach and it allows a significant reduction in the total energy consumption of selected USV_{CZ}.

We can also note that our approach was able to minimize the "Total Energy Consumption (TEC)" with an average gain of + 9.75% compared to the IDZ approach.

5.4. Average of Energy Consumption of Cleaning by USV_{CZ}

To analyze the mean energy consumption of cleaning by USV_{CZ}, the obtained results by the simulations are schematized by Figure 14.

Figure 14 presents the simulation which allows evaluating the behavior of the numbers of USV_{CZ} with respect to the average of the energy consumption of cleaning of zone 1 for our DDZ approach. We note that the mean "Cleaning Energy Consumption (CEC)" is decreased when the number of USV_{CZ} increases. So our proposal was able to decrease the energy consumption of cleaning by USV_{CZ}.

5.5. Number of Cleaned Cells by the Selected USV_{CZ}

To know and mark the number of cleaned cells by the selected USV_{CZ}, the obtained results by the simulations are shown in Figure 15.

The simulation in Figure 15 allows us to evaluate the behavior of a number of USV_{CZ} in relation to the number of cleaned cells in zone 1. We notice that the number of cleaned cells decreases when the number of USV_{CZ} increases. We notice that the number of cleaned cells decreases when the number of USV_{CZ} increases. We conclude that only one USV_{CZ} cleans many cells from the zone whereas a selected

Figure 13. Total energy consumption of selected USV_{CZ}

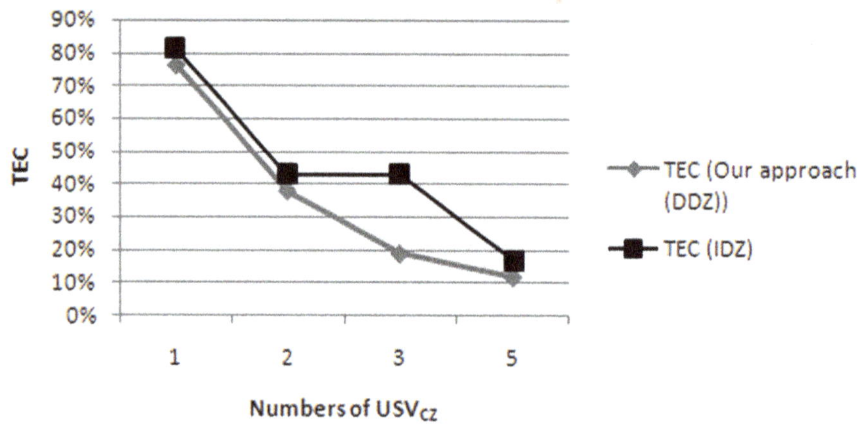

Table 4. Total energy consumption of selected USV_{CZ}

Numbers of USV_{cz}	1	2	3	5
TEC (DDZ)	76%	38%	19%	12%
TEC (IDZ)	81%	43%	43%	17%
Gain	5%	5%	24%	5%
Average gain	+9.75			

USV_{CZ} cleans a few cells as it belongs to a group of USV_{CZ} from the same zone. Thus the chosen USV_{CZ} has gained an energy consumption so that it can complete the cleaning of zone 2.

6. CONCLUSION

In this article, we presented a hierarchical decision-making architecture for the cooperation and the coordination air-sea. We defined and explained the characteristics and functioning of the system. This proposed approach is composed of a UAV for each region and a USV for a dirty zone. In the monitoring section, we chose the color as the degree of dirt so that the UAV can detect the dirty zones and initiate the cleaning. We suggested using the LEDs for each USV to measure the amount of energy. We formalized our proposal by means of a set-theoretic representation. The metrics measured are the total energy consumption of the selected USV, average of energy consumption of cleaning by USV and number of cleaned cells by the selected USV. The realized architecture shows that the proposed method provides encouraging results. For the practical experimentation of our approach, we plan to use a hybrid system including fixed wings and rotary wings for the UAV, and for the USV to use a helicopter-dropped system over the oil layers with embedded bio-sensors. As future work, we plan to start the following axes:

Figure 14. Average of energy consumption of cleaning by USV_{CZ}/zone1

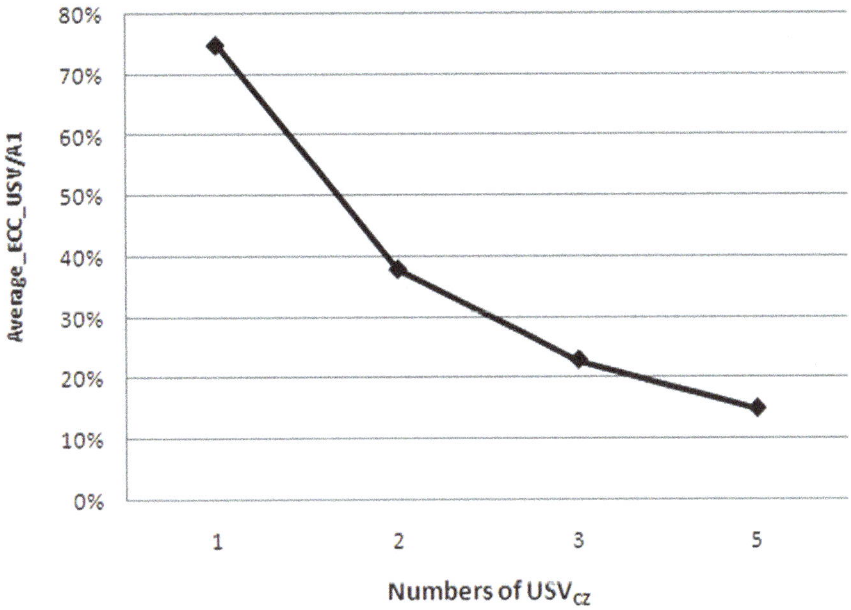

Figure 15. Number of cleaned cells by the selected USV_{CZ}

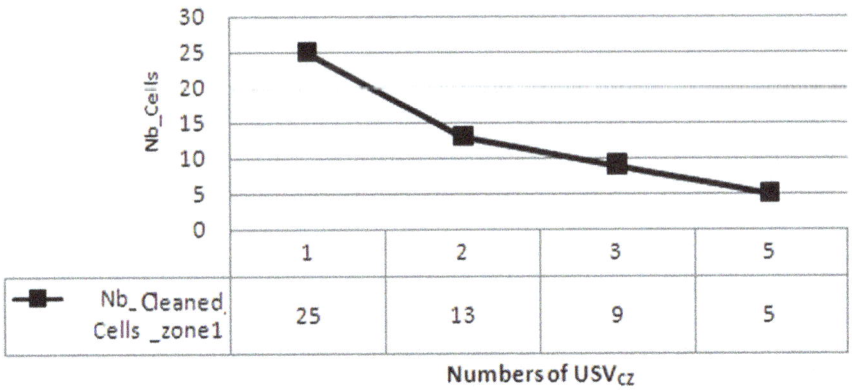

- Build a decentralized architecture for each UAV and USV with a task scheduler.
- Allocate a UAV cluster for each region in a complex environment through obstacles.
- Use other criteria to determine if zone is dirty such as chemical criteria in (Québec, 2003) and other indicators of water pollution as in (RAMP, 1997).
- Use the "Defense Advanced Research Projects Agency (DARPA) (Atherton, 2016)" system for the UAV, and the " Bio-Cleaner (Davies, 2012)" as a USV system.

REFERENCES

Atherton, K. D. (2016). DARPA Wants A Fast Cargo Drone That Can Land Anywhere. *Popular Science*. Retrieved February, 2017, from http://www.popsci.com/watch-darpas-new-vtol-concept-drone

Baba, A., & Chatila, R. (2008). Experiments with Simultaneous Environment Mapping and Multi-target Tracking. In O. Khatib, V. Kumar, & D. Rus (Eds.), *Experimental Robotics* (Vol. 39, pp. 201–210). Berlin, Heidelberg: Springer. doi:10.1007/978-3-540-77457-0_19

Batalin, M. A., Sukhatme, G. S., & Hattig, M. (2004). Mobile robot navigation using a sensor network. In *Proceedings of the International Conference on Robotics and Automation,* New Orleans, LA (Vol. 1, pp. 636-641). 10.1109/ROBOT.2004.1307220

Belbachir, A., & Escareno, J. (2016). Autonomous Decisional High-level Planning for UAVs-based Forest-fire Localization. In *Proceedings of the 13th International Conference on Informatics in Control, Automation and Robotic,* Lisbon, Portugal (Vol.1, pp.153-159). 10.5220/0005972501530159

Chatila, R., & Laumond, J. P. (1985). Position referencing and consistent world modeling for mobile robots. *In Proceedings of the IEEE International Conference on Robotics and Automation* (Vol. 2, pp. 138–145).

Cox, T. H., Nagy, C. J., Skoog, M. A., Somers, I. A., & Warner, R. (2004). Civil UAV Capability Assessment (Draft Version). *NASA*. Retrieved November 18, 2016, from https://www.nasa.gov/centers/dryden/research/civuav/civ_uav_doc-n-ref.html

Davies, A. (2012). Roomba-Like Robots Dropped by Helicopter Could Clean Up Oil Spills. *treehugger*. Retrieved March 2017, http://www.treehugger.com/clean-technology/floating-drone-cleans-oil-oceans.html

Finn, R. L., & Wright, D. (2012). Unmanned aircraft systems: Surveillance, ethics and privacy in civil applications. *Computer Law & Security Review*, *28*(2), 184–194. doi:10.1016/j.clsr.2012.01.005

Gal, O. (2011). Automatic Obstacle Detection for USV's Navigation Using Vision Sensors. In A. Schlaefer & O. Blaurock (Eds.), *Robotic Sailing: Proceedings of the 4th International Robotic Sailing Conference* (pp. 127-140). *Springer Berlin Heidelberg*.

Ghallab, M., Nau, D., & Traverso, P. (2004). *Automated planning: Theory & practice* (1st ed.). Elsevier.

Ghallab, M., Nau, D., & Traverso, P. (2016). *Automated Planning and Acting*. Cambridge University Press.

Hachour, O. (2008). Path planning of Autonomous Mobile robot. *International Journal of Systems Applications. Engineering & Development*, *2*(4), 178–190.

Hager, C. E., Kwon, H., Zarzhitsky, D., & Pack, D. (2011). A Cooperative Autonomous System for Heterogeneous Unmanned Aerial and Ground Vehicles. In Proceedings of the AIAA Infotech@ Aerospace, St. Louis, MI, March 29-31.

Harik, E. C., Guérin, F., Guinand, F., Brethé, J. F., & Pelvillain, H. (2014). Decentralized Control Architecture for UAV-UGV Cooperation. In *Proceedings of the 2nd AETOS International Conference On "Research challenges for future RPAS/UAV systems"*, BA106 Military Air Base, Bordeaux, France.

Ifremer. (2006). Etude des macro-déchets en milieu marin: une première synthèse. Retrieved November 01, 2016, from http://www.mddelcc.gouv.qc.ca/eau/interen.htm (In French)

Jakob, M., Semsch, E., Pavlıcek, D., & Pechoucek, M. (2010). Occlusion-aware multi-uav surveillance of multiple urban areas. In *Proceedings of 6th Workshop on Agents in Traffic and Transportation*. The 6th Workshop on Agents in Traffic and Transportation, Toronto, Canada (pp. 59-66).

Juliá, M., Gil, A., & Reinoso, O. (2012). A comparison of path planning strategies for autonomous exploration and mapping of unknown environments. *Autonomous Robots, 33*(4), 427–444.

Kulich, M., Juchelka, T., & Přeučil, L. (2015). Comparison of exploration strategies for multi-robot search. *Acta Polytechnica: Journal of Advanced Engineering, 55*(3), 162–168. doi:10.14311/AP.2015.55.0162

MacArthur, E. Z., MacArthur, D., & Crane, C. (2005). Use of cooperative unmanned air and ground vehicles for detection and disposal of mines. In *Proceedings of SPIE*—The International Society for Optical Engineering, Boston, MA (Vol. 5999). 10.1117/12.631314

Martinic, G. (2014). Unmanned maritime surveillance and weapons systems. *Journal of the Australian Naval Institute, 151*, 86–91.

O'Young, S., & Hubbard, P. (2007). RAVEN: A maritime surveillance project using small UAV. In *Proceedings of the IEEE Conference on Emerging Technologies and Factory Automation (ETFA)* (pp. 904-907). 10.1109/EFTA.2007.4416878

Québec. (2003). Québec's Water: A Resource To Be Protected. Retrieved November 01, 2016, from http://www.mddelcc.gouv.qc.ca/eau/inter_en.htm

Regional Aquatics Monitoring Program (RAMP). (1997). Water Quality Indicators. Retrieved November 08, 2016, from http://www.ramp-alberta.org/river/water+sediment+quality/chemical.aspx

Valada, A., Velagapudi, P., Kannan, B., Tomaszewski, C., Kantor, G., & Scerri, P. (2014). Development of a low cost multi-robot autonomous marine surface platform. In K. Yoshida & S. Tadokoro (Eds.), *Field and Service Robotics* (Vol. 92, pp. 643–658). Berlin, Heidelberg: Springer. doi:10.1007/978-3-642-40686-7_43

Zhang, Y. (2011). UCAV path planning based on FSCABC. *Information-An International Interdisciplinary Journal, 14*(3), 687–692.

This research was previously published in the International Journal of Software Innovation (IJSI), 6(2); edited by Roger Y. Lee and Lawrence Chung, pages 50-76, copyright year 2018 by IGI Publishing (an imprint of IGI Global).

Chapter 16
Models for Drone Delivery of Medications and Other Healthcare Items

Judy E. Scott
University of Colorado, USA

Carlton H. Scott
University of California, USA

ABSTRACT

This article describes how a healthcare delivery drone has the potential for developing countries to leapfrog the development of traditional transportation infrastructure. Inaccessible roads no longer will prevent urgent delivery of blood, medications or other healthcare items. This article reviews the current status of innovative drone delivery with a particular emphasis on healthcare. The leading companies in this field and their different strategies are studied. Further, this article reviews the latest decision models that facilitate management decision making for operating a drone fleet. The contribution in this article of two new models associated with the design of a drone healthcare delivery networks will facilitate a more timely, efficient, and economical drone healthcare delivery service to potentially save lives.

INTRODUCTION

There is a major new technology that is destined to be a disruptive force in the field of transportation: the drone. Just as the mobile phone allowed developing countries to leapfrog older technologies for personal communication, the drone has the potential to leapfrog traditional transportation infrastructure. Drones are not dependent on a well-developed road infrastructure because they fly to their destination.

Drones are commonly known as unmanned aerial vehicles (UAVs) (Choi-Fitzpatrick et al., 2016). Alternative terms include Unmanned Aircraft (UA), Remotely Operated Aircraft (ROA), Remotely Piloted Vehicle (RPV) and Remotely Piloted Aircraft (RPA) (Clarke, 2014).

DOI: 10.4018/978-1-5225-8365-3.ch016

Recent innovations have taken place in drone specific hardware, software and networks. For example, light composite materials and global positioning systems (GPS) enable efficient flight. Furthermore, lithium batteries are rapidly improving so drones can fly further on a charge (Raffaello, 2014). Drone software can use mobile phone or tablet apps for tracking and navigation (French, 2015; Khazan, 2016; Raptopoulos, 2013). The drone operating system manages the network by monitoring weather data from all the ground stations and optimizing the routes of the drones (Raptopoulos, 2013). The routes need to avoid adverse weather conditions and avoid other risk factors. Also, an on board webcam can enable communication with a control center (Prigg, 2014).

Although the definition of drones is complex because of the diversity of characteristics (Clarke, 2014; Villasenor, 2012), it is generally agreed that drones are devices which are capable of sustained flight, which do not have a human on board, and are under sufficient control to perform useful functions.

Worthwhile drone functions include delivery of small items that are urgently needed in locations with difficult access. However, delivery by drones is usually much more expensive than delivery by conventional transportation. Delivery costs per ton mile have been estimated at $2000 for drone delivery, versus $40 for same-day ground truck delivery, $1.37 for air freight and $0.20 for national truck (French, 2016). Furthermore, drones perform relatively poorly economically on route density and drop size (D. Wang, 2016). Nevertheless, analysis of "value density" versus urgency shows instances when the price premium makes sense (French, 2016). For example, an organ for transplant would have high value density and would be needed urgently to save a life. Other critical healthcare drone delivery applications include timely delivery of urgently needed medications, blood and vaccines. Despite the urgency, locations needing the delivery may have difficult access due to poor transportation infrastructure, or roads blocked by severe weather, disasters or traffic congestion. Since a drone can fly over an inaccessible road, innovative organizations have begun to use drones for healthcare delivery. In section three, we provide a review of documented examples of drone healthcare delivery.

To our knowledge, this study is one of the first academic papers published on drone delivery models for healthcare. While the limited academic literature published on modeling drone delivery has mainly focused on parcel delivery, this paper provides two new models for drone delivery in a healthcare context. A logistics network is designed to provide timely delivery of healthcare items using a tandem strategy involving land-based transportation and final delivery by drone. Both models focus on locating a warehouse with supplies and drone nests to complete final delivery. In the first model our objective is to minimize the total weighted time for delivery; the second model seeks to minimize the maximum weighted time to any delivery point thereby ensuring more equitable service to outlying regions. As such, the new models given are a contribution to continuous location theory with particular application to tandem truck drone delivery of healthcare products. For a review of location decisions see (Drezner & Hamacher, 2004).

The next section of the paper overviews drone applications and drone issues that need to be addressed. The third section describes drone applications in healthcare. In section 4, we review previous modelling work in drone package delivery and in the following section develop two new models with healthcare applications in mind. The final section has concluding remarks. An appendix provides data used in numerical examples.

BACKGROUND ON DRONE APPLICATIONS

Many drone applications involve surveillance using an onboard camera. However, drones also are capable of carrying devices other than cameras and capable of delivering small loads. Drones have been used extensively by the military in combat and for humanitarian aid. Useful non-military drone applications in different industries include agriculture surveillance and crop spraying, shark surveillance at beaches, monitoring wildlife for conservation, monitoring fires, scientific research and exploration, monitoring riots and international borders by police and governments, sports and entertainment event coverage, other media coverage, emergency services and disaster response (Choi-Fitzpatrick et al., 2016; Clarke, 2014; Esler, 2015).

Arguably, humanitarian drone applications are the most useful since human lives are at stake. For example, in Nepal after the 2015 earthquake, drones helped rescuers locate survivors (Choi-Fitzpatrick et al., 2016). Furthermore, a study used experiments to model automated response to a scenario of a San Francisco earthquake (Mosterman, Sanabria, Bilginc, Zhangd, & Zander, 2014). The automated system included fixed wing and rotorcraft drones integrated with autonomous ground vehicles.

In contrast to humanitarian drone applications, drones have been used for crimes, such as delivering contraband to prison inmates, firing weapons, terrorism and hacking (Choi-Fitzpatrick et al., 2016).

Also, drones have been criticized for inadequate regulation, safety issues and security and privacy abuse (Clarke, 2014; Welch, 2015). Celebrities and others are concerned that drones will spy on them. Furthermore, irresponsible drone owners have been a nuisance in cases related to photographing accidents or fires and interfering with emergency responders (Choi-Fitzpatrick et al., 2016). Other concerns are that people would shoot down drones and steal drone packages (Welch, 2015).

Regulatory bodies, such as the Federal Aviation Administration (FAA), usually ban commercial drones because they are wary of collisions in airline air space. Nevertheless, there have been reports of hundreds of collision close calls many due to non-commercial drones (Esler, 2015).

Since December 2015, drones weighing between .55 and 55 pounds need to be registered with the FAA and marked with the registration number so owners can be identified if there is an accident or criminal use of the drone (Choi-Fitzpatrick et al., 2016). Owners need to follow FAA restrictions, which currently only allow drone flights during daylight hours, within line of sight, not over people etc.

In the US, commercial drone use is illegal without a Section 333 exemption (French, 2015). FAA exemptions have been granted for surveillance applications, for example BP pipelines, rather than delivery. This has been frustrating to Amazon and other companies that are planning drone delivery (Welch, 2015). As discussed in the next section, Flirtey has had FAA approval for dropping medical supplies in rural Virginia with drones (Hackman & Nicas, 2015; Pepitone, 2015). However, the approval was only for a weekend. In contrast, countries with urgent healthcare needs, such as Rwanda, are more likely to quickly overcome regulatory hurdles (French, 2015).

DRONE APPLICATIONS IN HEALTHCARE

Drone applications in healthcare include delivery of medicine, defibrillators, blood samples, and vaccines (Agatz, Bouman, & Schmidt, 2015; Choi-Fitzpatrick et al., 2016; DHL, 2016; Hackman & Nicas, 2015; Haidari et al., 2016; Khazan, 2016; Pepitone, 2015; Preimesberger, 2016; Prigg, 2014; Raptopoulos, 2013; Tilley, 2016; Varnholt, 2016). Researchers have shown using simulation modeling of the Gaza,

Mozambique vaccine supply chain, that a system of drones could increase availability of vaccines and decrease costs in a wide range of situations (Haidari et al., 2016). Sensitivity analyses showed that the major drivers of costs savings are road speed of traditional land vehicles, the number of people needing to be vaccinated, and the distance that needs to be traveled.

Autonomous drones, such as those employed by Matternet, use GPS and other sensors to navigate between automated ground stations in order to deliver medications in remote locations that lack adequate roads (Raptopoulos, 2013). Matternet has delivered medications in Haiti following the 2010 earthquake and in the Dominican Republic (Choi-Fitzpatrick et al., 2016), as well as in Bhutan, New Guinea and Switzerland (French, 2015). The company works with UNICEF and Doctors without Borders.

Matternet's drones can carry one to two kilograms (2.2 to 4.4 pounds) and transport items about 10 km, traveling up to 40 km per hour, taking about 18 minutes including lift off and landing (French, 2015; Raffaello, 2014; Raptopoulos, 2013). A smartphone app enables senders to choose from a list of possible destinations. The drone then automatically generates a route based on the terrain, weather, and airspace and population density. The route will avoid airports, schools and public squares, as well as hills and buildings. A parachute will deploy in an emergency.

In Germany, DHL Parcel has researched three generations of medical drone delivery called Parcelcopter (DHL, 2016). The first generation travelled 1 km to deliver blood samples across the Rhine River at Bonn. The second generation tested drone delivery of medications and other urgently needed material for three months in 2014 to Juist, one of Germany's remote North Sea Islands (Agatz et al., 2015; Varnholt, 2016). The Parcelcopter traveled 12 km across open sea. From January to March 2016, DHL's third generation Parcelcopter tested delivery of over 130 parcels of urgently needed medicines or sporting goods between automated Skyports in two Bavarian Alpine villages. Drone delivery took 8 minutes compared to a 30-minute road trip in winter (DHL, 2016). The time difference could be significant in a medical emergency.

UPS and Zipline are working on a drone network to deliver vaccines and blood to 20 clinics in remote locations in Rwanda (Khazan, 2016; Preimesberger, 2016; Tilley, 2016). Malaria, infant deaths and mothers dying in childbirth are common in Rwanda. When rabies vaccine is needed urgently, drone delivery would not be hindered by washed out roads during rainy season. Only a third of Africans live within two kilometers of a road that functions year-round (Khazan, 2016). Zipline drones are launched from a nest and make a delivery by dropping items with a paper parachute. After the drone returns to the nest a sim card and new battery are inserted along with the blood or vaccines for the next delivery.

Zipline drones are the size of a large dog and can carry three pounds (Khazan, 2016). They can fly 45 miles in 30 minutes. Their route is tracked and changed with a tablet app.

Table 1. Comparison of drone healthcare delivery

Drone Company	Healthcare Items	Delivery Location
Matternet	Blood, medications	Haiti, Dominican Republic, Papua New Guinea, Switzerland
DHL Parcel	Blood, medications	Germany
Zipline	Vaccines, blood	Rwanda
Flirtey	Medications	Virginia, Nevada
Delft University	Defibrillators	Netherlands

Table 2. Comparison of healthcare delivery drones

Drone Company	Launching Pad	Delivery Method
Matternet	Automated ground station	Automated ground station
DHL Parcel	Automated Skyport	Automated Skyport
Zipline	Nest	Paper parachute
Flirtey	Airport	Dropped by rope
Delft University	Hospital, clinic?	Ground landing

The first FAA-approved drone delivery successfully dropped medical supplies to a health clinic in rural southwest Virginia (Pepitone, 2015). The clinic operates for about 3000 patients one weekend each year (Hackman & Nicas, 2015). Flirtey drones delivered prescription items from the Wise County Regional airport to the clinic in the remote fairgrounds in about 3 minutes. Delivery is usually 90 minutes along a winding bumpy road from the pharmacy in Oakwood 35 miles away. An experimental National Aeronautics and Space Administration (NASA) manned drone delivered from Oakwood to the Wise County airport since Flirtey drone batteries at that time were limited to about 20 miles. Flirtey drones have also delivered healthcare items in Nevada, Australia and New Zealand.

In the Netherlands, prototype ambulance drones have demonstrated delivery of defibrillators (Prigg, 2014). Traveling at speeds of up to 60 miles per hour, the drones can reach patients within a 4.6 square mile radius in a minute versus an average of 10 minutes for traditional emergency services. The faster drone response increases the chance of survival to 80% versus 8% for traditional emergency services. The drones track emergency mobile calls and use GPS to navigate. A paramedic, from a control room connected to a livestream web camera on the drone, can instruct a lay person assisting the patient. An emergency simulation study of drone defibrillator delivery shows the importance of device-specific customization of emergency operator instruction (Fleck, 2016). It is expected to take five years to develop an operational emergency drone network and address legal issues and improve the steering on the ambulance drones which are expected to cost $19,000 each (Prigg, 2014). Apart from defibrillators, there is potential for "flying medical toolbox" drones to carry other healthcare devices, such as oxygen masks or insulin injections for diabetes patients (Prigg, 2014).

Table 3. Comparison of healthcare delivery drone payload, range and speed

Drone Company	Payload	Range	Speed
Matternet	2 kg (4.4 lb)	10 km (6.25 miles)	40 kmph 25 mph
DHL Parcel	4.4 lb	12 km (7.5 miles)	>40 mph
Zipline	3 lb	45 miles	90 mph
Flirtey	2 kg (4.4 lb)	20 miles	?
Delft University	4 kg (8.8 lb)	12 km (7.5 miles)	60 mph

Table 1 shows that the healthcare items transported by the drones are predominantly medications, blood and vaccines. Potentially, defibrillators, oxygen and insulin could be transported by the ambulance drone, which is only a prototype currently.

In Table 2, we see that the launching pad is often automated and called a ground station, Skyport or nest. The delivery method can be the same, as in the case of Matternet and DHL Parcel or the load can be dropped as in the case of Zipline, by a paper parachute, and Flirtey dropped by rope.

Table 3 compares the payload, range and speed for the different drone delivery platforms. Matternet, DHL and Flirtey all carry about the same payload. Zipline carries a smaller payload but is much faster than the others because of its streamlined fixed wing plane design versus the helicopter-like design of the others. The range varies from Matternet's 6.5 miles to Zipline's 45 miles. Note that these numbers are based on reported information in articles and websites and may not reflect current capabilities. Also, as stated earlier, technology, such as the duration of the charge of lithium batteries, is rapidly improving.

MODELS FOR DRONE DELIVERY

Several researchers have developed various models for drone use primarily for parcel delivery inspired by Google, Amazon and DHL who are exploring this option. For example, Murray and Chu envisage that the primary delivery vehicle would be a truck with attendant drones and developed two models for delivery of parcels by drones to be launched from a truck (Murray & Chu, 2015). They model this as a mixed integer program of a travelling salesman type. Computational issues associated with this model have been discussed by others (X. Wang, Poikonen, & Golden, 2016). Related models use various meta-heuristic approaches to solve the resulting integer programs (Agatz et al., 2015; Ferrandez, Harbison, Weber, Sturges, & Rich, 2016; Ponza, 2015). These papers essentially conclude that there are cost savings to be realized when both trucks and drone work jointly to deliver parcels. Hong et.al. developed a sole drone based delivery system for urban areas by positioning drone recharging stations and routing delivery paths around obstacles (Hong, Kuby, & Murray, 2015). Mosterman et al. develop a model based emergency response system for a natural disaster involving a heterogeneous fleet of vehicles (Mosterman et al., 2014). Integration of the components of their response system is achieved with internet technology and a graphical interface. Dorling et al. extend vehicle routing problems to drone delivery by adding time and budget constraints to vehicle routing scenarios via integer programming formulations (Dorling, Heinrichs, Messier, & Magierowski, 2017). They explicitly recognize drone energy consumption and finite payload.

In the next section, we develop a particular type of model for location decisions that has application for the drone delivery of healthcare related items.

Model Development

We consider a scenario where emergency medical supplies need to be delivered to an outlying area that is not completely served by good roads but is too far for drone delivery alone. This is often the case in an undeveloped country. Consequently, a tandem strategy is adopted whereby land-based transportation is used to deliver from a warehouse (which could be a storage area at a hospital or clinic) where supplies are stored to a drone nest which is in sufficient proximity to the area of need. The problem at hand is to position both the warehouse that is served by road and the drone nests that are needed to serve the "last mile" delivery to the area of need.

We partition the area to be served into a number of demand clusters which could represent a village or group of villages, for example, and attempt to assign a drone nest to each demand cluster. A demand cluster would then consist of a number of demand points for emergency supplies (see Figure 1).

We provide two models to address the above location decisions. Our first model has as objective to minimize the total weighted delivery time; road plus air. A downside of this objective is that some demand clusters or demand points within a cluster may have unacceptably long delivery times. An alternative type of model is to minimize the maximum weighted time to deliver to all demand points in all demand clusters. Generally, this approach would lead to more equitable delivery times. Since there is a tradeoff between delivery time and cost, all models contain a budget constraint.

Model 1

For Model 1.

Minimize total weighted time for truck/drone delivery subject to a budget and drone travel constraint.

v_t, v_d are velocity of truck, drone $v_t < v_d$.

c_t, c_d are per unit costs of operation per mile for a truck and drone respectively, $c_t < c_d$.

Deliveries take place from a central depot at (*X, Y*) by truck to a point $(x_i, y_i), i = 1,...,n$ which are drone nests associated with a demand cluster. Final delivery is by drone to a demand point $(a_{ij}, b_{ij}), j = 1,...,m_i$. Here m_i is the number of demand points served by drone nest *i*.

Figure 1. Location of warehouse, drone nests and demand points

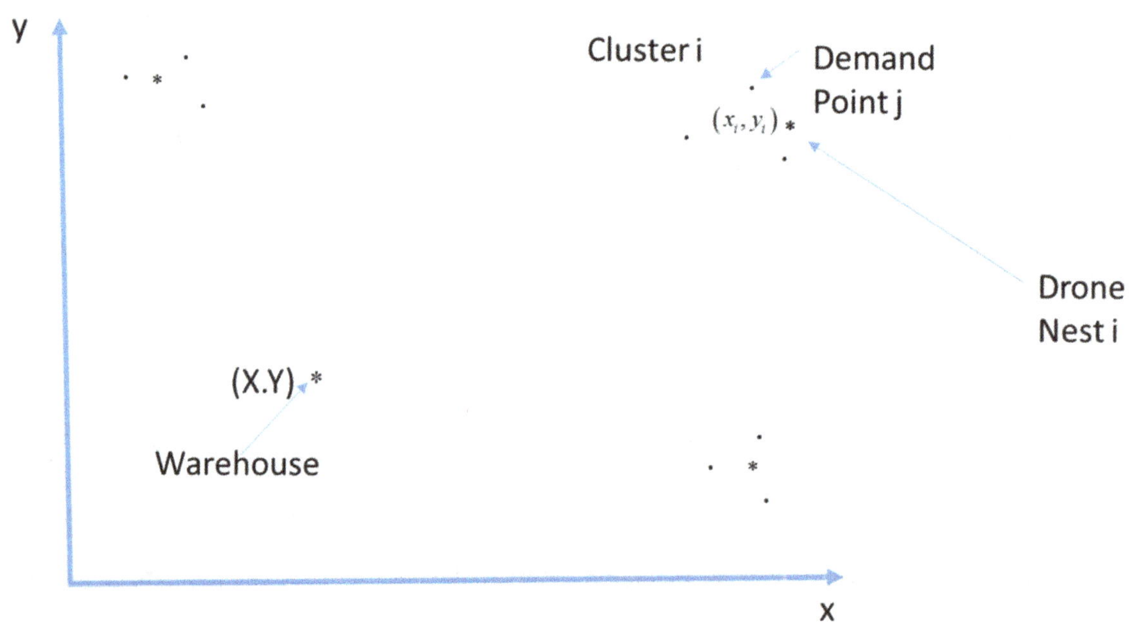

$d_i(\cdot,\cdot), d_{ij}(\cdot,\cdot)$ are distance functions and w_i, w_{ij} are weights with $w_i = \sum_j w_{ij}$. These are essentially the demand at each point (a_{ij}, b_{ij}), and conservation of demand flow at drone nest (x_i, y_i). B is budget, D is the distance limit on drone travel. We have the following mathematical program.

Minimize Total Weighted Time over X, Y and $x_i, y_i :\ i = 1,\ldots,n$

$$\sum_i \frac{w_i d_i(X - x_i, Y - y_i)}{v_t} + \sum_i \sum_j \frac{w_{ij} d_{ij}(x_i - a_{ij}, y_i - b_{ij})}{v_d}$$

subjcct to

$$\sum_i c_t w_i d_i(X - x_i, Y - y_i) + \sum_i \sum_j c_d w_{ij} d_{ij}(x_i - a_{ij}, y_i - b_{ij}) \le B$$

$$d_{ij}\left(x_i - a_{ij}, y_i - b_{ij}\right) \le D\ \forall (i,j)$$

Define $z_{i1} = X - x_i, z_{i2} = Y - y_i$ .The above program is then seen to be convex since d_i and d_{ij} are convex functions.

Model 2

For Model 2.

Minimize Maximum weighted time for truck/drone delivery subject to a budget and drone travel distance constraint.

The objective may be written as

$$\operatorname*{Min}_{X,x} \operatorname*{Max}_{i,j} \left[\frac{w_i d_i(X - x_i, Y - y_i)}{v_t} + \frac{w_{ij} d_i(x_{ij} - a_{ij}, y_{ij} - b_{ij})}{v_d} \right]$$

resulting in the following convex program (where we introduce a new variable T)

$$\text{Min } T$$

subject to

$$\frac{w_i d_i(X - x_i, Y - y_i)}{v_t} + \frac{w_{ij} d_{ij}(x_i - a_{ij}, y_i - b_{ij})}{v_d} \le T \quad \forall i,j$$

$$\sum_i c_t w_i d_i(X - x_i, Y - y_i) + \sum_i \sum_j c_d w_{ij} d_{ij}(x_i - a_{ij}, y_i - b_{ij}) \le B$$

$$d_{ij}\left(x_i - a_{ij}, y_i - b_{ij}\right) \le D\ \forall (i,j)$$

Here T is the maximum weighted time for delivery to any demand point. In the next section, we look at the use of these two models to both position assets (storage facilities and drone nests) and to assist managerial decision making with respect to budget allocation.

Numerical Results

We look at a simple example involving three demand clusters each serving three demand points. Drones can travel at 50 miles per hour and trucks at 40 miles per hour. Costs are very difficult to pinpoint. Welch looks at costs for Amazon Prime Air (Welch, 2015) but it is doubtful that these estimates would apply in an undeveloped country. Matternet estimates that drones would cost twenty-four cents for a six mile 15 minute trip by drone (Choi-Fitzpatrick et al., 2016). Since drones generally carry one item at a time and trucks many items we assume that the unit cost per mile to deliver by drone is twice that for a truck in our base model. Further we assume that drones are limited to 100 miles per delivery trip so the delivery point needs to be no more than 50 miles from the nest. This is actually more than current capability of a drone but we anticipate that battery life will significantly improve and extend trip times. In any case, the models are general and better estimates can be used as they become available. The overall budget will be chosen to enable feasible delivery times. Weights (demand estimates) are also used in the models. Detailed values are given in the Appendix.

Drones are assumed to fly in straight lines (no obstacles) so a Euclidean metric is appropriate; trucks on the other hand need to follow the road network. In this initial paper, we assume that the distance metric for trucks is a modified Euclidean metric that is empirically derived. In this case, we model the distance as 1.6 times the Euclidean metric. This is somewhat higher than empirical values used in the US (say $\approx$ 1.3) but reflects the fact that an undeveloped nation's road network is somewhat inferior.

Hence

$$d_i\left(X - x_i, Y - y_i\right) = 1.6\left(\left(X - x_i\right)^2 + \left(Y - y_i\right)^2\right)^{.5} \quad \forall i$$

and

$$d_{ij}\left(x_i - a_{ij}, y_i - b_{ij}\right) = \left(\left(x_i - a_{ij}\right)^2 + \left(y_i - b_{ij}\right)^2\right)^{.5} \quad \forall i, j.$$

The models are now solved using AMPL/MINOS (Fourer, Gay, & Kernigham, 2002). Since all models are convex, we are guaranteed global optima. We look at a few scenarios below.

Model 1: Budget=$9,000

Solving Model 1 we find that the warehouse is located at X=204, Y=148 with drone nests at (46, 207), (288, 197), (282, 55) serving clusters 1, 2 and 3 respectively. Hence, for example, a truck will travel from (204, 148) to (46, 207) where a drone will be loaded to fly to demand cluster 1. From the computer output, it is noted that village 2 in cluster 3 is maximally distanced from the drone nest 3 that serves it. Further we examine how budget changes affect the total weighted time. This is given in Figure 2 below.

Figure 2. Minimal total weighted time versus budget

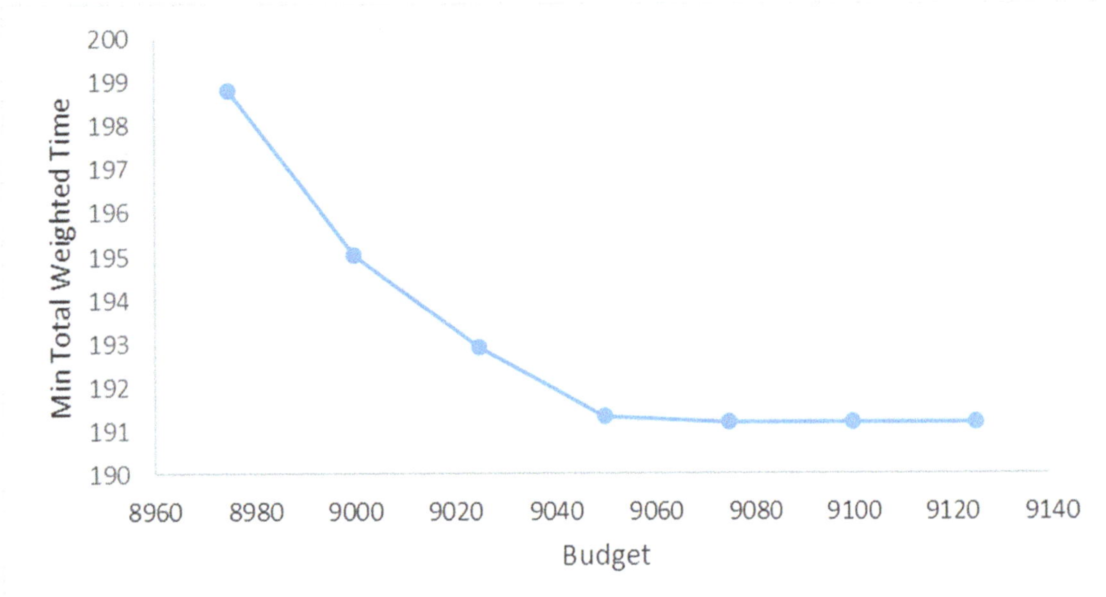

Increasing the budget up to about \$9075 will decrease the total weighted time but further increases do not help since the drone distance constraint becomes active. Hence in this scenario, there is no need for a budget allocation greater than \$9075. At the other extreme any budget allocation less than \$8980 will not meet service needs.

As drone technology improves, clearly the distance that a drone can fly from its' nest will increase. Below we give a table that that looks at the variation of maximal drone distance on warehouse and drone nest location. In order to get a richer set of results we have increased the budget by \$1000.

We note from Figure 3 that as the range *D* of a drone increases, the total weighted time decreases since more travel is done by faster drones than trucks. Note also that the minimal total weighted time is approximately linearly related to drone range with an average slope of 3.4. This implies that for each additional increase of one mile in range, the minimal total weighted time is reduced by 3.4 hours.

From table 4 above, we note that the location of the warehouse changes minimally (200,151) to (198,150). On the other hand, the drone nests can change location significantly; in the case of nest "1" from (48,205) to (66,197). However, since all the requirements for a drone nest can be carried in a container, this should not be of practical significance.

Model 2. Budget = \$9,000

In this case, we find that the warehouse is located at *X*=177, *Y*=157 with drone nests at (44, 209), (293, 202) and (287, 42). Note that the location of the drone nests is not too different from Model 1 which minimizes total weighted time. In this case, the computer output shows us that the worst weighted service time is to demand point 1 in cluster 1.

As above, we now look at the sensitivity of Model 2 to changes in budget and drone range.

Figure 3. Minimal total weighted time versus range

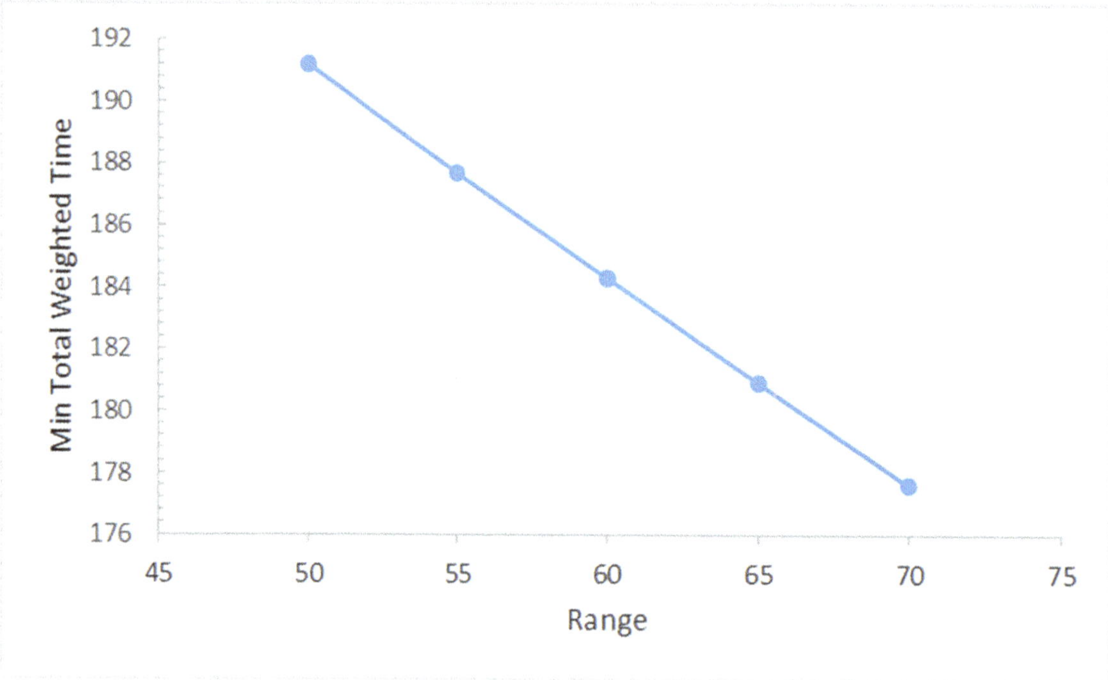

Table 4. Location versus drone range

D	(X,Y)	(x_1, y_1)	(x_2, y_2)	(x_3, y_3)
50	(200,151)	(48,205)	(267,191)	(283,56)
55	(200,151)	(52,203)	(262,189)	(279,60)
60	(199,151)	(57,201)	(258,186)	(275,63)
65	(198,150)	(62,199)	(253,184)	(271,67)
70	(198,150)	(66,197)	(149,181)	(266,69)

Once again, the objective MinMax Weighted Time decreases nonlinearly until it flattens out around a budget of $9075 where the drone range constraint becomes active. It is interesting this occurs around the same budget for both objectives. This change can be seen in Figure 4.

As the drone range increases, the MinMax Weighted Time decreases approximately linearly with an average slope of 2.01 (see Figure 5). Hence for each additional mile of range, the MinMax Weighted Time decreases by 2.01 hours.

Below we contrast the results from both models by comparing the total weighted times to service each demand point j from drone nest i.

We note that cluster 1 served by drone nest 1 gets better service using Model 2 (more equitable model) whereas clusters 2 and 3 get slightly worse service. The worst served demand point (point 1 in cluster 1) has total weighted service time reduced from 99 to 85. This data is shown in Table 5.

Figure 4. MinMax weighted time versus budget

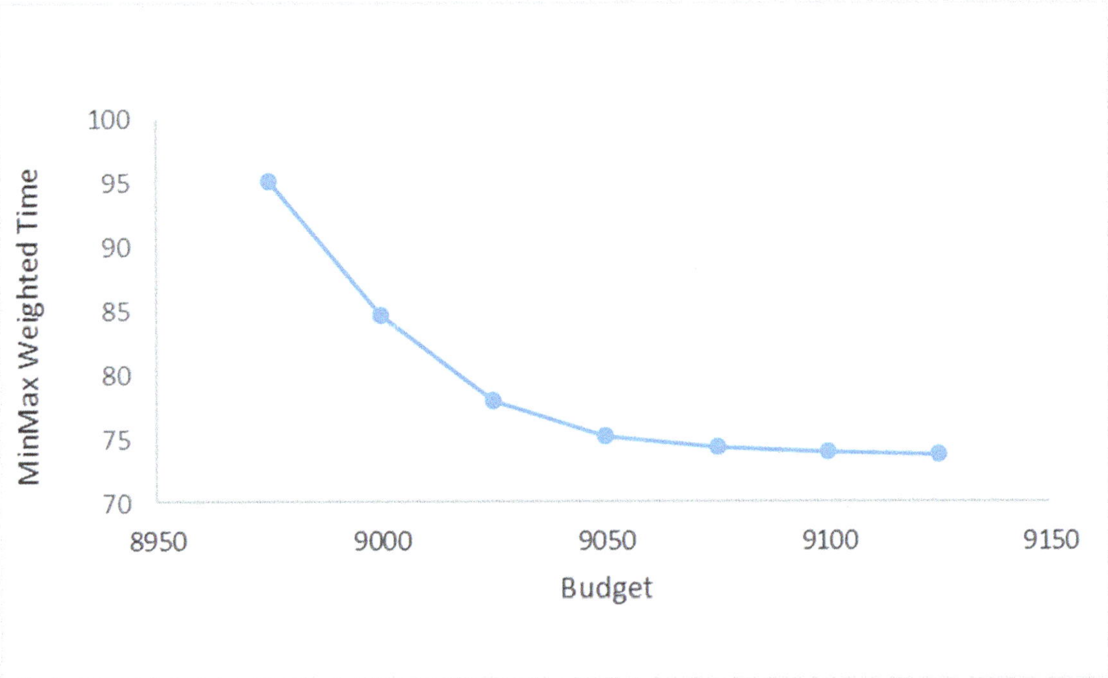

In an undeveloped country, it may be advisable to locate the warehouse at the international airport since this is the point of entry for medical supplies. This involves fixing the location of the warehouse say close to the international airport. Suppose then that the warehouse location is fixed at the origin so that $X = 0$ and $Y = 0$ is fixed. In this case, we find that the locations of the drone nests are at (21, 175), (283, 142) and (283, 24) for Model 1 for Budget = \$12,000 and at (26, 192), (297, 200), (281, 31) for Model 2 with Budget \$15,000. The extra budget is necessary due to the longer distance that trucks travel from the airport to reach a drone nest.

CONCLUSION

Drones, along with mobile technology, are enabling developing countries to leapfrog ahead with healthcare delivery to remote locations even with an unreliable road infrastructure. As stated earlier, only a third of Africans live within two kilometers of a road that functions year-round (Khazan, 2016). Even in developed countries, disasters, such as earthquakes and fires, can render roads inaccessible. Furthermore, extreme weather and city congestion can be an obstacle to emergency medical delivery.

Drone technology and its components, such as GPS and lithium batteries, are available and improving at a rapid pace. Despite issues related to privacy, security, safety and regulation, drones can provide beneficial and humanitarian applications, especially related to healthcare. Consequently, drone healthcare delivery to inaccessible locations is likely to become more ubiquitous in the near future.

Figure 5. MinMax weighted time versus range

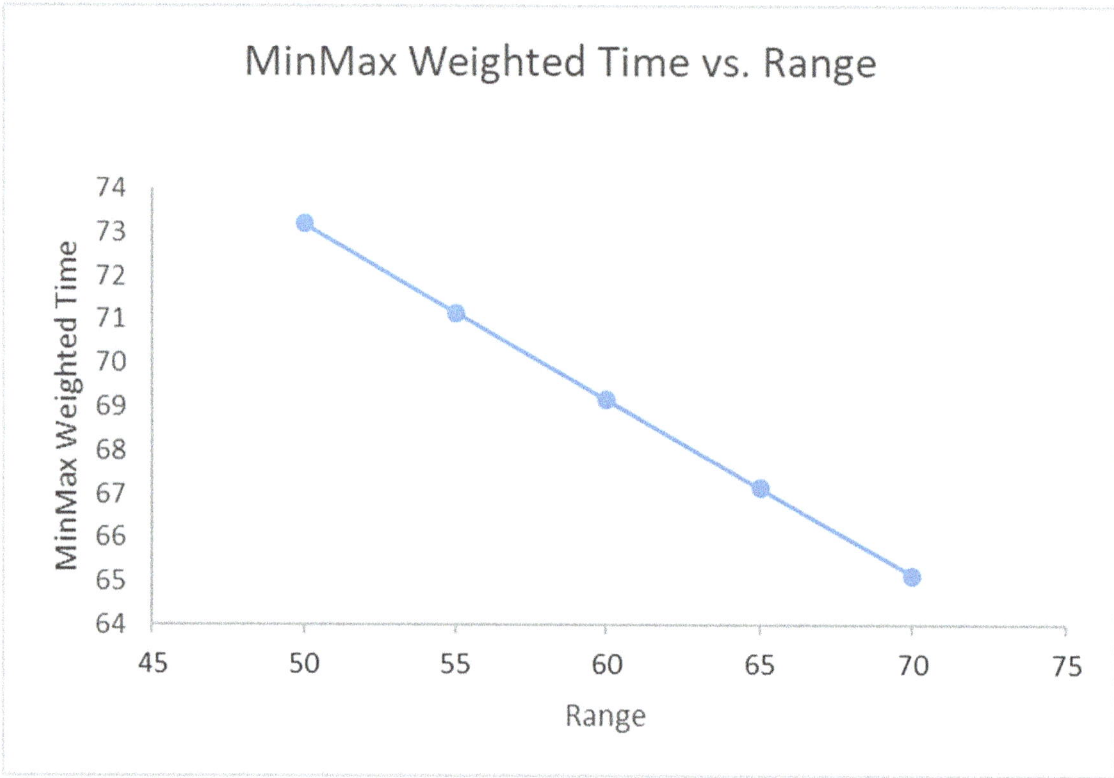

Table 5. Total weighted time comparison

Drone Nest/Demand Point	Total Weighted Time Model 1	Total Weighted Time Model 2
1,1	99	85
1,2	95	81
1,3	95	80
2,1	29	36
2,2	28	36
2,3	29	35
3,1	58	74
3,2	58	74
3,3	55	71

This paper discusses some innovative applications of drones in healthcare. Our contribution of two models, that use a tandem strategy involving traditional land transportation followed by drone delivery, will facilitate more timely, efficient and economical healthcare delivery. Healthcare costs are a major concern in both developed and developing countries. The models use a budget constraint while providing location decisions for warehouses and drone nests that enable timely delivery. Since time is of the essence in an emergency, faster response would prevent medical trauma and potentially save lives.

Drone companies, delivery companies, healthcare organizations, humanitarian organizations or governments could use our models to expedite delivery of urgently needed healthcare items. In some cases, governments could partner with drone companies, such as Rwanda does with Zipline to more economically deliver vaccines and blood to potentially save lives.

We are currently extending these models in a number of directions. Here we have used an empirically derived form of a distance metric for truck delivery. A better representation may be to explicitly model the actual road network. Future research can also investigate how many drones are needed at a particular drone nest and how well a network provides coverage for a region. Another issue is to account for the effect of obstacles such as a high mountain on a drone path and integrate it into the location model. For drone delivery in a metropolitan area, paths need to be routed around tall buildings and airports, for example. In conclusion, drone healthcare delivery is a new field with many opportunities for impactful future research.

REFERENCES

Agatz, N., Bouman, P., & Schmidt, M. (2015). Optimization Approaches for the Traveling Salesman Problem with Drone. *Erasmus University Rotterdam*. Retrieved from http://hdl.handle.net/1765/78472

Choi-Fitzpatrick, A., Chavarria, D., Cychosz, E., Dingens, J. P., Duffey, M., Koebel, K., . . . Almquist, L. (2016). Up in the Air: A Global Estimate of Non-Violent Drone Use 2009-2015. University of SanDiego: Joan B. Kroc School of Peace Studies at Digital@USanDiego. Retrieved from http://digital.sandiego.edu/gdl2016report/1

Clarke, R. (2014). Understanding the drone epidemic. *Computer Law & Security Review*, *30*(3), 230–246. doi:10.1016/j.clsr.2014.03.002

DHL. (2016). Successful Trial Integration of DHL Parcelcopter into Logistics Chain [Press release]. Retrieved from http://www.dhl.com/en/press/releases/releases_2016/all/parcel_ecommerce/successful_trial_integration_dhl_parcelcopter_logistics_chain.html

Dorling, K., Heinrichs, J., Messier, G. G., & Magierowski, S. (2017). Vehicle Routing Problems for Drone Delivery. *IEEE Transactions on Systems, Man, and Cybernetics. Systems*, *47*(1), 70–85. doi:10.1109/TSMC.2016.2582745

Drezner, Z., & Hamacher, H. (Eds.). (2004). *Facility Location: Applications and Theory*. Springer.

Esler, D. (2015, October). Drone Revolution. *Business & Commercial Aviation.*

Ferrandez, S. M., Harbison, T., Weber, T., Sturges, R., & Rich, R. (2016). Optimization of a Truck-drone in Tandem Delivery Network Using K-means and Genetic Algorithm. *Journal of Industrial Engineering and Management*, *9*(2), 374–388. doi:10.3926/jiem.1929

Fleck, M. (2016, May 7-12). Usability of Lightweight Defibrillators for UAV Delivery. *Paper presented at the 2016 CHI Conference Extended Abstracts on Human Factors in Computing Systems*, San Jose, CA.

Fourer, R., Gay, D. M., & Kernigham, B. W. (2002). *AMPL: A Modeling Language for Mathematical Programming* (2nd ed.). Cengage Learning.

French, S. (2015). Drone delivery is already here — and it works. *Marketwatch.* Retrieved from http://www.marketwatch.com/story/drone-delivery-is-already-here-and-it-works-2015-11-30

French, S. (2016). Amazon's first Prime Air delivery is just more drone hype *Marketwatch*. Retrieved from http://www.marketwatch.com/story/the-heavy-on-hype-light-on-substance-world-of-drone-delivery-2016-12-01

Hackman, M., & Nicas, J. (2015, July 17). Drone Delivers Medicine to Rural Virginia Clinic. *Wall Street Journal*. Retrieved from http://www.wsj.com/articles/drone-delivers-medicine-to-rural-virginia-clinic-1437155114

Haidari, L. A., Brown, S. T., Ferguson, M., Bancroft, E., Spiker, M., Wilcox, A., ... Lee, B. Y. (2016). The economic and operational value of using drones to transport vaccines. *Vaccine*, *34*(34), 4062–4067. doi:10.1016/j.vaccine.2016.06.022 PMID:27340098

Hong, I., Kuby, M., & Murray, A. (2015, May 20-23). Deviation flow refueling location model for continuous space: commercial drone delivery system for urban area. *Paper presented at the 13th International Conference on GeoComputation*, Richardson, TX.

Khazan, O. (2016, April 4). A Drone to Save the World. *The Atlantic*. Retrieved from http://www.theatlantic.com/technology/archive/2016/04/a-drone-to-save-the-world/476592/

Mosterman, P. J., Sanabria, D. E., Bilginc, E., Zhangd, K., & Zander, J. (2014). Automating Humanitarian Missions with a Heterogeneous Fleet of Vehicles. *Annual Reviews in Control*, *38*(2), 259–270. doi:10.1016/j.arcontrol.2014.09.008

Murray, C. C., & Chu, A. G. (2015). The flying sidekick traveling salesman problem: Optimization of drone-assisted parcel delivery. *Transportation Research Part C, Emerging Technologies*, *54*, 86–109. doi:10.1016/j.trc.2015.03.005

Pepitone, J. (2015). First FAA-Approved Drone Delivery Drops Medicine in Virginia. Retrieved from http://www.nbcnews.com/tech/tech-news/first-faa-approved-drone-delivery-drops-medicine-virginia-n393986

Ponza, A. (2015). *Optimization of Drone-Assisted Parcel Delivery* [Honors Thesis]. University of Padova. Retrieved from http://tesi.cab.unipd.it/51947/1/Andrea_Ponza_-_Analisi_e_ottimizzazione_nell'uso_di_droni_per_la_consegna_di_prodotti.pdf

Preimesberger, C. (2016, May 9). Drones Will Soon Be Dropping Medicines to Save Lives in Rwanda. *eWeek*.

Prigg, M. (2014, October 28). The ambulance drone that could save your life. *Daily Mail*. Retrieved from http://www.dailymail.co.uk/sciencetech/article-2811851/The-ambulance-drone-save-life-Flying-defibrillator-reach-speeds-60mph.html

Raffaello, D. A. (2014). Guest Editorial Can Drones Deliver? *IEEE Transactions on Automation Science and Engineering*, *11*(3).

Raptopoulos, A. (2013). No roads? There's a drone for that. *Paper presented at the TEDGlobal 2013*.

Tilley, A. (2016, May 9). UPS Experiments With Drone Delivery In Partnership With Zipline. *Forbes.* Retrieved from http://www.forbes.com/sites/aarontilley/2016/05/09/ups-experiments-with-drone-delivery-in-partnership-with-zipline/#13c4870f75da

Varnholt, H. (2016, January 19). DHL's Drone Demonstration Fails to Deliver. *Wall Street Journal.* Retrieved from http://www.wsj.com/articles/dhls-drone-demonstration-fails-to-deliver-1453226792

Villasenor, J. (2012). What Is a Drone, Anyway? *Scientific American.* Retrieved from http://blogs.scientificamerican.com/guest-blog/2012/04/12/what-is-a-drone-anyway/

Wang, D. (2016, January 5). The Economics of Drone Delivery. Retrieved from http://spectrum.ieee.org/automaton/robotics/drones/the-economics-of-drone-delivery

Wang, X., Poikonen, S., & Golden, B. (2016). The vehicle routing problem with drones: Several worst-case results. *Optimization Letters.*

Welch, A. (2015). *A cost-benefit analysis of Amazon Prime Air* [Honors Thesis]. University of Tennessee at Chattanooga.

This research was previously published in the International Journal of Healthcare Information Systems and Informatics (IJHISI), 13(3), edited by Joseph Tan, pages 20-34, copyright year 2018 by IGI Publishing (an imprint of IGI Global).

APPENDIX

Here we give data that was used in the numerical work on the two models.
For example, $w_{11} = 5$ is the weight associated with travel from drone nest 1 to demand point 1 in cluster 1.
For example, demand point 1 in cluster 1 is at (0,220).

Table 6. Weights associated with travel from drone nest i to demand point j

w	**1**	**2**	**3**
1	5	3	6
2	4	2	1
3	5	4	2

Table 7. x Coordinates of demand points

a	**1**	**2**	**3**
1	0	30	41.2
2	306	300	306
3	306	330	282

Table 8. y Coordinates of demand points

b	**1**	**2**	**3**
1	220	200	212
2	200	220	222
3	12	40	24

Section 6
Military Use and Ethics

Chapter 17
Lethal Military Robots:
Who Is Responsible When Things Go Wrong?

Lambèr Royakkers
Eindhoven University of Technology, The Netherlands

Peter Olsthoorn
Netherlands Defence Academy, The Netherlands

ABSTRACT

Although most unmanned systems that militaries use today are still unarmed and predominantly used for surveillance, it is especially the proliferation of armed military robots that raises some serious ethical questions. One of the most pressing concerns the moral responsibility in case a military robot uses violence in a way that would normally qualify as a war crime. In this chapter, the authors critically assess the chain of responsibility with respect to the deployment of both semi-autonomous and (learning) autonomous lethal military robots. They start by looking at military commanders because they are the ones with whom responsibility normally lies. The authors argue that this is typically still the case when lethal robots kill wrongly – even if these robots act autonomously. Nonetheless, they next look into the possible moral responsibility of the actors at the beginning and the end of the causal chain: those who design and manufacture armed military robots, and those who, far from the battlefield, remotely control them.

INTRODUCTION

Although the use of unmanned systems is still in its infancy in most armed forces, some militaries, especially those of the US and Israel, have developed and deployed highly advanced drones. Even though the majority of these unmanned systems used in operations today are unarmed and mainly used for reconnaissance and mine clearing, the increase of the number of armed military robots, especially airborne ones, is undeniable. Certainly, on the face of it, unmanned systems have some strong benefits that could reduce the number of 'unfortunate incidents' on the battlefield. To start with, the main causes of misconduct on

DOI: 10.4018/978-1-5225-8365-3.ch017

the battlefield: frustration, boredom, and anger are diminished.[1] What's more, these unmanned systems have no instinct of self-preservation, and are able to hold their fire in critical situations. On the other hand, the use of robots raises some serious ethical questions. For instance, under what circumstances, and to what extent, do we allow robots to act autonomously? What precautions should (and can) we take to prevent robots from running amok? Would the use of military robots not be counterproductive to winning the hearts and minds of occupied populations, or result in more desperate terrorist-tactics given an increasing asymmetry in warfare? (See for an overview Lin, Bekey, and Abney, 2008; Lichocki, Kahn & Billard. 2011; Olsthoorn & Royakkers 2011; Schwarz 2017). A particularly pressing question is what to do when things go wrong: who, if anyone, can be held morally accountable *in reason* for an act of violence that a) involves a military robot; and b) would normally be described as a war crime?

The answer to that latter question depends on the answer to a prior one: when is there reasonable ground to hold an agent morally responsible for a certain outcome in the first place? Following Fischer & Ravizza (1998) on moral responsibility, we will assume here that agents can only in reason be held responsible if they are *moral* agents, that is, persons (or organizations) who have *control* over their behavior and the resulting consequences. This means that agents can be held responsible for a certain decision only insofar as they have been able to make it *in freedom* and *knowingly*. The first term means that it is not reasonable to hold agents responsible for actions or their consequences if they were not coerced or under duress. The second term, 'knowingly,' has an important normative aspect in that it relates to what people should know, or can with reason be expected to know, with respect to the relevant facts surrounding their decision or action.[2]

According to some authors (Asaro, 2007; Sparrow, 2007; Sharkey, 2008), the use of armed military robots makes the attribution of responsibility problematic, as it is not sufficiently clear who can be held responsible for civilian casualties and other collateral damage that result from the use of military robots, whether by mechanical error or failing judgment. Is it the designer/programmer, the field commander, the robot manufacturer, the robot controller/supervisor, or the nation that commissioned the robot? The answer to that question depends on a number of factors. For instance, was the cause a programming error, a malfunctioning, an accident, or intentional misuse? Or did the procedure include a 'man-in-the-loop,' that is, an element of human control, or was the military robot a fully autonomous or even learning machine?

As to that last question, this paper distinguishes between semi-autonomous robots, autonomous (but non-learning) robots, and learning robots. To start with the latter, learning robots are able to develop new behavioral patterns without human intervention; these robots are able to go beyond the parameters they left the factory with, as the robot itself can change them in its interaction with the operating environment (Matthias, 2004). An example is a system that was developed for the automatic diagnosis of lung cancer, and which is able to learn to identify cancer cells on the basis of microscope images of specimens of needle biopsies obtained from the bodies of the persons to be diagnosed (Matthias, 2004). More learning systems are in development, and most of them will pose no moral problems. But this would be different in the case of lethal military learning robots: seeing that it will often be impossible to predict the future behavior of these robots, it is hard to see how one could have sufficient *control* over their actions, which in its turn makes it difficult to determine who can be held responsible with reason.

Autonomous robots are robots that are based on conclusions derived from gathered information and preprogrammed constraints, and are capable of independently selecting and engaging targets (Crootof 2015). This in contrast to semi-autonomous robots where a human must take some affirmative action before a specific target is elected or engaged, that is, they require a 'man-in-the-loop.' This means that the decision to open fire, or more in general, the taking of any action that could threaten human life, is to be considered and approved by a human agent. According to Sparrow (2007) and Singer (2009), this precondition is essential to avoid the problems with respect to the attribution of responsibility; it is not without reason that the "International Law of Armed Conflict dictates that unmanned systems cannot fire their weapons without a human operator in the loop" (Isenberg, 2007).

Yet, while it is certainly true that currently human agents are kept 'in-the-loop,' it is not certain, or even likely, that this will remain so. The logic that brought unmanned systems more or less leads naturally to the wish to take the human agent out of the system altogether (Sparrow, 2011, 121; see also Sullins, 2010), and it seems almost a given that the future will bring more autonomous, and possibly learning, robots. Illustrative for this direction is that The United States is investing heavily in autonomous robots, and the U.S. Department of Defence has described increasing autonomous capabilities as a high priority. One of the ultimate goals is to take the man out-of-the-loop (US Department of Defense 2013), which raises the essential moral question: should we relinquish the decision to kill a human to a non-human machine? (Johnson & Axinn, 2013). That a lethal decision should be made only by a human and not a machine was one of the main reasons that in 2009 four leading scientists (Noel Sharkey, Peter Asaro, Robert Sparrow and Jürgen Altmann) established the International Committee for Robot Arms Control, ICRAC. The committee wants to restrict the use of armed robots for military purposes as much as possible (Altmann et al., 2013). In 2010 in Berlin, the committee organised the expert workshop Limiting Armed Tele-Operated and Autonomous Systems and invited scientists, politicians and military delegates. Following the workshop, a statement was made – signed by a large majority of those present – which emphasised that a ban ought to be imposed on autonomous armed military robots because a human must always take life-and-death decisions. Part of the statement read,

It is unacceptable for machines to control, determine, or decide upon the application of force or violence in conflict or war. In all cases where such a decision must be made, at least one human being must be held personally responsible and legally accountable for the decision and its foreseeable consequences.[3]

In 2015, the Future of Life Institute gained considerable publicity in 2015 when it distributed an open letter signed by more than twenty thousand artificial intelligence and robotics researchers (and others), which called for a "ban on offensive autonomous weapons beyond meaningful human control"[4]. "Meaningful human control" has been identified as key for the responsible design of autonomous systems operating in circumstances where human life is at stake. By preserving meaningful human control human safety can be better protected and "responsibility gaps" can be avoided. However, we still lack a satisfactory definition of what meaningful human control precisely means in relation to killer robots (Ekelhof, forthcoming, see also Crootof, 2016). Sparrow (2007) was one of the first authors to discuss this question, and he claims that since no one can reasonably be held responsible for autonomous robots' behavior, robots should not be allowed to make life-and-death decisions. In Sparrow's reasoning, considerations of responsibility occupy centre stage. Mukerji (2016) agrees with Sparrow's conclusion that robots should not be allowed to make life-and-death decisions, but on a different argument. He argues that only subjects who are capable of responsibility should make morally significant choices. Since autonomous robots seem to lack that capability, it follows that they should not be entrusted to make life-and-death decisions and act on them. However, some researchers dispute these arguments and that the problem of determining responsibility for autonomous military robots can be solved by addressing it within the context of the military chain of commands (e.g., Schulzke, 2013, and Champagne and

Tonkens, 2015), or by developing responsibility practices that clearly establish lines of accountability (Noorman & Johnson, 2014, and Noorman, 2014).

The above raises plenty of issues to critically assess the attribution of responsibility with respect to the deployment of both semi-autonomous and (learning) autonomous lethal military robots. In this article, we will argue that, in contrast to what some other authors hold, that we can reasonably assign responsibility for the development and deployment of these robots to their developers and commanders. A neglected problem, however, regards the assignment of responsibility at the end of the causal chain: the human operators who remotely control armed military robots from behind a computer screen. The depersonalization of war, the dehumanization of the enemy, and the moralization of technology strengthened by the speed of decision-making, can disengage the human operator morally and emotionally. We will show that to make deliberate, and thus responsible, decisions it is essential that operators have control over their decisions, and a vivid awareness of what is at stake. We will start by looking at the role of those with whom responsibility normally lies, the commanders. After that, we will turn to those at the beginning and the end of the causal chain, respectively the manufacturers and designers, and the human operators. We end by drawing some conclusions.

ULTIMATELY RESPONSIBLE: THE COMMANDING OFFICER

Discrimination and proportionality are the two-main *jus in bello* principles of the Just War Tradition. The first forbids the intentional killing of 'the innocent,' and the underlying idea is that civilians should not be made to suffer in war; overall, this is a rights-based principle. The principle of proportionality is a consequentialist one: enemy combatants should not be subjected to unnecessary suffering and superfluous injury, and a mission is permitted only if the expected military gain outweighs the expected number of unintended (the intended killing of civilians being prohibited anyway) civilian casualties. These principles have materialized in international humanitarian law and, for instance, in international treaties banning, regulating, or limiting the possession and use of certain weaponry.

Sparrow has pointed out that these principles assume that "we can identify the persons responsible for the actions that these principles are intended to govern" (Sparrow, 2007; see also Fieser & Dowden, 2007). Although some authors (Sparrow, 2007; Asaro, 2007; Sharkey, 2008) hold that the use of military robots makes this attribution of responsibility complex, it seems evident that the ultimate responsibility for, for instance, a war crime that crucially involved a military robot lies in the hands of the operators, the commanding officers who ordered the deployment of that robot, the military organization on whose behalf they did so, and the state of which that military is an instrument of. So, at the heart of it, the use of unmanned systems is, as far as responsibility concerns at least, not *very* different from using an aircraft to drop a bomb from a high altitude.

Still, this straightforward approach only works as long as there is 'a human in the loop' – from a legal and ethical perspective that human element seems a precondition for the attribution of responsibility (cf. Singer, 2009). Some have therefore argued that the attribution of responsibility in the case of autonomous military robots is so problematic that for that reason alone their use should be forbidden (Asaro, 2007; Sparrow, 2007). One of the most outspoken advocates of this position on the use of autonomous military robots is Noel Sharkey. In his view, the problem is that armed autonomous robots simply cannot be held accountable for their actions, as "[t]here is no way to punish a robot" (2010, 380). According to Sharkey this "leaves the question about who is responsible somewhere along the long causal chain that includes: the manufacturer for the description they gave, the programmer, the designer, the Department of Defense, the generals or admirals in charge of the operation and the operator" (2010, 380-1). If it proves impossible to establish where the responsibility lies, the use of unmanned robots should be, according to Sharkey, illegal. Other experts in the field hold similar views: a majority of the participants of an Expert Workshop on "Arms Control for Robots – Limiting Armed Tele-Operated and Autonomous Systems" organized by the International Committee for Robot Arms Control (ICRAC) in 2010, for example, found it 'unacceptable for machines to control, determine, or decide upon the application of force or violence in conflict or war. In all cases in which such a decision must be made, at least one human being must be held personally responsible and legally accountable for the decision and its foreseeable consequences' (Altmann et al., 2013).[5] The general drift of these comments is that responsibility evaporates without a human in the loop.

But on second though it seems clear that, also without a human agent directly controlling the robot, it is still a human agent that has decided on whether or not to deploy this robot: 'even if a system is fully autonomous, it does not mean that no humans are involved. Someone has to plan the operation, define the parameters, prescribe the rules of engagement, and deploy the system' (Quintana, 2008). Krishnan therefore thinks that

The legal problems with regard to accountability might be smaller than some critics of military robots believe. (...) If the robot does not operate within the boundaries of its specified parameters, it is the manufacturer's fault. If the robot is used in circumstances that make its use illegal, then it is the commander's fault. (2009, 105).

The basis of this argument is the Doctrine of Command Responsibility, and although this doctrine, 'that goes back to the dawn of time' (Garraway 2009, 704), is interpreted differently by different authors (Garraway, 2009), it will usually cover the deployment of armed military robots. So, if such a robot is used in circumstances that make its use illegal, then it is the commander's fault. The question then arises: what degree of knowledge must a commander have about the workings of a military robot in order to be reasonably responsible? Commanders are expected to take "all necessary and reasonable measures in their power" to prevent war crimes (Heckaerst & Doswald-Beck, 2006), and that requires according to Dunlap (forthcoming) a reasonable understanding of the foreseeable consequences of the deployment of armed military robots, and that we cannot hold the commanding officer responsible if an armed military robot would perform unforeseeable consequences. The latter is in line with the argument of Sparrow (2007), that it is unfair to assign responsibility to the commanding officer of learning autonomous robots, since "[t]he use of autonomous weapons (…) involves a risk that military personnel will be held responsible for the actions of machines whose decisions they did not control". This problem. However, can be overcome by using a more explicit social contract or "blank check". According to Champagne and Tonkens (2015), Sparrow overlooks the possibility of a sufficient high-ranking commanding officer accepting responsibility for the robot's performance, and thus being responsible for any violation of the rules for the ethical conduct of warfare. So,

The moral blame and accompanying punishment could be placed squarely on a human agent (or agents) who, through her own volition, has traded a part of her freedoms for the prestige of occupying a high-ranking position in a given social hierarchy (…). If no one in willing to accept this responsibility, then they should not deploy autonomous killer robots in the first place. (Champagne & Tonkens, 2015, 136; see also Schulzke 2013).

That the future will hold more autonomous systems seems almost a given, and although renouncing certain types of autonomous robots might be a good idea for many reasons, a lack of clarity as to who is responsible for their use is probably not among them (see also Kershnar, 2013). Whether one would want to have that responsibility is a different question altogether.

RESPONSIBILITY OF DESIGNERS

Although in general commanders can be held responsible for the decisions they make, in particular cases the matter is not that clear cut; in some specific cases it might be unjust to hold a commanding officer responsible for the actions of military robots. For instance, it is dubious whether we can hold a commanding officer responsible if, for example, civilian casualties (or casualties among own military personnel) are the result of a malfunctioning military robot, or of a programming error. There have already been instances of such malfunctioning: in October 2007 in South Africa a malfunctioning robotic gun fired five hundred high explosive rounds, killing nine soldiers and seriously injuring fourteen (Shachtman, 2007), while in April 2008 a SWORDS, an armed unmanned ground vehicle, produced unintended movements (Magnuson, 2008). Although partly a result of their flying low and slow, the relatively high loss rate of military robots in military operations is perhaps also an indication of their unreliability. For example, some years ago the US Air Force reported that it had lost 50 percent of its then 90 Predators (Jordan, 2007). Although it is clear that both military personnel and local civilians should be protected from malfunctioning military robots, the question is whether designers can be reasonably held responsible for casualties caused by a malfunctioning system. This is a difficult question for more than one reason. Especially with regard to learning autonomous robots, since according to Sparrow (2007) the designers then lose their responsibility for how the robot will act in the future: "[t]he connection between programmers/designers and the results of the system, which would ground the attribution of responsibility, is broken by the autonomy of the system" (Sparrow 207, p. 70). Sparrow treats the autonomy of military robots as a 'black box', leaving its working opaque, and then drawing out implications as if the autonomy of military robots are the same as human autonomy (Noorman & Johnson, 2014). This is, however, misleading since although autonomous military robots "would be able to go beyond their programming by learning new information and developing novel solutions to problems, the range of possible actions open to them would still constrained by their initial programming" (Schulzke, 2013, 213), and do not exhibit random behavior. These robots are designed to perform particular tasks and the extent to which their actions can vary is constrained by the designers. These constrains for military robots should be based on International Humanitarian Law (including the principle of discrimination, the principle of proportionality, and the principle of superfluous injury or unnecessary suffering) and the rules of engagement.

The possibility that a military robot may engage the wrong targets – because it cannot adequately distinguish citizens from soldiers – could be an acknowledged limitation of the system. If the designers have made this clear to those who have purchased or deployed a system, then, Sparrow (2007) argues, they can no longer be held responsible should this occur; in that case the responsibility should be assumed by the nation or commanding officer who (willfully and knowingly) decided to send the robot into the battlefield despite its known limitations. Conversely, one could argue that designers have an (at least moral) obligation to make products that are 'safe' and satisfy the legal and ethical conditions, and that it poses an unacceptable risk to soldiers and civilians alike to introduce a robot with such 'an acknowledged limitation' onto the battlefield. Although it might seem somewhat inconsistent to require armed military robots to be safe, as their main purpose is to destroy, in reality there is of course ample reason to be extra cautious with systems that are designed to be lethal. As it stands, there is, regarding the responsibility of designers of military robots (and militaries that deploy them), no lack of relevant legal and ethical concepts we can turn to, with *liability* seeming to be the most appropriate one (see also Lucas, 2011). The advantage of responsibility as liability *in this case*, that is, of malfunctioning robots, is that it does not require the government to foresee the consequences of new technology but rather – under certain conditions – makes the ones developing the technologies legally liable for these consequences. This places the responsibility where it can be met best: in the hands of the ones developing the technology. They have the best knowledge of new innovations and their possible effects and they are in the best position to avert certain disadvantages. Moreover, a scheme of liability would stimulate them to take their (active) moral responsibility seriously (Van de Poel & Royakkers, 2011).

Regrettably, in an international environment liability issues are not always that straightforward; for starters, the conditions that have to be met for someone to be liable depend on national laws that differ from country to country. Generally speaking, however, in the Western world the main condition for liability is negligence. Negligence implies that a manufacturer failed to do something that was morally or legally required, and thus can be held responsible for certain harms produced by his product – in legal terminology this is called *reasonable care*. Reasonable care is based on the sometimes implicit moral responsibilities of programmers and designers – it is not something made explicit in law. In military practice, however, designers are rarely held responsible for accidents caused by poor design (Thompson, 2007). It is therefore important that there will be a regulatory framework that entails 'reasonable care' to be met by designers of armed military robots. A good candidate to offer some guidelines for a formal approach of reasonable care would be the *precautionary principle*.[6] This principle was initially proposed to deal with environmental problems, yet it has far wider applications. It is especially suited for situations in which it is not possible to calculate the risks because we have insufficient scientific knowledge. In general, the precautionary principle states that precautionary measures must be taken if there are indications that there are certain hazards despite the fact that the precise nature and extent of these hazards cannot be scientifically established (Raffensperger & Tickner, 1999). Applying the precautionary principle to armed military robots would imply that the potential hazards of military robots should first be properly assessed before they could be introduced into the armed forces on a large scale, and that these robots, for example, should design with some emergency stop mechanism in case of malfunctioning such as happened in South Africa with a SWORDS (see above). The present practice to test if a system is properly functioning is done through simulations and experiments in laboratories or in small-scale field tests. Although such simulations, experiments, and tests are very important with respect to technological innovations, they do not always provide complete and reliable knowledge of the functioning of technological products and the potential hazards and risks involved. For example, laboratory and field tests are not always representative of the dynamic and complex circumstances in which military robots have to function. Evidently, it is difficult to capture the fog of war in a test setting.

Nevertheless, it seems that regulation on risk assessment can provide us with a regulatory framework for the design of military robots by formulating a set of boundary conditions for the design, production, and use of military robot technologies, such as that a military robot follows the rules of engagement and the safety protocols. If such a regulatory framework meets certain criteria, it could be considered an adequate way of dealing with the ethical issues the design of a technology raises with regard to safety and responsibility. Such a regulation for risk management should be laid down in specific norms, such as the ISO 10218 requirements for industrial robots. This norm 'specifies requirements and guidelines for the inherent safe design, protective measures, and information for use of industrial robots. It describes basic hazards associated with robots, and provides requirements to eliminate or adequately reduce the risks associated with these hazards' (ISO 2011). Although this ISO norm was developed for industrial robots, it can be used outside that context: "Examples of non-industrial robot applications include, but are not limited to: undersea, military and space robots" (ISO 2011).

Until now, however, such a regulatory framework specifically for military robots has been lacking (see also Lin et al., 2008, and Lucas, 2011). The potential use of military robots in saving 'friendly' human lives possibly explains why most of today's governments refrain from regulation with respect to the design of military robots, and especially from regulation that could lead to a ban on certain designs of military robots. On the other hand, if no-one is legally responsible for the malfunction of a military robot, there is hardly any incentive to learn from mistakes or do better in future.

Driven by this faith in the rationalisation of warfare by constraining autonomous armed robots so that they meet certain criteria, Arkin (2009) investigates the introduction of an adequate artificial ethical conscience into the armed robots by programming international humanitarian law and the rules of engagement into the robot. An advantage of these robots is that they "have the potential capability of independently and objectively monitoring ethical behaviour in the battlefield by all parties and reporting infractions that might be observed. This presence alone might possibly lead to a reduction in human ethical infractions" (Arkin, 2007). The idea of the development of autonomous machines with an additional ethical dimension is quite a new emerging field of machine ethics. These robots, with the help of formal logic, are "able to calculate the best action in ethical dilemmas using ethical principles" (Anderson & Anderson, 2007). It is thus assumed that it is sufficient to represent ethical theory in terms of a logical theory and to deduce the consequences of that theory. Beavers (2012) sees machine ethics as a threat to *ethical nihilism*,

The doctrine that states that morality needs no internal sanctions, that ethics can get by without moral 'weight,' that is, without some type of psychological force that restrains the satisfaction of our desire and that makes us care about our moral condition in the first place. (Beavers, 2012, 343).

According to Johnson and Axinn (2013), robots with an artificial ethical conscience can only mimic moral actions with no human emotions and no feelings about the seriousness of killing a human. Therefore, they conclude that autonomous robots cannot be moral.

This view that ethics can be made computable also misunderstands the unique – non-reducible – nature of ethics. Arkin (2009) agrees that some ethical theories, such as virtue ethics, do not lend themselves well by definition to a model based on a strict ethical code. But he claims that the solution is simply to eliminate ethical approaches that refuse such reduction. However, the hallmark of ethics is its non-reducibility (Royakkers & Topolski, 2014). While many ethical situations may be reducible, it is the ability to act ethically in situations that call for a judgement that is distinctly human. A consequence of this approach is that ethical principles themselves will be modified to suit the needs of a technological imperative:

Technology perpetually threatens to co-opt ethics. Efficient means tend to become ends in themselves by means of the 'technological imperative' in which it becomes perceived as morally permissible to use a tool merely because we have it. (Kaagman & Kaufman, 2009).

Another problem concerning the attempt to program the rules of the law of war and engagement into robots is that these rules are subject to interpretation and they do not provide a ready answer in any given situation. Moreover, the application of these rules is context dependent and requires the ability to understand complex social situations (Asaro, 2009). This ability is necessary in order to make ethical decisions, and as long as robots lack this ability, they will certainly not be able to take better ethical decisions than humans (Sullins, 2010).

In conclusion, although risks are inherent to the use of military robots, there is a moral obligation to limit these risks to an acceptable level by procedures of risk assessment. To ensure the allocation of responsibility for malfunction or program errors of military robots, it should be specified what this acceptable level is. Still, the possibility of an accident remains, and Krishnan suggests that in such cases all the military robots of the type that was involved in the accident should be abandoned: "If the weapon is not withdrawn from service, it can only be interpreted as a failure of politics and maybe as a war crime or crime against humanity committed by the political leadership of a state" (2009, 105).

RESPONSIBILITY OF HUMAN OPERATORS

What is often underplayed in the literature on responsibility and military robots is the role of the human operators; an omission that is all the more regrettable since it is, in fact, especially there where a number of complexities lie. In this section, we will work out the position of the human operator, often working far away from the actual battlefield. Today's UAVs connect human operators with the war zone; they are the eyes – and sometimes hands – of the tele-soldier. Partly autonomous military robots such as the Predator or the Reaper, which are able to navigate independently to their objective but require a human operator to open fire, send GPS-coordinates and camera images back to an operator who then, based on the information projected on his computer screen, decides whether or not to engage a target.

The point is that the things they see on screen can emotionally and psychologically affect these human operators, influencing their decision-making. Although fighting from behind a computer is not as emotionally potent as being on the battlefield, killing from a distance remains stressful; various studies have reported physical and emotional fatigue and increased tensions in the private lives of military personnel operating the Predators in Iraq and Afghanistan (Donnelly, 2005; Kaplan, 2006; Lee, 2012). For example, a drone pilot may witness war crimes which he is unable to prevent, or he may even see how his own actions kill civilians; unfortunately no longer a hypothetical case.

The ensuing 'residual stress' of human operators has led to proposals to diminish these tensions. In particular, the visual interface can play an important role in reducing stress; interfaces that only show abstract and indirect images of the battlefield will probably cause less stress than the more advanced real images (Singer, 2009). From a technical perspective this proposal is a feasible one, since it will not be hard to digitally recode the war scene in such a way that it induces less psychological discomfort with the war operator. This cure may have some unwanted side-effects though. Showing abstract images would in fact dehumanize the enemy, and as a result could desensitize military personnel operating unmanned systems.

Today's operators already find it at times difficult to distinguish between playing a video war game and operating an actual drone (Singer, 2009). From a technological perspective it is only a minor step to let operators think they are playing a computer game, and destroying enemy 'avatars,' while they are actually killing real people at the other side of the globe. In that case, it would be no longer the real war that would numb the soldier, but the digital recoding of it. From a moral point of view this would mean that a soldier is likely to become even further physically and emotionally detached from his actions than is already the case at present (cf. Royakkers & Van Est, 2010). Thus, where increased detachment may reduce or even eliminate human operator stress, it could at the same time limit reflection; human operators only focused on the outcome – the targeting of the blips on a screen – might not be fully aware of the consequences of their decisions.

This last observation brings us to the important role dehumanization can play. Social psychologists point out that dehumanization, that is, seeing people for something less than human, can open the door to more serious forms of unethical conduct (see for instance Bandura, 1999, 200; Mastroianni, 2011; Moller & Deci, 2010, 43-4; see also Slim, 2007, 218). It is rather hard to imagine how to have respect for the local population, a vital element of the hearts and minds approach, from, for instance, a control room in Nevada (where the pilots of Predators and Reapers mostly work from). The most effective remedy for preventing unethical conduct is 'humanization,' a not so clearly defined concept that, however, includes the affirmation of common humanity, instead of distancing oneself 'from others or divesting them from human qualities' (Bandura, 1999, 202-3). Hugo Slim explains in *Killing Civilians* that, to be effective, civilian immunity "requires that armed people find a fundamental *identification* with those called civilians and not an excessive *distinction* from them" (2007, 34). As shown by the famous Milgram experiments on obedience, it is easier to be cruel when the other does not have a face (Milgram, 1974). At a time that unmanned aerial vehicles take out insurgents from afar, and a human operator thinks that his job is "like a video game. It can get a little bloodthirsty. But it's fucking cool" (Singer, 2009, 332), that 'face' is hardly ever there. With such a distance – physical, but also psychological – between a soldier and the horrors of war, it has to be feared that this can limit, or even eliminate, proper reflection of the human operators on their decisions to open fire (see for a different view Lee, 2012; Vincente, 2013).[7]

Unclear responsibility plays a role here: the social-psychologist Albert Bandura counts the displacement and diffusion of responsibility among the "many social and psychological manoeuvres by which moral self-sanctions can be disengaged from inhumane conduct" (1999, 194), which leaves the answer to the question of responsibility of the human operator far from straightforward.

Sullins (2010) believes that in the foreseeable future such inhumane conduct could be a thing of the past, as life-and-death decisions might then be mediated by a computer-aided diagnosis of the war situation, and that military robots may even have ethical constraints built into their design – a so-called 'ethical governor' that suppresses unethical lethal behavior. For example, Arkin (2007) has done research (sponsored by the US Army) to create a mathematical decision mechanism consisting of prohibitions and obligations derived directly from the laws of war. The idea is that future military robots might give a warning if orders, according to their ethical governor, are illegal or unethical. For example, a military robot might advise a human operator not to fire because from an evaluation of the camera images the robot informs the operator that he or she is about to attack non-combatants, that is, the software of the military robot diagnosing a war situation provides a human operator with ethical advice.

Achterhuis (1998) has called this development, in which for instance an ethical governor helps to shape moral decision-making, the 'moralization of technology.' Instead of only moralizing the *human operator* ('do not shoot non-combatants'), it is the *material environment* that should be moralized, Achterhuis claims. In our case, the military robot should then see to it that no rules of engagement are violated (cf. Verbeek, 2005). Achterhuis' plea for a moralization of technology has received severe criticism, with the main criticism being that human freedom is diminished when human actions are explicitly and deliberately steered with the help of technology (cf. Verbeek, 2009). A consequence may be that humans would simply show a type of behavior that was desired by the designers of the technology instead of explicitly choosing to act this way, diminishing the freedom of, and control over, their actions. According to Cummings (2006), this would also be the case with ethical governors, since they may form a 'moral buffer' between human operators and their actions, allowing human operators to tell themselves that it was the military robot that took the decision.

The moralization of military robots could blur the line between semi-autonomous and autonomous systems, as the decision of a human operator is not the result of human deliberation, but is mainly determined or even enforced by a military robot. Another possible effect is that human operators may come to over-rely on military robots (Cummings, 2006). This is bound to happen more often in future, due to an ongoing shift from controlling to monitoring.

In brief, the moralizing of a military robot may deprive the human operator from *controlling* the situation; his future role may be restricted to *monitoring*. For our case, this would imply that we could no longer hold a human operator in reason responsible for his decisions, since it would not really be the operator taking the decisions but a military robot. This could have consequences for the question of responsibility in another way too: Detert et al. (2008) have argued that people who believe that they have little personal control in certain situations – such as those who monitor – are more likely to go along with rules, decisions, and situations even if they are unethical or have harmful effects. However, Leveringshaus (2016) argues that this "lack of moral awareness" does not relieve the human operator of taking "adequate steps to attain the morally relevant facts that enabled him to assess the risks arising from automation", and that he can still be judged negligent for deploying and relying on a technology that put innocent lives at risks.

CONCLUSION

Niccolò Machiavelli held that in war nothing ever really changes, and hence thought that the invention of the firearm amounted to nothing more than just a new variety of the age-old catapult. It is tempting to think likewise about the progress of unmanned systems, that is, as a development that does not really raise issues different from those raised a long time ago by artillery, and more recently by highflying bombers. And in part, there is something to be said for this view. But the fact is that Machiavelli was, of course, wrong; the invention of the firearm proved as crucial for warfare as the spread of the stirrup some thousand years before. Possibly, the use of unmanned systems will prove to be equally significant, especially since the development of these systems has only just begun. That will raise a host of ethical issues that are truly new, and this article addressed one of them.

We have seen that most military robots currently find their applications in surveillance, reconnaissance, and the location and destruction of mines and IEDs. These robots are unarmed, harm no one, and save lives. But not all military robots are unarmed; different types of armed military robots are currently used on the battlefield. This development may cause an important ethical problem in the assignment of responsibility in the long causal chain involved in the design and deployment of armed military robots, stretching from the manufacturer, programmer, and designer, to the departments of defense, commanding commander, and human operator. According to Sparrow (2007) a necessary condition for fighting a just war, under the principle of *jus in bello*, is that there is someone that can reasonably be held responsible for an atrocity, such as killing non-combatants, in the course of war; this condition also applies when a military robot is employed. Especially with respect to armed learning military robots there are serious legal and ethical questions regarding responsibility due to the problematic attribution of responsibility: since no-one can have sufficient *control* over the actions of these robots, and no one can foresee the consequences of these actions.

The issue of responsibility is the main reason why many lawyers and ethicists argue that a human agent should remain in-the-loop, so that traditional accountability can be ensured, and that this would therefore avoid the problem of the allocation of responsibility. Yet, in section 2 we have argued that even in the case of learning autonomous robots the assignment of responsibility is still very well possible: it normally lies with the commanding officer. However, in the subsequent two sections we have shown that the existence of responsibility gaps cannot always be avoided. Müller (2016) states that possibility of this existence should not morally preclude the use of autonomous weapons, since in civil life we only expect 'due care' from those involved. He states that if technology produces rare cases of killings where no person is responsible, this does not by itself compel a ban of autonomous robots, but that we have to regulate autonomous robots is such a way that they reduce the frequency of responsibility gaps. Therefore, we should be focused on developing responsibility practices that work to minimize risk and that clearly establish lines of responsibility. This means, that we have to negotiate about how to best assign responsibility and what that entails. Noorman and Johnson conclude that "[d]elegation of responsibility to human and non-human components is a sociotechnical design choice, not an inevitable outcome of technological development. Robots for which no human actor can be held responsible are poorly designed sociotechnical systems (Noorman & Johnson 2014, 61; see also Noorman 2014). This will be a significant challenge, since this implies that this shared responsibility is distributed broadly and which can lead to 'the problem of many hands' (see Van de Poel et al. 2015), meaning that it can be very difficult to determine what share of responsibility each person has in this sociotechnical system when something goes wrong.

For example, developing responsibility practices is necessary for a lethal mishap due to a malfunctioning of a military robot. The application of the legal concept of 'product liability' is inadequate to deal with this problem since it misses a clear regulatory framework for the design of military robots about safety and risk assessment, and more specifically, what degrees of reliability should autonomous robots meet in terms of proportionality and discrimination in their decision to attack. Therefore, governments should have a moral obligation to determine the appropriate tolerance level for autonomous robots, and to licence them for use only after a thorough testing and inspection process (Simpson and Müller 2016).

As far as semi-autonomous robots are concerned, we have discussed the largely neglected problem of the responsibility of those at the other end of the causal chain, the human operators. After all, they are the ones who decide to use lethal force, and at first sight it might seem that the attribution of responsibility is clear-cut in their case. We have seen, however, that the criteria for attributing responsibility might not always be met. The depersonalization of war, the dehumanizing the enemy, and the moralization of technology, emotionally and morally disengages the human operator. The aim of this disengagement is to reduce the psychological stress among human operators being simultaneously 'present' in and absent from the battlefield. Unfortunately, this can also limit, or even eliminate, proper reflection on their decisions; human operators have partly lost control over their decisions. Hence, if we want to hold human operators in reason responsible for their decisions, it is essential that they have control over their decisions, and a clear awareness of what is at stake.

It is, in all, important to strike a proper balance between emotional and moral attachment and detachment. This requires a lot from the computer systems that human operators use to make their life-and-death decisions. On the one hand, such systems should communicate the moral reality of the consequences of the decisions of human operators, and on the other hand such systems should reduce the strong emotions human operators feel to reduce the number of war crimes (Sparrow 2009). To develop such systems is a real challenge, but necessary to solve the problem of the attribution of responsibility, and a precondition for fighting justly in the robotic era.

REFERENCES

Achterhuis, H. (1998). *De erfenis van de utopie*. Amsterdam: Ambo.

Alley, R. (2013). *The drone debate. Sudden bullet or slow boomerang?* New Zealand: Victoria University of Wellington.

Altmann, J., Asaro, P., Sharkey, N., & Sparrow, R. (2013). Armed military robots. *Ethics and Information Technology*, *15*(2), 73–76. doi:10.100710676-013-9318-1

Anderson, M., & Anderson, S. L. (2007). Machine ethics: Creating an ethical intelligent agent. *AI Magazine*, *28*(4), 15–26.

Arkin, R. C. (2007). *Governing lethal behavior: Embedding ethics in a hybrid deliberative/reactive robot architecture* (Technical report GIT-GVU-07-11). Atlanta, GA: Georgia Institute of Technology.

Arkin, R. C. (2009). Ethical robots in warfare. *IEEE Technology and Society Magazine*, *28*(1), 30–33. doi:10.1109/MTS.2009.931858

Asaro, P. (2007). Robots and responsibility from a legal perspective. In *Proceedings of the 8th IEEE 2007 International Conference on Robotics and Automation*. IEEE.

Asaro, P. M. (2008). How just could a robot war be? In A. Briggle, K. Waelbers, & Ph. Brey (Eds.), *Current issues in computing and philosophy* (pp. 50–64). Amsterdam: IOS Press.

Asaro, P. M. (2009). Modeling the moral user. *IEEE Technology and Society*, *28*(1), 20–24. doi:10.1109/MTS.2009.931863

Bandura, A. (1999). Moral disengagement in the perpetration of inhumanities. *Personality and Social Psychology Review*, *3*(3), 193–209. doi:10.120715327957pspr0303_3 PMID:15661671

Beavers, A. F. (2012). Moral machines and the threat of ethical nihilism. In P. Lin, K. Abney, & G. A. Bekey (Eds.), *Robot ethics. The ethical and social implications of robotics* (pp. 333–344). Cambridge, MA: The MIT Press.

Champagne, M., & Tonkens, R. (2015). Bridging the responsibility gap in automated warfare. *Philosophy & Technology*, *28*(1), 125–137. doi:10.100713347-013-0138-3

Coeckelbergh, M. (2013). Drones, information technology, and distance: Mapping the moral epistemology of remote fighting. *Ethics and Information Technology*, *15*(2), 87–98. doi:10.100710676-013-9313-6

Crootof, R. (2015). War, responsibility, and killer robots. *North Carolina Journal of International Law and Commercial Regulation*, *40*(4), 909–932.

Crootof, R. (2016). A meaningful floor for 'meaningful human control'. *Temple International and Comparative Law Journal*, *30*, 53–62.

Cummings, M. L. (2006). Automation and accountability in decision support system interface design. *Journal of Technology Studies*, *32*(1), 23–31. doi:10.21061/jots.v32i1.a.4

Detert, J. R., Treviño, L. K., & Sweitzer, V. L. (2008). Moral disengagement in ethical decision making: A study of antecedents and outcomes. *The Journal of Applied Psychology*, *93*(2), 374–391. doi:10.1037/0021-9010.93.2.374 PMID:18361639

Donnelly, S. B. (2005, December 4). Long-distance warriors. *Time Magazine*.

Dunlap, Ch. J. Jr. (n.d.). Accountability and autonomous weapons: Much ado about nothing. *Temple International and Comparative Law Journal*. (forthcoming)

Ekelhof, M. A. C. (n.d.). Complications of a common language: Why it is so hard to talk about autonomous weapons. *Journal of Conflict and Security Law*. (forthcoming)

Fielding, M. (2006). Robotics in future land warfare. *Australian Army Journal*, *3*(2), 99–108.

Fieser, J., & Dowden, B. (2007). Just war theory. *The Internet Encyclopedia of Philosophy*. Retrieved June 26, 2017, from http//www.iep.utm.edu/j/justwar.htm

Fischer, J. M., & Ravizza, M. (1998). *Responsibility and control: A theory of moral responsibility*. Cambridge, UK: Cambridge University Press. doi:10.1017/CBO9780511814594

Garraway, C. (2009). The doctrine of command responsibility. In J. Doria, H.-P. Gasser, & M. C. Bassiouni (Eds.), *The legal regime of the international criminal court* (pp. 703–725). Leiden: Nijhoff. doi:10.1163/ej.9789004163089.i-1122.194

Hellström, T. (2013). On the moral responsibility of military robots. *Ethics and Information Technology*, *15*(2), 99–107. doi:10.100710676-012-9301-2

Henckaerts, J.-M., & Doswald-Beck, L. (2009). Customary international humanitarian law: Vol. I. *Rules*. Cambridge, UK: ICRC and Cambridge University Press.

Human Right Council. (2013). *Report of the special rapporteur on extrajudicial,summary or arbitrary executions, Christof Heyns (A/HRC/23/47)*. United Nations.

International Organization for Standardization. (2011). *Robots for industrial environments - Safety requirements*. Retrieved June 26, 2017, from http://www.iso.org/iso/iso_catalogue/catalogue_tc/catalogue_detail.htm?csnumber=36322

Isenberg, D. (2007). Robots replace trigger fingers in Iraq. *Asia Times Online*. Retrieved June 26, 2017, from http://www.atimes.com/atimes/Middle_East/IH29Ak01.html

Johnson, A. M., & Axinn, S. (2013). The morality of autonomous robots. *Journal of Military Ethics, 12*(2), 129–141. doi:10.1080/15027570.2013.818399

Jordan, B. (2007). Half of predators fielded have been lost. *Air Force Times*. Retrieved June 26, 2017, from http://www.airforcetimes.com/news/2007/02/AFpredatorlosses070223

Kaag, J., & Kaufman, W. (2009). Military frameworks: Technological know-how and the legitimization of warfare. *Cambridge Review of International Affairs, 22*(4), 585–606. doi:10.1080/09557570903325496

Kaplan, R. D. (2006, September). Hunting the Taliban in Las Vegas. *Atlantic Monthly*, 81–84.

Kershnar, S. (2013). No moral problem. In B. J. Strawser (Ed.), *Killing by remote control: The ethics of an unnmanned military*. Oxford, UK: Oxford University Press. doi:10.1093/acprof:oso/9780199926121.003.0011

Krishnan, A. (2009). *Killer robots. Legality and ethicality of autonomous weapons*. Farnham, MA: Ashgate Publishing Limited.

Lee, P. (2012). Remoteness, risk and aircrew Ethos. *Air Power Review, 15*(1), 1–19.

Leveringhaus, A. (2016). Drones, automated targeting and moral responsibility. In E. D. Nucci & F. Santoni de Sio (Eds.), *Drones and responsibility. Legal, philosophical and socio-technical perspectives on remotely controlled weapons* (pp. 169–181). London: Routledge.

Lichocki, P., Kahn, P. Jr, & Billard, A. (2011). The Ethical Landscape of Robotics. *IEEE Robotics & Automation Magazine, 18*(1), 39–50. doi:10.1109/MRA.2011.940275

Lin, P., Bekey, G., & Abney, K. (2008). *Autonomous military robotics: Risk, ethics, and design*. San Luis Obispo: California Polytechnic State University. doi:10.21236/ADA534697

Lucas, G. (2011). *Industrial challenges of military robotics*. Paper presented at the conference The Ethics of Emerging Military Technologies of International Society for Military Ethics (ISME), San Diego, CA. Retrieved from http://isme.tamu.edu/ISME11/isme11.html

Magnuson, S. (2008, May). Armed robots sidelined in Iraq. *National Defense Magazine*.

Mastroianni, G. R. (2011). The person–situation debate: Implications for military leadership and civilian–military relations. *Journal of Military Ethics, 10*(1), 2–16. doi:10.1080/15027570.2011.561636

Matthias, A. (2004). The responsibility gap: Ascribing responsibility for the actions of learning automata. *Ethics and Information Technology, 6*(3), 175–183. doi:10.100710676-004-3422-1

Milgram, S. (1974). *Obedience to authority: An experimental view*. London: Tavistock.

Moller, A. C., & Deci, E. L. (2010). Interpersonal control, dehumanization, and violence: A self-determination theory perspective. *Group Processes & Intergroup Relations, 13*(1), 41–53. doi:10.1177/1368430209350318

Mukerji, N. (2016). Autonomous killer drones. In E. D. Nucci & F. Santoni de Sio (Eds.), *Drones and responsibility. Legal, philosophical and socio-technical perspectives on remotely controlled weapons* (pp. 197–214). London: Routledge.

Müller, V. C. (2016). Autonomous killer robots are probably good news. In E. D. Nucci & F. Santoni de Sio (Eds.), *Drones and responsibility. Legal, philosophical and socio-technical perspectives on remotely controlled weapons* (pp. 67–81). London: Routledge.

National Research Council. (2005). *Autonomous vehicles in support of naval operations.* Washington, DC: The National Academies Press.

Noorman, M. (2014). Responsibility practices and unmanned military technologies. *Science and Engineering Ethics, 20*(3), 809–826. doi:10.100711948-013-9484-x PMID:24142234

Noorman, M., & Johnson, D. G. (2014). Negotiating autonomy and responsibility in military robots. *Ethics and Information Technology, 16*(1), 51–62. doi:10.100710676-013-9335-0

Olsthoorn, P., & Royakkers, L. M. M. (2011). *Risks and robots – Some ethical issues.* Paper presented at the conference The Ethics of Emerging Military Technologies of International Society for Military Ethics (ISME), San Diego, CA. Retrieved from http://isme.tamu.edu/ISME11/isme11.html

Orend, B. (2006). *The morality of war.* Orchard Park, NY: Broadview Press.

Raffensperger, C., & Tickner, J. (Eds.). (1999). *Protecting public health and the environment: Implementing the precautionary principle.* Washington, DC: Island Press.

Royakkers, L. M. M., & Topolski, A. R. (2014). Military robotics & relationality: criteria for ethical decision-making. In J. van den Hoven, N. Doorn, T. Swierstra, B.-J. Koops, & H. Romijn (Eds.), *Responsible innovation 1. Innovative solutions for global issues* (pp. 351–367). Dordrecht: Springer. doi:10.1007/978-94-017-8956-1_20

Royakkers, L. M. M., & Van Est, Q. (2010). The cubicle warrior: The marionette of digitalized warfare. *Ethics and Information Technology, 12*(3), 289–296. doi:10.100710676-010-9240-8

Schulzke, M. (2013). Autonomous weapons and distributed responsibility. *Philosophy & Technology, 26*(2), 203–219. doi:10.100713347-012-0089-0

Schwarz, E. (2017). *Death machines: The ethics of violent technologies.* Manchester, UK: Manchester University Press.

Shachtman, N. (2007, October 18). *Robot cannon kills 9, wounds 14.* Wired.com.

Sharkey, N. (2008). Cassandra or false prophet of doom: AI robots and war. *IEEE Intelligent Systems, 23*(4), 14–17. doi:10.1109/MIS.2008.60

Sharkey, N. (2010). Saying 'no!' to lethal autonomous targeting. *Journal of Military Ethics, 9*(4), 369–383. doi:10.1080/15027570.2010.537903

Simpson, T. W., & Müller, V. C. (2016). Just war and robots' killings. *The Philosophical Quarterly, 66*(263), 302–322. doi:10.1093/pq/pqv075

Singer, P. W. (2009). *Wired for war: The robotics revolution and conflict in the twenty-first century.* New York: Penguin Books.

Slim, H. (2007). *Killing civilians: Method, madness and morality in war.* Hurst Company.

Sparrow, R. (2007). Killer robots. *Journal of Applied Philosophy*, *24*(1), 62–77. doi:10.1111/j.1468-5930.2007.00346.x

Sparrow, R. (2009). Building a better warbot: Ethical issues in the design of unmanned systems for military applications. *Science and Engineering Ethics*, *15*(2), 169–187. doi:10.100711948-008-9107-0 PMID:19048395

Sparrow, R. (2011). Robotic weapons and the future of war. In P. Tripodi & J. Wolfendale (Eds.), *New wars and new soldiers: Military ethics in the contemporary world* (pp. 117–133). Farnham, UK: Ashgate.

Sullins, J. (2010). RoboWarfare: Can robots be more ethical than humans on the battlefield? *Ethics and Information Technology*, *12*(3), 263–275. doi:10.100710676-010-9241-7

Thompson, M. (2007). V-22 Osprey: A flying shame. *Time*. Retrieved June 26, 2017, from http://www.time.com/time/politics/article/0,8599,1665835,00.html

US Department of Defense. (2013). *Unmanned systems integrated roadmap 2013-2038*. Washington, DC: Government Printing Office.

Van de Poel, I. R., & Royakkers, L. M. M. (2011). *Ethics, engineering and technology*. Oxford, UK: Blackwell.

Van de Poel, I. R., Royakkers, L. M. M., & Zwart, S. D. (2015). *Moral responsibility and the problem of many hands*. New York: Routledge.

Verbeek, P. P. (2005). *What things do: Philosophical reflections on technology, agency, and design*. University Park, PA: Pennsylvania State University Press.

Verbeek, P. P. (2009). Moralizing technology: On the morality of technological artifacts and their design. In D. M. Kaplan (Ed.), *Readings in the philosophy of technology* (pp. 226–242). Plymouth: Rowman & Littlefield Publishers.

Vicente, J. P. (2013). The American way of remote air warfare. *Journal of Military Studies*, *4*(1). doi:10.1515/jms-2016-0185

Welsh, A. (2008). *What is Honor? A Question of Moral Imperatives*. New Haven, CT: Yale University Press.

ENDNOTES

1. As the authors of a report on autonomous military robots, written for the US Navy, put it: unmanned systems are 'unaffected by the emotions, adrenaline, and stress that cause soldiers to overreact or deliberately overstep the Rules of Engagement and commit atrocities, that is to say, war crimes' (Lin, Bekey, and Abney 2008, 1).
2. There are more criteria to hold someone reasonably responsible, such as a *causal relationship* between the behavior and the consequences, but in this paper we focus on *control*.
3. For the statement see www.icrac.co.uk/Expert%20Workshop%20Statement.pdf.

4 https://futureoflife.org/open-letter-autonomous-weapons/ (retrieved July 3, 2017).

5 A footnote made clear to what kind of decisions this principle should apply to 'The decision to kill or use lethal force against a human being; The decision to use injurious or incapacitating force against a human being; The decision to initiate combat or violent engagement between military units; The decision to initiate war or warfare between states or against non-state actors.' The statement can be found at http://icrac.net/statements/ (retrieved July 4, 2017).

6 This principle originates from the Rio Declaration, the closing statement of the first conference of the United Nations on sustainable development, which was held in Rio de Janeiro in 1992.

7 According to Coeckelbergh (2013), the practice of using drones is illustrative of the claim that new technological practices that aim to bridge physical distance create more moral distance and make it difficult for people to exercise moral responsibility.

This research was previously published in The Changing Scope of Technoethics in Contemporary Society edited by Rocci Luppicini, pages 106-123, copyright year 2018 by Information Science Reference (an imprint of IGI Global).

Chapter 18
Autonomous Systems in a Military Context (Part 1):
A Survey of the Legal Issues

Tim McFarland
The University of Melbourne, Australia

Jai Galliott
The University of New South Wales, Australia

ABSTRACT

While some are reluctant to admit it, we are witnessing a fundamental shift in the way that advanced militaries conduct their core business of fighting. Increasingly autonomous 'unmanned' systems are taking on the 'dull, dirty and dangerous' roles in the military, leaving human war fighters to assume an oversight role or focus on what are often more cognitively demanding tasks. To this end, many military forces already hold unmanned systems that crawl, swim and fly, performing mine disposal, surveillance and more direct combat roles. Having found their way into the military force structure quite rapidly, especially in the United States, there has been extensive debate concerning the legality and ethicality of their use. These topics often converge, but what is legal will not necessarily be moral, and vice versa. The authors' contribution comes in clearly separating the two parts. In this paper, they provide a detailed survey of the legality of employing autonomous weapons systems in a military context.

INTRODUCTION

This paper, the first in a two-part series, surveys the expected legal implications of employing increasingly autonomous systems (including vehicles, weapons and other systems) in domestic and overseas operations, both in armed conflict and in military operations other than war. The intention is to present a broad overview touching on many areas of potential concern rather than detailed analyses of specific topics. The points raised cover the domestic law of Australia (our home nation), a number of other nations and relevant international legal regimes, including the law of armed conflict, arms control law, international human rights law, and others. The paper outlines the ways in which the nature of a military's

DOI: 10.4018/978-1-5225-8365-3.ch018

legal obligations, and its ability to meet those obligations, may be affected by the use of equipment, principally weapons and vehicles, with a significant capacity for autonomous operation.

The autonomous behaviours of interest in this discussion are those that are 'on the horizon' as well as those that have already been implemented in working systems. Specifically, consideration is given to the behaviours of systems currently in use, those that exist as prototypes and those that are the subject of a formal development program, such as a Defence Advanced Research Projects Agency (DARPA) program.

Due to the relative novelty of systems with a legally significant capacity for autonomous operation and the forward-looking nature of this discussion, the points made here come with an unavoidable margin of error. As of October 2015, there is no case law relating specifically to the types of autonomous systems under discussion, and thus no directly relevant precedent on which to base advice. Similarly, regulatory efforts by domestic legislatures and international treaty-making bodies are at very early stages and are subject to change. Areas of particular uncertainty are identified in the relevant sections below, and recommendations for further research are made.

Where there are no legal rules directed specifically at autonomous systems, regulatory efforts directed at remotely operated systems, particularly vehicles, are likely to be applicable to autonomous systems, as autonomous systems of the types under discussion would also typically operate without a crew. The effects of legal instruments relating to uncrewed vehicles and other systems are noted where relevant.

The legal discussion is presented in three main sections. The first covers some preliminary issues: how autonomous systems are understood and defined for legal purposes, and why they raise particular legal questions. The second discusses the legal implications of acquisition and use of autonomous weapon systems; broadly, issues arising under the law of armed conflict, arms control law and human rights law. The third discusses matters arising in relation to use of autonomous vehicles under various national laws and international law. The discussion covers both 'hard' law (legally binding, enforceable rules) and 'soft' law (undertakings or agreements that are not legally binding but may still carry significant moral or political weight), and the difference is noted where relevant. The reader should also note that points of law that apply to non-conflict applications would generally continue to apply during armed conflict unless specifically overridden by a rule of the law of armed conflict.

PRELIMINARY LEGAL ISSUES

Understanding and Defining Autonomous Systems

There are scant references to autonomous systems in the legal instruments which currently govern their use, and considerable uncertainty among legislators and others who must decide how to deal with weapons, vehicles and other systems with increasing abilities to operate autonomously. Given the pace of technical development, there is an urgent need for lawyers to understand the nature and operation of autonomous systems, but significant challenges face those who attempt to do so. Such systems utilise advanced technologies from well outside the typical field of expertise of lawyers, such as robotics and artificial intelligence, and with the state of the art in those areas advancing rapidly, any detailed legal analysis is at risk of becoming obsolete before it is complete. It is perhaps understandable, then, that discussions about legal issues relating to use of autonomous systems often suffer from a lack of clarity about what machine autonomy is and how it affects the operation of a weapon, vehicle or other device. There is a general understanding among legislators, legal authors and others that autonomous systems are those which can function to some significant extent without needing a human operator to actively control the system, but beyond that, differing understandings of the nature of machine autonomy and its implications in specific fields hinder discussions about regulation. This is a significant source of uncertainty for organisations seeking to develop or acquire increasingly autonomous systems.

When the prospect of military use of autonomous weapon systems and other autonomous machines began to draw public attention some years ago, legal practitioners and academics turned their attention to the matter and began speculating about the legal implications of increasing levels of machine autonomy. Many of those articles were based on popular images of autonomous machines with advanced capabilities which did not correspond closely with reality. Autonomous systems were often treated as a discrete class of futuristic machines, distinguishable from 'semi-autonomous' or 'automated' systems by virtue of being able to 'make their own decisions' and operate 'independently' of human control.

That view is still held by some legal authors today, but more sophisticated understandings are also emerging. Some authors treat machine autonomy as a matter of degree, a capacity for self-management that exists in weapons and other systems that are in use by armed forces today and will be present to a greater degree in future systems. In this view, weapon systems such as Close-In Weapon Systems (CIWS), cruise missiles and even anti-personnel mines are seen as having some capacity for autonomous operation, albeit not necessarily sufficient to raise novel legal issues. Other authors maintain the distinction between 'autonomous', 'semi-autonomous' and 'automated' systems and seek to define autonomous systems in a way that distinguishes them on the basis of the legal issues they raise.

Whichever approach is used, legal views of autonomous systems, and whether and how to regulate their use, are far from settled. It is likely that, for the foreseeable future, regulatory efforts will continue to respond in a piecemeal fashion to technological progress in specific areas, and legal definitions of autonomous systems will continue to differ depending on the application, such that they reflect the characteristics of concern in relation to each type of system.

For example, the definition of autonomous weapon systems used in this paper and most commonly used by legal writers is one put forward by the United States Department of Defense in 2012, in a policy directive relating to autonomy in weapon systems. According to that definition an autonomous weapon system is:

> *[a] weapon system that, once activated, can select and engage targets without further intervention by a human operator. This includes human-supervised autonomous weapon systems that are designed to allow human operators to override operation of the weapon system, but can select and engage targets without further human input after activation (pp. 13–14).*

That definition relies on the most legally fraught aspects of weapon system operation: selecting and engaging targets, and being within the direct control of the weapon system, even if human intervention is possible. As a counterpoint, albeit from the civilian context, a recent circular of the International Civil Aviation Organization (2011) defined an autonomous aircraft as '[a]n unmanned aircraft that does not allow pilot intervention in the management of the flight' (p. ix), thus excluding uncrewed air systems in which a remote pilot would be in a position to intervene (note that neither of these definitions is part of a legally binding instrument. They are mentioned here as indicators of the different conceptions of autonomous systems currently held by potential regulators in different fields).

How Autonomous Systems Raise Legal Issues

Legal systems regulate the decisions and actions of 'legal persons' (generally individuals and organisations). Only entities which are legal persons, whether human or otherwise, have legal rights and obligations. In relation to autonomous systems, those rights and obligations belong to the people and organisations which develop, sell, purchase, operate or otherwise interact with such systems. Perhaps unfortunately, it is common practice to anthropomorphise autonomous systems; people often speak of robots thinking, making decisions, watching, listening and so on. Even the definition of autonomous weapon systems given above states that it is the weapon system itself which makes selections and engages targets. This is a natural way to think of a device which operates autonomously, but it can be misleading in a legal discussion when it gives the impression that an autonomous device which replaces humans in an operational sense also does so in a legal sense.

Various legal authors, particularly those writing on autonomous weapon systems, have written in terms which suggest that it is the autonomous system which, for example, selects targets and is therefore responsible for distinguishing between combatants and civilians. Regarding robots and other devices in human terms, as decision-makers in particular, tends to cloud legal issues. While autonomous systems may relieve humans of the need to physically perform a task, they do not remove humans' legal obligations relating to how the task is performed, or even whether it is to be performed in the first place.

With that point in mind, it is suggested that two questions be asked when deciding which tasks or processes can be automated, and to what degree. First, one should ask whether the use of an autonomous system to perform some task is illegal *per se*, in the way that use of a weapon which is specifically prohibited by an arms control convention may be illegal. If use of the system is not inherently illegal, the second question is whether the people and organisations responsible for the autonomous system can meet their existing legal obligations while using it for that task. For example, while there may be no rule explicitly preventing the use of a highly autonomous weapon system in a complex urban environment, limitations of the system's hardware and software may make it impossible to meet the obligation to 'do everything feasible' to distinguish between combatants and civilians in that environment.

This paper uses these two questions as a framework for the legal discussion to follow.

How Autonomous Systems Raise Legal Issues

By their nature, autonomous systems will be able to operate unattended, so in many cases it is not only necessary to look for relevant legal rules under international law, but also in several jurisdictions: that to which the owner of the system is attached; that from which the system is controlled; that in which it is physically located; and possibly those through which it transits (in the case of vehicles); or those through which it transmits information.

Any military force looking to deploy autonomous systems will need to establish processes (or check the suitability of existing processes) to clarify the interaction between legal jurisdictions where an autonomous system activated in one jurisdiction operates in, or even just enters, another; where an autonomous system operating in one jurisdiction transmits information into another; and other novel interactions between legal jurisdictions.

AUTONOMOUS WEAPONS

Law of Armed Conflict

The law of armed conflict (LOAC) is the part of international law which regulates the conduct of armed hostilities between states, or between states and non-state armed groups. The rules of LOAC are contained in a range of treaties to which Australia is a party (most notably the four Geneva Conventions of 1949 and their two Additional Protocols of 1977) as well as a significant body of customary international law. Efforts by various states to develop military systems which utilise technologies of machine autonomy to assist, supplement and, to an extent, replace human soldiers in combat roles are well known. The controversy these efforts have caused has been considerable in recent years, including strong opposition from various states as well as a range of civil society groups. For example, in February 2014 the European Parliament adopted a non-binding resolution on uncrewed aerial vehicles (UAVs) in which it called on member states to 'ban the development, production and use of fully autonomous weapons which enable strikes to be carried out without human intervention.' Nevertheless, development efforts underway today have already produced machines that can replicate some of the functions of fighter pilots (Lockheed Martin, 2015) and sentries (Tim Hornyak, 2010), among others, and it is very likely that levels of autonomy in military systems, most notably weapon systems, will continue to expand.

Given how recent these technological developments are, it is no surprise that the core LOAC instruments do not contemplate significant levels of machine autonomy being used by military forces, and as of October 2015 there is no conventional or customary international law that specifically addresses the use of autonomous weapon systems in armed conflict. Unless or until overriding legal rules specifically targeted at autonomous weapon systems come into effect, generally applicable rules of LOAC and other relevant bodies of law will govern their use. Where rules are directed at particular types of weapon systems, it is likely that autonomous weapons will be subject to the legal rules directed at the equivalent devices which lack a capacity for autonomous operation. For example, an autonomous weapon system which launches a cluster munition is likely to be subject to the same legal restrictions as other cluster munitions; an autonomous system equipped with a machine gun would need to operate in a manner which complies with the same legal rules as a human soldier operating that machine gun; and so forth.

Despite the strong opposition from various quarters, it is an open question as to whether increasing levels of autonomy in weapon systems will have any significant effect on a state's ability to meet its obligations under LOAC. As the law does not refer specifically to autonomous weapon systems, the obligations which states bear in relation to their use are those which apply to use of any weapon system which is not, by its nature, illegal. There are weapon systems in use today that arguably conform to the definition of 'autonomous' given above, such as the CIWS systems commonly installed on warships for detecting and destroying incoming missiles and aircraft; or Israel's Iron Dome; and similar defensive systems which are all capable of identifying targets of interest and firing without needing human input. Not only are such weapon systems in use today, but there are no serious claims that such systems are illegal for use in armed conflict. It remains to be seen whether more advanced systems, capable of more complex behaviour in less predictable environments, might reach some threshold beyond which the law of armed conflict cannot adequately account for their use.

Essentially, the task for lawyers is to identify the places in which the current LOAC regime will first fail as levels of autonomy on the battlefield increase. The novel characteristic of autonomous weapon systems that may lead to such a failure is that the software-based controller governing the behaviour of such a weapon 'steps into the shoes' of a human operator in respect of specified decisions and actions. This is because the decision processes encoded into the system's control software replace the decision processes which would otherwise be followed by the human operator. Legal responsibilities relating to attacks and other operations involving weapons must therefore be met by the combination of the human decision-makers who are responsible for the use of the autonomous weapon, and the weapon's own controller software and sensors. This requirement acts as a constraint on the ways in which an autonomous weapon system with a given set of operational characteristics could be employed.

The following sections briefly summarise the legal principles that govern the use of weapons in armed conflict, and are most likely to be important in military decisions about whether autonomous weapons should be utilised and, if so, and how to utilise them. Many of the specific obligations discussed below are customary in nature, meaning they arise from general state practice that is accepted as law. They are also embodied in treaties to which Australia is a party, most notably *Additional Protocol 1 (AP1) to the 1949 Geneva Conventions* (United Nations, 1977).

Principle of Distinction

According to the principle of distinction, parties to an armed conflict must at all times distinguish between civilians and combatants; and between civilian objects and military objectives; and must direct attacks only against combatants and military objectives; never against non-combatants or civilian objects. These are customary legal obligations binding on all states.

The two most common objections expressed by opponents of autonomous weapon systems concern the weapons' ability to distinguish between military and civilian targets and their lack of accountability (the latter discussed below). Addressing the former objection, to uphold the principle of distinction, a state employing an autonomous weapon system in armed conflict must ensure that it is used only in situations in which its ability to distinguish between military and civilian targets is sufficient to accurately assess all potential targets to which it may reasonably be expected to be exposed. That ability might depend, for example, on the limitations of the sensors and software used in the weapon system. Thus, the need to observe the principle of distinction imposes requirements on the design of an autonomous weapon system intended to be used in some scenario as well as restrictions on how it may be legally utilised. Of course, a sufficiently capable autonomous system used in the correct manner may exceed the ability of a human soldier to distinguish between civilians and combatants, and so enable superior adherence to the principle of distinction (indeed, that would presumably be a design goal of many systems). Consideration of that possibility would form part of the precautions that commanders are obliged to take when planning an attack.

Principle of Proportionality

The principle of proportionality states that damage and injury resulting from military action should not be excessive in relation to the anticipated concrete and direct military advantage. This is a customary legal obligation binding on all states, and is also codified in AP1 (arts. 51.5b, 57). Like distinction and accountability, the concern about the ability of autonomous weapon systems to perform proportionality assessments is often cited as a concern by opponents, and indeed it is widely acknowledged that upholding the principle of proportionality requires expert human judgement.

However, as with the principle of distinction, this concern is best seen as a restriction on the circumstances in which a given autonomous weapon system could be used. Commanders bear the legal obligation to see that the incidental damage caused by an attack is not disproportionate to the expected military advantage. Consideration of the capabilities of the chosen weapon system is one factor to consider in making that judgement.

Precautions in Attack

In planning and conducting an attack, military commanders must take all feasible precautions to avoid, and in any event to minimise, incidental loss of civilian life, injury to civilians and damage to civilian objects. This, again, is a customary obligation which is also codified in AP1 (art. 57).

The notable part of this rule in relation to autonomous weapon systems is the feasibility of precautions. Given that part of the decision process leading to selecting and engaging a target would be encoded in the software which forms a part of the weapon system, under a natural reading of the law the obligation to take precautions imposes constraints on the design of the weapon system: the abilities of the controller software; and the sensors which provide information to the controller, must together be sufficient that their use constitutes taking 'all feasible' precautions in respect of whatever part of the targeting process is performed by the weapon system.

There are several conflicting considerations in relation to taking precautions in attack. The duty to take precautions rests with 'those who plan or decide upon an attack' (art. 57.2a), typically military commanders, while controller software would in general be written largely by developers, probably civilians. Given the expanded role to be played by autonomous system software in the decisions leading to a use of force, and the correspondingly reduced role to be played by human decision-makers in the chain of command, stringent obligations would exist in relation to testing of controller software and understanding of the behaviour of highly autonomous systems.

However, the feasibility of taking such precautions may be affected by the complexity of the software which will govern autonomous weapon systems. A further consideration is that, where it can be shown that replacing a human operator with an autonomous system results in superior adherence to the law, it is conceivable that there may be an obligation to utilise the autonomous system where it is feasible to do so in the circumstances. More research is needed to determine how these competing factors will be balanced.

Situational Factors

The situational factors referred to here are duties incumbent on attacking forces that are likely to be important in some attacks but not others. These include the duties to give '[e]ffective advanced warning … of attacks that may affect the civilian population' (art. 57.2c), to avoid attacking persons who are *hors de combat*, to give quarter and accept surrender, and others. These duties will operate as additional constraints on the design of autonomous systems and their use for specific attacks.

Weapons Reviews

All parties to AP are obliged to review each new weapon system in accordance with Article 36 of AP1 'to determine whether its employment would, in some or all circumstances, be prohibited by [international law].'

Autonomous weapon systems are likely to introduce some new challenges for reviewers. The more highly autonomous a weapon system is, the more of its behaviour is determined by its control system software rather than by the actions of its human operators. Thus it is reasonable to expect that a military's ability to meet its obligation to review new weapons would require the ability to test, validate and verify the behaviour of increasingly complex software systems, which is a difficult and specialised field. At the moment there is a lack of standard methodologies to test and validate complex autonomous systems, and given the rapid pace of technical change it may be some time before proven, standard methodologies emerge. It has been noted that states wishing to employ autonomous weapons could co-operate in testing them, exchange experiences on development and use, and so on (International Committee of the Red Cross [ICRC], 2014). While that is true in some respects, testing procedures to establish legal compliance must take account of each state's planned application of the system as well as differing legal obligations (for example, where a state wishing to purchase an autonomous weapon system is party to treaties which do not apply to the state which produced the weapon). Military forces may need to adopt or develop new review processes in order to adequately test highly autonomous weapon systems before they could be legally procured for use in conflict.

An additional degree of complexity may stem from the need to review autonomous weapon systems that can exchange information with other autonomous systems, such as in a swarming scenario. The number of possible interactions between individual systems would increase exponentially with the number of systems which can exchange information, greatly complicating the task of understanding the likely behaviour of the system as a whole.

The military would need to ensure that processes are in place that are effective in determining whether a proposed weapon system, or a proposed use of a weapon system, would be consistent with its host nation's obligations under applicable conventional and customary law.

Arms Control and Disarmament Law

An autonomous weapon system is essentially a weapon operating under computer control, and restrictions that would apply to the weapon part of the system would also apply to the autonomous system as a whole. Thus, a state which is prohibited from using a certain type of weapon would also be prohibited from using an autonomous system of which that prohibited weapon forms a part. These is a range of arms control treaties which prohibit the development, acquisition and use of specified types of weapons. These include:

- The *Convention on Cluster Munitions*, which prohibits munitions that are designed to disperse explosive submunitions
- The *Protocol on Blinding Laser Weapons (Protocol IV to the Convention on Certain Conventional Weapons)*, which prohibits laser weapons specifically designed to cause permanent blindness to unenhanced vision
- The *Mine-Ban Convention* ("*Ottawa Treaty*"), which prohibits anti-personnel mines

- The *Chemical Weapons Convention*, which prohibits weapons designed to cause harm through the properties of toxic chemicals
- The *Biological Weapons Convention*, which prohibits weapons using chemical or bacteriological (biological) agents
- *Protocol I to the Convention on Certain Conventional Weapons*, which prohibits weapons the primary effect of which is to injure by fragments not detectable by X-rays

Neither the import, export, possession or use of autonomous weapon systems *per se* is specifically prohibited by any existing arms control regime, but several other arms control instruments regulate dealings in classes of weapon systems which include, or are likely to soon include, systems with a significant capacity for autonomous operation. Notable instruments under which countries may have obligations are outlined as follows.

Arms Trade Treaty

The *Arms Trade Treaty* (2013) regulates international trade in conventional arms, including many types of conventional arms that are currently, or are likely to be, the subject of efforts to incorporate a capacity for autonomous operation into weapon systems: combat aircraft, missiles, warships, tanks, small arms and others (art 2.1).

Many weapon systems with significant capacity for autonomous operation may fall within the scope of the *Arms Trade Treaty* and, if so, would be subject to the obligations Australia bears under that treaty. Prominent among these obligations are a prohibition against exporting weapons where a country 'determines that there is an overriding risk' of the weapons being used to undermine peace and security or commit serious violations of the law, and, in relation to imports, to provide 'appropriate and relevant' information to the exporting state.

United Nations Register of Conventional Arms

Many countries have agreed to regularly report their arms imports and exports to the United Nations Office of Disarmament Affairs and that information is made public in the *United Nations Register of Conventional Arms* (United Nations Office for Disarmament Affairs). This is a voluntary and non-binding process intended to build confidence between states, and to help determine whether excessive or destabilising accumulations of arms take place. Countries submit reports on both imports and exports of arms. The categories of arms covered by the Register form the basis of the categories covered by the *Arms Trade Treaty* (arts. 5.3, 13.3), so autonomous weapon systems which would enliven obligations under that treaty are also likely to appear on the Register.

Wassenaar Arrangement

The *Wassenaar Arrangement on Export Controls for Conventional Arms and Dual-Use Goods and Technologies* is another voluntary, non-binding arms transfer control regime in which Australia participates (Wassenaar Arrangement Secretariat, 2015a). Under the terms of the Arrangement, each participating state is obliged to implement export controls in its domestic legislation that relate to a defined list of conventional armaments and dual-use goods (Wassenaar Arrangement Secretariat, 2015b). The aim of the controls is to ensure that exports of these items do not contribute to the development or enhancement of military capabilities which undermine international and regional security and stability.

The *Wassenaar Arrangement* operates by defining a list of dual-use goods, technologies and munitions which are to be subjected to these export controls. Some of the items in the list describe autonomous weapon systems or likely components thereof; for example, autonomous UAVs (2015b, list. 10), robot components (list. 17), software related to operation of sensors (p. 6) and navigation equipment (p. 7) that may be present in autonomous systems, and related items. The Arrangement also regulates exports of cyber-weapons, including those with some capacity for autonomous operation (p. 4).

Missile Technology Control Regime

The *Missile Technology Control Regime* (MTCR) is another voluntary, non-binding agreement between states which seeks to restrict the proliferation of unmanned delivery systems capable of delivering weapons of mass destruction by encouraging states to adhere to a set of export policy guidelines (Government of Canada).

Convention on Conventional Weapons

As of July 2015, States Parties to the *Convention on Certain Conventional Weapons* (CCW, 1983) are considering 'questions related to emerging technologies in the area of lethal autonomous weapons systems, in the context of the objectives and purposes of the Convention' (United Nations, 2014a, para. 36). Although the discussions are at a very early stage, military stakeholders should be aware that they may eventually result in autonomous weapon systems being regulated in some way under the CCW. Thus far, two informal Meetings of Experts have been held at the United Nations Office at Geneva, in May 2014 (United Nations, 2014b) and April 2015 (United Nations, 2015), at which States Parties to the Convention discussed technical, operational, legal and ethical issues relating to autonomous weapons. The outcomes of the latest meeting were reported to the Meeting of the High Contracting Parties to the Convention in November 2015.

Although no consensus about regulation of autonomous weapons has yet emerged from these discussions, it is notable that many participants in the meetings, both states and non-state participants, have called for states to ensure there is 'meaningful human control' over weapon systems and weapon effects. The phrase 'meaningful human control' has been in use in relation to autonomous weapons since 2011 and, while it originally referred to meaningful human control *over individual attacks*, has since been used by various commentators without that modifier as something of a framing concept in discussions about regulation of increasingly autonomous weapon systems. The phrase has no specific legal meaning and does not in any way describe a legal obligation but is nevertheless significant due to its prominence in the debate, and because various participants have argued that a corresponding legal norm should be developed (ICRC, 2014). The exact content of the concept varies from speaker to speaker, but generally it relates to the need for those persons responsible for the use of force to have adequate information about the circumstances and effects of the use of force; the controllability of autonomous systems; and the ability to hold somebody accountable for the consequences of an attack.

International Human Rights Law

Human rights are basic rights and fundamental freedoms that are inherent to all human beings. They are inalienable and equally applicable to everyone. International human rights law (IHRL) is the branch of international law that recognises and protects human rights. It is largely embodied in a set of treaties, each of which protects a specific set of rights. Most notable are the *International Covenant on Civil and Political Rights* (ICCPR) and the *International Covenant on Economic, Social and Cultural Rights* (ICESCR), both adopted by the United Nations General Assembly in 1966. Other, more specialised human rights treaties, include treaties prohibiting torture or various forms of discrimination, and treaties protecting rights of children, persons with disabilities and other groups. States Parties to these treaties are obliged to refrain from interfering with the enjoyment of protected rights, to take positive steps to protect their enjoyment, and to restore them when they have been infringed. States Parties also have a duty to ensure that non-state actors do not interfere with the enjoyment of protected rights. In addition to treaty law, protection of some of the most fundamental rights has become widely recognised as customary law, binding on all states.

Right to Life

Article 6 of the ICCPR protects one of the most fundamental rights, the right to life. It states in part: 'Every human being has the inherent right to life. This right shall be protected by law. No one shall be arbitrarily deprived of his life.' While some protected rights are considered derogable (that is, they can be suspended in defined circumstances, such as states of emergency), the right to life is non-derogable; it applies at all times, including in armed conflict. The right is not, however, absolute or unlimited. Specifically, it grants protection only from arbitrary deprivation of life, in recognition that circumstances may sometimes justify the taking of life, where necessary, reasonable and proportionate; that is the case when combatants are killed in an attack which complies with the law of armed conflict.

It has been argued that delegating to an autonomous system the decision to take a life would constitute arbitrary deprivation of life, and therefore a violation of a fundamental human rights obligation. Two such arguments have been made. The first is that deprivation of life by an autonomous system would necessarily be arbitrary because non-human decision-making regarding the use of lethal force is inherently arbitrary (Asaro, 2012, p. 687). The second argument is that the risk of an autonomous system making an incorrect decision about taking a life would be unacceptably high due to the technical difficulty they would have in meeting the strict requirements posed by IHRL (Heyns, 2013).

Neither of these arguments is conclusive. The first suffers from the common mistake of anthropomorphising autonomous systems: it places the legal decision to take a life with the system rather than with the person or organisation responsible for the behaviour of the system and for the decision to employ it in a given situation. The second amounts to conjecture about the technical capabilities of a system to correctly process information about potential targets.

Although both arguments can be rebutted, the possibility remains that use of an autonomous system to take a life may come to be seen as a violation of the right to life. That debate is still evolving, and further research would be required to reach a conclusive view.

Right to Dignity

A related objection is the use of autonomous systems to apply lethal force could violate the 'inherent dignity of the human person', from which, according to the ICCPR and the ICESCR, all human rights derive. The rationale of this argument is that '[a]s inanimate machines, fully autonomous weapons could truly comprehend neither the value of individual life nor the significance of its loss. Allowing them to make determinations to take life away would thus conflict with the principle of dignity' (Human Rights Watch, 201, p. 3). This is essentially a variation of the first argument made in relation to the right to life and may be rebutted in the same way, by noting that the decision to take a life would legally lie with humans. However, it is true that machine autonomy is a technology that 'allows increasing distance to be put between weapons users and the lethal force they project' (Heyns, 2013) and those weapon users may, arguably, 'become more detached from decisions to kill – and their execution.' Again, this argument is not settled.

Criminal Accountability for Violations of the Law

One of the most frequently heard objections to the use of increasingly autonomous weapon systems is that nobody could be held accountable for a criminal act committed by such a system operating 'independently' of human control. While it is certainly an exaggeration to say that nobody could be held accountable, a significant level of autonomy in a weapon system does raise some questions about criminal liability which, while far from unanswerable, have not been fully answered yet.

Generally speaking, criminal liability is based on the nature of an incident and the role one plays in relation to it: a soldier who uses a rifle to shoot a civilian may be liable for direct perpetration of the war crime of willful killing; another soldier who did not shoot but assisted the primary perpetrator in some way may be tried for co-perpetration, aiding or abetting; and their commanding officer may be held responsible for failing to prevent the incident according to the doctrine of superior responsibility. The definitions of offences and modes of responsibility that exist in international law today reflect deeds and organisational structures found to have contributed to past violations. If autonomous weapons are deployed in combat roles, they will partly or fully displace some human soldiers from the roles that the criminal law is designed to regulate, and possibly alter the roles played by others in the chain of command, such that the criminal offences and modes of responsibility that we have today may perhaps not be readily applied to future incidents.

It is therefore necessary to understand how use of autonomous systems in combat situations will alter the decisions and actions of combatants, and whether those changes can be fully accommodated within our existing criminal law. At the very least it will be necessary to clarify how the law operates in respect of autonomous systems, and it is quite possible that some development of the law will be needed. However, that understanding is difficult to achieve in relation to autonomous systems in general at this early stage of development. It will be necessary to gain more detailed knowledge of the capabilities and uses of specific autonomous systems as development progresses.

It seems reasonable to assume that the need to be able to properly investigate an incident involving an autonomous system would mandate that some facility for logging and/or reporting the system's activities be made available and be used. While some such facility would presumably be operated for maintenance purposes anyway in most complex systems, military forces may also need to ensure that it records sufficient information to allow investigators to establish all facts required to support a decision about whether or not a breach of the law had occurred following an incident. If that is the case, developers and procurers would need considerable certainty about the legal framework under which the system would be operated.

AUTONOMOUS VEHICLES

Many autonomous systems being developed for military use take the form of air, sea or land vehicles. As such, they are subject to bodies of law which govern the use of aircraft, ships and land vehicles which may possess some capacity for autonomous operation. This section outlines legal considerations that would apply to military use of autonomous vehicles. Such considerations generally relate to the use of autonomous vehicles in environments shared with civilian vehicles, rather than in purely military environments.

Despite vehicles being one of the most active areas of development in autonomous systems over recent years, many of the relevant legal instruments either do not mention autonomous systems at all, or do not distinguish between autonomous and remotely-operated vehicles, being concerned mainly with the fact that no crew is on board the vehicle. Those rules are discussed in this section along with rules specifically directed at autonomous systems.

Australian National Law

Various organisations already operate driverless (that is, autonomous) vehicles in Australia, although not yet on public roads. Mining companies with operations in remote areas have been operating fleets of autonomous LHD (load, haul, dump) vehicles and truck haulage systems in underground and open pit mines for several years: Rio Tinto, for example, has operated fleets of autonomous trucks in the Pilbara since at least 2008, and BHP and Fortescue Metals Group are both pursuing similar programs at their mining sites. Rio Tinto is also developing 'the world's first fully-autonomous heavy haul, long distance railway system' (Rio Tinto), dubbed 'AutoHaul', to transport iron ore from the Pilbara to ports, with up to 41 autonomous trains expected to be in operation by the end of 2015 (Ker, 2015).

Of course, remote mining sites are relatively well-controlled environments, isolated from the general public. Perhaps more ambitious in regulatory terms is the New South Wales Government's Sydney Metro Northwest rail link which, when it opens to customers in early 2019, will be 'the first fully-automated metro rail system in Australia' (Sydney Metro Northwest).

In each of these cases, while all the vehicles involved are driverless, staff in a remotely situated control room have oversight of the system and, in the case of Sydney Metro Northwest, staff will circulate through the system during its operation.

Operation of autonomous vehicles on public roads, by contrast, is at a much earlier stage both technically and legally. Although no organisation is currently operating autonomous vehicles on Australian public roads, there are well-known efforts by several companies (most famously Google) to produce self-driving cars for use by the general public, and such vehicles may be made available for use in Australia over the next few years. According to Google's 'Self-Driving Car Project Monthly Report' of May 2015, the company is currently testing 32 autonomous vehicles on public roads in Mountain View, California, including 23 modified Lexus SUVs and nine two-seater electric prototypes of its own design. These vehicles have collectively driven over one million miles autonomously.

No Australian law expressly forbids or permits the use of autonomous vehicles on Australian roads. While this may be interpreted to mean that autonomous vehicles are to be regarded in the same manner as manually operated vehicles, there are indications that their use may be at odds with today's law, and their legal status is currently uncertain.

Important elements of Australia's road laws assume that vehicles have a driver who is in control of the vehicle at all times. For example, the *Australian Road Rules* state that '[a] driver must not drive a vehicle unless the driver has proper control of the vehicle' (National Transport Commission, 2006, reg. 297.1), and there are many other references to obligations which apply to drivers, to the point where it is very difficult to see how the law could apply in its current form to a driverless vehicle.

There are other potential barriers too, such as legislative no-fault compensation schemes in some Australian jurisdictions which pay compensation for road injuries from government funds (Davis, 2011, p. 166). It is likely that autonomous vehicles would be subject to careful scrutiny as to their potential for causing injury, either to the driver or to others, before they would be registrable in Australia.

In the future, it may be possible for designers and deployers of autonomous systems to rely on the *Australian Defence Force Road Transport Exemption Framework* (2014) to allow operation on public roads of vehicles which do not conform to State and Territory road transport laws in certain circumstances. However, that framework does not currently contain any reference to autonomous vehicles.

The regulatory environment does, however, appear favourable to making appropriate changes to facilitate autonomous vehicles. The National Transport Commission, in a 2013 report discussing progress towards implementing Intelligent Transport Systems, commented that:

There is undoubtedly growing research and development in [commercial deployment of automated vehicles] and a variety of elements of the driving experience are becoming more supported and eventually more automated. The NTC believes that there would be benefit for Australia to examine the potential implications of more automated vehicles, moving towards full automation, on the Australian transport system (p. 36).

In addition, at the state level, South Australian Transport Minister Stephen Mullighan, when interviewed about the possibility of autonomous cars in Australia earlier this year, referred to an 'overhaul of the *Road Traffic Act* and *Motor Vehicles Act*' when he said that 'accommodating new technologies would be one part of renewed legislation which he hoped to have in Parliament before the end of this year' (Washington, 2015). A month prior, the South Australian Governor Hieu Van Le, in his speech opening Parliament, said:

We need to encourage the embrace of new industries and technologies that will change the way in which we move around our vibrant city and safe communities. Legislation must reflect and facilitate these changes. Our Motor Vehicles Act was written when the FB model Holden was being released to the market in 1959, and our Road Traffic Act two years later. My Government will reform both pieces of legislation, and also legislate for driverless vehicles which will revolutionise transportation in South Australia (Van Le, 2015).

National Law of Other Jurisdictions

Operating vehicles in the territory of another state is generally governed by the laws of that state. Road laws in domestic jurisdictions other than Australia are also relevant insofar as they provide indications about possible developments in Australia.

In the United States, preparation for autonomous vehicles has progressed somewhat further than it has in Australia. In 2013, the National Highway Traffic Safety Administration released a 'Preliminary Statement of Policy Concerning Automated Vehicles', 'to help states implement this technology safely so that its full benefits can be realized' (p. 1). That statement includes recommendations for testing procedures to be followed by manufacturers, including testing on public roads, although the Administration 'does not recommend that states authorize the operation of self-driving vehicles for purposes other than testing at this time' (p. 14). Accordingly, several states, including Nevada, California and Florida have enacted legislation that expressly permits operation of driverless vehicles under certain conditions, and those states as well as Virginia, Michigan and Washington, D.C. permit testing of automated vehicles on designated stretches of public roads.

In February 2015, the United Kingdom Department for Transport published a review of regulations related to automated vehicles which 'recognises the significant benefits that driverless and automated vehicles will bring. As a result, it is working to support their development and introduction' (p. 17).' The cities of Greenwich, Milton Keynes, Coventry and Bristol have been selected to host trials of autonomous cars and the UK Government will provide £19 million to support testing.

Germany has also moved towards permitting the use of autonomous cars, having announced a testing program in the city of Wuppertal (Hofsaehs, 2015). Japan has also allowed testing of autonomous cars (Boyd, 2015).

International Law

Two international legal instruments provide indications that autonomous vehicles are likely to be permitted on public roads in many jurisdictions in coming years.

Vienna Convention on Road Traffic

Following a resolution adopted by the Economic and Social Council of the United Nations, the *Convention on Road Traffic* was opened for signature in 1968 and today has 73 States Parties. Intended 'to facilitate international road traffic and to increase road safety through the adoption of uniform traffic rules' (UNTS. 17), the convention sets out a range of 'rules of the road', vehicle technical standards, rules relating to driving permits, and other rules which aim to minimise incompatibilities between the domestic road laws of States Parties and ease international travel.

The Convention as originally passed did not contemplate autonomous vehicles, and some language appears to inhibit adoption of vehicles which do not have a person who could reasonably be called a driver. In particular, Article 8(1) states that '[e]very moving vehicle or combination of vehicles shall have a driver' and Article 8(5) requires that '[e]very driver shall at all times be able to control his vehicle or to guide his animals.'

In 2014 though, at the 68th meeting of the Working Party on Road Traffic Safety, the United Nations Economic and Social Council (ECOSOC) adopted amendments proposed by the governments of Austria, Belgium, France, Germany and Italy which appear to pave the way to permitting use of 'vehicle systems which do – temporarily or constantly – not allow for overriding their interventions at any time or for switching them off completely' (United Nations Economic Commission for Europe [UNECE], 2014a, para. 5). The amendments state in part that:

[v]ehicle systems which influence the way vehicles are driven shall be deemed to be in conformity with [relevant parts of this Convention] when they are in conformity with the conditions of construction, fitting and utilization according to international legal instruments concerning wheeled vehicles ... (UNECE, 2014b, Annex.)

World Forum for Harmonization of Vehicle Regulations

Also under the umbrella of the United Nations Economic and Social Council is the World Forum for Harmonization of Vehicle Regulations, a working party responsible for formulating and promoting uniform vehicle regulations in order to facilitate international trade. One of its committees, the Informal Working Group on Intelligent Transport Systems / Automated Driving (ITS) (2015), is currently working on setting definitions of terms related to automated vehicles, identifying legal obstacles and horizontal issues including liability, and seeking exchange of views from interested parties, with the goal of the 'development of technical regulations aimed at permitting the advancement of automated driving' (p. 1). The Group views automated vehicle technology as occupying one of three levels of increasing autonomy: 'driver assistance systems'; 'automated (partial automation) assistance systems' and 'autonomous (full automation) systems' (p. 3). Although the Group states that 'at this time, no work is planned to accommodate autonomous technology in the technical regulations' (p. 1), it notes that 'discussions on Autonomous Driving Technologies that envisage autonomy beyond that permitted by those conventions, will remain as an exchange of views' (p. 2), suggesting further developments to come.

Aircraft

The legal status of any UAV capable of a significant degree of autonomous operation will depend very much on whether it is considered to be an aircraft or a munition. A report on a recent Australian inquiry (Australian Senate 2015) into the use of unmanned platforms notes that:

The line between unmanned platforms and guided munitions is also being blurred. For example, the Harop (IAI, Israel) is a long range 'loitering munition', controlled in-flight by a remote operator, designed to detect and attack air defence radar systems by self-destructing into them. Similarly, the Switchblade (AeroVironment, US) is an expendable small UAV equipped with an explosive warhead which has been used by US Marines in Afghanistan.

An autonomous UAV which is categorised as a munition may fall within the purview of one or more of the arms control regimes discussed above, while a system which is categorised as an aircraft would be subject to the regulations discussed in this section below. Such decisions have been made on a case-by-case basis in relation to remotely piloted UAVs on several occasions. Relevant considerations have included technical characteristics such as the vehicle's range, payload and the presence or absence of flight controls, as well as its purpose and whether or not it is intended to return to base (Hyder & Lee, 2006, p. 129). This should not be considered an exhaustive list, and it is likely that further research would be required to clarify the distinction.

International civil aviation treaties do not directly affect military aircraft, but are still noteworthy insofar as they are influential on national regulators, including CASA, and to the extent that they reflect states' views on autonomous UAVs. That law, as it stands today, indicates that integration of autonomous aircraft into civilian airspace is some considerable distance away.

The 1948 *Chicago Convention on International Civil Aviation* states in relation to pilotless aircraft, which includes autonomous UAVs:

No aircraft capable of being flown without a pilot shall be flown without a pilot over the territory of a contracting State without special authorization by that State and in accordance with the terms of such authorization. Each contracting State undertakes to insure that the flight of such aircraft without a pilot in regions open to civil aircraft shall be so controlled as to obviate danger to civil aircraft (art. 8).

Further, a circular published by the International Civil Aviation Organisation in 2011 on 'Unmanned Aircraft Systems (UAS)' states that:

All UA, whether remotely-piloted, fully autonomous or a combination thereof, are subject to [the Chicago Convention]. Only the remotely-piloted aircraft (RPA), however, will be able to integrate into the international civil aviation system in the foreseeable future. The functions and responsibilities of the remote pilot are essential to the safe and predictable operation of the aircraft as it interacts with other civil aircraft and the air traffic management (ATM) system. Fully autonomous aircraft operations are not being considered in this effort ... (sect. 2.2)

Maritime Vehicles

As with autonomous aircraft, the legal status of any given autonomous maritime vehicle will depend on whether it is classified as a 'ship' (or 'vessel') for treaty purposes. Those that are not may be treated as torpedoes, buoys, or other equipment. The difficulty of making that assessment in respect of a new and rapidly evolving class of vehicles is compounded by the fact that there is no precise and authoritative definition of what constitutes a 'ship' or 'vessel' in international law. There are dozens of treaties relating to the law of the sea, and it is likely that a number of them would be relevant to any given operation involving an autonomous uncrewed marine vehicle (UMV). Where different treaties use different definitions of the vehicles to which they apply, the legal status of that vehicle is unclear. For example, the definition of 'warship' given in Article 29 of the *United Nations Convention on the Law of the Sea* (1982), which prescribes widely accepted rules for operations on and under the water, assumes that a warship will have a crew:

For the purposes of this Convention, "warship" means a ship belonging to the armed forces of a State bearing the external marks distinguishing such ships of its nationality, under the command of an officer duly commissioned by the government of the State and whose name appears in the appropriate service list or its equivalent, and manned by a crew which is under regular armed forces discipline.

On the other hand, the *International Regulations for Preventing Collisions at Sea* (1972) apply to 'vessels', where '[t]he word "vessel" includes every description of water craft … used or capable of being used as a means of transportation on water' (reg. 3a). The size of the vehicle, its purpose and whether it is designed to be a means of transport are a few of the relevant considerations in determining its status, and a capacity for autonomous operation may become relevant in itself.

For autonomous maritime vehicles that are found to be vessels, a range of further questions would arise. What passage rights exist for UMVs in each maritime zone? When is a UMV required to show its flag?

The law of the sea is complex, with many treaties and other legal instruments applicable to a vast range of vessels and activities, and a long history of case law relating to many jurisdictions. It is difficult to make reliable, general statements about a novel class of vehicles which are likely to exist in a range of configurations and be used in many different types of operations. Further research on specific questions of interest would certainly be required.

CONCLUSION

This is not an exhaustive survey of all legal issues that will arise as military forces seek to utilise increasingly autonomous systems in its operations. Depending on the nature and location of the operations in which autonomous systems are used, many other areas of national and international law could potentially be relevant: privacy law, tort law, environmental law, space law and many others. The intention here has been to highlight the areas of most concern and identify the most significant unanswered questions.

There is obviously much uncertainty about how the law may eventually treat specific autonomous systems, frequently because the relevant legal instruments simply do not contemplate the use of systems which have some capacity to operate autonomously, and so rely on the assumption that a human will be directly responsible for actively controlling all equipment. In many fields relevant to the military, authorities are generally still in the process of understanding and defining machine autonomy, identifying the effects it will have on their areas of concern, and formulating policy. Only when that process nears completion will it be possible to ascertain the legal status of specific systems and applications.

ACKNOWLEDGMENT

The authors gratefully acknowledge the support of the Defence Science and Technology Group of the Australian Department of Defence in the preparation of this research paper, derived from an internal report prepared for the same and *Military Robots: Mapping the Moral Landscape* (Ashgate 2015). The views expressed are the responsibility of the authors and do not necessarily reflect those of the Australian Government.

REFERENCES

Asaro, P. (2012). On Banning Autonomous Weapon Systems: Human Rights, Automation, and the Dehumanization of Lethal Decision-making. *International Review of the Red Cross*, *94*(886), 687–709. doi:10.1017/S1816383112000768

Boyd, J. (2015). Japan's Plan to Speed Self-Driving Cars. *IEEE*. Retrieved from http://spectrum.ieee.org/cars-that-think/transportation/self-driving/japans-plan-to-speed-selfdriving-cars

Civil Aviation Safety Authority. (2002). *Unmanned Aircraft and Rockets: Unmanned Aerial Vehicle (UAV) Operations, Design Specification, Maintenance and Training of Human Resources*. Canberra, ACT: Australian Government Publishing Service.

Civil Aviation Safety Regulations (1998). (Cth) reg 101.240.

Davis, J. (2011). The (Common) Laws of Man Over Vehicles Unmanned. *Journal of Law. Information and Science*, *21*(2), 166–179.

Defence Trade Controls Act (2012). (Cth).

European Parliament (2014). *Resolution on the Use of Armed Drones* (2014/2567(RSP)).

European Parliament. (2014). Resolution on the use of armed drones (2014/2567(RSP)).

Google. (2015). *Google Self-Driving Car Project, Monthly Report: May 2015*. Retrieved from http://www.google.com/selfdrivingcar/files/reports/report-0515.pdf

Government of Canada. (n. d.). *Missile Technology Control Regime*. Retrieved from http://www.mtcr.info

Gulam, H., & Lee, S. W. (2006). Uninhabited Combat Aerial Vehicles and the Law of Armed Conflict. *Australian Army Journal*, *3*(2), 123–136.

Heyns, C. (2013). *Report of the Special Rapporteur on extrajudicial, summary or arbitrary executions. UN Doc A/HRC/23/47*. Geneva: United Nations.

Hofsaehs, U. (2015). *Driverless Cars to be Tested in German 'Rust-Belt' City*. Govtech.com. Retrieved from http://www.govtech.com/fs/Driverless-Cars-to-Be-Tested-in-German-Rust-Belt-City.html

Hornyak, T. (2010). *Korean Machine-gun Robots Start DMZ Duty*. Retrieved from http://www.cnet.com/news/korean-machine-gun-robots-start-dmz-duty/

Human Rights Watch. (2014). *Shaking the Foundations: The Human Rights Implications of Killer Robots*. New York, NY: Human Rights Watch.

International Civil Aviation Organisation (1947). *Convention on International Civil Aviation*. Article 8.

International Civil Aviation Organisation (2011). International Civil Aviation Organisation Unmanned Aircraft Systems (UAS). Circular No 328-AN/190[2.2].

International Civil Aviation Organization (2011). *Unmanned Aircraft Systems* (UAS) (Circular 328 AN/190, 2011).

International Committee of the Red Cross. (2014). Report (Paper presented at Expert Meeting – Autonomous Weapon Systems: Technical, Military, Legal and Humanitarian Aspects. Geneva: Switzerland.

International Maritime Organization. (1972). International Regulations for Preventing Collisions at Sea R3(a).

Ker, P. (2015, March 10). Rio Tinto pushes ahead with driverless trains in Pilbara. *Sydney Morning Herald*. Retrieved from http://www.smh.com.au/business/mining-and-resources/rio-tinto-pushes-ahead-with-driverless-trains-in-pilbara-20150309-13z28v.html

Lockheed Martin. (2015). *UCLASS*. Retrieved from http://www.lockheedmartin.com.au/us/products/uclass.html

National Highway Traffic Safety Administration. (n. d.). *Preliminary Statement of Policy Concerning Automated Vehicles*. Retrieved from http://www.nhtsa.gov/staticfiles/rulemaking/pdf/Automated_Vehicles_Policy.pdf

National Transport Commission. (2013). *Cooperative Intelligent Transport Systems Final Policy Paper*. Melbourne, VIC: National Transport Commission.

National Transport Commission. (2014), *Australian Defence Force Road Transport Exemption Framework*. Retrieved from http://www.ntc.gov.au/Media/Reports/(B14CBF25-3E5E-4BD2-BC3E-3B34041722E7).pdf

National Transport Commission (Road Transport Legislation — Australian Road Rules) Regulations (2006). (Cth) reg 297(1).

Rio Tinto. (2015). *Mine of the Future*. Retrieved from http://www.riotinto.com/ironore/mine-of-the-future-9603.aspx

Senate Foreign Affairs, Defence and Trade References Committee (2015). *Report on Use of Unmanned Air, Maritime and Land Platforms by the Australian Defence Force*. Canberra, ACT: Parliament of Australia.

United Kingdom Department for Transport. (2015). *The Pathway to Driverless Cars*. Retrieved from https://www.gov.uk

United Nations (1977). *Convention on Road Traffic*. 1042 UNTS 17.

United Nations (1977). *Protocol Additional to the Geneva Conventions of 12 August 1949, and Relating to the Protection of Victims of International Armed Conflicts (Protocol I)*. (1125 UNTS 3).

United Nations (1983). *Convention on Prohibitions or Restrictions on the Use of Certain Conventional Weapons which may be deemed to be Excessively Injurious or to have Indiscriminate Effects*. 1342 UNTS 137.

United Nations (1994). *Article 29 of the United Nations Convention on the Law of the Sea*. opened for signature 10 December 1982, 1833 UNTS 3.

United Nations. (2014a). *Meeting of the High Contracting Parties to the Convention on Prohibitions or Restrictions on the Use of Certain Conventional Weapons Which May Be Deemed to Be Excessively Injurious or to Have Indiscriminate Effects*. Geneva: United Nations.

United Nations. (2014b). *Meeting of Experts on LAWS (13–16 May 2014)*. Retrieved from http://www.unog.ch/80256EE600585943/%28httpPages%29/A038DEA1DA906F9DC1257DD90042E261?OpenDocument

United Nations Economic Commission for Europe (2014). Consistency Between the Convention on Road Traffic (1968) and Vehicle Technical Regulations.

United Nations Economic Commission for Europe. (2014). *Report of the 68th Session* (24-26 March 2014) (ECE/TRANS/WP.1/145).

United Nations Economic Commission for Europe. (2015). *Informal Working Group on Intelligent Transport Systems / Automated Driving, Working Paper: Guidance to GRs concerning Automated Driving Technology* (ITS/AD-05-06).

United Nations Economic Commission for Europe (2015). *Informal Working Group on Intelligent Transport Systems / Automated Driving, Draft Proposal for Terms of Reference of ITS/Automated Driving Informal Working Group* (IG-ITS-AD) (ITS/AD-03-05).

United Nations Office at Geneva. (2015). *2015 Meeting of Experts on LAWS (13–17 April 2015)*. Retrieved from http://www.unog.ch/80256EE600585943/%28httpPages%29/6CE049BE22EC75A2C1257C8D00513E26?OpenDocument

United Nations Office for Disarmament Affairs. (n. d.). *UN Register of Conventional Arms*. Retrieved from http://www.un.org/disarmament/convarms/Register/

US Department of Defense. (2012). *Autonomy in Weapon Systems (Directive No 3000.09)*. Washington, DC: Office of the Under Secretary of Defense for Policy.

Van Le, H. (2015). *The Fifty-Third Parliament at the Opening of its Second Session (Speech delivered at the Parliament of South Australia, 10 February 2015)*. Retrieved from http://www.governor.sa.gov.au/node/1711

Washington, D. (2015, March 19). Google visit spurs SA's driverless car push. *InDaily*. Retrieved from http://indaily.com.au/news/2015/03/19/google-visit-spurs-sas-driverless-car-push/

Wassenaar Arrangement Secretariat. (2015a). Wassenaar Arrangement on Export Controls for Conventional Arms and Dual-Use Goods and Technologies.

Wassenaar Arrangement Secretariat. (2015b). List of Dual-Use Goods and Technologies and Munitions List (WA-LIST 14(2), 25 March 2015).

This research was previously published in the International Journal of Robotics Applications and Technologies (IJRAT), 4(2); edited by Dan Zhang, pages 34-52, copyright year 2016 by IGI Publishing (an imprint of IGI Global).

Chapter 19
Autonomous Systems in a Military Context (Part 2):
A Survey of the Ethical Issues

Jai Galliott
The University of New South Wales, Australia

Tim McFarland
The University of Melbourne, Australia

ABSTRACT

This is the second paper of two on the role of autonomy in the unmanned systems revolution currently underway and affecting military forces around the globe. In the last paper, the authors considered the implications of autonomy on the legal obligations of military forces and their ability to meet these obligations, primarily through a survey of the domestic law of a number of drone wielding nations and relevant international legal regimes, including the law of armed conflict, arms control law, international human rights law, and others. However, the impact of autonomy in the military context extends well beyond the law and also encompasses philosophy and morality. Therefore, this paper addresses perennial problems concerning autonomous systems and their impact on what justifies the initial resort to war, who may be legitimately targeted in warfare, the collateral effects of military weaponry and the methods of determining and dealing with violations of the laws of just war theory.

INTRODUCTION

This paper surveys the broad range of ethical issues that apply to increasingly autonomous systems. It begins by assessing the appeal of said systems and provides a social contract account that places their operational benefits in a moral context and acknowledges that any fundamental shift in the character of warfighting involves a change in the relations between society, state and the military institutions charged with their defence. From within the just war framework established by philosophers to regulate the resort to war and its conduct, it then examines these relationships at the robot-warmaker, -enemy and -civilian levels, considering questions of risk, transparency, cognitive dissonance, fairness, terrorism and peace,

DOI: 10.4018/978-1-5225-8365-3.ch019

among others. All are described as traditional concerns exacerbated by autonomy, and which are not insurmountable or do not necessarily counter the moral advantages on offer. It concludes by noting that government, as the most potentially powerful agential entity, holds the ultimate responsibility for avoiding that the moral pitfalls of unmanned systems use, followed by designers, engineers and their commanders/operators.

THE MORAL APPEAL OF AUTONOMOUS SYSTEMS

In making a moral assessment of any technology, it is necessary to engage in a process of discovering and documenting the potential benefits that are thought to accompany its use, as the justification for the use of a technology should always be closely tied to the advantages it yields. One of the most frequently touted benefits of exploiting unmanned systems is that they hold potential to mitigate the human cost of war, assessed physically, psychologically or otherwise, with the level of mitigation largely dependent on the degree of separation between the operator and their system. In this regard, all unmanned systems share the same benefit, with increasing levels of autonomy simply facilitating the waging of war over greater distances and adding to the number of systems available to spare people's lives.

But aside from having a tragic human toll, war can also be staggeringly expensive and unmanned systems may be able to alleviate the financial burden. Military budget figures demonstrate that the largest contributor to the overall operational cost of a manned system, accounting for roughly one-third of the total, is the number of personnel directly and indirectly required to operate and support it (United States National Research Council, 2005, p. 97). Today, in terms of basic workforce requirements,[1] autonomous systems are not all that different from a manned aircraft, with the exception that unmanned vehicle operators are safer and those working part of a long mission can go home to their families after their shift, whereas traditional aircrew cannot. However, unmanned systems have a great deal of currently unrealised potential to reduce these personnel costs over the long term, and improved autonomous operation is the key with which it is to be unlocked. Mechanisms for achieving workforce reductions are already in development. Studies have been conducted demonstrating that one operator is capable of flying up to four unmanned aerial vehicles at any given time, so long as the tasks are not overly demanding and they are supported by the right technology (Goncalves et al, 2011, p. 535). Industry is therefore developing more autonomous systems and software, which will permit operators to spread their attention across a greater number of systems, potentially to a point whereby humans are virtually removed from the operating chain. This means that while forward capital costs cannot be avoided,[2] there will be fewer people to equip for traditional battlefield service,[3] drawing fewer military-funded deployment benefits and working fewer paid hours, all of which reduces overall operational costs. Research conducted for the US military also contends that more automated systems would, additionally, enhance human warfighting capability, because they can integrate more information from a variety of sources far faster when assisted by computers that are not vulnerable to human emotion, which can cloud judgement as well as lead to poor operational decision making (Arkin, 2009, pp. 29–30). Computerised systems do not forget orders, get tired or become afraid (Weiner, 2005).

Having outlined some of the likely benefits associated with the use of autonomous unmanned systems, it is now important to highlight why these are of moral importance. For the moment, we will focus on only one of these benefits: reducing the risk of harm to soldiers. While contentious for reasons that will be outlined in the following sections, most would presumably agree that, if the use of autonomous systems can limit the danger faced by soldiers through removing them from harm's way, we would seem to have been provided with good reason for using them. Yet few have elaborated on this point and, more generally, it is concerning that the justification is thought to be obvious and uncontentious. B. J. Strawser (2010) is to date the only academic to have provided a detailed account of why we should use unmanned systems. Rather than say that their employment is ethically permissible – Strawser concludes that a state with the means to employ unmanned weaponry as a force protection measure is, in fact, ethically obligated to do so. Strawser's contention stems from what he labels the 'principle of unnecessary risk'. This principle holds that if X gives Y an order to carry out some objectively good goal, then X has an obligation, all other things being equal, to select the means that makes it possible to achieve this goal without violating the demands of justice, making the world worse or exposing the executor to potentially lethal risk, unless the goal cannot be accomplished through other means that are less risky in nature. Strawser takes this principle of unnecessary risk to be uncontroversial and many will simply accept Strawser's account of why states should be considered obligated to employ unmanned systems, but there are a few criticisms to be made. The first is that it seems rather uncontroversial to suggest that if these technologies exclusively minimise harm to warfighters – that is, while imposing no additional 'costs' of another kind – it would be a morally good thing to use them. The worry here is that things may not be as clear-cut or as perfectly fair as some premises of Strawser's argument seem to suppose, making it difficult to determine when this obligation exists. For instance, there will often be cases where we will be inclined to make delicate trade-offs between military capability and the protection of life. It is also overly focused on saving life and less capable of explaining why financial efficiency and enhancing warfighting capacity is morally important.

For us to fully comprehend the duty to employ autonomous systems, we first need to understand precisely in what the state's protective duties are grounded. We all know that states currently have a monopoly on the waging of violence, even if non-state actors are becoming more pernicious; however, we need to explain why states do, or indeed should, lay claim to the monopoly on the use of armed force before we can really understand any obligation to employ an unmanned weapon. We must therefore return to the theoretical and ideological foundations of the state and consider the relationship between citizen, state and military force. In societies such as ours, this relationship rests somewhat precariously on the rather imprecise notion of shared or common values and manifests itself in the form of a not particularly well-defined network of rights and duties (Bryce, 2009). The mechanism through which our relationship to other individuals and the state is regulated, and our responsibilities defined, has been termed the 'social contract'. This is essentially an implicit agreement that outlines how everyone is to gain mutual benefit by escaping the 'state of nature', where competition is intense, and living together under the aegis of the state. Under the contract, individuals give up their natural right to the use of armed force for their private protection and entrust their security and wellbeing to the sovereign, otherwise known as the state (Krahmann, 201, p. 23). This leads us to the welfare state we have today but, most importantly, the social contract obliges the sovereign to attain security and justice for all, which, in the first instance, means ensuring that all those within its territory respect the rule of law. The sovereign will therefore monopolise the use of force internally, to encourage individuals to obey the covenants that are in place and respect the lives, liberties and property of others, threatening them with punishment if they do not (Hobbes, 2004, p. 84). According to Thomas Hobbes' account of the social contract, the sovereign state is also conceived as having a monopoly over the means and use of force against external threats, whether these are other sovereigns or informal collectives. It is responsible for raising armies and decides when and how to wage war. According to all of the most common social contract accounts, most notably Hobbes', it is also in the sovereign's purview to command people to serve as instruments of these armies (Baumgold, 1983, p. 43).

It is important to pay careful attention to these social contract theories because they form the most influential body of thought that addresses why states exist and what, exactly, they exist for. However, a standard social contract does not explicitly explain how the prudential benefits described earlier become moral benefits. We must, therefore, posit the existence of a second contract to explain the duty to employ unmanned systems. This can usefully be construed as an implicit contract that, despite being somewhat obscure, is often used implicitly by states and military forces to justify how and why they invest their limited resources in particular (and often controversial) ways. Let us label this the 'military-state contract'. The military-state contract regulates the relationship between the state and its military institutions as abstract entities. As Peter Feaver (2003) has argued, in any liberal democracy, this relationship is one of delegation (p. 57). The state delegates or contracts out much of its protective function to the military and its armed forces, just as the citizenry entrusts many of its responsibilities to the state (Baker, 2001, p. 186). In assigning these critical protective duties to the armed forces, the state is essentially using it as an agent. Together, they share the responsibility of protecting the citizenry. To ensure that society is adequately safeguarded under this arrangement, the military-state contract assigns mutual obligations and duties to the involved parties. The state's primary duty under the contract is to provide adequate funding for the armed forces and to enact law, which enables them to go about fulfilling their protective function. In return, the military has an obligation to effectively and efficiently provide for the safety and security of the state and its people, while ensuring that it does not work against the people's broader interests. In other words, prudential action in the form of ensuring effectiveness and efficiency in military operations, such as with the use of autonomous drones, becomes a *prima facie* moral duty. This is because a state's claim to legitimacy is grounded on it being the most efficient and well-equipped entity to achieve security from internal and external violence and harm.

Interestingly, the moral rules that govern war and which are commonly employed to judge the use of unmanned systems, can also be conceptually grounded in contract theory. The essence of the most basic form of the contractual theory of war can be stated fairly simply: it rejects more mainstream models of war in favour of one that views warfare as a contract or agreement between two or more independent states to settle their quarrel(s) by force of arms (Neff, 2005, p. 137). That is, it presents a model of war that in some respects resembles a duel writ large. Historically, contractual accounts of war have received little attention and have appeared in the legal and philosophical literature in what can, at best, be described as a somewhat disappointing piecemeal fashion. The idea of a 'war contract' seems to have as one of its earlier sources, the work of sixteenth-century philosopher and Spanish Jesuit theologian and just war theorist, Francisco Suárez (1994). The basic idea is that both sides agree to prosecute their rights in a particular manner, as if agreed upon in advance. In reality, the two sets of principles that just war theory has developed over time into: *jus ad bellum* and *jus in bello*. The former applies to the decision to go to war and includes the principles of just cause, right authority, right intent, likelihood of success, proportionality (in relation to the significance of the harm) and last resort (Johnson, 1984, p. 18). The latter is supposed to govern conduct in war and includes the principles of proportionality (in relation to the amount of force needed to achieve operational objectives) and discrimination. The law of armed conflict is, in many regards, merely a more practical but imperfect refinement of these moral principles. In any case, just war theory often proves useful in thinking about the challenges that accompany the use of what are essentially robots in war. This will be demonstrated in the next section.

THE ROBOT-WARMAKER RELATIONSHIP

Any fundamental shift in the way that a military conducts its core business will have an impact that flows from the highest elected officials down to those on the receiving end of armed force. This is not without moral significance, as will be demonstrated in this section with reference to key warmakers. To begin our survey of the ethical challenges associated with autonomous unmanned systems employment, it needs to be recognised that insofar as unmanned systems disguise the true costs of war, or otherwise appear to minimise these costs, they may alter public perception and enable high-level political and military decision-makers to engage their military forces in warfare without the full knowledge or procedural consent of those they govern. This is a problem because the power to give or withhold consent for war – that is, to authorise war – is an essential element of the freedom of citizens and one that should not be subverted and circumvented (Ignatieff, 2000, p. 176). To protect the rights of individuals, the principle of right authority exists to regulate and legitimate any necessary use of force as well as to ensure that the power to authorise the deployment of weapons (such as unmanned systems) rests only with certain authorities. The requirement holds that states are permitted to wage war only if they meet all other *jus ad bellum* criteria and the conflict has been approved through all legitimate channels. In practice, this authorisation is typically embedded in domestic legal codes and constitutions, international law and/or sought from the United Nations. But again, the relevant concern here is not with the right authority requirement itself or even the tendency for certain politicians to directly bypass democratic channels, but rather with the thought that unmanned systems further compound and complicate adherence to the requirement by altering the psychological mindset of the public. This is namely by making it easier for military and political leaders to attempt to garner public support for their chosen conflicts by engaging in a range of deceptive practices including what Michael Ignatieff calls 'linguistic subterfuge' (p. 177). The word 'war', he says, is the last word that a key political decision maker wants to use. Thus the term 'autonomous weapons', representing force only slightly short of war, is avoided (p. 177). Anthony Cordesman, another military expert, has noted as well that 'one of the lessons of modern war is that war can no longer be called war' (Cordesman, 1999, p. 9). Since citizens are the ones who have to fight wars under the social contract as well as bear the impact on the economy and their lives, they tend to be more sceptical of war than their leaders. For most, the word 'war' is associated with the prolonged and deadly conflicts of earlier generations, so instead, military and political leaders use words like 'hostilities', 'precision strike', 'coercive diplomacy' and other synonyms for war. This seems to disguise the moral reality of the planned or undertaken action/s, further altering the public mindset.

In truth, any kind of unmanned warfare that can realistically be conceived of today involves human operators in or on the loop, and even then, are likely to be supported by traditional ground forces. Nevertheless, Western leaders continue to sell unmanned warfare to the public as risk-reduced (or even risk-free), and utilise multiple transfers of risk (from politician to machine, for example) to manipulate the public into granting them authority to wage war, despite the fact that these transfers often only shift or introduce new forms of death, disaster and failure.[4] One might have thought that with the media's power and omnipresence, the anticipated or ongoing costs of any particular war would be realised quickly, thus allowing the public to make an informed decision. However, truth is often war's first casualty, even in the modern sensor-saturated era. As Phillip Knightley (2004) reveals in his book on the subject of the war correspondent as a myth-maker, states and their military forces have become much more adept at managing public opinion via the media since Vietnam. In fact, some theorists go so far as to argue that, in the modern age and with its new weapons, the media has 'indisputably become a weapon of war'

(Payne, 2005, p. 84). While a military's ability to control the media is limited to some extent by ethical and legal considerations, military forces have long recognised the role of the media in shaping the public mindset. As a result, military forces may exercise their ability to limit access to the battlefield during violent conflicts in order to exert control over what information is divulged and to whom it is released (limited by ease of transferring digital data) (Payne, 2005, pp. 84–88). For instance, experience gained in Iraq has taught the US military that graphic images can have a debilitating effect on the public's support for war such that they cannot carry out operations in the way they would like to (Shurtleff, 2002, p. 104). As such, military forces have sought to distance the public from the horrors of war for many years by banning photography of the coffins of deceased soldiers (Benjamin, 2012, p. 154). The apparent aim in attempting to control the media is to morally disengage the public from war's harsh realities in the hope that they will more readily support military action. In this sense, to engage in media and risk management is really to engage in yet another form of mindset-altering politico-moral subterfuge.

A possible objection to this line of argument, at least as it applies to unmanned systems, is that these weapons actually improve public understanding of the horrors of war. These systems, particularly those of highly autonomous nature, typically have high-resolution optics and other equipment that allow them to accurately and clearly record events and conditions. Whether it is the military forces themselves for propaganda reasons, or the soldiers and other whistle-blowers who feel the need to share their experiences, footage taken from these systems has often found its way into the public arena. The effect of this drone footage is compounded by the fact that throughout the Iraq war, amateur videos of combat produced and posted by soldiers on video sharing websites such as YouTube and MySpace proved to be a ready source of humiliation for the US military (Strangelove, 2010, p. 153). In 2007, the Defense Department forbade military personnel from uploading similar material to the Internet, but the prohibition was reversed two years later, when it came to realise that, while a military force is able restrict the media's access to the battlefield, it does not have full-spectrum control over the actions of its own troops. There are now a 'large number of video clips showing members of the US (and British) military engaged in extremely violent, anti-social activities' (Strangelove, 2010, p. 153). The argument could be put forward by the advocates of autonomous weaponry that the raw footage from unmanned systems, coupled with that from phones and other personal technologies, makes it significantly less likely that the public will underestimate the moral gravity of war or otherwise be swayed toward supporting or accepting a particular engagement.

However, it remains to be seen whether this is the case. Medea Benjamin (2012) vehemently opposes such claims, arguing that while the footage may well expose the public to the horrors of war, it somehow makes it look 'fun'. She says that the video clips from drones, which are often edited to various sound tracks and captions, turn war into a form of entertainment that has been termed 'war porn' (p. 155). It would seem that unmanned systems actually counter what David Whetham labels the 'CNN effect', whereby public support for war is expected to fall as civilian casualties eventually mount and friendly forces, as well as non-combatants, become casualties in increasing numbers (Whetham, 2012, p. 203). Of course, it may be that a large segment of the public is much more conflicted. David Betz and Sangho Lee (2006) argue that what is likely to emerge is a cognitive dissonance in the public mind. This, they say, is due to being morally and physically disengaged from the battlefield while simultaneously being able to view the harm and suffering caused in a level of detail that was rarely available to the public in earlier conflicts (pp. 220–21). On the one hand, the public is told that their armed forces are utilising unmanned systems to conduct relatively safe and peaceful operations. But, on the other hand, the very same public is presented with dreadful imagery of dead and wounded. Regardless of whether unmanned weaponry results in some sort of moral disengagement or a form of public cognitive dissonance, the risk comes in generating a liberal democratic citizenry that is in some sense and to some degree psychologically divested of its moral right or ability to confer authority to wage war in accordance with just war theory. A disengaged or conflicted citizenry is unlikely to be able to balance the need to protect the community without unjustly threatening other communities and violating certain *jus ad bellum* requirements.

It should also be remembered that while high-level decision makers are responsible for making the initial decision to engage in warfare, it is the individual soldier who defends his or her state and society who must be most unconditional in their exercising of moral restraint and in their adherence to just war theory. Ignatieff (1988) writes that it is the soldiers who have the most influence on the war's outcomes. Furthermore, they have the ability to introduce a moral component, because they actually conduct the war in question. In his words, 'the decisive restraint on inhuman practice on the battlefield lies within the warrior himself – in his conception of what is honourable or dishonourable for a man to do with weapons' (p. 118). The related concern is that by removing soldiers from the battlefield and training them to fight via a technologically mediated proxy, we may lower their ability or willingness to exercise restraint and compassion in warfare. This process of psycho-moral disengagement and desensitisation not dissimilar to that affecting the general public, would also threaten the soldiers' adherence to the *jus in bello* principles of discrimination and proportionality, again keeping in mind that most autonomous systems still have a human operator for the time being. The specific contention here is that there is no other tactical weapon currently on the battlefield that facilitates killing with such physical and psychological ease, such that killing becomes a rather clinical and dispassionate matter, easing any of their operators' or overseers' existing moral qualms. This problem is only exacerbated by humans moving from being 'in the loop' to 'on the loop'. Noel Sharkey (2010) offers support to this argument by drawing attention to reports collected by Peter Singer (2009, pp. 371–72). Amongst a variety of other disturbing statements, he cites one twenty-one-year-old soldier who talks about his acts of killing with casual indifference: 'the truth is, it wasn't all I thought it was cracked up to be. I mean, I thought killing somebody would be this life-changing experience. And then I did it, and I was like, 'alright, whatever' (Singer, 2009, p. 391). Later, he says that 'killing people is like squashing an ant. I mean, you kill somebody and it's like 'alright, let's go get some pizza' (p. 392). In this clinical killing environment, in which it seems reasonable to propose that some human targets are divested of their humanity when it comes to on-screen representation, respect for *jus in bello* norms is likely to wane. Here, videos are demonstrative. Many show raw footage of Predator strikes which represent people as little more than 'hot spots' or 'blips' on the screen and which show operators often failing to take the necessary precautions to ensure non-combatants are protected. It seems that obscuring 'targets' in this way and increasing the distance to the maximum possible, as well as automating some elements of system operation, makes it easier for unmanned systems operators to kill in an indiscriminate and disproportionate fashion (at least for those predisposed to doing so).

However, not all are convinced by the argument that unmanned systems present a moral problem by making it easier to indiscriminately and disproportionately kill. Daniel Brunstetter and Megan Braun (2011) argue that these systems are subject to the same *jus in bello* requirements as other weapons used in war. They add that the systems' technological advantages coupled with the removal of risk to soldiers means that they should, at the least theoretically, more easily satisfy the principles of discrimination and proportionality. They may also make operators and politicians more reluctant to kill in situations where there is doubt with regard to the legitimacy of the potential victim of aggression (p. 339). Brunstetter and Braun say that in the case of surveillance, at the very least, the distance or what they call 'separation factor', arguably offers an increased level of control over lethal targeting decisions and ought to actually reduce unnecessary killing and destruction (p. 349). They regard not being physically present in the battlespace as a critical factor in encouraging ethical decision making in war, given that this permits the relevant people more time to think and act judiciously. In some instances, this may be exactly right, but in others, it might be that this additional time merely facilitates potentially harmful micromanagement. Christian Enemark (2013) also questions some of the assumptions relied on in this 'lower barriers' argument. He says that there is reason to suppose that being physically absent from the battlefield is *more* conducive to discrimination. In his view, the removal of risk allows decisions to be made in a more deliberate manner by removing anger and emotion, which he thinks might otherwise lead to morally unsanctioned killings (p. 60). That is, if a drone operator working from a desk in Nevada encounters the enemy, adherence to *jus in bello* protocols should improve as the operator is at little or no personal risk. This may well be true in some instances and is obviously morally preferable. However, if the human is so far removed, this seems likely to lead us back to the sort of disengagement referred to above, or even a morally concerning callousness. Ultimately, only empirical research can detail the extent of this problem.

THE ROBOT-ENEMY RELATIONSHIP

The robotics revolution also involves a change in relations between those states that rely on said robotics and enemy forces, both current and potential. Critical to this change is the matter of technological asymmetry, which refers to any imbalance in strength that each side in a conflict can bring to bear against the other (Enemark, 2013, p. 59). Many highlight that technological asymmetry is nothing new, pointing the Battle of Agincourt, Horatio Kitchener's success at Omdurman, the Gulf War, Kosovo, and other conflicts. One might then wonder whether the technological asymmetry generated by unmanned systems, even with an increasing degree of autonomy, really represents a 'radical' departure from previous levels of asymmetry. Is firing a missile from an autonomous drone more ethically problematic than dropping a 500-pound bomb at 20,000ft or pressing a button that launches a cruise missile? Probably not to the person on the ground. However, at the moral level, the answer is certainly: 'yes'. What makes the former morally repugnant is the ability to target a specific individual, at great distance, with the knowledge that you are invulnerable to retaliation. Unmanned systems, especially autonomous unmanned systems, make it possible to remove the human actor from the area of conflict – all the while allowing the operators of these systems, or the system itself, to target people and infrastructure almost as if there were an operator actually in the field – taking us as close to the notion of a 'risk free' warfare as we are likely to get at any time in the foreseeable future. This is unlike the conflicts with Kosovo and Iraq, in which pilots remained in the air, thus somewhat putting themselves at risk. Objectors to this sort of risk free war say that an issue arises when unmanned systems are used against another force, which does not possess such technology, and when the level of life-threatening risk incurred between the warring parties becomes so imbalanced that we cross a symmetry threshold which makes the fight (and thus the use of the weapon) intrinsically unjust. To illustrate what people find problematic and to test our moral intuitions, consider the following scenario.

State X, holding what it thinks is a just cause, declares war on State Y. State X possesses an arsenal of unmanned systems. Rather than sending in many traditional human warfighters, State X sends in its unmanned weaponry, starting with autonomous reconnaissance drones (ranging from wasp sized micro-unmanned aerial vehicles to more traditional high altitude vehicles) to survey the area and provide intelligence. Once this information is transferred wirelessly across the military network and analysed by super-computers, State X sends in various drones armed with technologically sophisticated munitions to kill the combatants and destroy the infrastructure of State Y. State Y, for whatever reason, does not have an arsenal of unmanned weaponry and has no enemy combatants within range of its conventional weaponry, since its enemy has a purely technological presence. So, at the moment at which State X commences its remote attack, State Y essentially is doomed to fail in defending itself (Killmister, 2008, p. 121).

It is this kind of scenario – where one side can inflict damage on the other with virtual impunity – which ethicists find morally troubling. The issue is how such unmanned conflict could possibly be considered ethical. There are two grounds on which the legitimacy of asymmetrical unmanned warfare might be questioned and these will be considered separately. The first has to do with the notion of fairness and the question of proportionality between combatants (at *jus in bello* level), and how it might be undermined by radical technological asymmetry. Ardent drone defenders such as B.J. Strawser dispute this argument by saying that if one combatant is ethically justified in their effort and the other combatant is not, then it is good and just that the justified combatant has the asymmetrical advantage associated with the use of unmanned systems (McMahan, 2009).[5] That is equivalent to saying that an unjust combatant need not think about *jus in bello* proportionality at all. In his view, this is because combatants fighting for an objectively unjust cause have no moral right to engage in violent action, or even defensive action, against combatants fighting for an objectively just cause. Here, Strawser is invoking recent work by Jeff McMahan, which presents a fairly novel reinterpretation of classical just war theory. However, few are convinced by the idea that only the good can fight fairly and according to the rules of war. The typical line of thought, advanced by notables such as Michael Walzer (2006, p. 41–43), is that *jus ad bellum* and *jus in bello* should be thought of as logically independent. Combatants on both sides deserve the respect of *jus in bello* requirements because soldiers engaging in the business of war are not always in the position to question the reasons for which war is fought, and can only do their moral best on the battlefield.

The second (and deeper) problem for those advancing the fairness objection consists in pointing out that using unmanned systems in place of the manned equivalent removes an important element of the justice of resorting to war, which remains important regardless of whether *jus in bello* fails. More specifically, when the technological imbalance reaches a certain level, it may actually override any justification for war. To clarify the fairness objection, we therefore need to look at two particular *jus ad bellum* principles. Both principles are grounded in consequentialist considerations (maximising utility, i.e. pleasure, preferences or such), although one can equally well think of them in deontological terms (i.e., in terms of certain duties). The first is the principle of last resort. It is generally recognised that, once war is unleashed, it can often get out of control and take on its own destructive power, with devastating consequences. Therefore, war should only ever be fought when necessity demands it; that is, when there is no other alternative. However, waging war without any risk to one's own troops clearly calls into question one's adherence to this principle. Where there are such high levels of asymmetry created by technology, as in our scenario, war surely cannot be considered a last resort. In most cases, State X would presumably have other less lethal options available. For example, State X could make clear and obvious to its opponent the fact that it possesses significantly superior technology, perhaps by putting on a non-lethal display of its technological capabilities. This may result in both states reaching some sort of peaceful political negotiation and settlement. Second, there is the principle of proportionality. In the *jus ad bellum* sense, this principle asks us to look at the harm our proposed military action is likely to cause and weigh it against the harm that we hope to avoid. But, where the technological imbalance is so radical – that is, where those on the receiving side of an attack are virtually unable to retaliate – it seems that the harm that the technologically superior state hopes to thwart will in many cases be so insignificant that it would present dire problems for the proportionality calculus. In other words, unmanned war of the sort described in the earlier scenario is rendered unjust. Of course, the deployment of unmanned force can be more limited in nature and need not cross the symmetry threshold, but the escalation of war poses a constant risk.

For the reasons described above, war does not necessarily need to be a 'fair' or perfectly symmetrical fight – this would defeat the whole purpose of war. But war arguably needs to be a fight of some description if it is to have any hope of fulfilling the *jus ad bellum* principles of last resort and proportionality (Enemark, 2013, p. 60). If not, it is merely slaughter. On the other hand, the use of unmanned systems in increasing numbers and with greater levels of autonomy may also increase the weaker's propensity to use guerrilla-type tactics. For instance, in order to avoid being targeted and possibly killed by their opponents who have better surveillance and attack capacities, the combatants of State Y may embrace a common asymmetric tactic and decide that, rather than carry arms openly, they will try to conceal themselves and their weapons amongst the vulnerable civilian population. This tactic has the effect of making it much more difficult and potentially dangerous for State X to abide by the just war principles of discrimination and proportionality. For conventionally weak states, the aim is often to provoke the enemy and then comingle with the civilian population in the hope that the enemy will be driven to continue with attacks causing a large number of civilian casualties and other collateral damage. Of course, this is likely to violate the principle of proportionality, especially in the case of unmanned warfare. This means that with the engagement in unmanned warfare on any sort of significant scale, there is the possibility that the threat posed by the enemy has the potential to be both so great and so insignificant as to override the moral justification for war.

At this juncture, it is important to note that the conventionally weak state Y is not alone in utilising this asymmetric tactic and shifting the burden of risk onto its civilian population. Drone pilots and command centres are regularly comingled with civilians, particularly in America where cities often rise around military bases. The most well-known US drone command centre operates from the CIA's headquarters in Langley, Virginia; whilst Nevada's Creech and Nellis Air Force bases near Las Vegas house similar military operations. Domestic bases also operate in California, Arizona, New Mexico, North and South Dakota, Missouri, Ohio, New York, and perhaps elsewhere (Lendman, 2012). In early 2013, Fox News reported that new jobs and drone technology were coming to the former Willow Grove Naval Air Station in Horsham, Pennsylvania (O'Connell, 2013). But do drone pilots count as combatants and therefore as appropriate targets for their enemies? This is an important question because, while a good deal has been said about the psychological effects of comingling and protecting those who may be vulnerable to drone strikes, little has been said about protecting the civilians amongst the operators of unmanned platforms. A drone operator who actively tracks and kills enemies is a possible target, because s/he is a participant in combat. However, for a large number of drone operators, their civilian and military worlds are intertwined. Problems about their status arise when an operator finishes her/his shift and goes home. We need to think carefully about whether drone operators are targetable when they are eating, sleeping and picking up the children from school. Conventional combatants do not acquire immunity when they eat or sleep and cannot simply 'turn off' their combatant status. But, as drones proliferate around the world and wars become an increasingly part-time endeavour for those engaged in fighting them, we may be forced to reconsider the characteristics and nature of combatant status. Is the overseer of a task-autonomous drone technically a combatant, for instance? How about the engineer who services them? Whether or not they are ethically 'targetable', the use of unmanned systems might provoke homeland attacks, whether focused on drone pilots with semi-autonomous systems, or command centres, operational engineers and higher-level decision makers in the case of systems with higher-level autonomy.

THE POST-CONFLICT ROBOT-CIVILIAN RELATIONSHIP

Nearly every discussion of unmanned systems pays homage to classical just war theory, as has this one. Just war theory is known to be a coherent set of concepts and values designed to enable systematic moral judgment in wartime, with two main tenets: *jus ad bellum* and *jus in bello*. The principles within these categories are problematic, as highlighted in the preceding section. However, if given proper consideration, they stand to ensure that they are as humane as they can be for all involved, but especially innocent civilians. Yet it may be argued that relying on the above two tenets when assessing unmanned warfare does not do justice to the task and leaves out an important question, which is related to the nature of post-war responsibility and the well-being of civilians after war. Accordingly, some just war theorists have argued that we need to develop just war theory. They say that it should include an additional tenet dealing with principles covering demilitarisation, compensation, rehabilitation and so on. This third tenet has been labelled *jus post bellum* and is focused on the establishment of peace after conflict and is especially focused on civilians and civilian infrastructure. Hostilities in Afghanistan and Iraq continue despite the end of actual combat operations; the withdrawal of coalition combat troops and the many assurances that the focus is now on the reconstruction of these nations. Despite this recent extension of just war theory, there has been surprisingly little academic enquiry and virtually no scholarship linking it to unmanned systems technologies.

The principles vary depending on whether one thinks there should merely be restrictive criteria governing what can and cannot be done to the vanquished, or more maximalist criteria in accordance with the idea that victors acquire special obligations to those they have defeated in order to ensure that conflict does not resurface. The latter kind of account is clearly the more normatively appealing as it recognises that meaningful action is required to ensure that we surpass the *status quo ante*. One such maximalist account comes from Brian Orend, who is perhaps the foremost scholar and advocate of *jus post bellum*.[6] On the first count, he stipulates that a just settlement of a war should secure the basic rights that were violated and which justified the initial waging of war (Orend, 2008). The relevant rights could include those to life, liberty, sovereignty and, more controversially, territory. Securing these rights must, for Orend, be the substantive goal of any morally legitimate peace settlement. On the second count, drawing on the classical just war principle of proportionality, he says that the peace settlement must be measured and reasonable, as well as publicly proclaimed. The call for a public proclamation ensures that the victorious state very clearly communicates to the belligerent state that it has sincere intentions for a lasting post-war settlement, while the proportionality condition ensures that the victor does not utilise the settlement as an instrument of revenge, which would only lead to more instability. Drawing on yet another familiar principle, he says that a distinction needs to be made between the soldiers, statesmen and civilians of the defeated state involved in the peace settlement. The latter group is generally considered immune from the punitive aspect of post-war measures, ruling out the imposition of broad socio-economic sanctions such as reparation taxes and the like. The soldiers and statesmen, on the other hand, are far from immune. Orend's fourth principle states that these categories of persons from the vanquished state must be punished proportionally to the moral crimes they facilitated. Proportional punishment serves as a powerful deterrent against future moral crimes and may spur change and rehabilitation on the part of the aggressor. Of course, soldiers from all sides can commit war crimes and must be held accountable. Through his fifth principle, Orend further claims that the defeated state may also owe financial restitution to the victor, though as already stated, it falls to the latter to determine how this fits with the requirements of proportionality and discrimination. Finally, and most importantly, he states that the aggressor state may require rehabilitation and demilitarisation depending on the nature of the violence it waged and the reasons for this violence, along with some consideration of the threat posed in the absence of such action.

The first few points he makes effectively stipulate that states must adhere diligently to the laws of war during the conflict and any subsequent period of occupation (p. 46). Not only is this morally vital for its own sake, it also serves to reinforce a state's cause and the legitimacy of its occupation to the local population. The US ran afoul of this principle in Iraq, owing to the notorious abuse and mistreatment of prisoners of war at Abu-Ghraib prison. However, the focus has already started to shift from these concerns over torture and humiliation to those about drone warfare. As already highlighted throughout this discussion paper, the employment of unmanned systems can breach the rules of war for any number of reasons: it can lead to more wars, lower barriers to killing and be intrinsically unfair and unjust in cases of technological asymmetry. However, perhaps the biggest concern is that these systems might be *perceived* as being indiscriminate and/or disproportionate. The history of the targeted killing facilitated by unmanned systems technologies is marked by a long series of errors that are said to have resulted in large numbers of civilian casualties and mass destruction. According to Avery Plaw (2013), while accounts of the imprecision of these systems differ, many report effective discrimination or 'kill rates' ranging from ten per cent through to an abysmal two per cent; that is, according to some accounts, only one in fifty drone attacks are said to have resulted in the deaths of al-Qaeda combatants or other militant group operatives (pp. 2–3). While it goes without saying that these claims and figures are often exaggerated, and that autonomy and technological development will improve precision, there is a common perception that Western political leaders have lost the moral high ground as a result of the drone campaigns in which they have been directly or indirectly involved (Kreps, 2014, p. 1). This is problematic, because if both the local population and the international audience begin to have doubts about a state's moral basis for continuing to wage war, they may withdraw their popular support and resources, and this will either undermine the state's chance of success altogether or extend the time it takes to accomplish a state's set goals, and so will lead to longer, bloodier wars, regardless of the marvel of modern technology. Importantly, allowing society to have misgivings about the moral rectitude of such operations will also make the disarmament and demilitarisation of a society during war and in post-conflict phases of operations inordinately difficult, when this is critical to the establishment of a just and lasting peace (Orend, 2008, p. 46). All of this points to a need for greater transparency in future drone operations, especially as automation occurs.

The latter few of Orend's principles, concerning financial restitution, rehabilitation and so on, seem to require the establishment or re-establishment of state and non-state associations to ensure that society will eventually flourish and that a minimally just government will form. A recent report from Columbia Law School on drone warfare in Iraq and Afghanistan identifies a number of relevant factors that make the affected public reluctant to associate with others, enjoy life and show a commitment to their society. First, a number of neutral civilians have found themselves caught in a precarious and rather dangerous position between local militants belonging to groups such as al-Qaeda and the United States drone effort. They have been suspected of being spies or informants working for the foreign coalition (Center for Civilians in Conflict [CCC], 2012, p. 22). In one incident reported by the *Los Angeles Times*, a shop keeper was abducted, thrown into the back of a car and taken to a safe house where, along with other suspected 'traitors', and he was beaten for eight weeks (Rodriguez, 2011). Unable to determine whether he was in fact guilty, he was released to another militant group and later set free. While those suspected of colluding with foreign powers have been harmed and present the main concern, many of those who have been mistakenly targeted by drone missiles are also stigmatised and assumed to be connected in some way to the militancy (CCC, 2012, p. 22). In such cases, civilians living nearby those liable to be bombed are likely to undertake their own vigilante action against the suspected militants in the hope of ridding their community of potential danger, not knowing that apart from having a particular affinity for hitting large gatherings such as weddings and funerals, drones have also been known to mistake the homes of peace activists for Taliban hideouts. This worry about unmanned systems and negative association is of particular moral concern because it may inhibit or discourage the formation of civic groupings that connect citizens to their government (and often protect them from the same), take pressure off the state and serve to legitimise the state. Without the sort of robust civil societies that strong public participation supports, tyrannical governments can take reign, potentially leading to further violence and war.

Apart from the abovementioned, the unwillingness of locals to cooperate with American-led coalition forces and the transitional government initially implemented in Iraq can surely be attributed, more generally, to a failure to secure the oft-cited hearts and minds of locals.[7] Securing the emotional and intellectual support of this, or any other populace, is problematic for an unmanned military force, for it is clearly imaginable that sending out robots to fight one's wars and kill one's enemy does not sit well with those on the ground. Enemy combatants might take it to be indicative of the fact that their counterparts are not willing to settle their dispute in the customary, and somewhat more chivalric, face-to-face manner. It is known for a fact that the Taliban, for instance, accustomed to fighting in close proximity to their Middle Eastern enemies, sees the use of these distance weapons as extremely cowardly (van Wifferen. 2011). This may be the same for civilians too, for as already explained, they may only ever hear the powerfully alienating hum of the Predator drones flying overhead and see the destruction and side effects they cause, without ever seeing the faces of those operators behind them (CCC, 2012, p. 23). Furthermore, there are far fewer, if any, soldiers on the ground to take responsibility for any collateral damage and carry forward values like honour, honesty, duty, respect, integrity and personal courage, which are traditionally attributed to warriors who are considered to be representatives of their respective states. However, the Middle East and Central Asia has not diverged from traditional practice as quickly as the West, so it is precisely these sorts of values that the US and Australia will need if they are to win the approval of the populations they fight amongst. The question is how to advance these values in the pursuit of increasingly autonomous warfare that reduces the number of 'boots on the ground'.

Warfighting nations could begin by utilising unmanned systems technologies in future conflicts to enhance their public affairs mission and communicate their intentions and policies to indigenous populations throughout the spectrum of military operations. This would minimise civilian repercussions in the *post bellum* phase and counter any damaging disinformation that has been spread by the enemy. For instance, in remote areas of Iraq or Afghanistan, drones could be used throughout the spectrum of military operations to distribute messages written in the local dialect, explaining what these technologies are and why they are being used over local neighbourhoods. They could also be used to rapidly replace communications infrastructure damaged in battle in order to communicate with the local population via radio or mobile telephone. The Swarming Micro Air Vehicle Project, for example, utilises a dynamic swarm of high-flying drones that create a roving network capable of providing wireless communications over areas where the terrain would otherwise prevent such communication (Hauert 2015). This network might also be used to communicate truthful and factual unclassified information to the domestic population about the progress made towards achieving whatever goal initiated the conflict. All of these are elements of a successful *jus post bellum* programme. Therefore, even highly autonomous drones can be used for good and fulfil a crucial and complementary function in presenting the other side of the story and shaping what intelligence officers specialising in psychological operations would call the 'cognitive domain' of the target population/s (Didziulis, 2008, p. 9).

CONCLUSION: A BRIEF NOTE ON MORAL RESPONSIBILITY

Central to the abovementioned ethical concerns are issues of moral responsibility: Who should be charged with ensuring that Australia does not fall into the same moral pitfalls as the US? How should they conduct themselves in dealing with these challenges? And who and to what extent should we blame and/or punish them if things go wrong? However, despite the importance of these questions, robot arms control groups have recently engaged in much fear mongering concerning the rise of increasingly intelligent autonomous systems as part of their call for a moratorium on 'killer robots'. Conventionally, there are several loci of responsibility for the actions of a machine, but both Andreas Matthias and Robert Sparrow argue that these robots will bring about a class of actions for which nobody is responsible, because no individual or group has sufficient control of these systems (Sparrow, 2007, p. 177). That is to say that we will end up with a 'responsibility gap'.

A full discussion of this supposed gap cannot be had anywhere but on the pages of a large book, but it must here be acknowledged that a machine's path to anything near human-level autonomy is a long (if not impossible) one. To say otherwise is simply to distract us from what should be the core task: guiding the development of what are really semi-autonomous systems into an age where they can perform most tasks on their own, if only with the required programming and design. Machines will not suddenly 'wake up' as depicted in films about human-hating 'terminators'. Indeed, there is simply no way in which someone could deliberately create such an entity without a collective effort on the scale of the Manhattan project. All of the involved agents, from politicians, engineers, programmers, other defence scientists and indeed anyone associated with the use of unmanned systems (including the user in the case of semi-autonomous systems), retain a share of responsibility, even though they may claim that they were not in complete or absolute control. It would be foolhardy, or even dangerous, to conclude from the observation that responsibility is obscured by the use of unmanned weaponry that nobody is, or ought to have been, held to account and that it is impossible to deal with the case of autonomous systems. On the contrary (and as others have argued in relation to informatics more generally) (Manders-Huits, 2006), we are at such an important junction in the development of unmanned systems that we have reason to adjust and refine our conception of moral responsibility, leaving behind the idea that the imposition of moral responsibility relies on agents having full control over all aspects of the design or deployment of advanced robotics. This is because these systems are becoming so complex that few of the design, development or deployment related decisions are made on an individual basis. Why concentrate on the intentions and actions of humans alone in our moral evaluation of these systems when no human exerts full control over the relevant outcomes? We need to move away from the largely insufficient notion of individual responsibility, upon which we typically rely, and move towards a more complex notion of collective responsibility. That is, it must be a holistic approach which is capable of acknowledging the contribution of various human agents, systems and organisations or institutions. The need to update our moral values and associated notions of responsibility will become more important as the technology develops, the risks increase and the number of potential responsibility-related problems accumulate.

Who should take ultimate responsibility for responsible unmanned systems innovation and the issues raised by robots in relation to the warmakers, the enemy and civilians? Generally speaking, the more power an agent has and the greater the resources at their disposal – whether intellectual, economic or otherwise – the more obliged s/he/it is to take reasonable action when problems arise. The companies of the military-industrial complex are in a unique position, and have it well within in their power, to anticipate risks of harm and injury and theorise about the possible consequences of developing increasingly autonomous systems. The costs of doing so after the fact are great and many. Moreover, manufacturers are best positioned to create opportunities for engineers to do what is right without fear of reprimand, whether that would be going ahead as planned, designing certain limitations to the system or simply refusing to undertake certain projects. It therefore seems reasonable to impose forward-looking responsibility upon them. However, it is possible that profit trumps morality for some in the military-industrial complex, so concerned parties should also continue to look to the government institutions which oversee these manufacturers and set the regulatory framework for the development and deployment of their product, should manufacturers fail to self-regulate and establish appropriate industry standards.

ACKNOWLEDGMENT

The authors gratefully acknowledge the support of the Defence Science and Technology Group of the Australian Department of Defence in the preparation of this research paper, derived from an internal report prepared for the same and *Military Robots: Mapping the Moral Landscape* (Ashgate 2015). The views expressed are the responsibility of the authors and do not necessarily reflect those of the Australian Government.

REFERENCES

Arkin, R. (2009). *Governing Lethal Behavior in Autonomous Robots*. Boca Raton, FL: CRC Press. doi:10.1201/9781420085952

Baker, D. P. (2011). To Whom Does a Private Military Commander Owe Allegiance? In P. Tripodi & J. Wolfendale (Eds.), *New Wars and New Soldiers: Military Ethics in the Contemporary World*. Surrey, UK: Ashgate.

Baumgold, D. (1983). Subjects and Soldiers: Hobbes on Military Service. *History of Political Thought*, *1*(4), 43–64.

Benjamin, M. (2012). *Drone Warfare: Killing by Remote Control*. New York, NY: OR Books.

Betz, D. J., & Lee, S. (2006). Information in the Western Way of Warfare-Too Much of a Good Thing? *Pacific Focus*, *21*(2), 197–231. doi:10.1111/j.1976-5118.2006.tb00324.x

Brunstetter, D., & Braun, M. (2011). The Implications of Drones on the Just War Tradition. *Ethics & International Affairs*, *25*(3), 337–358. doi:10.1017/S0892679411000281

Bryce, M. (2009). The Unique Nature of Military Service. *Defence Force Welfare Association Forum*. Retrieved from <http://www.vvaansw.org/ESO%20Roundtable%20Nov%2009/Att%2005e%20Unique%20Nature%20of%20Military%20Service.pdf>

Center for Civilians in Conflict [CCC]. (2012). *The Civilian Impact of Drones: Unexamined Costs, Unanswered Questions*. New York, NY: Columbia Law School Human Rights Clinic.

Coady, C. A. J. (2011). The Jus Post Bellum. In P. Tripodi & J. Wolfendale (Eds.), *New Wars and New Soldiers: Military Ethics in the Contemporary World* (pp. 49–58). Burlington: Ashgate.

Cordesman, A. (1999). *The Lessons and Non-Lessons of the Air and Missile War in Kosovo*. Center for Stategic and International Studies.

Cummins, J. (2010). *Why Some Wars Never End: The Stories of the Longest Conflicts in History*. Massachusetts: Fair Winds Press.

Didziulis, J. R. (2008). *Winning the Battle for Hearts and Minds: Operationalizing Cultural Awareness During Stability Operations*. Montgomery, AL: Air Command and Staff College.

Enemark, C. (2013). *Armed Drones and the Ethics of War: Military Virtue in a Post-Heroic Age*. Oxon, UK; New York, NY: Routledge.

Feaver, P. D. (2003). *Armed Servants: Agency, Oversight, and Civil-Military Relations*. Boston, MA: Harvard University Press. doi:10.4159/9780674036772

Ferreria, S., Gonçalves, G., Goncalves, R., Pinto, J., & Sousa, J. (2011). Authority Sharing in Mixed Initiative Control of Multitple Uninhabited Aerial Vehicles. *Paper presented at the Engineering Psychology and Cognitive Ergonomics*, Orlando, Florida.

Hauert, S. (2015) *The Swarming Micro Air Vehicle Network (Smavnet) Project*. Retrieved from <http://smavnet.epfl.ch/ >

Hobbes, T. (2004). *Leviathan. Ebook*. Nuvision Publishers.

Ignatieff, M. (1998). *The Warrior's Honour: Ethnic War and the Modern Conscience*. New York, NY: Henry Holt and Company.

Ignatieff, M. (2000). *Virtual War: Kosovo and Beyond*. New York, NY: Metropolitan Books.

Johnson, J. T. (1984). *Can Modern War Be Just?* New Haven, CT: Yale University Press.

Killmister, S. (2008). Remote Weaponry: The Ethical Implications. *Journal of Applied Philosophy*, *25*(2), 121–133. doi:10.1111/j.1468-5930.2008.00400.x

Killmister, S. (2008). Remote Weaponry: The Ethical Implications. *Journal of Applied Philosophy*, *25*(2), 121–133. doi:10.1111/j.1468-5930.2008.00400.x

Knightley, P. (2004). *The First Casualty: The War Correspondent as Hero and Myth-Maker from the Crimea to Iraq*. Baltimore, MD: Johns Hopkins University Press.

Krahmann, E. (2010). *State, Citizens and the Privatization of Security*. Cambridge, UK: Cambridge University Press. doi:10.1017/CBO9781139042727

Kreps, S. (2014). Flying under the Radar: A Study of Public Attitudes towards Unmanned Aerial Vehicles. *Research & Politics*, *1*(1), 1–7. doi:10.1177/2053168014536533

Lendman, S. (2012). *America's Drone Command Centers: Remote Warriors Operate Computer Keyboards and Joysticks, Global Research: Centre for Research on Globalisation*. Retrieved from http://www.globalresearch.ca/america-s-drone-command-centers-remote-warriors-operate-computer-keyboards-and-joysticks/30590>

Manders-Huits, N. (2006). Moral Responsibility and It for Human Enhancement. *Paper presented at the Association for Computing Machinery Symposium on Applied Computing*, Dijon, France. 10.1145/1141277.1141340

Matthias, A. (2004). The Responsibility Gap: Ascribing Responsibility for the Actions of Learning Automata. *Ethics and Information Technology*, *6*(3), 175–183. doi:10.100710676-004-3422-1

McMahan, J. (2009). *Killing in War*. Oxford, UK: Oxford University Press. doi:10.1093/acprof:oso/9780199548668.001.0001

Neff, S. C. (2005). *War and the Law of Nations: A General History*. Cambridge, UK: Cambridge University Press. doi:10.1017/CBO9780511494253

O'Connell, C. (2013). Drone Command Center, 200-Plus Jobs Coming to Horsham. *Fox News*. Retrieved from http://www.myfoxphilly.com/story/21676061/drone-command-center-200-jobs-coming-to-horsham

Orend, B. (2008). Jus Post Bellum: A Just War Theory Perspective. In S. Carsten & J. K. Kleffner (Eds.), *Just Post Bellum: Towards a Law of Transition from Conflict to Peace* (pp. 31–52). The Hague, The Netherlands: TMC Aasser Press. doi:10.1007/978-90-6704-719-7_3

Payne, K. (2005). The Media as an Instrument of War. *Parameters: The US Army War College Quarterly Senior Professional Journal*, *35*(1), 81–93.

Plaw, A. (2013). Counting the Dead: The Proportionality of Predation in Pakistan. In B. J. Strawser (Ed.), *Killing by Remote Control: The Ethics of an Unmanned Military* (pp. 126–178). Oxford, UK: Oxford University Press. doi:10.1093/acprof:oso/9780199926121.003.0007

Quinn, B. (2013, September 26). Mod Study Sets out How to Sell Wars to the Public. *The Guardian*. Retrieved from http://www.theguardian.com/uk-news/2013/sep/26/mod-study-sell-wars-public

Robben, A. C. G. M. (2010). Losing Hearts and Minds in the War on Terrorism. In A. C. G. M. Robben (Ed.), *Iraq at a Distance: What Anthropologists Can Teach Us About the War* (pp. 106–132). Philadelphia: University of Pennsylvania Press.

Rodriguez, A. (2011, December 28). Pakistani Death Squads Go after Informants to US Drone Program. *Los Angeles Times*. Retrieved from http://articles.latimes.com/2011/dec/28/world/la-fg-pakistan-death-squads-20111228

Sharkey, N. (2010). Saying 'No!' To Lethal Autonomous Targeting. *Journal of Military Ethics*, *9*(4), 369–383. doi:10.1080/15027570.2010.537903

Shurtleff, K. (2002). The Effects of Technology on Our Humanity. *Parameters: The US Army War College Quarterly Senior Professional Journal, 32*(2), 100–112.

Singer, P. W. (2009). *Wired for War: The Robotics Revolution and Conflict in the 21st Century*. New York, NY: The Penguin Press.

Sparrow, R. (2007). Killer Robots. *Journal of Applied Philosophy, 24*(1), 62–77. doi:10.1111/j.1468-5930.2007.00346.x

Strangelove, M. (2010). *Watching YouTube: Extraordinary Videos by Ordinary People*. Toronto, Canada: University of Toronto Press.

Strawser, B. J. (2010). Moral Predators: The Duty to Employ Uninhabited Aerial Vehicles. *Journal of Military Ethics, 9*(4), 342–368. doi:10.1080/15027570.2010.536403

Suárez, F. (1944). On War. In G. L. Williams, A. Brown, & J. Waldron (Eds.), *Francisco Suárez, Selections from Three Works*. Oxford, UK: Clarendon Press.

United States National Research Council, Committee on Autonomous Vehicles in Support of Naval Operations. (2005). *Autonomous Vehicles in Support of Naval Operations*. Washington, D.C: The National Academies Press.

van Wifferen, L. (2011). *Alienation from the Battlefield: Ethical Considerations Concerning Remote Controlled Military Robotics*. Utrect, The Netherlands: Universiteit Utrecht.

Walzer, M. (2006). Response to McMahan's Paper. *Philosophia, 34*(1), 19–21. doi:10.100711406-006-9006-z

Weiner, T. (2005, February 17). "GI Robot" Rolls toward the Battlefield. *Herald Tribune*. Retrieved from http://iht.com/articles/2005/02/16/business/robot.php

Whetham, D. (2012). Remote Killing and Drive-by Wars. In D. W. Lovell & I. Primoratz (Eds.), *Protecting Civilians During Violent Conflict: Theoretical and Practical Issues for the 21st Century* (pp. 199–224). Surrey, UK: Ashgate.

ENDNOTES

1. The number of personnel required to support the operation of the aircraft would be comparable between manned and unmanned systems. Both require refueling, maintenance and a large number of data analysts.
2. The technological development of unmanned systems, itself, is unlikely to more expensive than the development of new manned platforms designed for twenty-first century combat, especially considering that there is no need for life support equipment.
3. The cost of putting a single soldier on the battlefield is approximately US$1,000,000.
4. (In fact, the British Ministry of Defence has even commissioned a study in which it is recommended that contractors employ drones rather than military personnel in an attempt to distort the public's perception of the risks associated with war. See: Quinn, 2013)

5 (McMahan, 2009 This recent monograph draws on many papers published over the last decade or so and represents some of McMahan's clearest thought on the topic.)

6 Tony Coady puts forward a very similar elucidation of *jus post bellum* principles; however, in some sense it is even more comprehensive. His *jus post bellum* is all-encompassing, covering the termination of war, how to deal with post-war conditions and how to deal with one's own troops justly after the war. See: (Coady, 2011). As his account seems to conflate the subtle differences between these three sub-areas, Orend's account is utilised for the sake of this chapter.

7 Numerous scholars make mention of this. For two good discussions, see: (Cummins, 2010; Robben, 2010)

This research was previously published in the International Journal of Robotics Applications and Technologies (IJRAT), 4(2); edited by Dan Zhang, pages 53-68, copyright year 2016 by IGI Publishing (an imprint of IGI Global).

Chapter 20
Drone Warfare:
Ethical and Psychological Issues

Robert Paul Churchill
George Washington University, USA

ABSTRACT

The United States is now relying on Reaper and Predator drone strikes as its primary strategy in the continuing War on Terrorism. This paper argues for the rational scrutiny drone warfare has yet to receive. It is argued that drone warfare is immoral as it fails both the jus in bello and the jus ad bellum conditions of Just War theory. Drone warfare cannot be accepted on utilitarian grounds either, as it is very probable that terrorists will acquire drones capable of lethal strikes and deploy them against defenseless civilians. Moreover, by examining the psychological bases for reliance on drone warfare, as well as the message the United States is sending adversaries, we need to be concerned that, rather than reduce the likelihood of terrorists strikes, the U.S. reliance on drones strikes threatens to institutionalize terrorism as the status quo for the foreseeable future.

INTRODUCTION

Drones have received considerable attention in the news lately as a promising but controversial technological innovation. Private corporations foresee new business possibilities and greater profits, small lightweight drones marketed as self-"piloted" gadgets have attracted recreational users, and police and law enforcement officials across America foresee the advantages of using the "unblinking stare" of surveillance drones to police us (Jayakumar, 2015). Of course, the most controversial use the new technology concerns the ways in which remotely-controlled, armed drones are now used as President Obama's preferred way of responding to the perceived threats posed by al-Qaeda, the Taliban, ISIS and other terrorist operatives. Not since public debates over President Regan's proposed Strategic Defense Initiative, popularly dubbed "star wars," has American military strategy been the subject for such widespread concern at home and abroad.

DOI: 10.4018/978-1-5225-8365-3.ch020

Remotely "piloted" drones, or "unmanned aerial vehicles" (UAVs) known as Predators received extensive use for surveillance purposes during the 1999 NATO intervention in Kosovo. Subsequently armed with Hellfire missiles Predators as "unmanned combat air vehicles" (UCAVs) were first tested in February 2001. The first targeted death, authorized by President George W. Bush in June 2004 and supervised by the Central Intelligence Agency (CIA), occurred when a Predator armed with a Hellfire missile killed a suspected Pakistani al-Qaeda operative in South Waziristan (Coll, 2014, p. 102). Still, the Bush administration's use of drones was rather restrained compared with President Obama's. Bush allowed 49 drone strikes to be launched between 2004 and 2008. According to figures assembled by the Bureau of Investigative Journalism, between 2009 and mid-January 2015, there have been 364 drone strikes, mostly in Pakistan, but also in Iraq, Libya, Somalia, Syria, Yemen, and at least 1 in the Philippines (Kreps & Zenko, 2014).

While Amnesty International is concerned that lethal drone strikes may be war crimes (Boone, 2013), speaking at the Naval Defense University in Washington, DC, President Obama declared that drone warfare met the requirements of Just War theory (BBC News, 2013). Apparently Obama was encouraged by the greater precision of strikes by the Reaper, the Predator's successor, which can fire a smaller two-foot missile known as a "Small Smart Weapon" (SSW). The Reaper, a larger and heavier drone the Predator has greater surveillance capabilities, longer flight times, and carries payloads than its predecessor. Allegedly, a SSW can "take out" a human target without harming persons in an adjoining room (Coll, 2014, p. 101), although no one has explained how drone surveillance can be certain which room a target is in or how many people are in it. In any case, lethal drone strikes are now not only the cornerstone of the U.S. response to the "war on Terror" Obama inherited from Bush, they are drone so widely regarded as the most promising way of eliminating the terrorists (and possibly dissidents) that they are avidly sought be other states, and the Obama administration has recently decided that manufactures could sell them abroad (Ryan, 2015).

What should we think about lethal drone strikes, and more broadly about the wisdom and morality of a technological "fix" to the problem of terrorism? These are the general questions I address in this paper. However, I will not be including in this discussion unmanned and non-piloted robotic warfare or cyber-warfare. Contrary to putting issues of robotic and cyber-warfare on the back burner, we badly need a more robust public debate about the possible (or likely) advent of these innovations. Such warfare would continue to explode our paradigms for ethical, and perhaps rational, thinking about the conduct of war. However, unless checked, armed drones are likely to presage the development of a more dangerous generation of robotic weapons; moreover, drone strikes are likely to set important precedents about moral beliefs or intuitions about distance-operated and unmanned weaponry. Consequently, although some of the issues considered here pertain to more technological innovations more advanced than drones, there is a compelling case for focusing at present on the ethics and psychology of drone warfare.

In what follows I will give relatively little attention to the technological features of unmanned (but piloted) combat air vehicles, or drones (UCAVs). My purpose will be to advance two related arguments. I shall argue that armed drone strikes as they are being used by the Obama administration are unethical. They are unethical for two major reasons. First, contrary to Obama's claims, their use does not conform to the principles of Just War theory; on the contrary, efforts to apply make drone warfare conform to the principles of Just War theory fail completely. Second, I argue that the use of UCAVs produces effects far worse than the problem for which they are the presumed 'solution.' As the history of weaponry shows, the technological innovations give one side only a temporary advantage; for this reason and others, the long-term consequences of this innovation are likely to be very bad. Indeed, UCAVs are likely to worsen the barbarities of terrorism in the future.

In the second argument, I consider what makes drone warfare attractive, but not in terms of the explicitly stated, "official" reasons presented by the Obama administration and others. Rather, my concern is with the way discourse about drones and circulated imagery short-circuit rational reflection, thus enabling us to rely on mythical thinking about American exceptionalism and myths about technological "progress" and technological "fixes" to otherwise seemingly unresolvable political problems. In this section I rely on work in cognitive science and political psychology far more than on ethical theory as traditionally construed, although my conclusions point to grave moral implications for present military strategy.

In effect, I shall argue that our self-righteous uses of UCAVs vindicates myths about us as a people and a nation that had been severely battered by the attacks of 9/11 and the long drawn out but inevitably unwinnable wars in Iraq and Afghanistan. However, the difficulty with myths, I maintain, is that they serve as frames—their purpose is to frame 'in' selectively, namely, what helps us feel successful and good about ourselves, but also to frame 'out' what we might need desperately to think about, that is, anything inconsistent with our privileged world view. The result is, I believe, a heady jingoistic "cocktail" mixing American exceptionalism, national pride, and jubilation over our military prowess, but an inability to think constructively about the dismal future UCAVs warfare is creating.

THE IMMORALITY OF DRONE WARFARE: JUS IN BELLO

To assess the morality of drone warfare, we should begin by noting the reasons why drone attacks have a fateful attraction for policy makers. (1) Drone attacks apparently offer an extremely efficient "solution" because they are said to have measurable, precise effects, and because their use costs less than plausible alternatives. (2) Because drones can be operated remotely from locations far removed from their targets, operating them carries no risk of casualties among American Armed Services personnel. (3) Lethal drone strikes obviate the need to put "boots on the ground," that is, to deploy armed forces to apprehend and arrest persons suspected of terrorist activity. Hence, drone strikes score higher on the scale of "politically feasibility" than troop deployments. (4) Because the use of drones facilitates a "kill-don't-capture response," drones enable the Obama administration to avoid the public relations nightmares Bush created with prisoners at Abu Ghraib and Guantanamo (Benjamin, 2012, p. 134), not to mention renditions of prisoners to "black sites." (5) Because the drone strike "kill list" is determined by a small number of persons in the CIA and the National Security Council (NSC) and an attack requires nothing more than presidential approval, the Obama administration is free to operate largely in secret (e.g. we do not know what the criteria is for inclusion on the kill list) and with little accountability.

Finally, (6) despite initial widespread news coverage, UCAV strikes now occur mostly out of sight, and out of mind for many, if not a majority, of Americans. This means there has been a significant shift in the burden of apprehension—valuable from the administration's perspective. Eyewitnesses in Pakistan (where the vast majority of UCAV strikes occur) report pervasive anxiety and psychiatric distress (Coll, 2014, p. 104). Civilians cannot know when or under what circumstances death will strike from the sky for they may not know who is targeted or when a Taliban or al Qaeda target' plans or movements will endanger them and their families. Moreover, drone warfare has resulted in Taliban and al Qaeda torture and murder of suspected Pakistani "spies." Consequently, while Americans enjoy relative normalcy those subjected to the "unblinking stare" live with a "siege mentality" (Derbyshire, 2015).

The ethical controversy over UCAVs began with concern about the indiscriminate and disproportionality of drone strikes. Discrimination in Just War theory relates to jus in bello, or the just means of waging war, and requires that military force be discriminate: enemy combatants engaging in a war effort are legitimate targets, but non-combatants, or civilians, engaging in activities they would perform whether or not a war is underway are not to be targeted. As a jus in bello principle, proportionality, requires that the harm caused not outweigh the tactical or strategic advantage gained through military force (. it is important to appreciate that, according to Just War theory, proportionality in jus in bello is to be assessed in tandem with discrimination (Frowe, 2011; McMahan, 2011; Orend, 2006; Walzer, 2006). This means that even when the doctrine of double effect applies (unintended and unforeseen civilian casualties are morally permissible), gaining a military advantage cannot be accepted if unintended and unforeseen casualties are disproportionate. Nor can the risk of casualties be shifted to civilians in order to spare American soldiers. Morality requires that we find an alternative means of neutralizing dangers, unless we face conditions of extreme emergency.

At least for the present, the United States does not face conditions of extreme emergency. We do not face the need to use UCAVs to "neutralize" terrorists within our country (at least not yet), nor in close proximity, e.g. lurking just across our borders in Canada and Mexico (assuming the consent of the Canadians and Mexicans), nor as terrorist-pirates threatening to attack us from the seas. Al-Qaeda, Taliban, ISIS and other terrorist operatives are far away in Afghanistan, Iraq, Pakistan, Somalia, Syria, and Yemen. Thankfully they do not (as of now) have the capacity to launch remote UCAV strikes against us.

Given the absence of a supreme emergency, Just War theory requires that our counterterrorism, if a bona fide war (and more on this later) be both proportionate and discriminate. Yet, from the beginning, there have been three serious problems with satisfying these jus in bello conditions. First, Hellfire missiles are far from precise as they have a "kill zone" with a 15-meter radius. This is a rather alarming area of about 706.5 square meters, or over 2,300 square feet. Second, limitations on the accuracy of surveillance and drone-guidance systems between 2001 and 2012 led to the targeting of presumed terrorists when they were immobile or before the dispersal of groups they might be in. Hence strikes often occurred when suspects were lodged as "guests" in civilian homes or guesthouses or when gathered in relatively large groups.

Due to these defects (and a third soon to be mentioned), the use of UCAVs in the first decade resulted in far too many civilian casualties. Estimates of civilian deaths have been controversial. According to journalist Steve Coll, an analysis of media reports by the New America Foundation concluded that that drones probably killed some 250 to 300 civilians up to 2014. Researchers, working at the London-based Bureau for Investigative Journalism and using field interviews as well as media reports, estimate that American drone strikes killed between 400 and 950 civilians during the same period (Coll, 2014, p. 100).

From the outset, the CIA has disputed nongovernmental counts, often claiming they reflect Pakistani or Taliban propaganda. At the same time, however, official estimates remain tightly guarded secrets. In 2014 Senator Dianne Feinstein, chair of the Select Committee on Intelligence, broke the silence by reporting that CIA classified documents showed civilian deaths per year as "typically in the single digits" (Coll, 100). Such figures are hardly believable, given the defects of UCAV strikes noted above, unless the CIA figures refer only to UCAV strikes made in 2013 and 2014. The lower reported kill ratios of 2013-14 might reflect the increased discrimination produced by the improved surveillance introduced with the Reaper and greater reliance on SSWs.

The likelihood that drone strikes were very probably indiscriminate and disproportionate at least between 2004-2012 can be inferred from U.S. government policies. In 2008 CIA director Michael Hayden approved "signature strikes," with President Bush's approval. This policy allowed drone operators to shoot at any male of military age whose behavior corresponded to a "significant suggestion of suspicious activity." Americans were never told how these words were being defined or why drone operators should be allowed the authority to make death sentences. Even more condemnatory, public defenses of "kills" occurring between 2008 and 2013 were justified by the CIA and John Brennan, President Obama's counterterrorism advisor (and CIA chief by early 2013), in a completely question begging way. In effect all military-age males in a strike zone were counted as combatants, "unless there is explicit intelligence posthumously proving them innocent" (Becker & Shane, 2012). Incredibly, in June of 2011, John Brennan, speaking bout UCAVs at a public seminar on counterterrorism actually said, "Nearly for the past year, there hasn't been a single collateral death, because of the exceptional proficiency, precision of the capabilities that we have been able to develop" (Coll, 2014, p.107).

Taking Brennan's claim at face value would require that we believe Obama's top counterterrorism adviser knew nothing about at least a dozen cases covering the period he mentioned in which there were documented civilian casualties. In the widely investigated UCAV strike at Datta Khel, Pakistan on March 17, 2011,just two months before Brennan's speech, a drone killed 41 tribal elders gathered for a jirga, or traditional council meeting, including local authorities. Coll avers "from the available evidence, it seems likely that C.I.A. targeting analysts tracked a suspect to the jirga and then decided to kill everyone present" (2014, p. 107). An anonymous American official told The New York Times that all the dead were "terrorists": "These people weren't gathering for a bake sale" (Coll, 2014, p. 107).

When he asserted in May 2013 that drone warfare met the conditions of Just War theory, Obama also announced some increased constraints on drone-targeting policies, including "near certainty" that civilians are in harms way (BBC News, 2013). However, Obama did not indicate what criteria he, the CIA and NSC, will use in determining whom to target. Nor did he indicate whether he had suspended signature strikes or restricted them to 'high value" targets. Nevertheless, even if we assume that increasing reliance on Predators and SSWs have ameliorated the first two problems of discrimination and proportionality, talk about "precision" in drone warfare obscures a more serious way in which this response to distant terrorists erodes jus in bello.

Consider the way Steve Coll, writing in The New Yorker (2014) compares estimates of UCAV civilian casualties in Pakistan with those resulting from jets in other past air wars. Coll makes an educated estimate that, for Pakistan, the total death toll from CUAVs since 2004 is between 2 and 4-thousand. He adds that even if one accepts a civilian death toll of 957—the highest nongovernment estimate—drones have probably spared more lives than civilians than jets have in past air wars (p. 108) Coll regards this as a "big jump in precision" and one of the war's "measurable successes" (p. 108). This comparison and the implied judgment that this is a "measurable success" invite confusion. So, one might ask, the UCAVs kill fewer civilians than do F-16 jets, does this mean that UCAV caused civilian deaths are acceptable? Coll does not say this; presumably he means only that UCAVs are now more successful in reducing civilian casualties than are jets.

Over reliance on technological capacities and quick fixes for pressing security problems can lull us into substituting a mechanical formula (X produces fewer units of Z-civilian deaths than Y) for a moral judgment. As French philosopher Grégoire Chamayou (2015) notes, talk of "precision" conflates two very different things: the technical precision of the weapon to destroy a target with the capacity to be discriminate in selecting targets. But the capacity to destroy whomever you wish does not mean that you are more capable of deciding who is and who is not a legitimate target. (p. 142). In this sense drone warfare is only as precise as your intelligence and the quality of your justifications for selecting human targets. Obviously, this second, critical decision cannot be made by technology; it is a result of human judgment that ought to be informed by careful moral reasoning. Somehow this critical distinction as missed at the highest echelons of command.

Finally, in connection with jus in bello, Bradley Jay Strawser (2010), a philosopher at the U.S. Naval Postgraduate School argues for a "morally obligatory" principle of combatant immunity (p. 342). Strawser derives this conclusion that soldiers fighting a just war must be provided with the maximum amount of protection. Not only does Strawser beg the question about the justice of drone warfare (see below), he believes incorrectly that a greater capacity to protect our soldiers permits us to override the principles of jus in bello. As I have indicated, however, according to the Just War theory, it is impermissible to shift the risks of death and injury onto civilians in this way. All the same, confusions of what we can do with what we ought to do, and infatuation with the advantages of technology, lead political commentator Waleed Aly (2013) to ask whether we reached "an era when civilian deaths are deemed an acceptable inevitability of war, rather than a reason to avoid it …"? (p.33)

THE IMMORALITY OF DRONE WARFARE: JUS AD BELLUM

The preceding discussion facilitates a transition to t consideration jus ad bellum. Does drone warfare meet the necessary condition of just cause? Certainly the terrorist operatives against whom we are "fighting," are far-flung, often well financed, able to gain and train recruits, and occasionally well organized. It is also true that recognized leaders plan to attack and kill Americans and our allies at home or abroad, they attempt to incite and train others to do the same. In addition, the United States has a moral obligation to protect civilians from mass atrocities whether citizens or others (Evans, 2008). Yet, we can concede these points but conclude that there is no real war against terrorists.

As we know, Bush chose to declare a "War on Terror" for political reasons, leveraging American anxiety about terrorism and humiliation over 9/11 to wage an ill-conceived war in Iraq. At most, jus ad bellum supported the hunt for Osama bin-Laden and the al-Qaeda terrorists who carried out the terror attacks on 9/11 and while they continued to pose risks of danger. The discourse about a "war on terrorism" (like the earlier wars on poverty and drugs) are convenient American myths to preclude alternative options, most prominently, the U.S. failure to ratify treaties establishing the International Criminal Court (ICC), and hence, complicating efforts to respond to the al-Qaeda terrorists more appropriately, namely as perpetrators of crimes against humanity and brought to trial before the ICC or ad hoc tribunals under ICC auspices.

The critical questions for just cause concern self-defense or the defense of innocents against aggression or the imminence of an impending and serious danger, and an inability to avert an impending attack except by attacking first. Thus Just War theory recognizes circumstances in which preemptive attacks are just (Walzer, 2006, pp.74-85), however, even morally permissible preemptive strikes are constrained by further principles of just ad bellum. The principle of proportionality in jus ad bellum requires that preemptive force cannot be more than is absolutely necessary to remove the an imminent danger. As soon as time and space have been opened for the consideration of alternatives, the principle of last resort requires that the use of deadly force cease. Jus as bellum does not sanction retaliation against an adversary no longer intending or capable of the infliction of harm. Most important for our purposes, jus ad bellum does not sanction preventive aggression.

Whether or not any of the Obama administration's UCAV strikes are preemptive or all are preventive depends on facts it is difficult to obtain. Evidence available to the CIA might have indicated that some "high value" targets were engaging in activities that, had they been carried to completion, would have resulted in the death of innocents (those not themselves engaged in harming). There is no justice in killing or maiming one for what he may be feeling, thinking or intending to commit aggression, in war or peace, and certainly many suspects' deaths were unjustified for this reason. It is far more difficult to believe drones trikes don't fail the test of last resort, except when they protect civilians in the regions in which terrorists are actually operating, such as Afghanistan and Iraq.

Yet, because of the global reach of UCAV warfare, we are entering a new phase in which it will be increasingly difficult to conceive of war according to the paradigm in which a power has a just cause because it is required to respond to unjust aggression initiated by an adversary. In fact, UCAVs make preventive warfare much more attractive, and for this reason constitute a moral hazard: the increased precision of drone technology and ability to strike with impunity is likely to tempt regimes to order drones into action more frequently than that they would commit conventional bombers or missiles. In other words, UCAVs may spare more civilians in any particular war than would conventional air force weapons, but possibly a greater number overall, if they also lead to more frequent wars (Kreps & Zenko, 2014).

We now see why, as I claimed earlier, reliance on UCAVs explodes the Just War paradigm. If UCAV warfare are so liable to fail the tests of just cause, proportionality, and last resort, what of the other rules of jus ad bellum? They surely fail the test of rightful intention, requiring that lethal force be used only in a just cause and to repel aggression or to vindicate rights wrongfully violated. Yet, the reasons most frequently for deploying UCAVs, as noted at the outset, have to do with political considerations, efficiency, and cost-effectiveness that have nothing to do with Just War theory. Indeed, UCAV warfare is extremely convenient because it preempts the need for an honest analysis of the roots of hostility and the effect of drone strikes in intensifying ant-American hostility. It is hardly surprising that a Pew Research Center (2010) found in surveys that 59-percent of Pakistanis regard the U.S. as an enemy while only 11-percent regard the U.S. as a partner.

It might be thought that, because the president orders UCAV strikes, their use satisfies the jus ad bellum condition of legitimate authority. There are several difficulties with this claim, however, and each increases in gravity. First, since the second half of the twentieth century legitimate authority has not been invested in states acting on their own, but with states acting with U.N. sanctions, as, for example, with the U.N. backed NATO intervention in Kosovo. Second, the president is not the individual Bush or Obama while in office, but the chief executive acting according to democratic principles and bound by constitutional limits. Given the opacity of strategic drone policy and the non-accountability of executive targeting and firing decisions, it is difficult to affirm that either Bush or Obama have acted with legitimate authority. Third, the chief executive, if anyone, should be scrupulous scrupulously in observing the rule of law. Yet, what are we to make of situations in which those condemned to death are not informed of any crimes they have committed, have no chance of surrender, and are not put on trial, given counsel and the opportunity to make a defense? There is no due process under such circumstances. Present UCAV targeting policies turn Fifth Amendment protections inside out: Human targets are presumed guilty and, assuming the accuracy of the UCAV strike, will never have the opportunity to prove their innocence. Such killings are extrajudicial in the extreme and should correctly be called assassinations.

I return to the significance of extrajudicial assassinations below to make the connection between this first argument—condemning drone warfare as a violation of Just War theory—and the further argument about the way our reliance on militarized technology abet the formation and propagation of dangerous myths. Before moving ahead, however, a few additional but brief points need to be made about proportionality, and reasonable chance of success, only remaining rule of jus ad bellum: proportionality and reasonable chance of success.

It is logically possible to conceive of circumstances in which limited UCAV warfare for limited objectives could satisfy the conditions of jus ad bellum. However, it is not possible to calculate whether use of UCAVs is proportionate when we are not using them to repel aggression, nor preemptively to avert a catastrophe, and not as a last resort. The U.S. preparation to use UCAVs strikes preventively to "win" a possibly interminable "war on terrorism" makes it virtually impossible to estimate proportionality. What counts as "winning" in such a war? Likewise, what can one make of a reasonable chance of success under these circumstances? We face nightmare scenarios in which completing the strikes on a "kill list" is interpreted as success similar to the way tallies of Vietcong and North Vietnamese dead supposedly indicated the U.S. was winning the Vietnam War. In effect, when our use of deadly force is not bounded by a just cause (including a reasonable end point) and not undertaken as last resort, then it is open-ended, and there is no "horizon" against which we can sensible measure proportionality or probable success.

A CRITICISM AND REJOINDER

At this juncture a detractor might respond that my demonstration of the immorality of drone warfare demonstrates nothing more than the "fact" that Just War theory has outlived its usefulness. It has been superseded by changes the theory was not designed to handle. (A detractor might claim as well that I have not shown that using limited and precise UCAVs, as now deployed under Obama, are indiscriminate and disproportionate. However, this charge misses the crucial point that an adversary failing to satisfy the rules of jus ad bellum cannot be using lethal force justifiably, however well it is observing the principles of jus in bello.) It might be claimed that Just War theory is simply a convention and current events have shown that states and other agents now have good reasons for not being restrained by these paper claims. Alternatively, my critic might claim that Obama's use of drones can be defended by ethical considerations not in or adequately represented by Just War theory.

In response it can be noted that the Just War theory developed as a long and continuous tradition of moral reflection going back to Augustine's City of God approximately 1,500 years ago. Thus insofar as provisions of Just War theory have the status of international law this is not a consequence of their having been drawn up and ratified by states since the close of World War II. More important, the principles of the theory can be justified independently on ethical grounds. To see this, consider both the ethics of philosopher Immanuel Kant and Human Rights theory. Kant (1785/2012) emphasized the necessity of applying the Categorical Imperative and treating all persons as possessing inherent dignity, and hence, worthy of our respect. Thus according to Kantian ethics it is always morally wrong to treat persons as mere means rather than as ends in-themselves. Yet this is precisely what we do when persons are used as means of achieving our political ends, when they are attacked based on principles that cannot be universalized, and when we use lethal force against persons who have not made this response necessary as the only way to protect innocents against their illegitimate violence. Moreover, because the use of lethal force against persons who are not endangering others constitutes a violation of such basic human rights as the right to life, bodily security, liberty, and health, such lethal force is condemned by Human Rights theories as well (Boylan, 2014; Churchill, 2006; Gewirth, 1996; Griffin, 2008).

My critic might say that so far I have relied on teleological (specifically deontological) moral theories, and not consequentialist theories. The charge is not entirely correct, for at least two rules of jus ad bellum are forward looking in the sense that they require considering probable outcomes. However, my critic is sure to claim that there are promising consequences of the U.S. UCAV warfare in addition to the efficiencies and political "benefits" I list at the outset of the paper. Fair enough, if we are to consider the morality of UCAV warfare on a consequentialist, or utilitarian, then we ought to consider all likely and foreseeable consequences. As we see below, however, this exercise suggests the long-term consequences of drone warfare are likely to be very bad.

SOME PROBABLE LONGER-TERM CONSEQUENCES OF DRONE WARFARE

The introduction of new and splashy technology tends to be evaluated with a disproportionate emphasis on in its proximate and short-term benefits. It is in the nature of a quick fix, that we think only of solving a problem immediately on hand. It is also in the nature of innovation that we tend to be aware only of the "benefits" promoted by designers, manufacturers, or marketers. For instance mobile phones are now regarded as a practical necessity, although research is just now indicating how much more stressed many people feel as a consequence of their existence or how they have undermined control over our time. New functions demanding our attention become "indispensable" functions before we fully understand how they will affect us. Likewise the recent surge of interest in distance learning and other technological innovations in the classroom far outstrips educators' abilities to find reliable research indicating that the latest platforms actually enhance student learning. Instead those who stand to profit by the adoption of these platforms pose as the experts sanctioning their use.

While we cannot know what the more remote causal effects of drone warfare will be, and this is a perpetual difficulty for ethicists wishing to rely on such calculations (Johnson, 2015), even hardcore utilitarians should consider that the long-term prospects of drone warfare are likely to be catastrophic. Presently the U.S. dominates the use of UCAVs for military purposes but 47 countries already possess some (Kreps & Zenko, 2014). It probably will not be long before rebels UCAVs (e.g. Russian-supported separatists in Ukraine), rogue states (e.g. Libya, North Korea), or non-state terrorist groups (e.g. ISIS or al-Qaeda in the Arabian Peninsula) deploy UCAVs.

We also need to consider the message our current use of UCAVs sends abroad as well as the effects produced by this message. Obviously, the U.S. wants to communicate that terrorists cannot succeed against us. But we also are projecting, the global reach of our power, to choose the time and place at which we strike, to induce constant insecurity, and to act with impunity. This is hardly surprising because aggressors seek to enhance their own power by denigrating and humiliating opponents, and because physical attacks always succeed, when they do, by producing a psychological change in the minds of adversaries (Dyer, 1985). Yet historian and psychologists who study war and have long known antagonists produce "mirror images" of each other (Bronfenbrenner, 1990; Keen, 1991; Mann, 2014, p. 192). Consequently, a belligerent's choice and use of weaponry send very effective weapons. Weapons are chosen to win, and hence, the logic of this will not be lost on the adversary. Moreover, because winning entails producing a significant psychological effect in the opponent, military tactics and tactics indicate not only what a belligerent believes its opponent will find most degrading and humiliating, but, a fortiori, what it itself would experience as most devastating should the opponent use them more effectively against one.

As we saw on 9/11, Bin–Laden and the al-Qaeda terrorists did not use the 4 seized jetliners to maximize total American deaths. Rather they attempted to humiliate the United States by striking targets vital to our national identity and symbolically and emotionally charged. Given our use of UCAVs to exterminate "high value" targets, the mirror image and psychology of reciprocal denigration will dictate that terrorists, if and when they are able to deploy UCAVs, will strike at "high value" targets on our side, possibly including the president.

Many Americans awoke on January 27, 2015 to news reports that the previous evening a two-foot-by-two-foot 2.8-pound DJI Phantom manufactured in China crashed into a tree on the South Lawn of the White House compound. As radar for White House security was calibrated to warn of bigger threats such as airplanes or missiles, the incoming "quad-copter" drone was not detected by the Secret Service until after it had landed (Shear & Schmidt, 2015). Although unarmed and piloted by a recreational user, the incident illustrates how easily terrorists could have carried out a strike. Airspace needs to be controlled only if drones, such as the 7,000 pound MQ-9 Reaper cruising at an average altitude of 20,000 feet, are airborne for days, weeks, or months at a time. Much smaller drones, such as the U.S.'s 4-pound RQ-HB drone could fall into the hands of terrorists. The Raven can be launched by hand and, although presently used only for surveillance it could be modified to carry explosives. Drones purchased at hoppy shops are relatively inexpensive, easily piloted, small in size, and highly maneuverable, and don't show up on radar and acoustic and visual sensors may have difficulty detecting them.

Small drones such as the "quad-copters" sold at hobby shops and online are limited by short battery life and are too light to carry even the smallest payload used by U.S.-military drones (recently Raytheon developed a 13-pound mini-'smart bomb' called Pyros), but small drones can be modified to mount attacks with light payloads of or chemical weapons. Kreps and Zenko (2014) aver that drones are perfectly suited to deliver biological and chemical agents. A January 2015 Wall Street Journal report described a summit by federal and international security officials to study how terrorists might "weaponize" drones purchased privately. Attendees heard how U.S. military and counterterrorism officials had commonly accessible drones to drop explosives on targets, how they can be equipped with guns and fired remotely, and how they could release deadly biochemical weapons (Nicas, 2015).

Criminals and terrorist are not waiting, however. A drone crashing in Mexico near the U.S. border was found to be carrying six pounds of methamphetamine (McVicker, 2013). Moreover, the same Wall Street Journal story about the summit reported that law enforcement authorities in the United States, Germany, Spain, and Egypt, have already foiled at least six potential terrorist attacks with drones since 2011. In the summer of 2013 German authorities raided Islamic militants and right-wing extremists and seized bomb-making materials and a drone. The group was allegedly planning to use the device to bomb a German summer camp. In France authorities were investigating a series of illegal drone flights over fourteen nuclear power plants (Nicas, 2015). Finally, given that plans to deploy thousand of small drones over civilian airspace are already being implemented (Howell, 2013), the odds that 'home-grown' terrorist groups will use the technology to kill and maim Americans.

COGNITIVE AND PSYCHOLOGICAL CONCERNS

Why do we really want to use UCAVs? One might think the reasonable answer just requires a re-enumeration of the advantages of drone warfare for the U.S. set out at the beginning of this paper (and perhaps other reasons I have missed). Of course, if these advantages are ethically justifiable, then they must be objectively and impartially moral for any other country that might seek to use UCAVs as America is doing now. As I have argued, this outcome would be disastrous. Here I wish to investigate the attractions of drones using a different approach. In effect, I consider possible motives for the use of UCAVs that are non-rational, and possibly irrational, but are missed because of the difficulties of bringing them into consciousness. It is my contention that before anyone can claim that using UCAVs is the best course, a fair effort to consider these issues needs to be made, despite the difficulties of revealing and putting into words what often occurs at the mythic, primal and pre-reflective levels.

From Roland Barthes (1982) we learn that myths have two important features. First, myths are immune from counterevidence: "they do not rely on structures of truth that undergird empirical claims" (Mann, 2014, p. 208). Myths evoke self-justificatory patterns of thought and feeling in the "national imaginary" that govern moral thought and action (Mann, 2014, p. 94). For instance, as philosopher Bonnie Mann (2014) points out, George W. Bush's Wild West language ("We'll smoke 'em out!" "Wanted: Dead or Alive!" "Bring 'em on!" and "Mission accomplished") animated the national imaginary of the white settler-heroes and invincible gunslingers who 'won the West' (p. 158). It is hardly surprising that the code name chosen to designate the Navy Seal raid to assassinate Osama bin laden was operation Geronimo, after the great Apache Indian.

After deciding on the "surge" in Iraq, in 2005 Bush gave a series of speeches to restore public support for the war and in the process raised his approval rating to 40 percent. Bush's strategy was to again to strike popular chords by appealing to American myths. While he acknowledged what he called a "failure" of prewar intelligence—appealing to popular distrust of big government—Bush re-circulated the same discursive strategies about the relation between al-Qaeda and the Iraqi state that had been so thoroughly discredited by news investigations. Bush continued to appeal to fear ("they hate us") and pride stimulate fear and pride ("because we're better"); but most importantly, Bush appealed to the myth that Americans always win wars when allowed to finish the job. As public opinion polls showed, a majority of Americans responded positively. As Scott Shane (2005) reported in the New York Times, Americans would support even with mounting casualties, but only on the condition that America would win.

Myths are effective, Mann (2014) points out, not only because they enable us to disempower experience (i.e. discount contrary evidence), but also because they preemptively structure our worldviews at the deepest levels of subjectivity (p. 141). As Mann claims, myths are what cognitive scientists and linguists refer to as "frames" (p. 204). Drawing on research in cognitive science, George Lakoff in The Political Mind (2008) understands a frame as "a word or set of words that activates a whole network of cognitive, affective, emotional, and moral associations" (pp. 12-13). Lakoff notes that, because "language is a matter of neural connections" (p. 232), a frame can be activated, in a literal sense, physiologically and as a consequence of neural pathways developed through repeated encounters with certain terms. Consequently, as Lakoff claims, words and images can evoke whole structures, or "neural bindings" (p. 25) without having to pass through conscious thought.

In *The Righteous Mind* (2013), psychologist Jonathan Haidt draws on a large body of experimental and evolutionary studies that support a conclusion very similar to Lakoff's. In effect, we usually decide what to believe very quickly and through a combination of intuitions, feelings, and bodily responses, and only afterwards use reason to explain our decisions. The intuitive dog wages the rational tail (Haidt, 2013, pp. 32-60); that is, we justify our beliefs only when called upon to do so. We easily see that many intuitive, common sense, beliefs about technological innovation are uncritical, mythical and metaphorical parts of what Lakoff (2008) calls the collective "cognitive unconscious" (p. 226), and therefore, short-circuit conscious cognitive processes. Among these frames are the technological imperative: "If we don't develop it, they will"; the "inevitable" innovation; the myth of the "technological fix," or the notion that we can always depend on technology to solve our problems; the myth of "technological progress," that is, the idea that innovation always makes life better; and the myth that invention is always value neutral. (E.g. "Guns don't kill people; only people kill people.")

None of these myths stands up under critical scrutiny, however. Consider the question whether technology makes our lives better; certainly few of us would agree that innovation always does so, although we will accept that there are many technological innovations we would not want to do without. However, the technology of a society has an enormous effect on the conceptual and experiential modes generally accepted within that society (McLuhan & Powers, 1992). Thus subjective preferences—what people believe they want—are determined in large part by the technological bases of social structure and function. But logically, if progress (whatever the metric) is the end that is to justify technology as the means, then this end itself cannot be the dependent function of the means in question. In other words, the standard of value must be independent of what we intend to value (Mage, 1974, p. 7).

For all of these myths it is difficult, if not impossible, to know what evidence could be regarded as disconfirming the claims; and therefore, they are not empirical hypotheses, and therefore have a dubious truth-value. Second, when put into statement form, some of these myths can be shown to be counterfactual conditionals that, because they make assertions about what did not happen (e.g. "It was inevitable"), are again unreliable. In addition, some of the myths serve as self-fulfilling prophecies. Consider the claim, "If we don't build it, then they will" for example. Because the myth has its effect in motivating us, we innovate and produce the weapon. But then, because our side has this new weapon, and because our plans to deploy it communicate our beliefs in its effectiveness, their side catches up by building and deploying it too. It doesn't stop there, of course. We then rely on the fact that the other side now has the weapon as proof that we were right all along. Hence these myths undergo further rounds of "neural binding" and their grip on our minds tightens correspondingly.

Of course, it is Mann's (2014) point that such myths are not supposed to withstand rational scrutiny. Myths, or frames, determine in large part, which facts and questions are addressed, or "framed in" and which are "framed out." As Mann says, "A frame puts certain moral questions into play as it puts others out of play. Once one world view is deployed, it inhibits the deployment of others" (p. 143). Of course, there are many myths about military power, masculinity, and American exceptionalism ("the best nation in the world" and the "home of the free and the brave"). Thus I am especially concerned about the ways myths about the U.S. combine with myths about technology producing potentially dangerous frames. Mann (2014) warns us that frames are deeply interpretive, put into play our feelings, and are apprehended as self-justifying (p. 143). In other words, many myths animate self-justificatory mini-narratives and images, such that behavior initiated or explained through the use of these frames acquires a similar justification.

In *I Love My Rifle More Than You*, Kayla Williams (2005), an intelligence specialist for the U.S Army deployed in Iraq in 2003, offers a clear interpretation of the mythic power of guns: "If the violent act is, among other things, a way of relocating the capacity to be violent (always) elsewhere, it produces the appearance that the subject who enacts violence [fires the weapon] is impermeable to violence. The accomplishment of this appearance becomes the aim of violence" (p. 178). We can see a similar self-justificatory move in the assertion "Of course, it's right, it works." Terrorists were obviously immoral; UCAVs were invented to solve this problem; UCAVs are working; and therefore, UCAV's are self-justifying. Conveniently, we are relieved from asking, "But are UCAVs themselves moral?" and "Is there not a better solution?"

Warfare is laced with images and myths which serve to short-circuit critical reasoning and to conceal the underlying and psychological reasons why, seemingly, the military establishment soldiers must continue the production of death. Carol Cohn (1990, 2004) was a pioneer during the period of the Cold War in investigating the ways in which cold war strategists spoke of missiles in terms of phallic imagery (e.g. "hardening," "penetration," "thrust") and of "patting" nuclear bombs. (Cohen was herself once invited to "pat the missile.") The situation is no different with the present drone warfare. The badge of the U.S. Air Force MQ-9 Reaper command pictures the grim reaper with a ghastly grin and scythe dripping with blood. This image is between the identification 'USAF MQ-9 Reaper' at the top and the bottom the motto: "That Others May Die" (Chamayou, 2015, p. 32).

As noted above, war is essentially an effort to effect a psychological change (of will and intention) in an adversary by inflicting unacceptable physical destruction. Unfortunately psychologists, philosophers, and others have given far too little attention the significance of collective shame and humiliation in disposing a state to go to war, and reliance on physical violence as redemption (Mann, 2014). Elaine Scarry (1985) broached this subject brilliantly with publication of The Body in Pain. In her landmark study Scarry demonstrates that torture and war frequently function as processes of "replacement and conversion" requiring victimization (p. 125). A regime unsure of its own legitimacy, or one having suffered humiliation, needs a magnification of its power in contrast to the weakness of its "enemies." So great is this need for the vulnerability, pain and humiliation of others, that if there are no readily available enemies, a regime is likely to create them. (Hitler is reported to have said, "If the Jews did not exist, it would be necessary to create them.") This process is based on sets of unconscious brain pathways and "neural bindings" giving rise to a real myth that enables delusional thinking: "If I can hurt them, then it means I am powerful." Sadly, as Scarry (1985) notes, destruction with impunity and torture have play an especially important role in the "translation" of the "objectified elements of pain" into the insignia or emblems of the regime's strength (p. 56).

Not only were Americans deeply traumatized by the events of 9/11, for over a decade our leaders have been frustrated and humiliated by prolonged, seemingly unwinnable wars without exit strategies, and, as indicated by the Boston Marathon massacre of 2013, the continuation of terrorist threats. Moreover, as I have suggested UCAV strikes illuminate and mirror our deepest fears about what unknown terrorists could do to us: striking without warning, striking at the time and place selected by the adversary, assassinating legally innocent persons (before they have committed criminal activity), killing with impunity (at least at present), singling out for death whomever desired, terrorizing a far larger populace, and in addition, obliterating human targets (leaving at most body parts, but no genuine corpses to traditionally wash, bury, and mourn).

Hence, I suggest that the projection of American military power after 9/11, starting with the infamous "Shock and Awe" in Iraq in 2003 and right through current drone warfare, reflects the need for the "conversion" of shame and humiliation and its "replacement" by turning the tables on the terrorists. The deployment of UCAVs has enabled the U.S. to up the ante, but the better hand we now play with will not be an advantage for long.

If this analysis is correct, then obviously, there is little if any conscious attention being given to answering the question, "Why do we really want to use drones?" This is tragic and potentially catastrophic. Our use of UCAVs, inasmuch as they replicate so many effects of terrorism, consequently show the world not just how recklessly we can be in vindicating our self-perception as the global power, but also what we regard as particularly effective, and therefore, fearful. As noted above, there are good reasons for believing that our "enemies" will attempt to acquire UCAV's and use them against us. It is horribly ironic that in attempts to protect Americans from terrorist attacks, U.S. leaders and military reliance on drone strikes, far from reducing the threat of terrorism, will succeed in recruiting a steady stream of enemies well into the future. Tragically, the U.S. is not acting constructively do away with the terrorist threat that has dogged Americas since 9/11. Instead, the U.S. is now in the process of institutionalizing the present reality and turning it into the status quo.

Finally, two additional and critical points should be mentioned before closing. First, we cannot detach our evaluation of new technologies of war from our efforts to understand relations between military force and the economic "substructures" of society (Klein, 2007; Mann, 2014). Second, what does the U.S. use of UCAVs say about Americans in terms of our humanity, both individually and collectively? The president's "kill don't capture" policy intentionally degrades and dehumanizes our victims. Many are nameless, faceless, and anonymous, and killing them is relatively easy and out of sight, thus cheapening the value of their lives in our eyes. We are in danger of disregarding the truth that taking human life can never be undertaken lightly, that it is always tragic and regrettable, and can be justified only when absolutely necessary. There are ways in which humans are killed, however, that are inherently degrading, and as a consequence, degrade those who execute them. Extrajudicial assassinations inflicted by UCAV strikes threaten to turn their victims into vermin to be exterminated, rather like Gregor who is transformed, in Kafka's "Metamorphosis," into a cockroach. The arrogance, callousness, and cruelty of UCAV warfare degrade and dehumanize us.

CONCLUSION

I have made two arguments to the effect that our present policy employing UCAVs is seriously flawed. The first argument draws consecutively on two different ethical traditions, Just War Theory and then Utilitarianism, and shows that both condemn drone warfare as immoral. The second argument draws on what we can learn from cognitive science about myths, or frames, and the deep but subconscious influences these cognitive and affective patterns have on our motives and choices. The second argument very strongly suggests that there may be unrecognized psychological needs and impulses, so far not entirely under conscious control, but that nevertheless have made armed drones appear to be the technological fix for the problem of terrorism.

Morally, the U.S. ought to suspend drone warfare immediately, and resume UCAV strikes only if, after robust and thorough moral scrutiny, the moral case for resumption is stronger than the argument made above. It is also urgent that we continue investigating the issues brought to light by my second argument, for only by understanding and coming to terms with the attractiveness of UCAVs can we avert both the institutionalization of a "permanent state of terror" and the dehumanization of our adversaries and of ourselves. If it is not too late, we need put an end to talk and of a "war on terror" and view terrorist acts for what they really are: criminals perpetrating crimes against humanity. In addition we need to reallocate the billions and billions of dollars now destined for more drones, and better drone technology to better address the underlying causes of terrorism. We need the government to support at least as many academics, scientists, and activists investigating those causes as the number of drone operators being trained today (more than all the pilots of fighter planes and bombers put together). Finally, as already proposed for military robotics programs (Iannone, 2014, pp. 295-308) we need to take the lead in international efforts to develop treaties and protocols leading to the effective control of UCAVs. This should be the first important step in ensuring that no one in the future will be victimized by globalized cyber-wars or robotic weaponry.

REFERENCES

Aly, W. (2013). When making war is so easy, Obama's doubts about morality of some drone strikes may be irrelevant. *Sydney Morning Herald*. 31 May. (p. 33)

Barthes, R. (1982). Myth today. In S. Sontag (Ed.), *A Barthes Reader* (pp. 93–149). New York, NY: Hill and Wang.

Becker, J., & Shane, S. (2012). Secret 'kill list' proves a test of Obama's principles and will.' *New York Times*. 29 May. http://www.nytimes.com/2012/05/29/world/obamas-leadership-in-war-on-al-qaeda.html?_r=2&pagewanted=all&

Benjamin, M. (2012). *Drone warfare: Killing by remote control*. New York, NY: OR Books.

Boone, J. (2013). US drone strikes could be classed as war crimes, says Amnesty International. *The Guardian*. 22 October. www.theguardian.com/world/oct/22/amnesty-us-officials-war-crimes-drones

Boylan, M. (2014). *Natural human rights: A theory*. New York, NY: Cambridge University Press. doi:10.1017/CBO9781139342650

Bronfenbrenner, U. The mirror image in Soviet-American relations. In F. M. Cancian & J. W. Gibson (Eds.), *Making war/making peace: The social foundations of violent conflict* (pp. 81–86). Belmont, CA: Wadsworth Publishing.

Chamayou, G. (2015). *A theory of the drone* (J. Lloyd, Trans.). New York, NY: New Press.

Churchill (2006). *Human rights and global diversity*. Upper Saddle River, NJ: Pearson Prentice-Hall.

Cohn, C. (1990). Nuclear language and how we learned to pat the bomb. In F. M. Cancian & J. W. Gibson (Eds.), *Making war/making peace: The social foundations of violent conflict* (pp. 111–121). Belmont, CA: Wadsworth Publishing.

Cohn, C. (2004). Sex and death in the rational world of defense intellectuals. In N. Scheper-Hughes & P. Bourgois (Eds.), *Violence in war and peace* (pp. 354–362). Malden, MA & London, UK: Blackwell.

Coll, S. (2014). *The unblinking stare: Taking stock of drone warfare*. The New Yorker, November 24. (pp. 98-109).

Derbyshire, J. (2015). A provocative investigation: Drone theory by Grégoire Chamayou. www.theguardian.com/books/2015/jan21/drone-theory-by-chamanyou

Dyer, G. (1985). *War*. New York, NY: Crown Publishers.

Evans, G. (2008). *The responsibility to protect: Ending mass atrocities once and for all.* Washington, DC: Brookings Institution.

Frowe, H. (2011). *The ethics of war and peace: An introduction.* Oxford, UK & New York, NY: Routledge.

Gewirth, A. (1996). *The community of rights.* Chicago, IL: University of Chicago Press.

Gregor, M., & Timmerman, J. (2012). *Kant: Groundwork for the metaphysics of morals.* Cambridge, UK: Cambridge University Press.

Griffin, J. (2008). *On human rights.* New York, NY & Oxford, UK: Oxford University Press. doi:10.1093/acprof:oso/9780199238781.001.0001

Haidt, J. (2013). *The righteous mind: Why good people are divided by politics and religion.* New York, NY: Vintage Books.

Howell, K. (2013). Invasion: 7,500 drones in U.S. airspace within 5 years FAA warns. Washington Times, November 7. www.washingtontimes.com/news/2013/nov/7/faa-chief-annonces-progress-drone-regs/?pages=all

Iannone, A. P. (2014). *Seeking balance: Philosophical issues in globalization and policy making.* New Brunswick, NJ & London, UK: Tansaction Publishers.

Jayakumar, A. (2015). Selling land from the air. Washington Post. February, 29 (p. A10).

Johnson, S. (2014). *How we got to now: Six innovations that made the modern world.* New York, NY: Riverhead Books.

Keen, S. (1991). *Faces of the enemy: Reflections on the hostile imagination.* New York, NY: Harper & Row.

Klein, N. (2007). *The shock doctrine: The rise of disaster capitalism.* New York, NY: Metropolitan Books.

Kreps, S., & Zenko, M. (2014) *The next drone wars. Foreign Affairs*, March/April. www.foreignaffairs.com/articles/140746/sarah-kreps-and-micah-zenko/the-next-drone-wars

Lakoff, G. (2008). *The political mind: Why you can't understand 21st century politics with an 18th century brain.* New York, NY: Viking Press.

Mage, S. (1974). *The myth of technological progress.* The GVSU Review. 1 (1), 2-14. http://scholarworks.gvsu.edu/gvsu_review/vol1/issi/4

Mann, B. (2014). *Sovereign masculinity: Gender lessons from the war on terror.* Oxford, UK: Oxford University Press. doi:10.1093/acprof:oso/9780199981649.001.0001

McLuhan, M., & Powers, B. R. (1992). The global village: Transformations in world life and media in the 21st Century. Reprint ed. Oxford, UK: Oxford University Press.

McMahan, J. (2011). *Killing in war.* Oxford, UK: Oxford University Press.

McVicker, L. (2013). Drone carrying meth crashes near San Ysidro port of entry. www.nbcsandiego.com/news/local/drone-carrying-meth-crashes-near-san-ysidro-porty-of-entry-289353601.html

News, B. B. C. (2013). Barak Obama defends 'just war' using drones. 24 May. http://www.bbcco.uk/news/world-us-canada-2263533

Nicas, J. (2015). Criminals, terrorists find uses for drones, raising concerns. *Wall Street Journal*. 29 January. www.wsj.com/articles/criminals-terrorists-find-uses-for-drones-raising-concerns

Orend, B. (2006). *The morality of war*. Peterborough, Canada: Broadview Press.

Pew Research Center. (2010). Pakistanis see U.S. as an enemy. www.pewresearch.org/daily-number/pakistanis-see-u-s-as-an-enemy

Ryan, S. (2015) New policy on armed drones: Obama to allow more sales to allies. *Washington Post*. 18 February. (p. A1, p. A4).

Scarry, E. (1985). *The body in pain: The making and unmaking of the world*. New York, NY & Oxford, UK: Oxford University Press.

Shane, S. (2005). Bush's speech on Iraq War echoes voice of an analyst. *New York Times*, December 4. http://www.nytimes.com/2005/1204/politics/04strategy.html?pagewanted=all

Shear, M. D., & Schmidt, M. S. (2015). White House drone crash described as a U.S. worker's drunken lark. *New York Times*, 28 January. www.nytimes.com/2015/01/28/us/white-house-drone.html

Strawser, B. J. (2010). Moral predators: The duty to employ uninhabited aerial vehicles. *Journal of Military Ethics*, *9*(4), 342–368. doi:10.1080/15027570.2010.536403

Walzer, M. (2006). *Just and unjust wars: A moral argument with historical illustrations* (4th ed.). New York, NY: Basic Books.

Williams, K., & Staub, M. (2005). *Love my rifle more than you*. New York, NY: W. W. Norton.

Zenko, M. & Kreps, S. (2014). Limiting armed drone proliferation. Center for Preventive Action. Council

This research was previously published in the International Journal of Technoethics (IJT), 6(2); edited by Rocci Luppicini, pages 31-46, copyright year 2015 by IGI Publishing (an imprint of IGI Global).

Chapter 21
War 2.0:
Drones, Distance and Death

Jai Galliott
The University of New South Wales, Australia

ABSTRACT

Technology has always allowed agents of war to separate themselves from the harm that they or their armed forces inflict, with spears, bows and arrows, trebuchets, cannons, firearms and other modern weaponry, all serving as examples of technologies that have increased the distance between belligerents and supposedly made warfare less sickening than the close-quarters combat of the past. However, this paper calls into question the extent to which new military technologies actually mitigate the savagery of war. It contends that with the introduction of technologies that eliminate the need for a human presence on the battlefield, we are the cusp of a major revolution in warfare that presents new challenges and questions for military technoethics, namely as to how soldiers should conduct themselves and fight justly, if they are to do so at all. Ultimately, it argues that only way to address these issues is through the design of the mediating technologies themselves, which is by no means an easy task.

INTRODUCTION

War is an all-too-human affair and will probably require the endangerment of human lives in some shape or form, but military robots known as 'drones' or 'unmanned systems' promise to significantly offset the human cost of war by removing warfighters from the physical dangers of the battle zone and facilitating the conduct of what is purported to be more precise killing. However, the use of these systems toward such ends is not without other implications for thinking at the intersection of military technoethics and just warfare. In this paper, I examine the efficacy of unmanned systems with a particular focus on the mindset-altering dimensions of unmanned warfare and their impact on principal warmaking agents, namely unmanned systems operators. This is because many of the unintended effects of this technology cannot be attributed to the machine, but to human psychology. I first examine some problems associated with technologically mediated fighting and suggest that through a process of moral disengagement and desensitisation, the barriers to immoral conduct *in* war may be reduced. Having considered the impact

DOI: 10.4018/978 1 5225-8365-3.ch021

on the long distance warrior's capacity or willingness to adhere to the rules/laws of war, the next section examines the impact on the personal wellbeing of the operators themselves. Here, among other things, the impact of being simultaneously present in contrasting environments is considered in arguing that this, if nothing else, may lead to serious transgressions of just war principles. Toward the end of the paper, I consider whether we can eliminate or relieve some of these technologically mediated but distinctly human moral problems by automating elements of the decision making process. It is concluded that while greater automation certainly has the potential to alleviate some moral concerns generated by these systems, there is a strong case for keeping humans in the decision making chain, even if it involves having to make a delicate moral tradeoff between maintaining and/or improving warfighting capability and limiting harm to noncombatants.

THE ROLE OF THE INDIVIDUAL SOLDIER

While many of the campaigns to halt the development of 'killer robots' focus on high-level decision makers, as they are central to the initial decision to develop said systems and engage them in warfare, it is the individual soldier who defends his state and society that must be most unconditional in exercising moral restraint and adhering to just war theory. Michael Ignatieff (1998) writes that more than any other of warmaking agential group, it is the soldiers who actually conduct war that have the most influence on its outcomes and the ability to introduce the moral component. In his words, 'the decisive restraint on inhuman practice on the battlefield lies within the warrior himself – in his conception of what is honourable or dishonourable for a man to do with weapons' (Ignatieff, 1998, p. 118). Ironically, soldiers are the primary agents of both physical violence and compassion and moral arbitration in war. As Darren Bowyer (1998) remarks, they deliver 'death and destruction one moment ... [and deal] out succour to the wounded (of both sides) and assistance to the unwittingly involved civilian population, the next' (p. 276). The specific concern examined here is whether by removing soldiers from the battlefield and training them to fight via a technologically mediated proxy we may, through a process of psycho-moral disengagement and emotional desensitisation, lower their ability or willingness to exercise restraint and compassion in warfare and adhere to the moral laws of war, namely the principles of discrimination and proportionality enshrined within just war theory, which respectively require that war be directly only at legitimate targets and and involve a morally appropriate level of force. It will be argued that the employment of unmanned systems tracks unethical decision-making and/or lowers barriers to killing, endangering the moral conduct of warfare and countering much of the benefit of using these systems.

Most human beings are born with what can only be described as a primitive survival instinct that, without unchecked force, would lead to a degree of violence and savagery. But in most societies, people are raised and socialised in such a way that typically leads them to hold an aversion to harming other human beings. In a military context, this socialised reluctance to kill is evidenced by recounts and statistics from earlier wars. David Grossman (1995), a self-proclaimed 'killogist' or military psychologist, writes of two World War veterans. The first confirms that many WWI infantrymen never fired their weapons and relied instead on artillery, while the second says that platoon sergeants in WWII had to move up and down the firing line kicking men to get them to fire and that they felt they were doing good if they could 'get two or three men out of a squad to fire' (Grossman, 1995, p. xiv). While some have criticised his methodology, S. L. A. Marshall gave further supporting evidence in arguing from personal experience and studies conducted on firing ratios, which revealed that 'on average not more than 15 per cent of the men had actually fired at the enemy' (Marshall, 2000, p. 54). He attributed this startling inhibition to kill to an 'ingrained fear of aggression' that was based on society's teaching that killing is fundamentally wrong (Marshall, 2000, p. 71). For Marshall, success in combat and the welfare of the state and its people demanded that action be taken to correct or overcome this problem.

In the years following publication of the first edition of Marshall's book – that is, in those following WWII – there is evidence that Marshall's calls for corrective action were answered. The claims of very low firing rates had been replaced by very high and morally concerning firing rates. By the time of the Korean War, the American firing rate was said to be up to fifty-five percent and, in Vietnam, it was reported to be up to ninety or ninety-five percent (Meagher, 2006). Some expressed doubts about these firing rates too, with some finding troops with unspent ammunition in the rear of troop formations, but they were generally satisfied that among those who actually sighted the enemy, there appeared to have been extraordinarily high and consistent firing rates (Grossman, 1995). From a strictly military or operational perspective, this is a remarkable success story. In order to overcome the hesitancy to fire and kill that most people develop over time, Russel Glenn says that staff sergeants and platoon commanders watched their troops to ensure that they were actually engaging with the adversary and that in Vietnam, they listened for the steady roar of machine gun fire which indicated to them that their soldiers were unhesitatingly firing their weapons (Marshall, 2000). However, this corrective action seems unlikely to account for such a radical shift in the firing ratios. The real cause for the difference in the firing rates, it could be argued, has much more to do with technology employed in later conflicts and changes in military training which, together, allowed and continues to allow, individuals to achieve a physical, emotional and/or moral distance from their enemies, thus enabling them to kill somewhat easier. It is these distances that need to be explored in more detail, as unmanned systems only further them and the disengagement and the accompanying desensitisation to a point that gives rise to unique problems affecting their operators' ability to wage discriminate and proportional warfare.

The link between physical and emotional distance, ease of aggression and waging warfare is in no way a new discovery. As Grossman (1995) writes, it has long been understood that there is a positive relationship between the empathetic and the spatial proximity of the victim and the resultant difficulty and personal trauma caused by the kill, or the morally problematic ease of killing, more generally. This relationship has been a cause for concern among anthropologists, philosophers, psychologists, theologians and, of course, soldiers themselves, who often struggle to understand their own actions. Jesse Glenn Gray (1959), an American philosophy professor whose career was interrupted by a period of service as a WWII counter-intelligence officer, wrote that unless one is caught in some sort of overwhelming murderous ecstasy where rage takes over, killing and destroying is much easier when done at a little remove and that with every foot of distance, there is a corresponding decrease in the accurate portrayal of reality. He argues that there is a point at which one's representation of the world begins to flag and another at which it fails altogether (Gray, 1959). Glenn put forward this argument over fifty years ago and the concern is that unmanned systems seem to increase the relevant distances to the flagging point that he referenced or that, at worse, the advancement of digital warfare may reach a point whereby the moral inhibitions of unmanned systems operators are almost totally overcome. That is, the worry is that we have reached a morally significant point in the history of weapons development. To really understand how these unmanned weapons make it easier to kill and dampen a soldier's capacity to act ethically in war, we have to think about unmanned systems in the wider context of weapons that increase physical distance to the target and lower resistance to killing.

UNMANNED SYSTEMS: KILLING AT MAXIMUM RANGE

In painting a mental picture of this conceptual phenomenon, one must imagine a spectrum. At one end we have close range, then middle range and long range at other end. Close range, for our sake, involves any easily attributable kill at 'point-blank' range, whether with one's bare hands, an edged weapon or even a projectile weapon. According to Grossman (1995), the key factor in close range killing is the undeniable responsibility one holds for the act. John Keegan and Richard Holmes (1986) cite the story of an Israeli paratrooper during the capture of Jerusalem in 1967: 'we looked up at each other for half a second and I knew that it was up to me, personally, to kill him' (p. 266). When a soldier kills at this range, more than any other, it's an intensely vivid and personal matter (Grossman 1995, p. 115). One can see the raw emotions on their enemy's face, hear their cries and smell the gunpowder. The Israeli paratrooper mentioned above goes on to say that having shot his enemy at close range, he could see the hate on his enemy's face and that 'there was so much blood...[he] vomited, until the rest of the boys came up' (Keegan and Holmes, 1986, p. 266). Combat at close proximity is an interpersonal affair, so much so that it is incredibly difficult to deny the enemy's humanity. For this reason, Grossman says, the resistance to close-range killing is 'tremendous' (Grossman, 1995, p. 118). At midrange – where you can still see and engage the enemy with hand grenades, sniper rifles and so on, but usually without being able to gauge the extent of the wounds inflicted – the experience of killing changes. At this range in the spectrum, the soldier can deny responsibility for the fatal shot or blow if there are others present and participating in the act of killing. One is still located on the battlefield and can hear the gunfire and feel the stress, but the distance between adversaries makes the act of killing both physically and psychologically easier, and thus more morally troubling (Grossman, 1995, p. 113).

At long range, at which one must use some sort of mechanical or electrical assistance to view potential victims (i.e., binoculars, cameras or radar), there is evidence to suggest that killing is made even easier. Among those who are least reluctant to kill are pilots, artillery numbers and missile silo attendees. Gwynne Dyer (2010, p. 57) writes that while being observed by their fellows puts pressure on them to kill (as was the case with the gunners in Vietnam), it has much more to do with the distance between them and their targets and how it acts as an emotional and moral buffer. She aptly notes that on the whole, 'gunners fire at grid references they cannot see; submarine crews fire torpedoes at 'ships' (and not, somehow, at the people in the ships); and pilots launch their missiles at 'targets'' (Dywer, 2010). Grossman (1995) also reports that in his extensive career researching and reading on the subject of killing in combat, he is not aware of a single instance in which an individual operating at such long range has refused to kill. We also have numerous examples of long distance killing made easy. Dyer (2010) reminds us that in the early nineteen forties, for instance, the British Royal Airforce 'firebombed' Hamburg. Using early bomber aircraft, munitions blew in windows and doors over four square miles and resulted in a firestorm which left seventy thousand people dead, mostly women, children and the elderly (Grossman, 1995). A further eighty thousand died in the firebombing of Dresden, two hundred and twenty-five thousand in Tokyo and many millions more in bombing conflicts since (Grossman, 1995). If the bomber crews had to kill each of these people with a flamethrower or, as Whetham (2012) writes, slit each of their throats with a knife, the majority would be unable to do it. The awfulness of killing people at such close proximity and the emotional trauma inherent to each act, and to the collective acts, would have been of such magnitude that they simply would not have happened.

Figure 1. Distance, technology and resistance to killing

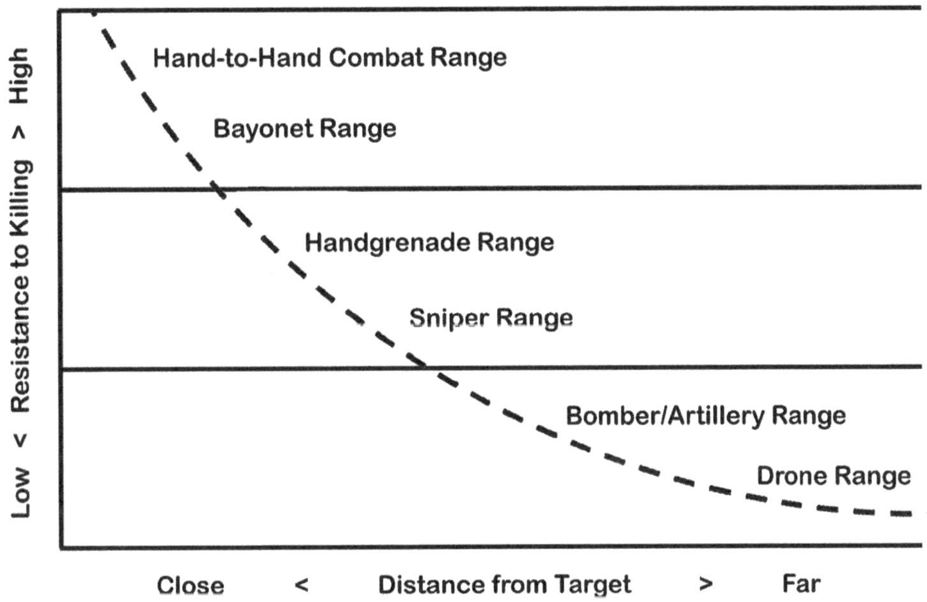

As indicated in Figure 1 above, killing conducted with unmanned systems belongs at the very end of the long-range killing spectrum and might even be worthy of its own designation: killing at maximum range. It is reasonable to expect that at this range, killing is made most easy and morally problematic. The contention here is that there is no other tactical weapon on the battlefield today that facilitates killing with such physical and psychological ease and that it becomes a rather clinical and dispassionate matter, easing any of their operators' existing moral qualms. Noel Sharkey (2010) offers support this argument in drawing attention to reports collected by Singer. Amongst a variety of other disturbing statements, he cites one twenty-one-year-old soldier who talks about his acts of killing with casual indifference: 'the truth is, it wasn't all I thought it was cracked up to be. I mean, I thought killing somebody would be this life-changing experience. And then I did it, and I was like, 'All right, whatever' (Singer, 2009, p. 391). Later, he says that 'killing people is like squashing an ant. I mean, you kill somebody and it's like 'allright, let's go get some pizza' (Singer, 2009, p. 392). In this clinical killing environment, in which it seems reasonable to propose that some human targets are divested of their humanity, respect for the laws of war wanes. Some of those videos mentioned in the previous section serve as a demonstration. Many show raw footage of Predator strikes with people being reduced to little more than 'hot spots' or 'blips' on the screen, with operators often failing to take the necessary precautions to ensure noncombatants are protected. Again, it seems that obscuring 'targets' in this way and increasing the distance to the maximum possible, makes it easier for unmanned systems operators to kill in an indiscriminate and disproportionate fashion.

While the distances involved show the powerful role of unmanned systems in overcoming moral-emotional qualms and the socialised inhibition to killing, there are a range of other mechanisms that further this and make it even easier for systems operators to kill without regard for the consequences of their actions or *jus in bello* norms. The first additional mechanism that Grossman (1995, p. 161) explicitly notes – and the one that is most relevant to America's use of unmanned platforms, but will only mention briefly here – is cultural difference, which perpetuates racial and ethnic differences and allows warfighters to further dehumanise the enemy. Military forces have long been trying to get their troops to think of the enemy/ies as 'inferior forms of life...as less than human' (Watson, 1978, p. 250). Put simply, the further one is technologically distanced from the enemy, the easier it is to think that they are distinctly different to you in some way and the easier it is to kill them. Unmanned systems separate soldiers from the cultural environment that they would operate in if they were in the field and, in that sense, permit them to racially demonise and 'other' the enemy. Connected to cultural distance is moral distance, which involves legitimising oneself and one's cause (Grossman, 1995, p. 164). Once it has been determined that the enemy is culturally inferior, it is not difficult for operators of unmanned platforms to incorrectly suppose that their counterparts are either misguided or share their leaders' moral guilt and think that this warrants waging unconstrained violence against those with the supposedly morally inferior cause (Grossman, 1995, pp. 164-7).

However, not all are convinced by the argument that unmanned systems present a moral problem in making it easier to indiscriminately and disproportionately kill. Daniel Brunstetter and Megan Braun (2011, p. 339) argue that these systems are subject to the same *jus in bello* requirements as other weapons used in war, but that their technological advantages coupled with the removal of risk to soldiers means that they should, at the least in theory, make satisfying the principles of discrimination and proportionality an easier task and perhaps make operators more reluctant to kill in situations where doubt exists as to the legitimacy of the potential victim of aggression. They say that the in the case of surveillance, at the very least, the distance or what they call 'separation factor', arguably offers an increased level of control over lethal targeting decisions and ought to actually reduce the emotional toll and unnecessary killing (Brunstetter and Braun, 2011, p. 339). They regard a drone operator's ability to confer with a superior officer as being a critical factor encouraging ethical decision making in war. In some instances, this may be the case, but in others, it might be that having a mission commander overseeing the operator's actions only places additional pressure on them to fire, just as the sergeants walking the trenches of WWI aimed to encourage reluctant soldiers to kill. Christian Enemark (2013) also questions some of the assumptions relied on here. He says that there is reason to suppose that being physically absent from the battlefield is *more* conducive to discrimination (Enemark, 2013). In his view, the removal of risk allows decisions to be made in a more deliberate manner and also removes anger and emotion that he thinks might otherwise lead to morally unsanctioned killings. That is, if a drone operator working from a desk in Nevada encounters the enemy, adherence to *jus in bello* protocol should improve as the operator is at little or no personal risk. It could be suggested, however, that if an operator is so emotionally removed, they are in fact likely to develop the sort disengagement referred to above or an even more morally concerning callousness.

What needs to be recognised is that the disengagement and depersonalisation of war can go so far that operators begin to lose sight of the fact that they are actually involved in real wars. It is becoming increasingly difficult to distinguish whether one is playing a video game or operating a real unmanned system with weapons that will also have very real consequences for those at which they are directed. As Nolen Gertz (2011) notes, while it should be rather clear that video games try to approximate the reality of combat insomuch as it is possible for such a medium to do so – with attempts at this reaching new levels with the designers of video games such as 'Medal of Honor' turning to former members of the military and its special forces for technical guidance – what is probably less well-known is that the military and its weapons designers often also try to approximate the reality of war gaming. Unmanned systems designers are now trying to model the controllers of these platforms to those of video game consoles or, in some cases, actually using the very same controllers, wired especially for use with these systems. Singer (2009) cites a project manager who was responsible for designing an unmanned ground vehicle and said that they modeled the controller after the PlayStation because that's what these 18-, 19-year-old Marines have been playing with pretty much all of their lives. The argument is not just that these systems create a militaristic mentality; that would be too reductive. What is being said is that recruits that are already familiar with the technology have a premediated and probably distorted idea of war. The related moral concern, of course, is that people often do things in video games that they would not do in normal life and that this may somehow carry over to the use of unmanned systems. That is, they may be unable to separate the reality from virtual reality, making them prone to launching the indiscriminate and disproportionate attacks that are characteristic of most violent video games, further supporting the argument that the increasing employment of unmanned systems tracks unethical conduct in war. As argued in the next section, this inability to separate the two worlds may also have deep psychological effects on the operators themselves, only compounding the problem posed by distancing, emotional disengagement and disindividualisation and the effect they have on adherence to *jus in bello* norms.

THE EMOTIONAL WELLBEING OF THE WARFIGHTER

In the previous section, it was argued that unmanned systems and the physical and emotional distances that they create magnify the challenges that warfighters face in applying and adhering to *jus in bello* norms. In this section, there is less focus on the mindset-altering effect or psychological impact that these systems have on the warfighting role of unmanned systems operators and more on the impact to their *personal* wellbeing, given that at both the *jus ad bellum* and *jus in bello* level, the harm incurred to one's own warfighters must be given some weight, even if it is not, or should not, be the overriding priority. Some suggest that killing via proxy does not have a significant adverse psychological or emotional effect on the distance warfighter. Lambèr Royakkers (2010), for instance, says that 'creating moral disengagement reduces or even eliminates the stress of cubicle warriors' (p. 291). Grossman (1995) also writes that in his years of reading and research on the subject of killing, he has never come across a single instance of psychiatric or psychological trauma associated with such long-range killing. Sharkey (2010, p. 372) gives further anecdotal evidence supporting these claims in citing the commander of 432nd Wing at Creech Air Force Base, who said that 'on only four or five occasions had sensor operators gone to see the chaplain or their supervisors and that this was a very small proportion of the total number of remote operators'. The reasoning here seems to follow from an idea examined in a study conducted by Michael Osofsky, Albert Bandura and Philip Zimbardo (2005), which hypothesised that the selective disengagement that similarly distancing systems or processes facilitate, allow them to perform their role and go about killing people and conducting themselves in an injurious fashion while maintaining their emotional wellbeing. However, Sparrow (2009) expresses the concern that while distancing unmanned systems operators from the risk of death and allowing them to become disengaged might be thought to reduce stress, killing by the push of a button can still be a stressful job and might actually expose operators to new, somewhat different or unexpected stressors. In this section, the argument is that something is missing from previous evaluations of what unmanned systems can reasonably achieve and that it is particularly important to understand these psychological stressors because they will also influence moral decision-making in war and guide public opinion as to whether unmanned warfare really is an improvement over earlier forms, which links to the discussion in the first section of this paper.

Before delving deeper into the psychological issues that affect unmanned systems operators, we must comment more generally on the burden of killing as it affects conventional warfighters in regular close and midrange warfare. As has already been mentioned, the resistance to killing at these distances can often be so extreme that it overcomes one's own innate self-preservation mechanism and the coercive force exerted by sergeants and platoon commanders in an attempt to encourage killing. Furthermore, it can override the expectations of one's peers and the more general obligation, that of defending the state and its otherwise vulnerable population. As such, the ordinary warfighter is put in a situation such that they cannot avoid some sort of psychological trauma because of conflicting feelings, stresses and moral obligations that are reinforced by society and a range of regulations and procedures that govern wartime conduct. It has been claimed that if the conventional soldier can overcome the socialised resistance to harm and kill individuals in close or midrange combat, he will forever be burdened by guilt. As an example, William Manchester (1979), novelist and United States Marine veteran of WWII, describes his feelings after killing a sniper in a fishing shack who was, one by one, picking off the Marines in his company: 'I shot him with a 45 and I felt remorse and shame. I can remember whispering foolishly, 'I'm sorry' and then just throwing up... I threw up all over myself. It was a betrayal of what I'd been taught since a child' (pp. 3-6). If for some reason the soldier does not kill, he has also failed to fulfil his duty to his comrades. John Early, a former mercenary and Vietnam veteran, explains that 'you're depending on him [the man next to you] for the most important thing you have, your life, and if he lets you down you're either maimed or killed. If you make a mistake, the same thing happens to him, so the bond of trust goes extremely close' (Dywer, 2010, p. 34). Furthermore, the reluctant soldier is subject to the shame that comes with failing his profession, fellow citizens and state. In Grossman's (1995) words, the soldier is 'damned if he does, and damned if he doesn't' (p. 87).

For conventional soldiers, it becomes a matter of balancing the obligation to kill that is generated by the society that one serves and the comrades that one serves alongside, with the toll of guilt that inevitably follows killing in war. From their comments above, Royakkers and others seem to imply that due to the moral and emotional buffers introduced by unmanned systems, this balancing act is made significantly easier for the soldiers- turned-unmanned systems operators. If we follow Grossman's line of thought, we may even conclude that this balancing act is totally insignificant or not even a factor for these distance warriors. To make a comparison to combat at close or mid-range, it might be said that the psychological stress profile of a drone operator is much more similar to that of a higher commander, who may have some operational function, but is much more strategically focused. Unlike the average soldier, but quite like an unmanned systems operator, high level military decision makers generally do not face the same sort of risk that burdens those troops that are physically present at the frontline (Evans & Ryan, 2000). Most of the time, they can go to sleep at night knowing that they are very unlikely to be bombed or mortared during their sleep; they do not have to worry that field communication lines might be cut or that resupply might not happen; and they generally are not faced with the task of retrieving the dog tags of those who have been killed in action beside them, or writing personal letters home to their loved ones (Evans & Ryan, 2000).

While this may well be true, both of higher commanders and unmanned systems operators, the more relevant similarity is arguably that they both face a range of other psychological stresses that, while qualitatively different, seem to equal or surmount those faced by conventional combatants and combatant commanders. For instance, higher commanders may not have to worry about being killed by the enemy, but they too have to be concerned about the enemy's movements and how they impact operations. They may not be worried about impediments to their own progress or dangers to their own life, but they are likely to be worried about ensuring that their troops have sufficient resources to conduct their campaign, their commanders' ability to conduct operations and the views of the public and the media (Evans & Ryan, 2000). Likewise, unmanned systems operators face their own unique problems. While some remote operators may be spared from the raw first-hand experience of seeing people die on the battlefield, in many ways, they may actually see more of war and its deadly outcomes than any typical soldier engaged in close or midrange warfare. There are two key reasons for this, one more convincing than the other. Firstly, while it has been argued that most of today's unmanned systems reduce targets to a 'blip' on the screen, many say that some systems have such high resolution optics that operators can see the impact of their actions with greater clarity than most conventional airforce pilots (Kaurin, 2013). Again, there is doubt about this because of the physical limitations with which optical technologies must operate, but technological developments may make this more of a concern as time passes. However, unmanned systems do improve loiter capability over conventional manned systems, meaning that an operator may observe a target for twelve or more hours before taking lethal action (Singer, 2009). Thus, even if the raw footage is not particularly clear and the operator is somewhat disengaged, this extended surveillance period in which many disturbing events may unfold, has potential to be psychologically damaging, if not at the time of operation, at a later stage when the operator has had time to come to a realisation concerning what has happened and what role they have played. This is the first sort of morally relevant stress we need to be concerned about: that of witnessing traumatic events.

Another associated stress rarely mentioned in the academic literature is that related to limited participation in these events. Like a higher commander who has to exert influence on the ground via troops that he is not directly in control of, with only that information conveyed by field-based combatant commanders, a drone operator has to contend with having limited capacity to intervene or influence events that are unfolding in the combat zone. Sparrow (2009) recognises that this general concern is more significant than it perhaps seems because despite the 'hype' surrounding them, many unmanned systems are limited in the information they can relay, the force they can actually exert and the range of actions that they make possible. It is not always the limitations of the systems themselves, but often the capacity of these systems to be utilised within particular environmental surroundings that limits adherence to just war theory and creates the psychological problem. In his first-hand account of piloting drones, Martin (with Sasser, 2010) details the anguish he experienced resulting from being unable to shape or rectify some injustices he witnessed on the battlefield. He describes how he had carefully planned to kill a group of enemy combatants and recounts the feelings he experienced when two children walked into the attack path post-missile release: 'there was nothing I could do to save them...we could only stare in abject horror as the silent missile bore down upon them' (Martin & Sasser, p. 21). The feeling of helplessness and shame that Martin tells of is compounded when it is the enemy waging the violence and one's own comrades that stand to be injured or killed. In much the same way as a missile release gone badly, an unmanned system may limit the distance warrior's moral agency in the sense of limiting their capacity to act in reference to right and wrong. In a combat scenario, this means they cannot aid fellow warfighters in the way that they could if they were physically present on the battlefield. Benjamin cites a young sensor operator who said, having witnessed the death of troops from afar, that 'it was a traumatic experience. It shocked me' (Benjamin, 2012, p. 97). This experience of sitting in a peaceful country in relative safety and taking indirect action, while seeing fellow military personnel die, is one that is unique to unmanned systems operators and one that operators need time to process and understand.

Unfortunately, this time for processing potentially traumatic experiences, or lack thereof, is another problem with consequences for ethical conduct in warfare. Unmanned systems operators are under increasing operational pressure driven by cost cutting and unrealistic expectations. A recent US Air Force study revealed that one of the biggest residual sources of stress for drone operators were the long hours and frequent shift changes that occurred due to staff shortages (Chappelle, McDonald & MacMilan, 2011). These staff shortages perhaps offer further anecdotal evidence confirming the concern about the stress to which these operators are exposed. Often-intense working hours and constant shift changes deny an operator time to critically reflect on their actions and prevents them from fully integrating into ordinary civilian life once they leave the base. Unmanned systems have fundamentally changed what the experience of 'going to war' is like. As Singer (2009) writes, it has given war a whole new meaning. For the last five-thousand years, when we described somebody as going to war in defence of the state – whether it be the ancient Greeks going to war against Troy or coalition forces fighting against the Iraqis in the Gulf War – we talked about them physically going to the battlefield, putting themselves in danger, knowing that they may never come home or see their family again (Singer 2009). That is what 'going to war' has involved for the last five millennia, but the experience of war has changed.

One US Air Force Pilot that Singer interviewed for his book describes how he would wake up in the morning, drive to the base and then spend twelve hours putting 'missiles on targets' (note the dehumanising language) before driving home again. Twenty minutes after he had been at war, 'he would be at the dinner table talking to his kids about their school work' (Singer, 2009, p. 80). However, this does not mean that modern unmanned warfighters are at less psychological risk than those physically present in the battlespace. This new experience of being at war – of being simultaneously at home and at war and having to balance the two – is itself a major source of stress that distance warfighters encounter every working day. As we are only in the early stages of understanding this problem, few studies detail the psychological effects of this juxtaposition between being at peace and being at war, but Martin & Sasser (2012) hint at the significance of this problem in a powerful example given on the final page of his book. Having just killed a number of enemy combatants and destroying the building they were housed in, he recalls how on the way home to see his wife he pulled his 'car into an In-and-Out Burger, changed out of uniform, and grabbed a burger and fries. Only after I was standing in line...did I have time to reflect on what had just happened, how unreal this war truly was' (Martin & Sasser, 2012, p. 306). This and the mounting civilian death toll seems to highlight that while there may be much that we do not know, the human warfighter may not yet be capable of effectively handling the dual experience of killing people at war and being at home, coping with the domestic demands of children, partners and society. Going home after a day of combat does not necessarily afford an operator time for reflection, for few will go home and talk over the dinner table about the combatantsor noncombatants that they have killed that day. Instead, they keep it to themselves and must then find some way of dealing with the constant disjunction between military and civilian life.

In many cases, it may prove morally preferable to expose operators to these unique stresses rather than those they would experience if they were in the field themselves (Sparrow, 2009). However, others seem to view it is as another eliminable problem concerning distance and that just as we have removed the warfighter from the battlefront, we should now remove the warfighter from the homefront (Gertz, 2011). Subscribing to this sort of thinking, one military commentator has put forward measures that he thinks will maximise efficiency while alleviating the military-civilian disjunction (Singer, 2009). These include establishing a 'communications bubble' that would see all personal phone calls to and from the control room banned and keeping units sequestered at a hotel or at barracks on a base before and after any large operation, keeping them isolated from their families and civilian lives in the hope of eliminating the accompanying troubles (Singer, 2009). This may seem good operational thinking but, if anything, it merely postpones the eventual meeting of the two opposing worlds (civilians and military) and may introduce further problems affecting ethical decision making if operators come to feel unnecessarily forced apart from their families and civilian lives. The attention is best focused elsewhere. It should be realised, for instance, that distance warfighters do not benefit from the same sense of camaraderie that conventional warfighters do (Singer, 2010). A squadron of drone operators is rarely together and never having the time to invest in the shared experience of the group or simply rest and recover. It may prove better to treat the stress as more of a manageable than eliminable problem, perhaps building comraderie or otherwise offering services that assist with management of the psychological stressors that come with unmanned warfare, improving operators' sense of wellbeing and ethical conduct in warfare.

Having this sort of support mechanism in place is particularly important in unmanned warfare as the proportion of inexperienced to experienced personnel is much greater than in conventional warfare and the former may not be able to cope with the emotional demands of unmanned warfighting on their own. Drone operators may be attracted to rendering military service in this form because they are familiar with the controls and the gaming experience that it replicates or simply because it seems less risky and allows them to maintain a connection with loved ones that they are reluctant to leave behind. However, many presumably do so not recognising that by fighting from the homefront, they actually cut themselves off from the support and comrade ship that one finds in a typical ground unit. The perception of the distance warrior is another factor that must be taken into account, for this has bearing on the level of support that one receives from both within the military and the public. The image of the 'top gun' combat pilot has been portrayed in novels and films and it is this image that likely resonates with the military leadership and the public. Unmanned systems operators do not seem to receive the same level of gratitude for rendering public service as conventional pilots do. They do not get the welcoming home parades or the medals. The American Distinguished Flying Cross, for instance, has never been awarded to an unmanned aircraft pilot and many think this for good cause (Fitzsimmonds & Mahnken, 2007). For this reason, it seems important for distance warfighters to initially establish support groups amongst themselves, particularly because our current understanding of the psychological attributes considered critical to performing well in unmanned warfare and coping with the results is at such an early stage.

All of the aspects of combat stress and trauma discussed here have the potential to profoundly impact not only the wellbeing of the warfighter, but the individual's contribution to the war effort and the aggregate contribution of arrangements of individuals that we call squadrons and military units. If we begin to understand the range of responses of unmanned systems operators, we can begin to understand how mediating warfighting technologies such as drones influence ethical conduct in warfare and then seek to make the necessary changes. As already mentioned, we do not yet have the psychological profiles that are necessary to select the most psychologically well-suited candidates for unmanned warfare and, even then, there is no guarantee that this will lead to more discriminate and proportional warfare. We also have a situation whereby political and military leaders may utilise unmanned to alter the public mindset regarding engagement in war and lower barriers to killing. For these reasons, many have suggested that we focus on technical solutions to what is a technologically mediated but largely human problem and I briefly detail some of these proposed solutions.

A TECHNOLOGICAL SOLUTION?

In the previous sections it was said that in practice, human warfighters often fail to discriminate between legitimate and illegitimate targets for at least two reasons. First, killing at maximum range temporarily reduces the humanity of one's enemy and detaches the distance warfighter from the consequences of the use of their weaponry. Second, this often results in traumatising emotional and psychological consequences for those warfighters when they eventually come to grasp their actions, which may then perpetuate further unethical conduct in wartime. Unfortunately, when it comes to trends in adherence to legal and ethical norms, the human record in adhering to the dictates of *jus in bello* principles is questionable, at best. In fact, Ronald Arkin dedicates an entire chapter in his book *Governing Lethal Behavior in Autonomous Robots* (2009) to human failings on the battlefield. He cites the US Surgeon General's report into battlefield ethics and the mental health of soldiers deployed in Operation Iraqi Freedom, which details some concerning findings. Among them were claims that: approximately ten percent of soldiers reported mistreating noncombatants; less than fifty percent agreed that noncombatants should be treated with dignity and respect; roughly fifteen percent agreed that all noncombatants should be treated as insurgents; and close to thirty percent reported facing ethical situations in which they did not know how to respond (Office of the Surgeon General 2006). Arkin (2009) suggests that with the increasing use of technology like semi-autonomous drones, the figures noted in these reports will become even more concerning and that atrocities in war will also become increasingly worse. His primary conclusion is that we cannot expect human warfighters, because of their fallible nature, to adhere to wartime standards when confronted by the horrors of the battlefield and that we ought to develop technology that extends beyond the limits of human warfighters (Arkin, 2009).

In other words, his solution is not to take the moderate step of tailoring the weapons control interface to encourage ethical decision making in war, but rather to eliminate the importance of the human and human-machine interface altogether and look at how one might actually provide unmanned systems with the capacity to make ethical decisions and carryout different kinds of actions, some of them lethal. He is deeply convinced of the possibility of ethical programming that would not only meet the normative requirements of international humanitarian law and just war theory, but would ultimately surpass their requirements, writing that: 'I am convinced that they [unmanned systems] can perform more ethically than human soldiers are capable of' (Arkin, 2009, p. 36). Arkin (2009) proposes that we equip unmanned systems with an 'ethical governor', which is based on Watt's mechanical governor for the steam engine. This governor acts as a transformer or suppressor of automatically generated lethal action, so in much the same way that the mechanical governor would shut down a steam engine running too hot, the ethical governor shuts down the autonomous machine when it is about to do something unethical, such as firing too closely to civilians. But this is just one component. The second is the 'ethical behavioral control' unit, which monitors all of the systems behaviours and ensures that they evolve from within set constraints (Arkin, 2009). The third is the 'ethical adaptor'. It allows the system to learn, adapt and overcome by updating the system's set of constraints (Arkin, 2009). It is presumably intended to act in place of human guilt or remorse in the case of unethical conduct managing to pass the previous two components. The final component is the responsibility advisor, which ensures that in the event of a lethal action occurring, responsibility can be clearly allocated between human and machine, in the case of joint interaction (Arkin, 2009).

Arkin believes that with these sub-systems working effectively together, autonomous machines would possess the capability to operate morally, ensuring that any force is delivered proportionally and with maximum discrimination between combatant and noncombatant targets. However, he does not seem to recognise that there is a difference between a machines that act in accordance with the dictates of morality and what Wendell Wallach and Colin Allen call a 'moral machine' (Wallach and Allen 2009). That is, he fails to recognise that there is a distinct difference between following rules or laws and actually being moral, in the full and proper sense of the word. While the technical acts of discriminating between targets and determining proportionality may be a necessary condition for moral conduct in war, they are not, alone, sufficient. For instance, Andreas Matthias (2011) has said that Arkin seems to view discrimination as a mere matter of technical classification between friend and foe and between legitimate and illegitimate targets of potentially lethal military action. He reduces discrimination down to a cognitive level operation that the machine mimics. It identifies all those targets within its specified radius of operation and sorts them into various categories, combatants and noncombatants, children, wounded persons and so on. While this is an important capacity for any moral agent in war, Arkin's system does not actually establish these categories, it presupposes them. 'Individual X standing beside weapon Y is Z' (an instance of discrimination), is a straightforward classification result and there is no sort of moral deliberation involved. To ask whether it is morally right for the machine to kill Individual X standing next to Weapon Y, on the other hand, involves genuine moral evaluation. For this very reason, if no other, it seems that Arkin is incorrect in saying that his proposed system can outperform human warfighters.

Part of the problem is that Arkin takes the laws of armed conflict, rules of engagement and just war theory to be mutually compatible and translatable into a set of rules that are also compatible with his control architecture and programming language. When the principles of just war theory are viewed in this way, it is easy to mistakenly think that what is left is a technical matter of ensuring that the machinery is capable of correctly and reliably carrying them out. In reality, while many military and political leaders also portray these sources of restraint as if they were a clear and steadfast set of rules amenable to automation, each is actually a set of highly context-dependent guidelines. The law of armed conflict is, in Asaro's (2009, p. 21) words, a 'menagerie of international laws, agreements and treaties' that are open to considerable interpretation. Rules of engagement, which are devised to instruct warfighters how to act in specific situations, can also be rather vague as they are an attempt to combine legal, political and strategic concerns on one small card and leave much up to the individual soldier. Likewise, just war theory is often presented as though it is an entirely coherent theory, when in most formulations it is necessarily, and by its very nature, a set of principles that, while setting certain values and rules of thumb, require a great deal of moral deliberation to be put to effective use. Take the *jus in bello* proportionality principle, for instance. There are a host of difficulties in operationalising this principle and in carrying out the necessary calculations in the field. As Lucas (2011) says, we do not have anything like a 'formal proportionality algorithm' and it requires weighing and comparing incommensurable goods under conditions of extreme epistemic uncertainty. How many men, women and children, for example, is it permissible to kill for a given military objective? Questions like these require a kind of moral evaluation of which Arkin's system is incapable. He might advance the more limited claim that his autonomous drones might outperform humans in classificatory tasks, but he wants to make the stronger claim that his proposed system will perform more ethically than humans.

Even if we put aside the fact that he glosses over the difference between representing morality and acting morally, both Lucas (2011) and Sparrow (2011) highlight that there is an additional conceptual problem associated with saying that an autonomous unmanned system might perform better than a human warfighter. As has already been said, one ought to be willing to concede that Arkin's proposed system might function better in classificatory tasks and thus might perform better under certain limited wartime conditions. We might discover through an empirical study, for example, that the failure rate of human warfighters confronting some specific sort of decision making scenario is ten percent or, alternatively, that they comply with the necessary legal and moral restraints ninety percent of the time. In this instance, it means that the human warfighters are fallible. Suppose that we can reliably track compliance with the relevant rules and principles. Arkin simply demands that his system perform as well or better than this ninety percent benchmark set by the human warfighter. When we send a human warfighter into battle, we expect one hundred percent compliance with the laws of war and just war theory, even though we know all to well that for those reasons stated earlier, soldiers are often statistically unlikely to meet these standards. But when human warfighters fail to do so, they are held accountable and punished. The problem with Arkin's argument is that he is effectively lowering the moral benchmark for conduct in war, which may result in further breaches of the law of armed conflict and the proportionality and discrimination principles. If we know in advance that in a particular scenario, Arkin's system will perform with less than one hundred percent compliance with the relevant laws and moral principles, it would be unethical to have it operate.

The solution may be to go beyond the sort of robotic architecture that Arkin proposes and develop systems that truly have full autonomy or what others call 'strong artificial intelligence'. That is, it might be morally preferable to develop a system that would, unlike Arkin's system that calls on a human to make certain decisions when it cannot, can create and complete its own tasks without the need for any human input, with the exception of the decision to build such a system (the human is so far removed from the loop that the level of direct influence on this sort of system is totally negligible). Such a system would display capacities that imitate or truly replicate the moral capacities of sentient human warfighters and would therefore be able to engage in the sort of moral deliberation that the laws of war and just war theory demand. There is considerable disagreement in the disciplines of artificial intelligence and philosophy as to when we can reasonably expect machines to become as capable as human warfighters on the battlefield or whether this 'strong artificial intelligence' or above human intelligence is at all possible. In 1965, Gordon Moore, Chairman Emeritus of Intel Corporation, described 'Moore's law', which predicted the doubling of transistors on silicon chips with each new generation leading to rapid acceleration in the power of computer, and it has held true to date (Moore, 1965). With this in mind, futurist Ray Kurzweil (2005) has prophesised that we will reach the 'technological singularity' – the point at which machine intelligence will surpass human intelligence – by 2045. Some are more reserved and say that while computers may exceed human intelligence in a few rather narrow domains, they will be never be as universally intelligent as is necessary for moral conduct in war. At the other end of the scale, there are those who hold no hope. Joseph Weizenbaum (1977) has famously stated that 'no other organism, and certainly no computer, can be made to confront human problems in human terms' (p. 233). The philosopher John Searle (1980) has also argued that a computer may be able to solve a complex problem, such as those in war, but without having any actual understanding of it.

Despite the fact that Weizenbaum and Searle seem to dismiss the idea of fully autonomous unmanned systems as the stuff of science fiction, it is wise to briefly consider whether they could even be a solution to immoral conduct both in resorting to war and conducting war these technological possibilities since a great deal of money is being invested toward the goal of creating them and it seems impossible to rule out their invention at this early stage. To be sure, the potential problems are many. One of the immediate concerns that surfaces about waging warfare with artificial moral agents is that they may eventually turn on human beings. Again, this might sound fantastical or highly unrealistic, but if we want to endow our unmanned systems with full moral agency (this cannot be assumed), this will involve giving them the capacity to contemplate both moral and immoral actions and it is a matter of conjecture as to what sorts of decisions autonomous military systems would make. A robot warfighter may not be motivated by sexual desire, power or dominance as human warfighters are, and may not have the same natural killer instinct, but they too may come to develop what some might claim are needs and aspirations that they will need to compete for with other agents for (Sullins, 2010). There is no guarantee that these systems will not evolve into the functional equivalent of today's unethical warfighters. What is more worrying is that these systems and the moral decisions they make may become so complex that humans are incapable of understanding them. Imagine that these systems eventually become so powerful that we come to cede power to one super system and entrust it to make *jus ad bellum* decisions in lieu of human political and military leaders, a possibility taken seriously by many technological optimists given that the latter often deceive the public. The variables it would take into consideration might be so many that no human could understand whether a particular war was just or unjust. Strawser (2011) argues that with machines filling the decision-making niche that humans once did, it could eventually get to a point where there is no point to humans making personal choices at all, a situation whereby they are better off surrendering their moral agency altogether. Both war and human life could be made meaningless. Of course, while this is logically conceivable, it does not mean it is worthy of any closer examination. What we do need to consider is whether might be some sort of more moderate technological solution, for those above do not overcome the emotional and mind-set altering effect that is linked to unethical conduct in unmanned warfare. That is, the solution to this technologically induced human problem is not necessarily to introduce more advanced autonomous technology, but work on adapting our existing technology to yield the best results. It may be that we need to find an appropriate tradeoff between military effectiveness and harm to warfighters and human beings in general.

CONCLUSION

What has been shown in this paper is that alongside the more technical issues that plague unmanned systems, there is a mindset-altering effect that impacts the ethical compass of our principal warmaking agents. First, it was demonstrated that we must be concerned with how technologically mediated fighting lowers barriers to conduct in war. It started by outlining the various killing ranges and located unmanned systems at 'maximum range', where killing is facilitated by physical and emotional distances that dehumanise and de-individualise the enemy, and concluded that this, coupled with the virtual nature of unmanned warfare, makes it significantly easier for soldiers to kill without proper respect for the *jus in bello* principles of discrimination and proportionality. Then it was argued that despite some arguments to the contrary, the constant disjuncture between being at war and being at peace, and having few psychological stress management tools, further magnifies the challenges of adhering to *jus in bello* norms. Finally, it examined some technological solutions to this largely human problem and argued that we really need humans to make some wartime decisions and that we must combine our best efforts in designing weapons control interfaces and judiciously introduce autonomy to ensure support for the aims of just war theory does not wane.

ACKNOWLEDGMENT

The author has received funding from the Australian Department of Defence, but the views expressed here are his alone. This paper is adapted from a chapter entitled 'Emotion and Military Robots: Challenges for Just War Theory' in the IGI *Handbook of Research on Synthesising Human Emotion in Intelligent Systems and Robotics* (2015).

REFERENCES

Arkin, R. (2009). *Governing lethal behavior in autonomous robots*. Boca Raton: CRC Press. doi:10.1201/9781420085952

Asaro, P. M. (2009). Modeling the moral user. *IEEE Technology and Society*, *28*(1), 20–24. doi:10.1109/MTS.2009.931863

Benjamin, M. (2012). *Drone warfare: Killing by remote control*. New York: OR Books.

Brunstetter, D., & Braun, M. (2011). The implications of drones on the just war tradition. *Ethics & International Affairs*, *25*(3), 337–358. doi:10.1017/S0892679411000281

Chappelle, W., McDonald, K., & McMillan, K. (2011). *Important and critical psychological attributes of USAF MQ-1 predator and MQ-9 reaper pilots according to subject matter experts*. Ohio: School of Aerospace Medicine.

Dyer, G. (2010). *War: The new edition*. New York: Random House.

Fitzsimonds, J. R., & Mahnken, T. G. (2007). Military officer attitudes toward UAV adoption: Exploring institutional expediments to innovation. *Joint Forces Quarterly*, *46*(3), 96–103.

Galliott, J. C. (Forthcoming). *Military robots: Mapping the moral landscape*. Farnham: Ashgate.

Gertz, N. (2011). Technology and suffering in war. *Paper presented at Technology and Security*, University of North Texas.

Gray, G. J. (1959). *The warriors: Reflections on men in battle*. New York: Harper & Row.

Grossman, D. (1995). *On killing: The psychological cost of learning to kill in war and society*. Boston: Little, Brown and Company.

Kaurin, P. (2013). Courage behind a screen. In B. J. Strawser (Ed.), *Killing by remote control: The Ethics of an unmanned military*. New York: Oxford University Press.

Keegan, J., & Holmes, R. (1986). *Soldiers: A history of men in battle*. New York: Viking.

Kurzweil, R. (2005). *The singularity is near: When humans transcend biology*. New York: Viking Penguin.

Lucas, G. Jr. (2011). Industrial challenges of military robotics. *Journal of Military Ethics*, *10*(4), 274–295. doi:10.1080/15027570.2011.639164

Manchester, W. (1979). *Goodbye darkness: A memoir of the pacific war*. New York: Dell Publishing Company.

Marshall, S. L. A. (2000). *Men against fire: The problem of battle command*. Norman: University of Oklahoma Press.

Martin, M. J., & Sasser, C. W. (2010). *Predator: The remote-control air war over Iraq and Afghanistan: A pilot's story*. Minneapolis: Zenith Press.

Matthias, A. (2011). Is the concept of an ethical governor philosophically sound? *Paper presented at the Tilting Perspectives: Technologies on the Stand: Legal and Ethical Questions in Neuroscience and Robotics*, The Netherlands.

Meadors, T. (2011). Virtual jus in bello: Teaching just war with video games. *Paper presented at the The Ethics of Emerging Military Technologies*, University of San Diego.

Meagher, R. E. (2006). *Herakles gone mad: Rethinking heroism in an age of endless war*. Northampton: Olive Branch Press.

Moore, G. E. (1965). Cramming more components onto integrated circuits. *Electronics*, *38*(8), 114–117.

Office of the Surgeon General. (2006). Mental health advisory team (MHAT) IV Operation Iraqi Freedom 05-07 (Final Report). Washington, DC: United States Department of the Army.

Osofsky, M. J., Bandura, A., & Zimbardo, P. G. (2005). The role of moral disengagement in the execution process. *Law and Human Behavior*, *29*(4), 371–393. doi:10.100710979-005-4930-1 PMID:16133946

Royakkers, L., & van Est, R. (2010). The cubicle warrior: The marionette of digitalized warfare. *Ethics and Information Technology*, *12*(3), 289–296. doi:10.100710676-010-9240-8

Searle, J. (1980). Minds, brains, and programs. *Behavioral and Brain Sciences*, *3*(3), 417–457. doi:10.1017/S0140525X00005756

Sharkey, N. (2010). Saying 'no!' to lethal autonomous targeting. *Journal of Military Ethics*, *9*(4), 369–383. doi:10.1080/15027570.2010.537903

Singer, P. W. (2009). *Wired for war: The robotics revolution and conflict in the 21st century*. New York: The Penguin Press.

Singer, P. W. (2010). The future of war. In G. Dabringer (Ed.), *Ethical and legal aspects of unmanned systems: Interviews* (pp. 71–84). Vienna: Institute for Religion and Peace.

Sparrow, R. (2009). Building a better warbot: Ethical issues in the design of unmanned systems for military applications. *Science and Engineering Ethics*, *15*(2), 169–187. doi:10.100711948-008-9107-0 PMID:19048395

Sparrow, R. (2011). Robotic weapons and the future of war. In J. Wolfendale & P. Tripodi (Eds.), *New wars and new soldiers* (pp. 117–133). Burlington: Ashgate.

Strawser, B. J. (2011). *Two bad arguments for the justification of autonomous weapons*. Paper presented at Technology and Security, University of North Texas.

Sullins, J. P. (2010). RoboWarfare: Can robots be more ethical than humans on the battlefield? *Ethics and Information Technology*, *12*(3), 263–275. doi:10.100710676-010-9241-7

Wallach, W., & Allen, C. (2009). *Moral machines: Teaching robots right from wrong*. Oxford: Oxford University Press. doi:10.1093/acprof:oso/9780195374049.001.0001

Watson, P. (1978). *War on the mind: The military uses and abuses of psychology*. New York: Basic Books.

Weizenbaum, J. (1977). *Computer power and human reason: From judgement to calculation*. New York: W.H. Freeman & Company.

Whetham, D. (2012). Remote killing and drive-by wars. In D. W. Lovell & I. Primoratz (Eds.), *Protecting civilians during violent conflict: Theoretical and practical issues for the 21st century* (pp. 199–214). Aldershot: Ashgate.

This research was previously published in the International Journal of Technoethics (IJT), 7(2); edited by Rocci Luppicini, pages 61-76, copyright year 2016 by IGI Publishing (an imprint of IGI Global).

Chapter 22
Robots in Warfare and the Occultation of the Existential Nature of Violence

Rick Searle
IEET, USA

ABSTRACT

We are at the cusp of a revolution in the development of autonomous weapons, yet current arguments both for and against such weapons are insufficient to the task at hand. In the context of Just war theory, arguments for and against the use of autonomous weapons focus on Jus in bello and in doing so miss addressing the implications of these weapons for the two other aspects of that theory- Jus ad bellum and Jus post bellum. This paper argues that fully autonomous weapons would likely undermine adherence to the Jus ad bellum and Jus post bellum prescriptions of Just war theory, but remote controlled weapons, if designed with ethical concerns in mind, might improve adherence to all of the theory's prescriptions compared to war as currently waged from a distance, as well as help to undo the occlusion of violence which has been a fundamental characteristic of all forms of modern war.

It is interesting… how weapons reflect the soul of their maker. – Don Delillo, The Underworld

INTRODUCTION

If current trends continue, sometime in the next few decades, or sooner, a machine will deliberately kill a human being. The idea that an autonomous weapon will be capable of making a decision to kill may seem like science-fiction, but it is a goal being pursued right now by the world's major militaries. The arrival of what the media has dubbed "killer robots" appears to be only a matter of time. (Markoff, 2014)

The following essay attempts to discern the possible implications of the rise of autonomous weapons for the waging of just wars. It is divided into four sections. The first section lays out the rise of remote controlled weapons over the past decade, and the evolutionary pressures pushing these weapons towards

DOI: 10.4018/978 1 5225-8365-3.ch022

greater levels of autonomy. The second section raises objections to the use of autonomous weapons on the basis of the *Jus ad bellum* and *Jus post bellum* prescriptions of Just war theory. The third section looks at two types of objections to both autonomous and remote controlled weapons (drones), and argues that both autonomous weapons and remote controlled weapons are merely a continuation of trends found in warfare in all advanced countries, which seeks to occlude the violent nature of war. The fourth and concluding section argues that remote controlled, though not autonomous weapons, may actually offer us a path away from this occlusion if they are designed and operated with this goal in mind.

THE RISE OF THE ROBOT WARRIORS: CRITICS AND DEFENDERS

The move towards autonomous weapons began with remotely operated armaments and in the US military. Since 2005 there has been a 1,200 percent increase of air combat patrols by drones. The US now has a fleet of 7,500 drones making up 31 percent of the Pentagon's air fleet. Its numbers are to be doubled by 2020. (Dowd, 2013) As of June 2012 the US had as many as 3,500 robots in landlocked Afghanistan doing everything from bomb removal to demolition (Hodge, 2012). Based on advanced robotics projects funded by the US military's advanced research arm, DARPA, the US is already experimenting with humanoid "infantrymen". Facing increasing budgetary constraints, the US army has plans under discussion to reduce the size of a brigade from 4,000 to 3,000 men with much of the difference to be made up by robots. (Ackerman, 2014)

The US may have begun the revolution in remote controlled and robotic warfare, but it is unlikely to end there. Innovation appears to be shifting to other countries that see in the new methods of war a way to overcome the asymmetric superiority of the American military in other forms of advanced weaponry. (Scharre, 2014) 75 countries now possess drone programs of their own with China to build 11 drone bases down its coast through 2015. Dowd (2013)

Given the superiority of US advanced weapons and its overwhelming military footprint, which comes at the cost of spending more on defense than the next ten countries combined, it is very likely that rival countries and groups will turn to much cheaper drones and eventually autonomous weapons to redress this asymmetry. (Fallows, 2014)

Drones and other remotely piloted vehicles may appear to be distinct from truly autonomous weapons, but the requirements of military competition are rapidly pushing the development of the former into the latter. Drones suffer vulnerabilities to jamming and problems with latency that truly autonomous weapons are seen to overcome. Piloting by individual human beings might be impossible should autonomous weapons in the form of "swarms" move from the drawing board to the battlefield. (Wong, 2013)

The specter of "killer robots" has led to international calls for a moratorium on the further development of such weapons, most notably in a 2012 paper issued by Human Rights Watch entitled "Losing Our Humanity". In that paper HRW drew a sharp line between remote controlled weapons, which believes are permissible, and autonomous weapons whose development it believes should be banned. (HRW, 2012)

HRW is not alone in its critique of autonomous weapons. The nascent rise of such systems has been met with a kind of widespread criticism that did not meet the growth of drone warfare. (Carpenter, 2014) These criticisms have in large part been based on the argument of the technology's prematurity.

The ethical superiority of remote controlled over autonomous weapons, it is argued, lies in the ability of human pilots and controllers to exercise discrimination on the battlefield aligned with the laws of war, and from the concern that autonomous weapons might prove especially dangerous in areas of combat with a large number of civilians. Self-directed armaments currently lack the capacity to properly attribute threat to human beings, and suffer AI's classic isotropy problem, where everything can be connected to everything else, so that a child's toy in a murky context might be confused with a weapon. (Guarini and Bello, 2012)

Noel Sharkey, perhaps the most prominent critic of autonomous weapons to date, has made a strong case that allowing weapons to act autonomously at this stage would be akin to the use of sub-munitions such as the BLU-108. The BLU-108 has no ability to distinguish between civilian and military targets merely unleashing its munitions on likely targets upon deployment. It has been this very inability to make targeting distinctions that has been the basis of the argument that autonomous war machines need to be banned, or at least a moratorium imposed on their use, similar to that of bans imposed on other indiscriminate weapons such as landmines. (Sharkey, 2012)

Autonomous weapons do, however, have their defenders. Ronald Arkin has been perhaps the most vocal and eloquent figure arguing that such systems might be a means of making warfare *more ethical.* In his view, properly programmed autonomous systems would not experience the kinds of negative emotions, such as fear, and the desire for revenge, that often lead human beings to violate the laws of war. (Arkin, 2009)

This idea that the application of robots to war could ultimately take some of the nastier parts of human nature out of the calculus of warfare is also touched upon by P.W. Singer in what was the first popular book on the new weapon systems, *Wired for War.* There, Singer brings up the case of Steven Green, a US soldier charged with the premeditated rape and murder of a 14 year old Iraqi girl. Singer contrast the young soldier "swirling with hormones" to the calm calculations of a robot lacking such sexual and murderous instincts. (Singer, 2009) Or as put by another defender of the idea of ethical robot warriors, Patrick Lin:

Robots wouldn't act with malice or hatred or other emotions that may lead to war crimes and other abuses, such as rape. They're unaffected by emotion and adrenaline and hunger. They're immune to sleep deprivation, low morale, fatigue, etc. that would cloud our judgment. They can see through the "fog of war", to reduce unlawful and accidental killings. And they can be objective, unblinking observers to ensure ethical conduct in wartime. So robots can do many of our jobs better than we can and maybe even act more ethically, at least in the high-stress environment of war. (Lin, 2011)

The problem with arguments of the type discussed above, both against and for autonomous weapons, is that they in a sense miss the essential ethical question by too strong a focus on technology. Arguments against the use of autonomous weapons would seem to collapse were it ever the case that such machines could make *better* discriminations on the battlefield than human beings. Whereas, arguments for the use of such machines would falter should battlefield discrimination at the level of human operators prove impossible for machines. The state of the technology decides the ethical questions.

Both sets of arguments focus on the *Jus in bello* aspect of Just war theory, rules which prescribe *how* warfare is to be conducted. Sharkey et al are basically arguing that autonomous weapons risk being unable to sustain *Jus in bello* prescriptions regarding *distinction* between military and civilian targets. The use of such weapons, as they currently exists, put civilians at risk due to the complexity of properly discriminating between combatants and noncombatants.

Arkins et al, on the other hand, argue that autonomous weapons (eventually) will prove *better* than human beings at discriminating between legitimate and illegitimate targets. The basis for this improved discrimination, in their view, will come through automated systems increased capacity over human beings to cut through the "fog of war". Such improved ability to adhere to the prescriptions of *Jus in bello* compared to human beings should also, as they see it, arise on account of such systems lacking the darker instincts of human beings such as revenge. Research should therefore continue to try to perfect such weapons.

If the position of those arguing against the use of autonomous weapons is based primarily on their violation of *Jus in bello* because these weapons are currently incapable of the kinds of distinctions required to wage war justly, then such an argument would collapse should technological projections such as Arkin's prove correct. Those who wish to develop a solid, non-technologically dependent, argument against the use of autonomous weapons will need to look elsewhere.

Perhaps they should look to Just war theory itself, for there is a case to be made that even if autonomous weapons can eventually be created that adhere to the prescriptions of *Jus in bellum,* they may fail to adhere to either or both of the theory's other two legs- *Jus ad bellum* and *Jus post bellum.*

AUTONOMOUS WEAPONS: *JUS AD BELLUM AND JUS POST BELLUM*

Let us imagine that Arkin's hopes for autonomous weapons are eventually proven correct, and that not only do such weapons adhere to *Jus in bello* they do so in a way that is ethically superior to the average human soldier. Are there any grounds, then, for still rejecting their use?

If one takes adherence to the prescriptions of Just war theory to be the primary way for us to establish whether the use of a weapon is ethical, then we need to explore how the use of autonomous weapons would affect the likelihood of wars being started that would be deemed just *(Jus ad bellum)* and how the use of such weapons might influence war's ending justly (*Jus post bellum).*

The argument that the use of autonomous weapons might somehow affect the decision of whether a war is just to wage in the first place, the prescriptions of *Jus ad bellum* might at first seem nonsensical. After all, no one is arguing whether we should use intelligent machines to *decide* whether or not we should wage war.

Tracing back to Augustine, the prescriptions of *Jus ad bellum* have been developed largely to morally inform political leadership in their decisions regarding whether to wage war. (Augustine, 1958) There is an argument, however, particularly well developed by the political philosopher Jeff McMahan, that the prescriptions of *Jus ad bellum* should also serve as the basis of the *individual's* willingness to fight. "Unjust wars can occur only if there are enough people willing to fight in them", he writes. (McMahan, 2009)

McMahan rejects the common view in Just war theory that the responsibility for ensuring a war follows the prescriptions of *Jus ad bellum* lies with political leaders alone, and that soldiers asked to engage in war need pay attention to the *Jus in bello* aspects of Just war theory only. For McMahan, any killing by the side waging an unjust war is on that account *already violating Jus in bello.* They are killing what are in effect innocent combatants who are engaged in a just war against them.

In McMahan's view, citizens in a democracy have a unique responsibility not to engage in unjust wars even when the decision to fight has been arrived at by legitimate democratic means. Soldiers should refuse to fight in, and civilians refuse to support, wars that do not meet the criteria of *Jus ad bellum.* McMahan's re-conceptualization of Just war theory would redefine the individual's relationship to war, and doubtless put pressure on statesmen to adhere to the prescriptions of *Jus ad bellum.* It is also a moral project the widespread adoption of autonomous weapons would likely make impossible.

One can be highly speculative and imagine a world in which autonomous systems are universally mandated to be given an activation or deactivation encryption code and could be turned on or off on the basis of an independent body's declaration of a war to be just or unjust, but it is almost inconceivable that states would agree to this.

The fact that wars could be waged largely using autonomous weapons might make political leaders more willing to engage in them no longer facing the risk of public opprobrium or the resistance of militaries reticent to engage in the risks of combat without strong cause.

The group in charge of running these weapons would very likely be far removed from the reality of war. They would be programmers, engineers, and defense contactors, rather than soldiers either at the front or even removed but still aware of the brutalities of the actual conduct of a war. Wholly different incentives between a military made up of human soldiers where the goal is to finish a conflict as quickly and with the least human damage (at least to one's own side) and the manufactures of military robots whose incentive would be to generate as much revenue as possible during a conflict and would even have a perverse incentive to encourage conflict.

There is thus a good argument that autonomous weapons might negatively impact our ability to follow the *Jus ad bellum* aspects of Just war theory constricting the range of actors who could effectively resist the onset or continuation of unjust conflicts, and perhaps even incentivize conflict for those groups who would be most responsible for how they were waged: no longer soldiers and their commanders, but programmers, engineers, and technology companies. Autonomous weapons might also negatively impact the adherence to the third leg of the Just war theory; namely, *Jus post bellum* or the justice that is to follow conflict.

As Andrew Rigby has pointed out, one of the primary components of a just peace (*Jus post bellum*) following war is *reconciliation* between the formerly warring parties. Of necessity this entails confronting past moral wrongs, and in the process generating *forgiveness* for those wrong doers whose actions were not in such violation of norms that they require punishment. Rigby sees three conditions that facilitate reconciliation and forgiveness:

1. Acknowledgement and apology by the wrongdoer
2. The promise not to repeat the wrong
3. Offer to make amends- reparations (Rigby, 2005).

In the case of autonomous weapons it is extremely unclear *who* would play any of these roles, though it could be the programmers and engineers of such machines, but it would certainly not be the machine that caused the unjustified killing or injury. Not being human an autonomous war machine cannot "apologize" for following its program, has no say in whether it will repeat a wrong, and has no capacity to make reparations, and though all of these things might be socialized, it is difficult to see how the world of common experience which those on opposite sides of a conflict share can be tapped as a bridge to reconciliation in cases where one side has waged war using robots instead of humans.

Autonomous weapons could make post war reconciliation increasingly difficult- a reconciliation that is essential in the most commonly wars fought today- civil wars, rather than those between states. It seems reasonable to assume that the population of the side in a war that utilized mainly autonomous weapons would find it difficult to take collective responsibilities for atrocities committed by machines they had little direct control over.

Wars waged by machines against humans will have little opportunity for the acts of conscious mercy such as that recorded by Ernst Jünger in his World War I memoir *Storm of Steel,* and would likely prevent the humanization of the enemy necessary for reconciliation and forgiveness. In his memoir Jünger records coming upon a lone enemy officer while alone on patrol.

I saw him jump as I approached, and stare at me with gaping eyes, while I, with my face behind my pistol, stalked up to him slowly and coldly. A bloody scene with no witnesses was about to happen. It was a relief to me, finally, to have the foe in front of me and within reach. I set the mouth of my pistol at the man's temple- he was too frightened to move- while my other fist grabbed hold of his tunic...

With a plaintive sound, he reached into his pocket, not to pull out a weapon, but a photograph which he held up to me. I saw him on it, surrounded by numerous family all standing on a terrace.

It was a plea from another world. Later, I thought it was blind chance that I had let him go and plunged onward. That one man of all often appeared in my dreams. I hope that meant he got to see his homeland again. (Jünger, 2004)

Heavy use of autonomous weapons in lieu of human soldiers would also likely lessen the extent to which societies grapple with the moral horrors of war. Human beings have struggled to understand war's terrible impact on individuals and societies since at least the Greek tragedians- a source soldiers continue to turn to in order to grapple with war's huge moral costs. (Healy, 2014) And in a society such as the United States where so small a percentage of citizens does the actual fighting, and war takes place at great distance from the country itself, the stories brought home by soldiers have become one of the few ways society can confront the ugly truth of "organized killing." (Grossman, 1995) Robots will not write such memoirs.

WAR "THAT HUMAN THING" AND REMOTE CONTROLLED WEAPONS

There are thus good arguments against the use of autonomous weapons in light of the *Jus ad bellum* and *Jus post bellum* criteria of Just war theory supporting a moratorium on these weapons even should technological innovations that make them better at discrimination during conflict occur. Yet is it the case that such a moratorium should apply to remote controlled weapons as well?

There are reasons to think that it should not. It was said that autonomous weapons might undermine new interpretations of *Jus ad bellum* because the moral consent of soldiers would no longer be required to engage in war. Drones and other remote controlled weapons that have a human pilot do not face these same risks. Pilots who find a war unjust are free to refrain/strike against using their skills to engage in combat. The fact that drone piloting facilities are often located in civilian areas might actually make them more susceptible to public protest and opposition to war.

Soldiers operating drones and remote controlled vehicles as opposed to the widespread use of autonomous weapons are likely to remain highly conscious of what combat looks like, that is, control and supervision as in the case of *Jus ad bellum* is, both better distributed and possess clearer chains of moral accountability in the case of remote controlled as it is with autonomous weapons.

Whether or not drones are also superior to autonomous weapons when it comes to *Jus post bellum* is somewhat more complicated and might best be illuminated by taking a look at arguments against the use of drones on grounds other than Just war theory. For the argument that these new forms of weapons make the adherence to the prescriptions of *Jus post bellum* more difficult can be reduced to the grounds that they make human recognition of the suffering caused by war and its risks less likely. And a good case has been made that both autonomous and remote controlled weapons share these flaws.

Perhaps the best arguments made against remote controlled weapons *and* autonomous weapons have to do with their effect on the human experience of war. Christopher Coker, for example, argues that technological innovation is resulting in a diminishment of our humanity, a humanity that ironically and tragically is often brought out by conditions of war. For Coker, combat seems to bring out the best and worst in human beings, the best being our bravery and willingness to sacrifice our lives for others and the worst our savagery. This strange amplification of our humanity is what led Thucydides to characterize war as "the human thing". (Coker, 2013)

Coker makes little distinction between the dehumanizing aspects of war by remote control and truly autonomous weapons. Persons fighting in wars from cubicles on the other side of the globe are barred from experiencing what Coker understands as the human characteristics of war no matter how good the simulation or resolution of battlefield conditions. Their lives are not at stake. This absence of the human reality of war, which has largely become limited to those who engage in close combat on land, would be doubly the case for any truly robotic warriors, which would have no sense of courage or fear or anger.

On at least some level, arguments against the use of technologies that remove soldiers from the battlefield are built on the idealization of human bravery in face of the mortal dangers of war. One of the defining features of Western civilization has been the honor it holds for those tasked with fighting its wars. (Hanson, 2001) This makes some sense given that the soldier since ancient Greece has been held in esteem for their willingness to sacrifice their life for the good of their political community. Yet such public perceptions of soldiers' innate heroism often leave veterans trapped between public mythology and the brutal reality of war itself. (Boudreau, 2008) War certainly offers opportunities for individual courage and heroism, but one should ask if such opportunities are worth their psychological and physical costs, not merely to enemies, but to one's own countrymen or oneself who has been asked to fight?

With this in mind the weaknesses of bio-conservative arguments against the use of not just autonomous, but also remotely operated weapons becomes clear. Unless motivated by some mythology of martyrdom, no one would argue that soldiers should be deliberately placed at risk of being maimed or killed in order to increase wars opportunity for heroic action. Following Coker's prescriptions would result in technologically fossilizing warfare to retain features that are more a matter of civilian myth than veterans' reality.

The weakness of the bio-conservative critique are further revealed the moment one starts to think about what war would look like if the rationale behind such an argument was applied to those areas where military exchange has long been instrumentalized. Should we return to the era when naval battles were fought through forced boarding followed by brutal hand to hand combat? (Paine, 2011) Air warfare, especially in the form of strategic bombings, has since its beginnings been waged at such distances that the humanity of a target was almost impossible to grasp. (Creveld, 2011) Coker's argument therefore, further boils down to petrifying only one form of warfare- land warfare- to retain the aspects of war he thinks make us more "human".

Another argument against remote weapons, regardless of their level of autonomy, is based on their effect on risk. In this view, efforts to avoid the risks of combat on the battlefield do not so much remove risk as relocate it. Risk is transferred to the other side, but doing so actually relocates risk onto *civilians,* whether because military targets now hide themselves among civilian population to make assaults more difficult, or because of the fact that the side attacked using remote controlled weapons needs to make the mortal risks of war apparent to the attacker and can only do so by attacking soft targets- namely civilians. Such is the argument made by counterinsurgency expert David Kilcullen:

If we consider it a legitimate act of war for a Predator to strike a target in, say, a city in Pakistan, killing militants in the houses where they live, but also potentially injuring or killing noncombatant civilians, is it legitimate for those same Pakistani militants to strike the city where that Predator's pilot lives? If it is legitimate to kill a militant attending a wedding in tribal areas, is it also legitimate to kill a Predator pilot at his kid's soccer game in Indiana Springs. (Kilcullen, 2013)

Yet like Coker's argument against both autonomous and remote controlled weapons on bio-conservative grounds, Kilcullen's position ignores the reality of advanced war as it is currently fought even when no such remote controlled or autonomous weapons are used. What difference is there for the attacked if the attacker is located in an unreachable location at the other side of the globe or flying 10,000 unreachable feet above them in a supersonic fighter they are powerless against? The technological superiority of advanced countries, where there airmen and seamen are largely safe from reprisal, has already shifted risk to soldiers on land and civilian populations completely independent of either drones or autonomous weapons.

In a very real sense Kilculen's scenario is merely another example of the situation we have had for quite some time absent drones. The same question of whether or not it is justifiable to attack, even at risk of injuring civilians, the area where a military assault on oneself originates would hold for the pilot of the long range bomber pilot based in Nevada flying sorties to and from a war on the other side of the world as it would for a drone pilot far away from the theater in which the effects of his actions occur.

Given that the piloted plane is impossible for the attacked to shoot down, the equation does not really change for the bombed if the pilot is actually in the plane or if he has never left the base.

It is true that drone- warfare has been experienced to be particularly terrifying to civilians who suffer under it. Yet this seems to be as much a matter of the way in which such warfare has been wagged rather than the fact that such weapons are operated remotely. The fact that in the quest to eliminate "targets" the US has used drones to attack civilian gatherings has led to a sense among the populations suffering such attacks that bombing can occur anywhere and at any moment for seemingly no reason with the effect of potentially killing them or their loved ones. The effect being something similar to that of the terrorism the drone strikes are supposedly being used to stop. (Ahmed, 2013)

What this response to criticisms of drones and autonomous weapons suggest is that these weapons are merely symptomatic of a problem with advanced warfare itself. Identifying that problem should place us more firmly on the path of deciding the ethical question of how these types of weapons should be developed and used or if they should be developed and used at all.

MODERN WARFARE AND THE OCCLUSION OF VIOLENCE

Most human beings seem to have an instinctive inhibition against the use of force directly against an individual that stems from our ability to exercise empathy. The army which has long recognized the dilemma confronted by the human aversion to kill in circumstances where the humanity of an opponent can be recognized has responded by trying to instrumentalize and automate the reactions of soldiers themselves so that empathy can, in a sense, be turned off. (Grossman, 1995) Automated weapons would remove this empathy entirely, and, contrary to Singer, such weapons would in fact perform with the cold detachment of psycho-paths, though without the risk they would use this coldness to pursue their own rather than group ends.

An important question to ask is why is it the army in which soldiers might be confronted with the command to kill one or perhaps a handful of enemy combatants that individuals have to be especially trained to overcome their aversion to kill rather than air forces or navies where a small number of airmen might be responsible for killing hundreds or even thousands of human beings that such aversion is so clearly an issue?

The answer might lie in the fact that the technologies of air and sea warfare foster the occlusion of violence given the greater distance from the granular human level against which their weapons are wielded. In a sense, warfare from a distance takes advantage of underlying features of human moral reasoning identified in the oft cited "Trolley Problem". In one of the most common versions of the Trolley Problem a person is faced with a stark moral choice in which a train is careening towards five people who will die: people who can only be saved by pushing an innocent fat man from a ledge unto the tracks.

Joshua Greene, most famously, has argued that those who choose to sacrifice one person to save five when presented in the Trolley Problem are exercising a higher "manual" mode of moral reasoning than those who refuse to push the hypothetically innocent man to his death who are guided by their instinctive "automatic" morality which is repulsed by the thought of the use of physical force against another. (Greene, 2013)

Yet Greene's privileging of "manual" over "automatic" means of moral reasoning is dependent upon how one defines the "greater good". In the Trolley problem that means sacrificing five people to save one, but for airmen and seamen who make a decision to kill by pulling a "switch" in the service of what they believe is the "greater good" seem to face much less of an instinctive human moral break on the use of force than soldier on land who can often see whom they intend to kill. Accessing their manual mode of morality through violence having been instrumentalized seems to result in an increased capacity to kill with impunity in the supposed name of some "greater good".

Indeed, individuals with diminished capacity for empathy, notably psychopaths, have been shown to score "better" i.e. more regularly chose to sacrifice one person to save five than their neuro-typical counterparts (Dutton, 2012), and the number of people who chose to sacrifice the fat man increases when subjects are given a version of the Trolley Problem where instead of having to push the fat man can merely let him fall by pulling a lever. (Greene, 2013)

Our occlusion of violence prevents us from seeing the reality of what we are doing, the brutality of the violence that it war. In that sense the horrific cruelty of a group like the Islamic State is like our mirror image glorifying in the unveiling of this occlusion as if accepting the inhumane barbarity of war were a form of strength. (Cottee, 2014)

The real reason for seeking to overcome our occlusion of violence is so that we can morally grapple with the reality of war. As Lawrence Keeley has noted, even in primitive societies where warfare was often endemic:

... warfare whether primitive or civilized involves losses, suffering, and terror even for the victors. Consequently, it was nowhere viewed as an unalloyed good, and the respect accorded to an accomplished warrior was often tinged with aversion.

For example, it was common the world over for the warrior who had just killed an enemy to be regarded by his own people as spiritually polluted or contaminated. He therefore had to undergo a magical cleansing to remove this pollution. Often he had to live for a time in seclusion, eat special food or fast, be excluded from participation in rituals and abstain from sexual intercourse.

It is really this occluding aspect of modern warfare that we need to escape, for remaining blind within it means that we will be unlikely to judge the true meaning of the human cost of war, and thus will be much less capable of following the prescriptions of *Jus ad bellum* to go to war only in the face of extreme circumstances. The occlusion of violence also makes the reconciliation between peoples required by *Jus post bellum* difficult if not impossible. How could we be prompted to make amends for deaths of innocents not even our soldiers would have witnessed which were ostensibly committed in our name?

The question we might ask is if there is any way technology might be designed to diminish this occlusion of the nature of violence rather than feed it? A distinctive feature of remote controlled warfare as it is currently fought is its use of cameras that not only allow the pilot or driver to see their interactions in intimate detail, but allow their engagements to be recorded for later viewing. As discussed by Coker, some pilots appear to view their actions as a sort of "video game" whereas others having watched their targets over long periods of time come to know them and have great difficulty with killing when the order to do so comes. What makes the latter remarkable is that for the first time in the history of warfare we have put eyes on the arrow, so to speak. Fighting from a distance no longer means that the enemy has been abstracted into a faceless target. Coker writes:

According to one drone pilot interviewed by a British newspaper, the modern warrior shares an intimacy with his targets that is unique in human history: the Reaper's sensors allow a pilot to 'an individual to a fidelity where you can perceive limb and arms, stuff that you are carrying'. (Coker, 2013 p. 122)

Weapons designed in order to minimize occlusion created by distance from the battlefield would aim at doing what even soldiers fighting on land, whose combat is the most intimate of all forms of war, are often unable to do; namely, to get close enough to their potential targets to actually be able to discriminate between combatant and non-combatant and to where possible injure, rather than kill, enemy soldiers.

Remote controlled weapons could turn the enemy into nothing more than simulacra to be killed with the same moral ease one has when dispatching with virtual enemies in a video game, or they could be used to reverse such occlusion and make the brutal reality of war more knowable both to civilians and to soldiers.

Imagine if we were to make footage from not just drones and remote controlled vehicles, but also soldiers themselves, open and available to the public for review, debate, or to enable humanitarian efforts at reconciliation for wrongdoing in the way the Wiki leaks tapes supplied by Chelsea Manning defied the government by giving us a searing glimpse of what it looks like when the laws of war are violated. (Jónsdóttir, 2014)

If we take democracy seriously then the ultimate responsibility for killing on the battlefield lies with the public that through their action or inaction have enabled such conflicts. Remote controlled weapons, if used properly, might allow us to see what exactly it is we have sanctioned. Were such records common and open we would be better able to hold to account those who had violated the laws of war in our name, and, it is hoped, be less willing to permit such violence in the first place.

Such might come to be the case in the US other Western countries, but developments elsewhere are likely to be different. It is very likely that much of the rest of the world will continue to pursue both remote controlled and autonomous weapons in order to redress the military superiority and spending prowess of the US. We might be able to craft regulation to limit the spread and use of such weapons, but that seems unlikely as long as this asymmetry is in place. In a way the specter of "killer robots" distracts us from addressing this unstable situation including what are likely the much more dangerous existential risk, and nuclear weapons, whose immorality is far more pronounced than even machines that can make their own decisions to kill.

Nevertheless, autonomous weapons do represent a real moral danger and technological pressures pushing in the direction of such weapons, the dangers from latency, jamming and especially the demand for *speed* will continue to exist. In a sense the weakness of both the supporters and opponents of autonomous weapons discussed initially, where ethics becomes dependent upon technological development might be inescapable.

As machines become more capable of autonomy, and especially if they improve in their capacity for discrimination between legitimate and illegitimate targets to the point of being better than human beings at such discrimination, then a hybrid solution combining some levels of machine autonomy and human oversight may prove the best way forward.

The question of ethical design lies in discriminating between military and non-military targets. This entails getting as close as possible to a target and may require something like "suicide" probes can get near enough to a potential target to establish with certainty whether it is, say, a fishing boat or a warship, a gathering of militants or a wedding party.

Even in such a scenario human beings should retain ultimate veto power over decisions to use lethal fire even if they do little else in terms of operating such weapons. Exercising such veto power would allow us to preserve the *Just ad bellum* and *Jus post bellum* prescriptions of Just war theory by both allowing such "vetoers" to prevent unjust wars in the first place, through refusal to pilot such weapons for a war that violates *Just ad bellum,* and in providing a human partner for reconciliation and responsibility after the conclusion of a war in adherence to *Jus post bellum.*

Humans having ultimate control over whether a machine kills would also prevent us from reaching a state of ultimate occlusion to which we have been heading that would leave us blind to the brutal reality of "organized killing" found in war. Only we are capable, and will likely remain the only ones capable of knowing, or cable of even caring about, what it means to kill another human being, to take from the world an absolutely unique manifestation of intelligence that will occur only once in the history of the universe. (Nicolelis, 2011) An intelligence that is part of the types of web of human relationships- a father, a husband, a son- such as those that convinced Jünger refrain from killing the enemy solider he stumbled across in the dead of night. In a more than metaphorical sense we are now tasked with being the "souls" of our machines, and should we fail in that task we will have created a world where intelligence has been rendered in a form that is largely soulless.

REFERENCES

Ackerman, E. (2014, January 22) U.S. Army Considers Replacing Thousands of Soldiers With Robots. *IEEE Spectrum*.

Ahmed, A. S. (2013) *The Thistle and the Drone: How America's War on Terror Became a Global War on Tribal Islam*. Harrisonburg: R.R. Donnelley.

Arkin, R. C. (2009). *Governing Lethal Behavior in Autonomous Robots*. Boca Raton: CRC Press. doi:10.1201/9781420085952

Augustine of Hippo. (426). De Civitate Dei Contra Paganos [The City of God Against the Pagans]. In G.C. Walsh, S.J. Demetrius, B. Zema, S.J. Grace, O.S.U. Monahan, & D.J. Hogan (Eds.), The City of God. Image Books/Doubleday.

Boudreau, T. E. (2008). *Packing Inferno: The Unmaking of a Marine*. Port Townsend, WA: Feral House.

Carpenter, C. (2014). *Lost Causes: Agenda Vetting in Global Issue Networks and the Shaping of Human Security*. Ithaca: Cornell University Press.

Coker, C. (2013). *Warrior Geeks: How 21st-century Technology Is Changing the Way We Fight and Think about War*. New York: Columbia University Press. doi:10.1093/acprof:oso/9780199327898.001.0001

Cottee, S. (2014, November 17). ISIS and the Intimate Kill. *Atlantic (Boston, Mass.), Retrieved from* http://www.theatlantic.com/international/archive/2014/11/isis-and-the-intimate-kill-peter-kassig/382861/

Creveld, M. (2011). *The Age of Airpower*. New York: Public Affairs.

DeLillo, D. (1997) Underworld. New York, NY: Scribner, 1997. (pp. 790).

Dowd, A. W. (2013). Drone Wars: Risks and Warnings. *Parameters*, *43*, 7–16. Retrieved from http://www.strategicstudiesinstitute.army.mil/pubs/Parameters/

Dutton, K. (2012). *The Wisdom of Psychopaths: What Saints, Spies, and Serial Killers Can Teach Us about Success*. New York: Scientific American/Farrar, Straus and Giroux.

Fallows, J. (2014, December 28). The Tragedy of the American Military. *Atlantic (Boston, Mass.), Retrieved from* http://www.theatlantic.com/features/archive/2014/12/the-tragedy-of-the-american-military/383516/

Greene, J. D. (2013). *Moral Tribes: Emotion, Reason, and the Gap between Us and Them*. New York: Penguin Press.

Grossman, D. (1995). *On Killing: The Psychological Cost of Learning to Kill in War and Society*. Boston: Little, Brown.

Guarini, M., & Bello, P. (2012). Robotic Warfare: Some Challenges in Moving from Non-civilian to Civilian Theaters. In P. Lin, K. Abney, & G. Bekey (Eds.), *Robot Ethics: The Ethical and Social Implications of Robotics* (pp. 129–145). Cambridge, MA: MIT Press.

Hanson, V. D. (2001). *Carnage and Culture: Landmark Battles in the Rise of Western Power*. New York: Doubleday.

Healy, P. (2009, November 11) The Anguish of War for Today's Soldiers, Explored by Sophocles. *The New York Times.* Retrieved from http://www.nytimes.com

Hodge, N. (2012) In the Afghan War, a Little Robot Can Be a Soldier's Best Friend. *The Wall Street Journal.* Retrieved from http://www.wsj.com

Jónsdóttir, B. (2013, August 21). Bradley Manning's Sentence: 35 Years for Exposing Us to the Truth. The Guardian. Retrieved from http://www.theguardian.com

Jünger, E., & Hofmann, M. (2004) Storms of Steel. (pp. 234).New York: Penguin Books.

Keeley, L. H. (1996). *War before Civilization* (p. 144). New York: Oxford University Press.

Kilcullen, D. (2013). *Out of the Mountains: The Coming Age of the Urban Guerrilla* (p. 174). New York: Oxford University Press.

Lin, P. (2011, December 15). Drone-Ethics Briefing: What a Leading Robot Expert Told the CIA. Atlantic (Boston, Mass.). Retrieved from http://www.theatlantic.com

Losing Humanity. (2012, November 19). *Human Rights Watch.* Retrieved from http://www.hrw.org/reports/2012/11/19/losing-humanity

Markoff, J. (2014, November 11) Fearing Bombs That Can Pick Whom to Kill. *The New York Times.* Retrieved from http://www.nytimes.com/2014/11/12/science/weapons-directed-by-robots-not-humans-raise-ethical-questions

McMahan, J. (2009). *Killing in War.* (pp. 6). Oxford: Clarendon Press. doi:10.1093/acprof:oso/9780199548668.001.0001

Nicolelis, M. (2011). *Beyond Boundaries: The New Neuroscience of Connecting Brains with Machines--and How It Will Change Our Lives.* New York: Times Books/Henry Holt and Co.

Paine, L. P. (2013). *The Sea and Civilization: A Maritime History of the World.* Alfred A. Knopf.

Rigby, A. (2005). Forgiveness and Reconciliation in Just Post Bellum. In M. Evans (Ed.), *Just War Theory A Reappraisal* (pp. 177–200). New York: Palgrave Macmillan. doi:10.3366/edinburgh/9780748620746.003.0009

Scharre, P. (2014, July 29) How to Lose the Robotics Revolution. *War on the Rocks.* Retrieved from http://warontherocks.com/2014/07/how-to-lose-the-robotics-revolution/#_

Sharkey, N. (2012) Killing Made Easy In P. Lin, K. Abney & G.A. Bekey (Eds.). Robot Ethics: The Ethical and Social Implications of Robotics. (pp. 111-128). Cambridge, MA: MIT Press.

Singer, P. W. (2009). *Wired for War: The Robotics Revolution and Conflict in the 21st Century* (pp. 391–392). New York: Penguin Press.

Wong, W. W. S. (2013). *Emerging Military Technologies: A Guide to the Issues Praeger Security International.* Santa Barbara: Praeger.

KEY TERMS AND DEFINITIONS

Autonomous Weapon: A weapon capable of making the decision of whether or not to engage in potentially lethal action.

Drone: A remotely piloted air weapon or observation platform.

Jus ad Bellum: Criteria emerging from Just War theory that help actors determine whether or not a war is just.

Jus in Bello: Criteria emerging from Just War theory that help actors determine whether or not a particular action preformed during war is just.

Jus post Bellum: Criteria emerging from Just War theory that help actors determine whether conditions following war are just.

Occlusion: The act of an object or phenomenon being hidden by something blocking its view.

Trolley Problem: A set of hypothetical scenarios used in moral philosophy and psychology in which participants are asked under what conditions they would sacrifice the life of an innocent person to save a greater number of innocent lives.

This research was previously published in Rethinking Machine Ethics in the Age of Ubiquitous Technology edited by Jeffrey White and Rick Searle, pages 225-238, copyright year 2015 by Information Science Reference (an imprint of IGI Global).

Section 7
Regulations and Privacy

Chapter 23
Drones in the U.S. National Airspace System

David Stuckenberg
United States Air Force (USAF), USA

Stephen Maddox
United States Air Force (USAF), USA

ABSTRACT

In 2012, the U.S. Congress passed the FAA Revitalization and Reform Act, which among other provisions called for the integration of drones into the U.S. national airspace by September 2015. While the statutory provision was an attempt to meet emerging industry needs which includes the defense sector, Congress inadvertently failed to examine many of the potential problems relating to domestic drone integration. In spite of industry and government efforts to mitigate these problems, four primary issues exist for UAS integration: the FAA is behind schedule, inadequate rule making, inadequate threat analyses, and an incomplete drone categorization system. Each problem is discussed in detail along with proposed solutions.

INTRODUCTION

The National Airspace System (NAS) is a highly integrated and complex network designed to provide safe and reliable air transportation over the contiguous and non-contiguous U.S. territories and waters. Although the regulatory scheme governing the NAS is intricate, it is sound. As a result, U.S. air operations are the safest in the world (Federal Aviation Administration [FAA], 2014d, subtitle). On average, the NAS handles more than 50,000 (manned) flights a day (Jones & Takemoto, 2011, para. 5). In 2014, domestic airlines transported in just one month more than 66.4 million passengers, or one-fifth of the U.S. population (FAA, 2014b). According to Federal Aviation Administration (FAA) estimates, the amount of air traffic will increase one percent per year for the next 20 years (Price, 2014, section IV).

DOI: 10.4018/978-1-5225-8365-3.ch023

In spite of this modest growth projection, Congress passed the FAA Modernization and Reform Act of 2012 (henceforth the Reform Act) as a measure to "improve aviation safety and capacity" (House of Representatives, 2012, p. 1). While such reforms are laudable, nested within the Reform Act's 300 pages is a statutory order that raises significant concerns. The Act stipulates, "The Secretary of Transportation ... shall develop a comprehensive plan to safely accelerate the integration of civil unmanned aircraft systems [UAS, also called unmanned aerial vehicles (UAVs) or remotely piloted aircraft (RPA)] into the national airspace system" (House of Representatives, 2012, p. 64). In addition, the Reform Act requires integration by September 30, 2015 (House of Representatives, 2012, p. 64).

However, there is a standing impediment to the drone industry. Drone operations in the NAS are restricted to below 400 feet above ground level (AGL) or Special Use Airspaces (SUAs) used by the government for testing and training. Access to airspace or flight corridors outside SUAs is granted only through a special FAA permit called a Certificate of Authorization (COA; FAA, 2014e). These operational restrictions exist because of the hazards drones pose to manned aircraft and the public.

As a consequence, the Association of Unmanned Vehicle Systems International (AUVSI), a leading robotics industry lobby, published a report in 2010 called "Unmanned Aerial Vehicle Integration in the National Air Space System: An Assessment of the Impact of Job Creation in the Aerospace Industry" (AUVSI, 2014; House of Representatives, 2013). The message of the report and other lobby efforts was predictable considering AUVSI's primary legislative goal: "To ensure that obstacles to advancing and fielding unmanned systems and robotics are removed" from airspace access (AUVSI, 2014, para. 1). In fact, the assessment clearly stated the U.S. economy would benefit enormously if drones had access to the NAS. Conversely, AUVSI warned a failure to integrate UAS would impede military readiness, limit the Department of Defense's (DOD) ability to stay at the forefront of new technology, and negatively impact jobs and the aviation industry for many years (House of Representatives, 2013).

In 2012, persuaded by industry efforts and a desire to ensure the viability of America's military capabilities, Congress supported the industry's drive for integration by incorporating statutory language into the Reform Act that ordered the FAA to include drones by 2015 (House of Representatives, 2012, p. 64). However, in passing the measure, Congress inadvertently failed to examine many of the problems involved in the use of domestic UAS.

Our government has made extensive investments to safeguard our most cherished monuments, critical infrastructure, and key government installations; unfortunately, much of it is easily bypassed with readily available small aircraft. In a Capitol Hill hearing about drones and national security, Congressman McCaul (2012) stated, "Now is the time to ensure these vulnerabilities are mitigated to protect our aviation system as the use of unmanned aerial systems continues to grow" (p. 3). Four primary problems exist for UAS integration: the FAA is behind schedule, inadequate rule making, inadequate threat analyses, and an incomplete drone categorization system. Each problem is discussed in detail along with proposed solutions.

THE FAA IS BEHIND SCHEDULE

Every generation encounters growing pains associated with new technology. Each leap forward in transportation and telecommunication technology came with its own set of problems and UAS are no exception. UAS technology presents both risks and benefits. Since 9/11, drones have become a tremendously versatile platform used to accomplish everything from crop dusting and aerial surveys to firefighting and border patrol (Homeland Security Committee Aide, phone interview, July 8, 2014).

However, three primary problems exist for the present UAS integration timetable. To start with, two enigmatic technological challenges hamper the UAS industry: sense and avoid (SAA) and control and communications (C2) links (Office of Inspector General, 2014, p. 5). Manned aircraft use a concept known as "see and avoid." When the pilot sees an obstacle, they will safely avoid it. A remotely piloted aircraft can not necessarily see obstacles and needs an alternative safety system. A major problem with C2 is the availability of frequency space. Just as the airspace is already filled with manned aircraft, the frequency spectrum is largely consumed by various users. A third challenge is addressing the general safety of UAS. It appears these challenges will not be resolved by the September 2015 deadline. The FAA's UAS roadmap anticipates SAA to be ready between 2016 and 2020 and C2 structure between 2016 and 2017 (2013, p. 57).

The Department of Transportation (DOT) Inspector General (2014) recently stated, "The [FAA] is behind schedule on most of the act's UAS provisions, and the magnitude of unresolved safety and privacy issues will prevent FAA from meeting Congress' September 2015 deadline for UAS integration" (p. 13). Dr. Mica Endsley (2014), Chief Scientist of the United States Air Force, explains, "Drones have scary implications, but the FAA has developed a plan. They are currently working on 'sense and avoid'—a new technology that will alert aircraft to the presence of drones" (phone interview, July 8). While this technology is both needed and desired by the FAA, Mr. Rory Paul (2014), the CEO of Volt Aerial Robotics Company, a firm that produces agricultural and specialty drones, notes, "There are still key safety components missing. At this time, there is no low-cost sense and avoid system applicable to small UAS and large aircraft… there is also no regulatory mandate for sense and avoid" (phone interview, June 26).

The need for SAA is underscored by an increasing number of FAA reports citing near collisions between aircraft and drones. In one case, an American Airlines regional jet nearly collided with a drone at 2,300 feet as it made its approach into Tallahassee, Florida. The UAS involved in the incident was supposed to remain within 400 feet of the surface (Nicas, 2014). A collision at that altitude might have ended in tragedy for the 23 passengers and crew. In a similar incident in 2011, an Air Force C-130 collided with an unmanned aircraft in Afghanistan. Fortunately, the unmanned aircraft did not strike any critical components on the cargo aircraft. No one was injured, although the incident resulted in costly repairs for the C-130 (Whitlock, 2014).

A glance at the number of accidents involving only military drones makes the need for SAA even more apparent. According to the *Washington Post* drone crash database, "About one-third of the crashes (since 9/11) occurred in Afghanistan, but nearly one-quarter happened in the United States" (Chow & Whitlock, 2014, para. 2). Under current regulations, these accidents occurred in special use airspace where risks to manned aviation are lower due to traffic segregation.

In a related concern, the DOT Inspector General (2014) stated:

FAA is not effectively collecting and analyzing UAS safety data to identify risks. This is because FAA has not developed procedures for ensuring that all UAS safety incidents are reported and tracked or a process for sharing UAS safety data with the U.S. Department of Defense (DOD), the largest user of UAS. (p. 2)

A great deal of information that could lead to the safe operation of UAS is still not available in a useful form to the FAA. Strangely, despite the FAA's overwhelming focus on aviation safety, their regional safety inspectors do not oversee UAS operations in person because the inspectors lack appropriate guidance and resources (Office of Inspector General, 2014, p. 13). These inspectors should be on the front lines of identifying NAS safety concerns. The FAA is retooling its process, but the delivery date for the new procedures is September 2015—rather late considering the importance of this issue and the congressional deadline (Office of Inspector General, 2014, p. 29).

Just as crowded airspace can be problematic for aircraft, a crowded frequency spectrum is problematic for unmanned aircraft. In the 2013 UAS roadmap, the FAA (2013) observed, "Government agencies and industry need to investigate link security requirements, such as protection against intended and unintended jamming, RF interference, unauthorized link takeover, and spoofing" (p. 29). As of June 2014, a DOT audit noted that frequency space still needs to be reserved for UAS and security vulnerabilities are unresolved (Office of Inspector General, 2014, p. 6).

To address these sorts of problems, the FAA, in coordination with states, universities, and industry, established six UAS test sites. However, research at the test sites is not well coordinated. According to the DOT Inspector General, the FAA needs to take a more proactive role in directing research to meet its data collection requirements (2014, p. 31). A good way to address the technological problems would be to sponsor competitions aimed at resolving the most vexing problems. For example, from 2004 to the present, the Defense Advanced Research Projects Agency (DARPA) successfully used the challenge model to generate solutions for robotic cars. Their Grand Challenge began in 2004 and resulted in significant advances in robotics, unmanned systems, and artificial intelligence (2014, para. 4). The same model can be applied to UAS. Because states, universities, and corporations have made considerable investment in the test sites it is likely they would be eager to compete for prizes and proprietary rights to SAA technology. At the same time, the FAA and UAS industry would benefit from multiple solutions to integration problems.

Today, the FAA finds itself bound by Congress on one hand and bound by a mandate to implement UAS integration on the other.

The FAA is responsible for developing plans and policies for the safe and efficient use of the United States' navigable airspace. This responsibility includes coordinating efforts with national security and privacy policies so that the integration of UAS into the NAS is done in a manner that supports and maintains the United States Government's ability to secure the airspace (FAA, 2013, p. 9).

However, the FAA (2013) also understands, "The safe integration of unmanned aircraft into the NAS is a significant challenge" (p. 42).

While a delay in the Act's ordered integration is both foreseeable and unavoidable, according to the DOT Inspector General, embracing this delay will afford the agencies concerned with much needed time to re-establish priorities and conduct needed research (Jansen, 2014, para. 4). Some might argue that integration will take place with or without federal direction; however, it's better to have fewer UAS operating while sound preparations are underway than thousands integrated under poorly conceived policy. While the current delay is a product of technical challenges, it must also be understood that a stay of the amendment will only delay the inevitable. The Reform Act does serve as a worthwhile first step to support a burgeoning industry. However, Congress should not demand integration before necessary technology can support it. The FAA agrees, noting:

Although aviation regulations have been developed generically for all aircraft, until recently these efforts were not done with UAS specifically in mind. This presents certain challenges because the underlying assumptions that existed during the previous efforts may not now fully accommodate UAS operations. (House of Representatives, 2012, sec. 332)

INADEQUATE RULE MAKING

Hobby aircraft represent another area where FAA policy is inadequate. In May 2014 a DJI Phantom II quadcopter was discovered crashed on top of the Metropolitan Square Building in St. Louis. It is the tallest building in downtown. The local media reported, "According to the FAA, 'civilians can have and operate drones, if you maintain a visual while flying and keep it under 400 feet above ground.' The drone that crashed Monday hit the 30th floor of the 593-foot-tall building" (Piper & Matthews, 2014, para. 5). Phantom quadcopters are entry level hobby aircraft—they can be purchased for less than $2,000. The current FAA rules for hobby aircraft do not provide adequate deterrent to reckless and criminal behavior.

Currently hobby aircraft, also known as model aircraft, are minimally regulated by the FAA. To be considered a model, an aircraft must meet the following criteria:

- Used only for recreational (i.e. non-commercial) purposes;
- Follows the safety guidelines of a community-based organization (such as the Academy of Model Aeronautics);
- Weighs 55 pounds or less (larger aircraft can be flown with approval from a community-based organization);
- Does not interfere with manned aircraft;
- Flown more than five miles from manned airports, unless coordinated with the control tower;
- Flown below 400 feet above ground;
- Flown within line of sight of the operator;
- Flown within sight of safety observers when necessary;
- Avoids populated areas;
- Not flown for aerial demonstrations or with spectators nearby until proven airworthy (FAA, 2014e, p. 6).

The FAA also applies some general aircraft prohibitions to model aircraft, such as prohibitions against reckless behavior and dropping hazardous objects.

Obviously these rules did not deter the drone operator in St. Louis. Local authorities searched for the drone's owner, but the investigation did not turn up any names (Piper & Matthews, 2014).

One of the most potentially devastating threats from drones is the risk they pose to people concentrated in open venues or settings such as sporting events. The U.S. Department of Interior recently banned the use of airborne drones in all national parks due in part to this type of hazard (Berman, 2014). Interestingly, such hazards are also occurring in the UK where drones also gained popularity. Late last year, "A man who flew a small unmanned aircraft over Alton Towers, endangering the safety of the theme park's customers, pleaded guilty to two separate charges at Stafford Magistrates Court," according to the UK Civil Aviation Authority (2014, para. 1). The operator involved was attempting to obtain video footage. Mr. Paul, CEO of Volt Aerial Robotics, was asked about the potential hazards to crowds at outdoor venues. He related his own story:

I recently confronted a phantom drone operator who was flying his quadcopter close to a crowd of 800. I went and spoke to the man controlling it and I said, "Look, what if this thing hits someone?" We got in a pretty heated argument about it. Looking down the road at these kinds of scenarios, all we need is one UAS to hit a person on the ground then it's all over [the industry will be hit hard]. (phone interview, June 26, 2014)

Where U.S. criminal code (U.S.C.) concerns aviation, there are very concrete penalties for certain disruptive acts. For instance, 49 U.S.C. covers air piracy, interference with crew members, destruction of aircraft, tampering with navigational aids, and other acts (Office of the United States Attorneys, 1999). What's more interesting is that federal law establishes a special jurisdiction for aircraft *in flight*. According to 49 U.S.C. § 46501, "An aircraft is 'in flight' from the moment when all external doors are closed following embarkation until the moment when one such door is opened" (Office of the United States Attorneys, 1999, para. 3). Legal descriptions such as this afford federal prosecutors the ability to criminally charge individuals in a uniform way when certain defined acts are committed. However, where drones are concerned, such acts have to be "construed" to fit existing laws, be tried based on unrelated precedents, or be judged based on case law. Such legal ambiguity could put individuals committing such acts at an advantage as it is often difficult to convict without specific legal prohibitions.

A possible counterargument to the establishment of UAS specific criminal code is that the Reform Act requires the FAA to develop rules and regulations pertaining to drones. However, there is a clear distinction to be made between administrative and criminal jurisdictions. The FAA has no authority to levy criminal penalties. In fact, the FAA is only able to bring administrative action. In the aviation industry, these actions are usually levied in the form of a fine or the revocation of airmen certificates (licensure); however, in the case of model aircraft, operators do not require certification. Since drone proliferation will doubtless continue to occur and operators may continue to operate them in a manner that creates a hazard to the public, Congress should establish statutory laws that will serve to deter those who would use drone technology in a reckless manner. However, by itself, criminal regulation cannot solve the security and safety problems arising from UAS integration.

Regulatory action should also explore the viability of point-of-sale limitations on certain common components needed to operate drones. It is conceivable that if certain components could be limited to those who have either received certification or background checks, regulators would have a much stronger chance of limiting drone capabilities to those who would use them in a legitimate way. Similar policies have had success in other industries. For instance, the Department of Justice, via the Bureau of Alcohol, Tobacco, Firearms and Explosives (ATF) restricts the availably of explosives to individuals or enterprises with a legitimate need. While explosives are obviously dangerous, the ATF describes their many legitimate uses:

Explosives industry members play an integral role in maintaining and improving our quality of life in the United States and work to bring countless benefits to our everyday lives in areas such as mining, oil and gas exploration; demolition; avalanche control; and the use of explosives in special industrial tools, fire extinguishers, air bag inflators, fireworks; and special effects in the entertainment industry. (n.d., para. 1)

The ATF also acknowledges that "because of the potential misuse of explosive materials, ATF plays a vital role in regulating and educating the explosives industry and in protecting the public from inadequate storage and security" (n.d., para. 1). Consequently, according to federal code, "Any person engaging in the business of importing, manufacturing, or dealing in explosive materials without a license issued under the Act, shall be fined not more than $10,000 or imprisoned not more than 10 years, or both" (Cornell University Law School, n.d., para. 1). The drone industry, like the explosives industry, has the potential to benefit society, but certain aspects should be closely regulated due to their capacity to cause damage or violate privacy. The FAA (2013) UAS roadmap acknowledges the need for "security and vetting requirements" (p. 28); however, the plan lacks any detail whatsoever. One way to augment criminal codes would be through point-of-sale restrictions.

A common argument against point-of-sale restrictions is that drones are easy to manufacture and build at home. Others, meanwhile, argue there are so many drones already available that attempting to limit certain technologies would be pointless. However, one could have made that same argument about explosives. In the past, high explosives were readily available. However, once regulation was brought to bear industry-wide, pre-regulation explosives eventually disappeared, leaving only licensed product. A similar outcome is likely with UAS. Over time, unregulated systems would become inoperable or obsolete. Amendment of federal law could facilitate criminal penalties for reckless acts, feasibly serve as a deterrent to criminal behavior, and could provide for accountability if implemented at the point-of-sale.

Hobby aircraft rules and the Reform Act provide a giant loophole for irresponsible individuals to fly through. It takes considerable talent, time, and money to build a traditional or large scale model aircraft, but these models still require a skilled pilot to operate. The situation has changed with the ready availability of automated aircraft. For less than $2,000 anyone can purchase a fully operational drone kit equipped with remote control, GPS, gyro stabilization, and a Wi-Fi controlled payload—typically a camera. While small, these aircraft are easy to purchase, modify, and operate. FAA rules and federal law should be updated to account for these changes in aviation technology.

INADEQUATE THREAT ANALYSIS

Some negligent behavior is inevitable, but it is also important to consider threats posed by intentional criminal behavior. There is mounting evidence that both domestic and international terror networks desire to utilize drones in their tactics (Chow, Cuadra, & Whitlock, 2014). As drones continue to proliferate, the government must begin to anticipate a rise in the types of dangerous scenarios that will be encountered. Presently, both civil and military targets are vulnerable. According to the Department of Justice:

On Sept. 28, 2011, (Rezwan) Ferdaus requested and instructed the undercover FBI employees (UCE) to deliver explosives and firearms for his attack plan. While inspecting the explosives and firearms in the UCEs' vehicle and inside his storage unit, Ferdaus placed some of the explosives inside a remote controlled aircraft that he had ordered and obtained for his attack plan. (The FBI: Boston Division, 2012, para. 5)

Ferdaus's plot involved using drones to ferry the explosives over barriers and into the U.S. Capitol and the Pentagon (The FBI: Boston Division, 2012). Because these kinds of conspiracies have not been isolated, concern in the drone community is mounting.

According to the National Priorities Project, "Adjusted for inflation, the United States has spent $635.9 billion on homeland security since FY2001 [as of 2012]" (2011). Much of this investment has been to shore up elements of our critical infrastructure and counter conventional threats. These threats include weapons such guns and knives, improvised explosive devices (IEDs), and vehicle-borne IEDs. However, one of the central issues arising from drones is that once integrated to the NAS, many of the security measures implemented after 9/11 may be easily bypassed via over flight. What's even more troubling is that even if the Reform Act's integration provisions were postponed or reversed, the availability of off-the-shelf technology already presents a tangible threat.

While most of the post-9/11 security investment is intact, the viability of these investments is put in question by the potential use of UAS to commit acts of terror. Today, the safety of America's key infrastructure, monuments, and vital interests is not just a matter of a ground strategy; much of the future threat to America will be from above. The FAA (2013) agrees:

Integrating public and civil UAS into the NAS carries certain national security implications, including security vetting for certification and training of UAS-related personnel, addressing cyber and communications vulnerabilities, and maintaining/enhancing air defense and air domain awareness capabilities in an increasingly complex and crowded airspace. In some cases, existing security frameworks applied to manned aircraft may be applicable. Other security concerns may require development of new frameworks altogether. (p. 12)

With the advent of widespread UAS, the United States must reframe its planning process for national security, air defense, and domain awareness.

According to Dr. Endsley (2014):

The Scientific Advisory Board began looking at this [threat] between 2005 and 2006. The threats were approached in phases: first bases, then civil infrastructure targets, [and so on]. It did cover defense against UAS. These [drones] are a very real and credible threat. (phone interview, July 8)

Major Michael (2014), Chief of Global Intelligence, Surveillance, and Reconnaissance (ISR) Operations for the USAF CENT (Central Command) agrees: "Expect the enemy is going to fly these [drones]." (phone interview, June 30)

When asked about the future security implications of drone operations in the NAS, Mr. Paul (2014) of Volt Aerial Robotics, stated, "I think they are very important. Recently Hezbollah and the Iranians attempted to fly a drone over a nuclear reactor in Israel. It went for the reactor in one of the most secure airspace systems in the world" (phone interview, June 26). According to Israeli officials, "[The Iranian drone] flew over sensitive installations inside southern Palestine and was shot down in an area near the Dimona nuclear reactor" (Karouny, 2012, para. 6). Mr. Paul went on to say, "Today, I could shut down Saint Louis Lambert International Airport from my basement; all I would have to do is fly it up and down the runway. No one would know who was controlling it or from where" (phone interview, June 26, 2014). Mr. Paul's notional scenario makes it clear that even non-military aircraft pose a significant security threat.

Lt. Col. (Ret.) Mitchell (2014), Chief of MQ-1 Predator Training for a USAF Special Operations Squadron agrees:

It is a very naive approach not to worry about non-weaponized drones. I'm not a chemist, but I'm pretty sure you can put a small amount of chemical in a water balloon and attach it to an RC plane, walk through on Friday afternoon taking coordinates with an iPhone GPS, then 24-hours later, launch an RPA from the parking lot into a full stadium using those coordinates. (phone interview, July 7)

Such a scenario provides a glimpse of what might be possible and, in fact, was already tried. In April 2014, the FBI foiled another terror plot involving the use of drones. The plot involved an attack that would use toy planes loaded with explosives against a school and a Connecticut federal building (Tacopino, 2014). The scheme was hatched over a period of three months, with children on the playground as the likely targets.

As previously mentioned, Hezbollah obtained several unmanned aircraft and used them to survey targets, violate the sovereign airspace of several nations, and incite terror. In July 2014, Hamas adapted Hezbollah's strategy and launched three drones against Israel (FOX News Network, 2014). According to the media report:

[The] strike against Israel introduces the use of drones with an offensive [terrorism] capacity, which could potentially inflict significant casualties—something the rockets [launched by Hamas] from Gaza have failed to do. Hamas released a video of what it said was the aircraft, about five yards long, with each wing holding two unidentified missiles. (FOX News Network, 2014, para. 12)

In terms of small drones, there is still debate both inside and outside of government on how much of a threat they truly pose. Thus, it is essential to conduct a credible impact assessment. Lt. Col. (Ret.) Mitchell, RPA Pilot, Evaluator, and Training Chief, USAF Special Operations Command (2014) posits:

Dangerous chemicals and substances available in ounces are numerous. I think a study should be considered where they hire someone and say—'Go buy [a drone] and see what you can do….' This kind of practical study will fill in gray areas very quickly. It's the little guy that's practicing—the smaller [the aircraft] is the more it's going to be available. (phone interview, July 7)

Lt. Col. (Ret.) Mitchell's comments underscore that many of the assumptions we are operating under could be wrong. According to a congressional aide (2014), "DHS did collaborate with the FAA on the privacy issues where test sites were concerned, but as far as any spoofing or hacking is concerned the DHS has yet to take any serious action. There should be more of a sense of urgency surrounding this issue, and there is not" (phone interview, July 7). Further proof can be seen in a study Congressman Michael McCaul (2012), Chairman of the House Subcommittee on Oversight, Investigation, and Management cited from the University of Texas at Austin. He noted:

Researchers from the Cockrell School of Engineering led by Dr. Todd Humphreys proved that civilian unmanned aerial systems can be hacked into and hijacked with a relatively small investment of money and time. These findings are alarming and have revealed a gaping hole in the security of using unmanned aerial systems domestically. (p. 3)

Presently, the USAF position on small drones is that if they are not designed to carry weapons, then they should not be a concern, according to Dr. Endsley (phone interview, July 8, 2014). Thus, while the Air Force focuses on drones designed to carry weapons and DHS focuses on ground threats, the FAA, a regulatory agency of the DOT, is left to fill the void. Simply put, the FAA is not equipped to safeguard our nation from this kind of threat. In fact, this aspect of drone integration is entirely incongruent with the FAA's mission, role, or resources. Consequently, America is becoming increasingly vulnerable to the whims of those have, in some cases, already attempted to do harm with an effective, anonymous, and easily acquired airborne delivery system.

With the advent of widespread UAS, the United States must reframe its plans for air defense. Some in leadership have begun to take notice. In recognition of the changes in UAS technology, 43 states introduced 130 pieces of legislation concerning privacy, law enforcement, research, and other topics (National Conference of State Legislators, 2014, para. 1). Recently Congressman McCaul (2012), made the following indictment:

The Department of Homeland Security mission is to protect the homeland. Unfortunately, DHS seems either disinterested or unprepared to step up to the plate to address the proliferation of unmanned aerial systems in U.S. air space [NAS], the potential threats they pose to our national security, and the concerns of our citizens of how drones flying over our cities will be used, including protecting civil liberties of individuals under the Constitution. For example, in discussions with my Subcommittee staff prior to this hearing, Department officials repeatedly stated the Department does not see this function (domestic use of drones) as part of their mission and has no role in domestic unmanned aerial systems. I strongly disagree. DHS's lack of attention about this issue is incomprehensible. It should not take a 9/11 style attack by a terrorist organization such as Hezbollah or a lone wolf inspired event to cause DHS to develop guidance addressing the security implications of domestic drones.... DHS lacks that initiative. I am concerned DHS is reverting back to a pre-9/11 mindset, which the 9/11 Commission described as a lack of imagination in identifying threats and protecting the homeland. (p. 3)

Even once the security gaps are understood, solutions are needed. One possibility is an air defense system specifically designed for small aircraft. When presented with the idea of deployable radar equipped with something to "knock down" drones—Major Michael (2014) noted, "We need to think along the lines of a 'kill switch' that could just turn them off. Another aspect would be bi-state radar, lasers, etc. Bi-state radar can see something as small as a bird flying" (phone interview, June 30). Lt. Col. (Ret.) Mitchell (2014) agrees:

Directed kill switches might be a good answer to this. The technology would be good—a deployable asset that would likely be high cost, but reasonable if portable to use at big events. Such technology could play a very important part in the future of an integrated homeland defense. (phone interview, July 7)

Consensus about the need for anti-drone technology also seems to exist among international defense experts. Dr. David Whetham (2014), Senior Lecturer at the UK Defense Academy and Kings College asserted: "The idea of deployable anti-drone countermeasures has got to be taken seriously.... There are technologies already developed that could be put to alternative uses which might be fairly easily adapted" (phone interview, July 8). He also postulated that the Phalanx system, a radar guided ship borne anti-aircraft system, could do the job (phone interview, July 8, 2014). While adapting and developing anti-drone technology is a long-term undertaking, both the authors and the experts interviewed agree—it presents the best approach if we are to maintain domestic security.

Developing anti-drone technologies suitable for domestic use will require a high degree defense and federal interagency cooperation. Interestingly, Congress established the Joint Planning and Development Office (JPDO), an office composed of members from the DOD, NASA, FAA, DOT, and DHS for similar purposes relating to UAS integration. However, the JPDO was defunded by Congress in 2014.

Notwithstanding, the FAA understood the value of this cooperative and re-established the JPDO as the Interagency Planning Office (Carey, 2014). Under the current framework, the DOD, in conjunction with DHS and the FAA, should immediately begin outlining the research requirements and seek support to develop anti-drone defense technologies.

INCOMPLETE DRONE CATEGORIZATION SYSTEM

The FAA (2013) states in the roadmap that "UAS pilots (the pilot-in-command [PIC]) will always have responsibility for the unmanned aircraft while it is operating" (p. 11). This is easy enough to apply to remotely operated aircraft where the PIC is simply located outside the aircraft, but it is somewhat difficult to apply to autonomous aircraft or "drones." In the case of an aircraft operated by artificial intelligence, who is the PIC and how are they held accountable for their actions? Automation is only hinted at in the FAA UAS roadmap; the near-term plan does not expect to authorize autonomous flights (2013).This is unfortunate, as some problems discussed earlier with C2 could be simplified if the aircraft do not require a C2 link.

Automation represents a significant growth area for unmanned aircraft. The examples from Volt Aerial Robotics and the DJI Phantom illustrate that the necessary technology already exists. Just as automation increases efficiency in industrial processes, aircraft are increasingly automated to improve safety and performance. Aircraft already exist—in fact they have existed for some time—that can takeoff, perform a mission, and return to the ground with no human input during flight.

Generally speaking, the FAA is aware of the wide variety of unmanned aircraft *possible*; unfortunately, the UAS roadmap lacks clarity in how to differentiate UAS. A high-level congressional staffer related:

Given the extraordinarily wide variety of aircraft known as "drones," starting to categorize the technology in terms of how it is operated, its range and capabilities, and what payloads it is able to carry may be a good way to distinguish among various models as we integrate drones into U.S. airspace. (Homeland Security Committee Aide, phone interview, July 8, 2014)

The current FAA system is designed to account for the wide variety of manned aircraft. A committee aide for Congress noted:

There is a sense of ambivalence about drones. While people recognize that drones can serve the public welfare in many ways, there is still some confusion over the term "drone" itself—whether we are talking about a five pound model or a border patrol aircraft. (Homeland Security Committee Aide, phone interview, July 8, 2014)

To resolve confusion and facilitate its licensing process, the FAA uses two primary descriptions to certify aircraft for operations in the NAS: category and type. Category describes the intended use of the aircraft and type describes the characteristics of the aircraft (FAA, 2009, 2014c). Additional descriptions are also used that differentiate aircraft by weight and complexity. The FAA has yet to announce the full certification plan for unmanned aircraft, but in the UAS roadmap the agency describes public, civil, and hobby UAS. Table 1 provides a partial listing of FAA category and class, along with two proposed additions.

Table 1. Partial listing of FAA categories and proposed additions

Category	Type
Normal	Airships
Utility	Balloons
Acrobatic	Rotorcraft
Commuter	Small Airplanes (< 19,000 lbs)
Transport	Transport Airplanes (>19,000 lbs)
Balloons	Unmanned Aircraft (public, civil, or hobby)
Special Classes	Small UAS (expected in 2014)
Proposed:	Proposed:
Automated/Autonomous	Multiple Aircraft (formation/swarm

In late 2014, The FAA released a number of interpretive legal requirements pertaining to small UAS based on the Reform Act's stipulations. However, these did not account for all possibilities. At least two other significant variations should be considered. Automated aircraft—true "drones"— are ideal for simple, repetitive, or dangerous work. Consequently, the final FAA classification system will need to account for autonomous systems.

Another UAS variant under development is the formation or "swarm," where multiple aircraft cooperate to perform a task. In a formation of manned aircraft, each plane has a pilot responsible for their aircraft. In the case of unmanned aircraft, one operator (or artificial intelligence) can direct the entire group. FAA regulations need to account for systems composed of multiple cooperative aircraft.

An audit by the DOT came to a similar conclusion. The inspector observed deficiencies in the FAA's current descriptions of unmanned aircraft. Their current system only describes UAS as weighing more than or under 55 pounds; by comparison, the DOD uses five categories to describe unmanned aircraft based on size, speed, weight, and capabilities (Office of Inspector General, 2014). The FAA needs to expand its categorization system as well as accelerate its plans for automated aircraft and swarms.

CONCLUSION

The question before us is not *if* drones will be a part of the U.S. NAS, but *how*? Will this endeavor unfold in a safe and reasonable manner, or will we repeat mistakes from aviation history? The DOT testified to the House Transportation subcommittee on aviation that the FAA would not be able to meet the deadline due to significant technological and safety concerns (Jansen, 2014). From a historical perspective, this is not surprising. Our current NAS took years to develop and resulted from reactive legislation that followed numerous accidents. The integration of UAS will require resolving multiple regulatory and legislative issues (Dulo, 2014).

The first problem is the schedule mandated by congress and laid out in the FAA UAS roadmap. According to an audit by the DOT:

Significant technological, regulatory, and management barriers exist to safely integrate UAS into the NAS. First, following many years of working with industry, FAA has not reached consensus on standards for technology that would enable UAS to detect and avoid other aircraft and ensure reliable data links between ground stations and the unmanned aircraft they control. (Office of Inspector General, 2014, p. 2)

To resolve these problems, FAA, along with its partner government organizations, need to lead the research and development planning at the UAS test sites in order to resolve technological and regulatory problems. Additionally, the challenge model should be considered as it provides a best problem solving practice for FAA to apply considering the time sensitive nature of the present technological issues.

In addition to the technological issues, current Federal and FAA rules for UAS are inadequate to deter reckless behavior or criminal acts. It's vital to address Congressman McCaul's (2012) warning: "The US government was concerned that these aerial vehicles could be modified and used to attack key assets and critical infrastructure in the United States" (p. 2). Additional threat analysis followed by focused research is needed now. According to Lt. Col. (Ret.) Mitchell (2014), "[Drones] will be one of the worst security, and potential security, issues that we have and that will continue to emerge" (phone interview, July 7).

Finally, the current and planned categorization system is inadequate. The system needs to account for fully automated systems and formations of UAS. These and other issues will delay the useful integration of UAS. The DOT Inspector General's office (2014) believes significant delays are inevitable:

It is important to note that the Act requires safe—not full—integration of UAS into the NAS by September 2015. Integrating UAS into the complicated U.S. airspace requires an incremental approach, and one that will allow the NAS processes currently in place, as well as those under development, to effectively ensure safety both in the air and on the ground (p. 26).

Delays will hinder the growing UAS industry and draw unwanted attention to the FAA; however, it will also afford the industry and government time needed for research and threat assessment.

REFERENCES

AUVSI. (2014). *About*. Retrieved from http://www.auvsi.org/new-item

AUVSI. (2014). *Issues and Policy*. Retrieved from http://www.auvsi.org/advocacy

Berman, M. (2014, June 20). National Park Service bans drone use in all national parks. *The Washington Post*. Retrieved from http://www.washingtonpost.com/news/post-nation/wp/2014/06/20/national-park-service-bans-drone-use-in-all-national-parks/

Carey, B. (2014, May 27). FAA's New 'Interagency Planning Office' Replaces JPDO. *AIN Online*. Retrieved from http://www.ainonline.com/aviation-news/air-transport/2014-05-27/faas-new-interagency-planning-office-replaces-jpdo

Chow, E., Cuadra, A., & Whitlock, C. (2014, June 20). Hazard Above: Drone Crash Database. *The Washington Post*. Retrieved from http://www.washingtonpost.com/wp-srv/special/national/drone-crashes/database/

Civil Aviation Authority. (2014, May 29). Guilty pleas for dangerous unmanned aircraft theme park flight. Retrieved from http://www.caa.co.uk/application.aspx?catid=14&pagetype=65&appid=7&mode=detail&nid=2364

Cornell University Law School. 27 CFR 555.161 - Engaging in business without a license. Retrieved from http://www.law.cornell.edu/cfr/text/27/555.161

DARPA. (2014, March 13). *The DARPA Grand Challenge: Ten Years Later*. Retrieved from http://www.darpa.mil/NewsEvents/Releases/2014/03/13.aspx

Department of Justice. Bureau of Alcohol, Tobacco, Firearms and Explosives. Explosives Industry. (n. d.). Retrieved from https://www.atf.gov/content/Explosives/explosives-industry

Dulo, D. (n. d.). UAS in the National Airspace: Aerial Goldmine or Legal Landmine. *Unmanned Systems: Association of Unmanned Vehicle Systems International 32*(6), 14-17. Retrieved from http://www.mckennalong.com/assets/attachments/uas_nationalairspace.pdf

Endsely, M. (2014, July 8, 12:47 CT). Phone interview.

Federal Aviation Administration. (2009). Standard Airworthiness Certificate. Retrieved from http://www.faa.gov/aircraft/air_cert/airworthiness_certification/std_awcert/

Federal Aviation Administration. (2013). *Integration of Civil Unmanned Aircraft Systems (UAS) in the National Airspace System (NAS) Roadmap*. Retrieved from http://www.faa.gov/uas/media/uas_roadmap_2013.pdf

Federal Aviation Administration. (2014a, February 26). *Busting Myths about the FAA and Unmanned Aircraft*. Retrieved from http://www.faa.gov/news/updates/?newsId=76240

Federal Aviation Administration. (2014b, March 13). Fact Sheet – FAA Forecast–Fiscal Years 2014-34. Retrieved from http://www.faa.gov/news/fact_sheets/news_story.cfm?newsId=15934

Federal Aviation Administration. (2014c, June 18). 14 CFR Part 91: Interpretation of the Special Rule for Model Aircraft. Retrieved from http://www.faa.gov/uas/media/model_aircraft_spec_rule.pdf

Federal Aviation Administration. (2014d, September 18). *Air Traffic NextGen Briefing*. Retrieved June 28, 2014, from http://www.faa.gov/air_traffic/briefing/

Federal Aviation Administration. (2014e, November 4). *Certificates of Waiver or Authorization (COA)*. Retrieved from http://www.faa.gov/about/office_org/headquarters_offices/ato/service_units/systemops/aaim/organizations/uas/coa/

FOX News Network. (2014, July 14). Israel military says it has shot down drone along southern coastline. *Fox News*. Retrieved from http://www.foxnews.com/world/2014/07/14/israel-military-says-it-has-shot-down-drone-along-southern-coastline/

General Atomics Aeronautical. (2013). MQ-1 Predator. Retrieved from http://www.ga-asi.com/products/aircraft/pdf/MQ-1_Predator.pdf

Homeland Security Committee Aide. (2014, July 8, 9:30 CT). Phone interview.

House of Representatives. (2012). *FAA Modernization and Reform Act of 2012* (Report 112-381). Integration of Civil Unmanned Aircraft Systems. Washington, DC: Government Printing Office. Retrieved from http://www.gpo.gov/fdsys/pkg/CRPT-112hrpt381/pdf/CRPT-112hrpt381.pdf

House of Representatives. (2013). Unmanned Aerial Vehicle Integration into the National Air Space System: An Assessment of the Impact of Job Creation in the Aerospace Industry. Washington, DC: Government Printing Office. Retrieved from http://www.gpo.gov/fdsys/pkg/CHRG-113hhrg78819/pdf/CHRG-113hhrg78819.pdf

Jansen, B. (2014, February 5). Watchdogs: FAA won't meet 2015 deadline for drone safety. *USA Today*. Retrieved from http://www.usatoday.com/story/news/nation/2014/02/05/faa-drones-inspector-general-gao/5226427

Jones, T., & Takemoto, P. (2011, July 6). *Fact Sheet*. Retrieved from http://www.faa.gov/news/press_releases/news_story.cfm?newsId=12903

Karouny, M. (2012, October 11). Hezbollah confirms it sent drone downed over Israel. *Reuters*. Retrieved from http://www.reuters.com/article/2012/10/11/us-lebanon-israel-drone-idUSBRE89A19J20121011

McCaul, M. (2012, July 19). Address, Using Unmanned Aerial Systems Within the Homeland: Security Game Changer? Washington DC. Retrieved from http://homeland.house.gov/sites/homeland.house.gov/files/07-19-12%20McCaul%20Open.pdf

Michael, N. (2014, June 30, 13:35 CT). Phone interview.

Mitchell, W. (2014, July 7, 17:10 CT). Phone interview.

National Conference of State Legislators. (2014). *2013 Unmanned Aircraft Systems (UAS) Legislation*. Retrieved from http://www.ncsl.org/research/civil-and-criminal-justice/unmanned-aerial-vehicles.aspx

National Priorities Project. (2011, May 26). *U.S. Security Spending Since 9/11*. Retrieved from https://www.nationalpriorities.org/analysis/2011/us-security-spending-since-911

Nicas, J. (2014, May 9). FAA: U.S. Airliner Nearly Collided With Drone in March: Incident Appears to be First Case of a Big U.S. Airliner Nearly Colliding With an Airborne Drone. *The Wall Street Journal*. Retrieved from http://online.wsj.com/articles/faa-u-s-airliner-nearly-collided-with-drone-in-march-1399659956

Office of Inspector General. (2014, June 26). *Audit Report*. FAA Faces Significant Barriers to Safely Integrate Unmanned Aircraft Systems into the National Airspace System. Retrieved from https://www.oig.dot.gov/sites/default/files/FAA%20Oversight%20of%20Unmanned%20Aircraft%20Systems%5E6-26-14.pdf

Office of the United States Attorneys. (1999, August). Special Aircraft Jurisdiction of the United States. *Criminal Resource Manual*. Retrieved from http://www.justice.gov/usao/eousa/foia_reading_room/usam/title9/crm01405.htm

Paul, R. (2014, June 26, 18:33 CT). Phone interview.

Piper, B., & Matthews, E. (2014, May 8). Drone crashes into Met Square building downtown. Retrieved from http://www.ksdk.com/story/news/local/2014/05/07/drone-crash-metropolitan-square-building/8828895/

Price, H. *Fact Sheet–FAA Forecast–Fiscal Years 2014-34*. (2014, March 12). Retrieved from http://www.faa.gov/news/fact_sheets/news_story.cfm?newsId=15934

Tacopino, J. (2014, April 8). Man planned to fly model planes packed with bombs into school: FBI. *The New York Post*. Retrieved from http://nypost.com/2014/04/08/man-busted-for-plotting-drone-attack-on-a-school/

The FBI Boston Division. (2012, July 10). Massachusetts Man Agrees to Plead Guilty to Plotting Attack on Pentagon and U.S. Capitol and Attempting to Provide Material Support to Terrorists. Retrieved from http://www.fbi.gov/boston/press-releases/2012/massachusetts-man-agrees-to-plead-guilty-to-plotting-attack-on-pentagon-and-u.s.-capitol-and-attempting-to-provide-material-support-to-terrorists

Airspace Operations Challenge.U. A. S. (2014). *Update*. Retrieved from http://uasaoc.org/index.htm

United States Department of Transportation. (2014, June 12). *March 2014 U.S. Airline Traffic Data*. Retrieved from http://www.rita.dot.gov/bts/press_releases/bts028_14

Whetham, D. (2014, July 8, 10:00 CT). Phone interview.

Whitlock, C. (2014, June 20). When Drones Fall From the Sky. *The Washington Post*. Retrieved from http://www.washingtonpost.com/sf/investigative/2014/06/20/when-drones-fall-from-the-sky/

This research was previously published in the International Journal of Aviation Systems, Operations and Training (IJASOT), 1(2); edited by Massoud Bazargan, pages 1-22, copyright year 2014 by IGI Publishing (an imprint of IGI Global).

APPENDIX

Expert Interviews and Transcripts

Overview

- Dr. Mica Endsley: Chief Scientist, United States Air Force;
- Committee Aide: U.S. House, Committee on Homeland Security;
- Major Michael: Intelligence, Surveillance, and Reconnaissance Division Chief, U.S. CENTCOM (last name withheld for security purposes);
- Rory Paul: CEO, Volt Aerial Robotics Company;
- Lt. Colonel (Ret.) Mitchell (last name withheld for security purposes): RPA Pilot, Evaluator, and Training Chief, USAF Special Operations Command;
- Dr. David Whetham: Professor and Senior Lecturer at the United Kingdom Defense Academy, King's College, London.

Dr. Mica Endsley: Chief Scientist, United States Air Force

Dr. Endsley is the scientific advisor to the Secretary of the Air Force and the department's chief of research and technology. Dr. Endsley also served on the Air Force Scientific Advisory Board as panel chair on Defense against remotely piloted vehicles and chaired an *S&T Review* of the Human Effectiveness Directorate at the Air Force Research Laboratory.

On the Record (recorded) phone interview conducted by author at 12:47 CT, July 8, 2014.

Starting with the discovery of an operational need, what is the process by which equipment is developed to meet that need?

Normally the Air Force's research needs are driven by the core competencies [Air Superiority, Global Precision Attack, Personnel Recovery, Special Operations, Global Integrated ISR, Command and Control, Cyber Security, Space Superiority, Agile Combat Support, Building Partnerships, Nuclear Deterrence Operations, and Rapid Global Mobility] in the master plan—these are handed down by the Major Commands [the purpose of a Major Command is to organize, train and equip forces for the Combatant Commands (COCOMS)]. Outside of this, the COCOMS hand down their needs. Needs also arise of war gaming, searching horizons, the Quadrennial Defense Review (QDR), and a host of other sources. However, the short term needs come from the Combatant Commands.

Authors note: Dr. Endsley declined to answer the remainder of the question citing the classification level of the requested information.

Considering the range of impacts that RPAs can have on security, what is the Air Force doing to examine this issue?

Many of the programs I can't go into, however, I can tell you we are looking at it. We have a group at the Air Force Academy working on this right now. [voice in- background—prompted Endsley] We also have the Joint Planning and Development Office (JPDO) working with the FAA, NASA, Air Force, et al. Drones have scary implications, but the FAA has developed a plan. They are currently working on "sense and avoid," a new technology that will alert aircraft to the presence of drones. In addition, there are six test sites around the U.S. that are there to help with appropriate R&D. As you know, commercial entities want to push everything from crop dusters to flying packages from Amazon.com.

Concerning RPAs, what are some of the major challenges going forward and how can we defend against drone related threats?

The Scientific Advisory Board began looking at this between 2005 and 2006. The threats were approached in phases: bases, then civil targets, et al. It did cover defense against UAS. They are a very real and credible threat. Umm, the amount of damage they can do is pretty much restricted to the amount of payload they can carry. So for the smaller ones that are harder to find, they can't carry very much. The large ones are easier to detect and can do more damage— are easier to shoot down basically.

What about the aspect that terrorism incites such a psychological dimension. For instance, everyone could feel that they are equally at risk? Thus, panic could do more damage than the actual payload?

That is certainly a possibility, but it has to be put into context.

Point of sale restrictions have been used with a degree of success with firearms, for instance, you can make any component on a weapon without a license except the upper. Could this be done with a similar component common to all drones?

I don't think you can do that at the aircraft component level because there are so many legitimate commercial and private uses for them, but you can do that at the weapons level. Because it's really only when you start weaponizing them that it becomes a concern.

So there is no common component that can be limited as far as OTS?

They are homebuilt, you can crank them out on a 3D printer if you want to; you can crank them out in your garage. And the actual vehicle, there's not any reason to restrict the building of the actual vehicle; it's only if you weaponize them it that it's a challenge from a security standpoint.

How do you assess the FAA's overall responsiveness to this emerging issue—are they incorporating future threats into their rationale?

Very prudent, very conservative, so as not to endanger the public, very cautious. However, the technology and safety and the security issues are being worked in a separate environment.

U.S. House of Representatives Committee on Homeland Security; Homeland Security Committee Aide

The aide is an advisor for the Chairman and Co-Chairman of the House Committee on Homeland Security. Office policy did not allow the interviewee to be identified by name. The quotes listed were approved for release.

What is the current disposition of the public and congress toward the use of drones in the NAS?

I think, generally, people accept that drones are here to stay. But there are still serious concerns about implications for privacy.

I think the UAV industry has done a good job of educating the public about the vast utility of drones, like tracking weather systems, fighting wildfires, protecting the environment, finding missing persons, etc.

What is being done to address the security issues associated with the use of drones in the NAS?

There is a sense of ambivalence about drones. While people recognize that drones can serve the public welfare in many ways, there is still some confusion over the term "drone" itself—whether we are talking about a five-pound model or a border patrol aircraft.

Should we try and limit certain aspects of drone technology at the point of sale, and how helpful would it be to have more definitive statutory language in the U.S.C. relating to drones?

As subcommittee Chairman, Rep. McCaul held a hearing on drones and urged DHS to take a larger role in ensuring that the technology would not be vulnerable to hacking, spoofing, or use as a weapon. To the best of my knowledge, however, DHS has not stepped up its efforts in that regard.

DHS did collaborate with the FAA on the privacy issues where test sites were concerned, but as far as any spoofing or hacking is concerned the DHS has yet to take any serious action. There should be more of a sense of urgency surrounding this issue, and there is not.

Given the extraordinarily wide variety of aircraft known as "drones," starting to categorize the technology in terms of how it is operated, its range and capabilities, and what payloads it is able to carry may be a good way to distinguish among various models as we integrate drones into U.S. airspace.

Major Michael: Intelligence, Surveillance, and Reconnaissance (ISR) Division Chief, USAF CENT (Central Command)

Major Michael is a former USAF RPA pilot and instructor. Today, he leads the global ISR operations program for the USAF CENT (Central Command). His full name was withheld from publication due to operational security requirements.

How will RPAs change the future security landscape within the U.S. and what is the DOD's disposition with respect to their use on U.S. soil?

The FAA has drug their feet; that is part of the reason the FAA is facing so much pressure. Expect the enemy is going to fly these [drones]—they will.

Regarding security and meeting the challenge of these threats to large venues with dense crowds, what technology could be used? Would portable equipment such as jammers be a viable option?

You are on to something with this line of thinking. However, if you jam the links they essentially become a cruise missile. Consequently, we need to think along the lines of a "kill switch" that could just turn them off. Another aspect would be bi-state radar, lasers, etc. Bi-state radar can see something flying as small as a bird. Other possibilities include a registry system for some aspect of technology common to all UAS. This could be registered to whoever buys that equipment. [Authors note: this is similar to what is done with assault weapons. Some might argue there are too many drones already out there. But similar legislation was brought to bear on dynamite and blasting caps. Eventually, the items sold pre-ban/registry went away leaving only licensed product.]

Rory Paul: CEO of Volt Aerial Robotics Company

Volt Aerial Robotics is a consultancy based in Saint Louis, Missouri, whose core business is the development, leasing, and sale of small autonomous aerial systems for academia, agriculture, environmental services, and industry.

What is your current assessment, from an industry standpoint, of the FAA's UAS integration plan?

There are still key safety components missing. At this time, there is no low-cost sense and avoid system applicable to small UAS and large aircraft. At just over one year away, there is also no regulatory component of sense and avoid. This is a vital element to safety where the operation of civil drones in the NAS are concerned. At this time, the onus is still on the UAS operator to keep the equipment clear. What I predict is that the systems will continue to proliferate and then the FAA will back-fill and play catch-up on all the requirements.

What is behind the push for UAS integration in the NAS?

The 2010 industry report published by AUVSI on the potential economic impacts of UAS systems in the U.S. was a large piece of it. However, AUVSI is a lobby group for the UAS industry and they were paid to write the report. In my opinion, the economic benefits were way over stated. Note: not the entire industry agrees with AUVSI.

How important are the security issues potentially arising of drone operations in the NAS?

I think they are very important. Recently Hezbollah and the Iranians attempted to fly a drone over a nuclear reactor in Israel. It went for the reactor in one of the most secure airspace systems in the world. Today, I could shut down Saint Louis Lambert International Airport from my basement; all I would have to do is fly up and down the runway. No one would know who was controlling it or from where.

Given drones are supposed to increase in number in the years ahead, do you think the hazards to crowds and other outdoor venues will increase?

I recently confronted a phantom drone operator who was flying his UAS 500 feet above a crowd of 800. I went and spoke to the man operating it and we got in a pretty heated argument. I said, "Look, what if this thing hits someone?" Looking down the road at these scenarios, all we need is one UAS to hit a person on the ground then it's all over [the industry will be hit hard].

Given the following solutions: (1) point of sale, (2) legal deterrence, or (3) technological inhibitors such as equipment deployed on site (similar to a TFR), which of these in your opinion presents the most viable solution to threats against the public?

Point of sale is very much worth exploring. Currently there are export regulations in place, but many companies are skirting these by developing technology here in the U.S. and then having the components manufactured out of the nation. 3D Robotics received approximately $35 million in funding and then began manufacturing in Mexico. From there the autopilot systems are being sold internationally. Putting someone in handcuffs and taking them away is useful, but it has limitations. However, portable radar units have a lot of potential. Jamming is also a viable option; however, many of the frequencies bands UAS operate on are the same as Wi-Fi (2.4 GHz), so there are some potential problems there, mainly, de-conflicting with other networks. So again, trade regulations are not being enforced. The premise of everything you are saying is right.

Lt Col. (Ret) Mitchell: Former Director of Training, Special Operations Squadron, USAF

Lt Col. (Ret) Mitchell is a former USAF Special Operations RPA pilot and instructor. He is also a member of the Steering Committee on the development of RPA programs at a major university. He requested if this study is publically published that his name be withheld for security reasons.

How difficult is it for the average person to build and employ RPAs? Should this ability be limited?

Very easily. It's so easy it's almost scary. I can go online or to RadioShack and buy most of the components I need very inexpensively. It's very cheap technology to purchase—and can be used as an indiscriminate and potent weapon.

What is domestic terrorism likely to resemble in the future and could it involve drones?

The nation must assume that there will always be some risk of terrorism. There is hardly a foolproof way to look at every aspect of what someone might do as new technology comes on line—thus, it's hard to defend against. Even if we were to deny links, it's possible that assets [UAS] would continue to their targets or have triggers that would detonate them if they lost link. It's very cheap technology to purchase and use as an indiscriminate potent weapon.

How useful would a technological resource be that could be employed in a manner similar to a TFR around high density crowds or other large events and venues?

Directed kill switches might be a good answer to this. The technology would be good—a deployable asset that would likely be high cost, but reasonable if portable to use at big events. Such technology could play a very important part in the future of an integrated homeland defense.

What will the future of drone operations look like and how will that change the domestic security picture?

When you get right down to it, what they are good for is very limited in scope. It's just a marketing hype. You can't take a quad copter and go land in someone's yard. I laugh when I see that kind of thought because of where we are at from a liability standpoint. When we have options like Amazon Prime Delivery—there is really no incentive to it [drone delivery]. We are not going to fertilize our lawns with drones when a walk-behind spreader is so easy. Going forward, it will be as difficult to manage the unlicensed user in the future as it is now. The nature of it is not too far off from that of video. In other words, not much that we do in public today can be done without being captured on video. The nature of the UAV is a similar mindset—you can't limit the proliferation of RC [remote controlled] aircraft, weapons, or balloons. It will be one of the worst security, potential security, issues that we have and that will continue to emerge.

Some say we should not worry about small drones and those not designed to carry weapons. What is your opinion on this?

It is a very naive approach not to worry about non-weaponized drones. I'm not a chemist but I'm pretty sure you can put a chemical in a water balloon and attach it to an RC plane, walk through Friday afternoon taking coordinates with an iPhone GPS, then 24-hours later, launch an RPA from the parking lot into a full stadium using those coordinates and be there 10 seconds later. They are very fast now.

Of the following three possible solutions, (1) federal regulation as a deterrence to use drones to do harm, (2) point of sale restrictions on technology, or (3) deployable directional technology, which would you recommend investing more time and resources exploring?

The first two items are more directly related to slowing down the proliferation of users. The question on the third is: how do you get the right people looking at it? I remember back closer to the events of 9/11 there was a lot of banter about why weren't we prepared for "this" or "that." Many can spend $50K on security systems for their home, systems that will monitor nearly aspect of the home. But, if you've never had a threat you don't spend money on it. But you and I understand that there is a standing goal to do harm to Western society. We understand the need. In fact, we may have such a detailed view that no one else is looking at. Someone who is serious about this like us [himself and the author], may very well have the capability to really do some serious harm. The difference is in the application of the technology.

Does the potentially small payload of drones weighing under 55 pounds diminish the overall threat?

Dangerous chemicals and substances available in ounces are numerous. I think a study should be considered where they hire someone and say, "Go buy one [a drone] and see what you can do." You then will know a small plane that can do a hundred miles an hour (with a lithium battery upgrade), launched from x-location that can go y-number of miles can accomplish "this." Then look at some of the past attacks on subways or some documented aspect of saran gas where this many ounces in a balloon or small condom causes this much harm in this environment (wind conditions, etc.) in a stadium. This kind of practical study will fill in gray areas very quickly. It's the little guy that's practicing—the smaller it is the more it's going to be available.

Dr. David Whetham: Professor at the United Kingdom Defense Academy and King's College, London

Dr. Whetham lectures and researches ethics in warfare, the laws of war, history of warfare, and strategy. His students include British and international officers in the Joint Services Command and Staff College. He holds a PhD in War Studies.

Security planners should not be concerned with drones not designed originally to carry weapons. In other words, small drones should not be a concern. In light of the facts we are discussing, do you have an opinion on that?

I'd say that's probably a bit naive considering that the first armed drone wasn't supposed to be an armed drone either; it was a Predator with effectively an ad hoc missile system strapped on to it. I think the question of whether it is purpose-built is basically a red herring because you can adapt any piece of hardware and provide it with a military purpose. I don't think that's a typical position.

How about the use of deployable anti-drone technologies?

The idea of deployable anti-drone countermeasures has got to be taken seriously. There are obviously alternatives out there—there are technologies already developed that could be put to alternative uses which might be fairly easily adapted, I would have thought. I'm starting to wonder if trying to get some more details of the security for the Olympics might be useful for you because there was certainly a lot of missile systems brought into the capitol. I wonder if something like the Phalanx technology might be adapted.... It's one thing using that at sea, but the amount of lead it puts out in the air you wouldn't want that ending up in the suburbs.

Chapter 24
An Analysis of Unmanned Aircraft Registration Effectiveness

Ronald Pentz
Eastern Michigan University, USA

He (Herman) Tang
Eastern Michigan University, USA

ABSTRACT

This article describes how small unmanned aircraft systems (sUAS) are growing at a rapid pace. They are inexpensive and widely available for both hobbyist and commercial use. However, with this rapid growth, regulations are having a difficult time keeping pace to safely incorporate them into the United States National Airspace. Recent regulations requiring the registration of all sUAS have been overturned by the United States Courts of Appeals. This research provides a statistical analysis of the effectiveness of the registration regulation in the reduction of unauthorized and careless sUAS operation prior to being overturned by the courts. Statistical analysis including descriptive statistics and chi square hypothesis tests were used to analyze more than 3,000 reported unauthorized and careless events. The findings show a significant difference in events pre-registration and post registration.

INTRODUCTION

Small Unmanned Aircraft Systems (sUAS), otherwise known as drones are growing rapidly for both hobbyist and commercial use. Research (Lee, 2016) shows economic projections of the civil sUAS market will continue strong growth over the next several years. This is based on the current consumer spending projections which remains consistent from multiple authors and industry practitioners. Additionally, the Federal Aviation Administration (FAA) Aerospace Forecasts for fiscal years 2017-2037 estimates that the hobbyist fleet will triple in size over the next 5 years, going from 1.1 million units in 2016 to more than 3.5 million units in operation by 2021. Similarly, the commercial sUAS fleet is expected to grow from 42,000 in 2016 to 420,000 by 2021. When both the hobbyist and commercial fleets are combined,

DOI: 10.4018/978-1-5225-8365-3.ch024

the number of unmanned aircraft operating in the National Airspace (NAS) reaches an estimated 4 million by 2021. With that estimated exponential increase in air traffic, it raises obvious concerns about the safe and effective integration of these drones into the NAS.

Unlike manned aircraft, part of the sUAS growth comes from simplicity to manufacture, wide availability both online and in-store, and inexpensive to purchase. There are approximately 89 unmanned aircraft manufacturing companies in the United States building these unmanned aircraft (Canis, 2015). The number of manufacturers worldwide is much larger. The product range offered is anything from a small unit that costs less than $200 and can fly on a single charge for around 10 minutes to a commercial grade unit costing upwards of $10,000 that can fly for a much longer period of time (Canis, 2015).

This rapid sUAS growth leads to concerns such as safety, reliability, and privacy. In turn, this has made it a challenge for the FAA to integrate the expanding sUAS population into the NAS (Branum, 2016). A fine balance must be achieved to avoid over regulation in the United States which could unnecessarily slow drone growth while under regulating could potentially induce unnecessarily high levels of risk into the NAS.

Statement of the Problem

In an attempt to safely integrate the expanding sUAS population into the NAS, the FAA has released recent regulations. Specific to this research was a regulation released on December 16, 2015 that requires all sUAS owners to register their unmanned aircraft prior to operation (Registration and Marking Requirements for Small Unmanned Aircraft, 2015). The intent of the registration requirement was to combat unsafe sUAS operations in the NAS. For example, in the case of careless or unauthorized sUAS operation, the registration markings on the side of the unmanned aircraft is one way to directly link it back to the owner allowing education and/or penalties to deter future occurrences of unsafe flight.

However, on May 19, 2017 the United States Court of Appeals overturned the registration requirement for hobbyists to register their sUAS by citing the FAA Modernization and Reform Act exclusion of model aircraft (On Petitions for Review of Orders of the Federal Aviation Administration). With the overturning of the registration requirements for hobbyists the sUAS owner is no longer required to register their drone which then becomes hard to identify.

The U.S. Court of Appeals may have made the correct decision based on the rule of law, but statistical analysis is needed to show if there is a need to change the laws that led to their decision. The problem remaining to solve is that there has been no statistical analysis regarding the effectiveness of sUAS registration requirements in reducing unauthorized and careless operation in the NAS. Once that problem is solved, decisions regarding the need to change laws could be made with the support of quantitative evidence.

Purpose of the Research

While recognizing and respecting the United States Court of Appeals decision to overturn the registration requirement based on their legal interpretation, this research will provide a statistical analysis of the effectiveness of sUAS registration in reducing careless and unauthorized operation within the boundaries of the United States of America. The proposed research will statistically analyze the effectiveness of sUAS registration in the reduction of unauthorized and careless drone operation.

Limitations

This research is not intended to interpret the law nor decide whether the United States Court of Appeals came to the correct decision with the order to exempt sUAS hobbyists from registration requirements. The research is limited to a quantitative analysis without rule of law or public opinion considered in the results.

The study is pulling data from publicly available information released by the FAA that does not require the report to include the exact size, make or model of the drone. Also, it does not differentiate whether it was being flown by a hobbyist or commercial operator. The population size of drone operators and forecast populations of drone operators are estimates with an uncertain confidence level obtained from the FAA.

Lastly, the main confounding variable that was not accounted for in this research that may influence the results were initiatives by the FAA, communities, and drone/aircraft associations to educate operators on the regulations applicable to unmanned aircraft flight.

Literature Review

A thorough literature review was conducted across the topic of drone operation in the NAS. Of all the literature reviewed, the common themes and most relevant to this research were (a) reliability (b) safety and (c) regulation.

Reliability

The topic of sUAS reliability was examined and found to be a popular theme in the majority of the literature reviewed. According to Van Blyenburgh (1999), sUAS reliability is relatively low when compared to manned aircraft. Besides sUAS having lower reliability levels, the causes of failure were found to be significantly different for failure types in a study of 152 cases of accidents over the last 10 years when compared to manned aircraft (Wild, Murray, & Baxter, 2016). Most failures were related to equipment problems rather than pilot error. It should be noted that Wilder, Murray, and Baxter (2016) was the most influential research found and appears to be the first of its kind to have compared unmanned aircraft failures to those of manned failures using statistical methods.

In a similar report, the primary causes of accidents were found to be electro-mechanical failure and human error (Weibel, & Hansman, 2006). What this means for the general population and operators of drones is that there will need to be more done to increase reliability levels in order to compete with manned aircraft reliability.

Safety

Because there are no crew or passengers onboard, the risk involved with sUAS operations are only to those flying onboard manned airplanes and people on the ground (Clothier & Walker, 2006). A small drone impacting a manned aircraft midair has the potential to cause a catastrophic accident. Additionally, a drone that crash lands or sheds parts during system failures and midair collisions have the potential to impact people on the ground (Weibel and Hansman, 2006).

As reported by Weibel and Hansman (2006) in a safety analysis, it is possible for drone operation with few restrictions over the majority of the United States without significant risk due to their small size and low population density. However, as the size of the drone increases and population density increases such as that of a large metropolitan city, the risk to the general public increases (Weibel and Hansman, 2006).

Regulation

A good point made in a conference paper by Clothier & Walker (2006) was that the full potential of drones will not be realized until they reach unencumbered access into the NAS, but that will not be made possible until regulatory bodies and the general public perceive that the level of safety remains the same without increasing risk to other airspace users and persons or property on the ground. Clothier and Walker (2006) also concluded that the class "A" mishap (destroyed aircraft and/or fatality) rate for UAVs is approximately 5-7 times higher in magnitude than that of manned aircraft which will require more rigorous regulations and operational restrictions to reach a higher safety level objective, but over-zealous regulation has the potential to impede the industry.

Branum (2016) conducted a mixed method quantitative analysis of sUAS growth and regulations. It was found that growth is outpacing current regulations. The survey consisted of 35 stakeholders in the local government. The conclusion of the research was that a significant support for policy to address unauthorized UAV flight into the national airspace that could potentially endanger manned aircraft or people.

An interesting find and counterpoint to regulation was a project in the form of a literature review by Madia (2016). The project critically analyzed the regulation and registration of drones for private use. The goal of the analysis was to understand the impact to users with the new registration requirements in place attempting to control drone mischief and trace any illegal activity back to its owner. The author believes that there are too many flaws with drone registration, making it ineffective. Additionally, criminals and mischievous owners have no incentive to register their drones despite fines and criminal penalties; they will simply not register their drones.

The two main regulations of interest in this research are found in the code of federal regulations as part 101 – Special Rule for Model Aircraft, and part 107 – Small Unmanned Aircraft Systems. However, part 107 does not apply for those operating under part 101, Subpart E. The focus of this research is the removal of the registration requirement for those operating under part 101, Subpart E which have the following general rules governing their operation applicable to this research:

- The aircraft is flown strictly of hobby or recreational use
- Flown within visual line of sight of the operator
- The aircraft is limited to less than 55 pounds
- The aircraft does not interfere with any manned aircraft operation
- The aircraft gives way to any manned aircraft
- Cannot fly within 5 miles of an airport without prior notice to the airport

There appears to be a catch all rule for part 101, Subpart E found in 101.43 stating that no person may operate model aircraft (drones) in such a manner as to endanger the safety of the NAS. This certainly can lead to some subjectivity as to whether or not the operator posed a danger to the NAS or manned aircraft operating within it.

RESEARCH DESIGN AND METHODOLOGY

The design methodology is a quasi-experimental simple time-series design that will introduce an independent variable after observations, followed by more observations to determine if there was a change in the dependent variable. This is followed by descriptive statistics and then hypothesis testing to make statistical research conclusions.

Research Design

A quasi-experimental type of design was used for this research. The main reason is that the data cannot take into account all possible confounding variables such as locations, hobbyist vs commercial, regulatory education of the operator, or size of the drone.

The specific design used is a simple time-series design as previously stated. More specifically, it was observations of the dependent variable (unauthorized and careless sUAS operation reports) for a period of time, then the introduction of the independent variable (registration requirement), followed by additional observations of the dependent variable (unauthorized and careless sUAS operation reports).

Data Collection and Categories

The sampling technique pulled reports of unauthorized and careless activity from publicly accessible information. Specifically, the FAA has provided detailed data available to the general public with over 3,000 UAS incident reports going back to November 2014. The data was extracted, sorted, and filtered. The target data was 365 days of events prior to the registration requirement and 356 days of events post registration requirement.

All 3,014 events recorded were reviewed and placed into one of the following categories (a) sightings, (b) airspace sightings, (c) midair, (d) ground collisions and fly aways, and (e) civil concerns. Each of the categories were then broken down into 3 sub categories for a closer analysis. The remaining part of this section will describe in detail those categories and sub categories ending in a summary table with examples.

Sightings

The sightings category represents drone sightings by (a) aircraft in flight, (b) people on the ground in the vicinity of the airport, and (c) aircraft flying in the vicinity of the airport. Note, in the vicinity of the airport means within a 5-mile radius of the airport.

Airspace Sightings

These are sightings of unauthorized drones within (a) class bravo airspace (b) class charlie airspace, or (c) class delta airspace.

Midair

This category captures events of (a) near midair collisions, (b) evasive action taken to prevent a midair collision, and (c) midair collisions between a manned aircraft and an unmanned aircraft. In this research, the term near midair or near miss is defined as a drone coming within 200 feet of a manned aircraft in flight.

Ground Collisions and Fly Aways

The sub categories in ground collisions are drones that crashed and were (a) found on the grounds of an airport (b) found crashed not in the vicinity of an airport and (c) fly aways from owners with unknown crash locations. For definition, fly away is the drone losing signal with the control link and flying away uncontrolled from the operator.

Civil Concerns

Lastly, the civil concerns category was made up of events that would concern the general public. This was broken down into the sub categories of (a) medical (b) firefighting and (c) other. The medical sub category is medical helicopters that reported sighting drones that had the potential to interfere with their operation and delay patient safety/transport. The firefighting category consisted of drones that were in the vicinity of a fire and interrupted aircraft firefighting operations. The other category was made up of drones that were reported to have flown over facilities and places of interest to the public such as stadiums, power plants, and national monuments causing safety concerns.

Table of Categories and Examples

All together there were 5 categories that were each sorted into 3 sub categories. This was accomplished by carefully reviewing each reported event and placing it into the appropriate category and sub category. The selection of categories and sub categories was organically composed as trends in the data were revealed. Examples of categories, sub categories, and actual event descriptions are listed in Table 1 below.

RESULTS

The data analysis includes both graphical and statistical methods. The two graphical analysis methods that were used were (a) time series plots and (b) histograms. The time series plots were designed to show the general trend of events over time and the histograms were created to describe the distribution of events.

As for the statistical methods, the primary methods used were the (a) regression model for the population growth of operators (b) descriptive statistics and (b) chi square goodness of fit testing. Regression was used to build a model to predict what the population would be in a given time period between the years 2016 and 2021 based on FAA aerospace forecasts. These forecasts should be considered as estimates as mentioned in the limitations. The descriptive statistics were used to describe the data's mean, mode, standard deviation, variance, and events per million operators. Lastly, chi square testing was used to determine if a significant shift occurred in any of the event categories.

Table 1. Categories and examples of unauthorized and careless operation

Sightings By	
Aircraft (pilot reports in flight)	Sightings of a drone by aircraft pilots while in flight.
Example	Acft C172 KHWO/CRG VFR at 045 Reported a white drone 3ft in diameter 17 miles north of VRB airport.
Airport (person on ground)	Sightings of a drone by a person on the ground within 5 miles of an airport.
Example	State police reported an individual flying a UAS by the airport perimeter fence.
Aircraft in Airport Vicinity	Sightings of a drone by aircraft pilots while in flight within 5 miles of an airport.
Example	Pilot reported passing a drone on his right side, at 500 feet, one mile final to runway 28R (HWO).
Airspace Sighting	
Class Bravo	Sightings of a drone in class bravo controlled airspace.
Example	E75L, heading eastbound at 5000' observed a 4 rotor UAS 500' below at 4500' in the class bravo.
Class Charlie	Sightings of a drone in class charlie controlled airspace.
Example	Robinson R44, reported a UAS hovering at 1,700' with red and green lights, 5 S Burbank in class charlie Airspace.
Class Delta	Sightings of a drone in class delta controlled airspace.
Example	Aircraft inbound to land at EWB on a 2-mile left base to RWY 14 observed a red quad copter drone at 1,600" well within the class delta Airspace.
Mid Air	
Near Miss	Pilots reporting drones coming within 200 feet of their aircraft.
Example	Piper PA27 reported near collision with a UAS, 2' in diameter, quad copter style black with red lights at 8,500' 10 N Van Nuys Airport.
Evasive Action Taken	Pilots reporting having to take evasive action to prevent a midair collision.
Example	Pilot observed a UAS, 3-4 feet off wing, 8,000' 30 SE HWD. Special maneuvering needed to keep safe distance.
Mid Air Collision	Pilots reporting having had a midair collision with a drone.
Example	C188, reported possibly striking a UAS at 1,400' 6 NE of Modesto. The pilot believes it was a drone that struck his tires.
Ground Collisions/ Fly Aways	
Found at Airport	Drones found crashed at or in the vicinity of an airport.
Example	ATCT reported a UAS, less than 300 LBS, crashed by the perimeter fence.
Off Airport	Drones crashed and found outside of the vicinity of an airport.
Example	Citizen driving reported a UAS striking his car while north bound on I-35W service road 2.5 S of AFW.
Fly Away	Reports of drones flying away from their operator.
Example	UAS operator was operating an unauthorized UAS within 1-mile radius of the PAH airport. The operator lost link with the UAS and ultimately lost the UAS.
Civil Concerns	
Medical	Reports of drones spotted by medical flights.
Example	B206 reported that a security guard witnessed a drone 100 yards to the west of the helicopter during a landing at the cook children's helipad.
Fire Fighting	Drones that were reported to interfere with firefighting operations.
Example	Calfire contacted the Oakland ARTCC to report a UAS that interfered with aerial firefighting activities forcing Calfire to temporarily ground all air tankers.
Other	Reports of drones flying over stadiums, power plants, etc.
Example	Notification from Everbank Field, they detained a man flying a drone over the Jaguar game.

Time Series Plot(s)

The first-time series plot can be seen in Figure 1 that shows a run of unauthorized and careless events reported that ranges from 365 days prior to the registration requirement to 365 days following the registration requirement. As can be seen, a decrease in events occurs at day 292 which is approaching the registration requirement on day 365, followed by an increase of events at day 435 following the registration requirements. This decrease as time nears the registration requirement could likely be attributed to an increase in promotional regulation awareness/education through the FAA and drone communities that drone registration will allow authorities to directly trace unauthorized and careless violations back to the owner.

The plot represents the period 365 days prior to the registration requirement and 365 days following the registration requirements. Note, the registration requirement went into effect on day 365.

To get a better picture of the events, a multiple variable time series plot was created. Figure 2 combines both pre and post registration events onto a single run of 365 days prior to registration and post registration. The quantity of events associated with post registration (red on Figure 2) requirements clearly appears to be higher than those events associated with pre-registration requirements. It should be kept in mind that the population of operators is also growing as estimated in the forecasts. One could logically expect events to increase as the population increases but remain proportionally consistent which will be discussed later in the research.

Figure 1. Time series plot of unauthorized and careless events

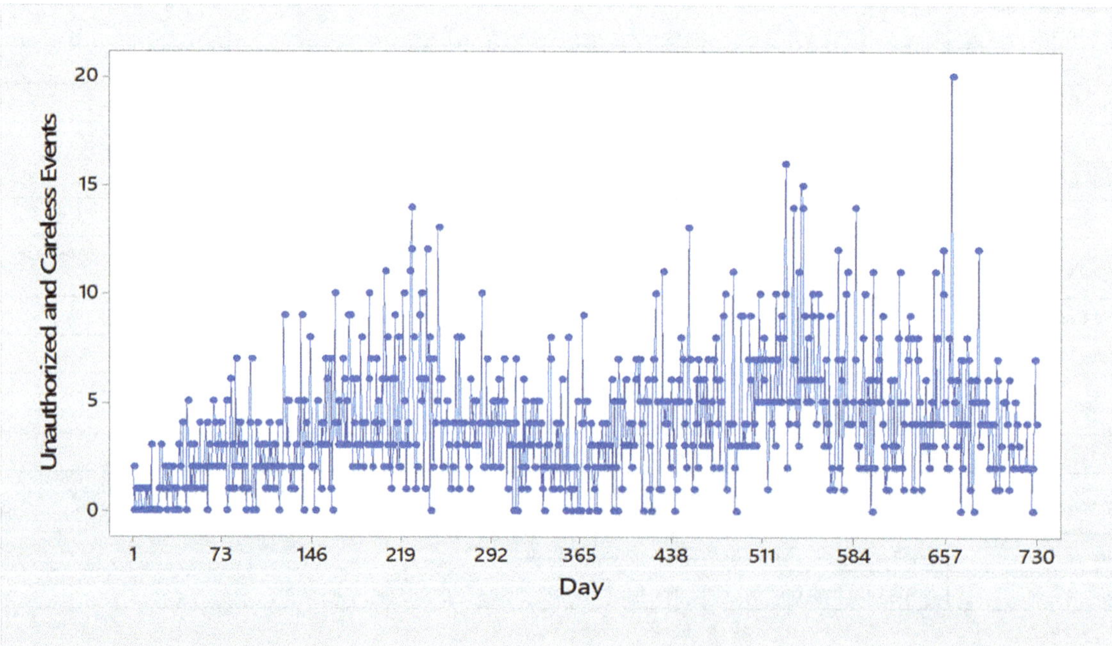

Figure 2. Time series plot overlaying both pre and post registration requirements during the 365 days prior to the registration requirement and 365 days following the registration requirements

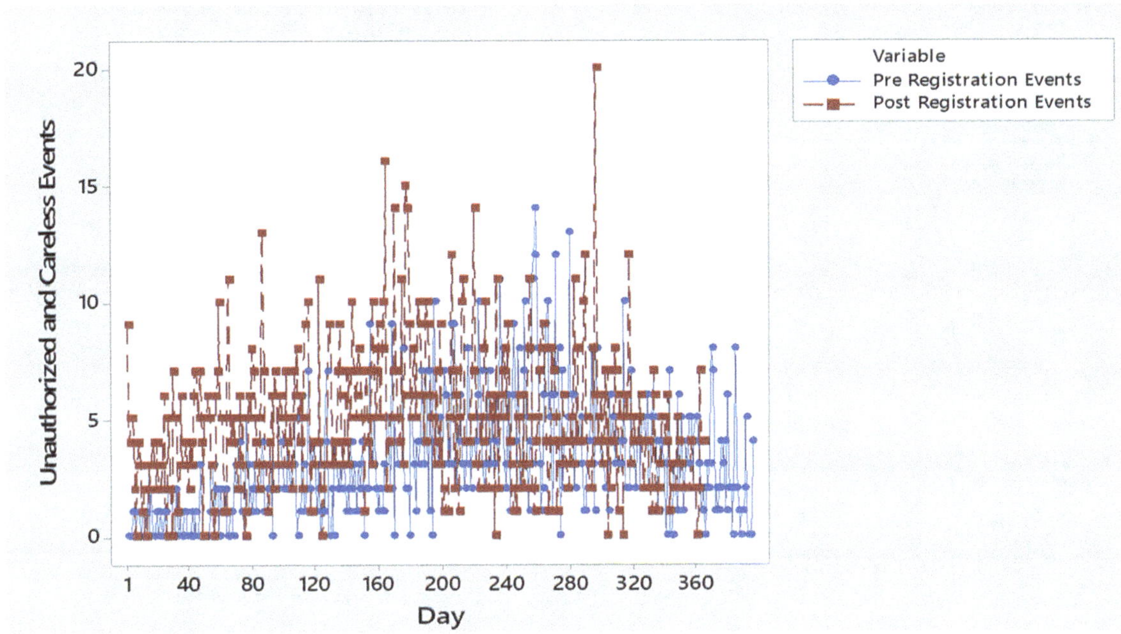

Regression Models for Population Growth

The regression models for the population growth of drone operators (hobbyist and commercial) developed in this research was based on published FAA Aerospace Forecasts for Fiscal Years 2017-2037. In Figure 3, the resulting model to predict population growth of drone operators in the United States is Y = - 441342 + 436.8 X - 0.1081 X^2. In this model, "Y" is the population and "X" is the year. As for accuracy, 99.61% of the variation can be explained by the regression model.

An additional fitted line plot of the quadratic regression is displayed in Figure 4 that shows the population growing by four times from the years 2016 to 2021. The quadratic regression model does have a high r-squared value of 99.36% indicating a good fit.

Lastly, Figure 5 shows the fitted line plot with 95% prediction intervals for the population growth. For example, in the year 2018 it can be expected to see the population of drone operators somewhere between 2.6170 and 3.2515 (millions) with a 95% prediction interval. Also, as can be seen in Figure 5 is that the prediction accuracy does decline as time increases.

Descriptive Statistics With Histogram(s)

Descriptive statistics and histograms were developed to provide an understanding of the event occurrence data. The number of days analyzed for unauthorized and careless events was 365 (N=365) for both pre-registration and post registration. As can be seen in Table 2 and Table 3, the mean has increased from 3.315 pre-registration to 4.811 events per day. Although the mean has increased, it should be kept

Figure 3. Regression summary report for population vs year

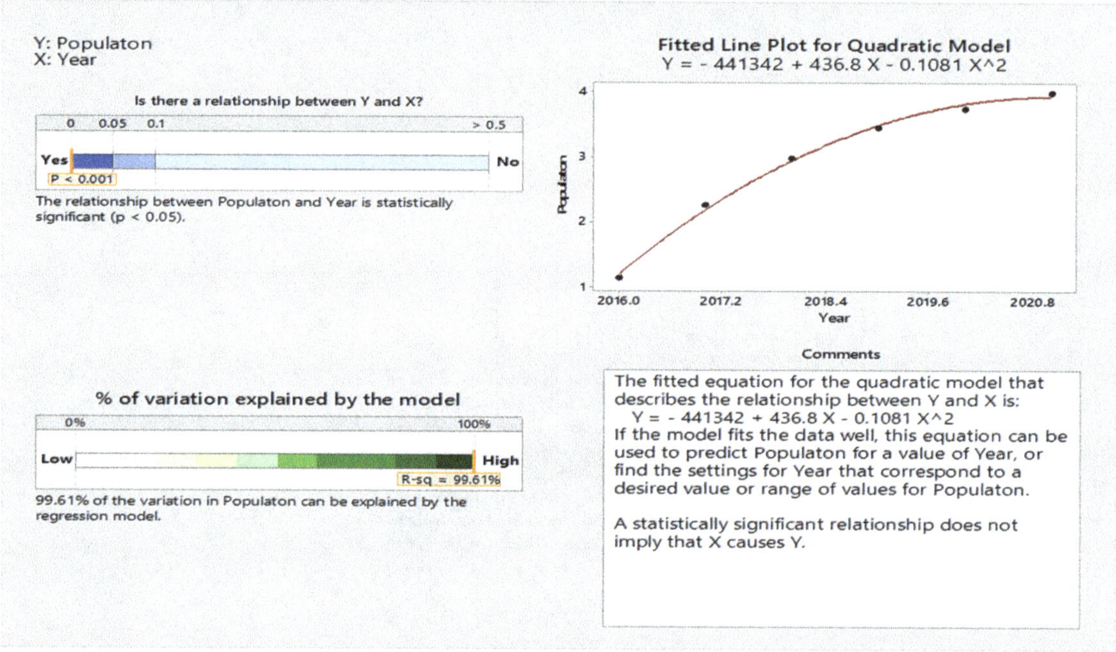

Figure 4. Fitted line plot for quadratic model

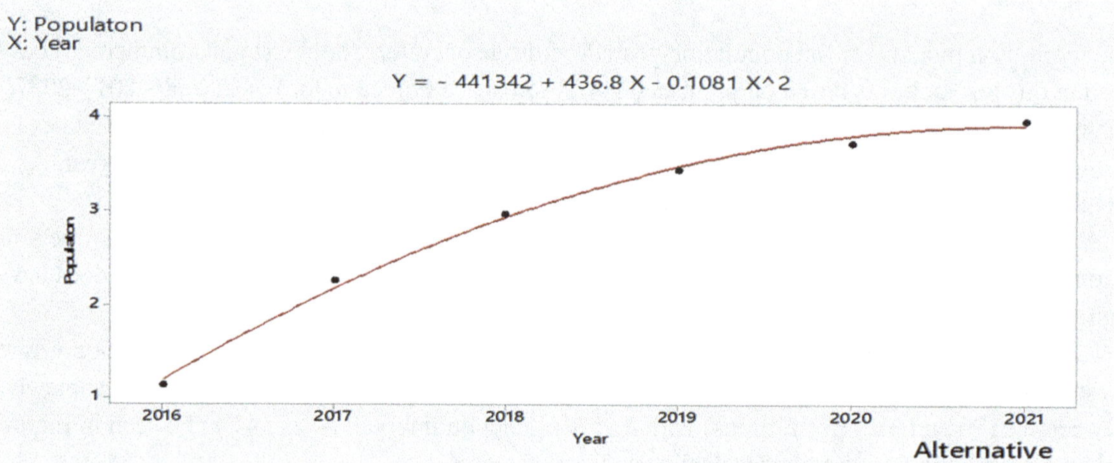

Statistics	Selected Model Quadratic	Alternative Model Linear
R-squared (adjusted)	99.36%	89.84%
P-value, model	<0.005*	0.003*
P-value, linear term	0.004*	0.003*
P-value, quadratic term	0.004*	—
Residual standard deviation	0.085	0.338

Figure 5. Regression for population vs year

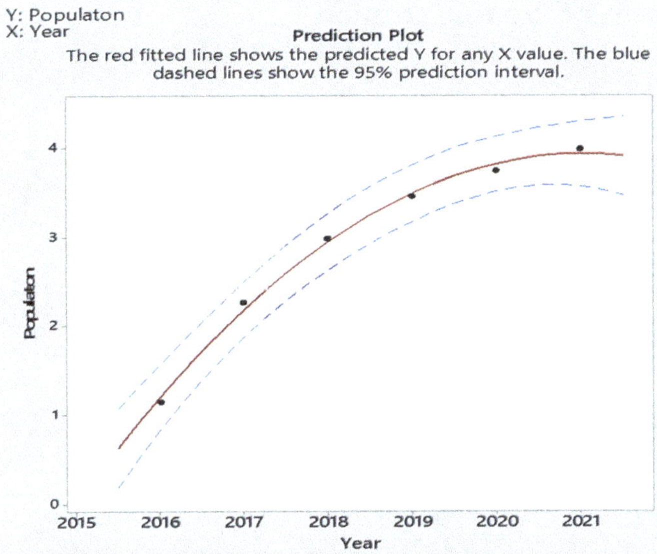

X	Predicted Y	95% PI
2015.5	0.62936	(0.18657, 1.0721)
2015.8	0.92063	(0.52098, 1.3203)
2016	1.1984	(0.83283, 1.5640)
2016.3	1.4626	(1.1222, 1.8031)
2016.5	1.7134	(1.3899, 2.0369)
2016.8	1.9506	(1.6368, 2.2645)
2017	2.1744	(1.8647, 2.4841)
2017.3	2.3846	(2.0752, 2.6940)
2017.5	2.5813	(2.2699, 2.8928)
2017.8	2.7645	(2.4501, 3.0789)
2018	2.9343	(2.6170, 3.2515)
2018.3	3.0905	(2.7713, 3.4096)
2018.5	3.2332	(2.9133, 3.5530)
2018.8	3.3623	(3.0432, 3.6815)
2019	3.4780	(3.1608, 3.7952)
2019.3	3.5802	(3.2658, 3.8946)
2019.5	3.6689	(3.3574, 3.9803)
2019.8	3.7440	(3.4346, 4.0535)
2020	3.8057	(3.4960, 4.1154)
2020.3	3.8538	(3.5400, 4.1677)
2020.5	3.8885	(3.5649, 4.2120)
2020.8	3.9096	(3.5692, 4.2500)
2021	3.9172	(3.5517, 4.2828)
2021.3	3.9114	(3.5117, 4.3110)
2021.5	3.8920	(3.4492, 4.3348)

in mind that the population of drone operators has also increased. Further examination of the remaining data shows an increase in the spread of events (standard deviation) and in increase in the quartiles.

For the distribution of events (Figure 6 and Figure 7), it can be seen that both pre-registration and post registration data exhibits a distribution that is skewed to the left. Of importance, it should be noted that there was a decrease in the frequency of days with zero events in the post registration group. Further examination reveals that there is a general shift to the right of the data after registration requirements were put in place. Population growth is likely the cause of the shifts in the histograms.

Table 2. Descriptive Statistics: Pre-Registration

Variable	N	Mean	StDev	Minimum	Q1	Median	Q3	Maximum
Pre Registration	365	3.315	2.566	0.000	1.500	3.000	4.500	14.000

Table 3. Descriptive Statistics: Post Registration

Variable	N	Mean	StDev	Minimum	Q1	Median	Q3	Maximum
Post Registration	365	4.811	2.985	0.000	3.000	4.000	6.000	20.000

Figure 6. Histogram of pre-registration

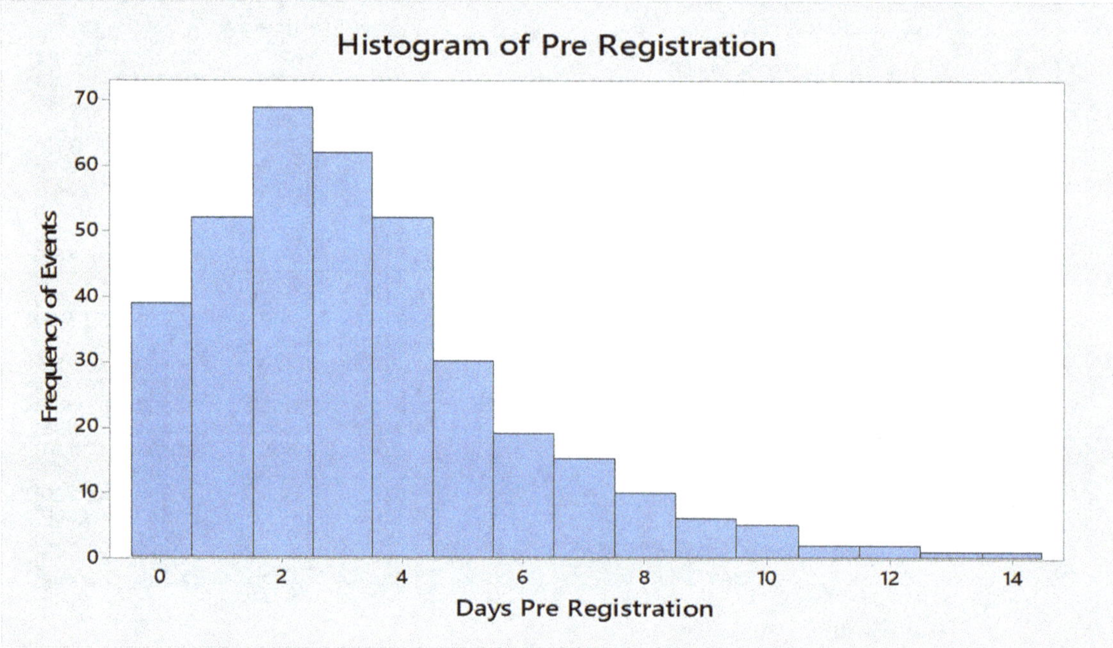

Figure 7. Histogram of post registration

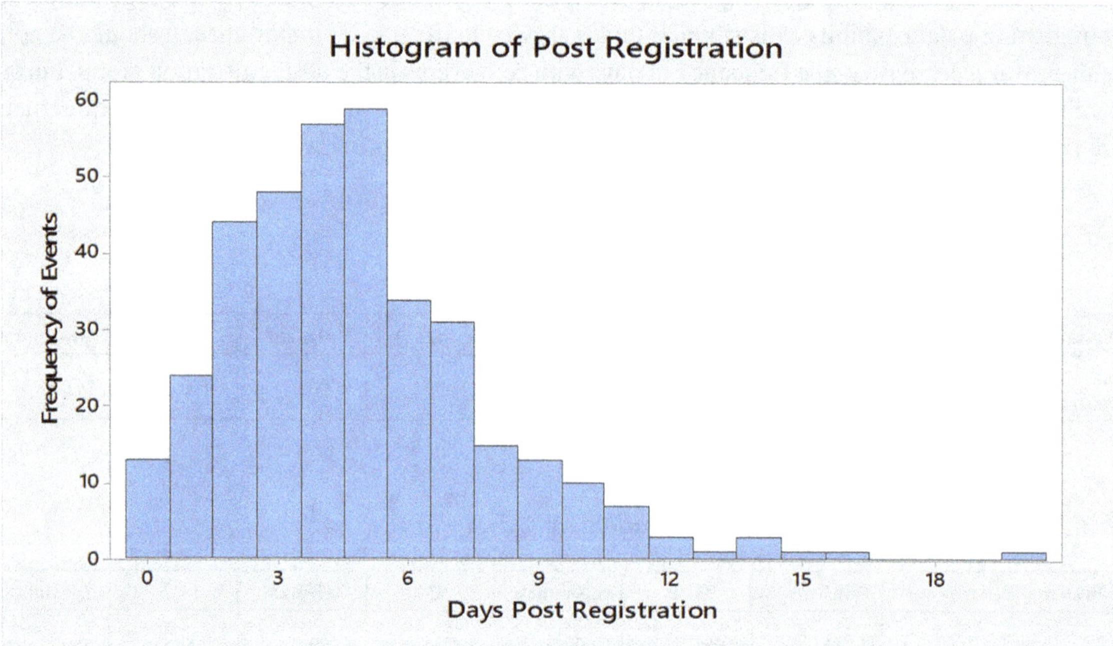

Events Per Million Operators Prior to Registration and Post Registration

Because there was a rapidly growing population of drone operators and an increase in events, consideration needs to be given when reporting an overall growth in events. To provide better clarity, events per million operators can be used to help provide context to the growth of events. Table 4 does show an increase of overall events, but when consideration is given to the population growth, events per million operators actually shows a decrease of events post registration requirements.

Chi Square Goodness of Fit Hypothesis Testing

The chi square goodness of fit test was conducted for each of the categories that resulted in 5 tests. The tests were used to determine if a significant difference existed in the proportion of events between registration requirements. The hypothesis below was either accepted or rejected for each test.

Null Hypothesis: The proportions of unauthorized and careless events do not differ with registration status.
Alternative Hypothesis: The proportions unauthorized and careless events differ with registration status.

The 5 chi square tests are presented in Table 5 below. For the calculation of the chi square goodness of fit:

- **Degrees of Freedom:** The degrees of freedom were equal to k -1 where k is the number of sub categories in each category.

DF = K - 1

- **Expected Frequency Counts:** The expected frequency counts of each sub category were equal to the sample counts times the expected proportion from the pre-registration reports.

Ei = NPi

- **Test Statistic:** The chi square (X2) test statistic was derived through the following equation.

$$X2 = \sum [(Oi - Ei)^2 / Ei]$$

Table 4. Events per million operators pre-registration and post registration requirement

Year	Events	Number of Operators	Events Per Million Operators
2015 (pre-registration)	1,210	630,000	1,921
2016 (post registration)	1,761	1,400,000	1,258

Table 5. Chi Square Tests

	Factor	Sightings By	Airspace Sightings	Mid Air Collisions	Ground Collisions	Civil Concerns
Registration Status	Chi Square Value	47.36562	6.65186	0.02252	4.02750	173.53848
	Degrees of Freedom	2	2	2	2	2
	Critical Value	5.99147	5.99147	5.99147	5.99147	5.99147
	Result	**Reject**	**Reject**	**Accept**	**Accept**	**Reject**

What can be seen in Table 5 is that the null hypothesis was rejected in the sightings category, airspace sightings category, and the civil concerns category. Both midair collisions and ground collisions had chi square values lower than the critical value and were accepted with no significant changes.

CONCLUSION

The population growth of drones operating in the NAS is growing rapidly. With this explosive growth, the average amount of unauthorized and careless events increased from 3.31 events per day prior to the registration requirements to 4.81 per day post registration requirements. We also saw a shift to the right in the distribution of events post registration. However, the amount of events per million operators decreased from 1,921 events per million operators to 1,258 events per million operators. The chi square goodness of fit tests showed significant differences exist in pre-registration versus post registration events in the categories of sightings, airspace sightings, and civil concerns.

Requiring drones to be registered appears to play some role in the proportionate reduction of unauthorized and careless operation in the NAS. The ability for authorities to trace a drone back to its owner is critical for education, accountability, and confidently knowing the number of drones in operation. The study has found no negative consequences when drones are required to be registered and supports the requirement of drone registration.

REFERENCES

Branum, M. (2016). Policy recommendation for integrating small unmanned aircraft systems with aviation facilities in North Central Texas [Doctoral dissertation].

Cameron, E. D. (2014). *Unmanned aircraft systems: Factors that affect the acceptance of unmanned aircraft usage within the United States national airspace system*. Eastern Michigan University.

Canis, B. (2015). *Unmanned aircraft systems (UAS): Commercial outlook for a new industry*. Washington: Congressional Research Service.

Clothier, R. A., & Walker, R. A. (2006). Determination and evaluation of UAV safety objectives.

DeGarmo, M. T. (2004). Issues concerning integration of unmanned aerial vehicles in civil airspace. Center for Advanced Aviation System Development. In *FAA Aerospace Forecasts Fiscal Years 2017-2037*. Retrieved from www.faa.gov/data_research/aviation/aerospace_forecasts/media/FY2017-37_FAA_Aerospace_Forecast.pdf

Federal Aviation Administration. (2017, July 31). Getting Started. Retrieved September 09, 2017, from https://www.faa.gov/uas/getting_started/

Federal Aviation Administration. (n.d.). ALC-42: Airspace, Special Use Airspace and TFRs. Retrieved September 09, 2017, from https://www.faasafety.gov/gslac/ALC/course_content.aspx?cID=42&sID=505&preview=true

Lee, B. R. (2016). Protection, profit, or privacy: exploring strategic solutions for integrating unmanned aerial systems (UAS) and the delicate balance between commercial opportunity and public safety [Doctoral dissertation]. Naval Postgraduate School, Monterey, CA.

Maida, N. (2016). The roles of unmanned aircraft systems in civilian, law enforcement, and military environments [Master of Science degree]. Utica College. Retrieved from http://search.proquest.com/docview/1836797492

United States Court of Appeals For The District Of Columbia Circuit. (2017, May 19). On Petitions for Review of Orders of the Federal Aviation Administration.

Operation and Certification of Small Unmanned Aircraft Systems; Final Rule, § 14 CFR Parts 21, 43, 61, 91, 101, 107, 119, 133, and 183 (2016).

Price, D. E. (2016). Unmanned aircraft systems for emergency management: a guide for policy makers and practitioners [Doctoral dissertation]. Naval Postgraduate School, Monterey, CA.

Registration and Marking Requirements for Small Unmanned Aircraft; Final Rule, § 14 CFR Parts 1, 45, 47, et al. (2015).

Van Blyenburgh, P. (1999). UAVs: An overview. *Air & Space Europe*, *1*(5-6), 43–47. doi:10.1016/S1290-0958(00)88869-3

Weibel, R. E., & Hansman, R. J. (2006). Safety considerations for operation of unmanned aerial vehicles in the national airspace system.

Wild, G., Murray, J., & Baxter, G. (2016). Exploring civil drone accidents and incidents to help prevent potential air disasters. *Aerospace*, *3*(3), 22. doi:10.3390/aerospace3030022

This research was previously published in the International Journal of Aviation Systems, Operations and Training (IJASOT), 4(1); edited by Massoud Bazargan, pages 54-66, copyright year 2017 by IGI Publishing (an imprint of IGI Global).

Chapter 25
Drones and Privacy

Nigel McKelvey
Letterkenny Institute of Technology, Ireland

Cathal Diver
Letterkenny Institute of Technology, Ireland

Kevin Curran
Ulster University, UK

ABSTRACT

Drones, also referred to as UAV's (Unmanned Aerial Vehicle), are an aircraft without a human pilot. Drones have been used by various military organisations for over a decade, but in recent years drones a have been emerging more and more in commercial and recreational capacity. The paper is aimed at drone and UAV technology capabilities and how they could and are currently effecting privacy laws globally in comparison to those currently in the Rep. of Ireland. Being investigated is the collection, retention and purpose of which civilian's information is being gathered. The authors also discuss the laws preventing the development and evolution of drone technology in the US in comparison to the Rep. of Ireland.

1. INTRODUCTION

In today's society the monitoring of flights has never been more regulated, and it would be difficult to argue the reasons why because in 2013 global terrorism rose by 43% alone (Ackerman, 2014). With terrorism on the rise and the use of drones in society, drones could become prime targets for terrorists and criminals, not necessarily to use the devices to crash into densely populated areas, injuring and potentially killing people, but to harvest the information on the drones to aid criminals and terrorists in cyber/digital crimes that could range from theft of personal bank details to the loss of highly sensitive military recon footage. With the FAA in the U.S. reporting that by 2020 there could be as many as 30,000 drones in the sky in the U.S. alone, concerns about privacy, security and safety could be justified (Waterman, 2012). If these projections are to be believed and drones are set to become a part of our everyday society then laws must be ratified for the use and policing of these machines specifically in civilian zones were human safety must be paramount.

DOI: 10.4018/978-1-5225-8365-3.ch025

Figure 1. Phantom advanced and pro-650-80 (TechRadar, 2015)

Companies like Google and Facebook started acquiring drone companies such as Titan Aerospace, an American based drone maker which specializes in high altitude long endurance flights, and Ascenta, a British based drone maker that also specializes in high altitude long endurance flights, in 2014 respectively, acquiring them in order to research develop drones that could potentially stay in the air for years without landing. These giants of the tech world are essentially internet companies, entirely dependent on we the consumer having access to the internet, which is the answer to the question. The more people they have using their systems, the more money they generate, as of the 2nd financial quarter in 2014 90% of Google's generated revenue came from Google websites or Google Member websites (invester.google.com, 2014), these companies are exploring the possibility of using these aforementioned drone companies to provide internet to regions in the world that currently have little to no access to the web. This could potentially be a great service even if it is just to increase profits for their companies, but the protection and confidentiality of customer information must be a priority for these projects, with drones from Titan Aerospace potentially having flight times as long as 5 years then user information privacy and security must take priority over even performance (Winter, 2014).

The track record of drones being hacked doesn't inspire confidence with Professor Todd Humphreys and his colleagues at the Radio Navigation Lab at the University of Texas demonstrated to the U.S. Department of Homeland Security how they could hack a drone, owned by the University, using a "spoofing" technique making the drone mistake their commands for those of the GPS satellite. Todd Humphreys and his colleagues were working off a budget of $1000, which is a nominal fee for something which could potentially cause catastrophic damage (BBC News, 2012). This same technique was used in late 2011 when a CIA guided recon drone had been hacked by enemy forces whilst flying over enemy territory in Iran. An engineer tasked with uncovering the drones stealth and security secrets claimed "The GPS navigation is the weakest point," when referring to the security of the drone (Peterson, 2011). The drone was hacked and landed by enemy forces with it containing very sensitive information. The DHS has since dismissed these accusations and claimed that it was a technical fault in the drone that forced it to crash land in enemy territory. If the accusations are true this would portray commercial drones in a bad light if state of the art military drones can be hacked so easily. Friday 26th of September 2014 is

Table 1. Different categories of drones on the basis of weight (European Parliament, 2015)

Uses	Type of operator	Targets, examples
Infrastructure protection, monitoring and safety / security inspections	Commercial State	Objects: - Transport (rail tracks, highways, bridges, traffic) - Energy (nuclear plants, dams, dykes, power grids, wind turbines, pipeline and power lines inspection) - Communications (mobile phone towers) - Industry (industrial installations) People: - monitoring unauthorised entry
Geo-spatial mapping	Commercial State	Objects: Mapping and surveying exploration, planning and crisis management
Environment monitoring	Commercial State	Objects: Air, water and other natural resources: pollution monitoring, hazardous material sensing, air/water quality testing, weather monitoring
Precision agriculture	Commercial Private individuals	Crop, animals: Crop and herd inspection, crop spraying to apply pesticides
Law enforcement, surveillance and monitoring of individuals and of people and of electronic communications	State (law enforcement) Commercial (sub-contractors)	Persons: Infrastructure protection against threats and illegal actions, targeted criminal investigation, crowd and public event monitoring, border control/protection, anti-social behaviour, supporting police response Geo-location, interception of communications and of electronic devices, profiling
Civil protection	State (law enforcement, civil protection authorities)	Objects, persons: Infrastructure monitoring, disaster relief and response, search and rescue, firefighting, hazard detection, crisis response
Regulatory enforcement	State (law enforcement, other authorities)	Pollution monitoring, fisheries monitoring, monitoring for illegal logging, wildlife protection and hunting regulations, etc
Journalism, media, film-makers	Journalists, camera-crews, film-makers	People and objects: Live journalistic reporting, investigative reporting, documentary filmmaking, promotional videos, fictional filmmaking
Electronic communications providers	Commercial (service providers)	Objects: Telecommunication and computing devices
Hobby, leisure	Private individuals	Objects and persons

the first time that a drone has been given clearance to be used in Europe in a commercial capacity, the goods delivery company DHL are using the drone to deliver medical supplies in Germany to an island that is only accessible by road (DHL, 2014).

The drone sector in Ireland is legislated by the Irish Aviation Authority, this authority is responsible for laws in terms of who, where and what people can fly planes, helicopters, drones, etc. in Ireland. The IAA regulations of drones state that drones cannot fly no higher than 150m or 500m from the drone control station, thus limiting the real application of long range drones in Ireland. In 2013 the IAA had issued 13 permits for the legal use of drones in Ireland, in total in as little as a seven month period this number almost doubled to 22 permits as of July 2014 highlighting the fact that the drone technology is becoming more and more apart of Irish society (McMahon, 2014). Currently in Europe the policing of drones is done by the state that the drone is present, but the European Aviation Safety Authority is in the process of drafting and introducing pan-European legislation for the legislating of drone use with a weight greater than 150kg with anything lower continuing to be legislated by individual states (McMahon, 2014). Depending on the type of drone, some drones must be compliant to the Irish Data Protection Act because of their data collection capabilities which include photography, video recording, thermal imaging etc. Table 2 endeavours to articulate some of the characteristics of drone technologies which must be considered by law makers:

The aim of this paper is to examine and research to what extent the Irish DPA act provides the right to privacy of Irish citizens in relation to the increasing introduction of drone technology into society. The Irish laws in relation to privacy will be examined in comparison to those currently in the United States of America and the United Kingdom. As the drone/UAV market rapidly expands and develops, many countries laws are out dated in relation to adequately protecting people's privacy in relation to the capabilities of these machines. As technologies advance, drones will become more and more part of everyday life and as this becomes the case, laws need to be ratified to prevent Ireland from gaining a big brother like society like the United States of America for example.

2. COLLECTION OF DATA

In the United States of America the drone market has expanded greatly in a military and a commercial capacity. Law enforcement agencies nationwide utilize drone technology to aid them in the prevention of crime and the apprehension of criminals. In the US state of Maryland traffic drones are in use to scan the licence plate number on vehicles travelling the major routes in the state. The information collected by these drones is used in relation to the apprehending of criminals possibly involved in crimes such as grand theft auto, but as this technology is being used for the benefit of the state and crime prevention, this technology also infringes on the privacy of innocents using those same routes as suspected criminals. With figures estimated that every 1,000,000 licence plates scanned in Maryland by a traffic drone, on average only 47 vehicles out of that 1,000,000 are suspected to be related to criminal activity. That makes the percentage of the data related to crime collected by the Maryland Police Department using traffic drones at an extreme low of 0.0047%, and those figures are reinforced with stats from Rhinebeck New York of 0.01% and 0.08 from High Point from North Carolina (Cushing, 2013). These figures would make this technology difficult to legally implement in the Rep. of Ireland because of the DPA, the stats from the use of this technology prove it to be excessive and that would infringe upon the Data Protection Act 1988 (amended 2014) Section 2(1)(c)(iii), this law dictates that the data being collected is "adequate, relevant and not excessive". The excessiveness does not relate to the scanning alone, the storing of the data collected is the most excessive part of the process. By the Irish Data Protection Act Section 2(5)(b) a data subject must be informed what s/he is being monitored for what purpose but when drones hover above highways recording video, collecting licence plate numbers etc. that no prompt is given to the data subjects they are not made aware as would be the requirement.

Table 2. Drones civil uses, operators, targets and examples of use (European Parliament, 2015)

Type on the basis of weight (MTOM)	Current uses and future potential uses	Description; Types; Price and diffusion	Regulation
Small (0-20 / 25 KG)	- Leisure use and commercial use (surveillance and inspection, photography)	- drones below 2 kg are also called micro-drones and are quickly developing - hundreds of different types; normally multi-rotor or fixed wing aircrafts, guided by GPS, live video streaming camera, - Price: 140 - 28.000 Euro Some available in shops (below 1 kg) - Take-up: those below 2 kg are very widespread	- Falls under MSs regulations
Light (20/25 - 150KG)	- geospatial surveying, wide-area surveillance - Potential to inspect pipelines/power cables, spray crops, search and rescue, border surveillance; forest fire monitoring	- Typically longer range, fixed-wing, B-VLOS, reaches altitudes of 3000 meters - ex: Luna, Hermes 90 - Price: 55.000 - 420.000 Euro	- Falls under MSs regulations
Large (>150KG)	- used by the military and defence - Potential for future cargo (and passenger) transport	NATO classifications: - Class II (150-600 kg): Sperwer, Hermes 450, Watchkeeper; - Class III (>600 KG): MALE - medium altitude, long endurance: Predator, Heron, Hermes 900[6] HALE - high altitude, long endurance: Global Hawk UACVs - strike or combat UAVs: MQ9-Reaper/Predator B Price: 670.000 Euro and above	- Falls under Regulation 216/2008/EC (EASA Regulation): EASA airworthiness certificate, unless operated by a State agency

2.1. Retention of Data

The Minnesota State Police Department delete the data collected within 48 hours of collection which is reasonable, but this is an exception. The majority of other agencies delete dates generally range from 90 days to 5 years. The most alarming fact wasn't the possible 5 yearlong delete date but the response from 3 Texas law enforcement agencies that they have no specified delete date, it could be presumed that the retention of the data gained from traffic drones in Texas is being stored indefinite (Cushing, 2013). This raises greater questions regarding all these agencies affiliation to the US-EU Safe Harbour Framework, with law enforcement agencies in the US subject to the Framework (Export.gov, 2014). In November 2013 the European Commission issued a press statement regarding its stance on the large scale U.S. intelligence agencies collection programmes that were being used to collect citizen's data without their knowledge, with the statement also stating that government agencies have been "*spying*" on U.S. citizens, companies and leaders. The Edward Snowden case was a factor in what prompted the European Commission's statement, with the revelations of that particular case seriously damaging the

Table 3. EASA Concept of Operations for Drones (European Parliament, 2015)

Categories of operations Regulatory regime	1st category: Open	2nd category: Specific	3rd category: Certified
Safety risks and risk level? taking "into account: mid-air collision with manned aircraft[15]; harm to people; and damage to property in particular critical and sensitive infrastructure"	Lower risk (Minimal aviation regulatory system, defining limits for operations, to be overseen by the police, "as for cars for instance")[16]	Medium risk	Higher risk (similar to normal manned aviation)
Operation authorisation by a Civil Aviation Authority for the flight?	No (Even commercial operations)	Yes - operator to perform a safety risk assessment with mitigation measures, addressing the airworthiness, operating procedures and environment, competence of personnel and organisations, airspace issues - to be reviewed and approved by the CAA through an "Operations Authorisation", with the support of the Qualified Entities as defined in the EASA	Yes - for the moment, for drones of 150 kg or more, but in the future kinetic energy, type of operation and complexity of the drone (notably autonomy), should be examined/defined - Certification required: Type Certificate (environmental, airworthiness, noise), design and production approvals,

trust between the E.U. and the U.S. in relation to data protection (European Commission, 2013). This is an example that could only enforce the argument that the data being collected by traffic drones is being stored indefinitely. The indefinite retention of the data generated by these traffic drones breaks the US-EU Safe Harbour principle Notice, that all individuals must be notified about the purpose for which the data is being collected. With traffic drones flying anonymously in the skies above highways, the Notice section in the US-EU Safe Harbour Principles not being adhered too. With the data being collected by drones being possibly stored indefinitely then the continuous monitoring of certain routes by drones or UAV's and the continuous use of similar routes by citizens then this could be used to track the location and travel habits of people without their consent or knowledge. The purpose of these traffic drones is to monitor traffic from above and process the data gained to aid law enforcement agencies with the locating and apprehension of people who have stolen cars, who have not paid there tax etc. but the continuous collection and retention of innocent citizens details, with the majority of them not aware that it is happening, may be legal in the USA but it would break Irish Data Protection laws should it be the implemented in Ireland. The purpose of this technology is to apprehend criminals, that is its purpose, meaning the retention of innocents details would infringe on the Data Protection Act 1988 (amended 2014) Section 2(1)(c)(i) "*the data shall have been obtained only for one or more specified, explicit and legitimate purposes*", this would make the technology being used difficult to implement in Ireland, the adjustment would need to be made that the only details being stored are the ones who are being flagged as crime related.

2.2. Privatized Collection of Data

In the USA a company called Vigilant has been collecting and retaining licence plate details. The company sells the data collected to over 2,200 law enforcement agencies in the country. The company has a shared database with another company called Digital Recognition Network, in this database there is estimated to be over 1.8 billion civilian records (Segar, 2014), one of the company's biggest contracts is with the Department of Homeland Security, providing citizens data straight to US government forces. The collection and distribution of this data in Ireland would break Irish Data Protection Laws, Data Protection Act 1988 (amended 2014) Section 2(1)(c)(i), prohibits the processing of citizens data without the without his or her consent. But in the U.S this is occurring on a daily basis mostly without the data subject having knowledge that they are being monitored. With companies like Vigilant and Digital Recognition Network having such a vast database of vehicle registrations and car owner details being collected in there thousands per day then the data being recorded and archived from months ago in fact becomes out dated. With figures estimated to be over 250 million used cars in the USA as of February 2014 and with approximately over 40 million of these cars titles being transferred that would make the data being stored by these companies out dated and in relation to this in Ireland it would break the Irish Data Protection Act Section 2(1)(b) that the data collected *"shall be accurate and, where necessary, kept up to date"*. The data could be run in conjunction with the US's DMV to keep track of car transactions and keep the data up to date and current but access to DMV databases are so restricted that law enforcement agency employees must sign contracts before they are allowed access so private company access could be difficult.

2.3. Anonymous Collection of Data

Drones are a useful tool in terms of being implemented in situations where a human's life would be at risk. But as people are identifiable by their own set of unique characteristics such as height, weight, ethnic origin, hair colour etc. Drones on the other hand are not as easy to identify. Without company markings and patented designs drones are notoriously hard to identify the user/operator of the drone. In France a spy drone was spotted flying in a restricted fly zone, the no fly zone contained a nuclear power plant. Originally the drones appearing in the area where originally dismissed as drone enthusiasts in the area who had simply strayed into a restricted zone by accident and without any malice intent but as time as passed it became apparent that these drones where not being controlled by simple drone enthusiasts but people with an agenda with over 13 confirmed drone sightings by Electricite de France SA, which runs the nuclear power grid in France, around French nuclear power stations and over 30 unconfirmed over a 2 months period beginning in October (Patel, 2014). Since the revelation of these drones have been spotted the media has been speculating to what purpose are these drones being used, with some media outlets speculating but none have any foundation. France is the most nuclear power dependant developed country in the world with nuclear power providing over 80% of the countries power (Lichfield, 2014). Greenpeace has voiced its opinions that the French government is trying to play down the risk of these events, even speculating that the drones could be implemented to perform acts of terrorism, stating "*A medium-sized drone could carry an explosive charge big enough to damage the pools [of cold water] in which spent nuclear fuels are stocked*" a spokesman said (Lichfield, 2014). Regardless of who or what these drones are being used for they appear to be collecting data for some purpose without permission, a drone being used in this capacity in Ireland would infringe upon the Irish DPA Section 2 (a)(1)

(a) and Section 2 (a)(1)(c)(iii). The example above is excessive but proves that the collection of data, video footage, photograph etc. by people implementing drone technology is extremely hard to prevent. Numerous drones have been flying around highly policed nuclear power plants in France but little can be done to prevent it, the plants have technology to detect UAV's in the plants airspace but beyond that there is little to nothing that law enforcement agencies can do without catching the pilots of the drones, which in the case of the drone power plant controversy has proofed difficult with three arrests made and zero convictions (Farivar, 2014).

2.4. Limitations

In terms of limitations to the research into how drone technology could potentially infringe on the Irish Data Protection Act there are many, globally drones are being implemented daily by companies and governments but outside of those organizations the exposure to what is being collected or generated by drone technology is very limited. If companies or governments are using drones in an unlawful or unconstitutional manner then these companies are not going to make the applications of these machines public knowledge in order to protect the interests, reputation and standing of an organization. These drones have the capabilities to collecting data from car speeds to human biometric data, but without the actual drone or control centre people have no way of knowing what is being collected.

Another limitation is that once the data has been collected that the user has no idea what is being there data is being used for ranging from the data being used to prevent crime to selling the data to companies to target people with advertising campaigns. The possibilities of what people's personal information and data is almost endless, but with the use of drone technology this data which can be so valuable can also be so accessed with such ease and anonymously. In terms of limitations of drone technology in Ireland, drones are still an emerging technology in Ireland. The drone sector is growing but in comparison to countries like the UK where as of October 2014 there was 359 drone permits issued up to that date (Quinn, 2014), when compared to the 22 that had been issued in Ireland as of July 2014, Ireland is lagging behind in terms of the use of drones with UK law enforcement agencies implementing drones and drones also being used more frequently in a recreational capacity. The population difference between the two countries has a factor but it also must be considered that in the UK drones are being used to their full potential of the technology using them in the police while in Ireland they are being implemented to make films and for enthusiasts flying them in a field. The industry in Ireland will grow, it will grow globally according to aerospace and defence research company Teal Group Corp. that the drone industry will become an \$87 billion industry by 2023 (Chow, 2014) but currently Ireland which is regarded as the silicon valley of Europe, is lagging behind in one of the most emerging technologies in the world currently (Weckler, 2014).

As the field of drone technology continues to expand so does the accessibility and availability of drones. Drones can be purchased for as little as £25 making them affordable to the vast majority of the population. The majority of adults in the Rep. of Ireland could afford to purchase the £25 drone, a drone can be next day delivered to a customer's door. Such accessibility raises concerns in regards to that anyone can, using the proper drone, collect information on another person of their choosing at their will, they can do it anonymously without the person even being aware that they are being monitored. As companies can be audited and are by law supposed to adhere to Irish DPA laws they are bound and restricted to what they can do, collect, use and keep because there drone usage must be documented but with civilians having the access to the same technology and not being monitored under the same scope as companies it exposes a frightening fact that someone could spy on another person without their knowledge. People are subject to the Irish Data Protection Act the same as companies and governments but individuals are a lot more difficult to police. An example of this would be the drones spotted around the French nuclear power stations. The identity of the controller of these drones as of December 2014, almost 3 months from when they were first spotted, is still unknown as is for what purpose these drones being used at those plants. The data gained from these drones is an example of the one of the biggest limitations to drone technology and that is transparency. With independent people implementing drone technology creates the massive problem of people's personal data being collected without people's knowledge and without rules or laws being enforced because the law enforcement doesn't know who it may need to be enforce.

2.5. Privacy and Security

"It is now established by case law that the personal rights of the citizen guaranteed by Article 40.3.1{o} of the Constitution include a right of privacy" (Lawreform.ie, 1996). The right to privacy is a civic right which is applicable to all citizens. In the USA, the supreme court has made various rulings promoting the right to human privacy, an example is the recent ruling that cell phones cannot be searched without a warrant (Liptak, 2014), all enforcing the right to human privacy. But as the US Supreme Court ruled to promote human privacy another bill that was filed was not even given a committee hearing, that bill was drafted by Senator Jim Tomes, a Republican from Wadesville, Indiana. The bill was aimed at making it illegal to use a drone covertly. By not even hearing the proposal for this bill the US government are maintaining their ability to survey citizens without their knowledge. Whether the government is watching citizens or not there want to have the ability to do so is a great cause for alarm, it is a great invasion of personal privacy. The advancement in commercial drones has been rapid with companies like the aforementioned DHL putting their new delivery drones into practice, companies in the US have had the door firmly shut with the banning of drone use for commercial purposes in 2007. The U.S. government are doing some back tracking now with new legislation on the use of commercial drones in the U.S. with them being possibly granted legality as early as September 2015 (IBNLive, 2013). Amazon, like DHL, have been developing their own delivery drones, with Amazon talking about the ability to have delivered orders to customers within 30 minutes after purchase according to Amazon CEO Jeff Bezos (IBNLive, 2013). This would effectively change the world of retail shopping for good, but will face its own problems also. The U.S. is notorious for it's no fly zones (Mapbox, 2014), for example in Washington D.C. large portions of the city are no fly zones, how could a company like Amazon operate or deliver to customers via drones who happen to live in these areas. There seems to be a lot hurdles for commercial drones before they can be made legal in the U.S. and there will have to be a lot legislation and laws for the use of these drones, there will be a lot of work to be done before they are legalized by the estimated time frame of September 2015. These drones being used for surveillance in a commercial capacity could become the next phone hacking scandal, these drones used for media purposes has limitless possibilities whether it be monitoring the situation in a war zone or stalking a celebrity, with the latter being very intrusive and potentially infringing on a person right to privacy. With drones likely to be put into commercial and more so in military use, all uses seem to infringe on all definitions of privacy whether that is with proper reasoning or justification and without.

3. THE FUTURE

Drones are here to stay in our society and are only going to become more involved with some countries around the world currently writing laws to legitimize the use of drones in society for example the aforementioned U.S expects legislation in place for September 2015 (IBNLive, 2013). According to British security technology company BAE Systems, drones may soon become invisible. BAE's project 'Taranis' is the development of a drone, which has been described as a 'Smart Multitasking Drone' that cannot be seen or tracked by radar thus making it invisible. Another feature of the drone being developed by the is the ability to develop its own flight patterns with the only human input being the destination, the drone then determines the how the drone flies, like speeds, routes and altitudes (BBC News, 2014). BAE has also made its concept of potential drone with a 3D printer in the drone. The 3D printer would be capable of producing parts, equipment and even mini drones as needed whilst on missions, the company predicts the first time we may see this drone could be the 2040's (BBC News, 2014). Drones enable war to be fought with minimal risk to your countries soldiers by putting drones rather than soldiers in the danger zones, the US government has predicted that by 2030 that 1 in 4 combat soldiers for the US military will be drones/robotic soldiers (Atherton, 2014). According to aerospace and defence research company Teal Group Corp. that the drone industry will become an $87 billion industry by 2023 (Chow, 2014).

The following is a partial list of U.S. companies that have received venture funding (McNeil, 2014):

- Airware - $40.4 M
- 3D Robotics - $35M
- Skycatch - $19.7M
- Crescent Unmanned Systems $250,000

The following is a partial list of U.S. companies that have received crowd-funding (McNeil, 2014):

- Flexbot - $500,000 with 4,670 funders
- Airdroids - $929,212 in pledged money
- PowerUp - $1.23M

Drones are currently being developed to help people in disaster zones, Harvard-MIT Division of Health Sciences has recently been awarded a $100,000 grant for its drone development project that will be capable of delivering vaccines and other critical medical supplies (Chow, 2014). Around 60% of the world currently does not have access to the internet (Internetlivestats.com, 2014). As mentioned before Google and Facebook are currently developing drones to deliver internet access to those who currently have not. Inaccessible terrain and underdeveloped technological infrastructure are some of the problems that these drone projects being developed by these companies, these Google and Facebook projects could potentially solve the internet access problem on a global scale (DeBord, 2014). Another application of drones is in archaeology, in April 2014 a group of archaeologists in New Mexico, USA, used a drone equipped with a thermal image camera to monitor a site from above. The drone images produced enabled the archaeologists to view a thermal image of the desert floor and in turn revealing never-before-seen structures in an ancient Native American settlement (Gannon, 2014). Drones have been in use in the field of agriculture for over 20 years now in Japan for seeding and crop dusting, but with further advancements drones they are expected to be used to monitor livestock like cattle and sheep for example and crops, detect deficiencies and diseases, determine plant ripeness and schedule harvesting. They can also spread poisoned bait for vermin which may be interfering with or harming crops, such as fire ants, mice and rats with minimal environmental impact as well as little crop disturbance (Szondy, 2013). Drones could even be in line to replace waiters, Japanese food chain Yo Sushi have begun tests on their quad-copter drone. The drone simply flies from kitchen to table delivering customers' orders, they have dubbed the quad-copter the "iTray". Yo Sushi is currently testing its drone in its Soho location in London but plans to apply the iTray to all its restaurants by the end of next year (Fincher, 2014).

Ireland's future in drone technology development is estimated to be a country at the head of the technology. Ireland, in comparison to the stringent drone laws in the US, has relatively relaxed laws in regards to the policing of the use of drones and this is one reason why entrepreneurs in drone technology are targeting and setting up drone companies in Ireland. Jay Bergman, a co-founder of the taxi ordering app company Hailo, stepped down in September 2014 from the taxi company and has since started a company called Cara which will develop and test drone technology (Weckler, 2014). Bergman was an integral part of a meeting in November with representatives of law firm Mason, Hayes and Curran, former junior minister Ciaran Cannon, the manager of the IAA and representatives from the IDA and more, the agenda of the meeting was *"to make Ireland the drone capital of Europe"* (Weckler, 2014). Taoiseach Enda Kenny has enlisted expert in drone technology including Bergman to prepare a template that could be used to inform new legislation for the policing of drone's in the future in Ireland whilst making Ireland an attractive location for drone companies. The key for Ireland according to strategists is to gain first-mover advantage in making Ireland attractive to companies with drone projects (Weckler, 2014). Plans to make Ireland attractive to drone technology companies are at such an advanced stage that County Mayo in the West of Ireland has been earmarked as one possible area for test sites for drone companies while another possible location is Galway Airport (Weckler, 2014). A recent commission study predicted that in the next 2 and a half years that industrial development in drones will reach 14 billion and this in turn will create 70,000 jobs, this is why the Enda Kenny and the Irish Government are trying to make Ireland the most attractive place for drone companies to invest this 14 billion and take advantage of the vast amounts of IT qualified people in the country (Weckler, 2014). The government's eagerness to capitalise on the lack of freedom in competitor countries when it comes to drone development has both positive and negative aspects. One such positive is the possible creation of thousands of new jobs but one aspect that may be being over looked is that with the lax attitude being taken to the development of drones in Ireland that citizens privacy may be being compromised to enable the creation of jobs and the enhancement of the already strong IT sector currently in Ireland.

In terms of further work on the protection of citizen's privacy in the United States, the US Government are currently drafting laws to legislate the use of drones in the US. these laws are scheduled to be complete and proposed by September 2015 but there are some reports indicating that the draft of these laws are behind schedules and will not meet the set date of September 2015 with news reports suggesting that the laws may slip from the mandate deadline until 2017 or even later because of the agency also faces significant technological, regulatory, and management challenges and these problems are inhibiting the progression of the proposed bill (O'Kane, 2014). Currently companies waiting for these laws to be ratified are growing frustrated with the slow progress as the bill was originally put into motion in 2012 (O'Kane, 2014). Global tech giant Amazon has in December threatened to move its drone development research abroad because of the current restrictions, the company is already testing delivery drones in Cambridge, England because of even test flights of drones are illegal in the US (Vincent, 2014).

5. CONCLUSION

In conclusion all evidence points to the fact that drones are only going to become more and more part of everyday life as we know it (Market Research Media, 2013), in order for these machines to be successfully integrated into our society, the transition must be gradual and well marshalled, with laws yet to be drawn up yet. Drones have many strong points and improve many fields of profession, whether that be archaeology, agriculture, retail etc. the possibilities are endless for the progression of drones in society. But the advancements of drones also has its negative effects with drones potentially doing the work that humans have been previously used to perform these tasks, then you also have the moral dilemmas of drones being used by the military for the assassination of terrorists. Also with drones possibly taking peoples jobs for cost and reliability reasons, drones may potentially be portrayed in a negative light but drones will work alongside people taking people out of harm's way whether that be in a military sense or a rescue sense. Drones doing the jobs we should not have to do, can only help society and lower job risks. As drones evolve and become an everyday part of our lives, drones will eventually greatly benefit the way we live and enhance our quality of life.

Technology never stands still, if you buy a laptop today in five years' time the odds are that the laptop's hardware and software would be out dated in comparison to the present times standards. Like technology the drone industry is and will continue to advance as it has done in the past. The drone industry has advanced in the past from military machine to commercial aid demonstrates the range of uses drones have and this expected to continue with drones now even being used to aid in the selling of homes and property in Canada (CBC News, 2014).

REFERENCES

Ackerman, S. (2014). Global terrorism rose 43% in 2013 despite al-Qaida splintering, US reports. The Guardian. http://www.theguardian.com/world/2014/apr/30/global-terrorism-rose-despite-al-qaida-splintering

Atherton, K. (2014). Robots May Replace One-Fourth Of U.S. Combat Soldiers By 2030, Says General. Popular Science. http://www.popsci.com/article/technology/robots-may-replace-one-fourth-us-combat-soldiers-2030-says-general

CBC News. (2014). Drones helping sell Canadian homes and property. Cbc.ca. http://www.cbc.ca/news/canada/windsor/drones-helping-sell-canadian-homes-and-property-1.2864042

Chow, D. (2014). The Future of Drones: Sky-High Hopes vs. Regulatory Realities. LiveScience.com. http://www.livescience.com/45242-future-of-drones.html

Chow, D. (2014). The Future of Drones: Sky-High Hopes vs. Regulatory Realities. LiveScience.com. http://www.livescience.com/45242-future-of-drones.html

Cushing, T. (2013). ACLU Presents Its Findings On License Plate Scanners, Most Of Which Is Bad News | Techdirt. Techdirt. https://www.techdirt.com/articles/20130717/16592323842/aclu-presents-its-findings-license-plate-scanners-most-which-is-bad-news.shtml

DeBord, M. (2014). You Won't Even Know These Facebook Drones Are In the Sky. Business Insider. http://www.businessinsider.com/you-wont-even-know-these-facebook-drones-are-in-the-sky-2014-9

Dhl.com. (2014). DHL. http://www.dhl.com/en/press/releases/releases_2014/group/dhl_parcelcopter_launches_initial_operations_for_research_purposes.html

European Commission. (2013). European Commission calls on the U.S. to restore trust in EU-U.S. data flows. http://europa.eu/rapid/press-release_IP-13-1166_en.htm

European Parliament. (2015). Privacy and Data Protection Implications of the Civil Use of Drones. http://www.europarl.europa.eu/RegData/etudes/IDAN/2015/519221/IPOL_IDA%282015%29519221_EN.pdf

Export.gov. (2014). Export.gov - U.S.-EU Safe Harbor Overview. http://www.export.gov/safeharbor/eu/eg_main_018476.asp

Farivar, C. (2014). Not the drone pilots you're looking for: Trio arrested, released. Ars Technica. http://arstechnica.com/tech-policy/2014/11/not-the-drone-pilots-youre-looking-for-trio-arrested-released/ [Accessed 8 Dec. 2014].

Fincher, J. (2014). Yo Sushi's "flying waiter" serves your meal via quadcopter. Gizmag.com. http://www.gizmag.com/yo-sushi-itray-quadcopter/27910/

Gannon, M. (2014). Drone Images Reveal Buried Ancient Village in New Mexico. LiveScience.com. http://www.livescience.com/44679-drone-images-reveal-buried-archaeological-ruins.html

IBNLive. (2013). Amazon's delivery drones raise security and privacy concerns, may face legal challenges. http://ibnlive.in.com/news/amazons-delivery-drones-raise-security-and-privacy-concerns-may-face-legal-challenges/437444-11.html

Independent News. (2011). Nowhere to run: drones, facial recognition, soft biometrics and threat assessments - End the Lie - Independent News. http://endthelie.com/2011/09/30/nowhere-to-run-drones-facial-recognition-soft-biometrics-and-threat-assessments/

Internetlivestats.com. (2014). Number of Internet Users (2014) - Internet Live Stats. http://www.internetlivestats.com/internet-users/

Investor.google.com. (2014). 2014 Financial Tables – Investor Relations – Google. https://investor.google.com/financial/tables.html

Lawreform.ie. (1996). The Law Reform Commission Consultation Paper Privacy: Surveillance and the Interception of Communications. http://www.lawreform.ie/_fileupload/consultation%20papers/cp-privacy.htm

Lichfield, J. (2014). French government on high alert after unexplained drone flights over nuclear power stations. The Independent. http://www.independent.co.uk/news/world/europe/french-government-on-high-alert-after-unexplained-drone-flights-over-nuclear-power-stations-9850138.html

Mapbox, (2014). Don't fly drones here. https://www.mapbox.com/drone/no-fly/

Market Research Media. (2013). U.S. Military Unmanned Aerial Vehicles (UAV) Market Forecast 2013-2018. http://www.marketresearchmedia.com/?p=509 [Accessed 22 Oct. 2014].

McMahon, A. (2014). Aviation body has issued 22 drone permits. Irish Times. http://www.irishtimes.com/news/technology/aviation-body-has-issued-22-drone-permits-1.1863750

McNeil, B. (2014). The top five things you need to know about Drones and GIS. http://www.directionsmag.com/entry/top-five-things-you-need-to-know-about-drones-and-gis/414810

News, B. B. C. (2012). Scientists 'hack' flying drone. http://www.bbc.com/news/technology-18643134

News, B. B. C. (2014). The smart multitasking drones of the future. http://www.bbc.com/news/technology-28742626

O'Kane, S. (2014). The FAA's drone regulations won't be ready until at least 2017. The Verge. http://www.theverge.com/2014/12/10/7370955/the-faas-drone-regulations-wont-be-ready-until-at-least-2017

Patel, T. (2014). France Probes Mystery Drones Flying Over Nuclear Reactors. Bloomberg. http://www.bloomberg.com/news/2014-10-29/edf-confirms-probe-into-mystery-drones-flying-over-nuclear-sites.html

Peterson, S. (2011). Exclusive: Iran hijacked US drone, says Iranian engineer (Video). The Christian Science Monitor. http://www.csmonitor.com/World/Middle-East/2011/1215/Exclusive-Iran-hijacked-US-drone-says-Iranian-engineer-Video

Quinn, B. (2014). Drone permits issued to UK operators increase by 80%. The Guardian. http://www.theguardian.com/world/2014/oct/26/drones-permit-uk-british-airline-pilots-association-unmanned-aircraft-house-of-lords

Segar, M. (2014). License plate reader makers sue Arkansas for banning their tech. Rt.com. http://rt.com/usa/166916-vigilant-drn-arkansas-alpr-lawsuit/

Szondy, D. (2013). Beyond military drones – the future of unmanned flight. Gizmag.com. http://www.gizmag.com/uav-future-of-unmanned-flight/27478/

TechRadar. (2015). Rule the skies with the best drones of 2015, from affordable to premium price. http://www.techradar.com/news/portable-devices/other-devices/the-best-drones-1281186

Vincent, J. (2014). Amazon tells FAA to change drone laws or it'll move research abroad. The Verge. http://www.theverge.com/2014/12/9/7359409/amazon-slams-FAA-for-restrictive-drone-laws

Waterman, S. (2012). Drones over U.S. get OK by Congress. The Washington Times, p.all. http://www.washingtontimes.com/news/2012/feb/7/coming-to-a-sky-near-you/?page=all

Weckler, A. (2014). Dublin to overtake London as Europe's technology hub - Independent.ie. Independent.ie. http://www.independent.ie/business/technology/dublin-to-overtake-london-as-europes-technology-hub-30025030.html

Weckler, A. (2014). Game of drones: Why Ireland could fly high in aviation's latest frontier - Independent.ie. Independent.ie. http://www.independent.ie/business/technology/game-of-drones-why-ireland-could-fly-high-in-aviations-latest-frontier-30818509.html

Winter, L. (2014). Solar-Powered Drone Could Fly Nonstop For Five Years. Iflscience.com. http://www.iflscience.com/technology/solar-powered-drone-could-fly-nonstop-five-years

This research was previously published in the International Journal of Handheld Computing Research (IJHCR), 6(1); edited by Mehdi Khosrow-Pour, D.B.A., pages 44-57, copyright year 2015 by IGI Publishing (an imprint of IGI Global).

Index

www.ingramcontent.com/pod-product-compliance
Lightning Source LLC
LaVergne TN
LVHW060253011026
849161LV00011B/2703